Ricardo Silvestre • Egídio Torrado
Editors

Metabolic Interaction in Infection

Springer

Editors
Ricardo Silvestre
Life and Health Sciences Research Institute
University of Minho
Braga, Portugal

Egídio Torrado
Life and Health Sciences Research Institute
University of Minho
Braga, Portugal

ISSN 1664-431X ISSN 2504-3692 (electronic)
Experientia Supplementum
ISBN 978-3-030-09108-8 ISBN 978-3-319-74932-7 (eBook)
https://doi.org/10.1007/978-3-319-74932-7

Printed on acid-free paper

This Springer imprint is published by the registered company Springer International Publishing AG part of Springer Nature.
The registered company address is: Gewerbestrasse 11, 6330 Cham, Switzerland

Preface

We are experiencing a renewed interest in cellular metabolism. Indeed, metabolism is now recognized to control multiple functions during cell development, differentiation, and effector function, thus playing a central role in homeostasis and disease development mechanisms. This revival has been compelled by impressive technical advances in our ability to measure metabolic processes in vitro and in vivo and to integrate distinct "omics" technologies to understand the multiple levels of metabolic regulation. The resurgence of cellular metabolism is particularly important when addressing microbial pathogenesis as it provides a new perspective on host–pathogen interactions. For too long, microbial pathogenesis and metabolism have been studied as separate entities. However, most metabolites used by pathogens and their hosts are chemically identical, and therefore, defining the metabolic cross talk and/or nutrient competition between pathogens and their host will be crucial to define microbial evasion strategies and the range of host niches that each pathogen can occupy. Therefore, it is crucial that we identify the metabolic players that control immune cell responses in health and disease.

Several seminal works have begun to delineate the intricate levels of connection between cell metabolism, immunity, and infection. The scientific community is beginning to understand the essential role for nutrient acquisition in pathogenesis and identifying pathogen-specific metabolic pathways that regulate the expression of virulence factors. Moreover, there is a growing understanding of pathogen-specific factors that interfere with the complex regulatory network controlling carbon and nitrogen metabolism of mammalian cells as well as host cell metabolic reprogramming critical to limit pathogenesis and infection progression. For example, the plasticity and polarization of immune responses is accompanied by a metabolic adaptation that is crucial to prevent and control infectious diseases. Therefore, a deeper knowledge on the molecular pathways underlying the immunological-metabolic cross talk may lead to effective breakthroughs in the identification of new drug targets and an improved understanding of the mechanism of action of existing drugs, which will provide opportunities for novel immunotherapeutic interventions.

In this book, we focus on host–pathogen interactions at the metabolic level. We will initially highlight the role of metabolism in cell signaling and immunity. We propose a thorough revision of current literature on nutrient-sensing mechanisms and pathways as well as the processes behind altered cellular functions and homeostasis upon nutrient modulation. This will provide the foundations for the understanding of host metabolism perturbation upon infection. Microbial auxotrophy and host cell metabolic rewiring by pathogens will also be dissected. We will explore the metabolic requirements of the infectious agents, the microbial metabolic pathways dedicated to circumventing host immune mechanisms, as well as the molecular mechanisms by which pathogens hijack host cell metabolism for their own benefit. Finally, to emphasize the importance of this emerging field, we will provide insights into the possible clinical and immunotherapeutic applications, as well as into the available experimental and analytical methods.

This book assembles a group of contributors who are breaking new ground in understanding the metabolic cross talk between host and pathogen to share their expert perspectives and outlook for the future.

Braga, Portugal

Ricardo Silvestre
Egídio Torrado

Contents

Part I
A Guide of Metabolism During Host–Pathogen Interactions

Chapter 1
Cellular Metabolism at a Glance

Inês Mesquita and Fernando Rodrigues

Abstract Metabolism is highly coordinated component of the cellular activity that involves sequential chemical transformations, within a so-called metabolic network. Through these coordinated actions, living organisms acquire energy and biosynthetic precursors to maintain cellular homeostasis and function. Metabolism relies on the breaking down of macromolecules to produce energy [catabolism] and/or intermediary metabolites that are then used to construct essential building blocks for macromolecule production [anabolism]. Overall, these metabolic processes are controlled by cellular energy status: when the energy released from catabolic processes exceeds the cellular demands the storage of metabolites in the form of lipids and glycogen takes place. These phenomena have been vastly associated with the genesis of metabolic disorders, such as obesity. In recent years, we have assisted to a rediscovery of metabolism through the identification of metabolic intermediaries that act as key players on differentiation, proliferation, and function of immune cells. This recent acknowledgement of the impact of metabolism in the overall immune response originated the ground-breaking field of immunometabolism. Here, we will provide a holistic view of metabolism highlighting the biochemical principles underlying its regulation.

Keywords Metabolic networks · Biosynthetic precursors · Energy · Homeostasis · Metabolic nodes · Regulation · Immunometabolism

---

I. Mesquita · F. Rodrigues (✉)
Life and Health Sciences Research Institute (ICVS), School of Medicine, University of Minho, Braga, Portugal

ICVS/3B's-PT Government Associate Laboratory, Braga/Guimarães, Portugal
e-mail: frodrigues@med.uminho.pt

R. Silvestre, E. Torrado (eds.), *Metabolic Interaction in Infection*, Experientia Supplementum 109, https://doi.org/10.1007/978-3-319-74932-7_1

1.1 Key Concepts on the Regulation of Cellular Metabolism

Metabolism (or the RoadMap of the metabolic reactions) relies on energy interconversion within the adenylate system [composed by adenosine triphosphate (ATP), adenosine diphosphate (ADP), and adenosine monophosphate (AMP)] and/or oxidation–reduction reactions, where the electrons are transiently stored [mainly within the nicotinamide adenine dinucleotide (NAD) and/or flavin adenine dinucleotide (FAD) systems, as well as nicotinamide adenine dinucleotide phosphate (NADP)] to govern the overall cellular metabolic fluxes. Despite few exceptions, these are the main players that dictate the dynamic flow of metabolites between the distinct metabolic pathways. These *metabolic interconnectors* are constantly linking and modulating specific pathways to maintain their balance. Although frequently forgotten, the Le Chatelier's principle is inevitably respected, and it is imperative for the maintenance of cellular homeostasis. The flow of metabolites between pathways relies on a sequence of reactions that are maintained in a state of dynamic equilibrium. For instance, it is known that the glycolytic flux is highly dependent on NAD^+ availability within the carbohydrate metabolism. Consequently, an increased rate of this pathway has to rely on increased rate of NAD^+ regeneration. This is commonly achieved through the utilization of pyruvate as final electron acceptor resulting in the production of lactate, in a process named lactic fermentation. Such process occurs due to the fact that the rate for aerobic respiration is unable to sustain the velocity of NAD^+ replenish to fulfill the high demand of this intermediate for glycolysis to accomplish the ATP requests. Although this process occurs mainly under anaerobic conditions, an impressive amount of recent data shows that oxygen is only an additional player in this crosstalk (Cheng et al. 2014; Déry et al. 2005; Jiang 2017; Jones and Bianchi 2015; Lunt and Vander Heiden 2011; Romero-Garcia et al. 2016).

Several factors strongly impact the metabolic route followed by a metabolite, namely: (1) the affinity of the enzyme for the substrate within a metabolic branching point, (2) the amount of enzyme (translated by the maximum velocity of the reaction), and (3) their post-translation activation/inactivation either by covalent modifications or by allosteric regulators. Having these concepts in consideration, to define a metabolic RoadMap for a certain substrate, it is essential to keep in mind that several metabolites have to cross cellular membranes and, for those that do not freely diffuse across the membranes, selective transporters are needed. For instances, NAD^+ is unable to cross membranes and, therefore, this molecule is stored in different cellular compartments, with the cytosol and the mitochondria representing the main NAD^+ pools (reviewed in Chap. 4). Therefore, the transport-mediated flux and counter flux of oxidized/reduced metabolites through cellular membranes (e.g., malate/aspartate shuttle) allow the maintenance of redox balance (NAD^+/NADH) in cellular compartments and are responsible for driving metabolic flow within specific pathways.

The overarching role of metabolism in cellular function has driven the scientific community to open access to detailed data regarding metabolites, cofactors, and enzymes involved in each metabolic process for different cellular condition/stimuli/

insult. However, the dynamic interplay between several metabolic nodes that regulate cellular function and metabolism cannot be overlooked when analyzing these data at the cellular functionality level. As such, the main key to integrate metabolism toward a better understanding of physiology and pathology is to comprehend the intricate connections between all of the metabolic pathways and their main sites of regulation.

1.2 An Introduction to Glycolysis

Glycolysis was the first identified and better studied metabolic pathway, which accounted for the contributions of several renowned scientists, among them was Eduard Buchner, Otto Warburg, and Hans von Euler-Chelpin. The development of molecular biology and biochemistry techniques allowed the purification of glycolytic enzymes and the discovery of important cofactors, such as NAD^+, which boosted the knowledge on this vital pathway. Glycolysis is a cytosolic pathway virtually present in all cells. During this process, one molecule of glucose is enzymatically broken in two molecules of pyruvate, with the consequent release of free energy in the form of two molecules of ATP and two NADH. The cellular rate for glucose metabolism depends mainly on the major cellular metabolic fluxes, as a cellular requirement of ATP or building blocks for the anabolic reactions.

1.2.1 Glucose Uptake

The first step on the metabolism of glucose is its entrance in the cell. Due to its biochemical properties, this molecule does not freely diffuse through cellular membranes. The majority of mammalian cells import glucose by facilitative diffusion, through the action of glucose transporters (GLUT) that belong to the (SLC2A) family of membrane transport proteins. In some cellular tissues, this transport defines the rate-determining step in glucose metabolism (Mueckler and Thorens 2013; Thorens and Mueckler 2010). The most studied members of this family range from GLUT1 to GLUT4, which display distinct kinetic properties and regulatory mechanisms. GLUT1, which is encoded by the gene *SLC2A1*, was one of the first membrane transporters to be purified and it is the most studied glucose transporter. Along with GLUT3, it mediates the basal and constant glucose uptake in several tissues, namely the brain, which is dependent on glucose catabolism. On the other hand, GLUT2 has a very low affinity for glucose, which implies that glucose uptake occurs only when high concentration of this metabolite is available. The *SLC2A4* gene encodes for GLUT4, which is mainly expressed in adipocytes, skeletal muscle, and cardiomyocytes. An interesting trait of this transporter is that, under basal conditions, GLUT4 resides in intracellular vesicles. However, upon insulin signaling, these intracellular vesicles containing transporters are fused to the plasma membrane exposing GLUT4 to the extracellular compartment, thus increasing glucose uptake and metabolism.

1.2.2 Glycolytic Process

The glycolytic process leads to the degradation of glucose [six-carbon (C6) molecule] in two pyruvates [three-carbon (C3) molecule]. Pyruvate may then follow different metabolic pathways (Fig. 1.1). The ten enzymatic steps that govern glycolysis occur in the cytosol, and some of them are tightly regulated to maintain an adequate glycolytic flux (Berg et al. 2002a, b). The first step relies on the phosphorylation of glucose by the rate-limiting enzyme hexokinase, which yields glucose-6-phosphate. The quick phosphorylation of intracellular glucose prevents its cellular efflux as well as the enzymatic conservation of high-energy bonds through the formation of phosphate esters. Additionally, phosphoryl groups destabilize glucose and facilitate its consequent metabolism. Hexokinase is regulated through a negative feedback loop by glucose-6-phosphate that when in high intracellular

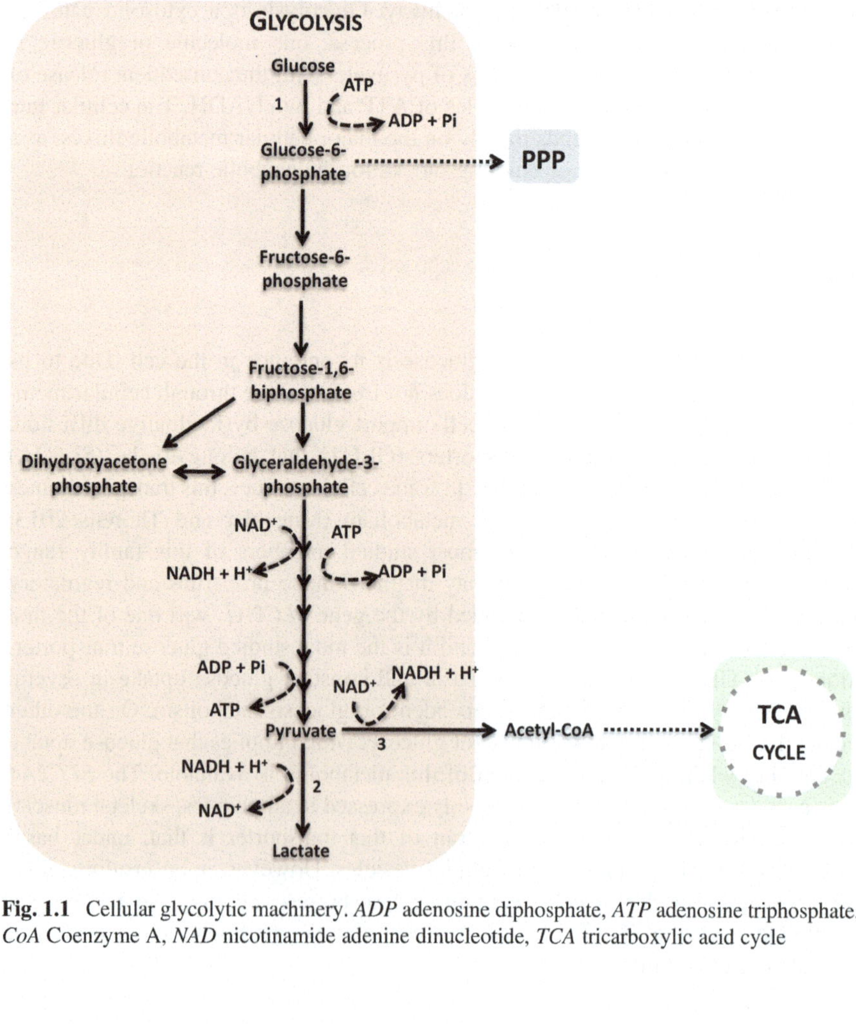

Fig. 1.1 Cellular glycolytic machinery. *ADP* adenosine diphosphate, *ATP* adenosine triphosphate, *CoA* Coenzyme A, *NAD* nicotinamide adenine dinucleotide, *TCA* tricarboxylic acid cycle

quantity signals for excess of cellular energy. Glucose-6-phosphate is then reversibly isomerized to fructose-6-phosphate, which can be further phosphorylated by phosphofructokinase (PFK), at the cost of one ATP molecule, to fructose-1,6-bisfosfate. The irreversible step mediated by PFK is a very important pacemaker of glycolysis being allosterically positively regulated by AMP and negatively by ATP, fructose-2,6-phosphate, and citrate. This balance points out the importance of the ATP/AMP ratio for the rate of glycolysis, as well as its regulation to channel carbon molecules through the oxidative branch of the pentose phosphate pathway (PPP) for maintaining the NADPH pool for fatty acid synthesis. Ultimately, FK is also capable of indirectly inhibiting hexokinase. If PFK is inactive, fructose-6-phosphate and glucose-6-phosphate accumulates, thus inhibiting hexokinase. However, this dynamic may be variable depending on the final destination of glucose-6-phosphate. Although the latter is an important intermediate, since it can modulate the glycolytic flux, it may also be shuttled to feed glycogenesis or the pentose phosphate pathway. The other irreversible step of glycolysis is mediated by pyruvate kinase where phosphoenolpyruvate is converted into the end glycolytic product, pyruvate. In fact, pyruvate kinase isoforms have been suggested to support divergent energetic and biosynthetic output and thus fulfilling differential requirements of cellular metabolism. As such, this catalytic activity plays a crucial role in energy metabolism and has recently been associated with cell proliferation and tumor growth (Israelsen and Vander Heiden 2015; Yang and Lu 2013).

Although glucose represents a high-energy substrate, only a small portion of free energy is released during glycolysis: most of it remains in the end product, pyruvate, that needs to be further catabolized.

1.2.3 The Fate of Glycolysis End Products: Pyruvate and NADH

In mammalian organisms, the end product of glycolysis, pyruvate, may then be further metabolized via two major catabolic pathways. In one hand, pyruvate may be oxidized in the mitochondria to yield acetyl-Coenzyme A, which will sustain the Tricarboxylic Acid (TCA) cycle (also named citric acid cycle or Krebs cycle). On the other, pyruvate may be reduced to lactate. Furthermore, in certain invertebrates, protists, and microorganisms, as the brewer's yeast, pyruvate may be redirected to the production of ethanol and carbon dioxide, in a process known ethanol fermentation (Boulton 1996).

To maintain glycolysis in perpetuation, cytosolic redox balance (NAD^+/NADH) has to be attained. One way to regenerate the cytosolic NAD^+ is through shuttles to the mitochondria (discussed below in this chapter) or by fermentation through the action of lactate dehydrogenase. In this process, pyruvate is reduced to lactate and the NAD^+ pool is reestablished. This process was thought to occur mainly at low oxygen tension (such as in hypoxic environments), but recent data show that it also occurs under normoxic conditions in specific cell types and within specific differential activations. Although the fermentation of pyruvate restores NAD^+ efficiently, only a small fraction of the glucose's free energy is released in this process. The yield of ATP per glucose consumed is lower under fermentative conditions when compared with respiration. However, the amount of ATP per time unit (velocity) is higher under fermentative conditions.

The mitochondrial oxidative decarboxylation of pyruvate by the pyruvate dehydrogenase complex, which is an irreversible reaction, produces acetyl-CoA, the carbon substrate of the TCA cycle (Berg et al. 2002a). Furthermore, pyruvate may also be shifted towards anabolism. Pyruvate may be transaminated into alanine, thus providing a direct link between amino acid and carbohydrate metabolisms. Also, pyruvate may also be carboxylated into oxaloacetate in the mitochondria, which yields a precursor for gluconeogenesis or for amino acid synthesis (aspartate).

1.2.4 The Pentose Phosphate Pathway, the Link to Glycolysis and Cellular Detoxification

The pentose phosphate pathway (PPP), also known as the hexose monophosphate shunt or phosphogluconate pathway, is a major glycolytic-divergent pathway that shifts metabolic reactions towards the production of precursors for nucleotide and amino acid synthesis as well as the mostly studied product, the reducing co-factor NADPH (Berg et al. 2002c) (Fig. 1.2). This pathway functions to support cell proliferation and survival (Buchakjian and Kornbluth 2010; Patra and Hay 2014). The reducing equivalent NADPH is a cofactor for the biosynthesis of fatty acids and sterols, and it also has a vital role in countering oxidative stress. The PPP comprises an oxidative and a non-oxidative branch. The rate limiting step of the oxidative branch is mediated by glucose-6-phosphate dehydrogenase (G6PD) that through the oxidative decarboxylation of glucose-6-phosphate, produces two molecules of NADPH and one ribose-5-phosphate. As expected, this branch relies on the availability of $NADP^+$, as the final electron acceptor, while excess of NADPH is a negative regulator. The non-oxidative branch includes several reversible reactions that interchange carbons from/to three-carbon and six-carbon molecules to/from five-carbon molecules, depending on the cellular concentration of NADPH and ribose-5-phosphate, being rate limited by transketolase and transaldolase.

The fine balance between the PPP and glycolysis is tightly regulated. Whereas ribose-5-phosphate can be produced for nucleotide synthesis, NADPH production mainly supports fatty acid synthesis and has an important action in counteracting oxidative stress. For the latter, six molecules of ribose-5-phosphate are shifted towards production of two molecules of glyceraldehyde-3-phosphate and four molecules of fructose-6-phosphate, thus fuelling glycolysis. The maintenance of redox homeostasis is a vital requisite for cell survival, particularly in aerobic organisms.

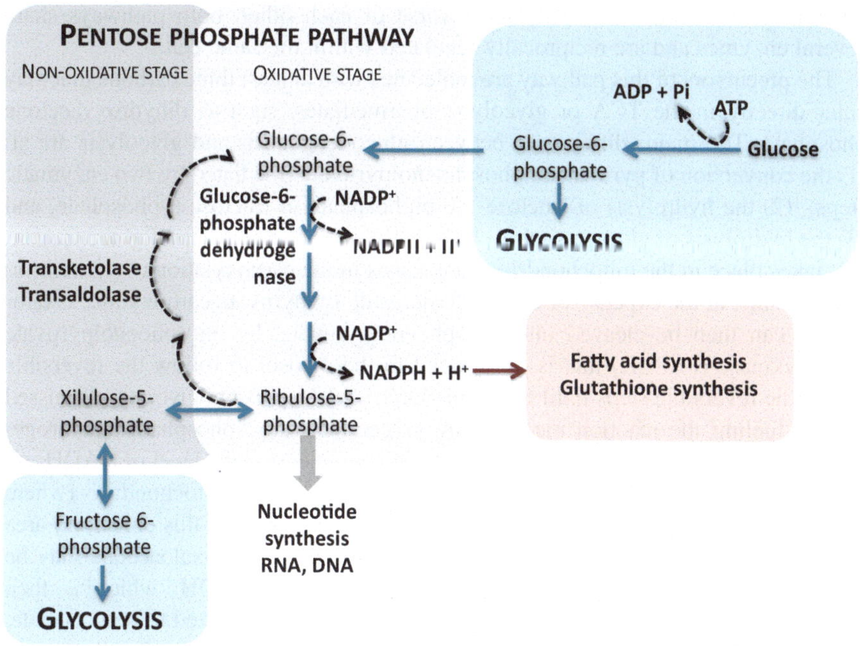

Fig. 1.2 The pentose phosphate pathway. *ADP* adenosine diphosphate, *ATP* adenosine triphosphate, *NADP* nicotinamide adenine dinucleotide phosphate

Reactive oxygen species (ROS) are continuously produced by the partial reduction of molecular oxygen, which release free radicals such as the superoxide anion or the hydroxyl radical. The exacerbated production of these molecules and/or decreased elimination by cellular detoxification systems originates molecular damage in several macromolecules, resulting in the development of mutations and eventually cell death (Dan Dunn et al. 2015; Ott et al. 2007). The steady-state control of oxidative stress is maintained by the action of several detoxification systems, which include glutathione reductase. This enzyme reduces oxidized glutathione (a tripeptide with a free sulfhydryl group) with the concerted action of NADPH, as electron donor.

When taken together, these data show that PPP is essential not only in the coordination of cellular metabolism (catabolism/anabolism) but also for fulfilling intermediaries that are needed for cellular defense and detoxification.

1.2.5 Gluconeogenesis

Gluconeogenesis is the anabolic process by which glucose is synthesized, occurring mainly at the liver and kidney. The noncarbohydrate biosynthetic precursors for this pathway are essentially lactate, amino acids, and glycerol, whose availability is responsible for determining the rate of gluconeogenesis. Although glycolysis and gluconeogenesis are not exactly the reverse of each other, both pathways share several enzymes and are reciprocally regulated within the same cell.

The precursors of this pathway are molecules with at least three carbons that may enter directly in the TCA or glycolytic intermediates, such as dihydroxyacetone phosphate. The major differences between gluconeogenesis and glycolysis are at: (1) the conversion of pyruvate to phosphoenolpyruvate (mediated by two enzymatic steps), (2) the hydrolysis of fructose-1-6-biphosphate in fructose 6-phosphate, and (3) the hydrolysis of glucose-6-phosphate in glucose. The first step of gluconeogenesis takes place in the mitochondria and consists in the carboxylation of pyruvate to oxaloacetate, at the expense of one ATP molecule by pyruvate carboxylase. Oxaloacetate can then be cleaved into phosphoenolpyruvate, by phosphoenolpyruvate carboxykinase (PEPCK) that is transported to the cytosol to follow the reversible glycolytic reversible steps until the formation of 1,3-bisphosphoglyerate (discussed below) fueling the reaction mediated by glyceraldehyde 3-phosphate dehydrogenase. However, this reaction requires reducing equivalents at the level of NADH. As such, translocation of reducing equivalents from the mitochondria (where β-oxidation of fatty acid is taking place leading to an excess of this cofactors) area is essential for gluconeogenesis. In this sense, mitochondrial oxaloacetate may be reduced to malate using the excess of mitochondrial NADH, which is then transported to the cytosol. Once in cytosolic, malate is re-oxidized to oxaloacetate, which, by the action of phosphoenolpyruvate carboxykinase, is simultaneously decarboxylated and phosphorylated to generate phosphoenolpyruvate at the expense of energy. Following these reactions, phosphoenolpyruvate is reversely converted in the upstream intermediates of the glycolytic pathway until the irreversible conversion of fructose-1-6-biphosphate into fructose-6-phosphate. The phosphate ester at C1 is hydrolyzed by fructose-1-6-biphosphatase, thus yielding fructose 6-phosphate. After isomerization by phosphoglucose isomerase, the obtained glucose-6-phosphate is cleaved by endoplasmic reticulum-bounded glucose-6-phosphatase, which originates glucose that is secreted to maintain the serum glucose levels. The ratio of mitochondrial transport of malate or phosphoenolpyruvate, or other source, to cytosol depends mainly on the availability of cytosolic NADH.

This pathway is energetically expensive, being, therefore, replenished via fatty acid β-oxidation. Indeed, fatty acids are rapidly mobilized by the hydrolytic action of lipases from triglycerides. The fatty acids products follow β-oxidation, whereas glycerol produced is mainly used to replenish gluconeogenic precursors.

1.3 The Tricarboxylic Acid Cycle and Oxidative Phosphorylation

The TCA cycle is the universal aerobic pathway for the complete oxidation of several macromolecules, including glucose, fatty acids, and some amino acids (Akram 2014; Citric and Cycle 2010; Kornberg 2000) (Fig. 1.3). In summary, the

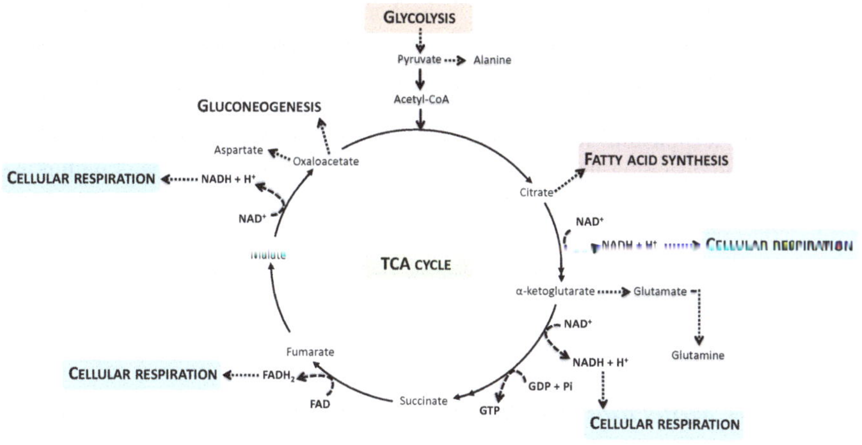

Fig. 1.3 The TCA cycle and the integration of multiple metabolic pathways. *CoA* Coenzyme A, *FAD* flavin adenine mononucleotide, *GDP* guanosine diphosphate, *GTP* guanosine triphosphate, *NAD* nicotinamide adenine dinucleotide

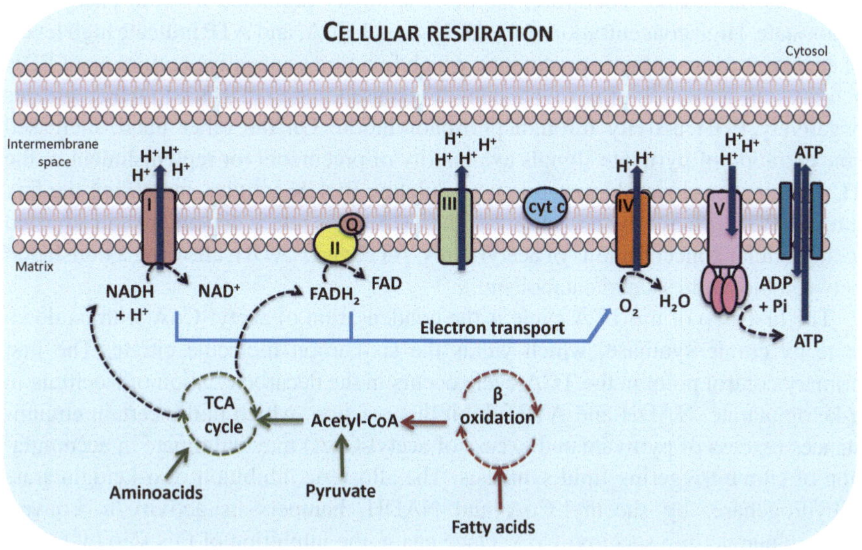

Fig. 1.4 Cellular respiration. *ADP* adenosine diphosphate, *ATP* adenosine triphosphate, *CoA* Coenzyme A, *FAD* flavin adenine mononucleotide, *NAD* nicotinamide adenine dinucleotide

organic carbon fuels are oxidized to acetyl-CoA, which is fed to TCA and completely oxidized to H_2O and CO_2. The released energy is conserved in the form of NADH and $FADH_2$, as well as GTP. The reduced molecules are highly energetic being used as substrates for the electron transport chain (ETC), which culminates with the reduction of molecular oxygen to H_2O and the buildup of a proton gradient along the internal mitochondrial membrane (Fig. 1.4). The energy of this electrochemical proton gradient is further used by ATP synthase for ATP production, in a process usually called oxidative phosphorylation. The respiratory control guarantees that the rate of the TCA cycle matches the cellular ATP utilization levels. These processes take place in the mitochondria, which are the cellular powerhouses responsible for the generation of energy in eukaryotic cells. While the TCA cycle occurs in the mitochondrial matrix, the ETC is localized within the mitochondrial inner membrane. Although the combination of these processes may result in the complete oxidative catabolism of organic carbon fuels, the TCA cycle is also a pathway used for the building blocks that will feed the de novo synthesis of several macromolecules (therefore designated an amphibolic pathway).

1.3.1 The Molecular Mechanisms Underlying the TCA Cycle

A mitochondrial complex named pyruvate dehydrogenase complex (PDH) is responsible for the decarboxylation of pyruvate in acetyl-CoA committing pyruvate carbons to be either converted into CO_2 or shifted to de novo lipid synthesis, via export of mitochondrial citrate. Therefore, the PDH is highly regulated by both energy and redox state. High concentrations of NADH, acetyl-CoA, and ATP indicate high levels of energy and biosynthetic precursors, which function as negative regulators of PDH activity, via activation of pyruvate dehydrogenase kinase (PDHK). PDHK regulates negatively PDH activity through phosphorylation. On the other hand, increased concentration of pyruvate signals availability of precursors for replenishment of the TCA cycle, consequently negatively modulating PDHK activity, increasing the flux through PDH. Therefore, it is clear that the TCA cycle rate is dependent and adjusted to the cellular concentrations of acetyl-CoA, ATP, and NADH, allowing a translation between biosynthesis and catabolism.

The first step of the TCA cycle is the condensation of acetyl-CoA with oxaloacetate by citrate synthase, which yields the six-carbon molecule citrate. The first primary control point in the TCA cycle occurs in the decarboxylation of isocitrate in α-ketoglutarate. NADH and ATP inhibit this reaction, which under certain circumstances (excess of pyruvate and excess of acetyl-CoA) may culminate in accumulation of citrate triggering lipid synthesis. The allosteric inhibition of α-ketoglutarate dehydrogenase, by succinyl-CoA and NADH, hampers its activity to convert α-ketoglutarate into succinyl-CoA. Once again, the inhibition of this step by reducing equivalents or end products may provide the metabolic intermediates for amino acid synthesis, via glutamate dehydrogenase. Succinyl-CoA is further converted into succinate, with the production of GTP, and then the carbon skeleton is oxidized by succinate dehydrogenase (complex II of the ETC) to form fumarate, which is again oxidized to malate. This final intermediate is the able to regenerate oxaloacetate, thus closing the cycle.

The result of one cycle turn of the TCA cycle is: three molecules of NADH, one of $FADH_2$, one of GTP, and two of CO_2. Each highly energetic reducing equivalent produced contains a pair of electrons that are substrates of the ETC.

1.3.2 TCA Cycle Intermediates Are Important Biosynthetic Precursors

TCA cycle is an amphibolic pathway that generates several biosynthetic precursors for de novo synthesis of macromolecules are produced. Oxaloacetate and α-ketoglutarate are precursors for aspartate and glutamate synthesis, which in turn may originate other amino acids, such as asparagine, arginine, proline, and glutamine. Furthermore, these intermediates can also contribute to the synthesis of pyrimidines and purines, which are the main building blocks of nucleic acids (Lane and Fan 2015). Oxaloacetate may also be converted into phosphoenolpyruvate, by phosphoenolpyruvate carboxykinase, thus completing the first step of gluconeogenesis. Succinyl-CoA withdrawal is responsible for the central metabolism that originates the synthesis of porphyrin rings, essential for hemoglobin production (Bonkovsky et al. 2013). Citrate production may have a vital role in de novo fatty acid synthesis. However, when these intermediates are withdrawn from the TCA cycle, the rate of the cycle is reduced and only their replenishment will allow the continuation of the oxidative capacity of this cycle. This exquisite and dynamic balance is mainly assured by pyruvate carboxylase that produces oxaloacetate via carboxylation of pyruvate, which guarantees the maintenance of homeostatic levels of TCA intermediates.

1.3.3 Oxidative Phosphorylation

Oxidative phosphorylation is the process where electrons from NADH or $FADH_2$ are transferred to the final electron acceptor oxygen, which is reduced to water originating a transmembrane electrochemical proton gradient (ΔΨm) (Fig. 1.4). The latter is further dissipated towards ATP formation through the ATP synthase (Balaban 1990; Hansford 2002; Nath and Villadsen 2015). This process is the major source of energy in aerobic organisms, resulting in the generation of 26 from the 30 molecules of ATP. These are produced upon complete oxidation of one molecule of glucose in water and carbon dioxide contrasting with those two from lactic acid fermentation. Oxidative phosphorylation takes place in the mitochondria, specifically in the inner membrane. This membrane is folded into *cristae*, and it is embedded with the mitochondrial respiratory complexes, which are responsible for the electron flux along respiration, as well as the proton extrusion from the mitochondrial matrix. During cellular respiration, electrons from NADH are transferred to the final electron acceptor through three major respiratory complexes: NADH-Q-oxidoreductase (complex I), Q-cytochrome c oxido-reductase (complex III), and cytochrome c oxidase (complex IV). On the other hand, the ETC II (complex II) is the succinate dehydrogenase that oxidizes succinate into fumarate linking directly the ETC and the TCA. Moreover, two mobile electron carriers, ubiquinone and cytochrome C, are responsible for the elaborated electron transport process from complex I and II to the complex III and from complex III to complex IV, respectively. As such, the final step of the ETC is the oxidation of the reduced cytochrome c by cytochrome oxidoreductase with the consequent reduction of oxygen to water. ATP production from ADP is achieved when protons flow back into the mitochondrial matrix, through the fifth complex, ATP synthase (or F_1F_0 ATPase). The F_1 subunit is responsible for the ATPase activity, while the F_0 one contains the proton channel that allows the reverse flux of H^+, pumping three protons back to mitochondria, resulting in the synthesis of one molecule of ATP. However, because of the electronegativity of the ATP/ADP translocator, the overall proton leakage per an ATP molecule is the equivalent to four protons. The rate of oxidative phosphorylation is dictated by the cellular energy flux, and it depends on the availability of a high-energy source of electrons (such as NADH and $FADH_2$), oxygen, ADP, and inorganic phosphate. However, intracellular ADP content is the most important and regulatory mechanism regarding the oxidative process.

Overall, the electron-motive force is coupled to the proton pumping activity of complexes I, III, and IV, which generates the mitochondrial proton motive force $\Delta\Psi m$ across the inner-mitochondrial membrane. The dissipation of energy of $\Delta\Psi m$ from the backflow of protons from the mitochondrial intermembrane space to the matrix is further used by ATP synthase to synthesize ATP from ADP and phosphate (Senior et al. 2002).

1.3.4 NAD⁺ Regeneration and Shuttles Across Mitochondrial Membranes

Although the mitochondrial membrane remains impermeable to most molecules, the exchange of certain molecules between the cytosol and the mitochondria (and vice versa) is mediated by transport systems essential for metabolic homeostasis. Some of these systems are used to translocate reducing equivalents from the cytosol to the mitochondrial matrix or vice versa (Fig. 1.5). In some cellular tissues, glucose is

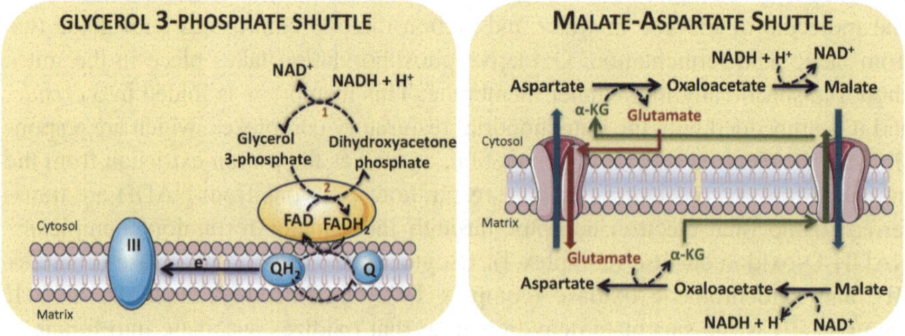

Fig. 1.5 Mitochondrial membrane shuttles. 1: Cytoplasmic glycerol 3-phosphate dehydrogenase; 2: Mitochondrial glycerol 3-phosphate dehydrogenase. *α-KG* alpha-ketoglutarate, *FAD* flavin adenine mononucleotide, *NAD* nicotinamide adenine dinucleotide

completely oxidized to CO_2, which requires a substantial transfer of reducing equivalents from cytosolic NADH to the mitochondria. As such, during glycolysis, the oxidation of glyceraldehyde-3-phosphate by glyceraldehyde-3-phosphate dehydrogenase originates NADH that needs to be oxidized to NAD^+ for endowing glycolysis. However, cytosolic NADH cannot freely diffuse through membranes. In accordance with this, the cytosolic and mitochondrial pools of NADH are conserved and independent as a whole. Therefore, the electrons are transferred from these reduced molecules to several oxidized carbon compounds that can be shuttled to the mitochondria and vice versa. The glycerol-3-phosphate shuttle ensures the introduction of electrons from NADH into the ETC, thus allowing the regeneration of cytosolic NAD^+ and consequently the maintenance of the glycolytic flux. Firstly, a pair of electrons is transferred from cytosolic NADH to dihydroxyacetone phosphate, thus originating glycerol-3-phosphate. This reaction is catalyzed by the soluble isoform of glycerol-3-phosphate dehydrogenase. Glycerol-3-phosphate is then reoxidized to dihydroxyacetone phosphate in the mitochondrial intermembrane space by the mitochondrial inner-membrane-bounded glycerol-3-phosphate dehydrogenase. This reaction yields $FADH_2$ that is further oxidized by ubiquinone responsible for transferring the electrons to the mitochondrial complex III. The oxidative/reductive state of each compartment cytosol versus inner-mitochondrial membrane is the driving force for this electric metabolic flux.

The malate-aspartate shuttle offers a distinctive way of regenerating NAD^+ with the simultaneous exchange of intermediates between the cytosol and the mitochondria. This shuttle is composed by two sets of enzymes, the malate dehydrogenase and the aspartate aminotransferase, each of which with mitochondrial and cytosolic localization, and two mitochondrial carriers, the α-ketoglutarate/malate and the aspartate/glutamate, with an antiport mechanism. As such, upon glycolysis, the excess of cytosolic NADH forces the equilibrium of cytosolic malate dehydrogenase to be shifted from oxaloacetate to malate, which allows the regeneration of NAD^+. Malate enters the mitochondria (with an export of mitochondrial α-ketoglutarate), and it is oxidized by malate dehydrogenase to oxaloacetate and NADH. This NADH is a substrate of complex I of the electron transport chain. Mitochondrial oxaloacetate transamination with glutamate gives rise to aspartate and α-ketoglutarate (used for the entrance of malate in the mitochondria). In the cytosol, aspartate transamination with in α-ketoglutarate gives rise to oxaloacetate and glutamate. The antiport aspartate/glutamate shuttles this cytosolic glutamate into the mitochondria with the counter flux of mitochondrial aspartate. As described above (Sect. 1.2.3), NAD^+ can also be regenerated through fermentation of pyruvate to lactate by the lactate dehydrogenase. Similar to NAD^+ and NADH, the mitochondrial membrane is also impermeable to ADP and ATP. Therefore, the carrier ATP-ADP translocase couples the flow of ADP and ATP through the mitochondrial membrane. This means that the mitochondrial efflux of ATP is accompanied by the influx of ADP.

Out of the scope of this chapter are other mitochondrial transporters that play crucial roles ensuring a correct functioning of cellular respiration.

1.4 Glycogen Metabolism

Glycogen is a multibranched polymer of glucose residues, which may quickly be degraded to supply glucose. Most of the residues are sequentially linked by α-1,4-glycosidic bonds, although branched chains are connected through α-1,6-glycosidic links. This energetic reservoir is capable of yielding large quantities of glucose, and consequently high amounts of energy upon catabolism. Contrary to what happens with fatty acids, glycogen-derived glucose may be used for fermentation, thus allowing the establishment of high glycolytic fluxes. The excess of serum glucose triggers insulin-driven regulatory processes that culminate in increased glucose uptake and consequent glycogen synthesis. The skeletal muscle and the liver are the main sites for glycogen storage, which is accumulated in the form of cytoplasmic granules. Although the muscle tissue uses glycogen for catabolism under intensive exercise for fulfilling high-energy requirements, the liver uses this reservoir to maintain glucose serum levels.

Overall, the net of cellular glycogen deposition depends on the coordinated suppression of glycogenolytic flux (glycogenolysis) and the stimulation of glycogen synthetic flux (glycogenesis). In spite of its importance, in the field of immunometabolism the knowledge of the effects of glycogen metabolism on immune (dys)function are scarce, and further studies are required to understand how the modulation of these processes impact cell response toward challenges.

1.4.1 Glycogenolysis

Glycogen degradation or glycogenolysis relies on the concerted activity of the following enzymes: (1) *glycogen phosphorylase* that degrades α-1,4-glycosidic bonds of glycogen with the release of glucose-1-phosphate; (2) a glycogen-debranching enzyme with a bifunctional activity, with *transferase* and *α-1,6-glucosidase* activity, that remodels glycogen within the boundaries of α-1,6-glycosidic bonds for further catabolism; and (3) *phosphoglucomutase* that converts the released intermediate (glucose-1-phosphate) in glucose-6-phosphate.

The key enzyme for glycogen breakdown by orthophosphate is glycogen phosphorylase. This phosphorolysis cleaves glycogen to yield a glucose-1-phosphate, in a bioenergetically advantageous reaction. Glucose-1-phosphate is readily converted into glucose-6-phosphate, the first intermediate of the glycolytic pathway or of gluconeogenic pathway in the liver. Produced glucose-6-phosphate has three possible outcomes: (1) glycolytic pathway yielding pyruvate; (2) enter the PPP yielding riboses and NADPH; or (3) re-converted into glucose through the last step of gluconeogenesis and released in the bloodstream (a process that takes place mainly in the liver).

As mentioned above, glycogen phosphorylase is responsible for cleaving α-1,4-glycosidic bonds and releasing glucose-1-phosphate. However, this enzyme is unable to recognize glycogen as substrate within a branch point with four glycosidic residues away from a branch point (after a α-1,6-glycosidic bond). Thus, to be further processed, this branch point of glycogen needs to be remodeled. As such, glycogen-debranching enzyme shifts three glycosidic residues from an outer branch to the linear chain of α-1,4-glycosidic substrate for action of phosphorylase. This reaction leaves an α-1,6-linked residue of glucose to an α-1,4-glycosidic linked chain that is released by a free molecule of glucose, by the action of α-1,6-glucosidase activity of the debranching enzyme. In the muscle, this glucose is prone to be phosphorylated by hexokinase, thus entering the conventional glycolytic pathway.

1.4.2 Glycogenesis

Glycogen is synthesized in a process that requires glucose-1-phosphate activation by uridine triphosphate (UTP) mediated by uridine diphosphate (UDP)-glucose pyrophosphorylase. This originates the ultimate glucose donor UDP-glucose. The initial priming is performed by the self-glucosylating protein glycogenin, which catalyzes the assembly of approximately the first ten glucose residues within a α-1,4-glycosidic linkage. The efficiency of transfer activity decreases pasting this length. This protein uses UDP-glucose for the polymerization of the initial glucose molecules. After junction of at least a minimal of four residues of glucose, glycogen synthase may begin the elongation steps. Using UDP-glucose, this enzyme is capable of adding these units to the nonreducing terminal residues of the growing glycogen molecule. However, since glycogen synthase is also capable of creating α-1,4-glycosidic linkages, another enzyme is required to form the α-1,6-glycosidic bonds that create the branched structure. This branching activity is catalyzed by a transferase that shifts a block of usually seven residues from a chain of at least eleven residues long to a more interior site to create an α-1,6-glycosidic branch point. This ramification process increases glycogen solubility, as well as the rate of synthesis versus degradation.

1.5 Fatty Acid Metabolism: The Balance Between Synthesis and Oxidation

Fatty acids are composed of long hydrocarbon chains and a terminal carboxylate group. These macromolecules have mainly four major biological roles as the building blocks for: (1) phospholipids and glycolipid synthesis, which are vital components of cellular membranes; (2) posttranslational modifications of proteins, as lipoprotein; and (3) the synthesis of the neutral lipids, triglycerides for energy storage.

In terms of metabolism, fatty acids can be synthesized or degraded, and these processes are not simply the reverse of each other and as such the fatty acid synthesis is not a reversal of the degradative pathway.

1.5.1 Fatty Acid β-Oxidation

Fatty acids are mainly stored in the form of triglycerides, which are constituted by three fatty acids and a glycerol molecule (Houten and Wanders 2010). Through the activity of a lipase, glycerol and the free fatty acids are released from the triglycerides. Following release of fatty acids, those with twelve or fewer carbons enter freely in the mitochondria where they are activated to acyl-CoA. Those with more than 14 and lower than 20 carbon are activated in the cytosol, which consists of the attachment of a CoA group to the fatty acid carboxyl group, in a process mediated by acyl-CoA synthase. After activation, the carnitinepalmitoyl transferase (CPT) I promotes an exchange of the CoA group of the acyl with a carnitine, giving rise to an acyl-carnitine. The acyl-carnitine is consequently transported into the mitochondrial matrix by the carnitine/acylcarnitine antiport system (translocase) (Fig. 1.6). Once inside the matrix, the CoA group is conjugated with the acyl group, through the activity of CPT II, giving rise to carnitine that fuels translocase for recycling in the cytosol. Inside the mitochondrial matrix, fatty acids are committed to β-oxidation. Overall, each round of oxidation consists of four reactions (oxidation, hydration, oxidation, and thiolysis) and culminates in the shortening of the acyl chain by two carbons, as well as the generation of $FADH_2$, NADH, and acetyl-CoA (Fig. 1.7). The

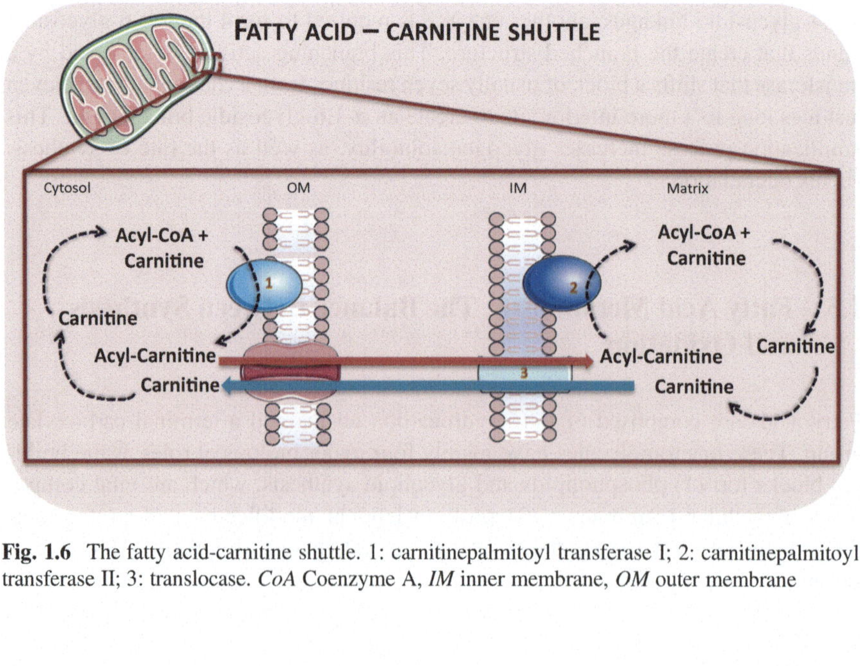

Fig. 1.6 The fatty acid-carnitine shuttle. 1: carnitinepalmitoyl transferase I; 2: carnitinepalmitoyl transferase II; 3: translocase. *CoA* Coenzyme A, *IM* inner membrane, *OM* outer membrane

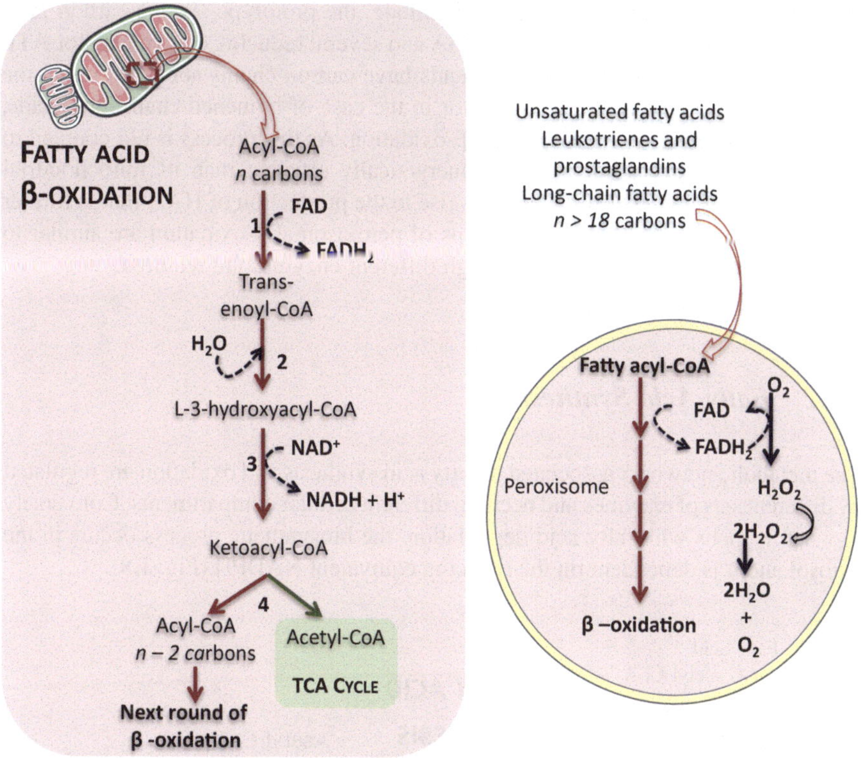

Fig. 1.7 Fatty acid β-oxidation. 1: Acyl-CoA dehydrogenase; 2: Enoyl-CoA hydratase; 3: 3-Hydroxy acyl-CoA dehydrogenase; 4: β-ketothiolase. *CoA* Coenzyme A, *FAD* flavin adenine mononucleotide, *NAD* nicotinamide adenine dinucleotide

first step consists of the oxidation of acyl-CoA by inner-mitochondrial membrane-bounded complex acyl-CoA dehydrogenase, which yields a trans-enoyl-CoA and one molecule of $FADH_2$. As mentioned above, this high-energy molecule immediately transfers the electrons to the mitochondrial respiratory chain, at the level of ubiquinone to produce ubiquinol. The acyl-CoA dehydrogenase displays length specificity. Acyl chains with 12–18 carbon atoms are oxidized by the long-chain enzyme while medium-chain fatty acids (four to fourteen carbon atoms) are metabolized by a medium-chain acyl-CoA dehydrogenase, and short-chain acyl-CoA dehydrogenase degrades only four to six carbon acyl chains. Following this first oxidation step, the double bond in the saturated enoyl-CoA is hydrated and L-3-hydroxyacyl-CoA is formed. This intermediate gives rise to one molecule of NADH, following another round of oxidation. In the final step of β-oxidation, catalyzed by a β-ketothiolase, an acetyl-CoA and an acyl-CoA shortened by two carbon atoms reenter the cycle for another round of β-oxidation. In the final step, the acetyl groups of the newly formed acetyl-CoA are prone to be oxidized in the TCA cycle, producing reducing equivalents used in the ETC for energy purposes. In terms of energy production, each molecule of palmitate, the prototype sixteen-carbon fatty acid, yields eight molecules of acetyl-CoA and several reducing equivalents for ATP production. Additionally, when fatty acids have carbon chains above 20, as in the case of hexacosanoic acid (26:0), and/or in the case of branched-chain fatty acids, they are shuttled into peroxisomes for β-oxidation. As this process is not coupled to oxidative phosphorylation, it is less energetically efficient than its mitochondrial counterpart. On the other hand, it gives rise to the production of H_2O_2 that is further degraded by catalases. Mainly, the steps of peroxisomal β-oxidation are similar to the mitochondrial counterparts, although different enzymes are required.

1.5.2 Fatty Acid Synthesis

The metabolic networks associated to fatty acid synthesis and oxidation are regulated by different sets of enzymes and occur in different cellular compartments. Conversely to what happens with fatty acid degradation, the biosynthetic process occurs in the cytosol and it is dependent on the reducing equivalent NADPH (Fig. 1.8).

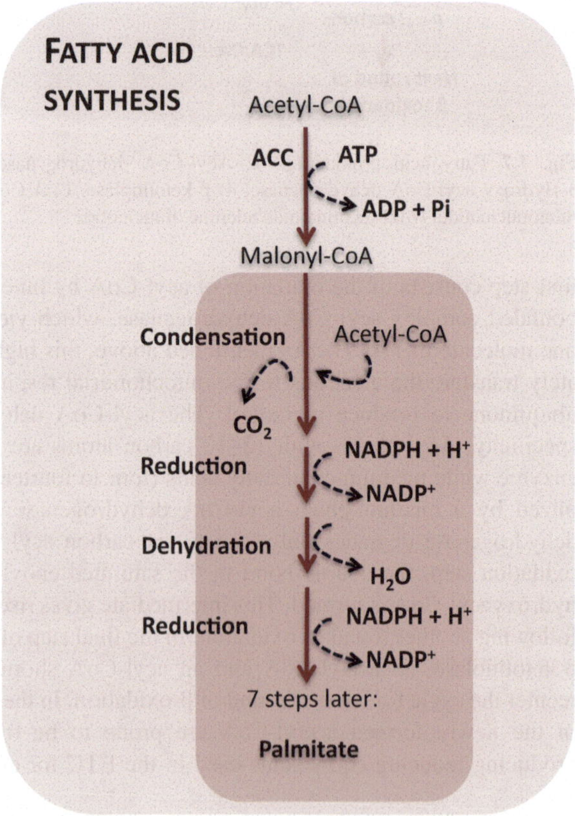

Fig. 1.8 Fatty acid synthesis. *ACC* acetyl-CoA carboxylase, *CoA* Coenzyme A, *NADP* nicotinamide adenine dinucleotide phosphate

The first step of fatty acid synthesis consists of the carboxylation of the cytosolic acetyl-CoA into malonyl-CoA. The major source of cytosolic acetyl-CoA is citrate that is synthesized from oxaloacetate and acetyl-CoA within the mitochondria. In fact, this important intermediate is shuttled through the citrate transporter from mitochondria to cytosol. In the cytosol, citrate is cleaved by acetyl-CoA and oxaloacetate by ATP-citrate lyase and through the malate-aspartate shuttle (described in Sect. 1.3.4). Oxaloacetate can then be reduced to malate (using the cytosolic excess of NADH) and further oxidized to pyruvate by malic enzyme with the production of NADPH. Pyruvate is then transported to mitochondrial matrix. On the other hand, the formation of malonyl-CoA is mediated by acetyl-CoA carboxylase (ACC) at the expense of energy. The metabolite produced by the activity of this enzyme plays a crucial role in controlling the rate of fatty acid synthesis and fatty acid degradation, as it is able to inhibit CPTI and, therefore, translocation of fatty acids into the mitochondria. Additionally, intracellular levels of other molecules involved in fatty acid anabolism and catabolism, as citrate, AMP, and palmitoyl-CoA, also dictate the flux of degradation versus synthesis. These negative feedback-regulatory mechanisms are particularly important since they represent clearly the global status of cellular metabolic needs. Cytosolic citrate is a two-carbon biosynthetic precursor for fatty acid synthesis, and, consequently, in the presence of high intracellular quantities, the cells are primed to initiate anabolic programs that culminate with the increased lipid content. Furthermore, citrate is an allosteric activator of ACC and inhibitor of the glycolytic enzyme phophofructokinase-1, thus diverting cellular metabolism from the catabolism of organic fuels to the storage of fatty acids and glycogen by promoting increased levels of glucose-6-phosphate, substrate of glucose-6-phosphate dehydrogenase, and phosphoglucomutase, respectively. Likewise, the presence of higher levels of palmitoyl-CoA inhibits by feedback the activity of ACC. Furthermore, palmitoyl-CoA is also responsible for inhibiting the transport of citrate to the cytosol, as well as glucose-6-phosphate dehydrogenase that provides NADPH for fatty acid synthesis. ACC activity can also be regulated by covalent modifications, as phosphorylation, mainly via AMP-associated protein kinase (AMPK) activation (Hardie and Pan 2002; O'Neill et al. 2013). AMPK is an important intracellular sensor and one of the central regulators of cellular metabolism. In the presence of a high AMP/ATP ratio, AMPK is activated and, consequently, the downstream phosphorylation of ACC inhibits fatty acid synthesis, thus promoting β-oxidation and ATP generation. Therefore, any stimuli capable of enhancing ATP production and precursor availability indicate that there is no need for an additional source of energy, thus priming anabolic processes.

Following malonyl-CoA synthesis, the elongation of acyl chains and consequent formation of fatty acids occur in a four-step sequence (condensation, reduction, dehydration, and reduction). As stated previously, in opposition to β-oxidation, which uses NAD^+ and FAD^+ as electron acceptors in fatty acid synthesis, one molecule of NADPH is oxidized to $NADP^+$ in each of the two reduction steps. Fatty acid synthase (FASN) is a multienzyme complex, which includes seven differentially functional catalytic sites. This complex is responsible for the catalysis of the final steps of the anabolic process originating palmitate. The synthesis of each molecule of palmitate requires eight molecules of acetyl-CoA as the primary biosynthetic precursor, as well as 14 molecules of NADPH and seven molecules of ATP. The NADPH precursors are the linkers between fatty acid synthesis and PPP pathway together with the action of malic enzyme. Further elongation and addition of saturated bonds are performed by different enzymes and are out of the scope of this chapter.

Most of newly synthesized fatty acids are either incorporated in triglycerides or in double-layer phospholipidic membranes. The chosen fate relies on current cellular requirements. In rapidly dividing cells, new membranes have to be promptly synthesized and, therefore, fatty acids are used as building blocks to produce phospholipids (Yao et al. 2016). However, in quiescent and in naïve cells, fatty acids may accumulate in the form of lipid bodies and constitute an energetic reservoir, available for degradation upon necessity (Knobloch et al. 2017; Lee et al. 2015; Yao et al. 2016).

1.6 Amino Acids

Amino acids are the building blocks of proteins and can be divided in essential, nonessential, and conditionally essential amino acids. Essential amino acids (histidine, isoleucine, leucine, lysine, methionine, phenylalanine, threonine, tryptophan, and valine) cannot be synthesized by the organism and, therefore, should be acquired through diet. On the contrary, dietary intake of nonessential amino acids such as alanine, aspartate, cysteine, and glutamate reduce the need of biosynthesis. Finally, some amino acids (as arginine, asparagine, glutamine, glycine, proline, serine, and tyrosine) are conditionally essential amino acids, which means that, in physiological conditions, these amino acids can be synthesized by the organisms and consequently dietary intake is not required. However, under certain situations, the net of synthesis may not fulfill the cellular/tissue/organ needs. When in excess, amino acids cannot be accumulated, as it happens with glucose (within glycogen) and fatty acids (within lipid bodies). Instead, amino acids are transformed into carbon skeletons and used as metabolic fuels or as biosynthetic precursor, depending on the cellular metabolic flux and regulation. A special highlight will be made for specific amino acids metabolism and their connection with the activation of intracellular signaling pathways.

1.6.1 Amino Acid Degradation: The Link Between the Urea Cycle and the TCA Cycle

Amino acid catabolism occurs mainly in liver, although in situations of fasting or prolonged exercise, the skeletal muscle may also use amino acids as a source of energy. In most cases, amino acid degradation starts with a transamination reaction that combines reversible amination and deamination. All transaminase reactions have the same mechanism and use pyridoxal phosphate (a derivative of vitamin B6) and transfer an amino group of an amino acid to a α-ketoacid to form a new amino acid. Aspartate aminotransferase (ASAT) and alanine aminotransferase (ALAT) are the most studied transferases, responsible for the metabolism of aspartate and alanine. The transamination of aspartate and alanine originates oxaloacetate and pyruvate, respectively, which are then shifted to feed the TCA cycle together with glutamate. Furthermore, the newly formed glutamate suffers oxidative deamination by glutamate dehydrogenase and yields the ammonium ion (NH_4^+). Carbamoyl-phosphate synthetase further catalyzes the production of carbamoyl phosphate from ammonia, bicarbonate (HCO_3^-; derived from the hydration of carbon dioxide), and ATP. This reaction is the first committed step of the urea cycle, which is important in the removal of excess ammonia. This cycle was the first cyclic metabolic pathway to be described, initially proposed in 1932 by Hans Krebs and Kurt Henseleit. Due to the presence of an anhydride bond, carbamoyl phosphate has a high transfer potential, originating the transference of the carbamoyl group to ornithine, thus forming citrulline, in a reaction catalyzed by ornithine transcarbamylase. Citrulline is then transported to the cytoplasm and, upon reaction with aspartate, which is the donor of the second α-amino group, argininosuccinate is formed. This step is mediated by argininosuccinate synthetase and driven by the cleavage of ATP in AMP. Argininosuccinate is then cleaved by argininosuccinase to yield arginine and fumarate, which represents the aspartate-driven carbon skeleton. Finally, arginine is hydrolyzed by arginase, thus originating ornithine, which is transported back to the mitochondria (within an antiport system ornithine/citrulline), and urea.

As mentioned above, the catabolism or anabolism of the resulting α-ketoacid provides metabolic intermediates that link the urea cycle and amino acid catabolism with the TCA cycle or gluconeogenesis. The metabolic integration of these pathways occurs upon the synthesis of fumarate that may be hydrated into malate and consequently oxidized to oxaloacetate. This precursor has four possible fates: (1) transaminated back to aspartate, which is again used in the urea cycle; (2) used in the gluconeogenesis, (Sect. 1.4.1); (3) condensed with acetyl-CoA to yield citrate, which may then feed the TCA cycle or contribute to fatty acid synthesis (Sect. 1.5.2); and (4) converted into pyruvate, which also has several different fates accordingly to cellular metabolic active networks (Sect. 1.2.3).

1.6.2 Amino Acids in the Reprogramming of Cellular Metabolism and Function

Despite the diversity of amino acids, upon degradation the carbon skeletons are transformed into seven major metabolic intermediates: pyruvate, acetyl-CoA, oxaloacetate, fumarate, succinyl-CoA, acetoacetyl-CoA, and α-ketoglutarate. Ultimately, the carbon skeletons can be converted into glucose or enter the TCA cycle. In terms of characterization, amino acids that are degraded to acetyl-CoA and acetoacetyl-CoA are named ketogenic amino acids, since their catabolism may culminate with ketone bodies of fatty acid synthesis. Conversely, amino acids degraded to pyruvate, fumarate, oxaloacetate, and α-ketoglutarate are considered glucogenic amino acids as they can feed gluconeogenic processes. Leucine and lysine are purely ketogenic amino acids, while tryptophan, phenylalanine, tyrosine, and isoleucine can be both ketogenic and glucogenic amino acids. The remaining 14 amino acids are considered only glucogenic ones.

Associated with their metabolic profiles, some amino acids have been vastly studied regarding their role in cell biology and functioning. The best-known example is glutamine as described in 1935 by Hans Krebs. This amino acid has a vital role in organism homeostasis, as well as in energy-generating and biosynthetic purposes. Glutamine is transported to the intracellular milieu by several transporters, with SLC1A5 being the most studied one. Once inside the cell, glutamine is converted into glutamate by mitochondrial glutaminases in a process termed glutaminolysis. Glutamate may then follow one of two fates: it is either transformed into α-ketoglutarate by glutamate dehydrogenases providing a direct link with the TCA cycle, or it is used by aminotransferases as an amino group acceptor. The role of glutamine in cellular metabolic rewiring during activation or oncogenesis has been vastly explored, thus emphasizing the role of this amino acid in several biological processes (Altman et al. 2016). Tryptophan catabolism is an interesting example of how a single amino acid may in one hand increase the anaplerotic reactions of the TCA cycle and, on the other, culminate in the de novo production of NAD^+ (Mesquita et al. 2016). Tryptophan degradation may yield acetoacetate, which is then converted into acetyl-CoA that directly enters the TCA cycle or is converted into citrate. However, tryptophan may be degraded by indoleamine 2,3-dioxygenase (IDO) or tryptophan 2,3-dioxygenase (TDO) to feed the de novo biosynthesis of NAD^+ (see Chap. 4 for more details). Arginine also has a vital role in the regulation of cellular functions and in the generation of immunomodulatory molecules that dictate the inflammatory state of the cell. As previously mentioned, this amino acid may be cleaved by arginase to yield ornithine and urea. Ornithine is important for the synthesis of polyamines, which have been vastly associated with tissue repair and anti-inflammatory phenotypes. On the other hand, arginine may be metabolized by nitric oxide synthase originating from nitric oxide (NO), an important microbicidal molecule associated with host defense mechanisms (Bronte and Zanovello 2005). In this reaction, citrulline is also produced, thus providing a direct link with the urea cycle.

Although amino acids are clearly involved in the metabolic choices and consequent reprogramming of cellular metabolism, it is known that through their action as signaling molecules, amino acids impact cell growth and proliferation. These effects are mainly mediated by amino acid sensing through the sensors mammalian target of rapamycin (mTOR) and general control non-derepressible 2 kinase (GCN2) (Gallinetti et al. 2013; Wolfson and Sabatini 2017; for more details, see Chaps. 3 and 12). Both sensors are regulated by nutrient availability and, more specifically, by amino acid starvation. Upon activation, these sensors signal to an increased uptake or synthesis of amino acids and proteins, which will support cell growth and proliferation. Branched chain amino acids (leucine, isoleucine, and valine) are capable of constitutively activating mTOR, which consequently activates the eukaryotic initiation factor 4E-binding protein 1 and the downstream target p70 S6 kinase. The subsequent cascades activate the transcription of specific mRNAs and signal for an initiation of protein synthesis and an arrest of its degradation. GCN2 activation during amino acid scarcity culminates in the phosphorylation of eIF2 and the adoption of specific transcriptional programs. Furthermore, GCN2 may also be activated while mTOR is inhibited and amino acid uptake is decreased (Averous et al. 2016), thus showing a crosstalk between these metabolic sensors.

Despite the diversity of the existing amino acids, recent works have highlighted the role of some of them in the regulation of cellular metabolism and its implications during the development of adequate responses (Ananieva 2015; Arts et al. 2016; Grohmann and Bronte 2010; O'Neill et al. 2016). However, additional studies are required to understand if the modulation of these central nodes can be used in distinct pathological contexts.

1.7 Conclusion

Metabolism comprises the biochemical basis of any biological function and as such spans all biological disciplines. The interconnection of metabolic networks from the cellular to the organismal level is essential for homeostatic function. Several metabolic pathways are highlighted as being the central of carbon or energy metabolic networks and as such have been extensive characterized in their metabolic as well as their regulatory constituents. In this chapter, we went through these central pathways emphasizing their co-regulatory aspects, in a simplified view for easy access and understanding. Therefore, only major metabolic events/pathways were covered.

References

Akram M (2014) Citric acid cycle and role of its intermediates in metabolism. Cell Biochem Biophys 68:475–478

Altman BJ, Stine ZE, Dang CV (2016) From Krebs to clinic: glutamine metabolism to cancer therapy. Nat Rev Cancer 16:619–634

Ananieva E (2015) Targeting amino acid metabolism in cancer growth and anti-tumor immune response. World J Biol Chem 6:281

Arts RJW, Novakovic B, ter Horst R, Carvalho A, Bekkering S, Lachmandas E, Rodrigues F, Silvestre R, Cheng SC, Wang SY et al (2016) Glutaminolysis and fumarate accumulation integrate immunometabolic and epigenetic programs in trained immunity. Cell Metab 24:807–819

Averous J, Lambert-Langlais S, Mesclon F, Carraro V, Parry L, Jousse C, Bruhat A, Maurin A-C, Pierre P, Proud CG et al (2016) GCN2 contributes to mTORC1 inhibition by leucine deprivation through an ATF4 independent mechanism. Sci Rep 6:27698

Balaban RS (1990) Regulation of oxidative phosphorylation in the mammalian cell. Am J Physiol 258:377–389
Berg JM, Tymoczko JL, Stryer L (2002a) The glycolytic pathway is tightly controlled. In: Berg JM, Tymoczko JL, Stryer L (eds) Biochemistry. W.H. Freeman, New York
Berg JM, Tymoczko JL, Stryer L (2002b) Glycolysis is an energy-conversion pathway in many organisms. In: Berg JM, Tymoczko JL, Stryer L (eds) Biochemistry. W.H. Freeman, New York. section 16.1
Berg JM, Tymoczko JL, Stryer L (2002c) The metabolism of glucose-6-phosphate by the pentose phosphate pathway is coordinated with glycolysis. In: Berg JM, Tymoczko JL, Stryer L (eds) Biochemistry. W.H. Freeman, New York
Bonkovsky HL, Guo JT, Hou W, Li T, Narang T, Thapar M (2013) Porphyrin and heme metabolism and the porphyrias. Compr Physiol 3:365–401
Boulton RB (1996) Yeast and biochemistry of ethanol fermentation. In: Boulton RB, Singleton VL, Bisson LF, Kunkee RE (eds) Principles and practices of winemaking. Springer, Boston, MA, pp 102–192
Bronte V, Zanovello P (2005) Regulation of immune responses by L-arginine metabolism. Nat Rev Immunol 5:641–654
Buchakjian MR, Kornbluth S (2010) The engine driving the ship: metabolic steering of cell proliferation and death. Nat Rev Mol Cell Biol 11:715–727
Cheng S-C, Quintin J, Cramer RA, Shepardson KM, Saeed S, Kumar V, Giamarellos-Bourboulis EJ, Martens JHA, Rao NA, Aghajanirefah A et al (2014) mTOR- and HIF-1-mediated aerobic glycolysis as metabolic basis for trained immunity. Science 345:1250684
Citric THE, Cycle A (2010) The citric acid cycle. In: Eastmond, PJ Graham (eds) pp 601–630
Dan Dunn J, Alvarez LAJ, Zhang X, Soldati T (2015) Reactive oxygen species and mitochondria: a nexus of cellular homeostasis. Redox Biol 6:472–485
Déry MAC, Michaud MD, Richard DE (2005) Hypoxia-inducible factor 1: regulation by hypoxic and non-hypoxic activators. Int J Biochem Cell Biol 37:535–540
Gallinetti J, Harputlugil E, Mitchell JR (2013) Amino acid sensing in dietary-restriction-mediated longevity: roles of signal-transducing kinases GCN2 and TOR. Biochem J 449:1–10
Grohmann U, Bronte V (2010) Control of immune response by amino acid metabolism. Immunol Rev 236:243–264
Hansford R (2002) Oxidative phosphorylation. Encycl Life Sci:1–8
Hardie DG, Pan DA (2002) Regulation of fatty acid synthesis and oxidation by the AMP-activated protein kinase. Biochem Soc Trans 30:1064–1070
Houten SM, Wanders RJA (2010) A general introduction to the biochemistry of mitochondrial fatty acid β-oxidation. J Inherted Metab Dis 33:469–477
Israelsen WJ, Vander Heiden MG (2015) Pyruvate kinase: function, regulation and role in cancer. Semin Cell Dev Biol 43:43–51
Jiang B (2017) Aerobic glycolysis and high level of lactate in cancer metabolism and microenvironment. Genes Dis 4:25–27
Jones W, Bianchi K (2015) Aerobic glycolysis: beyond proliferation. Front Immunol 6:227
Knobloch M, Pilz GA, Ghesquière B, Kovacs WJ, Wegleiter T, Moore DL, Hruzova M, Zamboni N, Carmeliet P, Jessberger S (2017) A fatty acid oxidation-dependent metabolic shift regulates adult neural stem cell activity. Cell Rep 20:2144–2155
Kornberg H (2000) Krebs and his trinity of cycles. Nat Rev Mol Cell Biol 1:225–228
Lane AN, Fan TWM (2015) Regulation of mammalian nucleotide metabolism and biosynthesis. Nucleic Acids Res 43:2466–2485
Lee J, Ellis JM, Wolfgang MJ (2015) Adipose fatty acid oxidation is required for thermogenesis and potentiates oxidative stress-induced inflammation. Cell Rep 10:266–279
Lunt SY, Vander Heiden MG (2011) Aerobic glycolysis: meeting the metabolic requirements of cell proliferation. Annu Rev Cell Dev Biol 27:441–464
Mesquita I, Varela P, Belinha A, Gaifem J, Laforge M, Vergnes B, Estaquier J, Silvestre R (2016) Exploring NAD^+ metabolism in host-pathogen interactions. Cell Mol Life Sci 73:1225–1236

Mueckler M, Thorens B (2013) The SLC2 (GLUT) family of membrane transporters. Mol Aspects Med 34:121–138

Nath S, Villadsen J (2015) Oxidative phosphorylation revisited. Biotechnol Bioeng 112:429–437

O'Neill HM, Holloway GP, Steinberg GR (2013) AMPK regulation of fatty acid metabolism and mitochondrial biogenesis: implications for obesity. Mol Cell Endocrinol 366:135–151

O'Neill LAJ, Kishton RJ, Rathmell J (2016) A guide to immunometabolism for immunologists. Nat Rev Immunol 16:553–565

Ott M, Gogvadze V, Orrenius S, Zhivotovsky B (2007) Mitochondria, oxidative stress and cell death. Apoptosis 12:913–922

Patra KC, Hay N (2014) The pentose phosphate pathway and cancer. Trends Biochem Sci 39:347–354

Romero-Garcia S, Moreno-Altamirano MMB, Prado-Garcia H, Sánchez-García FJ (2016) Lactate contribution to the tumor microenvironment: mechanisms, effects on immune cells and therapeutic relevance. Front Immunol 7:52

Senior AE, Nadanaciva S, Weber J (2002) The molecular mechanism of ATP synthesis by F1F0-ATP synthase. Biochim Biophys Acta – Bioenerg 1553:188–211

Thorens B, Mueckler M (2010) Glucose transporters in the 21st century. Am J Physiol Endocrinol Metab 298:E141–E145

Wolfson RL, Sabatini DM (2017) The Dawn of the age of amino acid sensors for the mTORC1 pathway. Cell Metab 26:301–309

Yang W, Lu Z (2013) Regulation and function of pyruvate kinase M2 in cancer. Cancer Lett 339:153–158

Yao CH, Fowle-Grider R, Mahieu NG, Liu GY, Chen YJ, Wang R, Singh M, Potter GS, Gross RW, Schaefer J et al (2016) Exogenous fatty acids are the preferred source of membrane lipids in proliferating fibroblasts. Cell Chem Biol 23:483–493

Chapter 2
Interaction Between Nutrition and Metabolism

Stella Maria Barrouin-Melo, Yadira Alejandra Morejón Terán, Johanna Anturaniemi, and Anna Katrina Hielm-Björkman

Abstract Diet plays a fundamental role in the nutritional status, in the homeostasis and in the capacity of an individual to adapt to the environment. A proper or an inadequate nutrition has an impact on the persistence, remission and incidence of various conditions, including the infectious diseases. Consequently, nutrition has a crucial importance on survival rates and health recovery of individuals or even populations around the globe. The synergistic relationship between nutritional needs and infectious processes has been demonstrated conclusively in diverse studies. This chapter will discuss the most important nutrients, their most common natural dietary sources, the different digestive processes for each one as well as the absorption, transport, storage, excretion and function of each of the nutrients within the organism. We also go through some concepts on the interaction between nutrition and the immune system, as well as examples on the influence of nutrition or specific nutrients on some infectious diseases, and their influence on the gene expression.

Keywords Metabolism · Infection · Disease · Nutrient · Immunonutrition · Nutrigenetic

S. M. Barrouin-Melo (✉)
Department of Anatomy, Pathology and Clinics, School of Veterinary Medicine and Zootechny, Federal University of Bahia, Salvador, Bahia, Brazil
e-mail: barrouin@ufba.br

Y. A. Morejón Terán
Institute of Collective Health, Federal University of Bahia, Salvador, Bahia, Brazil

J. Anturaniemi · A. K. Hielm-Björkman
Department of Equine and Small Animal Medicine, Faculty of Veterinary Medicine, University of Helsinki, Helsinki, Finland

R. Silvestre, E. Torrado (eds.), *Metabolic Interaction in Infection*, Experientia Supplementum 109, https://doi.org/10.1007/978-3-319-74932-7_2

2.1 Introduction

The World Health Organization defines *nutrition* as the ingestion of food in relation to the dietary needs of the body (WHO 2017). *Metabolism* is a combination of physical processes and chemical reactions that take place in the body to obtain energy and use it to synthetize, process, transform and eliminate substances, starting from the ingestion of nutrients, in order to maintain life (Smith and Morowitz 2004; Gil Hernández and Sánchez de Medina Contreras 2010; Lee 2013). The current concept of *infectious disease* is not only restricted to microbe invasion and host reaction but rather an orchestration of factors that include elements such as pathogen virulence, infectious burden, route of inoculation and the host's susceptibility to infection and previous pharmacological treatment (Traore et al. 2016). Susceptibility to the installation and development of an infection is intrinsically related to the host's immunological capacity, whose dependence on a proper nutrition has been extensively demonstrated (Cunningham-Rundles et al. 2005; Krawinkel 2012).

During an infection, apart from keeping the host's body structure and function, nutrition assumes an additional function. It provides the immune system with elements for pathogen neutralization and elimination while repairing any resulting tissue damage. Nutrition and the consequent metabolic processes should accomplish specific goals. Nutrition must sustain an effective immune response. It must provide an environment for an adequate protective and efficient self-limiting and self-resolving inflammatory response, without additional cell and tissue damage. It also must offer elements to facilitate the body in the handling and removal of catabolites and/or by-products resulting from the immune response itself, tissue injury or malfunction as well as medications given during disease treatment. Moreover, proper nutrition will enable quick and precise tissue regeneration without feeding or protecting the pathogen within the host's body or without helping it escape the host immune system.

This chapter will explore some basic concepts and current knowledge of nutrition and metabolism in the context of diseases caused by several pathogen agents.

2.2 Aspects of Nutritional Biochemistry

Balanced nutrition plays a special role in maintaining the health of an individual. The body uses nutrients from food to produce energy, maintain or repair body structures and regulate or modulate metabolism. In turn, every disease has a metabolic component that can lead to a depletion of reserves and the aggravation of the clinical condition. It is fundamental to understand the processes of digestion, absorption, transport and metabolism of each of the nutrients from natural sources and their most important functions within the body.

2.2.1 Carbohydrates

Carbohydrates are sources of energy for the body. The main dietary carbohydrates are the polysaccharides, oligosaccharides, disaccharides and monosaccharides (FAO/WHO 1998; Ochoa et al. 2014). Natural dietary sources of polysaccharides are amides of vegetal origin (amylose and amylopectin) and glycogen of animal origin (McCance and Lawrence 1929; Lovegrove et al. 2017). Sucrose and maltose are disaccharides of vegetal origin, while lactose is a disaccharide of animal origin. Fructose is the only natural monosaccharide that comes from dietary sources. Most monosaccharides are not free molecules, but they exist as basic components of disaccharides and polysaccharides (Cummings and Stephen 2007). The hexoses are the most important monosaccharides and include glucose, galactose and fructose (Berg et al. 2002a; Lee 2013). Amide digestion starts in the mouth with salivary amylase and continues in the intestine by the action of the pancreatic α-amylase (Lee 2013). Within the body, the end products of the digestion of dietary carbohydrates are hexoses and pentoses, which are rapidly absorbed by the intestine (Norton et al. 2015). Glucose molecules require specific transport mechanisms to enter the intestinal cell. Glucose uptake can occur by *facilitated diffusion*, which is a passive process that involves glucose transporters GLUT-1 to GLUT-5, a family of proteins present in cell membranes that bind and carry glucose into the cell (Carruthers et al. 2009; Mueckler and Thorens 2013). A secondary active transport requires energy by hydrolysis of ATP and consists of a *sodium-coupled glucose transporter* (SGLT) against a concentration gradient, from a low concentration outside the cell to a high concentration inside (Wright et al. 2011).

Monosaccharides pass into the portal vein and go directly into the liver, where they are oxidized to produce energy and are stored as glycogen, or pass into the general circulation for use by other tissues. Once glucose enters the hepatic cell with the help of insulin, it is phosphorylated into glucose 6-phosphate, which may follow one of the pathways depending on metabolic needs, namely, glycogenesis, glycogenolysis, glycolysis and gluconeogenesis as described in Chap. 1.

2.2.2 Lipids

Animal and plant triglycerides, membrane phospholipids and sterols comprise the principal dietary lipids for humans and other animals. *Triglycerides* are composed of three molecules of fatty acids bound to one molecule of glycerol. *Phospholipids* are composed of phosphoric acid, fatty acids and a nitrogen base. *Sterols* include cholesterol and Vit D. Fatty acids are categorized according to the number of carbon atoms in short-chain (2–4 carbon atoms), medium-chain (8–14 carbon atoms) or long-chain (16–20 or more carbon atoms) fatty acids. Considering the presence of carbon-carbon double bonds, fatty acids that have no double bond are categorized as saturated, while unsaturated fatty acids are those that have one or more double

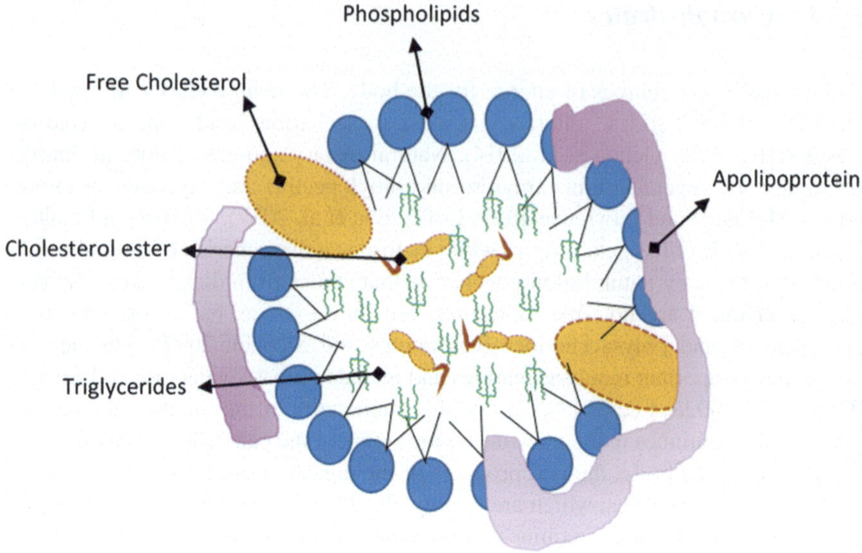

Fig. 2.1 Representation of the structure of a chylomicron, showing its nucleus of triglycerides and cholesterol esters, and the outer layer of apolipoproteins, cholesterol and phospholipids

bonds. The fatty acids obtained from land animals tend to be saturated, whereas the fatty acids of fish and plants are often polyunsaturated and therefore present as oils (Akoh and Min 2008; Berdanier and Zempleni 2009; Lee 2013).

Lipid digestion begins with lipase in gastric juice, but most fat digestion takes place in the duodenum, where fats are first emulsified by bile salts for optimal activity of lipases. Pancreatic lipase hydrolyses the triglyceride into mono- and diglycerides and free fatty acids, composing the digestion products, which together with liposoluble vitamins and cholesterol form micelles. Fatty acids containing less than 8–12 carbon atoms, thus being more soluble, pass through the intestinal mucosa and are transported linked by albumin directly into the portal vein, while the ones with more than 8–12 carbon atoms will form the micelles (Berdanier and Zempleni 2009; Lindquist and Hernell 2010; Lee 2013). Enterocytes take up the micellar compounds and resynthesize them into triglycerides that are packaged into chylomicrons, which are drained by lymphatic capillaries into the thoracic duct, and then discharged directly into the general circulation (Swift et al. 1990; Akoh and Min 2008). Chylomicrons are composed of a nucleus of triglycerides and cholesterol esters and an outer layer of lipoproteins, cholesterol and phospholipids (Fig. 2.1) (Akoh and Min 2008; Lindquist and Hernell 2010; Lee 2013). The chylomicrons circulate in blood vessels throughout the body to diverse tissues, such as the liver, adipose tissue and muscles. In the adipose tissue, the lipoprotein lipase (LPL) on the capillaries' endothelium partially digests the chylomicrons into free fatty acids, glycerol and chylomicron remnants (Lee 2013; Julve et al. 2016; Geldenhuys et al. 2017). Chylomicron remnants are lipoproteins rich in cholesterol and together with chylomicrons compose the transporting system for exogenous dietary lipids (Julve et al. 2016). Adipocytes take up free fatty acids, but not glycerol and chylomicron remnants, which are removed from the circulation and metabolized by the liver. Inside adipocytes, fatty acids are resynthesized into triglycerides, in a process that uses glycerol derived from glucose in the glycolytic pathway, to be stored as a source for metabolic energy and released on demand by other tissues (Akoh and Min 2008). Lipoproteins, including very low-density (VLDL), low-density (LDL) and high-density (HDL) lipoproteins, are the transport system for endogenously synthesized lipids (Beisiegel 1998; Rosa et al. 2015).

Free fatty acids transported by albumin are the main energy source for various organs. They are extensively used by the heart, the brain and other tissues to oxidize fatty acids which in the presence of O_2 are then catabolized into CO_2 and H_2O, producing energy. Nearly 40% of this energy is stored as ATP, and the remaining is released as heat (Frayn et al. 2006; Lee 2013). In the liver, the glycerol that is released by lipase is phosphorylated by glycerol kinase, and the resulting glycerol 3-phosphate is oxidized to dihydroxyacetone phosphate (DHAP) which participates in glycolysis among other metabolic pathways (Hagopian et al. 2008). The glycolytic enzyme triose phosphate isomerase converts DHAP to glyceraldehyde 3-phosphate, which is oxidized via glycolysis or converted to glucose via gluconeogenesis (Nye et al. 2008). After releasing triglycerides into the adipose tissue, the chylomicron remnants carry the cholesterol to the liver, where it becomes part of biliary fluid or is incorporated into VLDL (Julve et al. 2016).

2.2.3 *Proteins*

Proteins are the most important source of structural elements in the body, and they are distributed mainly in the muscles (40%), followed by the blood and skin (30%), and in the liver and intestine (10%) (Webb 1990; Gil Hernández and Sánchez de Medina Contreras 2010). Good protein sources include foods from animal origin such as red meat, poultry, eggs, milk and milk products and fish and seafood. Vegetal sources include nuts, seeds and legumes (Berdanier and Zempleni 2009). Protein digestion begins in the stomach where pepsins hydrolyse peptide bonds at the junctions between the aromatic amino acids—phenylalanine (Phe), tyrosine (Tyr), diiodotyrosine (Dit) and thyroxine (Thy) (Dabrowski 1983; Lee 2013; Liu et al. 2015). Some amino acids are released into the intestinal lumen, but others are digested on the surface of the cells of the intestinal mucosa by the aminopeptidases and dipeptidases from the villi. Some dipeptides and tripeptides are actively transported into intestinal cells and hydrolysed by intracellular peptidases. Final amino acid digestion therefore occurs at three sites: the intestinal lumen, the villi and the cytoplasm of the mucosal cells (Matthews 1975). Nucleic acids are fragmented into nucleotides within the intestine by pancreatic nucleases, and these nucleotides are themselves fragmented into nucleosides and phosphoric acid. Absorption of the amino acids is rapid in the duodenum and jejunum, but slow in the ileum.

Approximately 50% of digested proteins come from ingested foods, 25% from proteins of digestive juices and 25% from desquamated mucous cells. Some of the ingested proteins enter the colon and are finally digested by the action of intestinal bacteria (Matthews 1975; Webb 1990; Lee 2013).

Amino acids are categorized as essential when they must be provided by the diet, as the body cannot synthesize them. Essential amino acids include L-isoleucine, L-leucine, L-lysine, L-methionine, L-phenylalanine, L-threonine, L-tryptophan and L-valine. As infants are not fully capable of synthesizing L-arginine and L-histidine, these amino acids are also considered essential during early stages of body development (Fürst and Stehle 2004). On the other hand, L-alanine, L-aspartic acid, L-asparagine, L-glutamic acid, L-glycine, L-proline and L-serine, if not ingested in the diet in sufficient amounts, the body can synthesize them; they are therefore considered non-essential amino acids. They are also categorized as conditionally essential amino acids when their synthesis is reduced under certain pathophysiological conditions, such as prematurity in the infant or hypermetabolism or hypercatabolism. Conditionally essential amino acids are L-taurine, L-cystine, L-tyrosine, L-proline, L-serine and L-glycine (Reeds 2000; Obled et al. 2002; Lee 2013).

Amino acids are never stored in the body as proteins, and there is no reserve of free amino acids in the body; therefore they are synthesized de novo to satisfy demand (Lee 2013). The amino acid pool represents the amount of amino acids of dietary origin available for protein synthesis during a given time in the plasma and is adsorbed in cell surfaces after a meal (Picó et al. 1991; Proenza et al. 1994). During amino acid catabolism in the glycolytic pathways, the amino group is released in a process called deamination and is transformed mainly into ammonia in the liver. Most of the NH_3 formed is converted into urea and is secreted in the urine (Watford 2003).

Following the digestion of dietary nucleic acids, their purine and pyrimidine constituents are absorbed and metabolized, but most purines and pyrimidines in the body are synthesized from amino acids, especially in the liver (Liu et al. 2015). Purines or pyrimidines are combined with ribose to form nucleosides, which are components of a variety of coenzymes and related substances, such as uridine diphosphate glucose (UDPG), NAD, NADP and ATP. They also constitute ribonucleic acid (RNA) and deoxyribonucleic acid (DNA). The pyrimidines are catabolized to CO_2 and NH_3, and the purines are converted into uric acid, which is filtered into the kidneys, 90% of which is reabsorbed and 10% excreted (Moffatt and Ashihara 2002; Lee 2013; Maiuolo et al. 2016).

2.2.4 *Vitamins*

Vitamins (Vit) are organic components that cannot be synthesized by the body, and therefore they must come from the diet. The only exception is Vit D, which is synthesized by the skin (Zempleni 2007). Vits are not a source of energy, but they participate in the metabolism of macronutrients for energy production, usage and storage, as their most important function is to regulate and modulate metabolism. Vits are categorized according to the medium in which they are soluble. Vits A, D, E and K are liposoluble and stored in the liver with slow absorption and excretion processes. In excess, particularly Vit A and Vit D may become toxic. Hydrosoluble Vits include the ones in Group B and Vit C (Zempleni 2007), which are regularly required in the diet as they are not stored in large quantities, and generally any excess is non-toxic (Lee 2013).

2.2.4.1 Vitamin A: Retinol

Vit A is essential for all vertebrates and research on it spans about nine decades (Ross and Harrison 2007; Ross 2010). Vit A denomination includes a group of liposoluble compounds that exert retinol activity and exist as retinol, retinal and retinoic acid in dietary sources from animal origin, such as the liver, fish liver oils, milk and dairy, and eggs. Plant sources provide pro-Vit A carotenoids, found in green and yellow vegetables, such as carrots, sweet potatoes, pumpkin, kale, spinach, collards and squash (Tanumihardjo 2011; Lobo et al. 2012; Tanumihardjo et al. 2016). Animal products provide preformed Vit A in the form of an ester, called retinyl palmitate, while vegetables provide provitamin A mostly as β-carotene (Ross and Harrison 2007; Ross 2010). Hydrolysis of retinyl palmitate to become retinol requires at least two hydrolase enzymes, the pancreatic hydrolase retinyl ester in the proximal intestine and hydrolase on the surface of intestinal villi cells (Institute of Medicine 2001a; Ross and Harrison 2007). Free retinol then binds with the long-chain fatty acids, forming retinyl ester which is absorbed (Reboul 2013). Dietary β-carotene is absorbed by passive diffusion in the intestine and cleaved into two molecules of retinaldehyde within the cytoplasm of the cells of the mucosa (Hollander and Ruble 1978; Reboul 2013). Absorption of vegetable carotenoids in the small intestine is lower than that of animal previtamin A (Reboul 2013; Melse-Boonstra et al. 2017). In the digestive system, there is an efficient specific transporter system for retinol, whereas carotenoids are absorbed via non-specific transporters (Berdanier and Zempleni 2009). Retinaldehyde is esterified to form retinal and then becomes retinol, which couples with chylomicrons to be transported by the lymph into the blood and then to the liver (Nayak et al. 2001) where it is metabolized and stored as retinyl esters within stellate cells (Institute of Medicine 2001a; Ross and Harrison 2007). Lipoproteins carry carotenoids and retinyl esters to the fat portion of tissues that also store some Vit A. On the other hand, retinol, retinal and retinoic acid combine with specific retinoid-binding proteins within cells and in the plasma (Ross and Harrison 2007; Berdanier and Zempleni 2009). Studies have shown that more Vit A is synthesized from dietary β-carotene when subjects co-ingest β-carotene with Vit A, suggesting that retinoid status can influence carotenoid status (Ross 2006; Zempleni 2007). Either retinol or retinal can be converted to retinoic acid, but reverse reactions are not possible; moreover retinol and retinal are interconvertible only in the eye retina (Institute of Medicine 2001a). In the tissues, retinol is used either as retinol, retinal, or converted to retinoic acid, which is used faster than retinol (Ross 2010). Studies have shown that retinoic acid metabolites are recovered from expired air, but the structures of all the metabolites excreted in the urine and faeces are not known (Institute of Medicine 2001a; Ross 2006; Berdanier and Zempleni 2009; Bayer 2013). Excess Vit A intake is associated with pathological accumulation in the liver, inducing oxidative stress in the mitochondria (Lobo et al. 2012; de Oliveira 2015), and many clinical signs (Smith and Goodman 1976; Ross 2010). Retinal is the Vit A derivative that is the most toxic, due to its chemical reactivity, randomly modifying proteins through Schiff base formation (Zhong et al. 2012). Hypervitaminosis A appears to be due to abnormal transport and distribution of Vit A and retinoids caused by overloading of the plasma transport mechanisms (Smith and Goodman 1976; Zhong et al. 2012). Long-term ingestion of $\geq$10 times the recommended dietary allowance of Vit A is associated with toxicity (Hathcock et al. 1990; Penniston and Tanumihardjo 2006). There is no treatment or antidote for hypervitaminosis A, and its management is based on intake interruption (Penniston and Tanumihardjo 2006).

Vit A participates in many metabolic processes which are fundamental in the synthesis of protein for body development, the formation of epithelial cells, vision and the functioning of the immune system (Gerster 1997; McLaren and Kraemer 2012). In the eye, retinal combines with opsin to form rhodopsin, the molecule responsible for photoreception (Lee 2013). Retinoic acid is required to maintain normal gene expression and tissue differentiation (Ross and Harrison 2007; Chen et al. 2008; Aoto et al. 2008). Retinoic acid is an irreversibly oxidized form of retinol, which is an important hormone-like growth factor for epithelial and immune cells (Ross and Ternus 1993; Reichrath et al. 2007). Retinoid-dependent processes are required for the expression of many proteins of the extracellular matrix, such as collagen, laminin and proteoglycans (Lobo et al. 2012). In the brain, the synthesis of the calcium-binding protein, calbindin, is regulated by retinoic acid and not by Vit D as it is in other tissues, such as the intestine and kidneys (Wang and Christakos 1995; Berdanier and Zempleni 2009). Vit A also interacts with the metabolism of zinc and iron and is required for the synthesis of haemoglobin and erythropoiesis (Roodenburg et al. 2000). Retinal and retinol participate in many enzymatic reactions related to lipid metabolism, carbohydrate metabolism, protein metabolism and hormonal function (Chen and Chen 2014). Carotenoids and Vit A act as antioxidants, protecting cells against the effects of oxidants generated in aerobic metabolism or oxidative stress (Sies 1993; Valko et al. 2007; Omur et al. 2016).

2.2.4.2 Vitamin B1: Thiamine

Thiamine acts as a coenzyme, which means that it is required so that enzymes can perform normal physiological actions. Dietary animal sources of thiamine (also called thiamin or aneurin) include the liver, lean pork and dairy products; vegetable sources include wholegrains, yeast and legumes. Thiamine is absorbed by active transport in the proximal duodenum when dietary amounts are scarce, while passive diffusion occurs when excessive amounts are available in the diet. Uptake of Vit B1 depends on phosphorylation in both the intestinal lumen by the intestinal phosphatases and within cells by the enzyme thiamine pyrophosphokinase, which acts as a carrier (Lonsdale 2006; Berdanier and Zempleni 2009; Manzetti et al. 2014). Thiamine is delivered to the liver through the portal circulation bound to proteins, mainly albumin (Lonsdale 2006; Berdanier and Zempleni 2009; Makarchikov 2009). In the liver, thiamine is phosphorylated to thiamine phosphate derivatives that perform functions such as coenzymes; phosphorylation also takes place in other tissues, but to a lesser extent (Makarchikov 2009). The known natural thiamine phosphate derivatives are thiamine monophosphate; thiamine diphosphate, also called thiamine pyrophosphate (TPP) or cocarboxylase; thiamine triphosphate; and the recently discovered adenosine thiamine triphosphate and adenosine thiamine diphosphate (Lee 2013). The best-characterized form is TPP, which acts as a coenzyme in the catabolism of sugars and amino acids (Bettendorff and Wins 2013). Thiamine phosphate derivatives participate in many cellular processes, and their degradation takes place within the various biochemical cycles in the cells; after which it is excreted in the urine.

In the mitochondria, TPP is a coenzyme for pyruvate dehydrogenase (PDH), one of the key factors in carbohydrate metabolism in biochemical pathways that result in the generation of adenosine triphosphate (ATP), which is a major form of energy for the cell (Thurnham 2005). Thiamine is essential for the metabolism of carbohydrates; in its absence, carbohydrate metabolism is the first to deteriorate (Thurnham 2005; Lonsdale 2006); and therefore animals and humans must obtain Vit B1 from the diet (Thurnham 2005; Lee 2013). In the nervous system, PDH is also involved in the production of acetylcholine, a neurotransmitter, and for myelin synthesis, and insufficient intake of thiamine is associated with diverse neural conditions, such as degenerative polyneuropathies, Wernicke-Korsakoff syndrome or convulsions due to increased intracranial pressure (Bitsch 2003).

2.2.4.3 Vitamin B2: Riboflavin

The best dietary sources of Vit B2 are animal-derived foods, such as milk, meat and eggs, while the richest vegetable sources include green vegetables, mushrooms, legumes or almonds (Halsted 2003; Rivlin 2007; Berdanier and Zempleni 2009; Lee 2013). Dietary riboflavin is naturally present in free forms or along with flavoproteins, which are proteins that contain a nucleic acid derivative of riboflavin: the flavin adenine dinucleotide (FAD) or flavin mononucleotide (FMN), which is a FAD precursor (McCormick 2003; Halsted 2003; Lee 2013). Riboflavin is actively absorbed in the proximal small intestine, after being dephosphorylated by the action of hydrolases from the villi membrane cells, by a pyrophosphatase that cleaves FMN and FAD and by an alkaline phosphatase that liberates the vitamin from its coenzyme form (Berdanier and Zempleni 2009). In humans, absorption is also regulated by availability and becomes active when dietary amounts are scarce (McCormick 2003). Following absorption, the free Vit B2 is carried to the liver in the bloodstream of the portal system bounded to albumin (Bayer 2013), but riboflavin has more affinity with immunoglobulins (Gil Hernández and Sánchez de Medina Contreras 2010). In the cell, riboflavin enters as an important component of the intermediate metabolism in reactions that involve oxidation-reduction (McCormick 2003). Almost all tissues have the capacity to reconvert free riboflavin to FMN and FAD, which is particularly abundant in the liver, kidneys and myocardium, where flavoproteins are mainly located in the mitochondria because of their redox power (Lee 2013). Free riboflavin can be glycosylated, oxidized, demethylated and hydroxylated in the liver and excreted in the urine as a glycosylated metabolite or as free riboflavin. Vit B2 is not stored in the body for further uptake on demand, so it should come from the diet (Halsted 2003; Rivlin 2007; Berdanier and Zempleni 2009; Lee 2013).

Riboflavin has important functions as an antioxidant. Flavin adenine dinucleotide and flavin mononucleotide constitute the cofactors for flavoprotein enzymes, which catalyse oxidation-reduction reactions in cells and act as hydrogen transporters in the mitochondrial electron transport system (Walsh et al. 1978; McCormick 2003; Hühner et al. 2015). FAD has a more positive reduction potential than NAD+ and is a very strong oxidizing agent (Berg et al. 2002b; Mansoorabadi et al. 2007). FAD-linked proteins act in many metabolic pathways to execute functions such as DNA repair, nucleotide biosynthesis and synthesis of other cofactors such as coenzyme-A and coenzyme-Q (CoA, CoQ) and heme groups (Mansoorabadi et al. 2007). Other metabolic processes that require FAD include (1) the Krebs cycle (Berg et al. 2002b), (2) reduction of the oxidized form of glutathione (GSSG) to its reduced form (GSH) by glutathione reductase, (3) reduction of ubiquinone to ubiquinol (Xia et al. 2001; Berdanier and Zempleni 2009), (4) fatty acid oxidation by fatty acyl-CoA dehydrogenase (Powers 2003), (5) triglyceride synthesis by glycerol-3-phosphate dehydrogenase (Rivlin 2007) and (6) purine nucleotide catabolism by xanthine oxidase (McCormick 2003; Maiuolo et al. 2016). FAD also has important functions in the metabolism of other vitamins, such as the conversion of the amino acid tryptophan to niacin (Vit B3), production of 4-pyridoxic acid from pyridoxal (Vit B6), conversion of retinol (Vit A) to retinoic acid via cytosolic retinal dehydrogenase and synthesis of an active form of folic acid (McCormick 2003). Redox flavoproteins also act by non-covalent binding to FAD, for example, acetyl-CoA dehydrogenases, which function in the β-oxidation of fatty acids and in the catabolism of essential amino acids such as L-leucine, L-isoleucine, L-valine and L-lysine (Berdanier and Zempleni 2009).

2.2.4.4 Vitamin B3: Niacin

Niacin is the generic term for the two related and fast interconvertible chemical types, namely, nicotinamide (NAM) and nicotinic acid (NA). Niacin participates in metabolic cycles as a coenzyme to release energy from nutrients. Sources are wholegrain cereals and breads, milk, eggs, meats, liver, salmon, poultry, dry beans, dried fruit and green leafy vegetables (Institute of Medicine 1998a; Bender 2003a; Berdanier and Zempleni 2009; Lee 2013). In plants, niacin may be bound to macronutrients, such as peptides, hexoses and pentose. In animal sources, it comes in the form of NA and NAM. Dietary niacin of plant origin, particularly from dry grains, comes covalently bounded to small peptides and carbohydrates which are not released during digestion by humans; therefore they need to be previously exposed to alkaline treatment or boiling (Bates 1998; Institute of Medicine 1998a; Lee 2013). Niacin is found in foods of animal origin in the form of coenzymes NADH (nicotinamide adenine dinucleotide) and NADPH (nicotinamide dinucleotide phosphate), which are digested to release NA and NAM. In the intestine, the resulting NA or its hydrolysis product NAM crosses the intestinal cells by simple diffusion when abundant, or by a transporting molecule when scarce (Bechgaard and Jespersen 1977; Henderson and Gross 1979; Bates 1998). In the plasma, both molecules circulate freely and enter the cell tissues by passive diffusion. Within cells, NA and NAM can be resynthesized for NADH and NADPH from quinolinic acid, a metabolite of the amino acid tryptophan synthesis pathway. Niacin diverse metabolites are excreted in the urine (Institute of Medicine 1998a; Bender 2003a).

Like the mechanisms of Vit B2, Vit B3 acts as a precursor, and the resulting NADH and NADPH coenzymes have essential functions in more than 200 enzymes involved in the metabolism of carbohydrates, fatty acids and amino acids (Berdanier and Zempleni 2009; Bayer 2013). NADH and NADPH act as electric transporters because they serve as hydrogen receptors and perform different functions in the metabolism of the cell. Vit B3 has critical roles in the maintenance of the redox state of the cell (Institute of Medicine 1998a; Kirkland 2007). NADH functions with mitochondrial enzymes of the respiratory chains, while NADPH acts with cytosolic enzymes in mechanisms in which NADPH generated in the oxidation of different substrates is used in different biosynthetic processes. Examples of NADH/NADPH functions include fatty acid, cholesterol or steroid hormone synthesis, β-oxidation of fatty acids, amino acid deamination, oxidative decarboxylation and biosynthesis of polyalcohols, ethanol metabolism and some participation in diverse metabolic pathways such as the glycolytic pathways, Krebs cycle, respiratory chain or pentose phosphate pathway (Bender 2003a; Depeint et al. 2006; Berdanier and Zempleni 2009). The adipose tissue, spleen, immune cells and keratinocytes express high levels of protein G-coupled niacin receptors, which act by inhibiting cyclic adenosine monophosphate (cAMP) production and thus fat breakdown in adipose tissue, resulting in a decrease in VLDL and LDL (Gille et al. 2008; Wanders and Judd 2011). On the other hand, niacin increases the apolipoprotein A1 (apoA-1) and inhibits HDL uptake in the liver by downregulating the cholesterol ester transfer protein, which results in increasing HDL levels (Kamanna and Kashyap 2008; Berdanier and Zempleni 2009).

2.2.4.5 Vitamin B5: Pantothenic Acid

Humans and animals need to obtain pantothenic acid from the diet to synthesize CoA and to synthesize and metabolize carbohydrates, proteins and lipids. Vit B5 is present in small amounts in most foods. The main sources are organ meats, eggs, fish, mushrooms, avocados, broccoli and wholegrains. In foods, pantothenic acid is present mainly in the form of CoA and acyl carrier protein (ACP) that needs to be converted into free pantothenic acid to be absorbed by intestinal cells. The process requires hydrolysis of CoA and ACP into 4′-phosphopantetheine, which is dephosphorylated into pantetheine which undergoes hydrolysis by pantetheinase into free pantothenic acid (Bates 2005a; Trumbo 2006; Lee 2013). Free pantothenic acid is absorbed in the jejunum into intestinal cells by an active transport system that involves sodium and is saturable, which means that at high intake the absorption rate decreases (Institute of Medicine 1998b; Berdanier and Zempleni 2009). It is then transported in the form of free acid dissolved in the plasma and is captured by diffusion into the interior of the erythrocytes. Within the cell, Vit B5 becomes CoA, which is the predominant form in most tissues, particularly in the liver, adrenal glands, kidneys, brain, heart and testicles. Unlike the other B complex vitamers, pantothenic acid is not catabolized in the liver; it is excreted in its active form in the urine and faeces when in excess (Institute of Medicine 1998b; Bates 2005a; Berdanier and Zempleni 2009).

The main roles of Vit B5 are to act as a coenzyme in the metabolism of fatty acid and energy production. CoA binds to fatty acids in various reactions such as the synthesis of triglycerides, complex lipids, porphyrins and cholesterol or the catabolism of lipids and generation of ketone bodies. CoA also participates in other metabolic pathways in the synthesis and metabolism of carbohydrates and proteins in reactions such as acylation, fatty acid and pyruvate oxidation and the acetylation of choline. ACP groups act exclusively in the metabolism of fatty acid, where it is part of the synthetase multienzymatic complex (Institute of Medicine 1998b; Bates 2005a; Berdanier and Zempleni 2009; Lee 2013).

2.2.4.6 Vitamin B6: Pyridoxine

Vit B6 vitamers are also interconvertible chemical compounds in biological systems, similar to Vits B2 and B3. Vit B6 sources include meats, cereals, legumes, nuts, fruits and vegetables, as an aldehyde (pyridoxal), an alcohol (pyridoxine) or an amine (pyridoxamine). Most Vit B6 in foods of animal origin are in pyridoxal phosphate (PLP) or pyridoxamine phosphate forms, while plant foods contain mostly pyridoxine, which is more stable in cooking and other food processing (Merrill and Henderson 1990; Institute of Medicine 1998b; Berdanier and Zempleni 2009; Lee 2013). Depending on the ingested form, if free, glycosylated or bound with proteins, Vit B6 undergoes dephosphorylation by an alkaline phosphatase bound to intestinal cells (Merrill and Henderson 1990; Bender 2005). The vitamin is efficiently absorbed by passive diffusion in the jejunum and ileum after being converted into the active coenzyme form 5′-PLP by a pyridoxal kinase present in the cytoplasm of the cells (Merrill and Henderson 1990; Institute of Medicine 1998b; Berdanier and Zempleni 2009; Lee 2013). In the plasma, PLP circulates bound to albumin and haemoglobin to be distributed to the tissues, where it functions mainly in the metabolism of amino acids (Ink and Henderson 1984; Bender 2005; Lee 2013). In the liver, PLP is dephosphorylated and oxidized by the FAD and NAD enzymes in reactions that result in 4-pyridoxic acid and other inactive metabolites, which are excreted in the urine. A small amount of pyridoxine is also excreted in the faeces (Ink and Henderson 1984; Berdanier and Zempleni 2009).

Vit B6's active form PLP acts as a coenzyme in about 100 enzyme reactions in amino acid, as well as in glucose and lipid metabolism (Dakshinamurti and Dakshinamurti 2007). The majority of reactions involving PLP are transaminations, which are fundamental in the metabolism of amino acids, cell growth and cell division (Lichstein et al. 1945; Dakshinamurti and Dakshinamurti 2007). PLP is required as a coenzyme of glycogen phosphorylase, the enzyme that catalyses the release of hepatic and muscle glycogen as glucose 1-phosphate (Kohlmeier 2003; Bender 2005). It is fundamental for the metabolism of sphingolipids (Bourquin et al. 2011). Pyridoxine participates as a coenzyme in many reactions including racemization, elimination, replacement, decarboxylation and beta-group interconversion, which are required in pathways of macronutrient metabolism; as well, Vit B6 is required for gene expression and for the synthesis of haemoglobin, serotonin, histamine and niacin (Bender 2003b; Kohlmeier 2003; Dakshinamurti and Dakshinamurti 2007). PLP is also necessary for the absorption of other nutrients, such as Vit B12 and selenium (Institute of Medicine 1998b)

2.2.4.7 Vitamin B7: Biotin

Among all B complex vitamins that act as coenzymes, biotin is the only one that carries out its functions without the need for structural conversions (Gil Hernández and Sánchez de Medina Contreras 2010). Sources of biotin include brewer's yeast, organ meats, butter, some green legumes, eggs, royal jelly, soybeans, various ocean fish and wholegrains (Institute of Medicine 1998b; Lee 2013). Biotin is present in foods as a protein-bound coenzyme, linked to lysine residues (Mock 2007; Berdanier and Zempleni 2009). Gastrointestinal proteolytic digestion by proteases and peptidases releases biocytin (biotinyl-ε-lysine) and biotinylated peptides. The enzyme biotinidase, present in pancreatic and other intestinal secretions, intestinal flora and brush-border membranes, hydrolyses and releases biotin (Said 2008). In the colon, the intestinal microbiota synthesize biotin. In its free form, biotin is absorbed in the jejunum and proximal ileum by an active sodium-dependent multivitamin transporter (SMVT), which also has an affinity for pantothenic acid, and by a sodium-independent electrogenic system (Mock 2007; Berdanier and Zempleni 2009). Biotin circulates in the plasma in a free form and is bound to proteins, especially albumin (Berdanier and Zempleni 2009). Biotin uptake in the liver and peripheral tissues, as well as renal reabsorption, is mediated by the SMVT (Said 2008). Within cells, biotin is released from the proteins by the action of proteases and biotinidase, and this process is more intense in the liver, kidneys and adrenal gland and in the plasma (Mock 2007; Said 2008; Berdanier and Zempleni 2009). Biotin is catabolized by β-oxidation and sulphur oxidation and is mainly excreted with the urine together with its catabolites; a minor amount is excreted in the faeces (Mock 2007; Said 2008; Berdanier and Zempleni 2009; Lee 2013).

Biotin functions as a necessary cofactor for carboxylases which catalyse a critical step in the intermediary metabolism of amino acids, fatty acids and cholesterol, and are required for lipogenesis, gluconeogenesis and other processes (Mock 2007; Lee 2013). It bonds covalently to enzymes that catalyse carboxyl group transfer, acting as a carboxyl transporter (Mock 2007). Biotin carries carboxyl groups for pyruvate carboxylase, which converts pyruvate to oxaloacetate during gluconeogenesis; for acetyl-CoA carboxylase, which synthesizes malonyl CoA in the synthesis of fatty acids; for propionyl CoA carboxylase, which converts propionate into succinate allowing odd-chain fatty acid use; and for 3-methylcrotonyl-CoA carboxylase in the catabolism of leucine (Waldrop 2015). This cofactor property of biotin links its functions with the metabolic functions of folic acid, pantothenic acid and cobalamin (Institute of Medicine 1998b). Furthermore, biotin covalently binds to lysine residues in histones (Kothapalli et al. 2005), a vitamin-dependent modification which is critical in the regulation of gene transcription, mitotic condensation of chromatin and DNA repair (Institute of Medicine 1998b; Mock 2007; Said 2008; Berdanier and Zempleni 2009; Lee 2013).

2.2.4.8 Vitamin B9: Folic Acid

Folic acids—or folates—are members of the family of pteroylglutamates or gamma-glutamyl derivatives. Folate and folic acid are the usual synonyms for pteroylglutamate and pteroylglutamic acid, respectively (Daly et al. 1997). They are essentially provided by the diet, from a variety of sources such as the liver, mushrooms, green leafy vegetables, poultry, meat, seafood, potatoes and fruit. Folates are widely present in 150 different forms, most of which are labile and easily oxidized (Institute of Medicine 1998b; Shane 2008; Mataix and Sánchez de Medina 2009; Lee 2013), so 50 to 95% of the vitamin is lost during food preparation and processing. Folates are generally in the form of polyglutamate (also folylpolyglutamates, containing 2–8 glutamate residues) linked to proteins and are released by the action of digestive proteases. Following the release of its glutamic residues by the action of hydrolases, the vitamin is absorbed at the intestinal level as monoglutamyl folate. These molecules enter the enterocytes by an active transport mechanism that involves a receptor and a transporter (Bailey 2007), which is intensified by the presence of glucose and galactose (Mataix and Sánchez de Medina 2009). Once within the hepatic cell, folate is metabolized upon dihydrofolate reductase action to 5-methyl tetrahydrofolate (THF) in a reaction that depends on Vit B3 (Benkovic and Hammes-Schiffer 2003). The vitamin is then largely bound to albumin to circulate in the plasma, but a small fraction of circulating 5-methyl THF is bound to a high-affinity transporter, which is a soluble form of the transmembrane transporter (Bailey 2007). The vitamin diffuses especially to the tissues that have a high level of cellular division, such as the bone marrow, gastrointestinal mucosa and immune system (Bates 2003). The excess folate that is not used by tissues is excreted in the urine in different forms, and a very small amount is excreted in the faeces (Bates 2003; Lee 2013).

Folic acid is essential for the production and maintenance of cells in division, for DNA and RNA synthesis through methylation and for protecting DNA against modifications (Bates 2003; Bailey 2007). In the form of THF compounds, folates act as substrate in single-carbon transfers for methylation reactions (Shane 2008), which are required for the synthesis of purines and thymidylate and the remethylation of homocysteine to methionine (Fox and Stover 2008). Methyl group transfers also involve Vit B12 (Berdanier and Zempleni 2009). In the mitochondria, folates participate in the catabolism of choline, purines and histidine and in the interconversion of serine and glycine (Fox and Stover 2008). Because of its role in DNA and RNA synthesis, as well as in amino acid metabolism, folic acid is an essential nutrient in all processes that involve cell division, such as growth, gestation or immune responses (Bailey 2007; Lamers 2011).

2.2.4.9 Vitamin B12: Cobalamin

The structure of B12 vitamers has the element cobalt in the centre of a planar tetrapyrrole ring called a corrin ring, which makes it the most chemically complex of all the Vits (Institute of Medicine 1998b; FAO/WHO 2001). The natural form of Vit B12 is hydroxocobalamin, and conversion to the different forms of the vitamin occurs in the body after consumption (Institute of Medicine 1998b). The richest dietary sources of cobalamin include the liver and kidney, milk, eggs, fish, cheese and meats, but only bacteria and the microorganisms archaea are capable of synthetizing Vit B12 (Martens et al. 2002). Faeces are a rich source of Vit B12, and many animals, including rabbits, dogs and cats, eat faeces (Watanabe 2007). Ruminant animals absorb hydroxocobalamin of bacterial origin from the rumen, which precedes their small intestine, but they need to receive cobalt in their diet (McDowell 2000). Humans do not absorb bacteria-synthesized hydroxocobalamin, since absorption of Vit B12 occurs in the small intestine but bacterial synthesis occurs in the colon; thus humans must obtain Vit B12 from the diet. Dietary cobalamin comes bound to proteins and is released in the low pH of the stomach by the action of hydrochloric acid and pepsin. Still in the stomach, free forms of Vit B12 are rapidly bound to the salivary protein R, which is a haptocorrin (a family of B12 biding proteins). In the jejunum, the protein R is digested by pancreatic trypsin, and the free Vit B12 is bound to the gastric glycoprotein intrinsic factor (IF) to be absorbed (Green 2005; Zempleni 2007). The cobalamin-IF complex is absorbed via receptor binding in the presence of calcium on the cell membrane of the mucous cells of the terminal ileum (Seetharam 1999; Green 2005). Vit B12 is transported in the portal venous blood bound to one of the three transporter proteins called transcobalamins I, II and III. Transcobalamin II, a hepatic beta-globulin, is the main transporting form of Vit B12 to tissues such as the liver, bone marrow, reticulocytes, lymphoblasts, fibroblasts and kidneys. Transcobalamins I and III are granulocyte glycoproteins that have storage and lesser blood carriage functions

(Mataix and Sánchez de Medina 2009). In the tissues, transcobalamin II is degraded in the lysosomes, and the resulting mitochondrial and cytosolic free active forms, 5-deoxyadenosyl cobalamin and methylcobalamin, respectively, function as coenzymes for important cellular enzymes (FAO/WHO 2001; Green and Miller 2007). The main storage site for Vit B12 is the liver, and most is excreted via the bile. Most of the biliary B12 is recycled via enterohepatic circulation, and excess B12 beyond the blood's binding capacity is excreted in the urine.

Methylcobalamin interacts with folate in the metabolic pathways to form methionine, which is necessary for DNA synthesis in cells undergoing chromosomal replication and division, notably the bone marrow cells. It therefore plays a key role in the formation of erythrocytes and leukocytes (Martens et al. 2002). The folate-cobalamin interaction is critical for normal synthesis of purines and pyrimidines, as well as for amino acid metabolism in all tissues (Shane 2008). Furthermore Vit B12 has fundamental functions in the functioning of the brain and nervous system (Metz 1992). As a cofactor, cobalamins are essential for the mitochondrial methylmalonyl-CoA mutase conversion of methylmalonic acid to succinate, which is critical for fatty acid synthesis and myelin homeostasis, and thus brain and spinal cord functions (Mansoorabadi et al. 2005; Berdanier and Zempleni 2009). Methionine is also required in reactions for myelin sheath phospholipids and neurotransmitter metabolism, and methionine synthesis requires an adequate folate-cobalamin interaction, which means that Vit B12 plays an indirect role in the nervous system (Metz 1992; Mansoorabadi et al. 2005).

2.2.4.10 Vitamin C: Ascorbic Acid

Vit C is an essential nutrient, commonly ingested as ascorbic acid or its oxidized form, dehydroascorbic acid (Institute of Medicine 2000a; Kall 2003). Most animals and plants are able to synthesize Vit C by enzymatic reactions that convert monosaccharides to Vit C (Wheeler et al. 1998). Among the mammals that have lost the ability to synthesize Vit C are humans who need to obtain it from diet. Dietary sources of Vit C include raw vegetables and fruits such as papaya, rosehips, acerola, citrus fruit, guava, kiwi, broccoli, red peppers and other plants (Chatterjee et al. 1975; Ha et al. 2004; Lee 2013). Animal sources also contain Vit C, but it is broken down by cooking and other food processing (Bayer 2013). Ascorbic acid is easily absorbed into the small intestine by simple diffusion and by the active transport sodium vitamin co-transporters and glucose transporters (GLUTs) (Tsukaguchi et al. 1999; Daruwala et al. 1999; Li and Schellhorn 2007). The active form is rapidly oxidized to L-ascorbate acid and to stereoisomer L- and D-isoascorbate. In humans, the absorption rate varies between 70 and 95% with regular intake, but absorption decreases as intake increases. At high intake (1.25 g), human absorption of Vit C may be about 33%, while at low intake (<200 mg), the absorption rate can reach up to 98% (Levine et al. 1996; Wilson 2005). Ascorbic acid passes easily into the tissues and is found in high concentrations in the adrenals, kidneys, liver and spleen, due to the intense oxidative stress resulting from their metabolic processes (Mataix and Sánchez de

Medina 2009). Excess ascorbic acid is catabolized by oxidization by the action of L-ascorbate oxidase and excreted in the urine as Vit C metabolite oxalate, dehydroascorbic acid or ascorbate-2-sulphate; ascorbic acid is also excreted in the urine in its free form as a result of body saturation (Berdanier and Zempleni 2009).

Ascorbic acid is a powerful antioxidant and performs numerous physiological functions in the body. It deactivates free radicals that damage lipid membranes, proteins and DNA and protects other antioxidants such as Vit A and Vit E (Benzie 2003; Rahman 2007). It participates as a coenzyme in hydroxylation reactions in the synthesis of collagen (Carr and Frei 1999). Vit C also acts as a cofactor in the synthesis of carnitine, in the synthesis and catabolism of tyrosine and in the metabolism of microsome (Institute of Medicine 2000a). In reactions for biosynthesis, ascorbate donates electrons to keep iron and copper atoms in their reduced states, preventing oxidation (Jacob et al. 1987; Harris and Percival 1991; Hunt et al. 1994; Lee 2013). Throughout this reducing action to key enzymes, Vit C assures that the collagen molecule assumes its triple helix structure, essential to the development and maintenance of the scar tissue, blood vessels and cartilage (Institute of Medicine 2000a). The synthesis of carnitine, which is essential for the transport of fatty acids into the mitochondria for ATP generation, depends on the reducing action of Vit C (Dunn et al. 1984; Rebouche 1991). In the immune cells, Vit C is consumed in large quantities during infections, participating in the activities of phagocytes, the production of cytokines and lymphocytes (Shilotri and Bhat 1977; Delafuente et al. 1986; Ströhle and Hahn 2009).

2.2.4.11 Vitamin D: Cholecalciferol

Similarly to Vit A, Vit D is not a single compound, but a family of vitamers serving as provitamin D with a sterol structure that differ only in their side chains (Wolf 2004; Norman and Hentry 2007; Christakos et al. 2010; Lee 2013; Bikle 2014). Vit D2 comes from a common plant steroid, ergosterol (Bikle 2014). Most mammals can convert D2 to D3 and use it to make the active form (1,25-dihydroxycholecalciferol) which is responsible for its biological function (Berdanier and Zempleni 2009). Vit D3 is synthesized endogenously in the body, and its main dietary sources include fish, fish/cod liver oil, liver, buttermilk and egg yolks (Nair and Maseeh 2012; Lee 2013). Dietary Vit D is absorbed in the small intestine with long-chain fatty acids with the help of bile salts and becomes part of the chylomicrons in the lymphatic system (Norman and Hentry 2007; Nair and Maseeh 2012; Julve et al. 2016). Absorption takes place mainly in the jejunum and ileum. When there is more requirement, Vit D is reabsorbed in the duodenum from the bile, which is its principal excretory pathway (Berdanier and Zempleni 2009; Nair and Maseeh 2012). Absorbed Vit D is transported to the liver in its nonesterified form bound to the Vit D-binding protein, which is similar to the α-2 globulins and albumins (Berdanier and Zempleni 2009; Christakos et al. 2010; Bikle 2014). Dermal synthesis of Vit D also occurs when natural sunlight or ultraviolet irradiation acts on the 7-dehidrocholesterol to form cholecalciferol or D3, which is biologically inactive

(Webb et al. 1989; Holick 2005; Nair and Maseeh 2012). D3 undergoes two hydroxylation reactions; in the liver, cholecalciferol is converted to calcifediol (or 25-hydroxycholecalciferol or 25-hydroxyvitamin D), which is a biologically active metabolite. The other reaction occurs in the kidneys with the formation of calcitriol (1,25-dihydroxycholecalciferol or 1,25-dihydroxyvitamin D3), which is the hormonally active metabolite of Vit D. Calcitriol is capable of increasing calcium levels in the blood by inducing Ca uptake in the intestine, Ca reabsorption in the kidneys and Ca mobilization in the bone (Norman and Hentry 2007; Lee 2013; Hill and Aspray 2017). It is known that 25-hydroxycholecalciferol can accumulate in the heart, lungs, kidneys and liver (Christakos et al. 2010). Even though there are gaps in the knowledge of Vit D catabolism, the excretion of it and its metabolites occur primarily in the faeces with the aid of bile salts and small amounts in the urine (Norman and Hentry 2007; Lee 2013). Consumption of excess Vit D over extended periods of time causes toxic conditions resulting in excess calcification not only of bone but also of soft tissues, causing a variety of clinical alterations. The development of Vit D hypervitaminosis seems to depend on individual susceptibility (Morita et al. 1993; Holick 2005; Berdanier and Zempleni 2009).

The first understood Vit D function in the body was to regulate mineral homeostasis, mainly in calcium absorption and phosphate uptake (Holick 2005; Lee 2013; Bikle 2014). Calcitriol acts in conjunction with parathormone and calcitonin, to maintain serum calcium and phosphate levels by its actions on the intestine, kidneys, bone and parathyroid gland (Norman and Hentry 2007; Nair and Maseeh 2012). Calcitriol is also considered a hormone secosteroid (a subclass of steroids), which also promotes intestinal absorption of iron, magnesium, phosphate and zinc (Pérez-López 2007; Ghoneim et al. 2015). It also participates in neuromuscular and immune function (Lee 2013; Toussaint and Damasiewicz 2017). Muscle contraction is influenced by a direct effect on Ca^{2+} transport (Norman and Hentry 2007). Vit D influences insulin production by the endocrine pancreas (Nair and Maseeh 2012), skin and hair growth and regeneration (Bouillon et al. 2006). Concerning lipid metabolism, Vit D acts on the regulation of the glyoxylate cycle in the liver, stimulates cell proliferation in the adipose tissue via reactive oxygen species (ROS) (Sun and Zemel 2007) and modulates the adipocyte glucocorticoid function (Morris and Zemel 2005). Calcitriol regulates many genes (Omdahl et al. 2002) by binding to the Vit D-specific receptor in DNA sequences, following the typical mechanism of steroid hormone action (Pike and Meyer 2010). Thus, Vit D can be considered both a nutrient, particularly in conditions of little exposure to ultraviolet light, and a hormone because the body can completely synthesize the active form calcitriol from prohormone D2 or D3 (Norman 2008). Unlike herbivores and omnivores, dogs and cats are not able to synthesize Vit D3 adequately in the skin and are mainly dependent on dietary intake, which means that Vit D3 is an essential vitamin that should be present in dog and cat food (How et al. 1994).

2.2.4.12 Vitamin E: Tocopherol

Vit E includes two groups of compounds, tocopherols and tocotrienols, which come from animal sources such as fish oils and fats and vegetable sources such as wheat germ, rosehips, various nuts, corn, sunflower or soy oils. Four tocopherol and four tocotrienol molecules, identified by the prefixes alpha- (α), beta- (β), gamma- (γ) and delta- (δ), contribute to Vit E activity (Hacquebard and Carpentier 2005; Lee 2013). Alpha-tocopherol is the most biologically active form of the vitamin (Hacquebard and Carpentier 2005). During digestion, Vit E is included in micellar compounds together with lipids. Tocopherol favours the activity of Vit A to prevent its oxidation in the intestine. Additionally, other antioxidants affect Vit E absorption and organic levels; the intake of β-carotene, ascorbic acid and polyunsaturated fatty acid can markedly influence the rate of the use of α-tocopherol as an antioxidant (Omur et al. 2016). Low selenium intake increases the need for Vit E and vice versa, because selenium is a component of the glutathione peroxidase system, which suppresses free radical production (Berdanier and Zempleni 2009). Following absorption, Vit E is transported in the blood with the chylomicrons in the lymph to the liver. From the liver, tocopherol is transported with VLDL proteins in the blood to be stored in the adipose tissue (Drevon 1991; Lee 2013; Yang and McClements 2013; Goncalves et al. 2015). However, Vit E does not accumulate in the liver, suggesting that its catabolism and excretion are important (Brigelius-Flohé and Traber 1999; Herrera and Barbas 2001). Tocopherols are distributed to all cells in the body, and they are found in higher cell concentrations within adrenal, pituitary, testicular cells and platelets (Drevon 1991; Berdanier and Zempleni 2009). Tocopherol-specific binding proteins have been found within hepatic and heart cells, where α-tocopherol is transferred from liposomes to mitochondria (Stocker and Azzi 2000). The major excretion route is the intestine, via the elimination of hepatic Vit E into the bile, and a small part is excreted in the urine (Brigelius-Flohé and Traber 1999; Herrera and Barbas 2001).

Alpha-tocopherol is an important antioxidant that participates in the glutathione peroxidase pathway and protects cell membranes from oxidation by reacting with lipid free radicals produced by the peroxidation chain reaction, particularly on the phospholipids of the membranes (Omur et al. 2016). Besides removing the free radical intermediates as a peroxyl radical scavenger, α-tocopherol prevents oxidation from continuing in reactions that result in oxidized α-tocopheroxyl radicals (Hacquebard and Carpentier 2005). Other antioxidants, such as ascorbate, retinol or ubiquinol, act on the transformation of α-tocopheroxyl radicals into their active reduced form (Traber 2007; Omur et al. 2016). Thus, at the cellular level, α-tocopherol protects the membranes from oxidation resulting from the release of free radicals in many metabolic reactions, immune responses, inflammation or environmental toxins (Clarkson and Thompson 2000; Cases et al. 2005; Traber 2007; Choe and Min 2009). Glutathione peroxidase, a selenoenzyme, protects haemoglobin and the cell membranes by detoxifying lipid hydroperoxides to less toxic fatty acids and by preventing the formation of free radicals; Vit E potentiates its action by serving as a free radical scavenger to prevent the formation of lipid hydroperoxide (Rowe and Wills 1976; Krajcovicová-Kudlácková et al. 2004; Berdanier and Zempleni 2009; Birben et al. 2012). In addition, α-tocopherol protects against iron toxicity related to free radical formation (Berdanier and Zempleni 2009). Other functions of Vit E include (1) protection against DNA oxidation thus benefiting erythropoiesis, (2) steroid hormone synthesis and spermatogenesis (McCall and Frei 1999; Rahman 2007), (3) interaction with zinc in the metabolism and protection of skin lipids (Park 2015), (4) participation in the neuromuscular functions (Muller 1986) and (5) inhibition on thromboxane, prostaglandin and enhancing prostacyclin, thus regulating inflammation and the aggregation of platelets (Rimbach et al. 2002; Norman and Hentry 2007).

2.2.4.13 Vitamin K: Phylloquinone (Vit K1) and Menaquinone (Vit K2)

The menaquinone (Vit K2) family of liposoluble vitamins occurs in animal and human tissues as isomers, which are abbreviated as MK-n (M = menaquinone; K = Vit K; n = number of isoprenoid side chain residues, up to 20) (Klack and de Carvalho 2006; Shearer and Newman 2008; Bayer 2013). Dietary sources of Vit K2 are mostly in the form of menaquinone-4 or MK-4 (also menatetrenone), a short-chain menaquinone present in egg yolk, meat, liver or dairy products. Phylloquinone (Vit K1) is found in vegetables where it plays roles in the photosynthesis. Green vegetables and cereals are important dietary sources of phylloquinone, which is considered a precursor of Vit K for animals (Shearer and Newman 2008). Menakinone-4 is the most common type of Vit K2 in animal products because MK-4 is synthesized from Vit K1 in animal tissues. Within human and animal digestive tracts, a species-dependent amount of Vit K1 is absorbed by the cells of intestine segments (Berdanier and Zempleni 2009); the other amount of phylloquinone is converted to menaquinone-7 by bacteria, which also produce other isomers of Vit K2 that play a role in the metabolism of intestinal microbiota (Bentley and Meganathan 1982; Walther et al. 2013). Absorption of Vit K is an active energy-dependent process that requires bile and pancreatic juice; subsequently either phylloquinone or menaquinone-4 is incorporated into chylomicrons and transported through lymphatic vessels to the bloodstream reaching the liver (Suttie 2007; Card et al. 2014). The liver has an exclusive role in Vit K1 catabolism and excretion in a very fast turnover, and this is the reason why the diet should provide regularly adequate amounts of K1 (Suttie 2007; Lee 2013; Gonçalves et al. 2015). Vit K is distributed throughout the body tissues by the blood and transported by lipoproteins (Shearer and Newman 2008). However, most of Vit K1 and MK-4 are transported by triacylglycerol-rich lipoproteins to be metabolized in the liver (Dôres et al. 2001) from which only a small fraction is released into the circulation where Vit K forms are transported by LDL and HDL. The liver acts as a site for storage and redistribution of long-chain menaquinones (such as MK-7, MK-8 or MK-9) coupled to LDL. Because LDL has a long half-life in the circulation, long-chain menaquinones become available for extra-hepatic tissues, where they accumulate and perform Vit K functions not related to haemostasis, such as the suppression of inflammation, prevention of brain oxidative damage and a role in the synthesis of sphingolipids (Shearer and Newman 2008). On the other hand, tissues that accumulate high amounts of MK-4, such as the pancreas, arterial walls and testes, are the ones that have a significant capacity to convert the available Vit K1 into MK-4 (Thijssen and Drittij-Reijnders 1996; Suttie 2007). Conversion of phylloquinone or long-chain menaquinones to MK-4 is therefore a major metabolic pathway of Vit K use in tissues (Shearer et al. 2012; Lee 2013). Hepatic catabolism of Vit K involves oxidative degradation of the phytyl side chain, with the participation of enzymes also involved in fatty acids and steroids, and prostaglandin catabolism, and this results in the urinary excretion of catabolites (Bates 2005b; Shearer et al. 2012).

The functions of Vit K include those related to haemostasis and many others, more recently described (Shearer and Newman 2008), but in general terms, this vitamin serves as a co-substrate in the production of unique calcium-binding proteins in the blood, bone and kidneys (Berdanier and Zempleni 2009). In the liver, Vit K1 is reduced to hydronaphtoquinone (KH_2), which is the active cofactor for the enzyme carboxylase (Shearer and Newman 2008; Ferland 2012). Carboxylase acts on the carboxylation of glutamate residues in proteins to form gamma-carboxyglutamate (Gla), which has various functions. In the liver, Gla participates in the synthesis of coagulation factors prothrombin (factor II); factors VII, IX and X; and proteins C, S and Z (Suttie 2007). In bone metabolism, Gla forms osteocalcin, also called bone Gla protein which is secreted by osteoblasts and is essential for bone mineralization, periostin (Shanahan et al. 1998; Coutu et al. 2008) and the recently discovered Gla-rich protein (Viegas et al. 2008). Gla forms the matrix Gla protein, which is found in numerous body tissues and acts by inhibiting calcification, but its role is most pronounced in the cartilage, kidneys and arterial vessel walls (Cancela et al. 2012; Viegas et al. 2014). In the endothelial cells of arteries and leukocytes, Gla forms the growth arrest-specific protein-6 which acts for cell survival, proliferation, migration and adhesion (Hafizi and Dahlbäck 2006; Dihingia et al. 2017). Other non-cofactor functions of Vit K include modulation of inflammation, prevention of oxidative injury in the brain and a participation in the synthesis of sphingolipid (Shearer and Newman, 2008). Vit K2 acts in balance with Vit D3 to maintain calcium metabolism and homeostasis in the blood, bone and other tissues (Suttie 2007). A role for Vit K in the prevention of tumorigenesis has been described; synthetic menadione (K3) inhibits aryl hydrocarbon hydroxylase, thus reducing the levels of carcinogenic and mutagenic metabolites in the cell resulting in a reduction in tumour formation (Osada and Carr 2000; Osada et al. 2001).

2.2.5 *Minerals*

There are 103 known mineral elements. Living organisms are composed mainly of 11 of them, which are fundamental for maintaining key bodily functions. Minerals are categorized into macroelements, oligoelements and trace elements according to the daily dietary requirements. Macroelements must be supplied in more than 100 mg/day (Ca, Mg, P, Na, K, Cl and S). Oligoelements must be supplied between 1 and 100 mg/day (Zn, Fe, Mn, Cu and F). Trace elements must be supplied in less than 1 mg/day (Se, Mo, I, Cr, B and Co) (Freeland-Graves and Trotter 2003; Mataix and Sánchez de Medina 2009; Lee 2013). In this section, the main functions of minerals in the metabolic processes of the cells are discussed, as well as dietary natural sources and overall absorption processes.

2.2.5.1 Calcium: Ca

Calcium is the most abundant mineral in the body and regulates many metabolic functions. It comprises 1.5% to 2% of body weight and 39% of total body minerals; 99% of the body's total amount of Ca is in the bones and teeth, and 1% is in blood, extracellular fluids and inside soft tissue cells. Dietary sources of Ca are milk and dairy products, eggs, green leafy vegetables, almonds, molasses, soybeans, fish and seafood (L'Abbé 2003; Peacock 2010; Institute of Medicine 2011). Calcium absorption from the diet is variable; about 23 to 8% is absorbed in the duodenum whose content is predominantly acidic, but absorption is greatly reduced in the lower part of the intestinal tract where the contents are more alkaline (McCormick 2002). In the duodenum and proximal jejunum, absorption occurs by active transport, controlled by the action of Vit D3, which increases the uptake of Ca at the brush border of the intestinal mucosal cell by stimulating the production of a calcium-binding protein (calbindin) (Ebeling et al. 1992; Kojetin et al. 2006; Berdanier and Zempleni 2009). The role of calbindins in intestinal absorption cells is to temporarily store Ca ions after a meal and to transport them to the basolateral membrane for the final passage of absorption. A second mechanism for Ca absorption does not depend on Vit D, rather it is passive and non-saturable and occurs throughout the intestine, mainly in the jejunum but also in the colon; this system prevails if the diet is very rich in Ca (Wasserman and Fullmer 1989; Xue and Fleet 2009; Institute of Medicine 2011). Factors that promote higher absorption of Ca include Vit D and the presence of other nutrients such as lactose and proteins in the digestive content. Decreased absorption of Ca occurs during Vit D deficiency and excess fibre, oxalate and phytate in the diet (Pereira et al. 2009). Ageing and poor absorption of fats impair Ca absorption due to the saponification of fatty acid with Ca in the digestive tract (Gacs and Barltrop 1977; Wang et al. 2013). Half of blood calcium is bound to proteins, mainly to albumin and globulin; the rest circulates the form of free ions and diffusible complexes (Peacock 2010). Approximately 5 to 20% of the ingested Ca is excreted in the urine; although its intake may vary greatly, calciuria is slightly modified. Tubular reabsorption of Ca in the kidneys occurs by transport mechanisms similar to those of the intestine. The urine contains little Ca compared to the amount filtered in the glomerulus, because the proximal tubule, the loop of Henle and the distal tubule reabsorb most of it (approximately 85%). An average loss of 10% of dietary intake of Ca occurs by skin exfoliation and sweating in humans (Goldschmied et al. 1975; Heaney and Rafferty 2001; Institute of Medicine 2011). The faeces contain Ca from excess in diet, cell exfoliation of the whole digestive tract and Ca metabolites from bile (Abrams et al. 1991; Berdanier and Zempleni 2009).

Calcium functions as an integral part of all signalling systems (as a second messenger) between and within cells in the body, and metabolic regulation and interactions depend largely on cell communication (Berdanier and Zempleni 2009). The readily binding of Ca to proteins, changing the electrical charges on the protein chain, causes modification in the protein's tertiary structure, as occurs in the coagulation cascade that requires Ca as an activation factor. A large number of extracellular enzymes require Ca as a cofactor. Calcium participates in the contraction of the skeletal, smooth and heart muscles, neurotransmission, reactions during cellular division and various forms of endocrine and exocrine secretion (L'Abbé 2003; Peacock 2010; Institute of Medicine 2011). Calcium is involved in the transport of many nutrients in the cells of the digestive system and participates in many processes related to cell membrane homeostasis (Lee 2013).

2.2.5.2 Phosphorus: P

Phosphorus is one of the essential elements in nutrition, and it exists in the cells as phosphate (PO^{-}_{4}) compounds. In their free form, phosphates are called inorganic phosphate generally symbolized by Pi (Berdanier and Zempleni 2009). About 85% of phosphates in the body exist as calcium phosphate crystals in the bones and teeth, and the remaining is metabolically active and distributed in all cells of the body and in the extracellular fluid (Institute of Medicine 1997a; Anderson 2003; Lee 2013). Dietary sources of phosphates include meats, poultry, eggs, fish, dairy products, vegetables, cereals and legumes, where they are as organic and inorganic phosphates. Organic phosphates are bound to proteins, mostly at serine, threonine and tyrosine residues. Most inorganic phosphate is absorbed by the action of inorganic phosphatase. The presence of glucose in the diet enhances phosphate absorption. Absorbed phosphates are promptly distributed in the viscera, bones, skin and muscles (Wasserman 1981; Institute of Medicine 1997a; Anderson 2003; Lee 2013). The kidneys are the main site for phosphate regulation and excretion. Some phosphates are also excreted in the faeces and come from excess P in the diet and digestive tract secretions (Lee 2013).

Phosphorus plays an important role in organic and inorganic compounds and is involved in many aspects of metabolism and growth, including energy use, bone mineralization, blood buffering, cell membrane bipolarity and kinase-mediated signal transduction, and it has fundamental functions in immune responses (Kegley et al. 2001; Oster et al. 2016). Phosphorylation and dephosphorylation are important mechanisms for energy storage and are used in the regulation of metabolic processes involving ATP. Adenosine phosphates (AMP, ADP and ATP) release phosphates throughout hydrolysis by the action of phosphatases (Krebs and Beavo 1979; Kurosawa 1994; Berdanier and Zempleni 2009). Phosphorus is a component of adenine and guanine nucleotides and provides stability to DNA and RNA molecules, due to its capacity to retain a negative charge thus repelling other negatively charged molecules such as peroxides (Lodish et al. 2000a; Berdanier and Zempleni 2009). Phosphates are present in phospholipids, which are necessary for the functional characteristics of the membranes of cells and organelles, and phosphoproteins (Berdanier and Zempleni 2009). Phosphorus works in conjunction with Vit D3 and calcium, either in metabolic processes related to the formation of hydroxyapatite in bone mineralization or in the regulation of the homeostasis of cells and system as a second messenger (Gil Hernández and Sánchez de Medina Contreras 2010). Phosphates participate in all anabolic and catabolic pathways (Institute of Medicine 1997a; Berg et al. 2002c; Anderson 2003; Berdanier and Zempleni 2009; Lee 2013).

2.2.5.3 Magnesium: Mg

Magnesium is an essential mineral nutrient that occurs as the Mg^{2+} ion (Griffin 2003). The main dietary sources are brewer's yeast, green leafy vegetables, nuts, seeds, soybeans, oats or milk. An average of 20 to 50% of dietary Mg is absorbed throughout the small intestine, but most of the absorption occurs in the jejunum by a transporter-dependent diffusion and simple diffusion. At low intraluminal concentrations, Mg is absorbed via saturable active diffusion, whereas a non-saturable passive diffusion acts at high concentrations (Fine et al. 1991; Griffin 2003; Vormann 2003; Lee 2013). Magnesium absorption depends on the quantity of protein in the diet because a protein intake which is too low or too high inhibits Mg absorption, as does the amount of phosphate, phytate and fat in the gut (Fine et al. 1991; Vormann 2003). In the plasma, Mg is transported as free ions (55%) or in complexes with proteins and phosphate (45%) (Griffin 2003). Because membranes are impermeable to Mg, as they are to other ions, intracellular and extracellular concentrations of free Mg are sustained by a balance between buffering by binding to proteins and other molecules and transferring Mg ions to storage or extracellular spaces (DiSilvestro 2005). Thus, serum Mg levels may be normal even when intracellular magnesium is deficient (Jahnen-Dechent and Ketteler 2012). In the body, 60% Mg is distributed in the bones, 26% in the muscles and the rest inside soft tissue cells; 1% of Mg is found in extracellular space and body fluids (Institute of Medicine 1997b; Jahnen-Dechent and Ketteler 2012). Intracellular Mg correlates with intracellular potassium, and increased Mg lowers Ca (Whang and Whang 1990; Institute of Medicine 1997b; Jahnen-Dechent and Ketteler 2012). Unabsorbed dietary Mg is excreted in the faeces, while absorbed Mg is excreted in the urine and sweat (Wester 1987; Vormann 2003).

Magnesium interacts closely with phosphate and plays an essential role in the metabolism of nucleic acid. About 300 enzymes involved in nucleic acid metabolism require Mg as a cofactor, and one of its most important functions is to stabilize the structure of ATP in ATP-dependent enzymatic reactions (Vormann 2003), as it helps to maintain the double helical structure of DNA (Hartwig 2001). Magnesium-bound ATP is the substrate for enzymes such as kinases. Magnesium-calcium interactions are fundamental for the function of excitable membranes and neuromuscular transmission (Griffin 2003), as they are for the humoral and cellular immunocompetence by acting as an immunomodulatory agent and as an important element for thymic function (Keen et al. 2004).

2.2.5.4 Sodium Na, Chlorine Cl and Potassium K

Sodium, chlorine and potassium act together in the body being essential nutrients for metabolism. The body content of these three minerals is 2% sodium, 3% chlorine and 5% potassium. The main dietary source of Na and Cl is sodium chloride or common salt, while the main source of K are fruits and vegetables, fresh meat and milk (DiSilvestro 2005; Berdanier and Zempleni 2009; Lee 2013). All three elements are easily absorbed through the intestinal tract and distributed in all body fluids and tissues. The primary site of K absorption is the small intestine where most of the dietary K entry into the gastrointestinal tract occurs by passive diffusion (Stone et al. 2016). Sodium is the most prevalent metallic ion in extracellular fluid (Pohl et al. 2013), and ion transporters in the cell membrane maintain the balance between K and Na (Lodish et al. 2000b; Berdanier and Zempleni 2009; Lee 2013). A decrease in blood pressure and Na concentration triggers the kidneys' production of renin, which induces the liberation of aldosterone and angiotensin, which in turn promotes Na filtration from the urine. When the concentration of sodium increases, the receptors in the hypothalamus stimulate the sensation of thirst and the production of renin decreases, and excretion of Na in the urine by the kidneys returns to normal (Kurtzman et al. 1972; Greger 2000; Berdanier and Zempleni 2009). Ionized potassium is excreted in place of ionized sodium by the renal tubular exchange mechanism, which is also regulated by the renin-angiotensin system (Sealey et al. 1970; Laragh and Sealey 2011). Most Na excretion is via renal, but faeces and sweat excretion also occur. About 90% of dietary K is excreted in the urine and 10% in the faeces. Chlorine is excreted mostly as chloride acid in gastric secretion and the faeces (Klevay et al. 2007; Berdanier and Zempleni 2009).

Sodium regulates blood volume, blood pressure, osmotic equilibrium and pH throughout the renin-angiotensin system (Haddy 1991; Haddy et al. 2006; Lee 2013). Chlorine is needed for the production of hydrochloric acid in the stomach and in cellular pump functions and plays an important role in the acid-base balance (Linus Pauling Institute 2016). Potassium is the major cation inside cells, while Na is the major cation outside cells, and different concentrations of Na and K produce different intracellular and extracellular electric potential, which is called membrane potential. Electricity is fundamental for systems such as heart function, neurotransmission and muscle contraction (Alberts et al. 2002a; Berdanier and Zempleni 2009). This mechanism relies on Na^+/K^+-ATPase, an active transporter pumping ion against the gradient, and sodium/potassium channels (Lopina 2000). The transport of some amino acids requires the Na^+/K^+/Ca-ATPase pump system, which is also important in the regulation of the cell volume and the maintenance of membrane potential (Lopina 2000; Alberts et al. 2002b).

2.2.5.5 Sulphur: S

Sulphur is an essential element in nutrition. After calcium and phosphorus, it is the third most abundant element in the body and constitutes nucleic acids, amino acids, sulphate esters of polysaccharides, steroids, phenols and sulphur-containing coenzymes and cofactors (Nimni et al. 2007; Berdanier and Zempleni 2009; Lee 2013). Plant and animal dietary sources of sulphur are mostly proteins containing the amino acids cysteine and methionine, which are present in meats, fish, poultry, eggs, broccoli, cauliflower and beans. Methionine is an essential amino acid in diet, and apart from Vit B1 (thiamine), Vit B5 (pantothenic acid) and Vit B7 (biotin), cysteine and all sulphur-containing compounds in the body are synthesized from methionine (National Research Council 1989; Brosnan and Brosnan 2006; Berdanier and Zempleni 2009). Reduced sulphur is oxidized in the body by the enzyme sulphide oxidase, which is needed for the metabolism of methionine and cysteine (Florin et al. 1991; Sekowska et al. 2000; DiSilvestro 2005). As dietary sulphur comes with proteins, its digestion, absorption and transport are dependent on the processes that occur for protein metabolism in the body (see above). Cysteine and methionine are oxidized to sulphate by sulphide oxidase, eliminated in the urine or stored as glutathione, which can serve as a store for sulphur (Florin et al. 1991; Nimni et al. 2007). Moreover, the metabolism of other sulphur-containing amino acids such as homocysteine, a methionine catabolite, and taurine also generates inorganic acids, especially sulphate anions, in substantial amounts. Excess inorganic sulphur generated in the hepatic or renal metabolism is excreted in the urine, as sulphates (Florin et al. 1991).

Sulphur plays essential roles in amino acid and protein metabolism and functions. In cell metabolism, sulphur participates in reactions of oxidation-reduction as an antioxidant because the amino acids cysteine and methionine are used in the synthesis of glutathione. Reduced glutathione, a sulphur-containing tripeptide, is a reducing agent through its sulfhydryl (-SH) moiety derived from cysteine. Thioredoxins, a class of small proteins essential to all known life, use adjacent pairs of reduced cysteines to function as general protein reducing agents, with a redox similar function (Parcell 2002; Nimni et al. 2007; Berdanier and Zempleni 2009; Lee 2013; Mukwevho et al. 2014). In the metabolism of carbohydrates and fatty acids, important enzymes require sulphur-containing moieties in reactions involving acyl groups, such as CoA and alpha-lipoic acid (Oda 2006; Berdanier and Zempleni 2009). Sulphur is a component of heparin which acts as an anticoagulant and of chondroitin sulphate which constitutes cartilages and bones (Parcell 2002; Maeda et al. 2011; Sarrazin et al. 2011; Mikami and Kitagawa 2013).

2.2.5.6 Iron: Fe

Iron is distributed in the body as functional iron and as storage iron (DiSilvestro 2005; Lee 2013). Functional iron is included in haemoglobin, myoglobin, cytochromes and various proteins that work in the transport, storage and use of oxygen. Storage iron compounds include ferritin and hemosiderin within the liver, bone marrow, spleen and muscle cells (Lynch 2003). Most body iron exists as heme within haemoglobin (about 70%) and myoglobin (about 2.5%); the remaining is distributed in tissue macrophages (about 5%) and liver hepatocytes (about 20%) (Papanikolaou and Pantopoulos 2017), while enzymes contain about 0.5% of it (Harris 2013). Depending on the dietary origin, iron is categorized as heminic or non-heminic. Heminic iron is found in haemoglobin and myoglobin present in the liver, kidney and heart and meat, fish and poultry; non-heminic iron comes from plants, mainly from green leafy vegetables and beans, and from animal proteins that do not contain heme iron, such as milk and eggs (Lee 2013; Wessling-Resnick 2014). Iron is more easily absorbed in the duodenum in the ferrous (Fe^{2+}) state as it occurs in rich heme-containing meat products, but most dietary iron is in the form of inorganic ferric iron (Fe^{3+}), which comes from non-heminic dietary sources (Lynch 2003). Apoferritin on the membrane of duodenum cells binds some Fe^{3+} to form cytosolic ferritin, which stores iron; this iron will be eliminated with dead cells shed into the faeces (Lynch 2003; Wessling-Resnick 2014). Gastric and intestinal secretions containing duodenal cytochrome-b or other ferrireductases reduce intraluminal Fe^{3+} to Fe^{2+} (Papanikolaou and Pantopoulos 2017), a process that is facilitated by ascorbic acid, sugar and sulphur-containing amino acids (Lee 2013). Inhibition of iron absorption occurs by excess phytates, phosphates, oxalates, tannins, calcium and fibre, as well as by heightened intestinal motility or poor digestion of fats (Gillooly et al. 1983; Hurrell and Egli 2010). Ferrous iron is transported across the membrane of enterocytes by the divalent metal transporter-1 (DMT1), to be then exported to the circulation via ferroportin. Ferroportin transporting is helped by re-oxidation of Fe^{2+} to Fe^{3+} by a membrane-bound hephaestin, a cupper-dependent ferroxidase, or by ceruloplasmin, a soluble ferroxidase (Papanikolaou and Pantopoulos 2017). Iron transporting by ferroportin is inhibited by hepcidin, a 25-amino acid peptide hormone produced in the liver that regulates iron efflux to the plasma (Ganz and Nemeth 2015). Absorption of dietary heme iron requires uptake by mucosal apoferritin (Jacobs 1971), catabolism of heme within enterocytes and release of Fe^{2+} that from this point undergoes the same process as inorganic dietary iron (Papanikolaou and Pantopoulos 2017). Transferrin, the plasma iron carrier, takes up Fe^{3+} and transports it to reticuloendothelial cells in the bone marrow, spleen and liver and muscle cells (Klawe 2003). These cells have transferrin receptors, which facilitate the uptake of iron by endocytosis. Within the acidic endosomal cells, transferrin releases iron, which is reduced by a ferrireductase and then transported into intracellular compartments via DMT1. Membrane apo-transferrin releases iron back into the bloodstream, and transferrin recaptures it to engage in further cycles of iron delivery to cells (Papanikolaou and Pantopoulos 2017). The contribution of dietary iron released by enterocytes to maintain the circulating transferrin iron pool is very low because iron is not excreted by any physiological regulatory mechanism (Abbaspour et al. 2014). There is a steady and slight iron loss via gastrointestinal blood loss, sweating and shedding cells of the skin and the mucosal lining of the gastrointestinal tract (Drakesmith and Prentice 2012; Wessling-Resnick 2014). In physiological conditions, apo-transferrin is mostly reloaded with iron provided by tissue macrophages after erythrophagocytosis, the macrophage-mediated iron recycling system from senescent red blood cells into erythropoiesis (Knutson et al. 2005; Papanikolaou and Pantopoulos 2017). When plasma iron is limited, iron mobilization from ferritin contributes to erythropoiesis, via lysosomal ferritin degradation within the hepatocytes, bone marrow or spleen, and the liberated iron is exported to the bloodstream via ferroportin (Wang and Pantopoulos 2011; Ganz and Nemeth 2015). Hemosiderin is mostly found in macrophages, and it seems to result from the accumulation of ferritin or denatured ferritin following pathological conditions. The iron within the hemosiderin stores is much less bioavailable to supply iron on demand (Linder 2013; Abbaspour et al. 2014).

Several iron-containing proteins are involved in fundamental organic functions, such as cytochromes for energy production; ribonucleotide reductase, amino acid oxidases or fatty acid desaturases for intermediary metabolism and detoxification; tyrosine hydroxylase and thyroperoxidase for synthesis of hormones and neurotransmitters; and myeloperoxidase, nitric oxide synthases and lipoxygenases for host defence and inflammation (Ganz and Nemeth 2015). Iron acts as a cofactor for numerous metalloproteins, mainly as part of heme or iron-sulphur clusters, which are essential for oxygen transport and for electron transfer and catalytic reactions (Liu et al. 2014; Papanikolaou and Pantopoulos 2017). The chemical reactivity of iron makes it capable of interacting with proteins to function as an electron donor and acceptor in reduction and oxidation reactions (Berdanier and Zempleni 2009; Olson et al. 2013). However, this property makes iron extremely toxic in its free form because it acts as a catalyst of oxidative stress via the Fenton/Haber-Weiss reaction (Kanti Das et al. 2014; Papanikolaou and Pantopoulos 2017). The strict control of iron in the body assures its proper functioning while preventing its detrimental effects mainly by binding iron to proteins (DiSilvestro 2005; Gozzelino and Arosio 2016). The buffering capacity of apo-transferrin prevents the accumulation of free non-transferrin-bound iron, which is redox-active and toxic (Papanikolaou and Pantopoulos 2017). Controlling iron release in plasma is also an important mechanism of infection control by the body, because bacteria and other microorganisms have to obtain iron from the environment (DiSilvestro 2005; Drakesmith and Prentice 2012). The release of hepcidin as a response to signs of infection and inflammation causes intense hypoferremia due to iron retention in macrophages and inhibition of iron transport by ferroportin, resulting in the disruption of bacteria metabolic pathways and multiplication (Ganz and Nemeth 2015).

2.2.5.7 Zinc: Zn

Zinc acts as an intracellular ion and is an essential trace element for animals and humans. Zinc associates with more than 300 enzymes and is the only metal that appears in all enzyme classes (Sandstead 1994; Rink and Gabriel 2000; Institute of Medicine 2001b; Lee 2013). The body of an adult human contains about two grams of zinc, with higher concentrations in the brain, bones, liver, pancreas, kidneys, bones and muscles, followed by the eye, prostate, spermatozoa, skin, hair and nails (Berdanier and Zempleni 2009; Kaur et al. 2014). Dietary sources of zinc include shellfish, meat, fish, poultry eggs, dairy, seeds, alfalfa, celery, legumes, nuts and wholegrains (Institute of Medicine 2001b; Solomons 2003; Kaur et al. 2014). Zinc absorption is relatively poor, an average 10 to 40% of all the ingested minerals with food, and it occurs in the small intestine by passive diffusion or throughout binding to metallothionein protein, which is favoured by Vit D3 (Berdanier and Zempleni 2009; Roohani et al. 2013). Dietary proteins, glucose and lactose favour zinc absorption, while phytates and excess fibre, phosphates and iron or calcium inhibit it and favour zinc excretion in the faeces (Lönnerdal 2000; Lee 2013). After entering enterocytes, Zn is bound to a cysteine-rich intestinal protein that in turn transfers it to either metallothionein or albumin, which carries it in the plasma (Berdanier and Zempleni 2009). In the plasma, albumin-, transferrin- or α-2 macroglobulin-bound Zn is chelated to free amino acids or small peptides (Foote and Delves 1984; Solomons 2003). This Zn^{+2} ultrafiltrate is excreted in the urine or faeces through biliary excretion (Baek et al. 2012). Most of the absorbed zinc is taken up by the liver, where it is stored bound to metallothionein, before it is distributed to other tissues (Osredkar 2011).

Zinc has structural, catalytic and regulatory functions in the cells (Maret 2013). As a cofactor for enzymes, zinc has roles in cell replication, tissue growth and repair, bone formation, skin integrity and immunity (Freeland-Graves and Bavik 2003). It is a catalytic agent in hydroxylation and other enzymatic reactions, participating in reactions that lead to the synthesis or degradation of nutrients such as carbohydrates, proteins, lipids, nucleic acids and vitamins (Ruz 2003; Berdanier and Zempleni 2009). In the brain, zinc modulates neuronal excitability and is stored in specific synaptic vesicles by glutamatergic neurons (Huang 1997). Zinc also participates in intracellular signalling pathways to regulate cell functions that include salivary glands, the prostate, the immune system and intestine cell proliferation and survival, ion transport and hormone secretion (Hershfinkel et al. 2007).

2.2.5.8 Copper: Cu

Copper is an essential dietary trace element with biological importance in reactions of electron transport and oxygen transport (Osredkar 2011). In the body, copper is found mainly in muscles, where about 40% of it is found, followed by the liver, brain, kidneys and heart (Berdanier and Zempleni 2009; Lee 2013). It exists in most foods in very small quantities, but the richest dietary sources include animal products such as shellfish, meat, kidney, liver and a few vegetable foods such as legumes, nuts and seeds (Johnson 2003; Lee 2013). A certain amount of copper is absorbed from the stomach, but the greatest amount is absorbed in the duodenum by passive diffusion or active transport regulated by metallothionein (Institute of Medicine 2001c). Following metallothionein binding and transport within enterocytes, copper binds mostly to albumin and in lesser proportions to transcuprein and small peptides to be transported to the liver in a similar fashion to zinc (Institute of Medicine 2001c;

Berdanier and Zempleni 2009). About 90% of plasma copper is incorporated into ceruloplasmin, a glycoprotein synthesized in the liver, to be carried to tissues where it is used for the synthesis of other copper-containing enzymes (DiSilvestro 2005; Lutsenko 2010; Ramos et al. 2016). This metal is secreted from the liver as a component of bile for excretion in the faeces. Small amounts of copper are excreted in the urine and sweat (National Research Council 2000).

Copper is a component of many metalloproteins and acts in several enzymes involved in the aerobic respiration, such as cytochrome *c* oxidase in mitochondria proteins, or carboxypeptidase and carbonic anhydrase (Institute of Medicine 2001c; Berdanier and Zempleni 2009). It also participates in the structure of some transcription-regulating proteins, playing a critical role in DNA and RNA metabolism (Berdanier and Zempleni 2009). Copper is required in the function of many superoxide dismutases, protecting against oxidant agents and oxidative stress (Johnson 2003; Birben et al. 2012), and favours the synthesis of melanin and catecholamines (Institute of Medicine 2001c). Bound to ceruloplasmin, copper serves in the oxidation of iron before being transported into the plasma (Ramos et al. 2016).

2.2.5.9 Cobalt: Co

Cobalt is an essential trace element in nutrition as it is necessary to form cobalamin (Vit B12). Much of the cobalt in the body appears as Vit B12 deposits in the liver (MacPherson and Dixon 2003; Berdanier and Zempleni 2009); in its mineral form, it is found in the liver, heart, bones, kidney, trachea and spleen, with the highest levels in the liver and kidneys (Wehner and Craig 1972; Brune et al. 1980; Berdanier and Zempleni 2009). It is fundamental in the diet of ruminants, owing to the role of ruminal bacteria that produce Vit B12, which are transferred to meat products for human nutrition (Smith 1987). The main dietary sources of cobalt include green leafy vegetables and meat products, in which cobalt derives from Vit B12 (Cobalt Development Institute 2006; Lee 2013). Enterocytes absorb both B12-linked cobalt bound to the gastric glycoprotein IF and Co in its free form (Taylor 1962; Berdanier and Zempleni 2009). The main route of cobalt excretion is urine, and it is also eliminated in small amounts in the faeces, sweat and hair (Gregus and Klaassen 1986; DiSilvestro 2005).

Cobalt functions and metabolism are known as those pertaining to cobalamin, which is essential for carbon transfer reactions involved in the synthesis of DNA and is necessary for the maturation of erythrocytes and the normal functioning of all cells (Institute of Medicine 1998b; Lee 2013). The enzyme methionine aminopeptidase 2 (MetAP2) is a cobalt protein, which binds cobalt directly, instead of using the corrin ring; MetAP2 catalyses the removal of N-terminal methionine residues (Freeland-Graves and Bavik 2003; Garrabrant et al. 2004) and is critical for tissue repair and protein degradation and participates in endothelial cell proliferation (Griffith et al. 1997).

2.2.5.10 Iodine: I

Iodine is an essential trace element in nutrition, and its main role in the body is to constitute the thyroid hormone (Pennington 2003; National Research Council 2005). The human body normally contains 20 to 30 mg of iodine, with more than 75% in the thyroid gland and the rest distributed throughout the body, particularly in the mammary glands, gastric mucosa and blood (Institute of Medicine 2001d; Ahad and Ganie 2010). As the main source of iodine on earth are the oceans, most dietary iodine comes from salt, seafood (seaweed and animals) and plant foods cultured in iodine-rich soil (Ahad and Ganie 2010), particularly those from the Cruciferae family such as cabbage (Berdanier and Zempleni 2009). The ingested iodine is converted to iodide, a negatively charged ion, and in this form, it is easily absorbed (Institute of Medicine 2001d; Zimmermann 2009), as occurs with dietary sodium iodide or protein-bound iodide. Iodide circulates in the blood both in free form and linked to proteins (Ahad and Ganie 2010), and most dietary iodide (about 80%) is retained in the thyroid gland, which uses it for the synthesis of thyroxine (Zimmermann 2009). However, like the thyroid, many tissues can uptake albumin-bound iodide, such as the mammary gland, directly by sodium-iodide symporter, a transmembrane glycoprotein (Rousset et al. 2000). Thyroxine is synthesized through the iodination of tyrosine. Upon the action of the thyroid-stimulating hormone (TSH), the thyroid gland secretes thyroglobulin whose tyrosine residues are iodinated by the iodide peroxidase, and following this monoiodothyronine is produced and then diiodothyronine, triiodothyronine and thyroxine (Pennington 2003). The thyroid secretes triiodothyronine (T3) and thyroxine (T4), which are carried to all cells in the body by the thyroid-binding protein (Institute of Medicine 2001d). T3 and T4 are metabolized in the liver and other tissues, and some thyroid hormone derivatives are secreted into the bile (Pennington 2003; Berdanier and Zempleni 2009). In the target cells, thyroxine is deiodinated to triiodothyronine by the action of 5′-deiodinase, a selenium-containing enzyme (Institute of Medicine 2001d; National Research Council 2005). The released iodide in the plasma is either used by the thyroid or excreted as organic iodide in the urine (Institute of Medicine 2001d).

The main role of iodine in the body is to coordinate and maintain the metabolism of all cells, pregnancy and body growth and development through the thyroid hormones (Pennington 2003). In the mammary glands, iodine is either transferred to the newborn or acts as a potent antioxidant owing to its electron donor ability (Ahad and Ganie 2010). The antioxidant function is also performed in the other tissues that uptake iodine (Smyth 2003; Berdanier and Zempleni 2009; Lee 2013). Other roles for iodine have been described such as stabilizing the function of adrenal glands during stress or participating in immune regulation (Institute of Medicine 2001d; Ahad and Ganie 2010). Even though the body is efficient at excreting iodide, it can be toxic when there is a deficiency of selenium or excess dietary or supplement intake. Iron deficiency impairs iodine function in the thyroid (Institute of Medicine 2001d; Zimmermann and Köhrle 2002).

2.2.5.11 Manganese: Mn

Bones contain the highest concentration of manganese in the body, followed by the pituitary gland, liver, pancreas and digestive tissue. Very little is known about manganese intake, needs, bioavailability from different foods and supplement sources or the influence of changes in nutritional status in the metabolism of the body (Institute of Medicine 2001e; Keen and Zidenberg-Cherr 2003; DiSilvestro 2005). Some dietary sources such as green leafy vegetables, beetroot, blueberries, wholegrains, fruits or some kinds of teas have been mentioned (Lee 2013). Manganese is absorbed in the small intestine in a regulated way against excessive uptake, and cobalt and iron compete with Mn for common bonding sites for absorption (Leach and Lilburn 1978; Institute of Medicine 2001e; DiSilvestro 2005). In the bloodstream, most dietary absorbed Mn binds to albumin and rapidly enters the liver, where some of it goes into liver enzymes and some is transported to other tissues or is excreted in the bile (Institute of Medicine 2001e; DiSilvestro 2005; Lee 2013). Manganese transport to the tissues occurs by transferrin and α-2-macroglobulin, and it seems that bone accumulation is passive (Institute of Medicine 2001e). Excretion occurs in the faeces following secretion into the intestine via bile, and it is regulated by the body's Mn content (Britton and Cotzias 1966; Lee 2013).

Manganese is a constituent of enzymatic systems in the body and is an essential cofactor in several metalloenzymes, such as superoxide dismutase, arginase and phosphoenolpyruvate carboxykinase, and therefore plays a crucial role in the control of oxidative stress and the metabolism of proteins and carbohydrates (Institute of Medicine 2001e; Keen and Zidenberg-Cherr 2003). As such, Mn abounds in mitochondria of hepatic cells. Manganese is also important in activating the glycosyltransferases, which are necessary for the formation of glycoproteins, which are involved in many processes including bone formation and maintenance (Institute of Medicine 2001e; Keen and Zidenberg-Cherr 2003; DiSilvestro 2005). Sequestration of Mn by neutrophil calprotectin has been considered an important mechanism of host defence against fungal and bacterial infection (Kehl-Fie et al. 2011; Brophy and Nolan 2015).

2.2.5.12 Molybdenum: Mb

Molybdenum is a cofactor for oxidation enzymes and is important for several biological systems (Mendel 2013a) and flavoproteins (Garattini et al. 2003). Dietary sources of Mo include vegetables, cereals, grains, dark green leafy vegetables and animal viscera (Institute of Medicine 2001f; Berdanier and Zempleni 2009). Most ingested Mo is absorbed by the epithelial cells of the gastrointestinal system and circulates in the blood in a protein-bound complex (Institute of Medicine 2001f; DiSilvestro 2005; Berdanier and Zempleni 2009). It is found mostly in the liver, kidneys, bones and skin (Institute of Medicine 2001f). The principal route for the excretion of Mo is the urine, with small amounts excreted via the bile in the faeces (Berdanier and Zempleni 2009).

Molybdenum is considered an essential trace element that acts as a cofactor for molybdoenzymes (Schindelin et al. 1996; Romão et al. 1997; Mendel 2013b), in the mitochondrial amidoxime-reducing component, which performs important functions owing to its N-reductive activity (Ott et al. 2015). The enzymes xanthine oxidase, aldehyde oxidase and sulphate oxidase require a prosthetic group containing molybdenum (Rajagopalan 1988; Ott et al. 2015). Xanthine oxidase is necessary for the production of uric acid from purines; sulphite oxidase converts sulphite to sulphate; and aldehyde oxidase is involved in the hydroxylation of heterocyclic nitrogen compounds, such as nicotinic acid (Rajagopalan 1988; Garattini et al. 2003; Mendel 2013b). Thus, Mo is important for the metabolism of DNA and RNA, of iron and of energy (Maiuolo et al. 2016), as well as for sulphide detoxification (Wang 2012).

2.2.5.13 Selenium: Se

Selenium cooperates with Vit E as an antioxidant and has important functions in the immune system (Thomson 2003; Rayman 2012). In the body, selenium is distributed about 30% in the liver, 15% in the kidneys, 30% in muscles and 10% in plasma (Berdanier and Zempleni 2009; Lee 2013). Plant dietary Se sources reflect the quantity in the soil and include foods such as Brazil nuts, cereal grains, onions or soybeans, and animal sources are seafood and fish, meat, eggs and dairy (Institute of Medicine 2000b; Rayman 2012). Selenium is very well absorbed in the upper segment of the small intestine, with most ingested selenium absorbed readily from a variety of foods by mechanisms that include some overlap with amino acid absorption (Institute of Medicine 2000b; Berdanier and Zempleni 2009; Rayman 2012). Selenium absorption is not controlled by homeostatic mechanisms such as those for iron, so plasma selenium is proportional to selenium intake at both high and low levels (DiSilvestro 2005). Selenium is transported from the gut by LDL and VLDL to the tissues, and it is found in higher quantities in the liver, spleen, muscle, nails, hair and tooth enamel (Berdanier and Zempleni 2009). Most minerals that exist in metalloproteins are attached to selected amino acids, acting as a cofactor (Cook et al. 2008), or in the case of iron, by insertion into a heme or cytochrome structure, which can be inserted into a protein (Berdanier and Zempleni 2009; Jiang et al. 2009). Because selenium forms a selenium-sulphur bond using the SH group of a sulphur amino acid, methionine becomes a selenomethionine, or more typically cysteine becomes a selenocysteine amino acid (Thomson 2003; Schmidt and Simonović 2012). The particular synthesis mechanism for the incorporation of selenocysteine into protein requires Vit B6 as a coenzyme and results in selenoproteins (Soda et al. 1999; DiSilvestro 2005). Excess Se intake is toxic, but unlike iron and copper, whose excretion processes are not so efficient, it is actively excreted in the urine. The urinary system functions to maintain optimal selenium status, and under normal intake, equivalent amounts are excreted in both the urine and faeces (Institute of Medicine 2000b; Thomson 2003; Rayman 2012).

Selenium is important for almost all cell types in the body, owing to its critical role in glutathione peroxidase, which is the body's most powerful antioxidant and its isoenzymes (Thomson 2003; Rayman 2012). Glutathione peroxidase is fundamental for red blood cells because they lack mitochondria, an organelle that regulates the redox state. Erythrocytes contain haemoglobin; therefore they have to control the redox state so that haemoglobin can release oxygen in exchange for CO_2. Such a process favours the formation of hydrogen peroxide, which is inhibited by the action of glutathione peroxidase (DiSilvestro 2005; Berdanier and Zempleni 2009). Furthermore, glutathione peroxidase catalyses the reduction of various organic peroxides as well as hydrogen peroxides (Berdanier and Zempleni 2009). Membranes contain unsaturated fatty acids, which can be easily oxidized. The activity of glutathione peroxidase is critical to avoid stability impairment of these membranes through oxidation (Berdanier and Zempleni 2009; Lubos et al. 2011). Along with other enzymes of the antioxidant system, adequate provision of copper and vitamins A, E and C will assure that cell damage by free radicals is minimized, even if the diet is marginal in selenium (Clarkson and Thompson 2000; Choe and Min 2009; Omur et al. 2016). As an essential component of glutathione peroxidase, Vit E plays an important role in suppressing free radical production, but its site of action is distinct from that of Se (Berdanier and Zempleni 2009; Lubos et al. 2011). Different selenoproteins act as glutathione peroxidase isozymes, including a thioredoxin reductase (DiSilvestro 2005; Rayman 2012). Selenium is also an integral part of the enzyme iodothyronine deiodinase, which is necessary for thyroid hormone metabolism (DiSilvestro 2005; Schweizer et al. 2014). Adequate Se intake and metabolism have been associated with certain effects including (1) prevention of cancer (Naithani 2008; Brozmanová 2011; Chen et al. 2013), (2) infertility (Mistry et al. 2012; Pieczynska and Grajeta 2015), (3) cardiovascular diseases (Oster and Prellwitz 1990; Flores-Mateo et al. 2006) and (4) inflammatory conditions such as arthritis (Tarp et al. 1985; Huang et al. 2012; Mattmiller et al. 2013) or pancreatitis (Bowrey et al. 1999). The effects of selenium on the immune function have been extensively studied, and a variety of mechanisms have been identified, such as improving natural killer cell activity (Methenitou et al. 2001; Gershwin et al. 2004) or the production of lymphocytes and immunoglobulin (Keen et al. 2004; Rocha et al. 2016), as well as contributing to the maintenance of immune regulation (Kieliszek and Błażejak 2016).

2.2.5.14 Chromium: Cr

Chromium is associated with the metabolism of glucose (Mertz 1969, 1993, 1998). It has been found in bone and in breast milk of humans (DiSilvestro 2005; Berdanier and Zempleni 2009). Chromium dietary sources include corn oil, clams, grain cereals, yeast, meat and drinking water, and the amounts reflect its presence and quantity in soil (Anderson et al. 1992; Berdanier and Zempleni 2009). Organic chromium such as chromium picolinate is more efficiently absorbed than inorganic forms whose absorption is very low, under 2.5% in some studies (Ducros 1992;

Berdanier and Zempleni 2009). In blood, it can bind to a number of molecules, but it is mostly found in transferrin. Organic chromium is removed from the body in the urine (Mertz 1993; Institute of Medicine 2001g; Berdanier and Zempleni 2009).

Chromium enhances the action of insulin, which influences the metabolism of carbohydrates, lipids and proteins (Institute of Medicine 2001g; Berdanier and Zempleni 2009). It binds to DNA and has functions on the regulation of gene expression (Ye and Shi 2001; Hazane-Puch et al. 2010). Studies involving immune cells have demonstrated that Cr is required for lymphocyte proliferation (Keen et al. 2004). Some studies have also demonstrated that Cr may play a protective role against metabolic diseases such as obesity, hypertension and diabetes (Wiernsperger and Rapin 2010; Bai et al. 2015; Panchal et al. 2017).

2.3 Nutrition for Immune Response and Immunonutrition

Hypermetabolism and hypercatabolism characterize non-infectious conditions such as cancer (de Luis et al. 2014), burn injuries (Mendonça et al. 2011), post-surgery (Finnerty et al. 2013) and acute renal failure (Wooley et al. 2005) as well as infectious diseases such as sepsis (Longarela et al. 2000), tuberculosis (Gupta et al. 2009), malaria (Stettler et al. 1992) and HIV (Garcia-Lorda et al. 2000). These high stress conditions call for an increased need for nutrients, known as hypermetabolism—an abnormal increase in the basal metabolic rate—that will lead to a continual insufficiency of energy and/or protein to meet the metabolic demands of the body. This increased metabolic demand is for nutrients of all classes, such as carbohydrates, lipids and proteins, to provide energy and structure, as well as vitamins and minerals, which accomplish regulatory and modulatory functions in a number of metabolic processes, including hormonal homeostasis (Mehta and Duggan 2009).

Hypermetabolism is a process that if untreated evolves to hypercatabolism, which culminates in cachexia and fatal general organ failure. The concept of hypermetabolic response is well established (McClave and Snider 1994); however it is difficult to measure. In general, losses above 10% of total body weight in hospitalized patients are considered a threshold for diagnosing hypermetabolic response and equate to 15–20% losses of total body protein, a level beyond which significant deterioration in the clinical outcome is observed (Hansen et al. 2000; Gil Hernández and Sánchez de Medina Contreras 2010). The evaluation of the nutritional status is more appropriate to estimate the degree of hypercatabolism, which results from the consumption of structural proteins as cellular fuel for the maintenance of vital functions. Hypercatabolism is clinically assessed by using the Bistrian's index which determines the balance between ingested and excreted nitrogen (Bistrian 1979). A number of formulas have been designed to calculate needs for daily caloric requirements during illnesses associated with hypermetabolism, and Harries-Benedict's equation is the most used, so far (Harris and Benedict 1918; Japur et al. 2009). Based on an evaluation of the individual's nutritional status, factors representing activity level, intensity of stress, changes in body weight and temperature are added to the Harries-Benedict's equation (Long et al. 1979) to calculate the proper nutrient combination for a given case.

Studies on HIV have greatly contributed to the knowledge of nutritional aspects associated with hypermetabolism and hypercatabolism (Stein et al. 1990; Garcia-Lorda et al. 2000; Cunningham-Rundles et al. 2005; Kosmiski 2011; Serrano-Villar et al. 2017). Because such studies have explored a variety of events and systems in depth, they can be applied to other debilitating infectious and inflammatory conditions. A number of studies have addressed how co-infections, such as those between HIV and various other pathogens, are frequent and represent a critical association with malnutrition, resulting in cachexia and poor clinical outcome by several mechanisms (Van Lettow et al. 2003; Janssen et al. 2005; Semba et al. 2010; Werneck et al. 2011; Katz 2012; Hicks et al. 2014; Church and Maitland 2014; Cohen et al. 2015; Mkhize et al. 2017). Intracellular parasites, such as *Leishmania* spp. or *Plasmodium* spp., not only exploit nutrients already available in the cell and the cell's energy-yielding system, but they further induce the cell to provide actively for their nutrition (Trager 1974; Haanstra et al. 2016; Srivastava et al. 2016). Leishmaniasis in its visceral or tegumentary form is among the most studied debilitating infectious diseases in people and animals. Malnutrition is a risk factor for the establishment, maintenance and progression of *Leishmania* infection in the host. Studies in rodents have described that lack of sufficient protein in the diet (Pérez et al. 1979) or general malnutrition (Reithinger et al. 2001) clearly favours susceptibility to *Leishmania* infection. On the other hand, human visceral leishmaniasis has been considered a natural model of infection-induced cachexia (Pearson et al. 1992). Several studies have shown that undernutrition has a direct influence on the outcome of any clinical form of leishmaniasis caused by different *Leishmania* species in diverse parts of the globe, resulting in more severe presentations (Badaró et al. 1986; Harrison et al. 1986; Cerf et al. 1987; Dye and Williams 1993; Weigel et al. 1994; Hida et al. 1999; Cunha et al. 2001; Machado-Coelho et al. 2005; Baba et al. 2006; Alvar et al. 2006; Rukunuzzaman and Rahman 2008; Malafaia 2009; Rijal et al. 2010; Harhay et al. 2011; Gatto et al. 2013). In fact, evidence of an interaction between malnutrition, leishmaniasis and immunosuppression has been demonstrated (Harrison et al. 1986; Anstead et al. 2001; Kumar et al. 2014). Poorer nutritional statuses have been associated with poor response to chemotherapy with sodium stibogluconate and lower serum levels of matrix metalloproteinase 9 (MMP9), a cytokine that inversely correlates with disease progression (de Oliveira et al. 2013; Gadisa et al. 2017). In malaria, a similar dynamic between infection and nutrition also occurs (Méndez and Dobaño 2004). A consequence of the immunosuppression caused by HIV, *Plasmodium* or *Leishmania* infection is the augmented susceptibility of the host to acquire other infections and to develop metabolic imbalances or degenerative conditions, generating co-morbidities and deepening nutritional deprivation in a vicious circle. Indeed, the literature contains several examples of poorer clinical evolution and response to pharmacological treatment of malaria, leishmaniasis, tuberculosis, AIDS and a number of helminthic diseases in populations where malnutrition is endemic (Cegielski and McMurray 2004; Cunningham-Rundles et al. 2005;

Krawinkel 2012; Bhargava et al. 2014; McCuskee et al. 2014; Starr et al. 2015; Zacarias et al. 2017).

Immune responses are greatly responsible for heightened caloric requirements in the course of infectious diseases (Schaible and Kaufmann 2007; França et al. 2009). The immune system is prepared to respond intensely to antigenic stimuli and inflammatory signals, which require the acceleration of cellular metabolism and thus increased energy demands (Palmer et al. 2016). Nevertheless, the immune system depends on the activity and balance of the other systems. Nutrient sensitive hormones such as insulin, glucagon, corticosterone, growth hormone, thyroxin and catecholamine regulate the activity of leukocytes throughout hormone-receptor binding (Klasing 1994; Plummer and Deane 2016). Accordingly, an increase in neuroendocrine activity to ensure tissue repair and homeostasis characterizes hormonally induced increased glucose production, increased use of nitrogenous products and increased enzyme activity, whose consequences include the elevation of catabolism and reserve burning (Plank and Hill 2000; Simsek et al. 2014). Although such processes are not as well explored and described in infectious diseases that are not considered hypermetabolic, higher caloric expenditure should be expected as a result of an activated immune response to a given infection, in proportional levels.

Carbohydrates and lipids are the main sources of energy for the cells. Not only are their proportions in the diet important, but nowadays the quality of dietary carbohydrates and lipids with regard to metabolism and immune functions during disease is drawing attention (Vanderhoof 1998; Wolowczuk et al. 2008; Meckling 2009; Krawinkel 2012). Omega-3 fatty acids help reduce inflammation, while some omega-6 fatty acids tend to promote inflammation (Lachance et al. 2014). The caloric balance in the body as a whole has an impact on the progress and outcome of an infectious process (Schaible and Kaufmann 2007; França et al. 2009; Plummer and Deane 2016). Nutritional deficiency starting during intrauterine life interferes negatively with adult immunocompetence. Caloric overnutrition also negatively affects the immune response to infections (Krawinkel 2012). Extensive studies have demonstrated that glucose is not only the main energy source for the body, but its metabolic pathways participate in many processes related to immunity and its development (Vanderhoof 1998; Cunningham-Rundles et al. 2005; Norata et al. 2015), as well as in the pathophysiology of infectious diseases. Muscles are the primary source for glucose upon acute or immediate body demand, which explains why body losses occur at the expenses of muscles during intense disease processes and involves protein losses during hypercatabolism (Berg et al. 2013; Puthucheary et al. 2013; Preiser et al. 2015). Glucose from dietary sugars and body reserves associates with peptides to form glycoproteins, which are the molecular basis of a variety of functional elements of the immune response, such as antibodies, receptors, cytokines and chemokines, among others (Helderman 1981; Wolowczuk et al. 2008; Shi et al. 2011; Norata et al. 2015). Extrinsic signalling, such as MHC-TCR engagement, acts as a stimulation factor that regulates glucose uptake and metabolic activity in lymphocytes (Rathmell et al. 2000). Also, the glycolytic pathway plays a fundamental role in the differentiation of macrophages to the type 2 (M2) regulatory phenotype or to inflammatory type 1 (M1). Indeed, inhibition of glycolysis results in loss of the regulatory phenotype, including Il-10 production (Suzuki et al. 2016). Host-pathogen interactions and infection outcomes involve crucial aspects of glucose pathways; as an example, viruses are able to interfere (Palmer et al. 2014) and even disrupt the glucose metabolic machinery of their host T cells and phagocytes (Palmer et al. 2016). Sugar metabolism imbalances are not only the consequences, but they also participate in the progress of tissue damage, catabolism and dysfunction under infection and inflammatory processes. During sepsis, hyperglycaemia, hypoglycaemia and glycaemic variability are associated with adverse outcomes and death, and they occur due to the massive activation of pro-inflammatory mediators and release of counter-regulatory hormones, leading to excessive hepatic gluconeogenesis and peripheral insulin resistance (Plummer and Deane 2016). As shown by Palmer and collaborators (Palmer et al. 2016), during HIV infection a continuous activation of T lymphocytes, characterized by increased surface expression of glucose transporter 1 (GLUT-1) and high glycolysis activity, results in metabolic exhaustion and cell death.

The roles of lipids as components of cell membranes, as constituents of hormones, as the means by which energy is stored in the body and as maintainers of body temperature are all crucial during an immune response (Hwang 1989; Alvarez-Curto and Milligan 2016). In the healthy individual, adipose tissue is formed from dietary carbohydrates and lipids, and it is stable as fat storage. In the course of infection and immune response, the heightened metabolic demand induces lipid mobilization, and lipids are transformed into readily usable glucose for energy (Weinstein et al. 1994; Sjögren et al. 2007; Palmer et al. 2016). Liposoluble vitamins are also mobilized from the host's adipose tissue under hypermetabolism and hypercatabolism due to infection, immune response and inflammation conditions. Protection against oxidative stress is a crucial function of Vits A, E, D and K for the cells during the course of mobilization of the immune system against pathogens (Chandra and Chandra 1986; Rahman 2007; Valko et al. 2007; Albahrani and Greaves 2016; Omur et al. 2016). Metabolic and catabolic consequences of attempted caloric balance critically involve gluconeogenesis in a host under hypermetabolic demand. During glucose synthesis from lipid reserves, a great amount of free radicals is produced, leading to oxidation of cell membranes and the formation of ketone, which increase organic acidification (Simsek et al. 2014). Moreover, if the proportion of adipose tissue is above optimum before infection and disease onset, the host is more susceptible to developing higher levels of inflammation, more oxidative stress and more acidification (Krawinkel 2012; Cousin et al. 2016). Cumulative results of research have found the evidence that rather than quantity, the quality of dietary lipids is important to provide essential fatty acids that participate in several metabolic functions (Broadhurst 1997; Chandrasekharan 1999; Diekman and Malcolm 2009; Jacobson et al. 2012; Abumrad and Davidson 2012). More than in health, diet-related metabolic functions are vital to the outcome of diseases, including infectious and inflammatory processes. Prostaglandins, leukotrienes and acute phase protein cascades are key defence components that are directly involved in handling infectious processes, through first-line inflammatory response, and the proportion of end products of these inflammatory cascades is sharply influenced by the composition of available fatty acids from the diet (Calder 2013a; Alvarez-Curto and Milligan 2016; Marion-Letellier et al. 2016). On the other hand, it has been demonstrated in healthy individuals that there is an association between a lower grade of inflammation, as evaluated by inflammatory markers, such as IL-6, A1GP and haptoglobin, and the regulation of insulin sensitivity and lipid metabolism (Heliovaara et al. 2005). In a review of the role of infection in the development of atherosclerosis, Campbell and Rosenfeld (2015) discuss various studies that explore the interaction between hyperlipidemia, infection by several bacterial pathogens, and the development on inflammation.

The participation of proteins in the caloric balance also reflects the processes of hypermetabolism and hypercatabolism. This is because during gluconeogenesis, the muscular collagen, actin and myosin are mobilized and metabolized to produce glucose and thus energy (Gil Hernández and Sánchez de Medina Contreras 2010; Argilés et al. 2016). As muscle tissues function as storage of amino acids and glycogen for immediate release under elevated metabolic demand, during hypercatabolism the first losses occur in the musculature to preserve vital functions (Espinosa et al. 2016). When protein losses exceed 20% of total body protein, physiological systems begin to show functional deficiencies, especially in immunity, but also in cardiopulmonary and musculoskeletal systems. When protein losses surpass 30%, it means the glycogen and lipid reserves have been used, and then the body starts to catabolize its structure, and the survival rate reduces by 20% (Gil Hernández and Sánchez de Medina Contreras 2010). According to the Bistrian's index, the proportion of urea excreted in the urine helps to estimate the nitrogen balance in relation to the ingested amount of protein (Bistrian 1979). As proteins are made up of about 16% nitrogen, the excreted amount for a healthy individual should be proportional to that ingested. If the urinary losses of nitrogen are higher than the protein ingestion, this means that the protein reserves are being used or overcatabolized (Mahan et al. 2013; Ogunbileje et al. 2016). Thus, the protein balance should be re-established according to individual catabolic losses, which means that in diseases with greater structural losses, more proteins should be given to the patient. The nutrition of the individual under high demand will take into account not only maintenance of body mass but also the need to adapt to metabolic conditions associated with response to infections and recovery of the body (Mahan et al. 2013).

The concept of *immunonutrition* is somewhat different from the idea of general nutrition during a disease. The understanding of the role of malnutrition in the populations with a higher susceptibility to infectious agents, as well as in the development of more critical clinical pictures, together with the evolution of the knowledge in immunology, has led to the development of the concept of nutritional immunology or immunonutrition (Evoy et al. 1998; Grimble 2001; Mizock 2010; Chow and Barbul 2014; Mody et al. 2014; Murray et al. 2015; Vetvicka and Vetvickova 2016). Nutritional deficiencies underlie the impairment of immune functions in the course of diverse conditions, including infectious diseases (Keusch 2003; Zapatera et al. 2015). Immunonutrition is a growing area of interest. The initial publications pointing to the fact that interactions between infections and nutrition are synergistic date back to little before the 1960s (Keusch 2003; Scrimshaw 2007). The advance in immunological tools has supported extensive investigations to elucidate the specific mechanisms involved in the complex interactions between nutrition, immunity and disease in its various presentations (Scrimshaw 2007; Afacan et al. 2012). Immunonutrition as a concept is the therapeutic application of the beneficial effect of some nutrients on the immune response, owing to specific nutrient actions that regulate various activities of the immune system (Grimble 2001). In this sense, specific food items or nutrients are administered in higher quantities than those present in the regular diet to satisfy the particular demands of the body undergoing disease. The most commonly used nutrients for immunonutrition include amino acids, nucleotides, polyunsaturated fatty acids (PUFAs), vitamins and minerals, in the form of supplements or of foodstuffs which are very rich in some specific nutrient (Mainous and Deitch 1994; Redmond et al. 1998; Efron and Barbul 2000; Grimble 2001, 2005; Singh et al. 2002; Chan 2008; Chow and Barbul 2014). Other dietary components with immunomodulatory potential include polyphenols and nutrients with the ability of modulating the gut microbiota such as fibre, prebiotics and probiotics (Romeo et al. 2010; Pérez-Cano et al. 2012). Immunonutrition remains a challenge to researchers, despite the proliferation of research in the area due to the vastitude of its scope and the unavoidable controversy it raises because much has yet to be understood. Naturally, a substantial number of studies have been published on the use of diverse nutrients as *immunonutrients* (Table 2.1).

The activity of the immune system impacts the requirements for micronutrients in such a way that the immune status can be regarded as a sensitive indicator of micronutrient supply (Ströhle et al. 2011). All groups of micronutrients have been studied as direct participants in immune functions. For example, macrophages and dendritic cells (DC) in the skin and mucosal epithelia produce retinoic acid from the dietary provision of Vit A; this retinoic acid plays important roles in cell growth and differentiation during immune responses, preventing autoimmunity and immunopathogenic consequences (Mucida et al. 2007; Manicassamy et al. 2009; Cassani et al. 2012; Zhou et al. 2012; Reza Dorosty-Motlagh et al. 2016). Immunopathogeny is a common pathophysiological aspect of some infectious diseases, such as cutaneous or visceral leishmaniasis (Goto and Prianti 2009; Soong et al. 2012) or viral hepatitis (Bayraktar et al. 1997; Cassani et al. 1997), for example. Dendritic cells (DC) depend on retinoic acid to perform their antigen-presenting functions and to induce differentiation of naïve T cells into T regulatory cells (Mucida et al. 2007; Cassani et al. 2012). In the course of infections, DC-produced Vit A participates in metabolic pathways that favour the development of effector $CD4^+$ T cells into T helper (Th) type 1 and the production of pro-inflammatory cytokines (Raverdeau and Mills 2014).

Vit D participates in the network between the immune-brain-gut systems. It has been shown that many factors associated with the brain function, innate immunity and the gut microbiome are related to immune regulation by Vit D (Pandolfi et al. 2017; Chirumbolo et al. 2017). Calcitriol has immunomodulatory effects on the different cell types of the immune system by biding to the Vit D receptor, including regulating T-lymphocyte proliferation, immunoglobulin production, natural killer

Table 2.1 Immunonutrition: some relevant types, common natural sources and effects of nutrients

Nutrient	Type	Natural sources	Effects
Complex or slowly absorbed carbohydrates	Oligosaccharides and polysaccharides	Vegetables, fruits	Reduce metabolizable energy relative to sucrose; reduce insulin secretion; as a substrate that favours a healthy colonic microbiota, positively alter the intestinal flora (Vanderhoof 1998)
Eicosapentaenoic acid (EPA), docosahexaenoic acid (DHA), α-linolenic acid (ALA)	Omega-3 long-chain fatty acids	Deep sea fish, salmon, krill (EPA; DHA) Flaxseed or linseed oil (ALA)	Protect membranes against free radical damage; reduce pro-inflammatory cytokine production; increase the amounts of anti-inflammatory compounds such as resolvins and protectins in the site of inflammatory response; reduce systemic inflammation in diverse conditions, including sepsis (Alexander 1998; Grimble 2009; Pierre et al. 2013)
L-Glutamine	Non-essential amino acid	Meats, seafood, milk, nuts, eggs, cabbage, beans	Is a nutrient for immune cells and a substrate for glutathione formation; modulates the gut barrier function; precursor to the neurotransmitter γ-aminobutyric acid (GABA) (Grimble and Grimble 1998; Grimble 2009; Pierre et al. 2013)
L-Arginine	Dibasic conditionally essential amino acid	Meats, poultry, fish, seafood, eggs, milk, yogurt, cheese, nuts, seeds, oats, beans, wheat germ	Stimulates anabolic hormone release; improves nitrogen balance; augments lymphocyte mitogenesis in the thymus (Evoy et al. 1998; Grimble 2009; Pierre et al. 2013)
L-Leucine	Essential amino acid	Cheese, meats, fish, seafood, eggs, nuts, seeds, oats, beans, soybeans	Stimulates protein synthesis and is the only dietary amino acid that directly promotes cell growth in the body; slows down muscle

(continued)

Table 2.1 (continued)

Nutrient	Type	Natural sources	Effects
			proteolysis in the critically ill patient (Gershwin et al. 2004; Pierre et al. 2013)
L-Taurine	Conditionally essential, sulphonated β-amino acid	Shellfish, meat, beef, lamb, eggs	Stabilizes membranes and acts as an antioxidant for lymphocytes; stimulates glycolysis and glycogenesis; helps to preserve neutrophil phagocytic ability (Bachmann 2012)
L-Methionine and L-tryptophan	Essential sulphur amino acids	Fish, meat, eggs, milk, chicken, turkey, pork	Are substrates for glutathione synthesis; regulate T cell function, acute phase response, inflammatory response and cytokine production (Grimble and Grimble 1998)
Purines and pyrimidines	Nucleotides	Fish and seafood, organ meats, fresh vegetables	Participate in gut repair and mucosal defence, lipid metabolism and iron bioavailability (Gruenwedel 2003; Pierre et al. 2013)
Vit A	Liposoluble vitamin	Organ meats, fish liver, milk and dairy, eggs	Is necessary for epithelial metabolism and for vision and eye epithelia; stimulates collagen synthesis in wounds; promotes an efficient antiviral immune response; modulates innate tolerogenic dendritic cells and innate lymphoid cells in the gut; regulates the balance between Tregs and Th17 cells (Tanumihardjo et al. 2016; Elenius et al. 2017; Czarnewski et al. 2017; Erkelens and Mebius 2017)
Vit B6: pyridoxine	Hydrosoluble vitamin	Meats, cereals, legumes, nuts, fruits, vegetables	Is antioxidant, acting as a cofactor in the synthesis of L-cysteine which is a precursor for

(continued)

Table 2.1 (continued)

Nutrient	Type	Natural sources	Effects
			glutathione; required in the proliferation of lymphocytes and biosynthesis of antibodies (Grimble 1997)
Vit B9: folic acid	Hydrosoluble vitamin	Liver, mushrooms, green leafy vegetables, poultry, meat, seafood, potatoes, fruits	Has a protective function against cardiovascular disease, cancer and cognitive dysfunction; reduces hyperhomocysteinemia; regulates NK cell activities (Ebara 2017; Paniz et al. 2017)
Vit B12: cobalamin	Hydrosoluble vitamin	Liver and kidney, milk, eggs, fish, cheese, meats	Protects against haematological and neurological dysfunctions; has antioxidant functions and prevents hyperhomocysteinemia (Green et al. 2017)
Vit C	Hydrosoluble vitamin	Vegetables and fruits such as papaya, rosehips, acerola, citrus fruit, guava, kiwi, broccoli, red peppers and other plants	Is antioxidant, exerting free radical scavenging activity; enhances the neutrophil function; participates as a cofactor in the hydroxylation of L-proline and L-lysine residues in procollagen, favouring wound healing; protects against vascular leakage, epithelial barrier disruption and increased alveolar fluid levels during sepsis; reduce disease manifestations and prevent infections caused by bacteria, viruses and protozoa (Gershwin et al. 2004; Pierre et al. 2013; Hemilä 2017)
Vit D3	Liposoluble vitamin	Fish, fish/cod liver oil, liver, buttermilk, egg yolks	Regulates both adaptive and innate immunity and the inflammatory cascade; enhances lymphocyte responses by T cell activation, differentiation and migration;

(continued)

Table 2.1 (continued)

Nutrient	Type	Natural sources	Effects
			regulates gene expression involved in the interleukin signalling pathway, oxidative stress response, apoptosis signalling pathway and gonadotropin-releasing hormone receptor pathway (Jeffery et al. 2012; Szymczak and Pawliczak 2016; Pasing et al. 2017)
Vit E	Liposoluble vitamin	Wheat germ, rosehips, various nuts, corn, sunflower or soy oils	Is a potent antioxidant; protects the cell membranes against lipid peroxidation, particularly Vit E-rich immune cells; supports lymphocyte proliferation, IL-2 production and DTH response; restores an impaired response to infection by enhancing NK cell activity, IFN-γ production and neutrophil recruitment; has a role in the prevention against cardiovascular diseases, cancer, neurodegenerative conditions and pneumonia (Szymańska et al. 2017; Hemilä 2017; Pae and Wu 2017)
Vit K	Liposoluble vitamin	Egg yolk, meat, liver or dairy products	Inhibits calcification of soft tissues, participates in bone metabolism and protects against cardiovascular diseases, diabetes, arthritis and cancer; is necessary for brain metabolism, function and protection from oxidative stress (Ferland 2012; Schwalfenberg 2017; Suksomboon et al. 2017)

(continued)

Table 2.1 (continued)

Nutrient	Type	Natural sources	Effects
Magnesium	Macroelement	Brewer's yeast, green leafy vegetables, nuts, seeds, soybeans, oats or milk	Protects the DNA from oxidative damage; reduces insulin resistance when combined with chromium; has anti-inflammatory properties in chronicle inflammatory diseases (Petrović et al. 2016; Dou et al. 2016; Brenner et al. 2017)
Selenium	Essential trace element	Brazil nuts, cereal grains, onions or soybeans, and animal sources are seafood and fish, meat, eggs and dairy	Act as antioxidant; necessary for proliferation and differentiation of naïve CD4- T cells towards T-1 cells; directs macrophages towards the M2 phenotype (Pierre et al. 2013; Steinbrenner et al. 2015)
Zinc	Oligoelement	Shellfish, meat, fish, poultry eggs, dairy, seeds, alfalfa, celery, legumes, nuts and wholegrains	Regulates the innate and adaptive immune responses; enhances the defence against bacterial infection and sepsis; controls oxidative stress during inflammatory responses; necessary for pathogen elimination by neutrophil extracellular trap (NET) formation (Gammoh and Rink 2017)
β-Carotene (pro-Vit A)	Phytochemical	β-Carotene is found in green and yellow vegetables, such as carrots, sweet potatoes, pumpkin, kale, spinach, collards and squash	Enhances cell-mediated immune responses, expression of surface molecules on monocytes and secretion of the antitumoural cytokine and protects immune cells from oxidative damage by inhibiting lipid peroxidation on cell membranes (Hughes 2001)
Curcumin or diferuloylmethane	Phytochemical	It is a polyphenol found in the rhizome of *Curcuma longa*	Inhibit mediators of the inflammatory response; reduce the secretion of

(continued)

Table 2.1 (continued)

Nutrient	Type	Natural sources	Effects
		(turmeric) and in other *Curcuma* spp.	TNF-α and IL-1β and the production of COX-2-induced prostaglandin G2; potent antioxidant (Lestari and Indrayanto 2014; Ghosh et al. 2015; Pulido-Moran et al. 2016)
Flavonoids	Phytochemical Family of polyphenolic	*Flavonols* are found in onions, apples, berries, kale, leeks, broccoli, blueberries, red wine and tea; *flavones* are found in fruit skins, parsley and celery; *isoflavones* are present in leguminous plants, mainly soy and soy products; *flavanones* are exclusive of citrus fruits; *flavanols* are found in cocoa and tea; *anthocyanidins* are found in red wine and berry fruits	Act as anti-inflammatory and antioxidant; modulate gene expression and inflammatory response; activate the innate immune system and the immune system against tumours; regulate the activities of numerous cell types, as basophils, neutrophils, eosinophils, lymphocytes, macrophages, platelets, smooth muscle or hepatocytes (Middleton 1998; Pérez-Cano and Castell 2016)
Lycopene	Phytochemical	It is a red, lipophilic and naturally occurring carotenoid in many fruits and vegetables, such as tomatoes	Enhances the antioxidant response; inhibits tumour proliferation, induces apoptosis and decreases the metastatic capacity (Luo and Wu 2011; Holzapfel et al. 2013)
Resveratrol	Phytochemical Family of polyphenolic	Polyphenol is found in grapes and grape products such as red wine	Inhibits inflammation; modulates the development of cytokine-producing $CD4^+$ and $CD8^+$ T cells; regulates specific antigen-induced generation of cytotoxic T lymphocytes; modulates natural killer (NK) activity of PBMC (Falchetti et al. 2001; Kowalska et al. 2017)

(continued)

Table 2.1 (continued)

Nutrient	Type	Natural sources	Effects
Prebiotics, probiotics and synbiotics	Bacterial strains and metabolites	Selectively fermented foods, such as sauerkraut, kefir, natto, tempeh, kombucha, kimchi, pickles and yoghurt, containing vegetable fibres that cannot be digested by the host, but are metabolized by the colonic microbiome and live microorganisms. Synbiotics is the combination of probiotics and prebiotics	Modulate the metabolism of short-chain fatty acids, amino acids and plasma lipoproteins; regulate the balance between immune tolerance and inflammation, keeping a healthy intestinal microbiota and maintaining proper immune functions; prevent and treat immunological disorders (Romeo et al. 2010; Frei et al. 2015)

(NK) cell toxicity and cytokine production (White 2012; Pandolfi et al. 2017; Chirumbolo et al. 2017). As an example of the role of Vit D in host defence against infectious diseases, a recent study has demonstrated that Vit D3 supplementation protects against severe forms of tuberculosis in wild animals (Risco et al. 2016). Similar findings have been described in studies on domestic species and humans (Waters et al. 2001; Nursyam et al. 2006; Ströhle et al. 2011; Lalor et al. 2012). Furthermore, Vit D deficiency is consistently related to susceptibility to sepsis and mortality in people (de Haan et al. 2014).

Some of the roles for Vit E in the immune system include participation in the regulation of T cell functions, in the production of immunoglobulins, pro-inflammatory cytokines and chemokines, and in the control of oxidative stress (Grimble 1994; Field et al. 2002; Ramakrishnan et al. 2004; Molano and Meydani 2012; Azzi et al. 2016; Bou Ghanem et al. 2017; Pantelidou et al. 2017; Bouamama et al. 2017). Studies in animals have demonstrated that Vit E has effects such as (1) enhancing immunity against bacteria such as *Streptococcus pneumoniae* (Bou Ghanem et al. 2015, 2017) or *Listeria monocytogenes* (Wu et al. 2012), (2) reducing helminth *Haemonchus* parasite burden (De Wolf et al. 2014) and (3) modulating the immune response to herpes simplex virus (Sheridan and Beck 2008). In humans, Vit E supplementation has been considered helpful in the treatment of hepatitis B (Andreone et al. 2001; Dikici et al. 2007).

Vit K, like Vit D, also plays a role in preventing oxidative damage and inflammation-induced injury to the host (Shearer and Newman 2008). It has been demonstrated that the anti-inflammatory properties of Vit K include the downregulation of the pro-inflammatory cytokines TNF-α, IL-1β and NF-kB mediated by a Gla-rich protein (Viegas et al. 2017). Vit K derivatives produce an inhibition of the proliferative response and the induction of apoptosis in activated T cells (Hatanaka et al. 2014).

The criteria used to provide vitamins and other micronutrients to vulnerable patients and populations, however, raise concern (Ramakrishnan et al. 2004; Long et al. 2007; Mizock 2010; Chawla and Kvarnberg 2014; Mangin et al. 2014; Mundi et al. 2016; Smedberg and Wernerman 2016; Hemilä 2017; Balcells et al. 2017). Indeed, the indiscriminate use of vitamins has been criticized (Feleszko et al. 2014). The participation of hydrosoluble vitamins in reactions that influence the immune response and host protection against pathogens has been studied in animals and in the course of human conditions, and the subject has yielded a multitude of publications. Many studies have reported that supplementation with these micronutrients results in fewer infections (Selmi et al. 2004; Wintergerst et al. 2006; Elste et al. 2017). The beneficial effects of vitamins of the B complex have been reported for a number of conditions, but harmful consequences of overuse should also be considered (Chawla and Kvarnberg 2014). The Vitamin B family participates in the protection of mitochondria during processes that result in higher oxidative status, such as during inflammatory responses (Depeint et al. 2006; Zhang et al. 2016).

A recent review has discussed the outcome of Vit C studies in animals in infections by diverse types of pathogens, concluding that indeed Vit C does play a role in preventing, shortening and alleviating the manifestation of such diseases (Hemilä 2017). Vit C is considered a strong modulator of immunity, enhancing pathogen phagocytosis and antimicrobial activities by macrophages, and it contributes to the proper functioning of T and B cells (Hughes 1999; Ramakrishnan et al. 2004; Ströhle et al. 2011; Schwager et al. 2015; Ellulu 2017).

Among the minerals, calcium participates in events related to the recognition of pathogens by APCs, acting in intracellular signalling in immune cells such as macrophages, lymphocytes and polymorphonuclear and mast cells (Libako et al. 2015). Magnesium has a modulating function over immunoinflammatory responses and antagonizes the actions of Ca (Vormann 2003; Mazur et al. 2007; Libako et al. 2016). Because the intracellular Mg concentration is strictly regulated, Mg does not work as an intracellular second messenger, but acts on changing the direction of metabolism (Vormann 2003). This has been demonstrated by Libako et al. (2016) using Ca blockers which prevented lipopolysaccharide-induced transcription and release of IL-1β, IL-6 and TNF-α, while extracellular Mg showed a modulating action (Libako et al. 2016). Among the fundamental roles phosphorus plays in the immune responses, phosphorylation and dephosphorylation reactions take place during intracellular signalling which is essential for the accomplishment of all immune functions (Kegley et al. 2001; Oster et al. 2016). Several studies on different species suggest that dietary phosphorus is fundamental for an appropriate immune response during infections, particularly for the adaptive immune system, and to maintain a stable gut microbial ecosystem to act as a barrier against potential pathogens (Heyer et al. 2015). Sulphur is a component of sulphur amino acid glutathione (GSH), homocysteine (Hcy) and taurine (Tau), which play important roles in various mechanisms of the immune response (Grimble and Grimble 1998; Grimble 2006). Sulphate and taurine are the major end products of sulphur amino acid metabolism, and in animals, high taurine intakes are anti-inflammatory (Grimble 2006). Methionine and cysteine play a role in the immune surveillance of the intestinal epithelial layer and regulation of the mucosal response to foreign antigens (Fang et al. 2010).

In the group of the trace elements, studies have found evidence of their importance in the immune response to infectious diseases (Weiss and Carver 2017). For example, an intrinsic relationship has been demonstrated between regulatory cytokines and blood levels of iron, copper, zinc and selenium in cutaneous leishmaniasis (Kocyigit et al. 2002). Selenium is essential for all mammalian species because it is required in various physiological functions, including important mechanisms of the reproductive system and notably in the immune system in resistance to infectious diseases (Kumar et al. 2008). Selenoproteins participate in macrophage activities (Zamamiri-Davis et al. 2002) as well as in the immune regulation (Kieliszek and Błażejak 2016). Selenium supplementation has been associated with enhancing immune competence, leukocyte function and specific immunity of humans and animals (Steinbrenner et al. 2015). Zinc contributes to form the structure of many enzymes and transcription factors and acts in the synthesis of acute phase proteins by the liver (Rink and Gabriel 2000; Osada 2013). Zinc deficiency has been shown to impact on B-cell lymphopoiesis and to induce potent atrophy of the thymus, subsequently leading to a decline in the number of peripheral T lymphocytes, both in a murine model of zinc deficiency (Keen and Gershwin 1990; King et al. 1995) and in zinc-deficient humans (Wolowczuk et al. 2008). In fact, for example, zinc deficiency has been associated with higher susceptibility to severe forms of diseases such as leishmaniasis (Mishra et al. 2010). Supplementation with zinc has resulted in protective effects in conditions related to diverse pathogens (Lazzerini and Wanzira 2016; Weiss and Carver 2017; Darling et al. 2017; Wang and Song 2017).

Some nutrients are categorized as conditionally essential nutrients because they are critical factors for a variety of functions in the course of diseases or determined metabolic conditions. In this miscellaneous group, nutrients such as carnitine, glutamine, arginine, choline and phosphatidylcholine, inositol, homocysteine, cysteine and taurine and a variety of phytochemicals that include flavonoids, carotenoids, lycopene, dietary fibres and phytosterols are included (Institute of Medicine 1998c; Craig and Beck 1999; Liu 2003; The editors 2004; Ploder et al. 2010; Hakim et al. 2012; Rasool et al. 2012; Marcinkiewicz and Kontny 2014; Elmahallawy et al. 2014; Leermakers et al. 2015). For example, glutamine is a much studied amino acid owing to its many properties as an immunonutrient and is considered an essential amino acid in patients undergoing hypercatabolism (McRae 2017). Glutamine is the preferred source of energy of enterocytes, and for this reason, it facilitates the digestion and absorption of other nutrients and keeps intestinal cells in the course of infectious diseases and their treatment (Kim and Kim 2017). This amino acid has been investigated as a useful nutrient for the therapeutic approach to various conditions including bacterial (Stehle et al. 2017; Liu et al. 2017), viral (Serrano-Villar et al. 2016; Wang et al. 2017) and protozoal infections (Kempaiah and Dokladny 2016), as well as to other conditions that favour infection, such as multiple trauma or extensive surgery (Lorenz et al. 2015).

2.4 Nutrient Absorption and Residue Excretion During Diseases

Disease interferes with nutrition in a variety of ways. Infection and body mechanisms to control infection can produce an infinity of manifestations and diverse effects on the body systems, at either local or systemic levels, including fever, anorexia and taste aversion (Nilsson et al. 2017). Fever slows down digestion. Inappetence and anorexia are evolutionarily conserved clinical consequences induced by infection and the immune response, involving cytokine-mediated signalling to the hypothalamus (van Niekerk et al. 2016; Rao et al. 2017) and on amygdala neurons (Francesconi et al. 2016). Furthermore, anorexia induced by acute or chronic inflammation is dependent on prostaglandins, which are synthesized by the cyclooxygenase-2 (COX-2) produced by different cell types involved in host defences (Nilsson et al. 2017). The nutritional approach to lack of appetite, anorexia or even medically induced fasting during disease can be controversial. For example, there is a debate as to whether a certain level of—and to what extent—fasting is beneficial in different disease conditions, including infections in their myriad of aspects (Wang et al. 2016). Moreover, a great issue is what would be the point at which nutraceutical intervention, perhaps before or after a specific fasting management, would be most beneficial in therapeutic terms.

A variety of infections can also cause local injury to any segment of the digestive tract, consequently leading to the development of pathological alterations such as dysphagia, pain, sensory disturbances, epithelial lesions and destruction, disturbances in the production and release of digestive juices, alterations in the pH and composition of digestive secretions, interference with exocrine gland secretion, alterations in gut motility and gut microbiota, organ metabolism malfunctions and altered catabolite handling and excretion (Ramig 2004; Andrade et al. 2011; Said 2011; Oliveira et al. 2013; Delgado et al. 2016). In general terms, the consequences of digestive disturbances are malabsorption and lack of nutrients on the one hand and toxaemia on the other (Sukhotnik et al. 2003; Shils and Shike 2006; Mahan et al. 2013).

Gastric juice acidity, bile and pancreatic enzymes exert bacteriostatic and bactericidal actions within the small intestine. The physiological propelling action of peristalsis drives the bacteria to the distal intestine, while the ileocecal valve prevents the migration of large numbers of colonic bacteria to the small intestine (Mahan et al. 2013). When intestinal homeostasis is compromised, either by invasion by pathogenic microbes, by toxicosis or by other factors related with microbiota overgrowth or imbalance, alterations in the digestive tube develop and result in clinical consequences, diarrhoea being the most common (Quera et al. 2005; Bondarenko et al. 2006). Examples of pathogenic bacteria that invade and colonize the intestine include *S. typhimurium*, *L. monocytogenes*, *Campylobacter* spp., *Shigella* spp. or enterotoxigenic *E. coli* (Khoshoo et al. 1990; Barbuddhe and Chakraborty 2009; Pawlowski et al. 2009; Taylor et al. 2017; Liu et al., 2017). Botulism caused by ingestion of food contaminated by *Clostridium* spp. is a common reported cause of toxicosis and severe dysphagia (Burke et al. 2016). Overgrowth of the commensal bacteria of gut microbiota is associated with a number of predisposing causes, including diet, and results in clinical alterations associated with diarrhoea (Mahan et al. 2013; Davis et al. 2016). As well, enteral protozoan pathogens, helminths, viruses and fungus can cause several alterations in the digestive tract (Ali and Hill 2003; Donskey 2004; Hodges and Gill 2010). Systemic infectious diseases caused by viruses such as HIV (García et al. 2006; Dillingham et al. 2011); by protozoa such as *L. donovani*, *L. infantum* (Baba et al. 2006; Cota et al. 2012) or *P. falciparum*; or by bacteria such as *Leptospira* spp. or *Yersinia* spp. can also produce enteritis among several other pathophysiological alterations. Vomiting is another important cause of nutrient loss in the course of infections, and this frequently occurs together with diarrhoea (Karney and Tong 1972; Brasitus 1983; Uysal et al. 2016).

Nutrient malabsorption is an important pathophysiological element in the course of infections and their consequences on the body. Malnutrition is the direct sequela of malabsorption and comes together with the heightened requirements for the ongoing immune response, increased metabolic needs, increased losses of body reserves and reduced intake of food that all characterize illnesses. On the other hand, malfunction of organs responsible for metabolism of nutrients and excretion result in elevated catabolite and by-product accumulation, which intensify functional impairment, reducing the metabolic capacity of distinct systems leading to systemic failure. It is a vicious cycle. Thus, a normal diet is often not sufficient to meet the increased demands for micronutrients in infectious diseases (Steinbrenner et al. 2015).

A great number of studies have characterized malabsorption and nutrient deficiencies owing to infections. Rotaviruses are well-defined examples of infection-related malabsorption which may even culminate in growth impairment and death in children (Dennehy 2000; Estes et al. 2001; Ramig 2004; Shea-Donohue et al. 2017). Helminth infections are associated with multifactorial iron deficiency, from bleeding helminth adherence situ and from worm spoliation and inflammation of the intestine epithelia (Notari et al. 2014; Grencis et al. 2014; Shea-Donohue et al. 2017). Intestinal pathogens are common causes of decreased absorption of nutrients including amino acids, lipids, Vit A, Vit B12, thiamine, riboflavin and zinc (Shils and Shike 2006; Mahan et al. 2013; Shea-Donohue et al. 2017). Intolerance to carbohydrates is associated with sugar fermentation owing to overgrowth of intestinal bacteria, a common phenomenon during parasite infection (Ortiz et al. 2000; Born 2007). One of the consequences of inflammation of the hepatic parenchyma is the impairment of albumin synthesis, which in turn compromises the synthetic properties and functioning of the liver, including, for example, the absorption of calcium and many other minerals or fat-soluble vitamins (Zhou et al. 2015; Adinolfi et al. 2016; Afify et al. 2017; Wang and Feng 2017). Altered renal function has important effects on the metabolism of nutrients (Engel 2003; Boullata 2009). Renal inflammation and parenchyma losses are other common results of systemic infections (Chan 1983; Epstein et al. 1988; Mahan et al. 2013; Imig and Ryan 2013) and kidney dysfunction, whose direct consequences are the kidney-dependent absorption and metabolism of Vit D, calcium, phosphorus, potassium, iron and some hydrosoluble vitamins (Engel 2003; Méndez and Dobaño 2004).

The involvement of the liver in various infectious conditions is an important cause of several metabolic dysfunctions that result in profound nutrient deficiencies on the one hand and in the accumulation of metabolites, catabolites and by-products, on the other hand, leading to toxic conditions. Hepatitises caused by bacterial infections, such as leptospirosis (Alvarado-Esquivel et al. 2016), bacterial pylephlebitis or hepatic abscesses, or yet involving viruses, protozoa and nematodes, are examples of direct causes of hepatic impairment (Tekwani et al. 1987; Davis et al. 1993; Shils and Shike 2006; Wahib et al. 2006; Fernando et al. 2016; Afify et al. 2017; Arain et al. 2017). Encephalopathy due to ammonia accumulation is an example of toxaemia caused by a dysfunctional liver (Souto et al. 2016). Bacteria in the gut produce ammonia as a catabolite from dietary protein metabolism, and ammonia is absorbed into the circulation. Because of hepatic malfunction, it cannot be converted to urea, which results in encephalopathy as ammonia crosses the brain-blood barrier and is neurotoxic (Holecek 2014; Souto et al. 2016). Kidney dysfunction during the course of diseases is also a significant cause of a toxaemic outcome. Diseases caused by infections commonly compromise the function of the kidneys, a crucial organ for excretion, and eventually reabsorption, of many catabolites, resulting in toxin accumulation and worsening of the metabolic conditions (Tanaka et al. 2017; Li et al. 2017; Leem et al. 2017; Roveran Genga et al. 2017). Uremia is an example of the accumulation of toxic nitrogen compounds from protein catabolism during renal failure (Boullata 2009; Rhee 2015; Chen and Koyner 2015). Even hydrosoluble vitamins and minerals that are considered non-toxic because in normal conditions any excess is easily excreted by a healthy liver and kidneys, during organ failure, these nutrients can accumulate and become toxic.

Altogether, the onset of diseases and their course have an influence on and are influenced by nutrient availability, absorption, distribution, synthesis, catabolism and excretion. The body's struggle during the health-disease process will result in variable levels of metabolism alterations and imbalances. The outcomes of these include malnutrition, even in the most advanced health-care centres of well-developed countries, as it commonly happens in the fields of the most deprived areas of the globe. Inadequate nutrition is the leading cause of morbidity and mortality for both humans and animals, as it impairs immunity, increases susceptibility to disease and severely alters the vital functions. Assuring proper nutrition must therefore also be an inseparable step of any therapeutic or preventive approach to diseases, including the infectious ones. As important as early detection of deficiencies or excess nutrients during an illness might make the difference between aggravation and recovery is the emerging knowledge of the power of certain nutrients as therapeutic agents themselves.

2.5 Nutrigenetics and Nutrigenomics in Health and Disease

Nutrigenomics is the study of how nutrients or diets affect gene expression, and nutrigenetics studies how genetic background affects the response to a nutrient or diet (Phillips 2013). The traditional "one-size-fits-all" approach is not optimal for genetic subgroups that may differ critically. The goal is to match the nutrient intake combination with the current genome status (inherited and acquired genome). In nutrigenetics and nutrigenomics, the study topics include (1) if nutrition can have an impact on health outcomes by affecting gene expression directly; (2) if the health effects of nutrients depend on inherited genetic variants that, for example, alter the metabolism of nutrients; and (3) if nutritional requirements should be customized for each individual so that better health outcomes can be achieved (Fenech et al. 2011).

Dietary macronutrients can alter the gene expression in the body. For example, overfeeding of carbohydrates increases the expression of sterol regulatory element-binding protein-1c (SREBP-1c), fatty acid synthase (FAS) and acetyl-CoA carboxylase (ACC), the key lipogenic enzymes in subcutaneous adipose tissue, resulting a 296% increase in de novo lipogenesis (Minehira et al. 2003). In addition, high-carbohydrate feeding induces the expression of interleukin-1β (IL-1 β), tumour-necrosis factor (TNF)-α and monocyte chemoattractant protein-1 (MCP1) in the liver, indicating inflammatory response (Li et al. 2015). On the other hand, a high-protein diet has been shown to increase the expression of glucose transporter 4 (GLUT-4) improving insulin sensitivity (Freudenberg et al. 2012). Biologically important long-chain omega-6 fatty acids, dihomo-gamma linolenic acid (DGLA) and arachidonic acid (AA) can be synthesized from linoleic acid (LA) via gamma-linolenic acid (GLA) by different enzymes, such as fatty acid desaturases 1 and 2 (FADS 1 and FADS 2) (Sergeant et al. 2016). The efficiency of this pathway is highly impacted by genetic variations within the FADS genes (Chilton et al. 2014), ultimately affecting the plasma GLA, DGLA and AA levels and ratios. DGLA-derived eicosanoids have long been recognized as anti-inflammatory; however, AA produces pro-inflammatory metabolites. While GLA supplementation has been reported to attenuate various inflammatory responses, for example, in atopic dermatitis and rheumatoid arthritis (Zurier et al. 1996; Andreassi et al. 1997), some reviews have questioned the results (van Gool et al. 2004; Macfarlane et al. 2012). The genetic variation in genes coding for enzymes used in the fatty acid pathway probably has significant importance in how different individuals respond to GLA supplementation. Fatty acids can also have nutrigenomic effects. For example, NF-κB is a pro-inflammatory transcription factor which is activated by inflammatory stimuli, such as bacterial lipopolysaccharides (LPS) (Calder 2013b), and eicosapentaenoic acid and fish oil have been shown to decrease the gene expression of the pro-inflammatory cytokine TNF by lowering NF-κB activity (Lo et al. 1999; Zhao et al. 2004). Th1-type cytokines seem to be more sensitive to the effects of fatty acids in the diet than Th2-type cytokines, and fish oil has been reported to induce a shift away from Th1-type response (Wallace et al. 2001), possibly explaining the low incidence of inflammatory and autoimmune disorders among Greenlandic Inuit people (Kromann and Green 1980) and the benefits seen in patients with ulcerative colitis (Rodgers 1998), Crohn's disease (Belluzi and Miglio 1998) and psoriasis (Ziboh 1998) after the administration of fish oil.

Vit D receptor (VDR) is a protein that intermediates the effect of 1,25-dihydroxy-Vit D3 on a specific DNA segment. Through VDR, Vit D regulates the immune system, for example, downregulating inflammation-related genes such as IL-1, IL-6 and interferon-γ (IFN-γ) (Manolagas et al. 1994) and controlling macrophage and lymphocyte function (Wientroub et al. 1989). Vit D also activates many neurotrophic hormones, such as the glial cell-derived neurotrophic factor (Naveilhan et al. 1996), the leukaemia inhibitory factor (Furman et al. 1996), neurotrophin-3 (Neveu et al. 1994a) and the nerve growth factor (Neveu et al. 1994b), indicating a role of VDR in neural cell growth and differentiation.

Micronutrients such as zinc, copper and iron play an important role in maintaining and reinforcing the immune system and antioxidant performances. These micronutrients can affect many genes (nutrigenomic approach), and also genetic interindividual variability may affect the absorption and uptake of the micronutrients (nutrigenetic approach) (Mocchegiani et al. 2012). In addition to an adequate amount of zinc from the diet, the efficient functioning of the proteins which handle zinc ions is critical for maintaining good health (Maret and Sandstead 2006). Mutations in genes that code for zinc-related proteins are the basis for inborn errors of zinc metabolism, for example, mutation in zinc transporter Zip4 resulting in severe zinc deficiency (Küry et al. 2002), substitution in zinc transporter ZnT-8 resulting in the risk of type 2 diabetes (Sladek et al. 2007) and polymorphism in Zip2 associated with carotid artery disease (Giacconi et al. 2008). Age-related alterations in gene expression of zinc transporters Zip1, Zip2 and Zip3 are associated with reduced cellular zinc uptake (Giacconi et al. 2012). This is due to chronic inflammation and also DNA methylation, which can be linked to genomic instability in chronic inflammation, highlighting the pivotal role of zinc transporters in altering zinc homeostasis and metabolism (Cousins 2010; Mocchegiani et al. 2012). On the other hand, dietary zinc plays a key role in adaptive immunity, oxidative stress, DNA repair and protein degradation (Mocchegiani et al. 2012). For example, in zinc deficiency Th 1 cytokines (IFN-γ, IL-2) decrease and Th2 cytokines (IL-4, IL-10) increase leading to a shift towards Th2 production with chronic low-grade inflammation (Franceschi 2007). Furthermore, the expression of peroxisome proliferator-activated receptor-α (PPAR-α) is very sensitive to zinc signalling, and zinc deficiency leads to impaired reactive oxygen species (ROS) production (Meyer et al. 2002). Copper serves as an important cofactor for many proteins but is toxic in excessive amounts. Mutations in ATP7A/B genes are responsible for Wilson disease when copper accumulates in the liver (Mercer 2001), providing a nutrigenetic example of an individual's copper requirement. On the other hand, low copper intake has a nutrigenetic impact in adult humans, as it impairs IL-2 gene expression affecting innate immunity (Hopkins and Failla 1999). Iron affects many genes related to the inflammatory/immune response, cell functions and cell growth (Mocchegiani et al. 2012). Polymorphisms of the iron transporter Nramp1 influence the gene transcription causing vulnerability to infections and to altered inflammatory/immune responses (Bellamy 1999; Zaahl et al. 2004), and a mutation of *Nramp2* gene leads to increased inflammation and iron overload (Iolascon et al. 2005). The altered gene expression of these transporters can lead to storage of iron in various tissues and organs together with inflammatory degenerative pathologies such as cancer and autoimmune diseases (Cellier et al. 2007).

Nutrigenetics highlights the importance of the genetic make-up of the individual for recommendations of different kinds of nutrients and diets and the benefits or disadvantages of supplements. Nutrigenomics shows the effect that nutrients can have on gene expression, changing it for the better or worse and affecting individuals' health. It also introduces the possibility of one's own health outcome being affected by lifestyle choices, despite the gene variants that are present in the individual's DNA.

Funding Statement This work has not been funded by any public or private entity.

Acknowledgements We thank Denise Sara Key for proofreading the English of this manuscript.

References

Abbaspour N, Hurrell R, Kelishadi R (2014) Review on iron and its importance for human health. J Res Med Sci 19:164–174

Abrams SA, Sidbury JB, Muenzer J et al (1991) Stable isotopic measurement of endogenous fecal calcium excretion in children. J Pediatr Gastroenterol Nutr 12:469–473

Abumrad NA, Davidson NO (2012) Role of the gut in lipid homeostasis. Physiol Rev 92:1061–1085. https://doi.org/10.1152/physrev.00019.2011

Adinolfi LE, Rinaldi L, Guerrera B et al (2016) NAFLD and NASH in HCV infection: prevalence and significance in hepatic and extrahepatic manifestations. Int J Mol Sci. https://doi.org/10.3390/ijms17060803

Afacan NJ, Fjell CD, Hancock REW (2012) A systems biology approach to nutritional immunology? Focus on innate immunity. Mol Aspects Med 33:14–25. https://doi.org/10.1016/j.mam.2011.10.013

Afify M, Hamza AH, Alomari RA (2017) Correlation between serum cytokines, interferons, and liver functions in hepatitis C virus patients. J Interferon Cytokine Res 37:32–38. https://doi.org/10.1089/jir.2016.0044

Ahad F, Ganie SA (2010) Iodine, iodine metabolism and iodine deficiency disorders revisited. Indian J Endocrinol Metab 14:13–17

Akoh CC, Min DB (2008) Food lipids: chemistry, nutrition, and biotechnology. CRC Press/Taylor & Francis Group, Boca Raton, FL

Albahrani AA, Greaves RF (2016) Fat-soluble vitamins: clinical indications and current challenges for chromatographic measurement. Clin Biochem Rev 37:27–47

Alberts B, Johnson A, Lewis J et al (2002a) Ion channels and the electrical properties of membranes. In: Molecular biology of the cell. Garland Science, New York

Alberts B, Johnson A, Lewis J et al (2002b) Carrier proteins and active membrane transport. In: Molecular biology of the cell. Garland Science, New York

Alexander JW (1998) Immunonutrition: the role of omega-3 fatty acids. Nutrition 14:627–633

Ali SA, Hill DR (2003) Giardia intestinalis. Curr Opin Infect Dis 16:453–460. https://doi.org/10.1097/01.qco.0000092817.64370.ab

Alvar J, Yactayo S, Bern C (2006) Leishmaniasis and poverty. Trends Parasitol 22:552–557. https://doi.org/10.1016/j.pt.2006.09.004

Alvarado-Esquivel C, Sánchez-Anguiano LF, Hernández-Tinoco J et al (2016) Leptospira exposure and patients with liver diseases: a case-control seroprevalence study. Int J Biomed Sci 12:48–52

Alvarez-Curto E, Milligan G (2016) Metabolism meets immunity: the role of free fatty acid receptors in the immune system. Biochem Pharmacol 114:3–13. https://doi.org/10.1016/j.bcp.2016.03.017

Anderson JJB (2003) Phosphorus|Physiology. In: Encyclopedia of food sciences and nutrition. Elsevier, London, pp 4539–4546

Anderson RA, Bryden NA, Polansky MM (1992) Dietary chromium intake. Freely chosen diets, institutional diet, and individual foods. Biol Trace Elem Res 32:117–121

de Andrade JAB, Haapalainen EF, Fagundes-Neto U (2011) Enteroaggregative *Escherichia coli* as a cause of persistent diarrhea: an experimental model using light microscopy. Rev Paul Pediatr 29:60–66. https://doi.org/10.1590/S0103-05822011000100010

Andreassi M, Forleo P, Dilohjo A et al (1997) Efficacy of?-Linolenic acid in the treatment of patients with atopic dermatitis. J Int Med Res 25:266–274. https://doi.org/10.1177/030006059702500504

Andreone P, Fiorino S, Cursaro C et al (2001) Vitamin E as treatment for chronic hepatitis B: results of a randomized controlled pilot trial. Antiviral Res 49:75–81

Anstead GM, Chandrasekar B, Zhao W et al (2001) Malnutrition alters the innate immune response and increases early visceralization following *Leishmania donovani* infection. Infect Immun 69:4709–4718. https://doi.org/10.1128/IAI.69.8.4709-4718.2001

Aoto J, Nam CI, Poon MM et al (2008) Synaptic signaling by all-trans retinoic acid in homeostatic synaptic plasticity. Neuron 60:308–320. https://doi.org/10.1016/j.neuron.2008.08.012

Arain SQ, Talpur FN, Channa NA et al (2017) Serum lipid profile as a marker of liver impairment in hepatitis B Cirrhosis patients. Lipids Health Dis 16:51. https://doi.org/10.1186/s12944-017-0437-2

Argilés JM, Campos N, Lopez-Pedrosa JM et al (2016) Skeletal muscle regulates metabolism via interorgan crosstalk: roles in health and disease. J Am Med Dir Assoc 17:789–796. https://doi.org/10.1016/j.jamda.2016.04.019

Azzi A, Meydani SN, Meydani M, Zingg JM (2016) The rise, the fall and the renaissance of vitamin E. Arch Biochem Biophys 595:100–108. https://doi.org/10.1016/j.abb.2015.11.010

Baba CS, Makharia GK, Mathur P et al (2006) Chronic diarrhea and malabsorption caused by *Leishmania donovani*. Indian J Gastroenterol 25:309–310

Bachmann MF (2012) Taurine: energy drink for T cells. Eur J Immunol 42:819–821. https://doi.org/10.1002/eji.201242450

Badaró R, Carvalho EM, Rocha H et al (1986) *Leishmania donovani*: an opportunistic microbe associated with progressive disease in three immunocompromised patients. Lancet (London, England) 1:647–649. https://doi.org/10.1590/S0037-86821988000400001

Baek M, Chung H-E, Yu J et al (2012) Pharmacokinetics, tissue distribution, and excretion of zinc oxide nanoparticles. Int J Nanomedicine 7:3081–3097. https://doi.org/10.2147/IJN.S32593

Bai J, Xun P, Morris S et al (2015) Chromium exposure and incidence of metabolic syndrome among American young adults over a 23-year follow-up: the CARDIA trace element study. Sci Rep 5:15606. https://doi.org/10.1038/srep15606

Bailey LB (2007) Folic acid. In: Zempleni J (ed) Handbook of Vitamins, 4th edn. Taylor & Francis, Boca Raton, FL, pp 412–385

Balcells ME, García P, Tiznado C et al (2017) Association of vitamin D deficiency, season of the year, and latent tuberculosis infection among household contacts. PLoS One 12:e0175400. https://doi.org/10.1371/journal.pone.0175400

Barbuddhe SB, Chakraborty T (2009) Listeria as an enteroinvasive gastrointestinal pathogen. Curr Top Microbiol Immunol 337:173–195. https://doi.org/10.1007/978-3-642-01846-6_6

Bates CJ (1998) Niacin. In: Encyclopedia of human nutrition. Elsevier, Oxford, pp 253–259

Bates CJ (2005a) Pantothenic acid. In: Encyclopedia of human nutrition. Elsevier, Oxford, pp 467–472

Bates CJ (2003) Folic acid|Physiology. In: Encyclopedia of food sciences and nutrition. Academic Press, London, pp 2564–2569

Bates CJ (2005b) Vitamin K. In: Encyclopedia of human nutrition. Elsevier, Oxford, pp 398–405

Bayer P (2013) Ingesta: digestión, absorción, transporte y excreción de nutrientes. In: Mahan K, Escott-Stump S, Raymond J (eds) Krause's food and the nutrition care process, 13th edn. Elsevier España, S.L, España, pp 18–12

Bayraktar Y, Bayraktar M, Gurakar A et al (1997) A comparison of the prevalence of autoantibodies in individuals with chronic hepatitis C and those with autoimmune hepatitis: the role of interferon in the development of autoimmune diseases. Hepatogastroenterology 44:417–425

Bechgaard H, Jespersen S (1977) GI absorption of niacin in humans. J Pharm Sci 66:871–872

Beisiegel U (1998) Lipoprotein metabolism. Eur Heart J 19(Suppl A):A20–A23

Bellamy R (1999) The natural resistance-associated macrophage protein and susceptibility to intracellular pathogens. Microbes Infect 1:23–27

Belluzi A, Miglio F (1998) n-3 Fatty acids in the treatment of Crohn's disease. In: Kremer JM (ed) Medicinal fatty acids in inflammation. Birkhauser, Verlag Basel, pp 101–191

Bender DA (2003a) Niacin|Physiology. In: Encyclopedia of food sciences and nutrition. Elsevier, London, pp 4119–4128

Bender DA (2005) Vitamin B6. In: Encyclopedia of human nutrition. Elsevier, Oxford, pp 359–367

Bender DA (2003b) Vitamin B6|Physiology. In: Encyclopedia of food sciences and nutrition. Elsevier, London, pp 6020–6032

Benkovic SJ, Hammes-Schiffer S (2003) A perspective on enzyme catalysis. Science 301:1196–1202. https://doi.org/10.1126/science.1085515

Bentley R, Meganathan R (1982) Biosynthesis of vitamin K (menaquinone) in bacteria. Microbiol Rev 46:241–280

Benzie IFF (2003) Evolution of dietary antioxidants. Comp Biochem Physiol A Mol Integr Physiol 136:113–126

Berdanier CD, Zempleni J (2009) Advanced nutrition: macronutrients, micronutrients, and metabolism. CRC Press, Boca Raton, FL

Berg A, Rooyackers O, Bellander B-M, Wernerman J (2013) Whole body protein kinetics during hypocaloric and normocaloric feeding in critically ill patients. Crit Care 17:R158. https://doi.org/10.1186/cc12837

Berg JM, Tymoczko JL, Stryer L (2002a) Monosaccharides are aldehydes or ketones with multiple hydroxyl groups. In: Freeman WH (ed) Biochemistry. W H Freeman, New York

Berg JM, Tymoczko JL, Stryer L (2002b) The citric acid cycle oxidizes two-carbon units. In: Biochemistry. W H Freeman, New York

Berg JM, Tymoczko JL, Stryer L (2002c) Metabolism consist of highly interconnected pathways. In: Biochemistry. W H Freeman, New York

Bettendorff L, Wins P (2013) Biochemistry of thiamine and thiamine phosphate compounds. In: Encyclopedia of biological chemistry. Liège, Belgium, pp 202–209

Bhargava A, Benedetti A, Oxlade O et al (2014) Undernutrition and the incidence of tuberculosis in India: national and subnational estimates of the population-attributable fraction related to undernutrition. Natl Med J India 27:128–133

Bikle DD (2014) Vitamin D metabolism, mechanism of action, and clinical applications. Chem Biol 21:319–329. https://doi.org/10.1016/j.chembiol.2013.12.016

Birben E, Sahiner UM, Sackesen C et al (2012) Oxidative stress and antioxidant defense. World Allergy Organ J 5:9–19. https://doi.org/10.1097/WOX.0b013e3182439613

Bistrian BR (1979) A simple technique to estimate severity of stress. Surg Gynecol Obstet 148:675–678

Bitsch R (2003) THIAMIN|Physiology. In: Encyclopedia of food sciences and nutrition. Elsevier, London, pp 5772–5780

Bondarenko VM, Lykova EA, Matsulevich TV (2006) Microecological aspects of small intestinal bacterial overgrowth syndrome. Zh Mikrobiol Epidemiol Immunobiol 6:57–63

Born P (2007) Carbohydrate malabsorption in patients with non-specific abdominal complaints. World J Gastroenterol 13:5687. https://doi.org/10.3748/wjg.v13.i43.5687

Bou Ghanem EN, Clark S, Du X et al (2015) The α-tocopherol form of vitamin E reverses age-associated susceptibility to *Streptococcus pneumoniae* lung infection by modulating pulmonary neutrophil recruitment. J Immunol 194:1090–1099. https://doi.org/10.4049/jimmunol.1402401

Bou Ghanem EN, Lee JN, Joma BH et al (2017) The alpha-tocopherol form of Vitamin E boosts elastase activity of human PMNs and their ability to kill *Streptococcus pneumoniae*. Front Cell Infect Microbiol 7:161. https://doi.org/10.3389/fcimb.2017.00161

Bouamama S, Merzouk H, Medjdoub A et al (2017) Effects of exogenous vitamins A, C, and E and NADH supplementation on proliferation, cytokines release, and cell redox status of lymphocytes from healthy aged subjects. Appl Physiol Nutr Metab 42:579–587. https://doi.org/10.1139/apnm-2016-0201

Bouillon R, Verstuyf A, Mathieu C et al (2006) Vitamin D resistance. Best Pract Res Clin Endocrinol Metab 20:627–645. https://doi.org/10.1016/j.beem.2006.09.008

Boullata JI (2009) An introduction to drug–nutrient interactions. In: Handbook of drug-nutrient interactions. Humana Press, Totowa, NJ, pp 3–26

Bourquin F, Capitani G, Grütter MG (2011) PLP-dependent enzymes as entry and exit gates of sphingolipid metabolism. Protein Sci 20:1492–1508. https://doi.org/10.1002/pro.679

Bowrey DJ, Morris-Stiff GJ, Puntis MCA (1999) Selenium deficiency and chronic pancreatitis: disease mechanism and potential for therapy. HPB Surg 11:207–216. https://doi.org/10.1155/1999/97140

Brasitus TA (1983) Parasites and malabsorption. Clin Gastroenterol 12:495–510

Brenner M, Laragione T, Gulko PS (2017) Short-term low-magnesium diet reduces autoimmune arthritis severity and synovial tissue gene expression. Physiol Genomics 49:238–242. https://doi.org/10.1152/physiolgenomics.00003.2017

Brigelius-Flohé R, Traber MG (1999) Vitamin E: function and metabolism. FASEB J 13:1145–1155

Britton AA, Cotzias GC (1966) Dependence of manganese turnover on intake. Am J Physiol 211:203–206

Broadhurst CL (1997) Balanced intakes of natural triglycerides for optimum nutrition: an evolutionary and phytochemical perspective. Med Hypotheses 49:247–261

Brophy MB, Nolan EM (2015) Manganese and microbial pathogenesis: sequestration by the Mammalian immune system and utilization by microorganisms. ACS Chem Biol 10:641–651. https://doi.org/10.1021/cb500792b

Brosnan JT, Brosnan ME (2006) The sulfur-containing amino acids: an overview. J Nutr 136:1636S–1640S

Brozmanová J (2011) Selenium and cancer: from prevention to treatment. Klin Onkol 24:171–179

Brune D, Kjaerheim A, Paulsen G, Beltesbrekke H (1980) Pulmonary deposition following inhalation of chromium-cobalt grinding dust in rats and distribution in other tissues. Eur J Oral Sci 88:543–551. https://doi.org/10.1111/j.1600-0722.1980.tb01265.x

Burke P, Needham M, Jackson BR et al (2016) Outbreak of foodborne botulism associated with improperly Jarred Pesto--Ohio and California, 2014. MMWR Morb Mortal Wkly Rep 65:175–177. https://doi.org/10.15585/mmwr.mm6507a2

Calder P (2013a) Omega-3 polyunsaturated fatty acids and inflammatory processes: nutrition or pharmacology? Br J Clin Pharmacol 75:645–662. https://doi.org/10.1111/j.1365-2125.2012.04374.x

Calder P (2013b) n-3 fatty acids, inflammation and immunity: new mechanisms to explain old actions. Proc Nutr Soc 72:326–336. https://doi.org/10.1017/S0029665113001031

Campbell LA, Rosenfeld ME (2015) Infection and atherosclerosis development. Arch Med Res 46:339–350. https://doi.org/10.1016/j.arcmed.2015.05.006

Cancela ML, Conceição N, Laizé V (2012) Gla-rich protein, a new player in tissue calcification? Adv Nutr 3:174–181. https://doi.org/10.3945/an.111.001685

Card DJ, Gorska R, Cutler J, Harrington DJ (2014) Vitamin K metabolism: current knowledge and future research. Mol Nutr Food Res 58:1590–1600. https://doi.org/10.1002/mnfr.201300683

Carr AC, Frei B (1999) Toward a new recommended dietary allowance for vitamin C based on antioxidant and health effects in humans. Am J Clin Nutr 69:1086–1107

Carruthers A, DeZutter J, Ganguly A, Devaskar SU (2009) Will the original glucose transporter isoform please stand up! Am J Physiol Endocrinol Metab 297:E836–E848. https://doi.org/10.1152/ajpendo.00496.2009

Cases N, Aguiló A, Tauler P et al (2005) Differential response of plasma and immune cell's vitamin E levels to physical activity and antioxidant vitamin supplementation. Eur J Clin Nutr 59:781–788. https://doi.org/10.1038/sj.ejcn.1602143

Cassani B, Villablanca EJ, De Calisto J et al (2012) Vitamin A and immune regulation: role of retinoic acid in gut-associated dendritic cell education, immune protection and tolerance. Mol Aspects Med 33:63–76. https://doi.org/10.1016/j.mam.2011.11.001

Cassani F, Cataleta M, Valentini P et al (1997) Serum autoantibodies in chronic hepatitis C: comparison with autoimmune hepatitis and impact on the disease profile. Hepatology 26:561–566. https://doi.org/10.1002/hep.510260305

Cegielski JP, McMurray DN (2004) The relationship between malnutrition and tuberculosis: evidence from studies in humans and experimental animals. Int J Tuberc Lung Dis 8:286–298

Cellier MF, Courville P, Campion C (2007) Nramp1 phagocyte intracellular metal withdrawal defense. Microbes Infect 9:1662–1670. https://doi.org/10.1016/j.micinf.2007.09.006

Cerf BJ, Jones TC, Badaro R et al (1987) Malnutrition as a risk factor for severe visceral leishmaniasis. J Infect Dis 156:1030–1033

Chan DL (2008) The role of nutrients in modulating disease. J Small Anim Pract 49:266–271. https://doi.org/10.1111/j.1748-5827.2008.00589.x

Chan JC (1983) Acid-base disorders and the kidney. Adv Pediatr 30:401–471

Chandra S, Chandra RK (1986) Nutrition, immune response, and outcome. Prog Food Nutr Sci 10:1–65

Chandrasekharan N (1999) Changing concepts in lipid nutrition in health and disease. Med J Malaysia 54:408–427. quiz 428

Chatterjee IB, Majumder AK, Nandi BK, Subramanian N (1975) Synthesis and some major functions of vitamin C in animals. Ann N Y Acad Sci 258:24–47

Chawla J, Kvarnberg D (2014) Hydrosoluble vitamins. Handb Clin Neurol 120:891–914. https://doi.org/10.1016/B978-0-7020-4087-0.00059-0

Chen L-X, Koyner JL (2015) Biomarkers in acute kidney injury. Crit Care Clin 31:633–648. https://doi.org/10.1016/j.ccc.2015.06.002

Chen N, Onisko B, Napoli JL (2008) The nuclear transcription factor RARalpha associates with neuronal RNA granules and suppresses translation. J Biol Chem 283:20841–20847. https://doi.org/10.1074/jbc.M802314200

Chen W, Chen G (2014) The roles of vitamin A in the regulation of carbohydrate, lipid, and protein metabolism. J Clin Med 3:453–479. https://doi.org/10.3390/jcm3020453

Chen Y-C, Prabhu K, Mastro A (2013) Is selenium a potential treatment for cancer metastasis? Nutrients 5:1149–1168. https://doi.org/10.3390/nu5041149

Chilton FH, Murphy RC, Wilson BA et al (2014) Diet-gene interactions and PUFA metabolism: a potential contributor to health disparities and human diseases. Nutrients 6:1993–2022. https://doi.org/10.3390/nu6051993

Chirumbolo S, Bjørklund G, Sboarina A, Vella A (2017) The role of vitamin D in the immune system as a pro-survival molecule. Clin Ther 39(5):894–916. https://doi.org/10.1016/j.clinthera.2017.03.021

Choe E, Min DB (2009) Mechanisms of antioxidants in the oxidation of foods. Compr Rev Food Sci Food Saf 8:345–358. https://doi.org/10.1111/j.1541-4337.2009.00085.x

Chow O, Barbul A (2014) Immunonutrition: role in wound healing and tissue regeneration. Adv Wound Care 3:46–53. https://doi.org/10.1089/wound.2012.0415

Christakos S, Ajibade DV, Dhawan P et al (2010) Vitamin D: metabolism. Endocrinol Metab Clin North Am 39:243–253. https://doi.org/10.1016/j.ecl.2010.02.002

Church J, Maitland K (2014) Invasive bacterial co-infection in African children with Plasmodium falciparum malaria: a systematic review. BMC Med 12:31. https://doi.org/10.1186/1741-7015-12-31

Clarkson PM, Thompson HS (2000) Antioxidants: what role do they play in physical activity and health? Am J Clin Nutr 72:637S–646S

Cobalt Development Institute (2006) Cobalt in the environment. Cobalt Development Institute, England

Cohen AL, McMorrow M, Walaza S et al (2015) Potential impact of co-infections and co-morbidities prevalent in Africa on influenza severity and frequency: a systematic review. PLoS One 10:e0128580. https://doi.org/10.1371/journal.pone.0128580

Cook JD, Penner-Hahn JE, Stemmler TL (2008) Structure and dynamics of metalloproteins in live cells. Methods Cell Biol 90:199–216

Cota GF, Gomes LI, Pinto BF et al (2012) Dyarrheal syndrome in a patient co-infected with *Leishmania infantum* and *Schistosoma mansoni*. Case Rep Med 2012:1–4. https://doi.org/10.1155/2012/240512

Cousin B, Casteilla L, Laharrague P et al (2016) Immuno-metabolism and adipose tissue: the key role of hematopoietic stem cells. Biochimie 124:21–26. https://doi.org/10.1016/j.biochi.2015.06.012

Cousins RJ (2010) Gastrointestinal factors influencing zinc absorption and homeostasis. Int J Vitam Nutr Res 80:243–248. https://doi.org/10.1024/0300-9831/a000030

Coutu DL, Wu JH, Monette A et al (2008) Periostin, a member of a novel family of vitamin K-dependent proteins, is expressed by mesenchymal stromal cells. J Biol Chem 283:17991–18001. https://doi.org/10.1074/jbc.M708029200

Craig W, Beck L (1999) Phytochemicals: health protective effects. Can J Diet Pract Res 60:78–84

Cummings JH, Stephen AM (2007) Carbohydrate terminology and classification. Eur J Clin Nutr 61:S5–S18. https://doi.org/10.1038/sj.ejcn.1602936

Cunha DF, Lara VC, Monteiro JP et al (2001) Growth retardation in children with positive intradermic reaction for leishmaniasis: preliminary results. Rev Soc Bras Med Trop 34:25–27

Cunningham-Rundles S, McNeeley DF, Moon A (2005) Mechanisms of nutrient modulation of the immune response. J Allergy Clin Immunol 115:1119–1128.; quiz 1129. https://doi.org/10.1016/j.jaci.2005.04.036

Czarnewski P, Das S, Parigi SM, Villablanca EJ (2017) Retinoic acid and its role in modulating intestinal innate immunity. Nutrients 9(1):68. https://doi.org/10.3390/nu9010068

Dabrowski K (1983) Comparative aspects of protein digestion and amino acid absorption in fish and other animals. Comp Biochem Physiol A Comp Physiol 74:417–425

Dakshinamurti S, Dakshinamurti K (2007) Vitamin B6. In: Zempleni J (ed) Handbook of vitamins, 4th edn. Taylor & Francis, Boca Raton, FL, pp 360–315

Daly S, Mills JL, Molloy AM et al (1997) Minimum effective dose of folic acid for food fortification to prevent neural-tube defects. Lancet 350:1666–1669. https://doi.org/10.1016/S0140-6736(97)07247-4

Darling AM, Mugusi FM, Etheredge AJ et al (2017) Vitamin A and zinc supplementation among pregnant women to prevent placental malaria: a randomized, double-blind, placebo-controlled trial in Tanzania. Am J Trop Med Hyg 96:826–834. https://doi.org/10.4269/ajtmh.16-0599

Daruwala R, Song J, Koh WS et al (1999) Cloning and functional characterization of the human sodium-dependent vitamin C transporters hSVCT1 and hSVCT2. FEBS Lett 460:480–484

Davis MY, Zhang H, Brannan LE et al (2016) Rapid change of fecal microbiome and disappearance of Clostridium difficile in a colonized infant after transition from breast milk to cow milk. Microbiome 4:53. https://doi.org/10.1186/s40168-016-0198-6

Davis TM, Brown AE, Smith CD (1993) Metabolic disturbances in Plasmodium coatneyi-infected rhesus monkeys. Int J Parasitol 23:557–563

de Haan K, Groeneveld AJ, de Geus HR et al (2014) Vitamin D deficiency as a risk factor for infection, sepsis and mortality in the critically ill: systematic review and meta-analysis. Crit Care 18:660. https://doi.org/10.1186/s13054-014-0660-4

de Luis DA, de la Fuente B, Izaola O et al (2014) Clinical effects of a hypercaloric and hyperproteic oral suplemment enhanced with W3 fatty acids and dietary fiber in postsurgical ambulatory head and neck cancer patients. Nutr Hosp 31:759–763. https://doi.org/10.3305/nh.2015.31.2.8481

de Oliveira FA, Vanessa Oliveira Silva C, Damascena NP et al (2013) High levels of soluble CD40 ligand and matrix metalloproteinase-9 in serum are associated with favorable clinical evolution in human visceral leishmaniasis. BMC Infect Dis 13:331. https://doi.org/10.1186/1471-2334-13-331

de Oliveira MR (2015) Vitamin A and retinoids as mitochondrial toxicants. Oxid Med Cell Longev 2015:1–13. https://doi.org/10.1155/2015/140267

De Wolf BM, Zajac AM, Hoffer KA et al (2014) The effect of vitamin E supplementation on an experimental Haemonchus contortus infection in lambs. Vet Parasitol 205:140–149. https://doi.org/10.1016/j.vetpar.2014.07.013

Delafuente JC, Prendergast JM, Modigh A (1986) Immunologic modulation by vitamin C in the elderly. Int J Immunopharmacol 8:205–211

Delgado ME, Grabinger T, Brunner T (2016) Cell death at the intestinal epithelial front line. FEBS J 283:2701–2719. https://doi.org/10.1111/febs.13575

Dennehy PH (2000) Transmission of rotavirus and other enteric pathogens in the home. Pediatr Infect Dis J 19:S103–S105

Depeint F, Bruce WR, Shangari N et al (2006) Mitochondrial function and toxicity: role of the B vitamin family on mitochondrial energy metabolism. Chem Biol Interact 163:94–112. https://doi.org/10.1016/j.cbi.2006.04.014

Diekman C, Malcolm K (2009) Consumer perception and insights on fats and fatty acids: knowledge on the quality of diet fat. Ann Nutr Metab 54(Suppl 1):25–32. https://doi.org/10.1159/000220824

Dihingia A, Kalita J, Manna P (2017) Implication of a novel Gla-containing protein, Gas6 in the pathogenesis of insulin resistance, impaired glucose homeostasis, and inflammation: a review. Diabetes Res Clin Pract 128:74–82. https://doi.org/10.1016/j.diabres.2017.03.026

Dikici B, Dagli A, Ucmak H et al (2007) Efficacy of vitamin E in children with immunotolerant-phase chronic hepatitis B infection. Pediatr Int 49:603–607. https://doi.org/10.1111/j.1442-200X.2007.02419.x

Dillingham R, Leger P, Beauharnais C-A et al (2011) AIDS diarrhea and antiretroviral drug concentrations: a matched-pair cohort study in Port au Prince, Haiti. Am J Trop Med Hyg 84:878–882. https://doi.org/10.4269/ajtmh.2011.10-0541

DiSilvestro RA (2005) Handbook of minerals as nutritional supplements. CRC Press, Boca Raton, FL

Donskey CJ (2004) The role of the intestinal tract as a reservoir and source for transmission of nosocomial pathogens. Clin Infect Dis 39:219–226. https://doi.org/10.1086/422002

das Dôres SMC, de Paiva SAR, Campana ÁO (2001) Vitamina K: metabolismo e nutrição. Rev Nutr 14:207–218. https://doi.org/10.1590/S1415-52732001000300007

Dou M, Ma Y, Ma AG et al (2016) Combined chromium and magnesium decreases insulin resistance more effectively than either alone. Asia Pac J Clin Nutr 25:747–753

Drakesmith H, Prentice AM (2012) Hepcidin and the iron-infection axis. Science 338:768–772. https://doi.org/10.1126/science.1224577

Drevon CA (1991) Absorption, transport and metabolism of vitamin E. Free Radic Res Commun 14:229–246

Ducros V (1992) Chromium metabolism. A literature review. Biol Trace Elem Res 32:65–77

Dunn WA, Rettura G, Seifter E, Englard S (1984) Carnitine biosynthesis from gamma-butyrobetaine and from exogenous protein-bound 6-N-trimethyl-L-lysine by the perfused guinea pig liver. Effect of ascorbate deficiency on the in situ activity of gamma-butyrobetaine hydroxylase. J Biol Chem 259:10764–10770

Dye C, Williams BG (1993) Malnutrition, age and the risk of parasitic disease: visceral leishmaniasis revisited. Proceedings Biol Sci 254:33–39. https://doi.org/10.1098/rspb.1993.0123

Ebara S (2017) Nutritional role of folate. Congenit Anom (Kyoto) 57:138–141. https://doi.org/10.1111/cga.12233

Ebeling PR, Sandgren ME, DiMagno EP et al (1992) Evidence of an age-related decrease in intestinal responsiveness to vitamin D: relationship between serum 1,25-dihydroxyvitamin D3 and intestinal vitamin D receptor concentrations in normal women. J Clin Endocrinol Metab 75:176–182. https://doi.org/10.1210/jcem.75.1.1320048

Efron D, Barbul A (2000) Role of arginine in immunonutrition. J Gastroenterol 35(Suppl 1):20–23

Elenius V, Palomares O, Waris M et al (2017) The relationship of serum vitamins A, D, E and LL-37 levels with allergic status, tonsillar virus detection and immune response. PLoS One 12: e0172350. https://doi.org/10.1371/journal.pone.0172350

Ellulu MS (2017) Obesity, cardiovascular disease, and role of vitamin C on inflammation: a review of facts and underlying mechanisms. Inflammopharmacology 25:313–328. https://doi.org/10.1007/s10787-017-0314-7

Elmahallawy EK, Jiménez-Aranda A, Martínez AS et al (2014) Activity of melatonin against *Leishmania infantum* promastigotes by mitochondrial dependent pathway. Chem Biol Interact 220:84–93. https://doi.org/10.1016/j.cbi.2014.06.016

Elste V, Troesch B, Eggersdorfer M, Weber P (2017) Emerging evidence on neutrophil motility supporting its usefulness to define vitamin C intake requirements. Nutrients 9(5):503. https://doi.org/10.3390/nu9050503

Engel BE (2003) Renal function and disorders|Nutritional management of renal disorders. In: Encyclopedia of food sciences and nutrition. Elsevier, London, pp 4943–4951

Epstein FH, Badr KF, Ichikawa I (1988) Prerenal failure: a deleterious shift from renal compensation to decompensation. N Engl J Med 319:623–629. https://doi.org/10.1056/NEJM198809083191007

Erkelens MN, Mebius RE (2017) Retinoic acid and immune homeostasis: a balancing act. Trends Immunol 38:168–180. https://doi.org/10.1016/j.it.2016.12.006

Espinosa A, Henríquez-Olguín C, Jaimovich E (2016) Reactive oxygen species and calcium signals in skeletal muscle: a crosstalk involved in both normal signaling and disease. Cell Calcium 60:172–179. https://doi.org/10.1016/j.ceca.2016.02.010

Estes MK, Kang G, Zeng CQ et al (2001) Pathogenesis of rotavirus gastroenteritis. Novartis Found Symp 238:82–96–100

Evoy D, Lieberman MD, Fahey TJ, Daly JM (1998) Immunonutrition: the role of arginine. Nutrition 14:611–617

Falchetti R, Fuggetta MP, Lanzilli G et al (2001) Effects of resveratrol on human immune cell function. Life Sci 70:81–96

Fang Z, Yao K, Zhang X et al (2010) Nutrition and health relevant regulation of intestinal sulfur amino acid metabolism. Amino Acids 39:633–640. https://doi.org/10.1007/s00726-010-0502-x

FAO/WHO (1998) The role of carbohydrates in nutrition. In: Expert Consultation (ed) Carbohydrates in human nutrition: report of a joint FAO/WHO. World Health Organization, Roma, p 140

FAO/WHO (2001) Vitamin B12. In: Human vitamin and mineral requirements. FAO/WHO, Bangkok, Thailand

Feleszko W, Ruszczyński M, Zalewski BM (2014) Non-specific immune stimulation in respiratory tract infections. Separating the wheat from the chaff. Paediatr Respir Rev 15:200–206. https://doi.org/10.1016/j.prrv.2013.10.006

Fenech M, El-Sohemy A, Cahill L et al (2011) Nutrigenetics and nutrigenomics: viewpoints on the current status and applications in nutrition research and practice. J Nutrigenet Nutrigenomics 4:69–89. https://doi.org/10.1159/000327772

Ferland G (2012) The discovery of vitamin K and its clinical applications. Ann Nutr Metab 61:213–218. https://doi.org/10.1159/000343108

Fernando S, Wijewickrama A, Gomes L et al (2016) Patterns and causes of liver involvement in acute dengue infection. BMC Infect Dis 16:319. https://doi.org/10.1186/s12879-016-1656-2

Field CJ, Johnson IR, Schley PD (2002) Nutrients and their role in host resistance to infection. J Leukoc Biol 71:16–32

Fine KD, Santa Ana CA, Porter JL, Fordtran JS (1991) Intestinal absorption of magnesium from food and supplements. J Clin Invest 88:396–402. https://doi.org/10.1172/JCI115317

Finnerty CC, Mabvuure NT, Ali A et al (2013) The surgically induced stress response. JPEN J Parenter Enteral Nutr 37:21S–29S. https://doi.org/10.1177/0148607113496117

Flores-Mateo G, Navas-Acien A, Pastor-Barriuso R, Guallar E (2006) Selenium and coronary heart disease: a meta-analysis. Am J Clin Nutr 84:762–773

Florin T, Neale G, Gibson GR et al (1991) Metabolism of dietary sulphate: absorption and excretion in humans. Gut 32:766–773

Foote JW, Delves HT (1984) Albumin bound and alpha 2-macroglobulin bound zinc concentrations in the sera of healthy adults. J Clin Pathol 37:1050–1054

Fox JT, Stover PJ (2008) Folate-mediated one-carbon metabolism. Vitam Horm 79:1–44. https://doi.org/10.1016/S0083-6729(08)00401-9

França T, Ishikawa L, Zorzella-Pezavento S et al (2009) Impact of malnutrition on immunity and infection. J Venom Anim Toxins Incl Trop Dis 15:374–390. https://doi.org/10.1590/S1678-91992009000300003

Franceschi C (2007) Inflammaging as a major characteristic of old people: can it be prevented or cured? Nutr Rev 65:S173–S176

Francesconi W, Sánchez-Alavez M, Berton F et al (2016) The proinflammatory cytokine interleukin 18 regulates feeding by acting on the bed nucleus of the stria terminalis. J Neurosci 36:5170–5180. https://doi.org/10.1523/JNEUROSCI.3919-15.2016

Frayn KN, Arner P, Yki-Järvinen H (2006) Fatty acid metabolism in adipose tissue, muscle and liver in health and disease. Essays Biochem 42:89–103. https://doi.org/10.1042/bse0420089

Freeland-Graves JH, Bavik C (2003) Coenzymes. In: Encyclopedia of food sciences and nutrition. Elsevier, London, pp 1475–1481

Freeland-Graves JH, Trotter PJ (2003) Minerals–dietary importance. In: Encyclopedia of food sciences and nutrition. Elsevier, London, pp 4005–4012

Frei R, Akdis M, O'Mahony L (2015) Prebiotics, probiotics, synbiotics, and the immune system: experimental data and clinical evidence. Curr Opin Gastroenterol 31:153–158. https://doi.org/10.1097/MOG.0000000000000151

Freudenberg A, Petzke KJ, Klaus S (2012) Comparison of high-protein diets and leucine supplementation in the prevention of metabolic syndrome and related disorders in mice. J Nutr Biochem 23:1524–1530. https://doi.org/10.1016/j.jnutbio.2011.10.005

Furman I, Baudet C, Brachet P (1996) Differential expression of M-CSF, LIF, and TNF-alpha genes in normal and malignant rat glial cells: regulation by lipopolysaccharide and vitamin D. J Neurosci Res 46:360–366. https://doi.org/10.1002/(SICI)1097-4547(19961101)46:3<360::AID-JNR9>3.0.CO;2-I

Fürst P, Stehle P (2004) What are the essential elements needed for the determination of amino acid requirements in humans? J Nutr 134:1558S–1565S

Gacs G, Barltrop D (1977) Significance of Ca-soap formation for calcium absorption in the rat. Gut 18:64–68

Gadisa E, Tasew G, Abera A et al (2017) Serological signatures of clinical cure following successful treatment with sodium stibogluconate in Ethiopian visceral leishmaniasis. Cytokine 91:6–9. https://doi.org/10.1016/j.cyto.2016.11.016

Gammoh NZ, Rink L (2017) Zinc in infection and inflammation. Nutrients 9:624. https://doi.org/10.3390/nu9060624

Ganz T, Nemeth E (2015) Iron homeostasis in host defence and inflammation. Nat Rev Immunol 15:500–510. https://doi.org/10.1038/nri3863

Garattini E, Mendel R, Romão MJ et al (2003) Mammalian molybdo-flavoenzymes, an expanding family of proteins: structure, genetics, regulation, function and pathophysiology. Biochem J 372:15–32. https://doi.org/10.1042/BJ20030121

Garcia-Lorda P, Serrano P, Jiménez-Expósito MJ et al (2000) Cytokine-driven inflammatory response is associated with the hypermetabolism of AIDS patients with opportunistic infections. J Parenter Enter Nutr 24:317–322. https://doi.org/10.1177/0148607100024006317

García C, Rodríguez E, Do N et al (2006) Intestinal parasitosis in patients with HIV-AIDS. Rev Gastroenterol Peru 26:21–24

Garrabrant T, Tuman RW, Ludovici D et al (2004) Small molecule inhibitors of methionine aminopeptidase type 2 (MetAP-2). Angiogenesis 7:91–96. https://doi.org/10.1007/s10456-004-6089-7

Gatto M, de Abreu MM, Tasca KI et al (2013) Biochemical and nutritional evaluation of patients with visceral leishmaniasis before and after treatment with leishmanicidal drugs. Rev Soc Bras Med Trop 46:735–740. https://doi.org/10.1590/0037-8682-0198-2013

Geldenhuys WJ, Caporoso J, Leeper TC et al (2017) Structure-activity and in vivo evaluation of a novel lipoprotein lipase (LPL) activator. Bioorg Med Chem Lett 27:303–308. https://doi.org/10.1016/j.bmcl.2016.11.053

Gershwin ME, Nestel P, Keen CL (2004) Handbook of nutrition and immunity. Humana Press, Totowa, NJ

Gerster H (1997) Vitamin A-functions, dietary requirements and safety in humans. Int J Vitam Nutr Res 67:71–90

Ghoneim AH, Al-Azzawi MA, Elmasry SA et al (2015) Association of vitamin D status in the pathogenesis of chronic obstructive pulmonary disease. Egypt J Chest Dis Tuberc 64:805–812. https://doi.org/10.1016/j.ejcdt.2015.06.004

Ghosh S, Banerjee S, Sil PC (2015) The beneficial role of curcumin on inflammation, diabetes and neurodegenerative disease: a recent update. Food Chem Toxicol 83:111–124. https://doi.org/10.1016/j.fct.2015.05.022

Giacconi R, Malavolta M, Costarelli L et al (2012) Comparison of intracellular zinc signals in nonadherent lymphocytes from young-adult and elderly donors: role of zinc transporters (Zip family) and proinflammatory cytokines. J Nutr Biochem 23:1256–1263. https://doi.org/10.1016/j.jnutbio.2011.07.005

Giacconi R, Muti E, Malavolta M et al (2008) A novel Zip2 Gln/Arg/Leu codon 2 polymorphism is associated with carotid artery disease in aging. Rejuvenation Res 11:297–300. https://doi.org/10.1089/rej.2008.0671

Gil Hernández A, Sánchez de Medina Contreras F (2010) Capítulo 1.2 Funciones de los nutrientes. Metabolismo energético y metabolismo intermediario. Regulación metabólica. In: Tratado de Nutrición. Médica-Panamericana, p 3412

Gille A, Bodor ET, Ahmed K, Offermanns S (2008) Nicotinic acid: pharmacological effects and mechanisms of action. Annu Rev Pharmacol Toxicol 48:79–106. https://doi.org/10.1146/annurev.pharmtox.48.113006.094746

Gillooly M, Bothwell TH, Torrance JD et al (1983) The effects of organic acids, phytates and polyphenols on the absorption of iron from vegetables. Br J Nutr 49:331–342

Goldschmied A, Modan B, Greenberg RA et al (1975) Urinary calcium excretion in relation to kidney function in the adult. J Am Geriatr Soc 23:155–160

Goncalves A, Roi S, Nowicki M et al (2015) Fat-soluble vitamin intestinal absorption: absorption sites in the intestine and interactions for absorption. Food Chem 172:155–160. https://doi.org/10.1016/j.foodchem.2014.09.021

Goto H, Prianti M d G (2009) Immunoactivation and immunopathogeny during active visceral leishmaniasis. Rev Inst Med Trop Sao Paulo 51:241–246

Gozzelino R, Arosio P (2016) Iron homeostasis in health and disease. Int J Mol Sci 17:130. https://doi.org/10.3390/ijms17010130

Green R (2005) Cobalamins. In: Encyclopedia of human nutrition. Elsevier, Oxford, pp 401–407

Green R, Miller JW (2007) Cobalamin (vitamin B12). In: Zempleni J (ed) Handbook of vitamins, 4th edn. Taylor & Francis, Boca Raton, FL, pp 258–413
Green R, Allen LH, Bjørke-Monsen A-L et al (2017) Vitamin B12 deficiency. Nat Rev Dis Prim 3:17040. https://doi.org/10.1038/nrdp.2017.40
Greger R (2000) Physiology of renal sodium transport. Am J Med Sci 319:51–62
Gregus Z, Klaassen CD (1986) Disposition of metals in rats: a comparative study of fecal, urinary, and biliary excretion and tissue distribution of eighteen metals. Toxicol Appl Pharmacol 85:24–38
Grencis RK, Humphreys NE, Bancroft AJ (2014) Immunity to gastrointestinal nematodes: mechanisms and myths. Immunol Rev 260:183–205. https://doi.org/10.1111/imr.12188
Griffin IJ (2003) Magnesium. In: Encyclopedia of food sciences and nutrition. Elsevier, London, pp 3641–3646
Griffith EC, Su Z, Turk BE et al (1997) Methionine aminopeptidase (type 2) is the common target for angiogenesis inhibitors AGM-1470 and ovalicin. Chem Biol 4:461–471
Grimble RF (2001) Nutritional modulation of immune function. Proc Nutr Soc 60:389–397
Grimble RF (1994) Malnutrition and the immune response. 2. Impact of nutrients on cytokine biology in infection. Trans R Soc Trop Med Hyg 88:615–619
Grimble RF (1997) Effect of antioxidative vitamins on immune function with clinical applications. Int J Vitam Nutr Res 67:312–320
Grimble RF (2005) Immunonutrition. Curr Opin Gastroenterol 21:216–222
Grimble RF (2006) The effects of sulfur amino acid intake on immune function in humans. J Nutr 136:1660S–1665S
Grimble RF (2009) Basics in clinical nutrition: immunonutrition – nutrients which influence immunity: effect and mechanism of action. E Spen Eur E J Clin Nutr Metab 4:e10–e13. https://doi.org/10.1016/j.eclnm.2008.07.015
Grimble RF, Grimble GK (1998) Immunonutrition: role of sulfur amino acids, related amino acids, and polyamines. Nutrition 14:605–610
Gruenwedel DW (2003) NUCLEIC ACIDS|Properties and determination. In: Encyclopedia of food sciences and nutrition. Elsevier, London, pp 4147–4152
Gupta KB, Gupta R, Atreja A et al (2009) Tuberculosis and nutrition. Lung India 26:9–16. https://doi.org/10.4103/0970-2113.45198
Ha MN, Graham FL, D'Souza CK et al (2004) Functional rescue of vitamin C synthesis deficiency in human cells using adenoviral-based expression of murine l-gulono-gamma-lactone oxidase. Genomics 83:482–492. https://doi.org/10.1016/j.ygeno.2003.08.018
Haanstra JR, González-Marcano EB, Gualdrón-López M, Michels PAM (2016) Biogenesis, maintenance and dynamics of glycosomes in trypanosomatid parasites. Biochim Biophys Acta 1863:1038–1048. https://doi.org/10.1016/j.bbamcr.2015.09.015
Hacquebard M, Carpentier YA (2005) Vitamin E: absorption, plasma transport and cell uptake. Curr Opin Clin Nutr Metab Care 8:133–138
Haddy FJ (1991) Roles of sodium, potassium, calcium, and natriuretic factors in hypertension. Hypertens 18:III179–III183
Haddy FJ, Vanhoutte PM, Feletou M (2006) Role of potassium in regulating blood flow and blood pressure. Am J Physiol Regul Integr Comp Physiol 290:R546–R552. https://doi.org/10.1152/ajpregu.00491.2005
Hafizi S, Dahlbäck B (2006) Gas6 and protein S. Vitamin K-dependent ligands for the Axl receptor tyrosine kinase subfamily. FEBS J 273:5231–5244. https://doi.org/10.1111/j.1742-4658.2006.05529.x
Hagopian K, Ramsey JJ, Weindruch R (2008) Enzymes of glycerol and glyceraldehyde metabolism in mouse liver: effects of caloric restriction and age on activities. Biosci Rep 28:107–115. https://doi.org/10.1042/BSR20080015
Hakim S, Bertucci MC, Conduit SE et al (2012) Inositol polyphosphate phosphatases in human disease. Curr Top Microbiol Immunol 362:247–314. https://doi.org/10.1007/978-94-007-5025-8_12

Halsted CH (2003) Absorption of water-soluble vitamins. Curr Opin Gastroenterol 19:113–117
Hansen RD, Raja C, Allen BJ (2000) Total body protein in chronic diseases and in aging. Ann N Y Acad Sci 904:345–352
Harhay MO, Olliaro PL, Costa DL, Costa CHN (2011) Urban parasitology: visceral leishmaniasis in Brazil. Trends Parasitol 27:403–409. https://doi.org/10.1016/j.pt.2011.04.001
Harris ED (2013) Cofactors: Inorganic. In: Encyclopedia of human nutrition. Elsevier, Oxford, pp 357–365
Harris ED, Percival SS (1991) A role for ascorbic acid in copper transport. Am J Clin Nutr 54:1193S–1197S
Harris JA, Benedict FG (1918) A biometric study of human basal metabolism. Proc Natl Acad Sci USA 4:370–373
Harrison LH, Naidu TG, Drew JS et al (1986) Reciprocal relationships between undernutrition and the parasitic disease visceral leishmaniasis. Rev Infect Dis 8:447–453. https://doi.org/10.1016/0307-4412(86)90203-7
Hartwig A (2001) Role of magnesium in genomic stability. Mutat Res 475:113–121
Hatanaka H, Ishizawa H, Nakamura Y et al (2014) Effects of vitamin K3 and K5 on proliferation, cytokine production, and regulatory T cell-frequency in human peripheral-blood mononuclear cells. Life Sci 99:61–68. https://doi.org/10.1016/j.lfs.2014.01.068
Hathcock JN, Hattan DG, Jenkins MY et al (1990) Evaluation of vitamin A toxicity. Am J Clin Nutr 52:183–202
Hazane-Puch F, Benaraba R, Valenti K et al (2010) Chromium III histidinate exposure modulates gene expression in HaCaT human keratinocytes exposed to oxidative stress. Biol Trace Elem Res 137:23–39. https://doi.org/10.1007/s12011-009-8557-9
Heaney RP, Rafferty K (2001) Carbonated beverages and urinary calcium excretion. Am J Clin Nutr 74:343–347
Helderman JH (1981) Role of insulin in the intermediary metabolism of the activated thymic-derived lymphocyte. J Clin Invest 67:1636–1642
Heliovaara MK, Teppo A-M, Karonen SL et al (2005) Plasma IL-6 concentration is inversely related to insulin sensitivity, and acute-phase proteins associate with glucose and lipid metabolism in healthy subjects. Diabetes Obes Metab 7:729–736. https://doi.org/10.1111/j.1463-1326.2004.00463.x
Hemilä H (2017) Vitamin C and Infections. Nutrients 9(4):339. https://doi.org/10.3390/nu9040339
Henderson LM, Gross CJ (1979) Metabolism of niacin and niacinamide in perfused rat intestine. J Nutr 109:654–662
Herrera E, Barbas C (2001) Vitamin E: action, metabolism and perspectives. J Physiol Biochem 57:43–56
Hershfinkel M, Silverman WF, Sekler I (2007) The zinc sensing receptor, a link between zinc and cell signaling. Mol Med 13:331–336. https://doi.org/10.2119/2006-00038.Hershfinkel
Heyer CME, Weiss E, Schmucker S et al (2015) The impact of phosphorus on the immune system and the intestinal microbiota with special focus on the pig. Nutr Res Rev 28:67–82. https://doi.org/10.1017/S0954422415000049
Hicks RM, Padayatchi N, Shah NS et al (2014) Malnutrition associated with unfavorable outcome and death among South African MDR-TB and HIV co-infected children. Int J Tuberc Lung Dis 18:1074–1083. https://doi.org/10.5588/ijtld.14.0231
Hida M, Mouane N, Ettair S et al (1999) Visceral leishmaniasis and malnutrition: a case report. Arch Pediatr 6:290–292. https://doi.org/10.1016/S0929-693X(99)80268-1
Hill TR, Aspray TJ (2017) The role of vitamin D in maintaining bone health in older people. Ther Adv Musculoskelet Dis 9:89–95. https://doi.org/10.1177/1759720X17692502
Hodges K, Gill R (2010) Infectious diarrhea. Gut Microbes 1:4–21. https://doi.org/10.4161/gmic.1.1.11036
Holecek M (2014) Evidence of a vicious cycle in glutamine synthesis and breakdown in pathogenesis of hepatic encephalopathy-therapeutic perspectives. Metab Brain Dis 29:9–17. https://doi.org/10.1007/s11011-013-9428-9

Holick MF (2005) Vitamin D|Physiology, dietary sources and requirements. In: Encyclopedia of human nutrition. Elsevier, Oxford, pp 368–377

Hollander D, Ruble PE (1978) Beta-carotene intestinal absorption: bile, fatty acid, pH, and flow rate effects on transport. Am J Physiol 235:E686–E691

Holzapfel NP, Holzapfel BM, Champ S et al (2013) The potential role of lycopene for the prevention and therapy of prostate cancer: from molecular mechanisms to clinical evidence. Int J Mol Sci 14:14620–14646. https://doi.org/10.3390/ijms140714620

Hopkins RG, Failla ML (1999) Transcriptional regulation of interleukin-2 gene expression is impaired by copper deficiency in Jurkat human T lymphocytes. J Nutr 129:596–601

How KL, Hazewinkel HA, Mol JA (1994) Dietary vitamin D dependence of cat and dog due to inadequate cutaneous synthesis of vitamin D. Gen Comp Endocrinol 96:12–18. https://doi.org/10.1006/gcen.1994.1154

Huang EP (1997) Metal ions and synaptic transmission: think zinc. Proc Natl Acad Sci USA 94:13386–13387

Huang Z, Rose AH, Hoffmann PR (2012) The role of selenium in inflammation and immunity: from molecular mechanisms to therapeutic opportunities. Antioxid Redox Signal 16:705–743. https://doi.org/10.1089/ars.2011.4145

Hughes DA (1999) Effects of dietary antioxidants on the immune function of middle-aged adults. Proc Nutr Soc 58:79–84

Hughes DA (2001) Dietary carotenoids and human immune function. Nutrition 10:823–827

Hühner J, Ingles-Prieto Á, Neusüß C et al (2015) Quantification of riboflavin, flavin mononucleotide, and flavin adenine dinucleotide in mammalian model cells by CE with LED-induced fluorescence detection. Electrophoresis 36:518–525. https://doi.org/10.1002/elps.201400451

Hunt JR, Gallagher SK, Johnson LK (1994) Effect of ascorbic acid on apparent iron absorption by women with low iron stores. Am J Clin Nutr 59:1381–1385

Hurrell R, Egli I (2010) Iron bioavailability and dietary reference values. Am J Clin Nutr 91:1461S–1467S. https://doi.org/10.3945/ajcn.2010.28674F

Hwang D (1989) Essential fatty acids and immune response. FASEB J 3:2052–2061

Imig JD, Ryan MJ (2013) Immune and inflammatory role in renal disease. Compr Physiol 3:957–976. https://doi.org/10.1002/cphy.c120028

Ink SL, Henderson LM (1984) Vitamin B6 metabolism. Annu Rev Nutr 4:455–470. https://doi.org/10.1146/annurev.nu.04.070184.002323

Institute of Medicine (2001a) Vitamin A. In: Food and Nutrition Board (ed) Dietary reference intakes for vitamin A, vitamin K, arsenic, boron, chromium, copper, iodine, iron, manganese, molybdenum, nickel, silicon, vanadium, and zinc. National Academies Press (US), Washington, DC

Institute of Medicine (1998a) Niacin. In: Institute of Medicine (US) (ed) Dietary reference intakes for thiamin, riboflavin, niacin, vitamin B6, folate, vitamin B12, pantothenic acid, biotin, and choline. National Academies Press (US), Washington, DC

Institute of Medicine (1998b) Dietary reference intakes for thiamin, riboflavin, niacin, vitamin B6, folate, vitamin B12, pantothenic acid, biotin, and choline. National Academies Press, Washington, DC

Institute of Medicine (2000a) Vitamin C. In: Dietary reference intakes for vitamin C, vitamin E, selenium, and carotenoids. National Academies Press, Washington, DC

Institute of Medicine (2011) Overview of calcium. In: Ross AC, Taylor CL, Yaktine AL, Del Valle HB (eds) Dietary reference intakes for calcium and vitamin D. National Academies Press (US), Washington, DC

Institute of Medicine (1997a) Phosphorus. In: Dietary reference intakes for calcium, phosphorus, magnesium, vitamin D, and fluoride. National Academies Press (US), Washington, DC

Institute of Medicine (1997b) Magnesium. In: Dietary reference intakes for calcium, phosphorus, magnesium, vitamin D, and fluoride. National Academies Press, Washington, DC, pp 432–190

Institute of Medicine (2001b) Zinc. In: Dietary reference intakes for vitamin A, vitamin K, arsenic, boron, chromium, copper, iodine, iron, manganese, molybdenum, nickel, silicon, vanadium, and zinc. National Academies Press (US), Washington, DC

Institute of Medicine (2001c) Copper. In: Dietary reference intakes for vitamin A, vitamin K, arsenic, boron, chromium, copper, iodine, iron, manganese, molybdenum, nickel, silicon, vanadium, and zinc. National Academies Press (US), Washington, DC

Institute of Medicine (2001d) Iodine. In: Dietary reference intakes for vitamin A, vitamin K, arsenic, boron, chromium, copper, iodine, iron, manganese, molybdenum, nickel, silicon, vanadium, and zinc. National Academies Press, Washington, DC

Institute of Medicine (2001e) Manganese. In: Dietary reference intakes for vitamin A, vitamin K, arsenic, boron, chromium, copper, iodine, iron, manganese, molybdenum, nickel, silicon, vanadium, and zinc. National Academies Press, Washington, DC, pp 419–394

Institute of Medicine (2001f) Molybdenum. In: Dietary reference intakes for vitamin A, vitamin K, arsenic, boron, chromium, copper, iodine, iron, manganese, molybdenum, nickel, silicon, vanadium, and zinc. National Academies Press, Washington, DC, pp 441–420

Institute of Medicine (2000b) Selenium. In: Dietary reference intakes for vitamin C, vitamin E, selenium, and carotenoids. National Academies Press, Washington, DC, p 324, 284

Institute of Medicine (2001g) Chromium. In: Dietary reference intakes for vitamin A, vitamin K, arsenic, boron, chromium, copper, iodine, iron, manganese, molybdenum, nickel, silicon, vanadium, and zinc. National Academies Press (US), Washington, DC

Institute of Medicine (1998c) Choline. In: Dietary reference intakes for thiamin, riboflavin, niacin, vitamin B6, folate, vitamin B12, pantothenic acid, biotin, and choline. National Academies Press (US), Washington, DC

Iolascon A, d'Apolito M, Servedio V et al (2005) Microcytic anemia and hepatic iron overload in a child with compound heterozygous mutations in DMT1 (SCL11A2). Blood 107:354–349

Jacob RA, Skala JH, Omaye ST, Turnlund JR (1987) Effect of varying ascorbic acid intakes on copper absorption and ceruloplasmin levels of young men. J Nutr 117:2109–2115

Jacobs A (1971) Haemopoietic factors Iron absorption. J Clin Path (Roy Coll Path) 24:55–59

Jacobson TA, Glickstein SB, Rowe JD, Soni PN (2012) Effects of eicosapentaenoic acid and docosahexaenoic acid on low-density lipoprotein cholesterol and other lipids: a review. J Clin Lipidol 6:5–18. https://doi.org/10.1016/j.jacl.2011.10.018

Jahnen-Dechent W, Ketteler M (2012) Magnesium basics. Clin Kidney J 5:i3–i14. https://doi.org/10.1093/ndtplus/sfr163

Janssen I, Katzmarzyk PT, Srinivasan SR et al (2005) Combined influence of body mass index and waist circumference on coronary artery disease risk factors among children and adolescents. Pediatrics 115:1623–1630. https://doi.org/10.1542/peds.2004-2588

Japur CC, Penaforte FRO, Chiarello PG et al (2009) Harris-Benedict equation for critically ill patients: are there differences with indirect calorimetry? J Crit Care 24:628.e1–628.e5. https://doi.org/10.1016/j.jcrc.2008.12.007

Jeffery LE, Wood AM, Qureshi OS et al (2012) Availability of 25-hydroxyvitamin D(3) to APCs controls the balance between regulatory and inflammatory T cell responses. J Immunol 189:5155–5164. https://doi.org/10.4049/jimmunol.1200786

Jiang Y, Trnka MJ, Medzihradszky KF et al (2009) Covalent heme attachment to the protein in human heme oxygenase-1 with selenocysteine replacing the His25 proximal iron ligand. J Inorg Biochem 103:316–325. https://doi.org/10.1016/j.jinorgbio.2008.11.002

Johnson MA (2003) Copper| Physiology. In: Encyclopedia of food sciences and nutrition. Elsevier, London, pp 1640–1647

Julve J, Martín-Campos JM, Escolà-Gil JC, Blanco-Vaca F (2016) Chylomicrons: advances in biology, pathology, laboratory testing, and therapeutics. Clin Chim Acta 455:134–148. https://doi.org/10.1016/j.cca.2016.02.004

Kall MA (2003) Ascorbic acid|Properties and determination. In: Encyclopedia of food sciences and nutrition. Elsevier, London, pp 316–324

Kamanna VS, Kashyap ML (2008) Mechanism of action of niacin. Am J Cardiol 101:20B–26B. https://doi.org/10.1016/j.amjcard.2008.02.029
Kanti Das T, Wati MR, Fatima-Shad K (2014) Oxidative stress gated by Fenton and Haber Weiss reactions and its association with Alzheimer's disease. Arch Neurosci 2(3). https://doi.org/10.5812/archneurosci.20078
Karney WW, Tong MJ (1972) Malabsorption in plasmodium falciparum malaria. Am J Trop Med Hyg 21:1–5
Katz MH (2012) HIV infection among persons born outside the United States. JAMA 308:623–624. https://doi.org/10.1001/jama.2012.8670
Kaur K, Gupta R, Saraf SA, Saraf SK (2014) Zinc: the metal of life. Compr Rev Food Sci Food Saf 13:358–376. https://doi.org/10.1111/1541-4337.12067
Keen CL, Gershwin ME (1990) Zinc deficiency and immune function. Annu Rev Nutr 10:415–431. https://doi.org/10.1146/annurev.nu.10.070190.002215
Keen CL, Uriu-Adams JY, Ensunsa JL, Gershwin ME (2004) Trace elements/Minerals and immunity. In: Handbook of nutrition and immunity. Humana Press, Totowa, NJ, pp 140–117
Keen CL, Zidenberg-Cherr S (2003) Manganese. In: Encyclopedia of food sciences and nutrition. Elsevier, London, pp 3686–3691
Kegley EB, Spears JW, Auman SK (2001) Dietary phosphorus and an inflammatory challenge affect performance and immune function of weanling pigs. J Anim Sci 79:413–419
Kehl-Fie TE, Chitayat S, Hood MI et al (2011) Nutrient metal sequestration by calprotectin inhibits bacterial superoxide defense, enhancing neutrophil killing of *Staphylococcus aureus*. Cell Host Microbe 10:158–164. https://doi.org/10.1016/j.chom.2011.07.004
Kempaiah P, Dokladny K (2016) Reduced Hsp70 and glutamine in pediatric severe malaria anemia: role of hemozoin in suppressing Hsp70 and NF-jB activation. Mol Med 22:1. https://doi.org/10.2119/molmed.2016.00130
Keusch GT (2003) The history of nutrition: malnutrition, infection and immunity. J Nutr 133:336S–340S
Khoshoo V, Raj P, Srivastava R, Bhan MK (1990) Salmonella typhimurium-associated severe protracted diarrhea in infants and young children. J Pediatr Gastroenterol Nutr 10:33–36. https://doi.org/10.1007/978-3-642-01846-6_6
Kieliszek M, Błażejak S (2016) Current knowledge on the importance of selenium in food for living organisms: a review. Molecules 21:609. https://doi.org/10.3390/molecules21050609
Kim M-H, Kim H (2017) The roles of glutamine in the intestine and its implication in intestinal diseases. Int J Mol Sci. https://doi.org/10.3390/ijms18051051
King LE, Osati-Ashtiani F, Fraker PJ (1995) Depletion of cells of the B lineage in the bone marrow of zinc-deficient mice. Immunology 85:69–73
Kirkland JB (2007) Niacin. In: Zempleni J (ed) Handbook of vitamins. Taylor & Francis, Boca Raton, FL, pp 232–191
Klack K, de Carvalho JF (2006) Vitamina K: metabolismo, fontes e interação com o anticoagulante varfarina. Rev Bras Reumatol 46:398–406. https://doi.org/10.1590/S0482-50042006000600007
Klasing KC (1994) Avian leukocytic cytokines. Poult Sci 73:1035–1043
Klawe JJ (2003) Iron|Biosynthesis and significance of Heme (Haem). In: Encyclopedia of food sciences and nutrition. Elsevier, London, pp 3379–3381
Klevay LM, Bogden JD, Aladjem M et al (2007) Renal and gastrointestinal potassium excretion in humans: new insight based on new data and review and analysis of published studies. J Am Coll Nutr 26:103–110
Knutson MD, Oukka M, Koss LM et al (2005) Iron release from macrophages after erythrophagocytosis is up-regulated by ferroportin 1 overexpression and down-regulated by hepcidin. Proc Natl Acad Sci 102:1324–1328. https://doi.org/10.1073/pnas.0409409102
Kocyigit A, Gur S, Erel O, Gurel MS (2002) Associations among plasma selenium, zinc, copper, and iron concentrations and immunoregulatory cytokine levels in patients with cutaneous leishmaniasis. Biol Trace Elem Res 90:47–55. https://doi.org/10.1385/BTER:90:1-3:47

Kohlmeier M (2003) Vitamin B6. In: Nutrient metabolism. Elsevier, Amsterdam, pp 581–591
Kojetin DJ, Venters RA, Kordys DR et al (2006) Structure, binding interface and hydrophobic transitions of Ca^{2+}-loaded calbindin-D(28K). Nat Struct Mol Biol 13:641–647. https://doi.org/10.1038/nsmb1112
Kosmiski L (2011) Energy expenditure in HIV infection. Am J Clin Nutr 94:1677S–1682S. https://doi.org/10.3945/ajcn.111.012625
Kothapalli N, Camporeale G, Kueh A et al (2005) Biological functions of biotinylated histones. J Nutr Biochem 16:446–448. https://doi.org/10.1016/j.jnutbio.2005.03.025
Kowalska A, Siwicki AK, Kowalski RK (2017) Dietary resveratrol improves immunity but reduces reproduction of broodstock medaka Oryzias latipes (Temminck & Schlegel). Fish Physiol Biochem 43:27–37. https://doi.org/10.1007/s10695-016-0265-8
Krajcovicová-Kudlácková M, Pauková V, Baceková M, Dusinská M (2004) Lipid peroxidation in relation to vitamin C and vitamin E levels. Cent Eur J Public Health 12:46–48
Krawinkel MB (2012) Interaction of nutrition and infections globally: an overview. Ann Nutr Metab 61:39–45. https://doi.org/10.1159/000345162
Krebs EG, Beavo JA (1979) Phosphorylation-dephosphorylation of enzymes. Annu Rev Biochem 48:923–959. https://doi.org/10.1146/annurev.bi.48.070179.004423
Kromann N, Green A (1980) Epidemiological studies in the Upernavik district, Greenland. Incidence of some chronic diseases 1950–1974. Acta Med Scand 208:401–406
Kumar N, Garg AK, Mudgal V et al (2008) Effect of different levels of selenium supplementation on growth rate, nutrient utilization, blood metabolic profile, and immune response in lambs. Biol Trace Elem Res 126:44–56. https://doi.org/10.1007/s12011-008-8214-8
Kumar V, Bimal S, Singh SK et al (2014) *Leishmania donovani*: dynamics of *L. donovani* evasion of innate immune cell attack due to malnutrition in visceral leishmaniasis. Nutrition 30:449–458. https://doi.org/10.1016/j.nut.2013.10.003
Kurosawa M (1994) Phosphorylation and dephosphorylation of protein in regulating cellular function. J Pharmacol Toxicol Methods 31:135–139. https://doi.org/10.1016/1056-8719(94)90075-2
Kurtzman NA, White MG, Rogers PW, Flynn JJ (1972) Relationship of sodium reabsorption and glomerular filtration rate to renal glucose reabsorption. J Clin Invest 51:127–133. https://doi.org/10.1172/JCI106782
Küry S, Dréno B, Bézieau S et al (2002) Identification of SLC39A4, a gene involved in acrodermatitis enteropathica. Nat Genet 31:239–240. https://doi.org/10.1038/ng913
L'Abbé MR (2003) Calcium|Physiology. In: Encyclopedia of food sciences and nutrition. Elsevier, London, pp 771–779
Lachance C, Segura M, Dominguez-Punaro MC et al (2014) Deregulated balance of omega-6 and omega-3 polyunsaturated fatty acids following infection by the zoonotic pathogen Streptococcus suis. Infect Immun 82:1778–1785. https://doi.org/10.1128/IAI.01524-13
Lalor SM, Mellanby RJ, Friend EJ et al (2012) Domesticated cats with active mycobacteria infections have low serum vitamin D (25(OH)D) concentrations. Transbound Emerg Dis 59:279–281. https://doi.org/10.1111/j.1865-1682.2011.01265.x
Lamers Y (2011) Folate recommendations for pregnancy, lactation, and infancy. Ann Nutr Metab 59:32–37. https://doi.org/10.1159/000332073
Laragh JH, Sealey JE (2011) Renin-angiotensin-aldosterone system and the renal regulation of sodium, potassium, and blood pressure homeostasis. In: Comprehensive physiology. Wiley, Hoboken, NJ
Lazzerini M, Wanzira H (2016) Oral zinc for treating diarrhoea in children. Cochrane Database Syst Rev 12:CD005436. https://doi.org/10.1002/14651858.CD005436.pub5
Leach RM, Lilburn MS (1978) Manganese metabolism and its function. World Rev Nutr Diet 32:123–134
Lee M (2013) Ingesta: los nutrientes y su metabolismo. In: Mahan K, Escott-Stump S, Raymond J (eds) Krause's food and the nutrition care process, 13th edn. Elsevier España, S.L, España, pp 128–132

Leem AY, Park MS, Park BH et al (2017) Value of serum cystatin C measurement in the diagnosis of sepsis-induced kidney injury and prediction of renal function recovery. Yonsei Med J 58:604–612. https://doi.org/10.3349/ymj.2017.58.3.604

Leermakers ETM, Moreira EM, Kiefte-de Jong JC et al (2015) Effects of choline on health across the life course: a systematic review. Nutr Rev 73:500–522. https://doi.org/10.1093/nutrit/nuv010

Lestari MLAD, Indrayanto G (2014) Curcumin. In: Profiles of drug substances, excipients, and related methodology. Elsevier Academic Press, San Diego, pp 113–204

Levine M, Conry-Cantilena C, Wang Y et al (1996) Vitamin C pharmacokinetics in healthy volunteers: evidence for a recommended dietary allowance. Proc Natl Acad Sci USA 93:3704–3709

Li S, Siyuan T, Jiangmin F et al (2017) Analysis of the association between *Mycobacterium tuberculosis* infection and Immunoglobulin A nephropathy by early secreted antigenic target 6 detection in renal biopsies: a prospective study. Postgrad Med 129:307–311. https://doi.org/10.1080/00325481.2017.1289054

Li X, Lian F, Liu C et al (2015) Isocaloric pair-fed high-carbohydrate diet induced more hepatic steatosis and inflammation than high-fat diet mediated by miR-34a/SIRT1 axis in mice. Sci Rep 5:16774. https://doi.org/10.1038/srep16774

Li Y, Schellhorn HE (2007) New developments and novel therapeutic perspectives for vitamin C. J Nutr 137:2171–2184

Libako P, Miller J, Nowacki W et al (2015) Extracellular Mg concentration and Ca blockers modulate the initial steps of the response of Th2 lymphocytes in co-culture with macrophages and dendritic cells. Eur Cytokine Netw 26:1–9. https://doi.org/10.1684/ecn.2015.0361

Libako P, Nowacki W, Castiglioni S et al (2016) Extracellular magnesium and calcium blockers modulate macrophage activity. Magnes Res 29:11–21. https://doi.org/10.1684/mrh.2016.0398

Lichtstein HC, Gunsalus IC, Umbreit WW (1945) Function of the vitamin B6 group; pyridoxal phosphate (codecarboxylase) in transamination. J Biol Chem 161:311–320

Linder M (2013) Mobilization of stored iron in mammals: a review. Nutrients 5:4022–4050. https://doi.org/10.3390/nu5104022

Lindquist S, Hernell O (2010) Lipid digestion and absorption in early life: an update. Curr Opin Clin Nutr Metab Care 13:314–320. https://doi.org/10.1097/MCO.0b013e328337bbf0

Linus Pauling Institute (2016) Sodium (Chloride). In: Oregon State Univ. http://lpi.oregonstate.edu/mic/minerals/sodium#reference5. Accessed 6 Jun 2017

Liu G, Ren W, Fang J et al (2017) L-Glutamine and L-arginine protect against enterotoxigenic *Escherichia coli* infection via intestinal innate immunity in mice. Amino Acids 49 (12):1945–1954. https://doi.org/10.1007/s00726-017-2410-9

Liu J, Chakraborty S, Hosseinzadeh P et al (2014) Metalloproteins containing cytochrome, iron–sulfur, or copper redox centers. Chem Rev 114:4366–4469. https://doi.org/10.1021/cr400479b

Liu RH (2003) Health benefits of fruit and vegetables are from additive and synergistic combinations of phytochemicals. Am J Clin Nutr 78:517S–520S

Liu Y, Zhang Y, Dong P et al (2015) Digestion of nucleic acids starts in the stomach. Sci Rep 5:11936. https://doi.org/10.1038/srep11936

Lo CJ, Chiu KC, Fu M et al (1999) Fish oil decreases macrophage tumor necrosis factor gene transcription by altering the NF kappa B activity. J Surg Res 82:216–221. https://doi.org/10.1006/jsre.1998.5524

Lobo GP, Amengual J, Palczewski G et al (2012) Mammalian carotenoid-oxygenases: key players for carotenoid function and homeostasis. Biochim Biophys Acta 1821:78–87. https://doi.org/10.1016/j.bbalip.2011.04.010

Lodish H, Berk A, Zipursky SL et al (2000a) Structure of nucleic acids. In: Molecular cell biology, 4th edn. W. H. Freeman, New York

Lodish H, Berk A, Zipursky SL et al (2000b) Intracellular ion environment and membrane electric potential. In: Molecular cell biology, 4th edn. W. H. Freeman, New York

Long CL, Schaffel N, Geiger JW et al (1979) Metabolic response to injury and illness: estimation of energy and protein needs from indirect calorimetry and nitrogen balance. JPEN J Parenter Enteral Nutr 3:452–456. https://doi.org/10.1177/014860717900300609

Long KZ, Rosado JL, Fawzi W (2007) The comparative impact of iron, the B-complex vitamins, vitamins C and E, and selenium on diarrheal pathogen outcomes relative to the impact produced by vitamin A and zinc. Nutr Rev 65:218–232

Longarela A, Olarra J, Suárez L, García de Lorenzo A (2000) Metabolic response to stress, can we control it? Nutr Hosp 15:275–279

Lönnerdal B (2000) Dietary factors influencing zinc absorption. J Nutr 130:1378S–1383S

Lonsdale D (2006) A review of the biochemistry, metabolism and clinical benefits of Thiamin (e) and its derivatives. Evid Based Complement Alternat Med 3:49–59. https://doi.org/10.1093/ecam/nek009

Lopina OD (2000) Na^+, K^+-ATPase: structure, mechanism, and regulation. Membr Cell Biol 13:721–744

Lorenz KJ, Schallert R, Daniel V (2015) Immunonutrition – the influence of early postoperative glutamine supplementation in enteral/parenteral nutrition on immune response, wound healing and length of hospital stay in multiple trauma patients and patients after extensive surgery. GMS Interdiscip Plast Reconstr Surg DGPW 4.:Doc15. https://doi.org/10.3205/iprs000074

Lovegrove A, Edwards CH, De Noni I et al (2017) Role of polysaccharides in food, digestion, and health. Crit Rev Food Sci Nutr 57:237–253. https://doi.org/10.1080/10408398.2014.939263

Lubos E, Loscalzo J, Handy DE (2011) Glutathione Peroxidase-1 in health and disease: from molecular mechanisms to therapeutic opportunities. Antioxid Redox Signal 15:1957–1997. https://doi.org/10.1089/ars.2010.3586

Luo C, Wu X-G (2011) Lycopene enhances antioxidant enzyme activities and immunity function in N-Methyl-N′-nitro-N-nitrosoguanidine–induced gastric cancer rats. Int J Mol Sci 12:3340–3351. https://doi.org/10.3390/ijms12053340

Lutsenko S (2010) Human copper homeostasis: a network of interconnected pathways. Curr Opin Chem Biol 14:211–217. https://doi.org/10.1016/j.cbpa.2010.01.003

Lynch SR (2003) Iron|Physiology. In: Encyclopedia of food sciences and nutrition. Elsevier, London, pp 3373–3379

Macfarlane GJ, Paudyal P, Doherty M et al (2012) A systematic review of evidence for the effectiveness of practitioner-based complementary and alternative therapies in the management of rheumatic diseases: rheumatoid arthritis. Rheumatology (Oxford) 51:1707–1713. https://doi.org/10.1093/rheumatology/kes133

Machado-Coelho GLL, Caiaffa WT, Genaro O et al (2005) Risk factors for mucosal manifestation of American cutaneous leishmaniasis. Trans R Soc Trop Med Hyg 99:55–61. https://doi.org/10.1016/j.trstmh.2003.08.001

MacPherson A, Dixon J (2003) Cobalt. In: Encyclopedia of food sciences and nutrition. Elsevier, London, pp 1431–1436

Maeda N, Ishii M, Nishimura K, Kamimura K (2011) Functions of chondroitin sulfate and heparan sulfate in the developing brain. Neurochem Res 36:1228–1240. https://doi.org/10.1007/s11064-010-0324-y

Mahan LK, Escott-Stump S, Raymond JJ, Krause MV (2013) Krause dietoterapia. Elsevier, Amsterdam

Mainous MR, Deitch EA (1994) Nutrition and infection. Surg Clin North Am 74:659–676

Maiuolo J, Oppedisano F, Gratteri S et al (2016) Regulation of uric acid metabolism and excretion. Int J Cardiol 213:8–14. https://doi.org/10.1016/j.ijcard.2015.08.109

Makarchikov AF (2009) Vitamin B1: Metabolism and functions. Biochem Suppl Ser B Biomed Chem 3:116–128. https://doi.org/10.1134/S1990750809020024

Malafaia G (2009) Protein-energy malnutrition as a risk factor for visceral leishmaniasis: a review. Parasite Immunol 31:587–596. https://doi.org/10.1111/j.1365-3024.2009.01117.x

Mangin M, Sinha R, Fincher K (2014) Inflammation and vitamin D: the infection connection. Inflamm Res 63:803–819. https://doi.org/10.1007/s00011-014-0755-z

Manicassamy S, Ravindran R, Deng J et al (2009) Toll-like receptor 2-dependent induction of vitamin A-metabolizing enzymes in dendritic cells promotes T regulatory responses and inhibits autoimmunity. Nat Med 15:401–409. https://doi.org/10.1038/nm.1925

Manolagas SC, Yu XP, Girasole G, Bellido T (1994) Vitamin D and the hematolymphopoietic tissue: a 1994 update. Semin Nephrol 14:129–143

Mansoorabadi SO, Padmakumar R, Fazliddinova N et al (2005) Characterization of a succinyl-CoA radical-cob(II)alamin spin triplet intermediate in the reaction catalyzed by adenosylcobalamin-dependent methylmalonyl-CoA mutase. Biochemistry 44:3153–3158. https://doi.org/10.1021/bi0482102

Mansoorabadi SO, Thibodeaux CJ, Liu H (2007) The diverse roles of flavin coenzymes nature's most versatile thespians. J Org Chem 72:6329–6342. https://doi.org/10.1021/jo0703092

Manzetti S, Zhang J, van der Spoel D (2014) Thiamin function, metabolism, uptake, and transport. Biochemistry 53:821–835. https://doi.org/10.1021/bi401618y

Marcinkiewicz J, Kontny E (2014) Taurine and inflammatory diseases. Amino Acids 46:7–20. https://doi.org/10.1007/s00726-012-1361-4

Maret W (2013) Zinc biochemistry: from a single zinc enzyme to a key element of life. Adv Nutr 4:82–91. https://doi.org/10.3945/an.112.003038

Maret W, Sandstead HH (2006) Zinc requirements and the risks and benefits of zinc supplementation. J Trace Elem Med Biol 20:3–18. https://doi.org/10.1016/j.jtemb.2006.01.006

Marion-Letellier R, Savoye G, Ghosh S (2016) IBD: in food we trust. J Crohns Colitis 10:1351–1361. https://doi.org/10.1093/ecco-jcc/jjw106

Martens JH, Barg H, Warren MJ, Jahn D (2002) Microbial production of vitamin B12. Appl Microbiol Biotechnol 58:275–285. https://doi.org/10.1007/s00253-001-0902-7

Mataix J, Sánchez de Medina F (2009) Vitaminas. In: Mataix J (ed) Nutrición y Alimentación, Volumen 1. Oceáno/Ergón, Madrid, pp 330–185

Matthews DM (1975) Protein absorption. Bibl Nutr Dieta 5:28–41

Mattmiller SA, Carlson BA, Sordillo LM (2013) Regulation of inflammation by selenium and selenoproteins: impact on eicosanoid biosynthesis. J Nutr Sci 2:e28. https://doi.org/10.1017/jns.2013.17

Mazur A, Maier JAM, Rock E et al (2007) Magnesium and the inflammatory response: potential physiopathological implications. Arch Biochem Biophys 458:48–56. https://doi.org/10.1016/j.abb.2006.03.031

McCall MR, Frei B (1999) Can antioxidant vitamins materially reduce oxidative damage in humans? Free Radic Biol Med 26:1034–1053

McCance R, Lawrence R (1929) The carbohydrate content of foods. HMSO, London

McClave SA, Snider HL (1994) Understanding the metabolic response to critical illness: factors that cause patients to deviate from the expected pattern of hypermetabolism. New Horiz 2:139–146

McCormick CC (2002) Passive diffusion does not play a major role in the absorption of dietary calcium in normal adults. J Nutr 132:3428–3430

McCormick DB (2003) Riboflavin|Physiology. In: Encyclopedia of food sciences and nutrition. Elsevier, London, pp 4989–4995

McCuskee S, Brickley EB, Wood A, Mossialos E (2014) Malaria and macronutrient deficiency as correlates of anemia in young children: a systematic review of observational studies. Ann Glob Heal 80:458–465. https://doi.org/10.1016/j.aogh.2015.01.003

McDowell LR (2000) Vitamins in animal and human nutrition. In: Vitamins in animal and human nutrition, 2nd edn. Iowa State University Press, Ames, pp 564–523

McLaren DS, Kraemer K (2012) Vitamin A in health. World Rev Nutr Diet 103:33–51. https://doi.org/10.1159/000170954

McRae MP (2017) Therapeutic benefits of glutamine: an umbrella review of meta-analyses. Biomed Rep 6:576–584. https://doi.org/10.3892/br.2017.885

Meckling KA (2009) When good nutrients go bad: can we predict nutrient-drug interactions? Br J Nutr 102:334–336. https://doi.org/10.1017/S0007114508199494

Mehta NM, Duggan CP (2009) Nutritional deficiencies during critical illness. Pediatr Clin North Am 56:1143–1160. https://doi.org/10.1016/j.pcl.2009.06.007
Melse-Boonstra A, Vossenaar M, van Loo-Bouwman CA et al (2017) Dietary vitamin A intake recommendations revisited: global confusion requires alignment of the units of conversion and expression. Public Health Nutr:1–4. https://doi.org/10.1017/S1368980017000477
Mendel RR (2013a) Metabolism of molybdenum. Met Ions Life Sci 12:503–528. https://doi.org/10.1007/978-94-007-5561-1_15
Mendel RR (2013b) The molybdenum cofactor. J Biol Chem 288:13165–13172. https://doi.org/10.1074/jbc.R113.455311
Méndez C, Dobaño C (2004) Malaria and immunity. In: Handbook of nutrition and immunity. Humana Press, Totowa, NJ, pp 264–243
Mendonça N, Gragnani A, Masako Ferreira L (2011) Burns, metabolism and nutritional requirements. Nutr Hosp 26:692–700. https://doi.org/10.1590/S0212-16112011000400005
Mercer JF (2001) The molecular basis of copper-transport diseases. Trends Mol Med 7:64–69
Merrill AH, Henderson JM (1990) Vitamin B6 metabolism by human liver. Ann N Y Acad Sci 585:110–117
Mertz W (1969) Chromium occurrence and function in biological systems. Physiol Rev 49:163–239
Mertz W (1993) Chromium in human nutrition: a review. J Nutr 123:626–633
Mertz W (1998) Interaction of chromium with insulin: a progress report. Nutr Rev 56:174–177
Methenitou G, Maravelias C, Athanaselis S et al (2001) Immunomodulative effects of aflatoxins and selenium on human natural killer cells. Vet Hum Toxicol 43:232–234
Metz J (1992) Cobalamin deficiency and the pathogenesis of nervous system disease. Annu Rev Nutr 12:59–79. https://doi.org/10.1146/annurev.nu.12.070192.000423
Meyer K, Jia Y, Cao W-Q et al (2002) Expression of peroxisome proliferator-activated receptor alpha, and PPARalpha regulated genes in spontaneously developed hepatocellular carcinomas in fatty acyl-CoA oxidase null mice. Int J Oncol 21:1175–1180
Middleton E (1998) Effect of plant flavonoids on immune and inflammatory cell function. Adv Exp Med Biol 439:175–182
Mikami T, Kitagawa H (2013) Biosynthesis and function of chondroitin sulfate. Biochim Biophys Acta 1830:4719–4733. https://doi.org/10.1016/j.bbagen.2013.06.006
Minehira K, Bettschart V, Vidal H et al (2003) Effect of carbohydrate overfeeding on whole body and adipose tissue metabolism in humans. Obes Res 11:1096–1103. https://doi.org/10.1038/oby.2003.150
Mishra J, Carpenter S, Singh S (2010) Low serum zinc levels in an endemic area of visceral leishmaniasis in Bihar, India. Indian J Med Res 131:793–798
Mistry HD, Broughton Pipkin F, Redman CWG, Poston L (2012) Selenium in reproductive health. Am J Obstet Gynecol 206:21–30. https://doi.org/10.1016/j.ajog.2011.07.034
Mizock BA (2010) Immunonutrition and critical illness: an update. Nutrition 26:701–707. https://doi.org/10.1016/j.nut.2009.11.010
Mkhize BT, Mabaso M, Mamba T et al (2017) The interaction between HIV and intestinal helminth parasites coinfection with nutrition among adults in KwaZulu-Natal, South Africa. Biomed Res Int 2017:1–12. https://doi.org/10.1155/2017/9059523
Mocchegiani E, Costarelli L, Giacconi R et al (2012) Micronutrient (Zn, Cu, Fe)-gene interactions in ageing and inflammatory age-related diseases: implications for treatments. Ageing Res Rev 11:297–319. https://doi.org/10.1016/j.arr.2012.01.004
Mock DM (2007) Biotin. In: Zempleni J (ed) Handbook of vitamins. Taylor & Francis, Boca Raton, FL, pp 384–361
Mody A, Bartz S, Hornik CP et al (2014) Effects of HIV infection on the metabolic and hormonal status of children with severe acute malnutrition. PLoS One 9:e102233. https://doi.org/10.1371/journal.pone.0102233
Moffatt BA, Ashihara H (2002) Purine and pyrimidine nucleotide synthesis and metabolism. Arab B 1:e0018. https://doi.org/10.1199/tab.0018

Molano A, Meydani SN (2012) Vitamin E, signalosomes and gene expression in T cells. Mol Aspects Med 33:55–62. https://doi.org/10.1016/j.mam.2011.11.002
Morita R, Yamamoto I, Takada M et al (1993) Hypervitaminosis D. Nihon Rinsho 51:984–988
Morris KL, Zemel MB (2005) 1, 25-Dihydroxyvitamin D 3 modulation of adipocyte glucocorticoid function. Obes Res 13:670–677. https://doi.org/10.1038/oby.2005.75
Mucida D, Park Y, Kim G et al (2007) Reciprocal TH17 and regulatory T cell differentiation mediated by retinoic acid. Science 317:256–260. https://doi.org/10.1126/science.1145697
Mueckler M, Thorens B (2013) The SLC2 (GLUT) family of membrane transporters. Mol Aspects Med 34:121–138. https://doi.org/10.1016/j.mam.2012.07.001
Mukwevho E, Ferreira Z, Ayeleso A (2014) Potential role of sulfur-containing antioxidant systems in highly oxidative environments. Molecules 19:19376–19389. https://doi.org/10.3390/molecules191219376
Muller DP (1986) Vitamin E-its role in neurological function. Postgrad Med J 62:107–112
Mundi MS, Shah M, Hurt RT (2016) When is it appropriate to use glutamine in critical illness? Nutr Clin Pract 31:445–450. https://doi.org/10.1177/0884533616651318
Murray PJ, Rathmell J, Pearce E (2015) SnapShot: immunometabolism. Cell Metab 22:190–190.e1. https://doi.org/10.1016/j.cmet.2015.06.014
Nair R, Maseeh A (2012) Vitamin D: the "sunshine" vitamin. J Pharmacol Pharmacother 3:118–126. https://doi.org/10.4103/0976-500X.95506
Naithani R (2008) Organoselenium compounds in cancer chemoprevention. Mini Rev Med Chem 8:657–668
National Research Council (1989) Protein and amino acids. In: Subcommittee on the Tenth Edition of the Recommended Dietary Allowances (ed) Recommended dietary allowances, 10th edn. National Academies Press (US), Washington, DC
National Research Council (2000) Physiological role of copper. In: Copper in drinking water. National Academies Press (US), Washington, DC
National Research Council (2005) Health implications of perchlorate ingestion. National Academies Press, Washington, DC
Naveilhan P, Neveu I, Wion D, Brachet P (1996) 1,25-Dihydroxyvitamin D3, an inducer of glial cell line-derived neurotrophic factor. Neuroreport 7:2171–2175
Nayak N, Harrison EH, Hussain MM (2001) Retinyl ester secretion by intestinal cells: a specific and regulated process dependent on assembly and secretion of chylomicrons. J Lipid Res 42:272–280
Neveu I, Naveilhan P, Baudet C et al (1994a) 1,25-dihydroxyvitamin D3 regulates NT-3, NT-4 but not BDNF mRNA in astrocytes. Neuroreport 6:124–126
Neveu I, Naveilhan P, Jehan F et al (1994b) 1,25-dihydroxyvitamin D3 regulates the synthesis of nerve growth factor in primary cultures of glial cells. Brain Res Mol Brain Res 24:70–76
Nilsson A, Wilhelms DB, Mirrasekhian E et al (2017) Inflammation-induced anorexia and fever are elicited by distinct prostaglandin dependent mechanisms, whereas conditioned taste aversion is prostaglandin independent. Brain Behav Immun 61:236–243. https://doi.org/10.1016/j.bbi.2016.12.007
Nimni ME, Han B, Cordoba F (2007) Are we getting enough sulfur in our diet? Nutr Metab (Lond) 4:24. https://doi.org/10.1186/1743-7075-4-24
Norata GD, Caligiuri G, Chavakis T et al (2015) The cellular and molecular basis of translational immunometabolism. Immunity 43:421–434. https://doi.org/10.1016/j.immuni.2015.08.023
Norman AW (2008) From vitamin D to hormone D: fundamentals of the vitamin D endocrine system essential for good health. Am J Clin Nutr 88:491S–499S
Norman AW, Hentry HL (2007) Vitamin D. In: Zempleni J (ed) Handbook of vitamins, 4th edn. Taylor & Francis, Boca Raton, FL, pp 110–141
Norton JE, Gonzalez Espinosa Y, Watson RL et al (2015) Functional food microstructures for macronutrient release and delivery. Food Funct 6:663–678. https://doi.org/10.1039/c4fo00965g

Notari L, Riera DC, Sun R et al (2014) Role of macrophages in the altered epithelial function during a type 2 immune response induced by enteric nematode infection. PLoS One 9:e84763. https://doi.org/10.1371/journal.pone.0084763

Nursyam EW, Amin Z, Rumende CM (2006) The effect of vitamin D as supplementary treatment in patients with moderately advanced pulmonary tuberculous lesion. Acta Med Indones 38:3–5

Nye CK, Hanson RW, Kalhan SC (2008) Glyceroneogenesis is the dominant pathway for triglyceride glycerol synthesis in vivo in the rat. J Biol Chem 283:27565–27574. https://doi.org/10.1074/jbc.M804393200

Obled C, Papet I, Breuillé D (2002) Metabolic bases of amino acid requirements in acute diseases. Curr Opin Clin Nutr Metab Care 5:189–197

Ochoa L, de Paniagua Michel J, Olmos-Soto J (2014) Complex carbohydrates as a possible source of high energy to formulate functional feeds. Adv Food Nutr Res 73:259–288. https://doi.org/10.1016/B978-0-12-800268-1.00012-3

Oda H (2006) Functions of sulfur-containing amino acids in lipid metabolism. J Nutr 136:1666S–1669S

Ogunbileje JO, Porter C, Herndon DN et al (2016) Hypermetabolism and hypercatabolism of skeletal muscle accompany mitochondrial stress following severe burn trauma. Am J Physiol – Endocrinol Metab 311:E436–E448. https://doi.org/10.1152/ajpendo.00535.2015

Oliveira AGL, Brito PD, Schubach AO et al (2013) Influence of the nutritional status in the clinical and therapeutical evolution in adults and elderly with American Tegumentary Leishmaniasis. Acta Trop 128:36–40. https://doi.org/10.1016/j.actatropica.2013.06.005

Olson TL, Williams JC, Allen JP (2013) Influence of protein interactions on oxidation/reduction midpoint potentials of cofactors in natural and de novo metalloproteins. Biochim Biophys Acta – Bioenerg 1827:914–922. https://doi.org/10.1016/j.bbabio.2013.02.014

Omdahl JL, Morris HA, May BK (2002) Hydroxylase enzymes of the vitamin D pathway: expression, function, and regulation. Annu Rev Nutr 22:139–166. https://doi.org/10.1146/annurev.nutr.22.120501.150216

Omur A, Kirbas A, Aksu E et al (2016) Effects of antioxidant vitamins (A, D, E) and trace elements (Cu, Mn, Se, Zn) on some metabolic and reproductive profiles in dairy cows during transition period. Pol J Vet Sci 19:697–706. https://doi.org/10.1515/pjvs-2016-0088

Ortiz D, Afonso C, Hagel I et al (2000) Influencia de las infecciones helmínticas y el estado nutricional en la respuesta inmunitaria de niños venezolanos. Rev Panam Salud Pública 8:156–163. https://doi.org/10.1590/S1020-49892000000800002

Osada J (2013) The use of transcriptomics to unveil the role of nutrients in mammalian liver. ISRN Nutr 2013:1–19. https://doi.org/10.5402/2013/403792

Osada S, Carr BI (2000) Critical role of extracellular signal-regulated kinase (ERK) phosphorylation in novel vitamin K analog-induced cell death. Jpn J Cancer Res 91:1250–1257

Osada S, Osada K, Carr BI (2001) Tumor cell growth inhibition and extracellular signal-regulated kinase (ERK) phosphorylation by novel K vitamins. J Mol Biol 314:765–772. https://doi.org/10.1006/jmbi.2001.5171

Osredkar J (2011) Copper and zinc, biological role and significance of copper/zinc imbalance. J Clin Toxicol. https://doi.org/10.4172/2161-0495.s3-001

Oster M, Just F, Büsing K et al (2016) Toward improved phosphorus efficiency in monogastrics-interplay of serum, minerals, bone, and immune system after divergent dietary phosphorus supply in swine. Am J Physiol Regul Integr Comp Physiol 310:R917–R925. https://doi.org/10.1152/ajpregu.00215.2015

Oster O, Prellwitz W (1990) Selenium and cardiovascular disease. Biol Trace Elem Res 24:91–103

Ott G, Havemeyer A, Clement B (2015) The mammalian molybdenum enzymes of mARC. J Biol Inorg Chem 20:265–275. https://doi.org/10.1007/s00775-014-1216-4

Pae M, Wu D (2017) Nutritional modulation of age-related changes in the immune system and risk of infection. Nutr Res 41:14–35. https://doi.org/10.1016/j.nutres.2017.02.001

Palmer CS, Anzinger JJ, Zhou J et al (2014) Glucose transporter 1-expressing proinflammatory monocytes are elevated in combination antiretroviral therapy-treated and untreated HIV^+ subjects. J Immunol 193:5595–5603. https://doi.org/10.4049/jimmunol.1303092

Palmer CS, Cherry CL, Sada-Ovalle I et al (2016) Glucose metabolism in T cells and monocytes: new perspectives in HIV pathogenesis. EBioMedicine 6:31–41. https://doi.org/10.1016/j.ebiom.2016.02.012
Panchal SK, Wanyonyi S, Brown L (2017) Selenium, vanadium, and chromium as micronutrients to improve metabolic syndrome. Curr Hypertens Rep 19:10. https://doi.org/10.1007/s11906-017-0701-x
Pandolfi F, Franza L, Mandolini C, Conti P (2017) Immune modulation by vitamin D: special emphasis on its role in prevention and treatment of cancer. Clin Ther. https://doi.org/10.1016/j.clinthera.2017.03.012
Paniz C, Bertinato JF, Lucena MR et al (2017) A daily dose of 5 mg folic acid for 90 days is associated with increased serum unmetabolized folic acid and reduced natural killer cell cytotoxicity in healthy Brazilian adults. J Nutr 147:1677–1685. https://doi.org/10.3945/jn.117.247445
Pantelidou M, Tsiakitzis K, Rekka EA, Kourounakis PN (2017) Biologic stress, oxidative stress, and resistance to drugs: what is hidden behind. Molecules. https://doi.org/10.3390/molecules22020307
Papanikolaou G, Pantopoulos K (2017) Systemic iron homeostasis and erythropoiesis. IUBMB Life 69:399–413. https://doi.org/10.1002/iub.1629
Parcell S (2002) Sulfur in human nutrition and applications in medicine. Altern Med Rev 7:22–44
Park K (2015) Role of micronutrients in skin health and function. Biomol Ther (Seoul) 23:207–217. https://doi.org/10.4062/biomolther.2015.003
Pasing Y, Fenton CG, Jorde R, Paulssen RH (2017) Changes in the human transcriptome upon vitamin D supplementation. J Steroid Biochem Mol Biol 173:93–99. https://doi.org/10.1016/j.jsbmb.2017.03.016
Pawlowski SW, Warren CA, Guerrant R (2009) Diagnosis and treatment of acute or persistent diarrhea. Gastroenterology 136:1874–1886. https://doi.org/10.1053/j.gastro.2009.02.072
Peacock M (2010) Calcium metabolism in health and disease. Clin J Am Soc Nephrol 5(Suppl 1):S23–S30. https://doi.org/10.2215/CJN.05910809
Pearson RD, Cox G, Jeronimo SM et al (1992) Visceral leishmaniasis: a model for infection-induced cachexia. Am J Trop Med Hyg 47:8–15
Pennington JAT (2003) Iodine|Physiology. In: Encyclopedia of food sciences and nutrition. Elsevier, London, pp 3357–3360
Penniston KL, Tanumihardjo SA (2006) The acute and chronic toxic effects of vitamin A. Am J Clin Nutr 83:191–201
Pereira GAP, Genaro PS, Pinheiro MM et al (2009) Cálcio dietético: estratégias para otimizar o consumo. Rev Bras Reumatol 49:164–171. https://doi.org/10.1590/S0482-50042009000200008
Pérez-Cano FJ, Yaqoob P, Martín R et al (2012) Immunonutrition in early life: diet and immune development. Clin Dev Immunol 2012:1–2. https://doi.org/10.1155/2012/207509
Pérez-Cano F, Castell M (2016) Flavonoids, inflammation and immune system. Nutrients 8:659. https://doi.org/10.3390/nu8100659
Pérez-López FR (2007) Vitamin D: the secosteroid hormone and human reproduction. Gynecol Endocrinol 23:13–24
Pérez H, Malavé I, Arredondo B (1979) The effects of protein malnutrition on the course of Leishmania mexicana infection in C57Bl/6 mice: nutrition and susceptibility to leishmaniasis. Clin Exp Immunol 38:453–460
Petrović J, Stanić D, Dmitrašinović G et al (2016) Magnesium supplementation diminishes peripheral blood lymphocyte DNA oxidative damage in athletes and sedentary young man. Oxid Med Cell Longev 2016:2019643. https://doi.org/10.1155/2016/2019643
Phillips C (2013) Nutrigenetics and metabolic disease: current status and implications for personalised nutrition. Nutrients 5:32–57. https://doi.org/10.3390/nu5010032
Picó C, Pons A, Palou A (1991) A significant pool of amino acids is adsorbed on blood cell membranes. Biosci Rep 11:223–230

Pieczynska J, Grajeta H (2015) The role of selenium in human conception and pregnancy. J Trace Elem Med Biol 29:31–38. https://doi.org/10.1016/j.jtemb.2014.07.003

Pike JW, Meyer MB (2010) The vitamin D receptor: new paradigms for the regulation of gene expression by 1,25-Dihydroxyvitamin D3. Endocrinol Metab Clin North Am 39:255–269. https://doi.org/10.1016/j.ecl.2010.02.007

Pierre JF, Heneghan AF, Lawson CM et al (2013) Pharmaconutrition review: physiological mechanisms. JPEN J Parenter Enteral Nutr 37:51S–65S. https://doi.org/10.1177/0148607113493326

Plank LD, Hill GL (2000) Sequential metabolic changes following induction of systemic inflammatory response in patients with severe sepsis or major blunt trauma. World J Surg 24:630–638

Ploder M, Kurz K, Spittler A et al (2010) Early increase of plasma homocysteine in sepsis patients with poor outcome. Mol Med 16:498–504. https://doi.org/10.2119/molmed.2010.00008

Plummer MP, Deane AM (2016) Dysglycemia and glucose control during sepsis. Clin Chest Med 37:309–319. https://doi.org/10.1016/j.ccm.2016.01.010

Pohl HR, Wheeler JS, Murray HE (2013) Sodium and potassium in health and disease. In: Sigel A, Sigel H, Sigel RKO (eds) Interrelations between essential metal ions and human diseases. Springer Netherlands, Dordrecht, pp 29–47

Powers HJ (2003) Riboflavin (vitamin B-2) and health. Am J Clin Nutr 77:1352–1360

Preiser J-C, van Zanten A, Berger MM et al (2015) Metabolic and nutritional support of critically ill patients: consensus and controversies. Crit Care 19:35. https://doi.org/10.1186/s13054-015-0737-8

Proenza AM, Palou A, Roca P (1994) Amino acid distribution in human blood. A significant pool of amino acids is adsorbed onto blood cell membranes. Biochem Mol Biol Int 34:971–982

Pulido-Moran M, Moreno-Fernandez J, Ramirez-Tortosa C, Ramirez-Tortosa M (2016) Curcumin and health. Molecules 21:264. https://doi.org/10.3390/molecules21030264

Puthucheary ZA, Rawal J, McPhail M et al (2013) Acute skeletal muscle wasting in critical illness. JAMA 310:1591–1600. https://doi.org/10.1001/jama.2013.278481

Quera PR, Quigley EMM, Madrid SAM (2005) Small intestinal bacterial overgrowth. An update. Rev Med Chil 133:1361–1370. https://doi.org/10.4067/S0034-98872005001100013

Rahman K (2007) Studies on free radicals, antioxidants, and co-factors. Clin Interv Aging 2:219–236

Rajagopalan KV (1988) Molybdenum: an essential trace element in human nutrition. Annu Rev Nutr 8:401–427. https://doi.org/10.1146/annurev.nu.08.070188.002153

Ramakrishnan U, Webb AL, Ologoudou K (2004) Infection, immunity, and vitamins. In: Handbook of nutrition and immunity. Humana Press, Totowa, NJ, pp 116–193

Ramig RF (2004) Pathogenesis of intestinal and systemic rotavirus infection. J Virol 78:10213–10220. https://doi.org/10.1128/JVI.78.19.10213-10220.2004

Ramos D, Mar D, Ishida M et al (2016) Mechanism of copper uptake from blood plasma ceruloplasmin by mammalian cells. PLoS One 11:e0149516. https://doi.org/10.1371/journal.pone.0149516

Rao S, Schieber AMP, O'Connor CP et al (2017) Pathogen-mediated inhibition of anorexia promotes host survival and transmission. Cell 168:503–516.e12. https://doi.org/10.1016/j.cell.2017.01.006

Rasool S, Abid S, Iqbal M et al (2012) Relationship between vitamin B12, folate and homocysteine levels and *H. Pylori* infection in patients with functional dyspepsia: a cross-section study. BMC Res Notes 5:206. https://doi.org/10.1186/1756-0500-5-206

Rathmell JC, Vander Heiden MG, Harris MH et al (2000) In the absence of extrinsic signals, nutrient utilization by lymphocytes is insufficient to maintain either cell size or viability. Mol Cell 6:683–692

Raverdeau M, Mills KHG (2014) Modulation of T cell and innate immune responses by retinoic Acid. J Immunol 192:2953–2958. https://doi.org/10.4049/jimmunol.1303245

Rayman MP (2012) Selenium and human health. Lancet 379:1256–1268. https://doi.org/10.1016/S0140-6736(11)61452-9

Rebouche CJ (1991) Ascorbic acid and carnitine biosynthesis. Am J Clin Nutr 54:1147S–1152S

Reboul E (2013) Absorption of vitamin A and carotenoids by the enterocyte: focus on transport proteins. Nutrients 5:3563–3581. https://doi.org/10.3390/nu5093563

Redmond HP, Stapleton PP, Neary P, Bouchier-Hayes D (1998) Immunonutrition: the role of taurine. Nutrition 14:599–604

Reeds PJ (2000) Dispensable and indispensable amino acids for humans. J Nutr 130:1835S–1840S

Reichrath J, Lehmann B, Carlberg C et al (2007) Vitamins as hormones. Horm Metab Res 39:71–84. https://doi.org/10.1055/s-2007-958715

Reithinger R, Teodoro U, Davies CR (2001) Topical insecticide treatments to protect dogs from sand fly vectors of leishmaniasis. Emerg Infect Dis 7:872–876. https://doi.org/10.3201/eid0705.017516

Reza Dorosty-Motlagh A, Mohammadzadeh Honarvar N, Sedighiyan M, Abdolahi M (2016) The molecular mechanisms of vitamin A deficiency in multiple sclerosis. J Mol Neurosci 60:82–90. https://doi.org/10.1007/s12031-016-0781-0

Rhee EP (2015) Metabolomics and renal disease. Curr Opin Nephrol Hypertens 24:371–379. https://doi.org/10.1097/MNH.0000000000000136

Rijal S, Uranw S, Chappuis F et al (2010) Epidemiology of *Leishmania donovani* infection in high-transmission foci in Nepal. Trop Med Int Health 15(Suppl 2):21–28. https://doi.org/10.1111/j.1365-3156.2010.02518.x

Rimbach G, Minihane AM, Majewicz J et al (2002) Regulation of cell signalling by vitamin E. Proc Nutr Soc 61:415–425

Rink L, Gabriel P (2000) Zinc and the immune system. Proc Nutr Soc 59:541–552

Risco D, Salguero FJ, Cerrato R et al (2016) Association between vitamin D supplementation and severity of tuberculosis in wild boar and red deer. Res Vet Sci 108:116–119. https://doi.org/10.1016/j.rvsc.2016.08.003

Rivlin RS (2007) Riboflavin (vitamin B2). In: Zempleni J (ed) Handbook of vitamins. Taylor & Francis, Boca Raton, FL, pp 252–233

Rocha KC, Vieira ML, Beltrame RL et al (2016) Impact of selenium supplementation in neutropenia and immunoglobulin production in childhood cancer patients. J Med Food 19:560–568. https://doi.org/10.1089/jmf.2015.0145

Rodgers JB (1998) n-3 Fatty acids in the treatment of ulcerative colitis. In: Kremer JM (ed) Medicinal fatty acids in inflammation. Birkhauser Verlag, Basel, pp 1090–1103

Romão MJ, Rösch N, Huber R (1997) The molybdenum site in the xanthine oxidase-related aldehyde oxidoreductase from Desulfovibrio gigas and a catalytic mechanism for this class of enzymes. J Biol Inorg Chem 2:782–785. https://doi.org/10.1007/s007750050195

Romeo J, Nova E, Wärnberg J et al (2010) Immunomodulatory effect of fibres, probiotics and symbiotics in different life-stages. Nutr Hosp 25:341–349

Roodenburg AJ, West CE, Beguin Y et al (2000) Indicators of erythrocyte formation and degradation in rats with either vitamin A or iron deficiency. J Nutr Biochem 11:223–230

Roohani N, Hurrell R, Kelishadi R, Schulin R (2013) Zinc and its importance for human health: an integrative review. J Res Med Sci 18:144–157

Rosa C d OB, dos Santos CA, Leite JIA et al (2015) Impact of nutrients and food components on dyslipidemias: what is the evidence? Adv Nutr 6:703–711. https://doi.org/10.3945/an.115.009480

Ross A (2006) Vitamin A and carotenoids. In: Shils M, Shike M, Ross A et al (eds) Modern nutrition in health and disease. Lippincott Williams & Wilkins, Baltimore, MD, pp 351–375

Ross AC, Harrison EH (2007) Vitamin A: nutritional aspects of retinoids and carotenoids. In: Zempleni J (ed) Handbook of vitamins, 4th edn. Taylor & Francis, Boca Raton, FL, pp 41–41

Ross AC, Ternus ME (1993) Vitamin A as a hormone: recent advances in understanding the actions of retinol, retinoic acid, and beta carotene. J Am Diet Assoc 93:1285–1290

Ross C (2010) Vitamin A. In: Coates PM, Betz JM, Blackman MR et al (eds) Encyclopedia of dietary supplements, 2nd edn. Informa Healthcare, London/New York, pp 778–791

Rousset B, Dupuy C, Miot F, Dumont J (2000) Thyroid hormone synthesis and secretion. MDText.com, Inc, Dartmouth, MA. Endotext

Roveran Genga K, Lo C, Cirstea M et al (2017) Two-year follow-up of patients with septic shock presenting with low HDL: the effect upon acute kidney injury, death and estimated glomerular filtration rate. J Intern Med 281:518–529. https://doi.org/10.1111/joim.12601

Rowe L, Wills ED (1976) The effect of dietary lipids and vitamin E on lipid peroxide formation, cytochrome P-450 and oxidative demethylation in the endoplasmic reticulum. Biochem Pharmacol 25:175–179. https://doi.org/10.1016/0006-2952(76)90287-2

Rukunuzzaman M, Rahman M (2008) Epidemiological study of risk factors related to childhood visceral leishmaniasis. Mymensingh Med J 17:46–50

Ruz M (2003) Zinc|Properties and determination. In: Encyclopedia of food sciences and nutrition. Elsevier, London, pp 6267–6272

Said HM (2008) Cell and molecular aspects of human intestinal biotin absorption. J Nutr 139:158–162. https://doi.org/10.3945/jn.108.092023

Said HM (2011) Intestinal absorption of water-soluble vitamins in health and disease. Biochem J 437:357–372. https://doi.org/10.1042/BJ20110326

Sandstead HH (1994) Understanding zinc: recent observations and interpretations. J Lab Clin Med 124:322–327

Sarrazin S, Lamanna WC, Esko JD (2011) Heparan sulfate proteoglycans. Cold Spring Harb Perspect Biol 3:a004952–a004952. https://doi.org/10.1101/cshperspect.a004952

Schaible UE, Kaufmann SHE (2007) Malnutrition and infection: complex mechanisms and global impacts. PLoS Med 4:e115. https://doi.org/10.1371/journal.pmed.0040115

Schindelin H, Kisker C, Hilton J et al (1996) Crystal structure of DMSO reductase: redox-linked changes in molybdopterin coordination. Science 272:1615–1621

Schmidt RL, Simonović M (2012) Synthesis and decoding of selenocysteine and human health. Croat Med J 53:535–550. https://doi.org/10.3325/cmj.2012.53.535

Schwager J, Bompard A, Weber P, Raederstorff D (2015) Ascorbic acid modulates cell migration in differentiated HL-60 cells and peripheral blood leukocytes. Mol Nutr Food Res 59:1513–1523. https://doi.org/10.1002/mnfr.201400893

Schwalfenberg GK (2017) Vitamins K1 and K2: the emerging group of vitamins required for human health. J Nutr Metab 2017:6254836. https://doi.org/10.1155/2017/6254836

Schweizer U, Schlicker C, Braun D et al (2014) Crystal structure of mammalian selenocysteine-dependent iodothyronine deiodinase suggests a peroxiredoxin-like catalytic mechanism. Proc Natl Acad Sci USA 111:10526–10531. https://doi.org/10.1073/pnas.1323873111

Scrimshaw NS (2007) Prologue: historical introduction. Immunonutrition in health and disease. Br J Nutr 98(Suppl 1):S3–S4. https://doi.org/10.1017/S0007114507833034

Sealey JE, Clark I, Bull MB, Laragh JH (1970) Potassium balance and the control of renin secretion. J Clin Invest 49:2119–2127. https://doi.org/10.1172/JCI106429

Seetharam B (1999) Receptor-mediated endocytosis of cobalamin (vitamin B12). Annu Rev Nutr 19:173–195. https://doi.org/10.1146/annurev.nutr.19.1.173

Sekowska A, Kung HF, Danchin A (2000) Sulfur metabolism in *Escherichia coli* and related bacteria: facts and fiction. J Mol Microbiol Biotechnol 2:145–177

Selmi C, Invernizzi P, Zuin M et al (2004) Evaluation of the immune function in the nutritionally at-risk patient. In: Handbook of nutrition and immunity. Humana Press, Totowa, NJ, pp 1–19

Semba RD, Darnton-Hill I, de Pee S (2010) Addressing tuberculosis in the context of malnutrition and HIV coinfection. Food Nutr Bull 31:S345–S364

Sergeant S, Rahbar E, Chilton FH (2016) Gamma-linolenic acid, dihommo-gamma linolenic, eicosanoids and inflammatory processes. Eur J Pharmacol 785:77–86. https://doi.org/10.1016/j.ejphar.2016.04.020

Serrano-Villar S, Vásquez-Domínguez E, Pérez-Molina JA et al (2017) HIV, HPV, and microbiota: partners in crime? AIDS 31:591–594. https://doi.org/10.1097/QAD.0000000000001352

Serrano-Villar S, Vázquez-Castellanos JF, Vallejo A et al (2016) The effects of prebiotics on microbial dysbiosis, butyrate production and immunity in HIV-infected subjects. Mucosal Immunol. https://doi.org/10.1038/mi.2016.122
Shanahan CM, Proudfoot D, Farzaneh-Far A, Weissberg PL (1998) The role of Gla proteins in vascular calcification. Crit Rev Eukaryot Gene Expr 8:357–375
Shane B (2008) Folate and vitamin B12 metabolism: overview and interaction with riboflavin, vitamin B6, and polymorphisms. Food Nutr Bull 29:S5–16–9. https://doi.org/10.1177/15648265080292S103
Shea-Donohue T, Qin B, Smith A (2017) Parasites, nutrition, immune responses and biology of metabolic tissues. Parasite Immunol 39:e12422. https://doi.org/10.1111/pim.12422
Shearer MJ, Fu X, Booth SL (2012) Vitamin K nutrition, metabolism, and requirements: current concepts and future research. Adv Nutr 3:182–195. https://doi.org/10.3945/an.111.001800
Shearer MJ, Newman P (2008) Metabolism and cell biology of vitamin K. Thromb Haemost 100:530–547
Sheridan PA, Beck MA (2008) The immune response to herpes simplex virus encephalitis in mice is modulated by dietary vitamin E. J Nutr 138:130–137
Shi LZ, Wang R, Huang G et al (2011) HIF1alpha-dependent glycolytic pathway orchestrates a metabolic checkpoint for the differentiation of TH17 and Treg cells. J Exp Med 208:1367–1376. https://doi.org/10.1084/jem.20110278
Shilotri PG, Bhat KS (1977) Effect of mega doses of vitamin C on bactericidal ativity of leukocytes. Am J Clin Nutr 30:1077–1081
Shils ME, Shike M (2006) Modern nutrition in health and disease, 10th edn. Lippincott Williams & Wilkins, Philadelphia, PA
Sies H (1993) Strategies of antioxidant defense. Eur J Biochem 215:213–219
Simsek T, Uzelli Simsek H, Canturk NZ (2014) Response to trauma and metabolic changes: posttraumatic metabolism. Turkish J Surg 30:153–159. https://doi.org/10.5152/UCD.2014.2653
Singh R, Gopalan S, Sibal A (2002) Immunonutrition. Indian J Pediatr 69:417–419
Sjögren M, Alkemade A, Mittag J et al (2007) Hypermetabolism in mice caused by the central action of an unliganded thyroid hormone receptor ?1. EMBO J 26:4535–4545. https://doi.org/10.1038/sj.emboj.7601882
Sladek R, Rocheleau G, Rung J et al (2007) A genome-wide association study identifies novel risk loci for type 2 diabetes. Nature 445:881–885. https://doi.org/10.1038/nature05616
Smedberg M, Wernerman J (2016) Is the glutamine story over? Crit Care 20:361. https://doi.org/10.1186/s13054-016-1531-y
Smith E, Morowitz HJ (2004) Universality in intermediary metabolism. Proc Natl Acad Sci 101:13168–13173. https://doi.org/10.1073/pnas.0404922101
Smith FR, Goodman DS (1976) Vitamin A transport in human vitamin A toxicity. N Engl J Med 294:805–808. https://doi.org/10.1056/NEJM197604082941503
Smith RM (1987) Cobalt. In: Trace elements in human and animal nutrition. Elsevier, Burlington, pp 143–183
Smyth PPA (2003) Role of iodine in antioxidant defence in thyroid and breast disease. Biofactors 19:121–130
Soda K, Oikawa T, Esaki N (1999) Vitamin B6 enzymes participating in selenium amino acid metabolism. Biofactors 10:257–262
Solomons NW (2003) Zinc|Physiology. In: Encyclopedia of food sciences and nutrition. Elsevier, London, pp 6272–6277
Soong L, Henard CA, Melby PC (2012) Immunopathogenesis of non-healing American cutaneous leishmaniasis and progressive visceral leishmaniasis. Semin Immunopathol 34:735–751. https://doi.org/10.1007/s00281-012-0350-8
Souto PA, Marcotegui AR, Orbea L et al (2016) Hepatic encephalopathy: ever closer to its big bang. World J Gastroenterol 22:9251–9256. https://doi.org/10.3748/wjg.v22.i42.9251

Srivastava A, Philip N, Hughes KR et al (2016) Stage-specific changes in plasmodium metabolism required for differentiation and adaptation to different host and vector environments. PLoS Pathog 12:e1006094. https://doi.org/10.1371/journal.ppat.1006094

Starr LM, Scott ME, Koski KG (2015) Protein deficiency and intestinal nematode infection in pregnant mice differentially impact fetal growth through specific stress hormones, growth factors, and cytokines. J Nutr 145:41–50. https://doi.org/10.3945/jn.114.202630

Stehle P, Ellger B, Kojic D et al (2017) Glutamine dipeptide-supplemented parenteral nutrition improves the clinical outcomes of critically ill patients: a systematic evaluation of randomised controlled trials. Clin Nutr ESPEN 17:75–85. https://doi.org/10.1016/j.clnesp.2016.09.007

Stein TP, Nutinsky C, Condoluci D et al (1990) Protein and energy substrate metabolism in AIDS patients. Metabolism 39:876–881

Steinbrenner H, Al-Quraishy S, Dkhil MA et al (2015) Dietary selenium in adjuvant therapy of viral and bacterial infections. Adv Nutr 6:73–82. https://doi.org/10.3945/an.114.007575

Stettler N, Schutz Y, Whitehead R, Jéquier E (1992) Effect of malaria and fever on energy metabolism in Gambian children. Pediatr Res 31:102–106. https://doi.org/10.1203/00006450-199202000-00002

Stocker A, Azzi A (2000) Tocopherol-binding proteins: their function and physiological significance. Antioxid Redox Signal 2:397–404. https://doi.org/10.1089/15230860050192170

Stone MS, Martyn L, Weaver CM (2016) Potassium intake, bioavailability, hypertension, and glucose control. Nutrients 8(7):444. https://doi.org/10.3390/nu8070444

Ströhle A, Hahn A (2009) Vitamin C and immune function. Med Monatsschr Pharm 32:49–54–6

Ströhle A, Wolters M, Hahn A (2011) Micronutrients at the interface between inflammation and infection-ascorbic acid and calciferol. Part 2: calciferol and the significance of nutrient supplements. Inflamm Allergy Drug Targets 10:64–74

Sukhotnik I, Krausz MM, Sabo E et al (2003) Endotoxemia inhibits intestinal adaptation in a rat model of short bowel syndrome. Shock 19:66–70

Suksomboon N, Poolsup N, Darli Ko Ko H (2017) Effect of vitamin K supplementation on insulin sensitivity: a meta-analysis. Diabetes Metab Syndr Obes 10:169–177. https://doi.org/10.2147/DMSO.S137571

Sun X, Zemel MB (2007) 1Alpha,25-dihydroxyvitamin D3 modulation of adipocyte reactive oxygen species production. Obesity (Silver Spring) 15:1944–1953. https://doi.org/10.1038/oby.2007.232

Suttie JW (2007) Vitamin K. In: Zempleni J (ed) Handbook of vitamins. Taylor & Francis, Boca Raton, FL, pp 152–111

Suzuki H, Hisamatsu T, Chiba S et al (2016) Glycolytic pathway affects differentiation of human monocytes to regulatory macrophages. Immunol Lett 176:18–27. https://doi.org/10.1016/j.imlet.2016.05.009

Swift LL, Hill JO, Peters JC, Greene HL (1990) Medium-chain fatty acids: evidence for incorporation into chylomicron triglycerides in humans. Am J Clin Nutr 52:834–836

Szymańska R, Nowicka B, Kruk J (2017) Vitamin E – occurrence, biosynthesis by plants and functions in human nutrition. Mini Rev Med Chem 17:1039–1052. https://doi.org/10.2174/1389557516666160725094819

Szymczak I, Pawliczak R (2016) The active metabolite of vitamin D3 as a potential immunomodulator. Scand J Immunol 83:83–91. https://doi.org/10.1111/sji.12403

Tanaka K, Tanabe K, Nishii N et al (2017) Sustained tubulointerstitial inflammation in kidney with severe leptospirosis. Intern Med 56:1179–1184. https://doi.org/10.2169/internalmedicine.56.8084

Tanumihardjo SA (2011) Vitamin A: biomarkers of nutrition for development. Am J Clin Nutr 94:658S–665S. https://doi.org/10.3945/ajcn.110.005777

Tanumihardjo SA, Russell RM, Stephensen CB et al (2016) Biomarkers of Nutrition for Development (BOND)-vitamin A review. J Nutr 146:1816S–1848S. https://doi.org/10.3945/jn.115.229708

Tarp U, Overvad K, Thorling EB et al (1985) Selenium treatment in rheumatoid arthritis. Scand J Rheumatol 14:364–368

Taylor DM (1962) The absorption of cobalt from the gastrointestinal tract of the rat. Phys Med Biol 6:445–451

Taylor DN, Hamer DH, Shlim DR (2017) Medications for the prevention and treatment of travellers' diarrhea. J Travel Med 24:S17–S22. https://doi.org/10.1093/jtm/taw097

Tekwani BL, Tripathi LM, Mukerjee S et al (1987) Impairment of the hepatic microsomal drug-metabolizing system in rats parasitized with *Nippostrongylus brasiliensis*. Biochem Pharmacol 36:1383–1386

The editors (2004) Preface: carnitine: lessons from one hundred years of research. Ann N Y Acad Sci 1033:ix–xi. https://doi.org/10.1196/annals.1320.019

Thijssen HH, Drittij-Reijnders MJ (1996) Vitamin K status in human tissues: tissue-specific accumulation of phylloquinone and menaquinone-4. Br J Nutr 75:121–127

Thomson CD (2003) Selenium|Physiology. In: Encyclopedia of food sciences and nutrition. Elsevier, London, pp 5117–5124

Thurnham DI (2005) Thiamin|Physiology. In: Encyclopedia of human nutrition. Elsevier, London, pp 263–269

Toussaint ND, Damasiewicz MJ (2017) Do the benefits of using calcitriol and other vitamin D receptor activators in patients with chronic kidney disease outweigh the harms? Nephrology (Carlton) 22(Suppl 2):51–56. https://doi.org/10.1111/nep.13026

Traber MG (2007) Vitamin E. In: Zempleni J (ed) Handbook of vitamins. Taylor & Francis, Boca Raton, FL, pp 174–153

Trager W (1974) Some aspects of intracellular parasitism. Science 183:269–273

Traore FA, Sako FB, Sylla D et al (2016) Tetanus in women of childbearing age in the infectious disease department in the national hospital of Conakry (Guinea). Med Sante Trop 26:323–325. https://doi.org/10.1684/mst.2016.0594

Trumbo PR (2006) Pantothenic Acid. In: Shils ME, Shike M, Ross AC et al (eds) Modern nutrition in health and disease. Lippincott Williams & Wilkins, Philadelphia, PA, pp 462–467

Tsukaguchi H, Tokui T, Mackenzie B et al (1999) A family of mammalian Na^+-dependent L-ascorbic acid transporters. Nature 399:70–75. https://doi.org/10.1038/19986

Uysal S, Tunalı V, Akdur Öztürk E et al (2016) Incidence of parasitic diarrhea in patients with common variable immune deficiency. Turkiye Parazitol Derg 40:67–71. https://doi.org/10.5152/tpd.2016.4687

Valko M, Leibfritz D, Moncol J et al (2007) Free radicals and antioxidants in normal physiological functions and human disease. Int J Biochem Cell Biol 39:44–84. https://doi.org/10.1016/j.biocel.2006.07.001

van Gool CJAW, Zeegers MPA, Thijs C (2004) Oral essential fatty acid supplementation in atopic dermatitis-a meta-analysis of placebo-controlled trials. Br J Dermatol 150:728–740. https://doi.org/10.1111/j.0007-0963.2004.05851.x

Van Lettow M, Fawzi WW, Semba PH, Semba RD (2003) Triple trouble: the role of malnutrition in tuberculosis and human immunodeficiency virus co-infection. Nutr Rev 61:81–90. https://doi.org/10.1301/nr.2003.marr.81-90

van Niekerk G, Isaacs AW, Nell T, Engelbrecht A-M (2016) Sickness-associated anorexia: mother nature's idea of immunonutrition? Mediators Inflamm 2016:1–12. https://doi.org/10.1155/2016/8071539

Vanderhoof JA (1998) Immunonutrition: the role of carbohydrates. Nutrition 14:595–598

Vetvicka V, Vetvickova J (2016) Concept of immuno-nutrition. J Nutr Food Sci. https://doi.org/10.4172/2155-9600.1000500

Viegas CSB, Costa RM, Santos L et al (2017) Gla-rich protein function as an anti-inflammatory agent in monocytes/macrophages: Implications for calcification-related chronic inflammatory diseases. PLoS One 12:e0177829. https://doi.org/10.1371/journal.pone.0177829

Viegas CSB, Herfs M, Rafael MS et al (2014) Gla-rich protein is a potential new vitamin K target in cancer: evidences for a direct GRP-mineral interaction. Biomed Res Int 2014:1–14. https://doi.org/10.1155/2014/340216

Viegas CSB, Simes DC, Laizé V et al (2008) Gla-rich protein (GRP), a new vitamin K-dependent protein identified from sturgeon cartilage and highly conserved in vertebrates. J Biol Chem 283:36655–36664. https://doi.org/10.1074/jbc.M802761200

Vormann J (2003) Magnesium: nutrition and metabolism. Mol Aspects Med 24:27–37

Wahib AA, El-Nasr MSS, Mangoud AM et al (2006) The liver profile in patients with hepatitis C virus and/or fascioliasis. J Egypt Soc Parasitol 36:405–440

Waldrop GL (2015) Biotin. In: eLS. Wiley, Chichester, UK

Wallace FA, Miles EA, Evans C et al (2001) Dietary fatty acids influence the production of Th1- but not Th2-type cytokines. J Leukoc Biol 69:449–457

Walsh C, Fisher J, Spencer R et al (1978) Chemical and enzymatic properties of riboflavin analogues. Biochemistry 17:1942–1951

Walther B, Karl JP, Booth SL, Boyaval P (2013) Menaquinones, bacteria, and the food supply: the relevance of dairy and fermented food products to vitamin K requirements. Adv Nutr 4:463–473. https://doi.org/10.3945/an.113.003855

Wanders D, Judd RL (2011) Future of GPR109A agonists in the treatment of dyslipidaemia. Diabetes Obes Metab 13:685–691. https://doi.org/10.1111/j.1463-1326.2011.01400.x

Wang A, Huen SC, Luan HH et al (2016) Opposing effects of fasting metabolism on tissue tolerance in bacterial and viral inflammation. Cell 166:1512–1525.e12. https://doi.org/10.1016/j.cell.2016.07.026

Wang D (2012) Redox chemistry of molybdenum in natural waters and its involvement in biological evolution. Front Microbiol. https://doi.org/10.3389/fmicb.2012.00427

Wang G, Feng D (2017) Dynamic relationship between infantile hepatitis syndrome and cytomegalovirus infection. Exp Ther Med 13:3443–3447. https://doi.org/10.3892/etm.2017.4375

Wang J, Pantopoulos K (2011) Regulation of cellular iron metabolism. Biochem J 434:365–381. https://doi.org/10.1042/BJ20101825

Wang K, Hoshino Y, Dowdell K et al (2017) Glutamine supplementation suppresses herpes simplex virus reactivation. J Clin Invest. https://doi.org/10.1172/JCI88990

Wang L, Song Y (2017) Efficacy of zinc given as an adjunct to the treatment of severe pneumonia: a meta-analysis of randomized, double-blind and placebo-controlled trials. Clin Respir J. https://doi.org/10.1111/crj.12646

Wang TY, Liu M, Portincasa P, Wang DQ-H (2013) New insights into the molecular mechanism of intestinal fatty acid absorption. Eur J Clin Invest 10. https://doi.org/10.1111/eci.12161

Wang YZ, Christakos S (1995) Retinoic acid regulates the expression of the calcium binding protein, calbindin-D28K. Mol Endocrinol 9:1510–1521. https://doi.org/10.1210/mend.9.11.8584029

Wasserman RH (1981) Intestinal absorption of calcium and phosphorus. Fed Proc 40:68–72

Wasserman RH, Fullmer CS (1989) On the molecular mechanism of intestinal calcium transport. Adv Exp Med Biol 249:45–65

Watanabe F (2007) Vitamin B12 sources and bioavailability. Exp Biol Med (Maywood) 232:1266–1274. https://doi.org/10.3181/0703-MR-67

Waters WR, Nonnecke BJ, Rahner TE et al (2001) Modulation of mycobacterium bovis-specific responses of bovine peripheral blood mononuclear cells by 1,25-dihydroxyvitamin D3. Clin Vaccine Immunol 8:1204–1212. https://doi.org/10.1128/CDLI.8.6.1204-1212.2001

Watford M (2003) The urea cycle: teaching intermediary metabolism in a physiological setting. Biochem Mol Biol Educ 31:289–297. https://doi.org/10.1002/bmb.2003.494031050249

Webb AR, DeCosta BR, Holick MF (1989) Sunlight regulates the cutaneous production of vitamin D3 by causing its photodegradation. J Clin Endocrinol Metab 68:882–887. https://doi.org/10.1210/jcem-68-5-882

Webb KE (1990) Intestinal absorption of protein hydrolysis products: a review. J Anim Sci 68:3011–3022

Wehner AP, Craig DK (1972) Toxicology of inhaled NiO and CoO in Syrian golden hamsters. Am Ind Hyg Assoc J 33:146–155. https://doi.org/10.1080/0002889728506624

Weigel MM, Armijos RX, Racines RJ et al (1994) Cutaneous leishmaniasis in subtropical ecuador: popular perceptions, knowledge, and treatment. Bull Pan Am Health Organ 28:142–155

Weinstein SP, O'Boyle E, Fisher M, Haber RS (1994) Regulation of GLUT2 glucose transporter expression in liver by thyroid hormone: evidence for hormonal regulation of the hepatic glucose transport system. Endocrinology 135:649–654. https://doi.org/10.1210/endo.135.2.8033812

Weiss G, Carver PL (2017) Role of divalent metals in infectious disease susceptibility and outcome. Clin Microbiol Infect. https://doi.org/10.1016/j.cmi.2017.01.018

Werneck GL, Hasselmann MH, Gouvea TG (2011) An overview of studies on nutrition and neglected diseases in Brazil. Cien Saude Colet 16:39–62

Wessling-Resnick M (2014) Iron. In: Ross A, Caballero B, Cousins RJ et al (eds) Modern nutrition in health and disease, 11th edn. Wolters Kluwer Health/Lippincott Williams & Wilkins, Philadelphia, PA, pp 176–188

Wester PO (1987) Magnesium. Am J Clin Nutr 45:1305–1312

Whang R, Whang DD (1990) Update: mechanisms by which magnesium modulates intracellular potassium. J Am Coll Nutr 9:84–85. https://doi.org/10.1080/07315724.1990.10720354

Wheeler GL, Jones MA, Smirnoff N (1998) The biosynthetic pathway of vitamin C in higher plants. Nature 393:365–369. https://doi.org/10.1038/30728

White JH (2012) Vitamin D metabolism and signaling in the immune system. Rev Endocr Metab Disord 13:21–29. https://doi.org/10.1007/s11154-011-9195-z

WHO (2017) Nutrition. In: World Heal. Organ. Heal. Top. http://www.who.int/topics/nutrition/en/. Accessed 7 Apr 2017

Wientroub S, Winter CC, Wahl SM, Wahl LM (1989) Effect of vitamin D deficiency on macrophage and lymphocyte function in the rat. Calcif Tissue Int 44:125–130

Wiernsperger N, Rapin J (2010) Trace elements in glucometabolic disorders: an update. Diabetol Metab Syndr 2:70. https://doi.org/10.1186/1758-5996-2-70

Wilson JX (2005) Regulation of vitamin C transport. Annu Rev Nutr 25:105–125. https://doi.org/10.1146/annurev.nutr.25.050304.092647

Wintergerst ES, Maggini S, Hornig DH (2006) Immune-enhancing role of vitamin C and zinc and effect on clinical conditions. Ann Nutr Metab 50:85–94. https://doi.org/10.1159/000090495

Wolf G (2004) The discovery of vitamin D: the contribution of adolf windaus. J Nutr 134:1299–1302

Wolowczuk I, Verwaerde C, Viltart O et al (2008) Feeding our immune system: impact on metabolism. Clin Dev Immunol 2008:639803. https://doi.org/10.1155/2008/639803

Wooley JA, Btaiche IF, Good KL (2005) Metabolic and nutritional aspects of acute renal failure in critically ill patients requiring continuous renal replacement therapy. Nutr Clin Pract 20:176–191. https://doi.org/10.1177/0115426505020002176

Wright EM, Loo DDF, Hirayama BA (2011) Biology of human sodium glucose transporters. Physiol Rev 91:733–794. https://doi.org/10.1152/physrev.00055.2009

Wu W-H, Pang H-JE, Matthews KR (2012) Immune status and the development of Listeria monocytogenes infection in aged and young guinea pigs. Clin Invest Med 35:E309

Xia L, Björnstedt M, Nordman T et al (2001) Reduction of ubiquinone by lipoamide dehydrogenase. Eur J Biochem 268:1486–1490. https://doi.org/10.1046/j.1432-1327.2001.02013.x

Xue Y, Fleet JC (2009) Intestinal vitamin D receptor is required for normal calcium and bone metabolism in mice. Gastroenterology 136(1317–27):e1–e2. https://doi.org/10.1053/j.gastro.2008.12.051

Yang Y, McClements DJ (2013) Vitamin E and vitamin E acetate solubilization in mixed micelles: physicochemical basis of bioaccessibility. J Colloid Interface Sci 405:312–321. https://doi.org/10.1016/j.jcis.2013.05.018

Ye J, Shi X (2001) Gene expression profile in response to chromium-induced cell stress in A549 cells. Mol Cell Biochem 222:189–197

Zaahl MG, Robson KJ, Warnich L, Kotze MJ (2004) Expression of the SLC11A1 (NRAMP1) 5′-(GT)n repeat: Opposite effect in the presence of −237C→T. Blood Cells, Mol Dis 33:45–50. https://doi.org/10.1016/j.bcmd.2004.04.003

Zacarias DA, Rolão N, de Pinho FA et al (2017) Causes and consequences of higher *Leishmania infantum* burden in patients with kala-azar: a study of 625 patients. Trop Med Int Health 22:679–687. https://doi.org/10.1111/tmi.12877

Zamamiri-Davis F, Lu Y, Thompson JT et al (2002) Nuclear factor-kappaB mediates over-expression of cyclooxygenase-2 during activation of RAW 264.7 macrophages in selenium deficiency. Free Radic Biol Med 32:890–897

Zapatera B, Prados A, Gómez-Martínez S, Marcos A (2015) Immunonutrition: methodology and applications. Nutr Hosp 31(Suppl 3):145–154. https://doi.org/10.3305/nh.2015.31.sup3.8762

Zempleni J (2007) Handbook of vitamins, 4th edn. Taylor & Francis, Boca Raton, FL

Zhang P, Tsuchiya K, Kinoshita T et al (2016) Vitamin B6 prevents IL-1β protein production by inhibiting NLRP3 inflammasome activation. J Biol Chem 291:24517–24527. https://doi.org/10.1074/jbc.M116.743815

Zhao Y, Joshi-Barve S, Barve S, Chen LH (2004) Eicosapentaenoic acid prevents LPS-induced TNF-alpha expression by preventing NF-kappaB activation. J Am Coll Nutr 23:71–78

Zhong M, Kawaguchi R, Kassai M, Sun H (2012) Retina, retinol, retinal and the natural history of vitamin A as a light sensor. Nutrients 4:2069–2096. https://doi.org/10.3390/nu4122069

Zhou R, Horai R, Silver PB et al (2012) The living eye "disarms" uncommitted autoreactive T cells by converting them to Foxp3(+) regulatory cells following local antigen recognition. J Immunol 188:1742–1750. https://doi.org/10.4049/jimmunol.1102415

Zhou X, Liu W, Gu M et al (2015) Helicobacter pylori infection causes hepatic insulin resistance by the c-Jun/miR-203/SOCS3 signaling pathway. J Gastroenterol 50:1027–1040. https://doi.org/10.1007/s00535-015-1051-6

Ziboh VA (1998) The role of n-3 fatty acids in psoriasis. In: Kremer JM (ed) Medicinal fatty acids in inflammation. Birkhauser Verlag, Basel, pp 53–45

Zimmermann MB (2009) Iodine deficiency. Endocr Rev 30:376–408. https://doi.org/10.1210/er.2009-0011

Zimmermann MB, Köhrle J (2002) The impact of iron and selenium deficiencies on iodine and thyroid metabolism: biochemistry and relevance to public health. Thyroid Off J Am Thyroid Assoc 12:867–878. https://doi.org/10.1089/105072502761016494

Zurier RB, Rossetti RG, Jacobson EW et al (1996) gamma-Linolenic acid treatment of rheumatoid arthritis. A randomized, placebo-controlled trial. Arthritis Rheum 39:1808–1817

Chapter 3
Interplay Between Metabolic Sensors and Immune Cell Signaling

Prashant Chauhan, Arup Sarkar, and Bhaskar Saha

Abstract The immune system, like all other systems, responds to perturbations of the baseline, homeostatic functioning of immune cells. These perturbations come in the form of infection, tumors, autoantigens, and can occur after mismatched transplantation. During response, immune cells alter their metabolic activities. However, the subsets of the same cell type differ to distinctively associate specific immune function to a particular metabolic profile. The response is mounted as a joint function of metabolic receptor and immune receptor signaling that target various metabolic pathways: glycolysis the pentose phosphate pathway; oxidative phosphorylation; beta-oxidation of fatty acids and transamination. The products from these cycles are integrated in the tricarboxylic acid cycle. However, many more pathways lead to many secondary metabolites that are not directly related to energy derivation or maintaining structure of the cells. These secondary metabolites can again work in an autocrine manner to re-tune the immune cells to optimize their restorative effector functions.

Keywords *Leishmania* · Macrophage metabolism · Immunometabolomic · Neutrophil metabolism · Lymphocyte metabolism · Metabolic regulation of immune response

3.1 Prelude: Immunometabolomics, A New Sequel in the 'Omics' Quadrology

The quartet of 'omics'—genomics including transcriptomics, proteomics, bioinformatics, and metabolomics—has grown rapidly during the past three decades. The result is the merging of the regulatory systems that used to be considered unique for each organ system. One such example is the rise of immunometabolomics that

P. Chauhan
National Centre for Cell Science, Pune, India

A. Sarkar · B. Saha (✉)
Trident Academy of Creative Technology, Bhubaneswar, Odisha, India

R. Silvestre, E. Torrado (eds.), *Metabolic Interaction in Infection*, Experientia Supplementum 109, https://doi.org/10.1007/978-3-319-74932-7_3

deciphers the redundancy and interdependencies among innumerable metabolic pathways in immune cells. The convention holds that metabolic changes during pathogen invasion result from immune cell signaling. This view has been significantly remolded as a results of the increasing reports favor metabolism as the ‘first responder’ that can be decisive in dictating the robustness of immune response produced and determining the outcome based on the metabolic potential and fitness of the responding immune cells. This new insight has prompted the ‘reprogramming’ of cellular metabolism as an approach to direct immune cell fate and functions in order to improve disease outcome. In this chapter, we discuss how metabolic pathways and metabolites affect immune cell subsets and how this knowledge can be applied to better understand disease and to reveal novel therapeutic approaches.

Networks of various metabolic pathways, their downstream targets, adaptor proteins, enzymes of biosynthetic anabolic and catabolic pathways (McCarthy and Esser 2010; Hancock et al. 2010), and signaling intermediates reveal the multifocal nature of these pathways (Ghesquiere et al. 2014), which operate differently in immune cells during homeostatic and perturbed states. During the perturbed state, abnormal patterns or oscillations are observed in the synthesis of transcription factors, enzymes, and catabolic or anabolic product pathways (Pearce et al. 2013). Such aberrant patterns predicted via mathematical modeling can be used to identify sensitive nodes or an ‘Achilles heel’ that could be targeted as palpable checkpoints to develop therapeutic intervention strategies in many clinical scenarios, including chronic diseases, inflammation, pathogenic infections, autoimmunity, and oncogenic transformations. The consensus among several reports (Pearce et al. 2013; MacIver et al. 2013; Ganeshan and Chawla 2014; Biswas 2015; Watson et al. 2015) indicates that multifaceted interactions are involved within the pathways that drive the immune cells to develop into either infection-controlling phenotypes or disease-promoting phenotypes. To explain the rewiring of the metabolic circuitry in perturbed states, the dynamic changes need to be considered in the architecture of the metabolic and cellular signaling networks in the presence of a ‘perturbator’ (Fig. 3.1).

Deciphering how a cell’s metabolic grid—an interconnected, coordinated network of several metabolic pathways within a cell—is instrumental in producing phenotypically distinct immune cell subsets is key to understanding the role of metabolic pathways in immunoregulation. In this chapter, we review the six major metabolic pathways in immune cells: glycolysis, tricarboxylic acid (TCA) cycle, the pentose phosphate pathway, fatty acid synthesis, fatty acid oxidation, and amino acid metabolism.

That metabolism plays a deterministic role in the function of immune cells was first revealed by Alonso and Nungester in their 1956 seminal report ‘The effect of pneumococcal products on glycolysis and oxygen uptake by polymorphonuclear leucocytes’, which aimed at decrypting the factors responsible for the observed discrepancies between resistant and susceptible hosts. It was followed by reports on metabolic changes in lymphocytes in response to pathogenic invasions (Beisel 1975). Collectively, these reports described changes in carbohydrate metabolism and oxidative phosphorylation upon pathogen-derived stimuli, establishing the role of

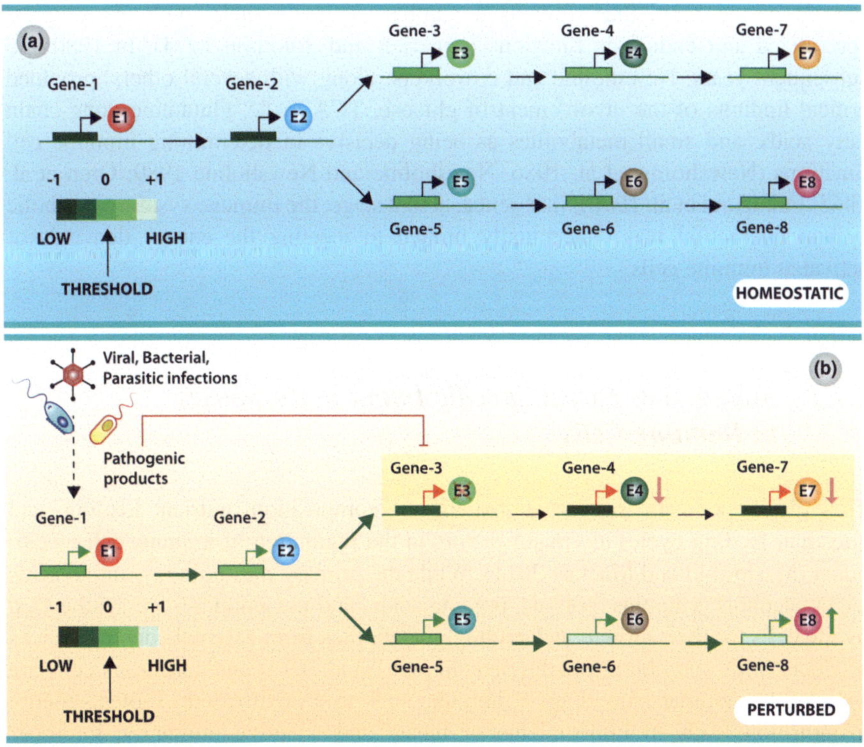

Fig. 3.1 A simple scheme of a hypothetical biochemical pathway of an immune cell. (**a**) The basal levels of synthesis of proteins/enzymes under non-perturbed conditions, where cellular homeostasis is maintained. Genes responsible for the synthesis of proteins/enzymes are translated at a basal rate. Hypothetically, proteins E3, E5, and E7 could be important enzymes of some biochemical pathways which help the cells to maintain a homeostatic intracellular atmosphere. Proteins E3 and E5 are synthesized via enzyme E2 at the same rate, whereas, as shown in (**b**), in the presence of a perturbator, genes E1 and E2 become activated, producing more enzyme E2 that drives the synthesis of proteins E3 and E5, but the pathogen-derived product may interfere with the protein E3 catalytic potential or may even inhibit production at the gene level by inhibiting gene 3. In such a scenario, the downstream gene 3 targets, i.e., gene 4 and gene 7, are not transcribed efficiently and the production of proteins E4 and E7 is hampered. E4 and E7 may be important drivers of antimicrobial responses such as proinflammatory cytokines or even enzymes like NADPH oxidase (Nphox), which is crucial for free-radical-mediated pathogen clearance. Thus, befitting the fundamental principle of host–pathogen interaction, altered transcription of crucial genes by pathogen or pathogenic products, even metabolites, represents an effective immune-evasion strategy deployed by pathogens. It effectively dampens the host immunogenic response and ensures pathogen survival

altered metabolism in reciprocating the elevated energy demands and optimally fit effector responses in activated lymphocytes (Roos and Loos 1973a, b). The concept of nutrients affecting immune system functions—the causal relationship between infection, nutrients, and immunodeficiency—also emerged in the 1970s (Chandra 1990; Meydani et al. 1990). At the same time, the concept of neuroendocrine regulation of metabolism arose and, in particular, the role of the hypothalamus, a portion of the brain that contains a number of small nuclei with neurological specialized and endocrine functions (Amkraut and Solomon 1974). In 1980 and subsequent years, Newsholme and coworkers, along with several others, provided critical findings of the involvement of glucose, TCA cycle, glutamine, long-chain fatty acids, and small metabolites as being decisive in determining immune cell functions (Newsholme et al. 1986, Newsholme and Newsholme 1989; Oren et al. 1963; Fukuzumi et al. 1996). In essence, at that stage, the immune system–metabolic system interaction was conceptually limited to meeting the energy demands of activated immune cells.

3.1.1 Metabolites Direct Specific Immune Responses in Immune Cells

The 'diet hypothesis' suggested that dietary composition, nutrient loading, and circadian feeding cycles are major factors in the regulation of immune cell metabolism (Maslowski and Mackay 2011). While the primary metabolites—amino acids (AA), alcohols, vitamins, polyols, organic acids, and nucleotides—are synthesized by the cells for their growth and cycling, metabolites from external sources such as products from microbial growth (Belkaid and Hand 2014) and cancerous cells are reported to modulate functions of immune cells to a great extent. Protein–energy malnutrition leads to impediments of innate and adaptive immunity, improper phagocyte functions, complement system, humoral immunity, and cytokine production (Chandra 1997). Overnutrition and obesity lead to immune malfunction, increased adipose tissue inflammation, increased insulin resistance (Wensveen et al. 2015), increased risk of cardiovascular diseases (Eckel 1997), and cancer (De Pergola and Silvestris 2013). The pathologies associated with activation of inflammation, obesity, and ageing-related health changes together indicate some commonalities interrelated with immune cells, such as increased TNF-α, IL-6 production, increased NF-κB transcription factor, C-reactive protein (CRP), hyperactive B-cell responses, and insulin resistance. Some major examples of metabolites directing specific metabolic responses in immune cells are highlighted below.

3.1.1.1 Vitamin A

Active metabolite of vitamin A, retinoic acid (RA) controls the presence of $ROR\gamma^+$ innate lymphoid cells (ILCs) in the small intestine (Goverse et al. 2016). RA promotes ILC3; its deprivation results in the expansion of IL-13, producing ILC2, proving ILC to be a primary sensor of RA dietary stress and in helminthic infections (Mortha et al. 2014). ILC1 and ILC3 show RA-dependent increase in CCR9 in the gut, whereas ILC2 expresses gut-homing receptors during development in the bone marrow (Kim et al. 2015).

The T-helper cells (T_H) bifurcate into two functional subsets, T_H1 and T_H2, characterized by their discrete secretory cytokines and surface cytokine receptor expression (Mosmann et al. 1986). Naïve or non-committed $CD4^+$ T cells, when activated via T-cell receptors (TCR) and high levels of IL-12, derived from macrophages and dendritic cells (DCs) in response to antigenic stimuli, inflammation, or even IFN-γ produced by natural killer (NK) cells, tend to polarize towards a T_H1 phenotype. By contrast, naïve T cells enriched in a milieu containing high levels of IL-4 become polarized towards T_H2 (Mosmann and Coffman 1989). The fact that RA is a T_H cell polarizing factor (Mora et al. 2008), as its deficiency is found to be conducive for T_H1 development (Cantorna et al. 1995), is now well supported experimentally. Conversely, the supplementation of naïve T cells with abundant RA differentiates them to IL-4-secreting T_H2 type (Hoag et al. 2002).

All-trans RA (ATRA) significantly reduces the capability of $CD4^+$ T memory cells and naïve T cells to undergo T_H17 cell differentiation (Mottaghi et al. 2014). ATRA inhibits type 1 diabetes by regulatory T (Treg) cell–dependent suppression of IFN-γ, producing T cells without involvement of T_H17 cells (Van et al. 2009). TGF-β alone converts naïve T cells into Treg cells that suppress autoimmunity but, with IL-6, it promotes IL-17-secreting T_H17 cells. RA regulates TGF-β-dependent immune responses, which allows the alteration of T_H17 cells to anti-inflammatory Treg cell generation (Mucida et al. 2007). RA promotes the proliferation and maturation of splenic B cells in mice. Enhancing the surface expression of IgG1 and CD138 ultimately augments humoral immunity in vivo (Chen and Ross 2007).

Like vitamin A, calcitriol or vitamin D exhibit immunomodulatory properties, but the underlying molecular mechanism remains poorly understood. As vitamin D receptor expression was induced in a T_H17 subset, the treatment of mice with vitamin D receptor agonist 1,25-dihydroxyvitamin D (D_3) was found to improve experimental autoimmune encephalitis (EAE), corroborative with reduced IL-17 and IL-17F expression. The expression of CEBP homologous protein was implicated in reduced cytokine production by T_H17 cells (Chang et al. 2010). Antagonistic or inhibitory effects of vitamin D3 on DCs are reported in patients with systemic lupus erythematosus (SLE) (Barragan et al. 2015; Piemonti et al. 2000; Ben-Zvi et al. 2010). Patients with severe SLE and overproduction of autoreactive antinuclear antibodies (Mok and Lau 2003) had reduced vitamin D3 levels. Therefore, hyperactive B-cell proliferation, differentiation, and antibody secretion could be controlled by exogenous vitamin D3 administration (Heine et al. 2002; Chen et al. 2007).

3.1.1.2 Metabolites of Fatty Acid Oxidation

Short-chain fatty acids (SCFAs), acetate C_2, propionate C_3, and butyrate C_4, are liberated through dietary fiber fermentation by the action of endosymbiotic microbiota of the gut (den Besten et al. 2013; Masui et al. 2013). A diet lacking fiber induces systemic endotoxemia or increased insulin resistance due to the loss of structural integrity of mucosal gut epithelia and exacerbated inflammation of the surrounding adipose tissue (Byrne et al. 2015). SCFAs can act on leukocytes either by G protein-coupled receptors (GPCRs) activation or by histone deacetylase (HDAC) inhibition to regulate cytokine (TNF-α, IL-1, IL-6, and IL-10), eicosanoid, and chemokine production (Vinolo et al. 2011; Mosser and Edwards 2008). By inducing IL-10, SCFAs diminish lipopolysaccharide LPS-induced NF-κB activation and inflammatory cytokine production, promoting the anti-inflammatory milieu conductive for microbial growth (Liu et al. 2012). Comparative nuclear mangetic resonance metabolomics showed a positive correlation between ambient SCFA levels and Treg cell numbers; butyrate specifically induced the differentiation of Treg cells whose adoptive transfer into $Rag1^{-/-}$ mice ameliorated the experimentally induced colitis (Furusawa et al. 2013). SCFAs induce both effector and Treg cells by suppressing HDACs and regulating the mTOR-S6K pathway (Park et al. 2015). The SFCA effects on Treg cells were mediated through GPCR43, a receptor for SFCAs prominent on sentinel colonic Treg cells. Mice fed an SCFA-rich diet were protected against experimentally induced colitis because SCFAs promoted GPCR43-expressing Treg cells (Smith et al. 2013). SCFAs also contribute to inflammation via the regulation of adipokines; for example, leptin increases T_H1 and suppresses T_H2 cells (Lord et al. 1998).

3.1.1.3 Long-chain Fatty Acids

Metabolites such as unsaturated fatty acids, particularly ω-3 fatty acids, prevent the activation of NLRP3 inflammasome, a molecular platform that is a culprit in many obesity-associated diseases in humans and that mediates the processing of IL-1β, caspase-1, and IL-18 in response to infections and stress (L'homme et al. 2013). Saturated fatty acids like palmitate, on the other hand, have been proposed to trigger NLRP3 inflammasome. Inflammasome activation in hematopoietic cells impairs insulin signaling in several target tissues to reduce glucose tolerance and insulin sensitivity (Wen et al. 2011). Transcendence of one of the most prevalent metabolic syndromes in the Western world, non-alcoholic fatty liver disease (NAFLD) to the chronic form non-alcoholic steatohepatitis (NASH), is suspected to be regulated by NLRP3 and IL-18. NAFLD-to-NASH progression is also promoted by TNF-α and the ligands of TLR4 and TLR9 (Rinella and Sanyal 2016; Henao-Mejia et al. 2012).

Polyunsaturated fatty acids (PUFAs) act as polarization factors for M1 and M2 types of macrophages (Lawrence and Natoli 2011; Murray et al. 2014). These distinctly polarized states in macrophages are directed by pathways that control epigenetic modifications and cell survival in response to microenvironmental or extrinsic factors, including microbial products and cytokines released in inflammation and infections or high-fat diet-induced chronic inflammation and metabolic instabilities (Song et al. 2016). Saturated and unsaturated fatty acids bring about opposite macrophage polarization. When macrophages from low fat-fed mice were exposed to palmitate, the expression of proinflammatory genes (IL-12 and inducible nitric oxide synthase [iNOS]) was upregulated but palmitoleate (PO) treatment reversed the inflammatory condition in high fat-fed mice by inducing anti-inflammatory genes (*TGF-β1* and *IL-10*) and oxidative metabolism (Chan et al. 2015).

3.1.1.4 Aryl Hydrocarbon Receptor

Aryl hydrocarbon receptor (AhR) and its endogenous ligand kynurenine, a tryptophan-derived catabolite, regulate T cell differentiation in tumors and inflammatory foci (Quintana et al. 2008; Veldhoen et al. 2009; Mezrich et al. 2010; Opitz et al. 2011). An AhR is expressed on lymphoid tissue inducer (LTi)-like cells and $NKp46^+$ subsets of ILC22, a distinct IL-22 secreting and NK cell natural cytotoxicity receptor ($NKp46^+$, $IL\text{-}22^+$) expressing an intestinal lymphocyte subset that provides protection and integrity to intestinal mucosa (Satoh-Takayama et al. 2008). AhR induces transcription factor Notch, which is required for $NKp46^+$ ILCs (Lee et al. 2012). AhR modulates the development of two phenotypically distinct Treg cells, the induced Treg (iTreg) and the IL-10-secreting type 1 Treg (T_R1) cells, through different signaling pathways (Pot Kreis 2012). The functions of AhR in T cells depend on the specific ligand bound to the receptor (the AhR induces detoxifying enzymes and modulates immune cell differentiation). For example, the binding of 2,3,7,8-tetrachlorodibenzo-p-dioxin to AhR suppresses EAE by promoting the development of $Foxp3^+$ Treg cells, whereas 6-formylindolo(3,2-β)carbazole enhances EAE by inducing the differentiation of IL-17, producing T cells, suggesting the importance of these receptors in the initiation of reciprocal developmental programs upon different ligand binding (Quintana et al. 2008; Singh et al. 2011; Nakahama et al. 2013). AhR deficiency or lack of AhR ligands compromises the maintenance of intraepithelial lymphocytes (IELs) in the intestine and the control of the microbial load and diversity, resulting in an unrestrictive immune response owing to the absence of Treg cells and increased vulnerability to epithelial damage (Li et al. 2011a, b).

3.1.1.5 Other Minor Metabolites

Taurine (2-aminoethanesulfunoic acid, a major organic component of bile) is known for its biovolumizing effects (Jeon et al. 2007). Activation of T lymphocytes and subsequent transformation of activated T cells into T cell blasts requires profound changes in cells' volumetric index. Indeed, the T cell volume regulating taurine transporter Taut is crucial for T cells survival and T cell-mediated immune responses. In $taut^{-/-}$ mice, T cells-mediated B16 melanoma rejection and hapten-induced hypersensitivity reactions were severely impaired (Kaesler et al. 2012).

3.2 Immune Regulation via Glucose Metabolism

3.2.1 Energy and Nutrient-sensing in Immune Cells: A Road to Directed Evolution?

Evolution of the primitive immune system to a highly complex sensors-coupled immune system is accompanied by the development of a signaling–metabolism–immune effector axis that is dynamically altered by a pathogenic perturbator (Hotamisligil 2006). Such alterations result in the transition from a healthy state to a plethora of clinical manifestations, such as inflammation, attenuated immune response, and metabolic stress (Frieden 2015; Niccoli and Partridge 2012; Vita et al. 1998). As the mechanisms of such alterations are as yet unknown, several questions on immunometabolomics continue to haunt researchers: (1) Does the immune system dysfunction precede the metabolic dysfunction or vice versa occurs in a diseased state? (2) What are the immunometabolomic differences between the aged and the young people? (3) Does a pathogen signal the host cells to alter metabolism to ensure pathogen survival? (4) Do hosts assess the pathogen-triggered signal and alter the metabolic pathways to protect themselves? (5) Are those metabolomic alterations different in case of genetically resistant or susceptible hosts? (6) Do nutritional deficiencies have any causal relationship with various immunocompromised or immunodeficient states? (7) How do immune cells regulate their growth and behavior in nutrient-abundant and nutrient-deprived conditions? (8) How is the metabolic sensor–nutrient crosstalk influenced by the presence of a perturbator such as a pathogen? (9) What are the metabolic sensors-associated signaling pathways that direct the immune cell phenotype and function under perturbed conditions? (10) Can a perturbator or a pathogen regulate the threshold of the sensors such that the pathogen-derived signals can be selectively picked up by the system and help the pathogen survive? (11) Can nutrients and immune effectors work on the same pathogen-perturbed cell to moderate the anti-pathogen response? (12) Will the answers to these questions lead to the clinical translation in the form of therapeutic intervention and control of the major diseases caused by infectious agents or by metabolic dysregulation? The field of immunometabolomics is rapidly evolving as an interdisciplinary science (Fig. 3.2) and is likely to provide interesting leads for therapeutic interventions.

3.2.2 Glucose Sensing

As both non-proliferative DCs and proliferative T cells use the glycolytic pathway, it is suggested that aerobic glycolysis is not always linked to cellular proliferation but plays crucial roles in immune activation (Chang et al. 2013; Jones and Thompson 2007; Gerriets and Rathmell 2012; Buck et al. 2015; O'Sullivan and Pearce 2015). Glut2 receptors (SLC2A2), a high-capacity low-affinity glucose transporter on

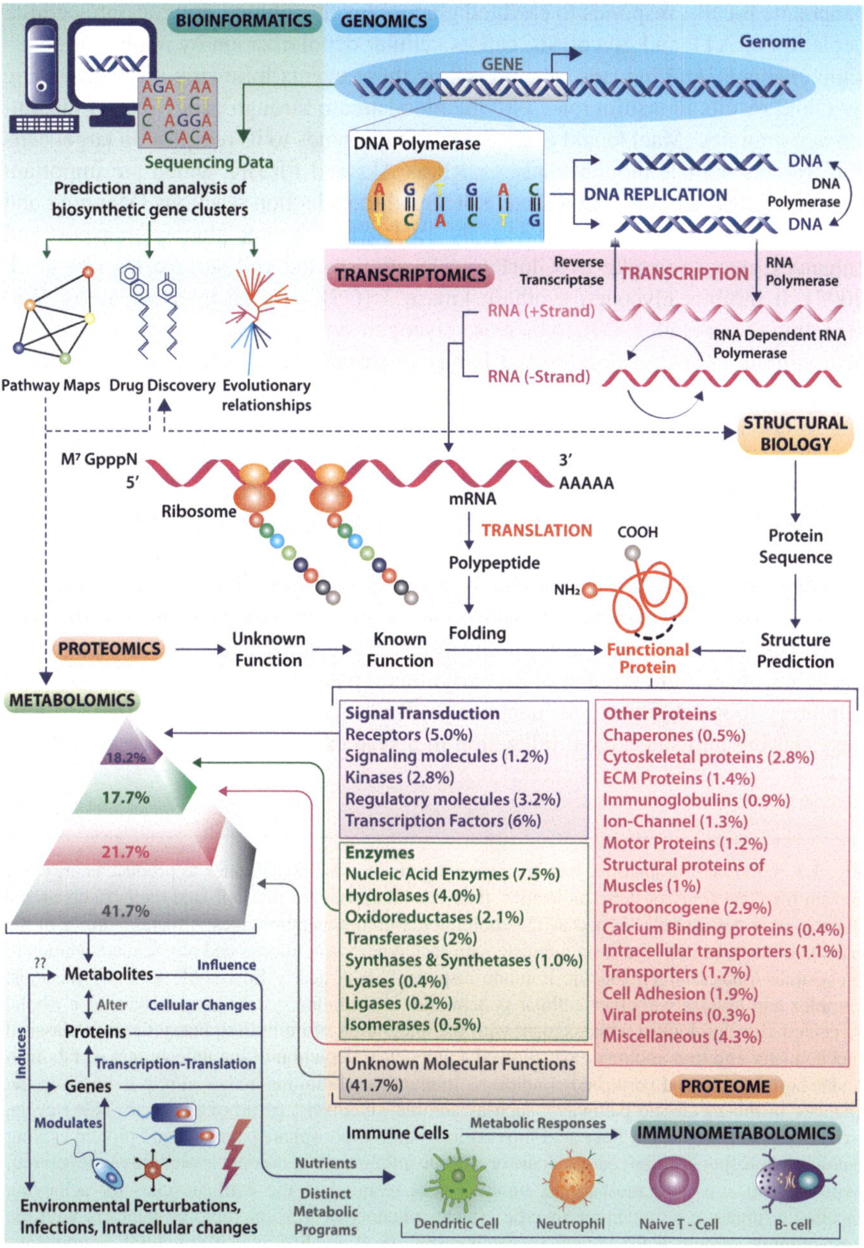

Fig. 3.2 What is 'immunometabolomics'? Immunometabolomics' functional importance can be realized by understanding the extended central dogma. The flow of genetic information from potentially the most complex naturally existing polymer of nucleotides DNA to protein in virtually every living cell ranging from simple prokaryotes to complex vertebrate life forms occurs via an RNA intermediate. Metabolomics is the study of metabolites, which are the products of chemical reactions occurring within a cell. Metabolomics is derived from 'metaballein' in Greek, meaning to

pancreatic β-cells, responds to elevated glucose levels in blood, triggers intracellular depletion of ATP and glycolysis, causes cellular depolarization by inhibiting potassium channels, and induces calcium influx; these events following glucose sensing by Glut2 results in insulin release to the bloodstream through exocytosis of insulin-storage granules (MacDonald et al. 2005). Insulin binds to its receptor on target cells and triggers signals though kinases such as Akt and PI(3)K, which are important upstream kinases regulating a variety of signal transduction reactions (Manning and Cantley 2007). In glucose-starved cells, PKB/Akt activates glucose transport and enhances glucose uptake by Glut1 translocation to the cell surface (Jacobs et al. 2008). It inhibits glycogen synthase kinase 3 (GSK-3β), an intermediate of Wnt signaling (Fang et al. 2000), to increase glycogen synthesis and glucose storage. Akt also inhibits FOXO, a Forkhead O family of transcription factors that responds to extrinsic signals such as pathogens, growth factors, inflammation, oxidative stress, and nutrient availability to halt transcriptional activation of gluconeogenesis enzymes (Puigserver et al. 2003; Daitoku et al. 2011). Facets of glycolysis include interlinked bioenergetic and anabolic roles of glucose metabolism, shaping transcriptional programs, post-transcriptional regulation, and signaling events in immune cells through glycolytic metabolites (Fig. 3.3). Glycolytic intermediates provide precursors for important anabolic reactions in proliferating cells, allowing cells to produce appreciable amounts of nucleic acids, AA, and lipids to undergo active cell division. Glucose metabolism through aerobic and anaerobic pathways thus controls cell growth and other biosynthetic pathways via both stimulatory and inhibitory loops. Multi-organ 'negative feedback loops' provide robustness to glucose sensing and subsequent utilization in a systemic manner. Glucose availability

---

Fig. 3.2 (continued) transform. Nutrients are converted to small catabolites to produce energy and sustain life. The study of these molecules, their chemical changes in a cell, and their effects on the phenotype is collectively termed as metabolomics. Immunometabolomics is metabolomics in the context of immune cells that communicate with each other via cytokines and physical association to orchestrate innate/antigen-specific immune responses. It is nearly impossible to study the most complex and diversely evolved cellular system, like the immune system in metazoans as a whole. Accelerated technological advancement with the integration of multidisciplinary fields to classical biochemistry and immunology have allowed a more clear depiction of immune system metabolism under homeostatic and perturbed conditions. Immunometabolomic studies aim at delineating the changes in the metabolic pathways, such as aerobic glycolysis, tricarboxylic acid (TCA) cycle, oxidative phosphorylation, fatty acid β-oxidation, pentose phosphate pathway, and protein cycling pathways in response to infections, acute or chronic inflammation, nutrient availability/competition, hypoxia, and stress. In most cases, immune cells respond to the extrinsic cues for achieving metabolic fitness, e.g., attaining specific effector phenotype. The microbiome interacts with the host immune system via a variety of mechanisms, such as shedding of structural components, production of metabolites, and direct physical activation. These interactions result in the disruption of immune homeostasis in local infection sites as well as distantly located metabolic organs, such as the liver, muscles, draining lymph nodes, and adipose tissues, which is, in turn, responsible for the initiation of different metabolic reprogramming in immune cells for producing optimal effector responses

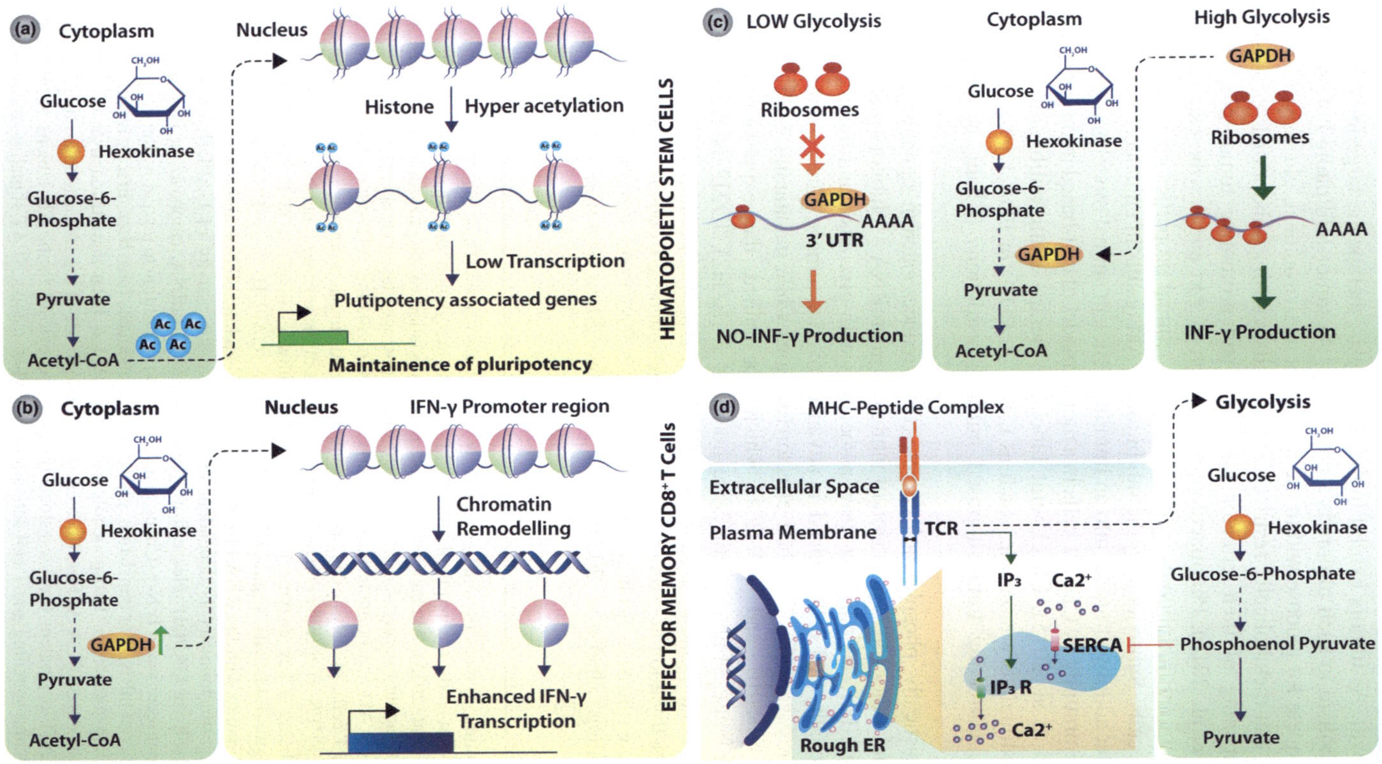

Fig. 3.3 Facets of glycolysis in immune cell metabolism regulation. Glycolysis not only provides energy to cells and fuels anabolic reactions in a cell, but also controls genetic programs and immune response. **(a)** Using a high-resolution nuclear magnetic resonance technique, distinct metabolic transitions occurring during early cytoplasmic differentiation were elucidated. The end product of glycolysis of acetyl-CoA is revealed as a decisive factor in regulating

affects the outcome of T cell-dependent processes such as immunological restricted tumor growth. As tumors exhibit higher glycolysis, therapeutic targeting of glycolytic metabolism in such tumors is a possibility. As tumor-infiltrating lymphocytes (TILs) compete for glucose within the tumor microenvironment, glucose utilization by tumors can metabolically restrict $CD8^+$ T cells, causing suboptimal responsiveness, dampening of their effector functions, and promoting tumor progression (Sukumar et al. 2015). Pharmacologic inhibition of Akt enables the expansion of TILs with the transcriptional, metabolic, and functional properties characteristic of memory T cells. Subsequently, the inhibition of Akt results in enhanced persistence of TILs after adoptive transfer into an immune-deficient animal model and augments antitumor immunity by $CD8^+$ T cells in a mouse model of cell-based immunotherapy (Crompton et al. 2015). Checkpoint blockade therapy using antibodies directed against costimulatory molecules such as CTLA-4, PD-1, or PD-L1 leads to tumor regression, correcting tumor-imposed nutrient restrictions (Chang et al. 2015) (Fig. 3.4).

Mechanistically, phosphoenolpyruvate (PEP) reduces calcium signaling or IFN-γ mRNA availability and suppression of inhibitory microRNAs. Inhibitory microRNAs such as miR-101 and miR-26a show association with glycolysis and tumor-specific polyfunctionality of T cells. Indeed, in advanced ovarian tumors, glucose restriction is imposed on T cells, limiting their effector functions while maintaining high expressions of miR-101 and miR-26a, which constrained the expression of the methyltransferase—Enhancer of zeste homolog 2 (EZH2). EZH2 is involved in activating Notch signaling by suppressing Notch repressors Numb and Fbxw7 via histone trimethylation at position lys27. As a consequence, T cells acquire versatility in cytokine expression (granzyme B, IFN-γ, IL-2, TNF) and sustained survival via Bcl-2 signaling (Zhao et al. 2015; Bantug and Hess 2016). The importance of Notch

---

Fig. 3.3 (continued) transcriptional programs by histone acetylation. Thus, high glycolytic rates in embryonic stem cells are required to uphold histone acetylation and maintain pluripotency (Moussaieff et al. 2015). Akt can also promote phosphorylation of the enzyme that catalyzes the conversion of citrate to acetyl-CoA, a substrate for histone acetylase that epigenetically modifies histones. Thus, Akt can promote histone acetylation by regulating ATP-citrate lyase (Lee et al. 2014). (**b**) In T cells, glycolysis can drive chromatin remodeling at promoter regions of inflammatory cytokines (IFN-γ), thereby enhancing the translation of these genes (Gubser et al. 2013). (**c**) In addition to being a key enzyme in the glycolytic pathway (involved in the payoff phase of glycolysis), glyceraldehyde 3-phosphate dehydrogenase (GAPDH) is an RNA binding protein, involved in binding with the RNA of the inflammatory cytokines IFN-γ and IL-2. The enzyme's RNA binding stoichiometry is dependent on the enzymatic activity of GAPDH; engaging in glycolysis results in GAPDH release from bound interferon mRNA, emphasizing its translation and subsequent secretion of IFN-γ (Chang et al. 2013). (**d**) In addition to regulating the transcription and availability of mRNA for translation, glycolytic intermediates have been directly reported to act as the signaling molecules. Phosphoenolpyruvate (PEP) has been demonstrated to bind to the calcium-dependent SERCA ATPase (located at the luminal side of the sarcoendoplasmic reticulum involved in ATP-dependent Ca^{2+} translocation from the cytosol to the ER lumen), resulting in sustained calcium signaling and T cell activation. PEP also sustains NFAT signaling and, thereby, acts as an integrator of glycolytic activity and TCR signaling (Ho et al. 2015)

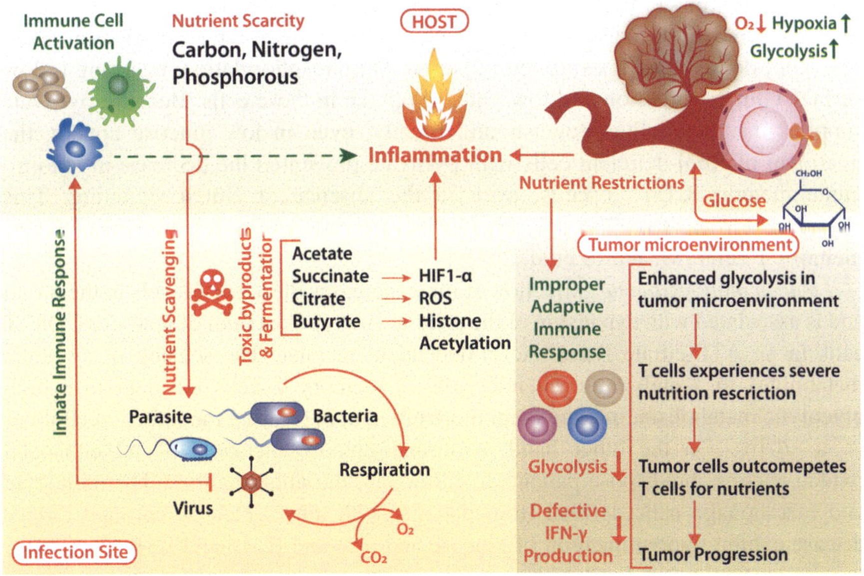

Fig. 3.4 Nutrient availability affects the outcome of several immune cell-dependent processes. During pathogenic encounters, tissue homeostasis is disrupted, resulting in infiltration of a variety of immune cells within inflammatory foci. The dynamic interactions between host-immune cells and pathobiota are predominant at infection sites because of the rapid production of a huge arsenal of metabolic derivatives, organic acids and byproducts, within immune cells. Organic acids produced as direct microbial growth or fermentation perform several functions: acetate contributes to inflammation, succinate induces HIF-1α and activates HIF-responsive genes, citrate derived reactive oxygen species (ROS) production, etc. Pathogens outcompete host immune cells and scavenge freely available micro-/macronutrients and electron acceptors for respiration, self-sufficing the needs of replication and mediating immune-evasion strategies; this also simultaneously skews the host-protective polyfunctionality of immune cells. These pathogens and their associated products activate several immune sensors, such as Toll-like receptors (TLRs) found on innate cells. TLRs recognize the evolutionarily conserved pathogenic moieties known as pathogen-associated molecular patterns (PAMPs). PAMPs activate several innate immune cells (neutrophils, dendritic cells, monocytes, and macrophages) for the initiation of non-specific inherent defense mechanisms against pathogenic infections initially, until the antigen-specific T cells are deployed. The failure of T cell-based protection and potent antigen recognition within tumors can be attributed to the chronic suppression of T cell metabolism by tumor cells and immune suppression by cytokines from tumor-associated immune cells. Interestingly, in tumor microenvironments, T cells are metabolically compromised due to severe nutrient deficiency, particularly glucose and amino acids (Gajewski et al. 2013; Chang et al. 2015). This leads to the inactivation of mTOR and other kinases controlling T cell effector functions and their proliferation (Guertin and Sabatini 2007). IFN-γ production is severely impaired during these conditions. As IFN-γ is regarded as the spearhead of proinflammatory cytokines and a potent antitumorigenic molecule, its dampened production ultimately results in tumor progression

signaling has been correlated with increased glucose uptake and recall responses by $CD4^+$ memory T cells, which maintain the essential signaling component of Notch signaling protein, known as recombination signal binding protein for the immunoglobulin-κJ region (Rbpj). Treatment of mice with a Notch inhibitor reduced numbers of memory $CD4^+$ T cells and prevented the recurrence of EAE. The Rbpj-deficient cells exhibited impaired Akt phosphorylation, resulting in low surface Glut1 expression and low glucose uptake in these cells. Because pyruvate supplies the metabolites downstream directly, even in low glucose uptake, the treatment of Rbpj-deficient cells with pyruvate prevented the decrease in autoimmune memory $CD4^+$ T cells, even in the absence of Notch signaling. This observation suggests a key role for glucose metabolism in the survival of $CD4^+$ memory T cells (Maekawa et al. 2015).

Listeria monocytogenes infection of mice increases the acetate levels in the blood and is associated with expansion of the cellular Acetyl-CoA pool of memory $CD8^+$ T cells in an ATP-citrate lyase (Acly)-dependent manner. The sensing of systemic metabolites in acutely infected mice allows memory T cells to adapt to a high glycolytic metabolism, making them more efficient in clearing the infection (Balmer et al. 2016). On the other hand, abnormal glucose metabolism and enhanced oxidative stress predispose patients to cardiovascular abnormalities. The monocyte and macrophage cells derived from patients with atherosclerotic coronary artery disease exhibit hyperutilization of glucose and promote IL-1β and IL-6 production, driving systemic tissue inflammation. This increased glycolytic flux, in turn, fuels the production of reactive oxygen species (ROS) from mitochondria, consequently promoting dimerization of the glycolytic enzyme pyruvate kinase M2 (PKM2), accentuating IL-1β and IL-6 production in a STAT-3-dependent manner ultimately exacerbating disease condition (Shirai et al. 2016).

As the genetic mutations altering the glucose metabolism in T cells are clinically associated with abnormal pathologies, the central importance of glucose metabolism in T cell immunity is evident. Class I PI(3)K promotes the phosphorylation of phosphatidylinositol-(4,5)-biphosphate (PIP_2) to phosphatidylinositol-(3,4,5)-triphosphate (PIP_3), which activates several downstream signaling molecules via membrane recruitment to signaling complexes (Walsh and Fruman 2014). Class IA PI(3)Ks are heterodimeric proteins, comprising of a p110 catalytic subunit and a p85 regulatory subunit. There are three isoforms of PI(3)K, p110α, p110β, and p110δ, encoded by PIK3CA, PIK3CB, and PIK3CD, respectively. p110α and p110β are ubiquitously expressed but p110δ is restricted to immune cells (Okkenhaug et al. 2007). Hyperactive PI(3)K activity is associated with chronic active glycolysis in primary immunodeficiency diseases. In many patients, infrequent dominant-activating germline mutations in the gene encoding the PI(3)K catalytic subunit p110δ are responsible for human immunodeficiencies. p110δ mutations from 14 patients from seven families revealed that these patients suffered sinopulmonary infections, lymphadenopathy, nodular lymphoid hyperplasia, and viremia due to cytomegalovirus (CMV) and/or Epstein–Barr virus (EBV) infection and deficiency of available naïve T cells but with the presence of exhausted or senescent-like T cell phenotype. The T cells from these patients displayed increased Akt phosphorylation and hyperactivation of metabolic checkpoint mTOR, enhanced glucose uptake, and terminal effector differentiation; these effects were prevented by treatment with the mTORC1 inhibitor rapamycin (Lucas et al. 2014). Likewise, owing to p110δ deficiency, $CD8^+$ T cells possess lower levels of metabolic activity, including the impaired ability to trigger aerobic glycolysis after TCR-specific activation, contributing to proliferative impairment, as well as diminished cytokine production (Gracias et al. 2016). Thus, precise and rapid regulation of glycolysis is required for maintaining proper T cell function; constitutive PI(3)K activity drives T cells towards an exhausted phenotype (Fig. 3.5).

In immune cells, the increased glucose uptake for glycolysis is achieved by promoting surface trafficking of glucose transporter Glut1 and this process is dependent upon the PI(3)K signaling pathway (Frauwirth et al. 2002). In this manner, the Glut1 glucose transporter allows maximal glucose uptake under minimal glucose availability. T cells activation requires costimulation via signals from CD28 following TCR stimulation and increased cellular trafficking of Glut1 (Jacobs et al. 2008). IL-7 also promotes Glut1 trafficking and glucose uptake via STAT5-mediated activation of Akt to sustain T cell lifespan (Wofford et al. 2008). In the absence of costimulation, T cells show a reduced ability to proliferate owing to failure to activate PI(3)K and increased glycolysis. Furthermore, T cells with constitutive Akt activation have increased glycolytic activity in cancer cells (Elstrom et al. 2004). ICOS and OX40-like inducible costimulatory molecules also induce robust PI(3)K signaling on activated T cells (Gigoux et al. 2009; Croft 2009). It is, therefore, possible that along with the constitutive Akt activation, these costimulatory molecules might be responsible for promoting even stronger glycolytic activity in antigen-exposed T cells. Thus, the primary immunodeficiency diseases have an immunometabolic regulatory backbone that may guide therapeutic

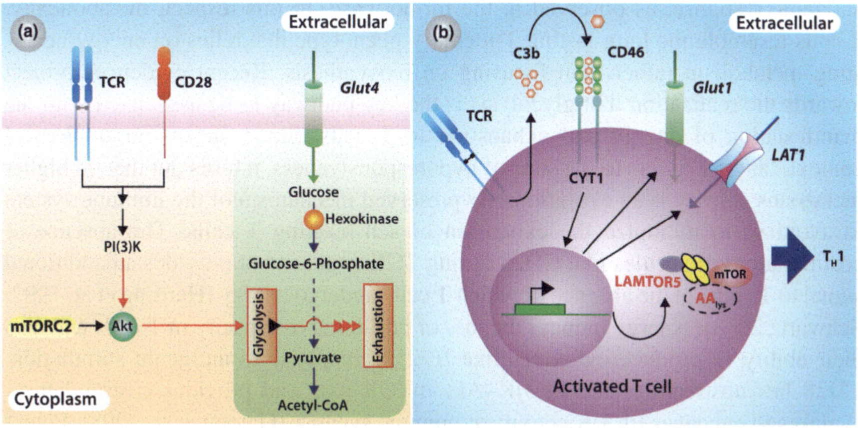

Fig. 3.5 (**a**) Precise and rapid signals from T cell receptors (TCR) (signal 1) and CD28 (signal 2) are required for the regulation of glycolysis and maintenance of proper T cell function. Constitutive activity of PI(3)K drives T cells towards an exhausted phenotype. (**b**) The in vivo signals that initiate and coordinate metabolic reprogramming in activated T cells remains highly elusive. Complement component C3b enhances nutrient sensing and uptake in human T cells, enabling increased glycolytic turnover, restraining exhaustive state and promoting T_H1 type responses (B-adapted from Kolev et al. 2015)

trials using the drugs classically identified as immunosuppressive in nature, such as inhibitors of mTOR or PI(3)-kinase.

Mutations in CD46, a complement regulatory protein and an effective costimulatory molecule, modulate T_H1 cells by orchestrating immunometabolic signals from several receptors, TCR, Glut1, and LAT1, involved in T cells reprogramming (Fig. 3.5b). CD46 ligation by C3b is required for the induction of glucose sensing, increased glycolysis, and enhanced amino acid uptake, which are essential for T_H1 polarization. CD46 activation drives the expression of LAMTOR5, which mediates assembly of the amino acid-sensing Ragulator–Rag–mTORC1 complex, resulting in increased glycolysis and oxidative phosphorylation, necessary for cytokine production. In patients harboring CD46 mutations, suboptimal metabolism is highlighted as a phenomenon behind the cellular immune defect. These results establish a critical link between the complement system and immunometabolic adaptations driving human $CD4^+$ T cell effector function (Kolev et al. 2015).

3.2.3 T Cells Anergy and T cell Exhaustion Are Controlled by Glycolysis

Quick and rapid antigenic stimulation to T cell results in their activation, while chronic stimulation with self-peptides or, in the case of autoimmunity, however, results in a self-tolerant or exhausted phenotype that prevents quick and rapid antigenic response, as observed in the former case. In this respect, metabolically, T cells resemble the long-lasting T memory phenotype that relies on energy generating metabolism rather than focusing on biosynthesis. Recent evidences project towards the realization that glycolysis is the key pathway to be associated with the manifestation of anergy and exhaustion in T cells. Since, in an immunological context, 'anergy' refers to the state of hyporesponsiveness, it turns out that, in higher metazoans, anergy is an evolutionarily preserved mechanism of the immune system to avoid or to minimize the expansion of self-reacting T cells. The absence of costimulatory molecules like CD28 during TCR engagement provides an additional signal to T cells, in the absence of which T cells undergo anergy (Harding et al. 1992; Schwartz 2003). Anergy induces the loss of proliferative capacity in T cells and also their ability to produce cytokines like IL-2, in response to antigenic stimulation. CD28 lies upstream of the PI(3)K–Akt–mTOR axis and provides costimulatory-signals and enhances PI(3)K activity at immune synapse (Garçon et al. 2008). Potent immunosuppressant macrolide-rapamycin blocks T cells proliferation in the late G1 phase but does not affect costimulation-dependent IL-2 production. The role of mTORC1 activation and T cell stimulation was highlighted in one of the studies (Powell et al. 1999), wherein the authors suggested that rapamycin is sufficient to induce T cells anergy, even in the presence of costimulatory signals. Also, in rapamycin-induced anergic T cells, decreased expression of MAPK was reported

(Powell and Delgoffe 2010). As discussed in several reports, the PI(3)K–Akt–mTOR pathway promotes Akt- and mTOR- dependent trafficking of Glut1 and amino acid receptors on the cell surface (Zheng et al. 2007; Chi 2012; Macintyre et al. 2014). Thus, costimulation plays a deterministic role in the expression of the full activation potential of T cells and evading anergy in T cells failure, to which will result in the prevention of upregulation of glucose uptake by activated T cells via Glut1, amino acids by CD98/Lat1 (solute carrier family 3 member 2-SLC3A2 activators of dibasic and neutral amino acid transport), and, also, transferrin by CD71 (Zheng et al. 2009).

While T cell anergy is due to inadequate TCR stimulation in the absence of costimulatory inputs (Harding et al. 1992), T cell exhaustion is due to the progressive loss of T cell functions. In fact, acquisition of a 'worn-out state' by T cells during viral infections, manifested as reduced functionality, and is termed as 'exhaustion' (Day et al. 2006; Barber et al. 2006; Bengsch et al. 2016). Several receptors suppress T cell inflammatory functions to promote self-tolerance. Activation of programmed cell death protein-1 (PD-1) has been shown to upregulate the enzyme ATGL (adipose triglyceride lipase) that cleaves intracellular triacylglycerol (TAG) into glycerol-3-phosphate and fatty acids for their utilization in FAO in activated T cells. Other major receptors from this category are: cytotoxic T lymphocyte antigen-4 (CTLA-4), lymphocyte activation protein 3 (LAG-3), T cell immunoglobulin, and mucin domain-containing-3 (TIM-3) (Blackburn et al. 2009). Great loss in the functionality of T cells is observed upon exhaustion pertaining to their ability to undergo proliferation: self-renewal, capacity to protect from secondary infection, and diminished cytokine production and cytolytic potential (Wherry 2011). So, reactivation of exhausted or chronic refractory T cells holds potential for the amelioration of many abnormal pathologies and clinical situations.

Ligation of CTLA-4 or PD-1 alone blocked CD3-/CD28-mediated upregulation of glucose metabolism and Akt activity synergistically, but via distinct molecular mechanisms (Parry et al. 2005). In patients with chronic infections and cancer, PD-1 blockade can lead to disease amelioration and rejuvenation by promoting fatty acid oxidation in T cells. As PD-1 promotes endogenous fatty acid pool utilization via carnitine palmitoyl transferase-1A (CPT1A) expression, mitochondrial β-FA oxidation, lipolysis by elevation of adipose triglyceride lipase (ATGL), and by inhibiting glycolytic and amino acid metabolism, these effects promote longevity in T cells (Patsoukis et al. 2015). PD-1 may also inhibit Ras and its subsequent downstream targets ERK-1 and ERK-2 through an SHP1- and SHP2-independent manner to inhibit cell cycle progression and proliferation of T cells (Patsoukis et al. 2012). CD274/PD-L1 (programmed death ligand 1) is reported to downregulate phospho-Akt, S6K, and ERK-2, along with mTOR complexes (with concomitant upregulation of PIP_3 to PIP_2 cycling enzyme phosphatase and tensin homolog [PTEN]) to promote inducible T regulatory (iTreg) development and help in the maintenance of peripheral tolerance. Other molecules like LGA-3 can also be tweaked to promote differentiation of naïve T cells towards the T_H1 phenotype and LAG-3 expression on T cells makes them more susceptible to Treg-mediated immunosuppression (Durham et al. 2014). Not just the inhibitory receptors but also the downregulation of costimulatory molecule CD28 within the $CD8^+$ T cells subset is correlated as a reduction factor in the recognition repertoire of antigen diversity, defective antigen-induced signalling, and shorter replicative lifespan but increased cytotoxicity and regulatory functions in CD28-negative T cells with age-associated decline of immune function (Weng et al. 2009). Akt-mediated evasion of cell death is dependent on glucose; Akt relies on glycolysis to inhibit induction of PUMA (p53 upregulated modulator of apoptosis), which is sensitive to the metabolic status of the cell. Cytokines can be inducers of antiproliferative effects in lymphocytes, as IFN-α inhibits protein synthesis, reduces amino acid incorporation, impairs glucose and glutamine metabolism, and arrest cell cycle (Navarro et al. 2010).

3.3 Metabolic Pathways in Immune Cells: Beyond Glycolysis

3.3.1 Amino Acid Sensing

Mammalian target of rapamycin (mTOR) is an evolutionally conserved kinase that orchestrates protein translation (Wang and Proud 2006), cell growth and proliferation (Schmelzle and Hall 2000; Fingar et al. 2004), autophagy (Jung et al. 2010), and also metabolism in cancerous settings (Shaw and Cantley 2006) in response to various environmental cues in eukaryotes (Wullschleger et al. 2006). Under the non-perturbed state, several mechanisms enforce inhibitory control over mTOR and maintain cellular homoeostasis. However, antigenic exposure triggers mTOR activation in $CD4^+$ and $CD8^+$ T cells and drives their differentiation towards specific effector lineage. mTOR instructive signals are not only restrictive to amino acid sensing; rather, signals transduced via DCs through antigens (MHC–Peptide–TCR complex), costimulatory molecules such as CD28, and proinflammatory cytokines can be collectively integrated to activate mTOR signaling in T cells. The other two types of instructive signals are mediated by environmental sensing of growth and immunomodulatory factors; also, metabolic cues from nutrients such as amino acids and glucose provide non-restrictive activation of mTOR signaling. mTORC1 activity in most eukaryotic cells is predominantly dependent on amino acid availability. Despite extensive research in amino acid sensing and mTOR kinases, the underlying mechanism through which mTORC1 senses amino acids remains an enigma, with the unraveling of new potential mechanisms experimentally, AA–mTOR sensing, and its regulation considered to be an ever-evolving story (Kim et al. 2013a, b; Bonfils et al. 2012; Han et al. 2012).

Cytoplasmic kinase mTOR responds to a variety of stimuli (Laplante and Sabatini 2009), but immune activation signals to T cells are very exclusive, as environmental and metabolic stimuli are speculated to act on almost all eukaryotic cells. A close positive correlation is reported between amino acid availability and translation rates (Kim et al. 2013a, b), as mTOR activation by nutrient abundance leads to the repression of translation, eukaryotic initiation factor 4E-binding protein 1 (eIF4E-BP1), and activation of translation activator S6 kinase (S6K1) (Hay and Sonenberg 2004; Inoki et al. 2002). The amino acids, in particular leucine and arginine depravation phenocopies, complete amino acid starvation conditions, making them the most potent activators of the mTORC1 complex. It becomes important to understand the function of the tuberous sclerosis complex (TSC1-TSC2) as a gatekeeper of the mTOR complex by acting as a GTPase-activating protein (GAP). The amino acids regulate mTORC1 through a TSC Rhob independent pathway, but, interestingly, growth factors have a diminished mTORC1-activating effect under amino acid-deprived conditions and can even suppress mTORC1 signaling, even in TSC2$^{-/-}$ cells, where mTORC1 is unrestrictedly activated (Smith et al. 2005).

mTORC1 is one of the two major mTOR complexes, regulated by lysosomal-bound small GTPase Rheb. Under amino acid-abundant conditions, the translocation of mTORC1 takes place where the Rag protein is localized. Rag GTPase recruits proteins from genes MAPKSP1, ROBLD3, and c11orf59, known as the 'Ragulator complex', which are essential for mTORC1 activation. Akt and ERK, as well as RSK downstream of ERK, phosphorylate the TSC2 and, thereby, allow Rheb to activate mTORC1. Unlike growth factors, amino acids communicate to mTORC1 through the Rag GTPase complex and the Ragulator complex that anchors the Rags to the lysosome in a spatiotemporal-dependent manner (Zhou et al. 2015a, b). Finally, the targeting of mTORC1 to the lysosomal surface eliminates its amino acid sensitivity (Sancak et al. 2008; Sancak et al. 2010). The amino acids leucine, arginine, and glutamine are reported to be the most potent activators of mTORC1 activity, each activating mTOR by distinct mechanisms. It has now been shown that, in addition to amino acid sensing by lysosomal activation, intracellular leucine sensor leucyl-tRNA synthetase (LRS) plays a crucial role in amino acid-induced mTORC1 activation by sensing intracellular leucine concentration and initiating a molecular cascade that culminates in mTORC1 activation. Mutations of LRS amino acid residues important for leucine binding render the mTORC1 pathway non-respondent to even high intracellular leucine concentration. LRS directly binds to Rag GTPase, the mediator of amino acid signaling to mTORC1, and functions as GAP for Rag GTPase (Han et al. 2012). Environmental and metabolic tension, such as DNA damage and nutrient scarcity, induces the expression of evolutionarily conserved proteins known as sestrins that negatively regulate nutrient-responsive protein kinase complexes in eukaryotic cells, such as mTORC1 (Lee et al. 2016; Chantranupong et al. 2014). Sestrin-mediated inhibition of mTORC1 activity takes place in both direct and indirect modes via interaction with mTORC1-regulating complex GATOR (in particular, the GATOR2 protein) and activation of AMP-activated protein kinase (AMPK), respectively. Glutamine sensing is mediated by adenosine diphosphate ribosylation factor-1 (Arf1) GTPase (Jewell et al. 2015). Recently, SLC38A9 transporter was identified as a putative lysosomal arginine sensor required for arginine to activate mTORC1 (Wang et al. 2015). The mechanism by which arginine deprivation represses mTORC1 activity has recently been shown. GATOR2 interacts with the cellular arginine sensor for mTORC1 (CASTOR1) and is required for arginine deprivation-dependent inhibition of mTORC1 activity (Chantranupong et al. 2014). When subjected to external perturbations such as starvation, stress, or viral infections, mammalian cells downregulate global protein synthesis by phosphorylating the eIF2α subunit (Hershey 1991). T cells, after priming via signals from TCR, express cytokine mRNA but retain it until any additional TCR stimulation is acquired. Phosphorylation of eIF2α increased the expression of stress-associated genes and accumulation of cytoplasmic granules associated with RNA-binding proteins, restimulation of cells resulted in rapid eIF2α dephosphorylating, ribosomal mRNA loading, and cytokine secretion (Scheu et al. 2006).

The highly conserved Gen2–eIF2α–Gcn4/ATF system plays an important role in achieving amino acid homeostasis in several cells, including immune cells. Serine–threonine kinase Gcn2 is activated by the accumulation of uncharged t-RNA (and binds Gcn2 through a domain with homology to histidyl-tRNA synthetase; this domain has higher affinity for uncharged tRNA than charged tRNA, and binding causes a conformational change that activates the kinase domain of Gcn2), which, in the case of amino acid scarcity, binds to Gcn2 and activates its kinase activity. External amino acid depletion triggers a cascade involving proximal sensor Gcn2, which is involved in the inhibition of T cell functions (Harding et al. 2000; Wek et al. 2006). Subsequently, eIF2-α phosphorylation results in global halting of the protein translation while enhancing the translation of the Gcn4-encoding gene that initiates the transcription of genes responsible for the biosynthesis of amino acids (Hinnebusch 2005). In addition to regulating the adaptation response to amino acid starvation, Gcn2 activation in T cells inhibits T_H17 differentiation and promotes Treg cell development (Sundrud et al. 2009), proliferation arrest, and anergy induction (Munn et al. 2005). Thus, a low-protein diet would potentially reduce the availability of circulating amino acids in the blood, leading to reduced proliferation of $CD8^+$ memory T cells and impaired memory-based recall responses (Iyer et al. 2012). The activating transcription factor 4 (ATF4) pathway canonically drives the autophagy gene expression program and is involved in the phosphorylation of Ser^{51} on the α subunit of eIF2α (Sood et al. 2000) and suppression of global translation rates, including a paradoxical increase for certain mRNAs, such as those encoding the transcription factors Gcn4 and ATF4 themselves (Vattem and Wek 2004). Most cellular RNAs are regulated via this system; activated ATF4 serves as the master regulator of transcription of specific target genes to salvage amino acids from existing proteins, leading to metabolic stress adaptation (Harding et al. 2003; B'chir et al. 2014).

3.3.2 Lipid Sensing

Sterol metabolism integrates cellular and humoral responses (Li et al. 2017). Cholesterol and free fatty acids are two types of lipids essential for a variety of cellular reactions. Cholesterol provides fluidity and dynamicity to plasma membranes and is used as a precursor for the synthesis of many steroidal hormones, like testosterone. It is also used for the synthesis of bile acids. Cholesterol sensing begins at the endoplasmic reticulum, via dynamic interactions between sterol regulatory element-binding proteins (SREBPs) (Horton et al. 2002), SREBP cleavage-activating protein (SCAP), 3-hydroxy-3-methylglutaryl coenzyme A reductase (HMG-CoA reductase), and insulin-induced genes (INSIGs) (Dong et al. 2012). SREBPs cannot exert their activities alone; therefore, they form a complex with SCAP protein and translocate to Golgi. Cholesterol sensing initiates when the membrane cholesterol levels fall below a certain threshold. Sensor and mediator protein Insig1 dissociates and the SCAP–SREBP complex translocates to Golgi apparatus, where SCAP-activated S1P and S2P initiate the cleavage of SREBP, liberating SREBP for its nuclear translocation (Sakai et al. 1996). SREBP1 is a member of the basic helix-loop-helix-leucine-zipper (bHLH-ZIP) family of transcription factors of 125 kDa that remains associated with the nuclear envelope and endoplasmic reticulum. In case of sterol depletion, the membrane-bound precursor is cleaved to generate a soluble NH_2 terminal fragment (~68 kDa) that translocates to the nucleus (Wang et al. 1994). Studies from pioneering lipid biologists suggest that the nuclear translocation of SREBP results in the transcription of genes encoding HMG-CoA reductase, a prime moderator enzyme in the synthesis of cholesterol and lipids (DeBose-Boyd 2008; Horton et al. 2002). Therapeutic targeting of HMG-CoA reductase is achieved by the various cholesterol-lowering statins class of drugs (Gazzerro et al. 2012). As a primary component of cell membranes or triglycerides in adipose tissues, fatty acids play structural or energy-providing roles in cells. The precise mechanism of sensing of different fatty acids is not clear, but the conjecture points towards peroxisome proliferator-activated receptor (PPAR) (Nunn et al. 2007; Ravnskjaer et al. 2010). PPARγ promotes monocyte/macrophage differentiation and uptake of oxidized LDL and other lipids (Tontonoz et al. 1998; Chawla et al. 2001). Cholesterol efflux PPARs (currently recognized PPARα, PPARδ, and PPARγ isoforms) (Greene et al. 1995; Berger and Moller 2002) are ligand-activated transcription factors that belong to the category of the superfamily of nuclear hormone receptors and regulate the expression of several genes involved in metabolic processes (Desvergne and Wahli 1999) that, when dysregulated, result in metabolic abnormalities like hyperlipidemia, diabetes, and obesity (Contreras et al. 2013). Essentially, PPAR binds to DNA as a heterodimer with retinoid X receptor (RXR) and recognize short sequences of DNA around target genes known as PPAR response elements (Kliewer et al. 1992). PPAR isoforms unequivocally sense the fatty acids milieu of cells through loose or weak binding of diverse FA derivatives (Chakravarthy et al. 2009; Kliewer et al. 1994). The low binding affinities associated with PPARα binding and free fatty acids (found to be in the order of 2–50 μM) in vitro suggest that only high concentrations of free fatty acids optimally activate PPARs (Tzameli et al. 2004). The physiological significance of this effect could be correlated with the PPARs involvement in cellular proliferation (Fajas et al. 2003) and differentiation (Jump 2004; Berod et al. 2014). PPARs regulate the expression of enzymes involved in fatty acid oxidation (Széles et al. 2007). PPARα is evidently considered as the nexus of a regulatory hub impacting an array of fatty acids-associated cellular processes, such as: FA uptake, FA activation, intracellular FA binding, mitochondrial and peroxisomal FA oxidation, ketogenesis, and triglyceride turnover (Kersten 2014). Leukotriene B4, a strong inflammatory mediator, also an inducer of PPARα, regulates the oxidative degradation of fatty acids and their derivatives (Devchand et al. 1996). PPARs are reported to exhibit anti-inflammatory effects, as observed in the case of PPARα$^{-/-}$ mice. The PPARγ isoform dominates in myeloid cells, whereas PPARα is expressed abundantly in T and B lymphocytes (Jones et al. 2002). PPARγ gene expression is upregulated in T cells by mitogenic activation. PPARγ ligands have been reported to inhibit IFN-γ and IL-2 production in murine mitogen-activated splenocytes and human T cells (Cunard et al. 2002; Yang et al. 2000), induce apoptosis in murine T helper cells (Harris and Phipps 2001), and inhibit proliferation of human T cells (Clark et al. 2000). These findings raise many unanswered questions in the lipid-sensing biology of immune cells. Some of which are: (1) To what extent do the cellular pools of fatty acids modify the sensing and subsequent signal transduction events in immune cells? (2) Do dietary polyunsaturated fatty acids (PUFAs) have any deterministic effect on immune functions in infectious and non-infectious diseases? (3) Can different classes of dietary fatty acids alter the immune response through differential modulation of PPARs? (4) To what extent can PPARs be targeted for controlling diseases with immune system involvement?

3.3.3 Nucleotide Sensing

Nucleotide reservoirs of cells are carefully managed, as excessive production or deficiency is associated with abnormal pathologies, such as increased mutagenesis and, often, genomic instability (Traut 1994; Fajardo et al. 2014; Pai and Kearsey 2017). During the S-phase of the cell cycle, the demand for nucleotides surges in lymphocytes to support active DNA replication. Enzymes involved in de novo DNA synthesis and others, such as ribonucleotide reductase (RNR) and catalyzing the rate-limiting step in deoxynucleotide triphosphates (dNTPs) production, are upregulated, which mediates the conversion of nucleoside diphosphate to nucleotide triphosphates, the precursors for dNTPs, and maintains subcellular pools of dNTPs for DNA replication (Eriksson et al. 1984). The control of dNTP concentrations is critical to the fidelity of DNA synthesis and repair. One level of regulation of intracellular dNTPs is through subcellular localization of ribonucleotide reductase (Wu and Huang 2008). Spatiotemporal localization of RNR, e.g., subcellular localization of RNR at the DNA damage site, adds an additional layer of its regulation in the cell (Lee et al. 2006; Yao et al. 2003) and proteolytic degradation (Chabes and Thelander 2000). RNR and other nucleotide biosynthesis enzymes are direct targets of c-Myc, a master regulator of the cell cycle (Dang 2013). The oncoprotein c-Myc can act as an illegitimate replication-licensing factor, which may amplify the translation of crucial rate-limiting enzyme-encoding genes, such as RNR (mouse R2 gene) that, in turn, permits de novo replication initiation and illegitimate replication, conferring genomic instability in cells (Kuschak et al. 2002).

Adenylates (ADP, AMP) are important factors, as their sensing regulates nucleotide biosynthesis and AMPK activation (Inoki et al. 2012). Exogenous and endogenous elevations in cAMP concentration result in diminished mitogen stimulation, cell cycle arrest, and decreased RNR mRNA concentrations in peripheral blood mononuclear cells (PBMCs) (Albert et al. 1990). Under oxidative stress, such as anoxia or exercise, the ADP:ATP ratio and intracellular AMP increase, indicating cellular energy exhaustion (Hardie and Hawley 2001). A number of groups (Hawley et al. 1996; Hamilton et al. 2002; Woods et al. 2003) have confirmed that elevated cellular AMP:ATP ratios activate AMPK, a cellular energy sensor, by phosphorylation of the Thr^{172} residue of the AMPKα1 catalytic subunit of the AMPK α, β, and γ heterodimeric assembly (Xiao et al. 2011).

3.3.4 Energy Sensing

Cellular homeostasis is maintained by the constitutive expression of housekeeping genes and maintenance of substantial amounts of ATP, which are used for biosynthetic reactions and by the ion transporters P-type ATPase (sodium potassium pump, calcium pump, and proton pump), F-type ATPase (mitochondrial ATP synthase, chloroplast ATP synthase), V-type ATPase (vacuolar ATPase), and ABC transporters (ATP binding cassettes MDR, CFTR, etc.) (Fox et al. 2005). Functions distinctly limited to immune cells, such as motor functions, antigen processing and processing, activation, and effector functions like mediating cytotoxicity, antibody, cytokine secretion, and regulatory functions also require a significant portion of the cellular ATP pool. In chronic inflammatory diseases (CIDs), lymphocyte activation requires huge amounts of energy (around 2000 kJ/day). Calculations indicate that active leukocytes require $\sim 8 \times 10^{-11}$ mol ATP/day, which is about 3.3 times that needed by quiescent leukocytes, i.e., 2.4×10^{-11} mol ATP/day (Straub et al. 2010). Immune cells sense and respond to this urgent need for replenishing the depleted ATP pools by modulating glucose metabolism (Vander Heiden et al. 2009; Wang et al. 2011). Activated T cells do not increase the TCA cycle but, instead, the TCA cycle serves as an intermediate or junction for 'reprogramming' β-oxidation of fatty acid and pyruvate oxidation to glycolysis, pentose phosphate, and glutaminolysis pathways (Guppy et al. 1993; Wang et al. 2011).

The heterotrimeric AMPK pathway serves as an ancient metabolic checkpoint that is associated with energy sensing in mammals. Across all eukaryotic species ranging from *Arabidopsis thaliana* (SnRK1) (Emanuelle et al. 2015), *Saccharomyces cerevisiae* (snf1) (Hong et al. 2003), *Caenorhabditis elegans* (AAK-2) (Lee et al. 2008), and *Drosophila* (Hardie 2002) to mammals, an ortholog of AMPK and its associated subunits can be found, and in all of these species, this enzyme mediates regulatory features of metabolism (both glucose and lipid metabolism, specifically in the context of uptake, utilization, and storage). LKB1, an upstream kinase, directly phosphorylates AMPK at Thr^{172} when cellular ATP levels are low (Shaw et al. 2004). This happens under low glucose and oxygen (hypoxia) or mitochondrial insults or under the influence of inhibitors of oxidative phosphorylation, electron transport chain (inhibitors such as antimycin, rotenone, and uncouplers), dinitrophenol, CCCP (Carbonyl cyanide m-chlorophenyl hydrazone- an oxidative phosphorylation inhibitor). Therefore, decreased intracellular ATP levels results in activation of the LKB1 kinases cascade. In the case of immune cells, cellular functions like cytotoxicity, antibody, cytokine secretion, and regulatory functions require significant portions of cellular ATP pools; therefore, LKB1 kinase gets activated. Evidently, in LKB1-deficient T cells, AMPK activation was decreased, thereby increasing the production of inflammatory cytokines by $CD4^+$ and $CD8^+$ T cells. In both LKB1 and AMPKα1 deficits, T cells mTORC1 activity was found to be upregulated and, thus, also, IFN-γ production by these cells; rapamycin treatment reversed this effect (MacIver et al. 2011). There are evidences now available that activators of AMPK (AICAR [5-aminoimidazole-4-carboxamide-1-β-D-ribofuranoside] and metformin) (Zhou et al. 2001) show potent anti-inflammatory activities and several studies support AMPK as an NF-κB signaling inhibitor (Huang et al. 2015; Guo et al. 2014), as it modulates SIRT1, FOXO3, PGC-1α, and p53 (Salminen et al. 2011). The antidiabetic drug metformin and related compounds, complex I inhibitors of oxidative phosphorylation (Wheaton et al. 2014; Ferrannini and DeFronzo 2015), decrease ATP output (Shaw et al. 2005). Pearce and coworkers showed that tumor necrosis factor (TNF) receptor-associated factor-6 (TRAF6), an important adaptor in TNF receptor and IL-1R/Toll-like receptor signaling, regulates $CD8^+$ T memory cell development after infection by modulating the fatty acid metabolism of the cell. TRAF6-deficient T cells display a defective AMPK activation and fatty acid oxidation in response to growth factor withdrawal. Metformin restored fatty acid oxidation and $CD8^+$ T memory cell generation in the absence of TRAF6 (Pearce et al. 2009). Similarly, AICAR, a synthetic AMPK mimetic, binds to the AMPK heterotrimeric complex, favoring its phosphorylation by an ATM-dependent and LKB1-independent manner (Sun et al. 2007). AICAR also exerts immunosuppressive effects by inhibiting IL-2 expression through the inhibition of GSK-3β phosphorylation and NFAT activation in the T cell line (Jhun et al. 2005). AMPK thus acts as a master switch that T cells push to 'shut down' all the major anabolic biosynthetic processes occurring at that particular time point when ATP availability is limited. The LKB1–AMPK axis can, therefore, be regarded as a crucial checkpoint to restrict intracellular ATP usage (Wang and Green 2012). This intriguing mechanism is still not sufficient to conserve the bleeding ATP to normalize intracellular ATP levels. AMPK also simulates fatty acid oxidation (Lee et al. 2006) and autophagy (Pauly et al. 2012) to achieve considerable intracellular ATP replenishment. Genetic and pharmacological studies from many pioneering laboratories have helped researchers enormously to delineate the tissues and cell-specific roles of AMPK more profoundly in the regulation of cellular energetics in different organisms.

3.3.5 Oxygen Sensing

Oxygen sensing becomes of utmost importance to cells because, in hypoxic or anoxic conditions, the deficiencies are overcome by enhanced activity of O_2-regulated ion channels. Chronic effects can be reversible or irreversible, depending on the modulation of hypoxia-induced transcription factors, which determine the expression of numerous genes encoding enzymes, transporters, and growth factors. Hypoxia-induced factor (HIF) is found in mammalian cells cultured under reduced oxygen tension, and is necessary for transcription activation mediated by the erythropoietin (EPO) gene enhancer in hypoxic cells (Semenza and Wang 1992; Wang et al. 1995). It is suggested that ROS generated by mitochondrial complex III are required for hypoxic activation of HIF (Klimova and Chandel 2008).

HIF-1α, in particular, is a binding partner of HIF-1β, which is constitutively expressed in cells. Binding of these two proteins, along with other co-factors (p300/CBP) (Arany et al. 1996), modulates the transcription of target genes whose promoters contain a hypoxia response element. Under normoxic conditions, HIF-1α is hydroxylated by iron-dependent prolyl hydroxylases (PHD) enzymes at proline-402 and -564 residues. These prolyl hydroxylation events are essential for the binding of HIF-1α and von Hippel–Lindau tumor suppressor protein (VHL)-dependent ubiquitination and subsequent proteasomal degradation of HIF-1α (Ohh et al. 2000). VHL targets HIF for oxygen-dependent proteolysis (Maxwell et al. 1999). Under hypoxic conditions, PHD becomes inactive, resulting in cytosolic accumulation of HIF-1α, ensuring its nuclear translocation. Angiogenic genes, such as those for vascular endothelial growth factor (VEGF), besides classically documented genes like phosphoglycerate kinase-1 (PGK), Glut1, and erythropoietin (EPO), are well-documented targets of the HIF downstream signaling network (Palazon et al. 2014; Forsythe et al. 1996). Not only VEGF but also HIF-1α promotes Glut1 expression, which directly correlates with increased glucose uptake by T cells, to encourage their activation during initial exposure to pathogens (Chen et al. 2001; Wang et al. 2011). Pathogen-associated molecular patterns (PAMPs) induce signaling via TLRs on macrophages. The signal culminates in activation of the inflammatory mediator NF-κB (Akira et al. 2006), which, in turn, leads to the increase of HIF-1α transcription (Rius et al. 2008). Factors inhibiting HIF add another layer of regulation of HIF by hydroxylating asparagine residues in HIF-1α and HIF-2α, impairing the complex formation between HIF-1α and CBP/p300 (Freedman et al. 2002). HIF-1α (but not HIF-2α) stimulates glycolytic gene expression in engineered HEK293 cells (Hu et al. 2003). Blouin et al. reported that LPSs from Gram-negative bacterial cell walls stimulates HIF-1α transcription in proportions comparable to that of hypoxia-induced HIF-1α expression (Blouin et al. 2004). As LPS is a natural ligand for TLR4, TLR4-MyD88 induced NF-κB activation or TLR4-TRIF induced IRF-3 activation may play an important role in HIF-1α production.

The two major transcription factors c-Myc and HIF-1α potentially responsible for activation-induced T cell metabolic transcriptome are the result of TCR-induced activation. c-Myc and HIF-1α are readily upregulated as soon as oxygen deprivation is initiated, and Myc deletion remarkably inhibited activation-induced glycolysis and glutaminolysis in T cells (Wang et al. 2011). Supposedly, Myc is involved in the stabilization of tumor suppressor p53 via inhibition of its negative regulator MDM2 in an ARF-dependent manner. In mouse embryo fibroblasts, Myc rapidly activates ARF and p53 gene expression to circumvent replicative crisis by induction of apoptosis (Zindy et al. 1998). Sensors such as ion channels mediate hyperventilation in carotid bodies and increase cardiac output to enhance glucose uptake and breathing rate to revert to normoxic conditions (Peng et al. 2011). Pioneering work emerged in the field of HIF1α and hypoxia from Gregg Semenza's group, indicating that hypoxia promotes the availability of heterodimeric (α/β) HIFs, which are stimulated by the EPO enhancer (Wang et al. 1995). This phenomenon is implicated in systemic in vivo response to hypoxia, where circulating concentrations of EPO increase greatly initially in response to anemic or hypoxemic stimulus to promote hematopoiesis (Jelkmann 2011). In a complex diabetic cardiomyopathy model induced by streptozotocin (STZ) in $HIF1\alpha^{-/-}$ homozygous mice, HIF-1α activated in T lymphocytes provides a protective function to cardiac tissue (Lin et al. 2016), indicating the complex nature of its operability under diverse cellular conditions. Dang et al. elegantly dissected the role of HIF-1α in maintaining the delicate balance between T_H17/Treg cells differentiation. ROR-γT, an obligatory transcription factor required for the differentiation of T_H17 cells, is directly activated by HIF-1α. A similar tertiary complex formation between ROR-γT and p300 is mediated by HIF-1α, regulating the translation of T_H17 signature genes. HIF-1α attenuates Treg development by binding Foxp3, a transcription factor required for naïve T cells differentiation to Treg cells (Dang et al. 2011). There are indications of depressed functions of cytotoxic T cell and blockade in B cell development in bone marrow in the case of HIF-1α absence (Kojima et al. 2003). Anoxia or hypoxia can induce $CD4^+$ T cells to convert to the Treg phenotype (Clambey et al. 2012; Ben-Shoshan et al. 2008; McNamee et al. 2013). HIF-1α is reported to be selectively expressed in differentiated T_H17 cells (Shi et al. 2011). HIF-1α is also implicated in maintaining Treg suppressive capacity and stability. HIF-1α knockdown could reverse the increased IFN-γ by VHL-deficient Treg cells and restore their immunosuppressive capacity in vivo (Lee et al. 2015).

During rapid proliferation, tumor cells extravagantly spend nutrients and self-limit O_2 availability. HIF-α subunits thereby stabilize as a result of diminished activity of PHD. HIF1-α/ARNT dimers (aryl hydrocarbon receptor nuclear translocator) potentially enhance glucose transport through LDH-A (Ryan et al. 1998). The HIF specifically blocks the access of glycolytic end products to mitochondria in a PDK1-dependent manner (Kim et al. 2006a, b). Disruption of pyruvate kinase M2 suppressed oncogenic mTOR-mediated tumorigenesis in kidney tumors. Synergistic targeting of glycolysis and mTOR blunted the proliferation and tumor development of mTOR hyperactive cells (Sun et al. 2011a, b). As discussed earlier, amino acid sensing 4E-BP1 hypophosphorylation is downstream of mTORC1 inhibition, resulting from AMPK activation under energy-depleted conditions (Liu et al. 2006) and HIF-induced activation of REDD1 (Brugarolas et al. 2004; Reiling and Hafen 2004). REDD1 and AMPK both inhibit mTORC1 activity via TSC2, but it is still unclear how HIF seems to integrate within this pathway (Fig. 3.6).

3.4 Signals from Immune Sensors: The Initiation of Metabolic Communication in Adaptive and Innate Immune Cells

The innate immune cells encompass those cells which are involved in initiating broad, non-specific immune responses against pathogens and serve as the first line of defense against invading microorganisms (Medzhitov and Janeway 2000). Crosstalks between innate immunity, inflammation, and metabolic changes reveal novel regulatory frameworks of the immune system. The role of TLRs, NOD-like receptors (NLRs), and RIG-like receptors (RLRs) in initiating immune responses is now established (Kawai and Akira 2009; Mogensen 2009; Brubaker et al. 2015). IL-1β seems to be a central regulator of proinflammatory responses exacerbating diabetes, multiple sclerosis, inflammatory bowel diseases, among other conditions (Netea et al. 2010; Masters et al. 2010; Wen et al. 2012). Metabolites such as NAD^+

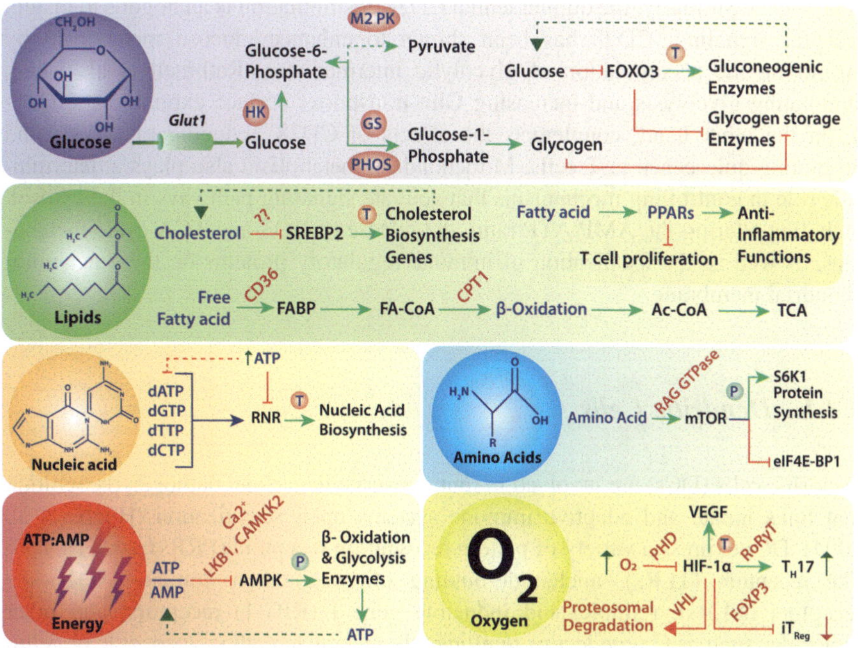

Fig. 3.6 The iterative theme of regulatory feedback and feed-forward motifs in the major metabolic pathways of immune cells, as discussed previously. (*Abbreviations: Please refer to the 'Abbreviations' section for details.)

(acting via deacetylases such as SIRT1 and SIRT2) (Kong et al. 2013) and succinate (regulating HIF-1α) (Misawa et al. 2013) are shown to regulate innate immunity. NLRP3 inflammasomes sense signals delivered by uric acid and cholesterol crystals and induce IL-1β production (Ghaemi-Oskouie and Shi 2011; Rajamäki et al. 2010; Duewell et al. 2010). A complex molecular platform comprising caspase-1, caspase-5, ASC (apoptosis-associated speck-like protein containing a CARD), and NLRP1 (NLR family, pyrin domain-containing 1), a macromolecular complex now recognized as NLRP1 inflammasome, is involved in innate immune response (Martinon et al. 2002; Mariathasan et al. 2004). The NLRP3 inflammasome is activated by LPS and ATP (Mariathasan et al. 2006), resulting in secretion of the interleukin-1 (IL-1) family of cytokines that mediate inflammatory abnormalities, such as gout and rheumatoid arthritis (Sims and Smith 2010). Inflammation has profound effects on aerobic glycolysis (the 'Warburg effect') and T cell activation at the inflamed foci, causing dramatic shifts in T cell metabolism (Wen et al. 2012). The metabolic status of the adaptive arm of the immune system is dictated by the type of invading pathogen, antigenic exposure and reinfection, infection severity, and nutritional status. It is well appreciated that stimulation of T cells via TCR requires proper engagement of the MHC–peptide complex on APC (signal 1), and an additional costimulatory signal (signal 2) is critical for perpetuating the downstream signaling and effector functions (Chauhan et al. 2017). Conversely, the absence of a costimulatory signal can lead to T cell anergy and subsequent deletion (Harding et al. 1992). Similarly, the quintessential CD28 costimulation is analogous to insulin receptor signaling. CD28 has been shown to enhance glucose metabolism by promoting the accumulation of glycolytic intermediates (Rathmell et al. 2003), stimulating glycolysis and increasing Glut transporter surface expression. CTLA-4, on the other hand, counteracts the effects of CD28, reducing glycolysis and promoting quiescence in T cells. Mitochondrial metabolism also plays a deterministic role in controlling mechanisms that activate signaling pathways in the cytosol, including altering the AMP:ATP ratio, the release of ROS and TCA cycle metabolites, as well as the localization of immune regulatory proteins on the outer mitochondrial membrane.

3.4.1 Dendritic Cells

Dendritic cells (DCs) are involved in immunosurveillance and pathogen recognition that links innate and adaptive immune systems upon sensitization (Hoebe et al. 2004). DCs express a variety of pattern recognition receptors (PRRs), such as Toll-like receptors (TLRs), nucleotide-binding oligomerization domain (NOD)-like receptors (NLRs), retinoic acid inducible gene I (RIG-1) receptors, and other molecules such as C-type lectins to efficiently recognize pathogen-specific or cellular damage-associated conserved motifs known as PAMPs and DAMPs, respectively (Hemmi and Akira 2005; Kawai and Akira 2011). These two factors play important roles in DCs maturation and acquisition of properties of an antigen-presenting cell.

The development of DCs from progenitor cells (Geissmann et al. 2010) is associated with PGC1α-driven mitochondrial biogenesis, which is controlled by transcription factors PPARγ, which negatively regulates DCs maturation and function (Klotz et al. 2007) and c-Myc. Mature DCs express L-Myc, which is required for optimal T cell priming (Wumesh et al. 2014).

Activation of DCs by antigens or by TLRs leads to increased glycolytic flux (Pearce and Everts 2015). Increased requirement for glucose in activated DCs is associated with increased synthesis of more fatty acids to support the expansion of the ER and Golgi for performing inflammatory activities (Everts et al. 2012; O'Neill 2014). As TLR signaling converges at the downstream kinase TANK-binding kinase 1 (TBK-1) and inhibitor of nuclear factor-κB kinase subunit-ε (IKKε), which phosphorylates Akt (Xie et al. 2011), the TLR-activated cells undergo Akt-mediated cellular growth and activation (Everts et al. 2014). After activation, DCs sustain the glycolytic pathway and decrease oxidative phosphorylation necessary for their sustenance (Everts et al. 2012). Because of increased Warburg metabolism in TLR-activated BMDCs, increased iNOS expression and NO production inhibit oxidative phosphorylation. Thus, a major function of increased glycolysis seems to be to an increased production of pyruvate to fuel the TCA cycle in activated DCs and to maintain cellular ATP levels when oxidative phosphorylation is inhibited. This metabolic switch, which is crucial for DC maturation, relates to the PI(3)K/Akt pathway and antagonism by energy sensor AMPK (Krawczyk et al. 2010). mTOR inhibition in TLR-activated DCs enhances their lifespan by suppressing NO production and by allowing DCs to use fatty acids or glucose as nutrients to fuel core metabolism and mitochondrial ATP generation (Amiel et al. 2014).

3.4.2 *Mast Cells*

Mast cells are critical in regulating the development of allergic diseases and in many immunoglobulin E-mediated immune responses. These cells are sensitized and activated by surface FcεR1–IgE–allergen cross-linking, TLR ligands, sphingosine-1-phosphate and its receptor S1PR, IL-3/IL-3R interaction, IL-33R/IL-33 interaction, and so on, to execute their effector functions, such as degranulation releasing histamine, mast cell-specific proteases chymase, tryptase, and other chemotactic factors (eosinophil chemotactic factors, ECF; neutrophil chemotactic factors, NCF), synthesis of eicosanoids, de novo synthesis of an array of cytokines such as TNF-α, IL-3, GM-CSF, and IL-5 (Stone et al. 2010; Galli et al. 2005; Sayed et al. 2008). In murine mast cells, inhibition of glycolysis by 2-deoxy-glucose impairs mast cells degranulation, suggesting that cellular ATP generation and histamine release in mast cells are dependent on glycolysis (Chakravarty 1968; Johansen 1979). Murine bone marrow-derived mast cells rapidly alter their metabolism in response to stimulation through the FcεRI, and the metabolic shift appears to be differentially required for the control of mast cell function (Phong et al. 2017). In humans and mice, the white adipose tissue-derived mast cells express minimal leptin, an adipokine that is primarily secreted by adipose tissue and, to some extent, also by mast cells (Taildeman et al. 2009). Leptin acts as a potent activator of immune cells (Fernández-Riejos et al. 2010). Mast cells are reported to be an important direct contributor in obesity-induced insulin resistance and type 2 diabetes (Liu et al. 2009). Leptin-deficient mast cells protect mice from obesity and diabetes by M2 type macrophage polarization and anti-inflammatory response, without affecting T cell differentiation (Zhou et al. 2015a, b). However, the mechanisms by which mast cells participate in metabolic diseases are yet to be fully uncovered. The importance of these innate cells in metabolic disorders is already realized. Effective mast cell inhibition-based medications, however, need to be therapeutically evaluated further in obesity and diabetes experimental models and further experimentation needs to be performed before offering optimism to patients with chronic metabolic inflammatory diseases.

3.4.3 *Natural Killer Cells*

Natural killer (NK) cells specialize in antibody-dependent cell-mediated cytotoxicity (ADCC) and proinflammatory cytokine secretion (Wherry et al. 1991; Seidel et al. 2013; Moretta and Moretta 2004) to selectively eliminate virus-infected cells (Topham and Hewitt 2009). Besides mediating synthesis and secretion of potent antiviral agents such as IFN-γ, these cells are also proactive against active oncogenic cells (Woo et al. 2015). Glycolytic restriction results in decreased NK cell effector function (Keppel et al. 2015). NK cells operate in nutrient-depleted, low oxygen conditions, such as tumor microenvironments; thus, hypoxic instructive signals induce the expression of HIF-1α in these cells (Balsamo et al. 2013). Intratumoral transfer of antisense HIF-1α reduces HIF-1α-dependent VEGF production and tumor angiogenesis but induces NK cell-dependent, and not T cell-mediated, rejection of tumors (Sun et al. 2001). NK cells proliferation is controlled by IL-2 and IL-15, which increase glycolysis and oxidative phosphorylation. Metabolic checkpoint kinase mTOR in NK cells was found to be activated by high IL-15 induction. mTOR stimulated the growth and nutrient uptake of NK cells and provided positive feedback for upregulation of IL-15R (Marçais et al. 2013). mTOR inhibition by rapamycin inhibited NK cell cytotoxicity in both mice and humans, justifying the immunosuppressive nature of rapamycin in the context of NK cells. TGF-β blocked IL-15-induced activation of mTOR in vitro. TGF-β and the mTOR inhibitor rapamycin both reduced the metabolic activity and proliferation of NK cells (Viel et al. 2016). IL-15 controls both the homeostasis and the peripheral activation of NK cells. The checkpoint kinase mTOR activates and boosts bioenergetic metabolism after exposure of NK cells to high concentrations of IL-15, whereas low doses of IL-15 triggered only phosphorylation of the transcription factor STAT5 (Marçais et al. 2014). IFN-γ enhances NK cell cytotoxicity by causing an increase in the levels of expression of TRAIL and Fas ligand. The production of IFN-γ, one of the NF-κB target genes in NK cells, was also found to be suppressed by adiponectin (Kim et al. 2006a, b). The CD56(dim) subset of human NK cells is more cytotoxic and a better IFN-γ producer than the CD56(bright) cells, which are metabolically more active than CD56(dim) cells (Keating et al. 2016).

3.4.4 Neutrophils

Neutrophils are innate immune cells that have ephemeral life span (4–6 days if not otherwise stimulated with GM-CSF [granulocyte macrophage colony-stimulating factor], which is known to delay the apoptosis in neutrophils under inflammatory pressure) (Brach et al. 1992), confront the pathogenic microorganism to clear them, kill phagocytosed parasites, and remove debris and apoptotic cells from the tissue cavity (Amulic et al. 2012). Sophisticated techniques like intra-vital live imaging and two-photon microscopy have allowed researchers to visualize neutrophils infiltration or recruitment in vivo, as these cells initially sense disruption of the tissue microenvironment and produce signals for immune activation. These cells arrive at the site of infection as early as 30 mins after pathogenic exposure (Peters et al. 2008). Neutrophils influence cellular processes like inflammatory responses, antiviral defense, hematopoiesis, fibrogenesis, and angiogenesis. Highly purified neutrophils in vitro are known to secrete IL-1β, IL-6, IL-8, and TNF-α (Cassatella 1999; Fujishima et al. 1993). The microbicidal action of neutrophils is attributed to autolytic release of chromatin exudates rich in histones and decorated with antimicrobial peptides, known as neutrophil extracellular traps (NETs) (Brinkmann et al. 2004) or to phagosome-mediated lysis of the target (Hampton et al. 1998; Van Kessel et al. 2014). NETs help in ensnaring and killing of trypanosomatid parasites, such as *Leishmania*, by the process known as NETosis (Guimarães-Costa et al. 2009). High glucose impedes and delays the NETosis response. IL-6 is a potent inducer of energy-dependent NET formation. Therefore, a hyperglycemic state in diabetic subjects mimic a state of a constitutively active proinflammatory condition that causes neutrophils to behave refractorily towards external stimuli, making diabetic subjects more susceptible and prone towards the acquisition of external infections due to such metabolic immunocompromise (Joshi et al. 2013).

Since harnessing the antimicrobial potential of neutrophils mandates a shift in their energy and metabolism, particularly glycolytic and glutaminolysis pathways, failure to achieve this might lead to the improper functioning of these cells, as observed in the case of diabetic state (Alba-Loureiro et al. 2007). Neutrophils have been reported to have little or no detectable oxidative phosphorylation (Kramer et al. 2014). Their mitochondria maintain a membrane potential (Δψm) which is usually generated by the respiratory chain complexes. Neutrophils can maintain Δψm via the glycerol-3-phosphate shuttle system, whereby their mitochondria play a supportive role in the regulation of aerobic glycolysis, rather than driving energy production by themselves (Van Raam et al. 2008). Neutrophils possess the HIF-1α and factor inhibiting HIF-1 (FIH) hydroxylase oxygen-sensing pathway. Using HIF-1$\alpha^{-/-}$

myeloid cells, it has been shown that this important transcription factor is directly involved in regulating neutrophil survival under hypoxic conditions (Walmsley et al. 2005). HIF-1α regulates the bactericidal capacity of myeloid cells greatly. HIF-1α regulates the critical effector molecules produced by neutrophils, which directs pathogen containment/elimination; some major ones are granule proteases, cathepsin G, TNF-α, nitric oxide, among others (Peyssonnaux et al. 2005). Collectively, this information indicates that metabolism acts as the spine of the innate immune system, reinforcing the structure of immunity and the synergistic regulation of glycolysis and transcription factor HIF-1α, along with various other mediators providing accessory supportive capacities to this metabolic structuring.

3.4.5 Macrophages

Tissue-resident macrophages or phagocytes are resultant cells of heterogeneous progeny and are not necessarily seeded by bone marrow-derived monocytes. Microglial cells are primarily derived from cells of the yolk sac (Ginhoux et al. 2010), whereas the skin-resident Langerhans cells originate predominantly from fetal liver, but with a minor contribution from the yolk sac (Hoeffel et al. 2012). Two reciprocally interacting subtypes of macrophages, classically activated macrophages (CAMs) and alternatively activated macrophages (AAMs), define macrophage dichotomy (Mills et al. 2000). Macrophages are the mediators of initial antipathogenic proinflammatory response by metabolic switching from oxidative phosphorylation to active glycolysis. Succinate dehydrogenase, the bifunctional enzyme of the TCA cycle and ETS, mediates metabolic repurposing of mitochondria to drive inflammatory macrophages (Mills et al. 2016). The withdrawal of citrate from the TCA cycle has been implicated in lipid biosynthesis in macrophages and DCs (Kelly and O'Neill 2015). In activated macrophages, Glut1-mediated glucose metabolism drives a proinflammatory phenotype, facilitating enhanced nucleotide flux through the pentose phosphate pathway, which fuels the production of purines and pyrimidines (Freemerman et al. 2014). NADPH generation via the PP pathway facilitates ROS generation and, thus, increases the antimicrobial potential of macrophages (West et al. 2011; Flannagan et al. 2016). Activation of macrophages can be achieved via a range of different stimuli, including LPS (Krawczyk et al. 2010), TLR3 synthetic ligand (poly I:C) (Reimer et al. 2008), and type 1 interferons (Uematsu and Akira 2007). Classically activated macrophages (M1 type) engage Warburg metabolism (Garedew et al. 2010). M1 macrophages promote inflammation, while alternatively activated (M2 type) macrophages are immunosuppressive by nature. It has been documented that polarization stimulus for M1 macrophages can be IFN-γ, LPS, or the presence of both (Gordon and Martinez 2010; Martinez and Gordon, 2014), while M1 macrophages express markers and products like TNFα, IL-1β, IL-6, IL-12, IL-23, CXCL10, pSTAT1, and MMP-9 (Jablonski et al. 2015). These cells express high iNOS activity and metabolism of arginine to NO (MacMicking et al. 1997), but M2 macrophages evolve under IL-4, IL-13, IL-10, glucocorticoids, and glucocorticoid derivatives, along with TGF-β (Gordon 2003). Major markers and products of M2 macrophages are IL-10, TGF-β, CCL17, CCL22, CD163, CD206, and pSTAT3/6 (Van Dyken and Locksley 2013), while arginine is metabolized to ornithine in these cells (Odegaard and Chawla 2011). Ornithine serves as a substrate for ornithine decarboxylase (ODC), a rate-limiting enzyme in the synthesis of polyamines, which are small, polycationic molecules regulating a multitude of cellular processes, such as DNA replication, protein translation, cell growth, and differentiation (Tabor and Tabor 1984). TSC1 is a critical regulator of the M1 and M2 phenotypes of macrophages. Molecular studies indicate that TSC1 inhibits M1 polarization by suppressing the Ras GTPase–Raf1–MEK–ERK pathway in an mTOR-independent manner, whereas TSC1 promotes M2 cells by the mTOR-dependent CCAAT/EBP-β pathways (Zhu et al. 2014).

M2 macrophages are polarized by the rewiring of Akt and IL-4 receptor signaling (Covarrubias et al. 2016) by two distinct pathways: (1) The Akt and mTOR pathway promotes the expression of enzyme Acly (ATP-citrate lyase) that supplies acetyl groups for epigenetic modifications of histones and mediates changes in the gene expression of those controlling M2 macrophage polarization, concurrently promoting glucose uptake, glycolysis, and production of Acetyl-CoA. (2) Increased fatty acid oxidation in M2 macrophages also lead to increased Acetyl-CoA turnover that feeds forward citrate conversion via the TCA cycle and exports to the cytoplasm. M2 macrophages indicate protective functions in insulin resistance (Sun et al. 2011a, b) but detrimental in parasitic infections (Sica et al. 2006; Mukhopadhyay et al. 2015), as they highly express the anti-inflammatory cytokine IL-10 and reduce the expression of NO and TNF-α, creating an anti-inflammatory milieu (Doyle et al. 1994).

CARKL (a carbohydrate kinase-like protein)-dependent metabolic reprogramming is required for proper M1- and M2-like macrophage polarization and uncovers a rate-limiting requirement for appropriate glucose flux in macrophage polarization. CARKL was identified via cellular kinome screening as a kinase that is rapidly downregulated (both in vivo and in vitro, in mice as well as humans) during LPS-driven M1 macrophage activation. Because M1 macrophages are highly dependent on glycolysis and PPP-derived NADPH for their effector needs, overexpression of CARKL metabolically drives M1 differentiation and oxidative phosphorylation, whereas an siRNA-mediated targeting of CARKL increases glycolysis but attenuates effector functions (Haschemi et al. 2012). M1 macrophages exhibit increased glycolysis, while oxidative phosphorylation promotes M2 polarization. Besides glycolysis, M1 macrophages also have higher levels of iNOS, HIF-1α, and u-PFK2 (Galván-Peña and O'Neill 2014). M2 macrophages have high AMPK and PFKB1 activity (Kelly and O'Neill 2015; Johnson et al. 2012). Arginase expression is increased in M2 macrophages, while iNOS expression is decreased (Morris 2007). Stimulation of macrophages with LPS resulted in increased IL-1β mRNA and HIF-1α expression, glycolysis, and TCA cycle intermediates accumulation, indicating a glutaminolysis-derived α-ketoglutarate, which leads to the accumulation of cytosolic succinate (Tannahill et al. 2013). Tumor-associated macrophages (TAMs) acquire angiogenesis and immunosuppressive properties detrimental for the host. TAMs upregulate the expression of REDD-1, a negative regulator of mTOR. REDD1 deletion enhances glucose uptake and glycolysis in TAMs experiencing hypoxia via mTOR (Wenes et al. 2016). TAMs show a 'skewed' inflammatory phenotype that promotes cancer-related inflammation (Coussens and Werb 2002; Mantovani et al. 2008) (Fig. 3.7).

3.5 Metabolic Pathways in Adaptive Immune Cells and Their Integrative Network

Research in the field of immunometabolomics is continuously providing new insights of intricate, vast, and interlinked networks of biochemical processes in immune cells at a particular point in time. These networks encompass the major receptors implicated in sensing nutrients in local extracellular milieu and invoke specific transcriptional and translational changes in immune cells. These metabolic receptors do not just sense nutrients but are also involved in directing the evolution of T cells into different subsets, such as T_H1, T_H2, Treg, and T_H17. Each of these subtypes of T cells is functionally distinct but they display overlapping cytokine signatures. The distinct cytokine produced by each of these subsets thus determines their dissimilarity from each other: predominantly, T_H1 cells produce IL-12, IFN-γ, and TNF; T_H2 cells produce IL-4 and IL-5. The T_H17 subset is now known to secrete cytokines IL-17, IL-21, IL-22, TNFα, and IL-6. The lineage commitment of activated T cells is determined by the integration of multiple environmental signals, some of which are provided by the external pathogenic microorganism, apoptotic cells, and nutrients perceptibly. Energy sensors or receptors such as nutrient-sensing serine–threonine kinases, AMPK, mTOR, and the reciprocal relationship between AMPK and mTOR mutually exert binary control in maintaining metabolically active (On) or quiescent (Off) states in a cell (Inoki et al. 2012).

3.5.1 Energy Sensor AMPK Is a Central Regulator of Maintaining Cellular Energetics

AMPK is now identified as a master regulator to modulate versatile cellular activities, (Hardie et al. 2003) while mTOR regulates cell growth by monitoring levels of amino acids and growth-stimulating signals. The cellular metabolism encompasses biosynthetic reactions of catabolic-dissimilatory and anabolic-assimilatory nature. It is the critical balance between these two polar processes that intricately automates cellular proliferation, differentiation, and the entire functional dynamics of a cell. AMPK is one of the most crucial cellular sensors, as it adjusts the energy status of cells (Rutter et al. 2003). It promotes ATP production by increasing the proteins involved in catabolic pathways, while conserving ATP by switching off anabolic/biosynthetic pathways (Fig. 3.8b). In vertebrates, AMPK also regulates metabolic energy homeostasis at the

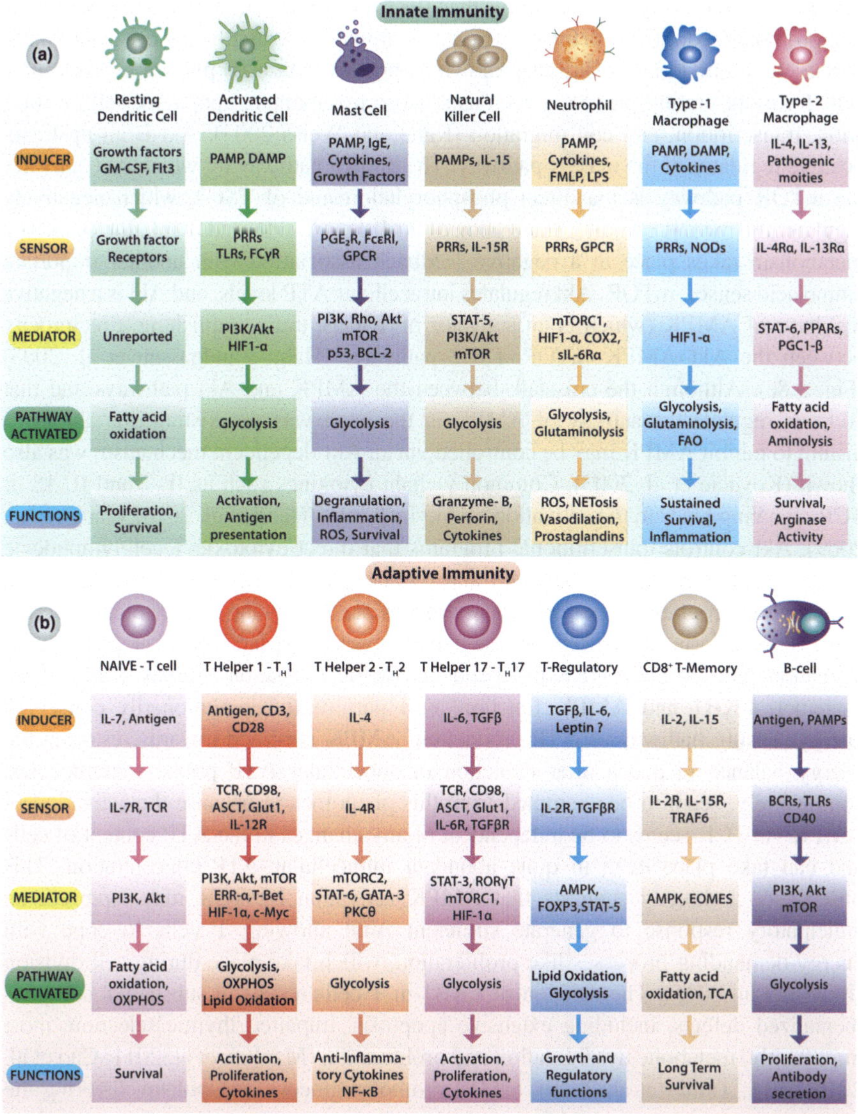

Fig. 3.7 From an immunological-metabolic standpoint, the activation of immune cells relies on four major constituents: (1) *Inducers* are the environmental cues relating to perturbations; (2) *Sensors* are transmembrane-spanning proteins composed of the extracellular sensor domain and the intracellular signal relay domain that acts as a scaffold for the recruitment of many other adaptor proteins. Immunological sensors are different from the metabolic sensors: the former senses pathogenic moieties, while the latter senses the nutrients and pathogenic byproducts; (3) *Mediators* are the scaffold, relay, or adaptor proteins that are recruited to the complex following inducer–sensor interaction and transduce the signals from sensors to its downstream targets; (4) *Effectors* are the molecules—transcription factors, enzymes of crucial biosynthetic pathways—that initiate the metabolic or the effector responses in immune cells. The signals transduced via inducer–sensor

systemic level; for instance, it controls the effects of agents acting on the hypothalamus that promote feeding and defines circadian rhythms of feeding behavior (Kola 2008). Serine–threonine kinase–Akt (also known as protein kinase B) plays a key role in T cells for orchestrating processes such as glucose metabolism, apoptosis, cell proliferation, transcription, and cell migration (Kane and Weiss 2003). Akt is an upstream positive regulator of the mTOR pathway. The known pathway by which Akt activates the mTOR pathway is via direct phosphorylation and of TSC2, which negatively regulates the mTOR complex. Activation of AMPK in T cells and regulation of energy metabolism takes place in a negative feedback association with another important amino acid sensor, mTOR. Akt regulates intracellular ATP levels, and Akt is a negative regulator of AMPK (which is an activator of TSC2), thus establishing a reciprocity between the Akt–AMPK–mTOR–TSC2 pathways (Hahn-Windgassen et al. 2005) (Fig. 3.8c). Although the crosstalk between the AMPK and Akt pathways and that Akt can regulate the activity of AMPK in the heart was preexisting, the ability of insulin to inhibit AMPK may be controlled via an Akt-dependent mechanism was also shown (Kovacic et al. 2003). Common γ-chain cytokines such as IL-2 and IL-15 or TCR signaling leads to the activation of Akt in T cells (Hand et al. 2010; Waugh et al. 2009). Akt controls transcriptional programs that direct cytotoxic T cell/lymphocyte (CTL) fate. Essentially, the strength and duration of Akt activity dictates the CTL transcription programs of perforin and cytokines like IFN-γ and also cytokine receptor expression perforin, and determines the fate of these cells (Macintyre et al. 2011). TCR activation promoted CAMKK-β-dependent AMPK activation (Tamás et al. 2006). Together, LKB1 and AMPK function as a part of the evolutionarily conserved energy-sensing pathway of ATP production. AMPK is biased towards restoring the energy balance in a cell after depletion of subcellular ATP pools. Lymphocytes, contrastingly, seem to be exempted from this phenomenon because the activation of AMPK via TCR seems to be independent of any changes in the ATP content of cells and can take place even in quite abundant intracellular ATP concentration. This observation indirectly suggests that AMPK activation by TCR may represent an anticipatory response to generate sufficient ATP allowing T cells to cope with energy-demanding processes like proliferation, which take place during cell division (Blagih et al. 2015). The deletion of LKB1 in T cells results in substantial compartmentalized defects, including extensive apoptosis, impaired thymic selection, most importantly, metabolic dysregulation and proliferation (MacIver et al. 2011; Cao et al. 2010, 2011; Tamás et al. 2010). mTOR promotes glucose metabolism, favoring the

Fig. 3.7 (continued) ligation results in the synthesis of transcription factors, transcriptional regulators, post-transcriptional and post-translational modifications, and also form the code of metabolic adaptation of immune cells. Such signals also affect allosterically regulated loops with metabolites and their cellular intake via respective sensors, regulating the stoichiometry of enzyme-catalyzed reactions involved in metabolic pathways and the rate-limiting steps catalyzed by crucial enzymes (Wang et al. 2011). (*Abbreviations: Please refer to the 'Abbreviations' section for details.)

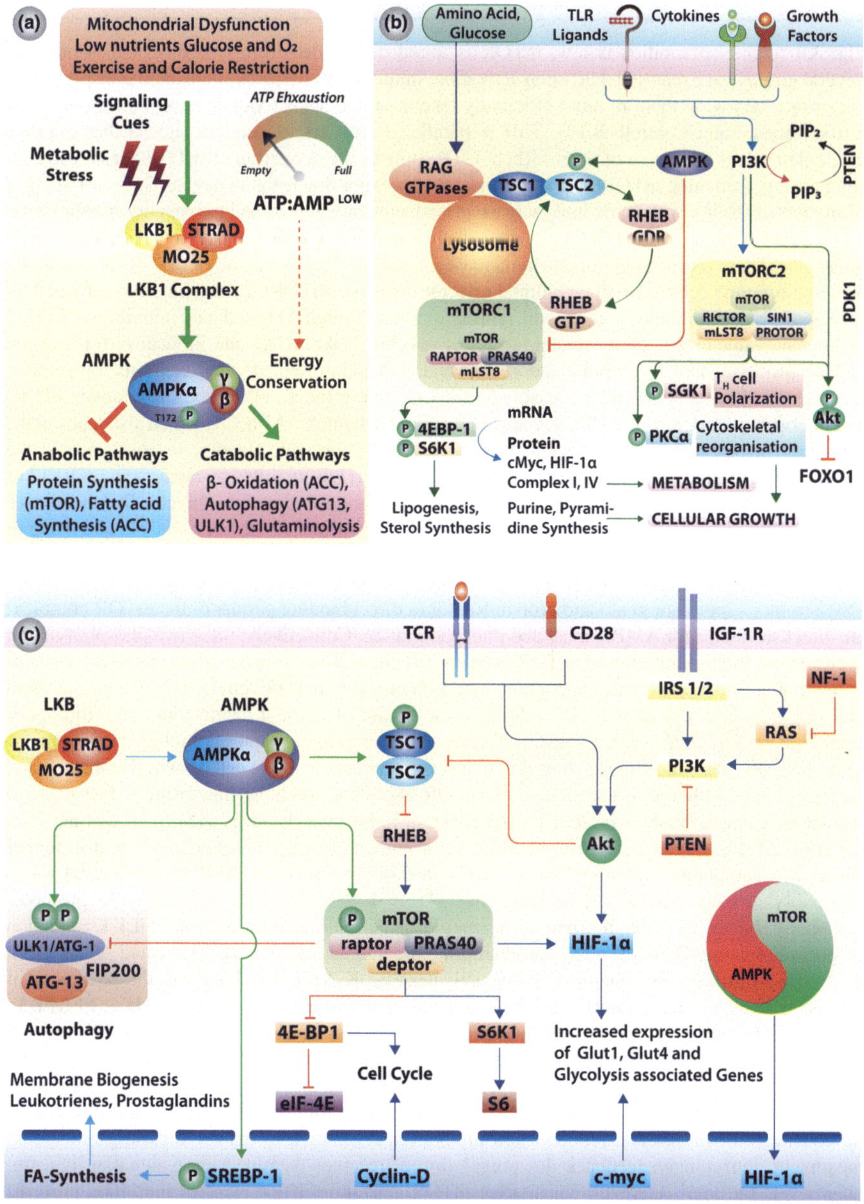

Fig. 3.8 (**a**) Cellular energy sensor AMPK responds to low ATP:AMP ratios. How do T cells detect and respond to the nutrients? When cellular energy levels are low, the AMPK becomes activated. Glucose or glutamine starvation in case of in vitro-cultured T cells exhibit the same type of effect. The general function of AMPK is energy conservation. This is achieved in two ways: (1) By switching off the anabolic/energy-consuming pathways; (2) Turning on catabolic pathways to generate energy. (**b**) The mTOR complex controls autonomous cell growth by integrating at least four major signals. Immune cells in quiescent states or when nutrients availability in extracellular

Fig. 3.8 (continued) milieu is low, mTOR is found dissociated from lysosome and is inactive. Several groups have reported that, upon activation, immune cells express amino acid transporters on the surface, allowing them to more efficiently take in amino acids (Nicklin et al. 2009; Carr et al. 2010; Ramsay and Cantrell 2015). This is paralleled with Akt-dependent signals that improve TSC2-dependent inhibition of Rheb. Rheb–GTP controls the activation of mTOR for its catalytic competency. Activated mTORC1 phosphorylates two major downstream targets, S6K and 4EBP-1 (Raught et al. 2001), with an overall increase in ribosome biogenesis, and polypeptide synthesis via the translation of important mRNAs coding for biosynthetic enzymes and immune system-related molecules. These reactions culminate in increased metabolic flux as a result of mTORC1 activation, enabling immune cells to produce optimal effector response. mTORC2, which is also activated by PI(3)K directly, contributes to metabolic reprogramming through Akt-mediated inhibition of FOXO to a considerable extent. Downstream targets of mTORC2 like SGK1 and PKC govern important cellular processes like T cell polarization, survival, and actin-oriented cytoskeletal rearrangement, respectively (Brunet et al. 2001; Larsson 2006). Ligation of the T cell receptor for antigen (TCR) and/or costimulatory receptor CD28 results in the rapid activation of PI(3)K, which also leads to the activation of Akt-dependent inducible transcription, cell survival, glucose metabolism, and the cellular translational machinery (Kane and Weiss 2003). mTOR inhibited by rapamycin FKBP12 or leucine starvation more potently leads to the disruption of the coordination of cyclin-D-dependent T cell cycle control in human embryonic kidney cells (Averous et al. 2008). (**c**) The Yin and Yang of cellular metabolism, the AMPK and mTOR pathways, antagonistically converge on a small number of rate-limiting substrates: AMPK regulates different aspects of cell growth and cell metabolism. TSC2 (tumor suppressor gene) and protein Raptor (a direct binding partner of the mTOR complex) are direct substrates for AMPK-mediated phosphorylation. Collectively, these two phosphorylation events result in the inhibition of mTOR activity. mTOR activation occurs only in the presence of nutrient abundance; in glucose- and amino acid-rich medium, mTOR activity is 'On' and cells will undergo growth and proliferate. In budding yeast in case of nutrient depravation, an orthologous version of AMPK-*Snf1* gets activated; it directly phosphorylates a Raptor homolog, a core component of mTOR, and immediately shuts down the central engine of entire cell growth (Hedbacker and Carlson 2008). Other direct effectors of mTOR signaling, such as transcription factor sterol regulatory element binding protein-1 (SREBP1), are phosphorylated by AMPK at distinct sites (Li et al. 2011a, b). A kinase from budding yeast known as Atg-1 (Autophagy-1) and in higher eukaryotes autophagy initiating kinase ULK-1 are direct targets of AMPK activity. ULK1 is phosphorylated by AMPK (at residues Ser-371 and Ser-777) and, thus, AMPK promotes autophagy; however, under nutrient sufficiency, high mTOR activity prevents ULK1 activation by phosphorylating ULK1 Ser-757 and disrupting the interaction between ULK1 and ATG1 (Kim et al. 2011). Insulin/growth factor signaling activates PI(3)K-AKT, which induces the expression of glucose transporters (e.g., Glut1) and the activation of glycolytic enzymes (e.g., HK2, PFKFB3). More so, IGF-1 receptor, upon stimulation with insulin and insulin-like growth factors, activates the phosphoinositide-3-kinase/Akt/mTOR pathway, causing redundant cellular effects like mTOR-dependent loss in insulin receptor substrate-1 expression, leading to feedback downregulation of signaling through the pathway. Mutational/oncogenic activation of PI(3)K/Akt kinase, a not so infrequent event in cancers, renders such tumors hypersensitive to mTOR inhibitors, including rapamycin. Intriguingly, feedback downregulation of receptor tyrosine kinase signaling is a frequent event in tumor cells with constitutive mTOR activation. IGF-I receptor inhibition prevents rapamycin-induced Akt activation and sensitizes tumor cells to the inhibition of mTOR (O'Reilly et al. 2006). PI(3)K-Akt signaling activates mTOR, which, in turn, activates the transcription factor HIF-1, which clinically correlates with HIF-1 and cooperates with other transcription factors or oncogenes, such as c-Myc, p53, or Oct4, to induce the expression of glycolytic genes, including Glut1, HK2, PFKFB3, LDHA, and suppressors of the TCA cycle, such as PDK (Cairns et al. 2011; Ward and Thompson 2012). (*Abbreviations: Please refer to the 'Abbreviations' section for details.)

development of effector T_H1 and T_H17 populations while inhibiting Treg generation (Sauer et al. 2008; Haxhinasto et al. 2008; Düvel et al. 2010). Mice with a T cell-specific disruption of mTOR do not support differentiation of T_H17 cells, which end up as Treg cells (Delgoffe et al. 2009). Macrolide rapamycin, a weak antibiotic but a strong irreversible inhibitor of mTOR signaling, promotes Treg development via fatty acid oxidation (Battaglia et al. 2005). Other reports also indicate the rapamycin–FKBP12 and cyclosporine A (CsA) potently inhibited the TGF-β and IL-6-induced generation of IL-17-producing cells. Rapamycin selectively binds to mTOR (Brown et al. 1994) and inhibits T_H17 cell differentiation but promotes immunosuppressive $FoxP3^+$ T regulatory cell differentiation (Kopf et al. 2007); a critical role of mTOR in T cell differentiation was deduced from this study. Rapamycin administration potentially induced memory T cell generation in the presence of antigen (Araki et al. 2009; Delgoffe et al. 2009). HIF-1α signaling seems to promote T_H17 cell development by enhancing glycolytic state and promotes the action of ROR-γt, an essential transcription factor for T_H17 cell development. HIF-1α also limits FoxP3 expression by targeting it for proteasomal degradation, thereby controlling the T_H17/Treg equilibrium (Dang et al. 2011). Considering the antagonistic influence of AMPK on the mTOR and HIF-1α signaling pathways (Bolster et al. 2002; Inoki et al. 2002; Faubert et al. 2013), and AMPK-driven shift towards FAO metabolism, it is rational to hypothesize that AMPK is an important node of metabolic circuitry acting as a checkpoint for regulating T effector/Treg plasticity (O'Neill and Hardie 2013). Through various genetics-involving knockout mouse models, it has now been shown that mTOR plays an important role in the regulation of T cell differentiation. The mTOR complex controls autonomous cell growth by integrating at least four growth regulatory inputs: (a) nutrient availability; (b) growth factor signaling; (c) cellular energy inputs; (d) cellular stress monitoring (Sengupta et al. 2010; Shimobayashi and Hall 2016). Tumor suppressor (TSC-1, TSC-2) functions as an integral complex that stringently regulates mTOR activity. TSC1 enforces a quiescent state in T cells to promote immune homeostasis and survival (Wu et al. 2011); therefore, it is but obvious that TSC would exert strong restraining signals on the mTOR complex to regulate crucial cellular processes like growth, cell cycle entry, and activation of metabolic machinery (Yang et al. 2011; O'Brien et al. 2011). Evidently, T_H1 and T_H17 cell development form T cells devoid of RHEB, and, hence, mTOR1 signaling, even when polarization signals were abundant. This failure of execution of the differentiation program in $RHEB^{-/-}$ T cells is associated with a decrease in STAT activation and decreased expression of lineage commitment transcription factors such as T_H1, T-bet, STAT-1, STAT-3, and T_H17-RORγT. Similarly, T cells lacking RICTOR and mTORC2 activity exhibit decreased STAT-6 activation, PKC activity, and GATA-3 expression, and, consequently, fail to direct towards T_H2 cells under polarizing conditions (Delgoffe et al. 2011). Alternatively, Boothby's group revealed that diminished T_H1 polarization can also occur in the absence of mTORC2 signaling (Lee et al. 2010). The cells lacking mTOR activity demonstrate increased SMAD3 activation and become $Foxp3^+$ Treg under normal conditions (Chi 2012). The PI(3)K/Akt/mTOR signaling network also crucially regulates Foxp3 expression and, thus, the development of Treg cells (Sauer et al. 2008).

The existence of redundancy, cross-regulation, feed-forward, and feedback motifs amongst metabolic pathways is a utilitarian method for an organism to drive metabolism in parallel and growth in order to accommodate the periods of feeding and starvation comfortably, and also to maintain proper cellular responses in homeostatic and perturbed states (Fig. 3.9). Some major examples of such interactions exhibiting cross-regulatory yet flexible nature of metabolic pathways are: (1) Growth factor signaling pathway activates mTOR through Akt-mediated inhibition of tumor suppressors TSC1/TSC2, the major repressors of mTOR signaling. Mechanistically, TSC2 is inactivated by Akt-dependent phosphorylation, which destabilizes TSC2 and disrupts its interaction with TSC1 (Inoki et al. 2002). (2) mTOR is directly inhibited by AMPK, via direct phosphorylation of Raptor on two well-conserved serine residues, and this phosphorylation induces 14-3-3 binding to Raptor. This crucial phosphorylation event of Raptor by AMPK causes mTORC1 inhibition and cell cycle arrest in energy-stressed conditions (Gwinn et al. 2008). (3) The energy-sensing AMPK-FOXO pathway mediates the lifespan extension induced by a novel method of dietary restriction in *Caenorhabditis elegans*. AMPK activates FOXO (Greer et al. 2007), which upregulates genes involved in the gluconeogenetic pathway, along with others, such as lipid metabolism and autophagy. (4) AMPK also inhibits glycogen synthase thereby leading to a decrease in glycogen synthesis (Wojtaszewski et al. 2002). (5) The activated monocytes express inducible 6-phosphofructo-2-kinase (iPFK-2), which synthesizes fructose 2,6-bisphosphate, a stimulator of glycolysis. Stimulation of glycolysis by hypoxia in activated monocytes requires the phosphorylation and activation of iPFK-2 by AMPK (Marsin et al. 2002). (6) Activation of AMPK by metformin or an adenosine analog suppresses the expression of SREBP-1, a key lipogenic transcription factor (Zhou et al. 2001). (7) AMPK directly phosphorylates SREBP-1 to inhibit its cleavage and translocation into the nucleus, thus repressing SREBP-1-mediated transcriptional activation of lipid synthesis in mammalian liver (Li et al. 2011a, b). (8) AMPK activation has also been reported to phosphorylate and inhibit HMG-CoA reductase activity, attenuating isoprenoid synthesis, and reduce cholesterol levels, thus imparting vaso- and cardioprotective effects (Henin et al. 1995). (9) Seminal reports from several groups (Egan et al. 2011; Kim et al. 2011; Lee et al. 2010; Shang et al. 2011) have shown that AMPK controls mammalian microautophagy. AMPK interacts and activates ULK1 protein kinase via its phosphorylation, a key initiator of the autophagic process in yeast (Alers et al. 2012). (10) mTOR can activate HIF-1α under normoxic conditions, promoting aerobic glycolysis and enhanced Warburg metabolism*, an effect predominantly observed in metabolically active macrophages (Sakamoto and Seiki 2009) and hyperactive cancerous cells (Harada et al. 2009; Hudson et al. 2002). (11) mTOR has also been highlighted in T cell activation and control of autophagy. Macroautophagy is a tightly regulated process in T cells and can be induced in response to TCR engagement to accommodate the bioenergetic requirements of activated T cells (Hubbard et al. 2010; Bronietzki et al. 2015). (12) A direct evidence of mTORC1 activating SREBP-1

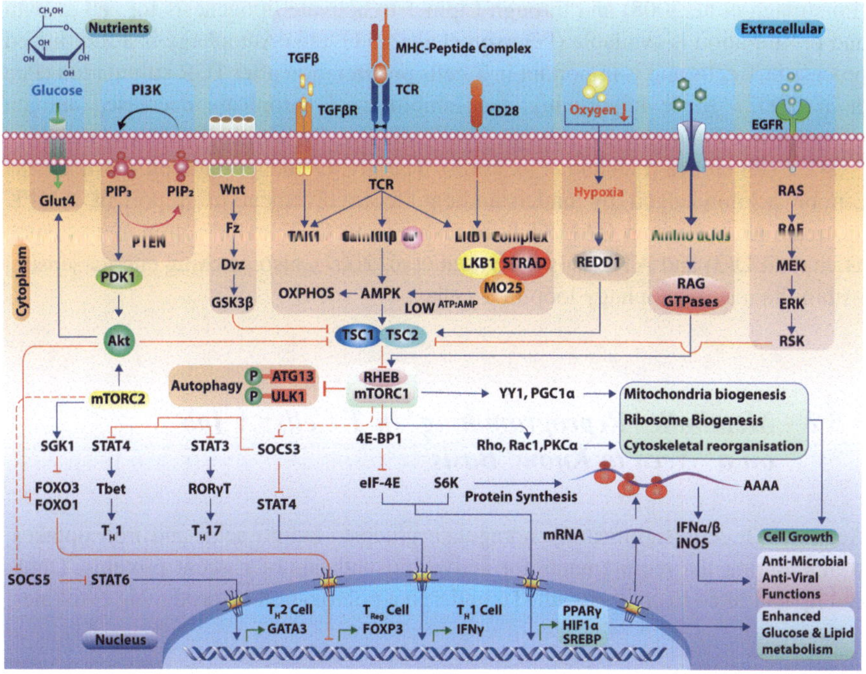

Fig. 3.9 A pathway map of metabolic reprogramming of T cells depicting AMPK and mTOR as key components. (**a**) Upper panel: mTOR as a metabolic sensor literally responds to cumulative stimulus provided by glucose, amino acids, growth factors, energy, and microorganisms to an immune cell. Immediately upstream of mTORC1 lies the Rheb. Rheb is a small GTPase that is essential for mTOR signaling (Avruch et al. 2009). Phosphorylation of TSC2 by membrane-localized Akt leads to dissociation of the TSC1–TSC2 complex and loss of GAP activity for Rheb-GTP. This leads to the accumulation of Rheb-GTP, which promotes mTORC1 activation. Insulin and insulin-like growth factors lead to the activation of PI(3)K that, via the production of PIP_3, leads to the activation of PDK1, which activates Akt and subsequent activation of mTORC1 by phosphorylation of TSC1 and TSC2. Signaling through mTORC1 versus mTORC2 selectively differentiates $CD4^+$ T cells into the T_H1 and T_H2 lineages, respectively (Lee et al. 2010; Delgoffe et al. 2011). CD28 is a potent activator of PI(3)K and plays an important role in the activation of mTORC1 in T lymphocytes. Upstream of the mTORC1 complex lies a heterotrimeric protein kinase (composed of α, β, γ subunits), AMPK. T cell receptor and CD28 provides signals to the LKB1 kinase complex comprising LKB1, STRAD, and MO25. LKB1 requires two adaptor proteins, STRAD and MO25, to be a functional kinase in the phosphorylation of AMPK (Hawley et al. 2003; Lizcano et al. 2004), which phosphorylates AMPK at the Thr-172 residue of the α subunit. LKB1 is activated when the cytoplasmic ATP:AMP ratio is low. Upstream kinases LKB1, CAMKKβ, and TAK1 promote AMPK phosphorylation at the Thr-172 residue in response to TCR-mediated $Ca2^+$ influx and signals from cytokine receptor TGF-βR and TGF-β interactions, respectively (Momcilovic et al. 2006; Warden et al. 2001; Hong et al. 2003). TRAF6 is also a positive regulator of AMPK in $CD8^+$ T cells and regulate AMPK via an unknown mechanism (not shown in the figure). Under homeostatic non-perturbed conditions, molecules that inhibit the mTORC1 complex maintain thymic and peripheral tolerance by preventing the T cells from undergoing any unnecessary fate commitment. Molecules like LKB1, PTEN, and TSC1–TSC2 exert enforcing control over T cell fate commitment via regulating mTOR complex activity. Alternatively, the pathways lead to the phosphorylation of TSC2:REDD1, in response to hypoxic conditions (Brugarolas et al. 2004)

(Porstmann et al. 2008) and through Lipin-1 to activate lipogenesis for cell growth and proliferation is available (Peterson et al. 2011). (13) Autophagy is a scrutinized process in T cells and is important in T cells proliferation after TCR stimulation (Pua et al. 2007). More importantly, the inhibition of autophagy decreased antigen recognition by endogenous Epstein–Barr virus nuclear antigen 1 (EBNA1)-specific $CD4^+$ T cell clones (Paludan et al. 2005). The involvement of autophagy in T cells can be acknowledged by understanding the involvement of important AMPK upstream kinases, such CAMKKβ, in response to $Ca2^+$ mobilization agents, such as vitamin D(3) and ATP (Høyer-Hansen et al. 2007), also patching energy sensing within the macroautophagy loop in T cells.

3.5.2 Metabolic 'Reprogramming' in T Cells: A Job on a 'Need to Know' Basis

Activated T cells significantly upregulate glucose, amino acid, and iron uptakes, compensating the requirements for active biosynthesis of nascent proteins, DNA, and lipids linked to cells growth. To undergo a polarization by producing cytokines,

---

Fig. 3.9 (continued) and GSK-3β, which is regulated by Wnt signaling and Akt (Inoki et al. 2006). Under amino acid sufficiency, Rag proteins bind to Raptor and promote the activation of mTORC1 via Rheb-GTP activity (Sancak et al. 2008; Long et al. 2005). Coinhibitory molecule PD-1 inhibits PI(3)K function, while positive costimulation TCR/CD28 signaling but not anti-CD28 alone (Sauer et al. 2008) and cytokine signaling are reported to enhance mTORC1 activation. The growth factors also seem to activate the mTORC1 complex. (**b**) Lower panel: downstream of mTORC1 and mTORC2 signaling. The activated mTORC1 complex phosphorylates S6K1, leading to the phosphorylation of ribosome-associated protein S6, which augments protein translation. The phosphorylation of 4E-BP1 by mTORC1 allows the release of eIF-4E to participate in mRNA-associated translation initiation complexes (Gingras et al. 2001; Zoncu et al. 2011). mTORC1 also enhances the gene expression of metabolic transcription factors important for vital cellular energetics and biosynthetic pathways, such as glucose and lipid metabolism, mitochondrial metabolism and biogenesis, ribosome biogenesis, and suppression of autophagy. In an immunological context, mTORC1 activity leads to the inhibition of SOCS3 and increased activation of STAT-3 and STAT-4. Consequently, this leads to increased expression of lineage commitment transcription factors of T-bet and RORγT, which promote the development of T_H1 and T_H17 cells, respectively. The counterpart of mTORC1–mTORC2 contains mTOR, mLST8, and mAVO3, but not Raptor, and is rapamycin–insensitive and seems to function upstream of Rho-GTPases to regulate the actin cytoskeletal rearrangement (Jacinto et al. 2004). mTORC2 is not upstream of the mTORC1 effector S6K. mTORC2 activity leads to the phosphorylation of Akt, resulting in its activation. This leads to cytoplasmic sequestration and phosphorylation of FOXO proteins (Dijkers et al. 2000), preventing FOXO proteins from activating the transcription of target genes like KLF2 (Ouyang et al. 2009), which is involved in chemokine CCR7 expression (Kerdiles et al. 2009). mTORC2 activity also inhibits SOCS5 and, thereby, enhances the STAT6 phosphorylation induced by IL-4 and subsequent expression of GATA-3 expression and development of T_H2 cells (Delgoffe et al. 2011). The mTOR2 complex can activate PKCθ, which indirectly promotes T_H2 cells differentiation (Chi 2012). (*Abbreviations: Please refer to the 'Abbreviations' section for details.)

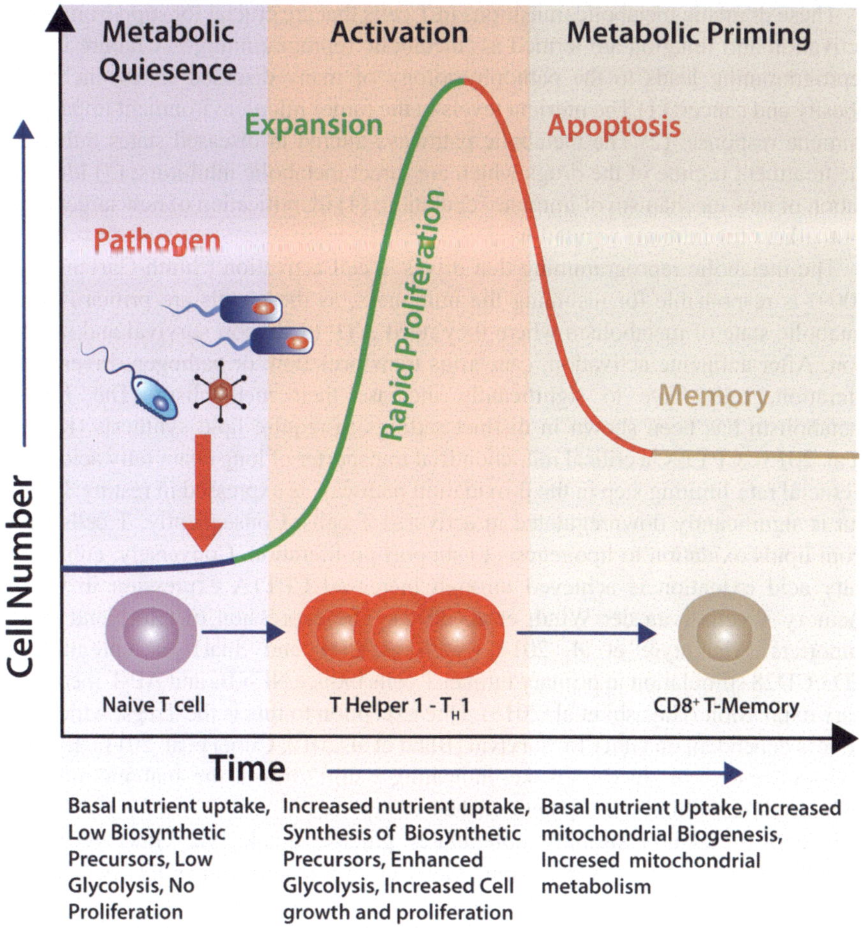

Fig. 3.10 ATP generation in resting lymphocytes facilitates their survival and migration. Following interaction with pathogens, signals transduced via different sensors/receptors initiate proliferation and differentiation programs. The activation is observed 6–10 h after antigenic exposure and cells divide after every 40–46 h. So, metabolic pathways are essential in dictating the antigen-specific expansion of T cells after pathogen exposure. Upsurge in supply:demand ratios of ATP and biosynthetic precursors for other pathways (e.g., ribose for nucleotide synthesis, lipids for membranes, and amino acids for protein synthesis) obligate T cells to a metabolic shift. Upon resolution of infection, the T cells stop proliferation, undergo apoptosis, and form memory cells; all these changes force T cells to adopt catabolic activities. In case a similar pathogen is encountered again, a rapid and robust immune response is initiated by the memory T cells as a counter-recall response. (Adapted form Pearce et al. 2013)

inflammatory mediators, or engaging foreign micro-entities, it is mandatory for immune cells to significantly rewire their metabolic network by developing exquisite mechanisms to sense energy and nutrient availability in the environment (Fig. 3.10).

These dramatic metabolic transitions in T cells that are crucial for supporting their activation and function are termed as 'metabolic reprogramming'. A failure in this reprogramming leads to the pathophysiology of many diseased states, including obesity and cancer: (1) The nutrient levels in the tumor microenvironment impact the immune response; (2) The metabolic pathways altered in diseased states influence the treatment regime of the drugs which are direct metabolic inhibitors; (3) Identification of new mechanism of immune regulation; (4) Identification of new targets and biomarkers for immune regulation.

The metabolic reprogramming that drives T cell activation (Smith-Garvin et al. 2009) is responsible for resuming the numbness, as these cells are primarily in a catabolic state of metabolism where they need ATP to support survival and migration. After antigenic activation, cancerous transformation, or pathogen-driven proliferation, cells have to significantly increase their metabolism. The T cell metabolism has been shown in distinct settings to require lipid synthesis (Kidani et al. 2013). CPT1A, a critical mitochondrial transporter of long chain fatty acids and a crucial rate-limiting step in the β-oxidation pathway, is expressed in resting T cells but is significantly downregulated in activated T cells. Consequently, T cells shift from lipids oxidation to lipogenesis to support proliferation. Conversely, enhanced fatty acid oxidation is achieved through increased CPT1A expression in $CD8^+$ memory T cells (van der Windt et al. 2012). The expression of Glut1 and other importers (Macintyre et al. 2014), such as Lat1 and Snat1, is upregulated. CD3/CD28 stimulation in primary human T cells induce NF-κB- and AP-1-mediated Lat1 expression (Hayashi et al. 2013). One exception to this is the Tregs, which are far less dependent on Glut1 for survival (Bhatt et al. 2012; Curiel et al. 2011). Lymph nodes were high in glucose uptake, indicating a shift towards the anabolic mode in cells residing within hematopoietic niches.

T cell stimulation strongly upregulates glucose uptake via Glut1 receptors (Macintyre et al. 2014), T cells require Glut1 upon activation, and Glut1 is essential for high rates of glucose uptake after activation to support proliferation. Naïve $CD4^+$ T cells bifurcate into effector T_H1 and T_H2 subsets under upon IFN-γ and IL-4 stimulation, respectively, while cytokines such as TGF-β and IL-6 promote the development of T_H17 subsets and TGF-β alone promotes the Treg phenotype in T cells (Zhu et al. 2009). Evidently, mass spectrometry studies have shown that each of these subsets is metabolically distinct. Broadly, T_H1 and T_H17 cells have comparably higher rates of glycolysis and glutaminolysis. Treg performs glycolysis at a lower rate and have higher mitochondrial metabolism (Fig. 3.11). This phenomenon has been elegantly demonstrated by Rathmell's group, where distinct T_H1, T_H2, T_H17, and Treg cells differentiated in vitro were treated with low doses of 2-DG (2-deoxy-glucose-hexokinase inhibitor) and rotenone (inhibitor of complex I of the electron transport chain). Interestingly, 2-DG did not seem to affect Treg differentiation, while proliferation of other subtypes T_H1, T_H2, and T_H17 was effectively dampened. Conversely, Treg proliferation was successfully dampened using rotenone (since the Treg have different metabolic demands), and T_H1, T_H2, and T_H17 proliferation was not affected (Cao et al. 2014). T cells require Glut1 to induce acute GvHD (graft versus host disease) rejection upon allogenic transplantation. In

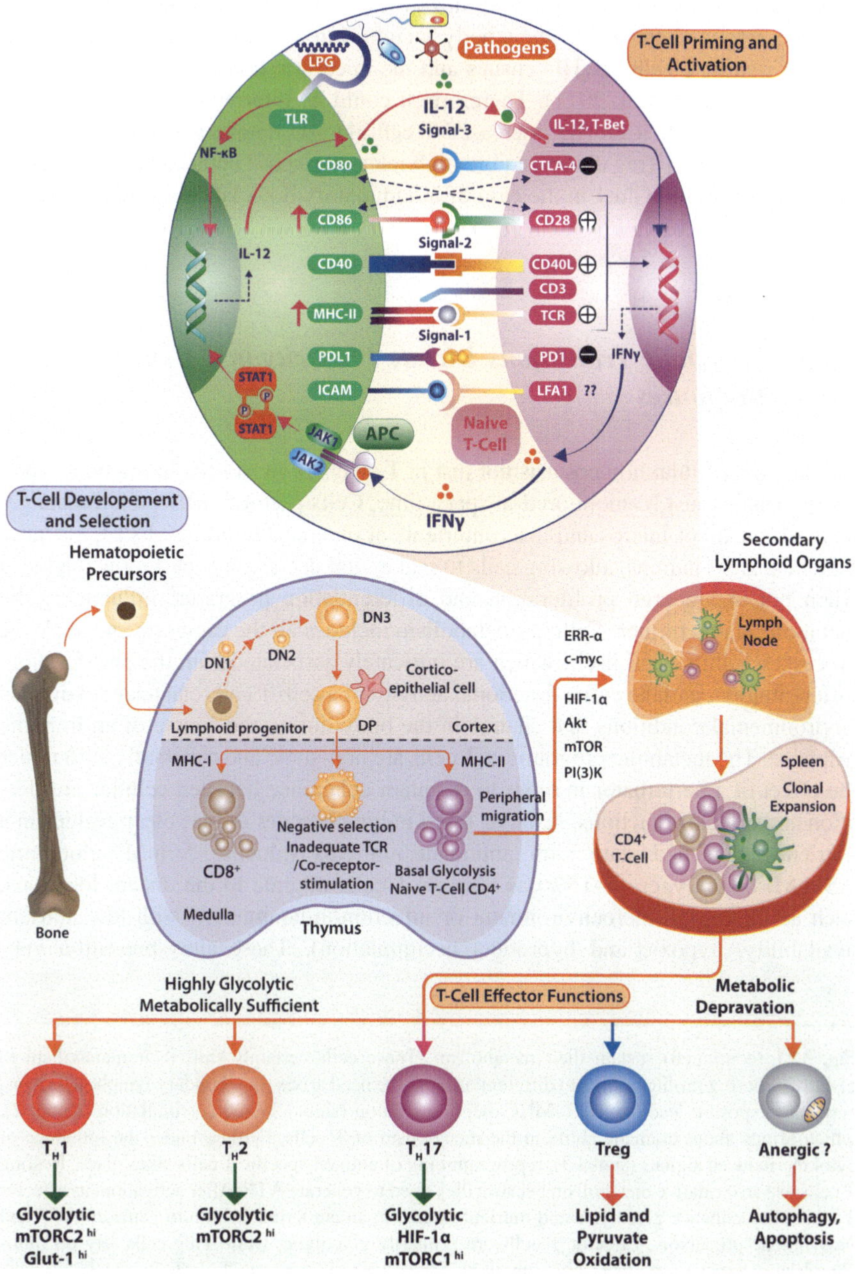

Fig. 3.11 A model for metabolic distinction in T effector subsets. T cells experience dynamic changes throughout their lifespan, in the context of their development surface receptor expression, migration receptors (chemokine and their receptors), lineage commitment, and antigen sensitization. Naïve T cells migrate from the thymus upon development, primarily relying on OXPHOS to

allogenic bone marrow transplantation, devoid of T cells, the mice will show no immune response and retain their body weight, but when transplanted with bone marrow rich in T cells, GvHD ensues and the mice lose body weight (Byersdorfer et al. 2013; Gatza et al. 2011). Therefore, it could be inferred that T effector cells require Glut1 to elicit GvHD, increased T cell glucose uptake is sufficient to elicit autoimmunity, and Treg do not require high rates of glucose uptake and can function to suppress without Glut1 in the case of inflammatory bowel disease (IBD) (Mayne and Williams 2013).

3.5.3 Metabolic Transitions Dictate Plasticity in T Cell Responses

To maintain cellular homeostasis not just in T cells but most cells in the body, their energy status must be monitored in space time. Cells perform active micromanagement of uptake of micro- and macronutrients or chemical building blocks, essential minerals, and vitamins, allowing cells to make vital decisions regarding when to, or when not to, undergo proliferation and differentiation programs, influencing the metabolic status further. Cellular metabolism includes all the catabolic and anabolic processes within a cell itself, which are intricately associated with the proliferation, differentiation, and the entire functional activity of a cell. T cells can adapt to diverse environmental conditions that change in the body during the course of an immune response. The metabolic programs in T cells are non-static and constantly shift under the effect of a perturbator in order to maintain coherence between cellular proliferation and effector functions. A naïve T cell in lymph nodes or the spleen resides in a nutrient-replenished state with abundant nutrients (glucose 5 mM, glutamine 0.5 mM) and oxygen (5–13%) as effector T cells migrate to the distant locations, such as the tumor microenvironment or infection/inflammation (with low nutrient availability, hypoxic and byproduct accumulation). These sites present a very

---

Fig. 3.11 (continued) sustain their metabolism. These cells certainly shift their metabolism to glycolysis during proliferation and differentiation as the need arises. In secondary lymphoid organs, antigenic exposure leads to TCR–MHC–peptide ligation (signal 1) and costimulation (signal 2), which brings about dramatic shifts in the metabolism of T cells. Further, under the influence of cytokine-induced signals (signal 3), reprogramming of antigen-specific T cells takes place. Resting T cells use an oxidative metabolism because they need to generate ATP. After activation, as effector T cells, they enhance glycolysis and nutrient uptake in an autocrine fashion to generate ATP and biosynthetic precursors. Effector T cells are primarily glycolytic, while Treg cells rely on mitochondrial pathways for suppressive functions. Glycolysis gives rise to T_H1, T_H2, and T_H17 cells, while robust fatty acid oxidation and catabolic cellular reactions give rise to Treg and, possibly, T memory cells. Treg cells adopt an entirely different program, because they do not require Glut1 and are able to survive on the mitochondrial oxidation pathway. Metabolic pathways, intermediates, are therefore attractive targets of cancer and transplantation immunotherapy. (*Abbreviations: Please refer to the 'Abbreviations' section for details.)

different environment than nutrient-replete zones within lymph nodes or the spleen (Pearce and Pearce 2013). Tumor microenvironment poses a very competitive atmosphere since the T cells compete with tumor cells directly for nutrient and oxygen availability, with tumor cells undergoing the Warburg effect and having elevated demands of glucose. Stable radioisotope tracer-assisted (SITA) metabolomics allied with GC-MS greatly adds to the existing knowledge of T cell metabolome (Buescher et al. 2015). T cells shift their metabolic program to mitochondrial oxidative phosphorylation when glucose is limited. Glucose and glutamine are two important fuel sources of T cells. Using SITA and metabolomic approaches, it has been shown that glucose is readily metabolized in T cells. Isotopic ^{13}C used for analyzing glucose to lactate conversion indicates that most of the lactate pools (~80%) come from ^{13}C labeled glucose, which indicates the high rate of Warburg metabolism in effector T cells. However, this case was not observed with the TCA cycle (which promotes ATP production by oxidative phosphorylation) metabolites, as TCA intermediates were not derived from ^{13}C labeled glucose (less than 20% of the fumarate and malate were derived from ^{13}C glucose). Instead, when the cells were fed with ^{13}C labeled glutamine, most metabolites of the TCA cycle were derived from ^{13}C labeled glutamine. This observation indicates that, for ATP production via the TCA cycle and oxidative phosphorylation, glutamine is more important than glucose in effector T cells. Glutamine depletion from the medium of cultured effector T cells reduces their ability to perform oxidative phosphorylation, while T cells increase glutamine metabolism under limiting glucose concentrations. This phenomenon relates to 'metabolic plasticity' that allows T cells to use different fuels, thereby enhancing their ability to adapt to diverse environmental conditions. Glutamine transporters such as Snat1 and Snat2 and the glutamine processing enzymes glutaminase (Gls), Got1 and Gpt1were found to be upregulated within a few hours after acute glucose starvation (Blagih et al. 2015; Pearce and Pearce 2013). Even the nutrient cycling in quiescent states of immune cells requires maintenance of baseline ATP levels and, to achieve this energy, sensing becomes vital by sensors such as AMPK. This mechanism is crucially important for the cells in order to enable them to sustain the homeostatic equilibrium if no perturbation is present or to rapidly transit from the inactive to active state in case any pathogen is encountered. Therefore, immune cells show a sensible flexibility in the execution of metabolic programs as per the requirement of body, depending upon the availability of glucose, glutamine, fatty acids, and, also, oxygen to utilize oxidative phosphorylation. In the case of the resting phase of lymphocytes, catabolic autophagy is employed to derive precursors for protein translation and energy production. Interestingly, the expression of transcription factors like Kruppel-like factor 2 (Klf-2) and FOXOs (both inhibited by mTORC2 activation) promotes the expression of inhibitory proteins and maintain the quiescent state in lymphocytes (Buckley et al. 2001; Hua and Thompson 2001). Other transcription factors, including Myc (Wang et al. 2011), estrogen-related receptor-α (ERRα) (Michalek et al. 2011a, b), and liver X receptors (LXRs) (Bensinger et al. 2008) facilitate T cell metabolic reprogramming by enhancing the expression of metabolic genes, which allows T cells to meet the bioenergetic and biosynthetic requirements of cell division and effector functions.

Consequently, upon activation, T cells switch to anabolic mode and activate the glycolytic pathway and oxidative phosphorylation to generate energy and create biosynthetic precursors. This metabolic shift is closely related to the immunological activation signals and costimulatory signals provided by TCR and costimulatory molecules like CD28. Diverse metabolic states in an immune cell are important for maintaining a conservatory balance between activation and dormancy in the biochemical flux of an immune cell in a time-dependent manner, which means that these cells will not be active all the time but can, and will, rapidly transit from an inactive to an active state to exert their effector functions in case a perturbation in immune homeostasis is encountered or in response to severe inflammation pathogenic invasion, oncogenic transformation, and autoimmunity. This involves T cell activation, via peptide–MHC–TCR co-ligation, stimulatory amplification signal, and, subsequently, clonal expansion/differentiation to an effector population having distinct cytokine, chemokine, and cytolytic molecular signatures. This process consumes energy in the form of ATP and NADH and, therefore, causes substantial upregulation in cellular energetics. It becomes necessary for a cell to rewire its intracellular biochemical circuitry to meet this increased energy expenditure and maintain cellular homeostasis simultaneously, without bringing about any disease like transformations. Naïve peripheral T cells are metabolically quiescent, with low cycling rates in both nutrient uptake and protein translation. It becomes important in this regard to understand that T cells migrate to relatively nutrient-depleted environments (e.g., tumor microenvironment and infection site) under chemotactic influence of cytokines and chemokines from nutrient-replete lymphoid niches. Thus, T cells must metabolically acclimatize to these antagonistic environments to perform or operate efficiently.

Insights into T lymphocytes function on the basis of their metabolic operability reveals that these cells highly are distinct from most other cells of the body when contextualized in terms of proliferative capacity. Upon activation, these cells are uniquely capable of adopting environmental cues and integrating these signals to reprogram their metabolism as per the requirements (Michalek and Rathmell 2010). For example, antigenic stimulation causes these cells to engage in Warburg metabolism for meeting increasing energy demands required for proliferation (Jones and Thompson 2007; Vander Heiden et al. 2009). Also, it is the choice of the selection of different metabolic pathways that dictate their effector phenotype, as explained in earlier sections. Briefly, effector T cells utilize aerobic glycolysis and have ephemeral lifespan, whereas long-lived memory T cells rely on mitochondrial fatty acid β-oxidation for their development and persistence (Vander Heiden et al. 2011). Moreover, this distinctive variation in the selection of metabolic pathways by these cells is not by their own choice to make; rather, these effector phenotypes evolve under the selection pressure exerted by perturbators (cumulative factors envisaging the presence of pathogenic moieties or PAMPs, cytokines working in tandem with pattern recognition receptors, the byproducts of microbial growth, nutrient- and oxygen-limiting conditions, immune complement system, and others) or environmental cues. Treg cells adopt oxidative phosphorylation and fatty acid β-oxidation for their development and proliferation (Michalek et al. 2011a, b), while

T_H17 cells are observed to require glycolysis (Shi et al. 2011). Nevertheless, the specific factors contributing to the attainment of metabolic distinction in immune cells under dynamic/perturbed versus homeostatic states add an element of mystery in immune cell immunometabolism and make it so fascinating to delve into deeply. According to Cohen et al., biological ageing correlates with progressive cellular loss and, in yeast, caloric restriction (CR) delays aging by activating the SIRT2 deacetylase (Cohen et al. 2004). Interestingly, in the case of 3T3-L1 cells, overexpression of SIRT2 inhibits differentiation, whereas reducing SIRT2 expression promotes adipogenesis (owing to increased FOXO1 acetylation leading to direction interaction between SIRT2 and FOXO1). These bimodal effects of SIRT2 are accompanied by corresponding changes in the expression of PPARγ, C/EBPα, and genes marking terminal adipocyte differentiation, including Glut4, aP2, and fatty acid synthase (Jing et al. 2007). It has also been reported that SIRT1 (a class III histone deacetylase) targeting promoted the expression of Foxp3, a key transcription factor in Treg cells, and increased Treg suppressive functions in vitro and in vivo (Beier et al. 2011). Similarly, genetic deletion of SIRT1 in DCs restrained the generation of Treg cells while promoting T_H1 cell development, resulting in enhanced proinflammatory response against microbial infections. Mechanistically Sirt1 signals through a HIF-1α-dependent pathway directing T_H1 and Treg generation reciprocally via DCs derived IL-12 and TGF-β production (Liu et al. 2015). Wnt signaling negatively modulates the crucial transcription factor Foxp3 required for Treg cell development. Therefore, Treg-mediated immunosuppression at the inflammatory loci is reduced due to increased Wnt production within this milieu (van Loosdregt et al. 2013).

van der Windt et al. showed that the level of $CD8^+$ T cells possesses high mitochondrial spare respiratory capacity (SRC) and IL-15 seems to be playing a critical role in promoting mitochondrial biogenesis and CPT1A expression, thus regulating SRC. The authors showed that SRC is dependent on fatty acid oxidation, which contributes to long-term survival and also their development of T memory cells. This observation strongly advocates that mitochondrial fatty acid β-oxidation supplies the fuel for the survival and maintenance of T memory cells (van der Windt et al. 2012). Interestingly, the differentiation of $CD8^+$ memory T cells depends on FOXO1 (Michelini et al. 2013; Kim et al. 2013a, b). However, more detailed studies are required to acquire precise insights into possibly many more factors contributing to the metabolic basis of long-term persistence of both central and peripheral T memory cells vis-a-vis their peripheral migration from lymph nodes upon antigenic restimulation (Sallusto et al. 1999). Cytokines like IL-2 and the costimulatory signaling molecules trigger the glycolytic pathway through the enhancement of nutrient transporter expression and the key regulator mTOR (Frauwirth et al. 2002; Jones and Thompson 2007; Wieman et al. 2007; Kolev et al. 2015). Critical intrinsic transcription factors such as HIF-1α, ERRα, and c-Myc, along with others, assists in the proliferation of T cells during the clonal expansion phase (Doedens et al. 2013).

Recently, in an interesting study, the non-redundant functions of Glut1 receptors in activated T cells have been described. $CD4^+$ T_H1 and T_H17 cells selectively require Glut1 in vivo to regulate immunological malignancies. In the absence of

Glut1, CD4$^+$ T cells fail to become activated; therefore, this study shows that glucose is a key metabolic substrate for T cells (Macintyre et al. 2014). Under glucose-limiting conditions, there is a loss of CD8$^+$ T cell functions, in particular IFN-γ production, mechanistically correlated with decreased phosphorylation of p70S6 kinase and eIF4E-BP1, which are required for the translation process (Cham and Gajewski 2005). Conversely, decreased diacylglycerol, an important regulator activating levels of Ras-GRP in cells, enhances ERK activation and augments CD8$^+$ T cell functional responses (Riese et al. 2011).

Besides glucose, glutamine is an essential metabolic substrate required for optimal T cell activation (Frauwirth et al. 2002). Fluctuations in the concentration of amino acid in the extracellular environment, as well as in the intracellular metabolic products, results in the alteration of T cell activity and polarization. Deletion of glutamine transporters leads to impairment in the differentiation of effector T cells; corroboratively, depletion of glutamine impedes cellular proliferation and cytokine production, which could be rescued by supplementing biosynthetic precursors of glutamine (Carr et al. 2010). In conjunction with this SLC7A5 (large neutral amino acid transporter small subunit-1), null T cells were unable to undergo metabolic reprogramming, also nullifying their capacity to undergo clonal expansion (Sinclair et al. 2013). Intracellular leucine concentration can also regulate metabolic reprogramming in the activation of T cells (Ananieva et al. 2014). Again, the expression of alanine serine and cysteine transporter system (ASCT2), another transporter of glutamine, becomes upregulated after T cell activation (Levring et al. 2012). The loss of ASCT2 renders activation impaired in CD4$^+$ T cells (Nakaya et al. 2014). Depletion of arginine in the extracellular environment is associated with impaired T cell proliferation and aerobic glycolysis, but not mitochondrial oxidative phosphorylation (Fletcher et al. 2015). It has been well documented that depletion of tryptophan in the cellular microenvironment inhibits proliferation of the effector T cell population and induces anergy (Munn et al. 2002; Uyttenhove et al. 2003). A recent study also revealed that, in addition to glucose and glutamine, ROS also play a critical role in T cell activation (Murphy and Siegel 2013; Sena et al. 2013). Mice with reduction in mitochondrial electron transport chain complex III showed reduced clonal expansion of T cells due to the lack of ROS generation, which was crucial for the activation of NFAT and production of IL-2 (Chow et al. 1999).

3.5.4 Role of Transcriptional Regulators in Shaping T Cells Metabolism

An increasing body of evidence shows that several transcription factors and signaling pathways coordinate the metabolic programming in T cells after activation. Transcription factors are the proteins that bind to specific DNA sequences in order to regulate the expression of certain genes. Acute metabolic programs are achieved by post-translational modifications of enzymes. The long-term metabolic programming of immune cells requires a much more complex system including activation of transcriptional regulators, along with coactivators that facilitate the necessary changes in gene expression which stabilize the changeable metabolomics during perturbed states (Ganeshan and Chawla 2014). In the light of rejuvenated interest of the Warburg phenomenon in cancer cells, it is reported that not just increased glycolysis cancerous cells depend upon optimal mitochondrial function for metabolic sustenance. Specifically, glutaminolysis plays an indispensable role in generating ATP and lactate in these cells besides glycolysis; therefore, glutamine intake and its cataplerotic conversion to generate ATP, nitrogen for biosynthesis, and carbon substrates for fueling cells' anabolic demands, as it turns out, is crucial for cancerous cell survival and proliferation. Gao et al. reported that an oncogenic transcription factor c-Myc regulates the miRNA expression transcriptionally (Gao et al. 2009). In T cells, c-Myc is induced upon activation and promotes T cell proliferation (Wang et al. 2011). The in silico analysis of gene expression reveals that overexpression of c-Myc is a characteristic feature of B cells (and, thus, the Myc induced downstream, gene expression) and underwent oncogenic transformation (Schuhmacher et al. 2001). c-Myc transcriptionally represses miR-23A and miR-23b to enhance the gene expression of mitochondrial glutaminase. In human B cell lymphoma cells P-493, this leads to upregulation of glutamine catabolism, thus establishing a critical link between c-Myc-controlled regulation of glutaminase activity (Gao et al. 2009). Thus, Myc regulates a transcriptional program that stimulates mitochondrial glutaminolysis and leads to glutamine dependence. c-Myc-dependent glutaminolysis is the reprogramming of mitochondrial metabolism that depends upon glutamine catabolism to sustain cellular viability and contribute to anapleorotic reaction of the TCA cycle via its reentry into TCA as an intermediary substrate α-ketoglutarate (Wise et al. 2008).

Receptors of the estrogen-related family are also implicated in directing a critical transitioning during development. The *Drosophila* ortholog of estrogen-related receptor (ERR) is known as dERR. Interestingly, a study shows that dERR plays a key role in driving critical switching in the gene expression program during *Drosophila* embryogenesis. ERRs acts as key regulators of energy metabolism by promoting mitochondrial biogenesis, fatty acid oxidation, and oxidative phosphorylation. In particular in this study, the role of dERR in orchestrating normal growth and development via carbohydrate metabolism was highlighted (Tennessen et al. 2011). ERRs act as cofactors that stimulate enhanced HIF-dependent transcription of glycolytic genes under hypoxia. It is now known that ERR also interacts with oncoprotein Myc, and these two transcription factors synergistically activate the transcription of glycolytic genes. In cancerous cells, depletion of ERR is detrimental for cell growth and aerobic glycolytic commitment; thus, it can be a potential target for anticancer therapy (Cai et al. 2013). The role of ERRα as a functional regulator was elucidated with mice lacking ERRα (ERRα$^{-/-}$ phenotype). In silico DNA microarray analysis of adipose tissue in these animals shows altered expression of genes regulating adipogenesis (particularly that of lipid, eicosanoid, and steroid synthesis) and energy metabolism. However, no lethality or gross anatomical alterations were observed in these animals besides reduced weight loss and peripheral fat deposits (Luo et al. 2003). Quiescent $CD4^+$ T cells express low levels of ERRα protein that subsequently upregulated upon activation. $ERR\alpha^{-/-}$ T cells and T cells treated with shRNA displayed suppressed upregulation of Glut1 protein glucose uptake and metabolism, indicating its broad effect but not specific effects on metabolism essential for T effector functions. Furthermore, in vivo inhibition of ERRα reduced T cells proliferation and T effector cells formation in both immunization and EAE models (Michalek et al. 2011a, b).

Through many reports, it is now known that invasive cancer cells use the transcriptional coactivator peroxisome proliferator-activated receptor gamma, coactivator 1 alpha (PGC-1α) to enhance oxidative phosphorylation and mitochondrial biogenesis and the oxygen consumption rate (LeBleu et al. 2014). Conversely, T cells infiltrating tumors (TILs) show a decrease in mitochondrial function and mass, loss of oxidative metabolism specific to activation in the tumor microenvironment, and, also, an additive effect of strong acidosis in the tumor microenvironment contributes to T cell dysfunction (Fischer et al. 2007; Mendler et al. 2012). T cell mitochondrial biogenesis is repressed via Akt-mediated inhibition of PGC-1α and subsequent reinforced expression of PGC-1α results in the augmentation of antitumor T cell function (Scharping et al. 2016). The T cell metabolism is greatly influenced by operational lipid pathways (Lochner et al. 2015). In particular, highlighting the crucial roles of de novo fatty acid synthesis is necessary for optimal functions of T effector cells, dependence of $CD8^+$ T memory cells on FA synthesis, and oxidation for their development (O'Sullivan et al. 2014), and distinct glycolytic and lipid oxidative metabolic programs are essential for $CD4^+$ effector and Treg subsets (Michalek et al. 2011a, b). Cellular growth requires a shift from fatty acid oxidation pathways towards lipogenesis pathways. As evident from a huge number of reports on cancer biology, cancer cells require fatty acids to replicate membranes like others and Akt is known to upregulate fatty acid synthesis via an oxygen-consuming enzyme stearoyl-CoA desaturase (Kamphorst et al. 2013). Cancer cells differentially activate and thrive on de novo lipid synthesis pathways in a low-lipid environment (Daniëls et al. 2014). Adoption of this lipogenic mode is conducive for cell division and is nearly applicable to both malignant transformed cells versus normal cells in the same manner. Ligand-dependent receptors, such as liver X receptors (LXRs) belonging to the nuclear receptor family, are of particular interest in regards to the immunometabolism of T cells, since these receptors are ligand (endogenous oxysterols)-activated and, therefore, druggable. This can serve as a basis for the development of novel therapeutic molecules targeting LXRs, thus, simultaneously inhibiting crucial cellular processes like cholesterol transport, glucose metabolism, and modulation of inflammatory responses. An important study based on the aforementioned theme justifies the importance of LXRs-based targeting via specific antagonist suppressed IL-17A expression from splenocytes derived from Vα2.3/Vβ8.2 TCR transgenic mice and from MOG (35–55)-immunized C57BL/6 mice, since cytokines IL-21 and IL-22 are produced by T_H17 cells and modulate immune responses. LXR antagonist suppressed MOG (35–55)-induced expression of IL-21 and IL-22 mRNA in splenocytes, which collectively provides an effective treatment strategy for multiple sclerosis (Xu et al. 2009) LXR activation interferes with IL-2- and IL-7-induced proliferation and cell cycle progression of human T cells, suggesting that LXR is a potent modulator of cytokine-dependent proliferation and survival of normal and transformed T and B lymphocytes (Geyeregger et al. 2009). LXR, along with RXRs, are also documented to inhibit apoptosis in murine macrophages upon activation and provide protection from infections with *Bacillus anthracis*, *Escherichia coli*, and *Salmonella typhimurium* (Valledor et al. 2004). Another interesting study reported that LXRs inhibit basal cytokine-inducible expression of macrophage matrix metalloproteinase-9 (MMP 9), since MMPs are implicated in extracellular matrix remodeling, and inappropriate expression of MMPs contribute to vascular pathological conditions, including atherosclerotic lesions. The treatment of murine peritoneal macrophages by LXR antagonists such as GW3965 or T1317 reduces MMP-9 expression at the transcriptional level and diminishes its induction in response to activation by lipopolysaccharides, IL-1β and TNF-α (Castrillo et al. 2003). LXR being an intracellular sterol sensor also links to SREPB transcription factors to maintain cellular cholesterol levels in dividing T cells. SULT2B1 is an oxysterol-metabolizing enzyme, which is upregulated in activated T cells. SULT2B1 activation leads to the suppression of LXR pathways for cholesterol transport and also encourages SREBP pathway-mediated cholesterol synthesis (Bensinger et al. 2008).

3.6 Conclusion

The mainstay of cellular metabolism is glycolysis. The TCA cycle provides the burning of 2-carbon units arising from glycolysis and fatty acid oxidation and of the intermediates originating from amino acids. While these reactions represent the catabolic processes to derive energy, the biosynthetic pathways represent the restoration mechanisms. When perturbed by pathogens, tumors, autoantigens, and transplant antigens, various cells of the immune system follow these same processes to not only respond to the perturbations but also restore normalcy. Studying these metabolic pathways can reveal a novel threshold-based shift in the dynamic network of metabolic processes. Identification of a key node in these networks may lead to therapeutic interventions to many diseases.

Funding The work has been recommended by Infect-ERA (Project INLEISH) and was funded by the Department of Biotechnology, New Delhi, Government of India. The grant number is BT/IN/Infect-ERA/33/BS/2016-17.

Glossary

Anergy T cells anergy is a tolerance mechanism by which lymphocytes become intrinsically refractory to antigenic sensitization, but do not undergo apoptosis readily. Instead, they exhibit loss in functionality and a state of hyporesponsiveness in their effector functions. Anergy affects both $CD4^+$ and $CD8^+$ lymphocytes, and two major mechanisms underlying this obscure phenomenon are documented: (1) Clonal anergy, which is predominantly characterized as growth arrest. This type of anergy arises from incomplete T cell activation and is observed in T cells which have undergone prior activation. Mechanistically, blockade in the Ras/MAPK pathway might be responsible for such observed numbness, and, evidently, it has been suggested that IL-2 or anti-OX40 signaling could reverse such a phenotype. (2) Adaptive tolerance, which represents a more generalized inhibition of proliferation and effector functions. This type of anergy is initiated in naïve T cell in vivo by an inadequate environmental stimulus of deficit costimulatory signals provided by CD28 or dominant inhibitory signals from coinhibitory molecules such as PD-1 and PD-L1 or even both. It has been shown that blockade of the interaction between PD-1 and its ligands PD-L1 and PD-L2 prevents anergy induction in iNKT cells. Adaptive tolerance can be induced in the thymus as well as in peripheral lymphoid organs. When persistent antigen is encountered, T cells block receptor tyrosine kinase activity, which ablates calcium mobilization, and also another independent mechanism involving the IL-2 signaling cascade.

First responders In the metazoan body, the immune system carries out specialized functions that coordinate several responses when encounters with a foreign microorganism/their derivatives take place. A myriad of such responses includes the induction of an inflammatory environment and coordinated execution of genes responsible for the activation and mobilization of cells of the immune system. Such highly tailored responses to external stimulus are mediated by, neutrophils, dendritic cells, macrophages, T cells, and B cells, along with other immune cells. These immune cells experiences dramatic shifts in their metabolic profiles upon activation, making them adequate to combat the infection directly (by synthesizing antimicrobial peptides, nitric oxide, and ROS) and indirectly (pleiotropic factors such as chemokines and cytokines, which attract and activate cells of the adaptive immune system), and, thus, making them first responders to any exogenous pathogenic stimulus.

Glycolysis The cellular economy is monetized by adenosine triphosphate (ATP). ATP is generated in a cell by two major pathways, glycolysis and oxidative phosphorylation. Depending upon the requirement of a cell, metabolism can shift towards the anabolic module (the de novo or salvage construction of molecules required for growth and biomass production) or the catabolic module (the breakdown of macromolecules into yet smaller molecules for the production of energy). The glycolytic pathway converts glucose into pyruvate via a series of intermediate metabolites that can enter other pathways (such as the pentose phosphate pathway PPP) and contribute to biosynthesis and cell growth. Glycolysis also leads to the conversion of two molecules of coenzyme NAD^+ to NADH, and the overall process can be summarized as:

$$Glucose + 2NAD^+ + 2ADP + 2P_i \rightarrow Pyruvate + 2NADH + 2ATP + 2H^+ + 2H_2O$$

Histone deacetylases Histone tails are positively charged, owing to amine groups present on their lysine and arginine AA residues. These positive charges help the histone tails to interact with and bind to the negatively charged phosphate groups on the DNA backbone. HDACs are a class of enzymes that remove acetyl groups (O—C CH3) from an ε-N-acetyl lysine amino acid on a histone, allowing the histones to wrap the DNA more tightly. This decreased binding allows chromatin expansion, permitting genetic transcription to take place.

Insulin resistance Insulin is secreted by β cells in the pancreas contained within islets clusters. Insulin is synthesized and secreted by these cells. The blood sugar level spikes after meals; in response, this provides a neuroendocrine stimulus to the pancreas (by the hypothalamus) to secrete insulin into the blood. There, insulin helps muscle, fat, and the liver absorb glucose from the bloodstream, lowering blood glucose levels. Insulin stimulates hepatocytes and muscle cells to store excess glucose in a process known as gluconeogenesis. Insulin also provides a positive feedback to hepatocytes to reduce glucose production in order to maintain glucose homeostasis in the body. In case of insulin resistance, muscle, adipose cells, and hepatocytes do not respond optimally to insulin and, thus, fail to absorb glucose from the bloodstream. The reasons for insulin resistance can be numerous, such as high fat intake causes obesity and obesity-induced inflammation, in case of cardiovascular diseases, insufficient physical exercise, excess FA-induced mitochondrial dysfunction, and many other factors, such as lifestyle, etc. As a consequence of this, β cells in the pancreas try to meet the elevated demand for insulin by producing more insulin but, consequentially over time, insulin resistance can lead to type 2 diabetes because, ultimately, β cells fail to supply enough insulin to the body. As a result, blood glucose levels build up, leading to serious metabolic syndromes.

Metabolic reprogramming As the concept of metabolic regulation of the immune system gained wider acceptance over years, the importance of bidirectional regulation of the immune system and their effector functions dictated by their metabolic status became clearer. Activated T cells significantly upregulate glucose, amino acid, and iron uptakes, compensating the requirements for active biosynthesis of nascent proteins, DNA, and lipids that are linked to proliferation and growth. To undergo a polarization under signals like cytokines, inflammatory mediators, or direct physical engagement via PAMP-TLR or the MHC–Peptide–TCR complex with pathogenic moieties, it is mandatory for immune cells to rewire their metabolic network. These dramatic metabolic transitions in T cells that are crucial for supporting their activation and function are termed as 'metabolic reprogramming'. The effects of immune cells include cytokine activation of cells such as adipocytes and hepatocytes, which are the central hub of mediating systemic metabolic changes (for example, in metabolic disorders such as type

2 diabetes and obesity-induced inflammation). In addition, the concept of reciprocal regulation of immune cells by host-commensal microbiota-derived metabolites and nutritive status of the body influencing the immune cells is now very well documented. Such multifaceted chemical interactions between different cells of the body with lymphocytes and macrophages eventually shape the immune responses significantly.

Oxidative phosphorylation Glycolysis and the TCA cycle help maintain the NAD^+ –NADH redox balance in a cell and the NADH and $FADH_2$ molecules are used to donate electrons to the electron transport chain on the inner membrane of a mitochondrion for oxidative phosphorylation. For each molecule of $FADH_2$ or NADH oxidized, two or three molecules of ATP, respectively, are generated. Redox reactions in the oxidative phosphorylation sequentially transfer electrons to generate a proton (H^+) gradient across the inner membrane of mitochondria, which drives the synthesis of ATP. Although immune cells utilize other metabolic pathways such as the pentose phosphate pathway, glutaminolysis, and fatty acid oxidation, the common molecule integrating these pathways is glucose.

Pentose phosphate pathway Generates ribose for nucleotide synthesis. During this process, $NADH^+$ is reduced to NADPH, forming the cofactor for ROS generation via the NADPH oxidase (NPhox) system in phagocytic cells.

Primary metabolites A primary metabolite is a kind of metabolite that is directly involved in normal growth, development, and reproduction. It usually performs a physiological function in the organism (i.e., an intrinsic function). A primary metabolite is typically present in many organisms or cells. Major examples include amino acids (AA), alcohols, vitamins, polyols, organic acids, and nucleotides. Metabolites from external sources such as products from microbial growth are also primary metabolites.

TCA cycle Acetyl-CoA can also undergo alternate fates, i.e., its conversion to lactate, which is excreted from cells. Acetyl-CoA (2C) enters the TCA cycle to combine with oxaloacetic acid (4C) and yield citrate (6C). Through a sequential set of reactions, citrate is completely oxidized to CO_2 and oxaloacetic acid is regenerated. The TCA cycle completes the oxidation of glucose to six molecules of CO_2, generating two molecules of GTP (guanosine triphosphate), ten molecules of NADH, and two molecules of reduced FAD ($FADH_2$). The TCA cycle provides intermediates that can be exported to cytosol for fatty acid synthesis. The NADH thus formed as a product in the above step needs to be recycled. In anaerobic organisms or hypoxic mammalian cells, NAD^+ is regenerated from NADH by the conversion of pyruvate to lactate. In aerobic organisms, the NADH and pyruvate that are generated by glycolysis are transported to mitochondria. One carbon of each pyruvate molecule is released as carbon dioxide (CO_2) and one molecule of NADH is generated; the remaining two carbons are added to a carrier, coenzyme A, to yield acetyl-CoA, which acts as a connecting link between glycolysis and the TCA cycle.

Transamination A chemical reaction that transfers an amino group to a ketoacid to form new amino acids. This pathway is responsible for the deamination of most amino acids. During glutaminolysis, α-ketoglutarate is generated from glutamine. Glutamine thus plays dual functions: replenishing the exhausted intermediates of the TCA cycle and complete oxidation to generate ATP. The choice that immune cells make to divert glutaminolysis in either direction depends on the energy requirements during states like infection and homeostasis. The β-oxidation of fatty acids also yields acetyl-CoA, which is burnt through the TCA cycle.

Warburg effect Under oxygen-plentiful conditions, non-proliferative tissues preferentially metabolize glucose to pyruvate via glycolysis, and then completely oxidize most of the pyruvate in the mitochondria to CO_2 and generate energy via a process known as oxidative phosphorylation (OxPhos). Because the oxygen is required as the final electron acceptor to completely oxidize glucose, this process is obligatorily aerobic. Thus, when hypoxic conditions are encountered, cells divert the pyruvate generated via glycolysis away from OxPhos by generating lactose. Lactate production allows glycolysis to continue (since NADH cycles back to NAD^+), but limits ATP production in comparison to OxPhos. Sir Otto Warburg made a pioneering discovery that cancer cells are inherently skewed towards converting most glucose to lactate in spite of utopic oxygenation. This aerobic glycolysis is also reported to occur in normal proliferative cells besides transformed malignant cells. While this mode of generating energy is far less efficient as compared to OxPhos, it justifies the explanation for faster glucose utilization rates exhibited by cancerous cells. Another plausible explanation for enhanced conversion of glucose to lactate by tumor cells is that HIF-1α is stabilized and accumulates in cytosol upon hypoxic induction. HIF-1α transactivates the glycolytic genes, as well as activating the PDK1 gene, which is responsible for the conversion of pyruvate to acetyl-CoA. Acetyl-CoA enters the TCA cycle, which donates electrons to mitochondrial respiratory chain complexes I to IV. Increased activity of transcription factors such as myc and HIF-1α coordinates with the loss of tumor suppressive function of p53. A loss of function of p53 correlates with the increase of Glut transporters (transcription and surface translocation) via NF-κB. In addition, mitochondrial biogenesis promoted by myc results in the overproduction of ROS. ROS can cause mtDNA spontaneous mutations, rendering mitochondrial enzymes dysfunctional and, thus, OxPhos, leaving only aerobic glycolysis, a probable mode of energy production in transformed cells.

References

Akira S, Uematsu S, Takeuchi O (2006) Pathogen recognition and innate immunity. Cell 124:783–801

Alba-Loureiro TC, Munhoz CD, Martins JO, Cerchiaro GA, Scavone C, Curi R, Sannomiya P (2007) Neutrophil function and metabolism in individuals with diabetes mellitus. Braz J Med Biol Res 40:1037–1044

Albert D, Kowalski J, Nodzenski E, Micek M, Wu P (1990) The dose dependent effect of cyclic AMP on ribonucleotide reductase in mitogen stimulated mononuclear cells. Biochem Biophys Res Commun 167:383–390

Alers S, Löffler AS, Wesselborg S, Stork B (2012) Role of AMPK-mTOR-Ulk1/2 in the regulation of autophagy: cross talk, shortcuts, and feedbacks. Mol Cell Biol 32:2–11

Alonso D, Nungester WJ (1956) Comparative study of host resistance of guinea pigs and rats: V. The effect of pneumococcal products on glycolysis and oxygen uptake by polymorphonuclear leucocytes. J Infect Dis 99:174–181

Amiel E, Everts B, Fritz D, Beauchamp S, Ge B, Pearce EL, Pearce EJ (2014) Mechanistic target of rapamycin inhibition extends cellular lifespan in dendritic cells by preserving mitochondrial function. J Immunol 193:2821–2830

Amkraut A, Solomon GF (1974) From the symbolic stimulus to the pathophysiologic response: Immune mechanisms. Int J Psych Med 5:541–563

Amulic B, Cazalet C, Hayes GL, Metzler KD, Zychlinsky A (2012) Neutrophil function: from mechanisms to disease. Annu Rev Immunol 30:459–489

Ananieva EA, Patel CH, Drake CH, Powell JD, Hutson SM (2014) Cytosolic branched chain aminotransferase (BCATc) regulates mTORC1 signaling and glycolytic metabolism in $CD4^+$ T cells. J Biol Chem 289:18793–18804

Araki K, Turner AP, Shaffer VO, Gangappa S, Keller SA, Bachmann MF, Larsen CP, Ahmed R (2009) mTOR regulates memory CD8 T-cell differentiation. Nature 460:108–112

Arany Z, Huang LE, Eckner R, Bhattacharya S, Jiang C, Goldberg MA, Bunn HF, Livingston DM (1996) An essential role for p300/CBP in the cellular response to hypoxia. Proc Natl Acad Sci USA 93:12969–12973

Averous J, Fonseca BD, Proud CG (2008) Regulation of cyclin D1 expression by mTORC1 signaling requires eukaryotic initiation factor 4E-binding protein 1. Oncogene 27:1106–1113

Avruch J, Long X, Lin Y, Ortiz-Vega S, Rapley J, Papageorgiou A, Oshiro N, Kikkawa U (2009) Activation of mTORC1 in two steps: rheb-GTP activation of catalytic function and increased binding of substrates to raptor1. Biochem Soc Trans 37:223–226

Balmer ML, Ma EH, Bantug GR, Grählert J, Pfister S, Glatter T, Jauch A, Dimeloe S, Slack E, Dehio P, Krzyzaniak MA (2016) Memory $CD8^+$ T cells require increased concentrations of acetate induced by stress for optimal function. Immunity 44:1312–1324

Balsamo M, Manzini C, Pietra G, Raggi F, Blengio F, Mingari MC, Varesio L, Moretta L, Bosco MC, Vitale M (2013) Hypoxia downregulates the expression of activating receptors involved in NK-cell-mediated target cell killing without affecting ADCC. Eur J Immunol 43:2756–2764

Bantug GR, Hess C (2016) Glycolysis and EZH2 boost T cell weaponry against tumors. Nat Immunol 17:41–42

Barber DL, Wherry EJ, Masopust D, Zhu B, Allison JP, Sharpe AH, Freeman GJ, Ahmed R (2006) Restoring function in exhausted CD8 T cells during chronic viral infection. Nature 439:682–687

Barragan M, Good M, Kolls JK (2015) Regulation of dendritic cell function by vitamin D. Nutrients 7:8127–8151

Battaglia M, Stabilini A, Roncarolo MG (2005) Rapamycin selectively expands $CD4^+$ $CD25^+$ $FoxP3^+$ regulatory T cells. Blood 105:4743–4748

B'chir W, Chaveroux C, Carraro V, Averous J, Maurin AC, Jousse C, Muranishi Y, Parry L, Fafournoux P, Bruhat A (2014) Dual role for CHOP in the crosstalk between autophagy and apoptosis to determine cell fate in response to amino acid deprivation. Cell Signal 26:1385–1391

Beier UH, Wang L, Bhatti TR, Liu Y, Han R, Ge G, Hancock WW (2011) Sirtuin-1 targeting promotes Foxp3+ T-regulatory cell function and prolongs allograft survival. Mol Cell Biol 31:1022–1029

Beisel WR (1975) Metabolic response to infection. Annu Rev Med 26:9–20

Belkaid Y, Hand TW (2014) Role of the microbiota in immunity and inflammation. Cell 157:121–141

Bengsch B, Johnson AL, Kurachi M, Odorizzi PM, Pauken KE, Attanasio J, Stelekati E, McLane LM, Paley MA, Delgoffe GM, Wherry EJ (2016) Bioenergetic insufficiencies due to metabolic alterations regulated by the inhibitory receptor PD-1 are an early driver of $CD8^+$ T cell exhaustion. Immunity 45:358–373

Ben-Shoshan J, Maysel-Auslender S, Mor A, Keren G, George J (2008) Hypoxia controls $CD4^+$ CD25+ regulatory T-cell homeostasis via hypoxia-inducible factor-1α. Eur J Immunol 38:2412–2418

Bensinger SJ, Bradley MN, Joseph SB, Zelcer N, Janssen EM, Hausner MA, Shih R, Parks JS, Edwards PA, Jamieson BD, Tontonoz P (2008) LXR signaling couples sterol metabolism to proliferation in the acquired immune response. Cell 134:97–111

Ben-Zvi I, Aranow C, Mackay M, Stanevsky A, Kamen DL, Marinescu LM, Collins CE, Gilkeson GS, Diamond B, Hardin JA (2010) The impact of vitamin D on dendritic cell function in patients with systemic lupus erythematosus. PLoS One 5:e9193

Berger J, Moller DE (2002) The mechanisms of action of PPARs. Annu Rev Med 53:409–435

Berod L, Friedrich C, Nandan A, Freitag J, Hagemann S, Harmrolfs K, Sandouk A, Hesse C, Castro CN, Bähre H, Tschirner SK (2014) De novo fatty acid synthesis controls the fate between regulatory T and T helper 17 cells. Nat Med 20:1327–1333

Bhatt AP, Jacobs SR, Freemerman AJ, Makowski L, Rathmell JC, Dittmer DP, Damania B (2012) Dysregulation of fatty acid synthesis and glycolysis in non-Hodgkin lymphoma. Proc Natl Acad Sci USA 109:11818–11823

Biswas SK (2015) Metabolic reprogramming of immune cells in cancer progression. Immunity 43:435–449

Blackburn SD, Shin H, Haining WN, Zou T, Workman CJ, Polley A, Betts MR, Freeman GJ, Vignali DA, Wherry EJ (2009) Coregulation of $CD8^+$ T cell exhaustion by multiple inhibitory receptors during chronic viral infection. Nat Immunol 10:29–37

Blagih J, Coulombe F, Vincent EE, Dupuy F, Galicia-Vázquez G, Yurchenko E, Raissi TC, van der Windt GJW, Viollet B, Pearce EL, Pelletier J, Piccirillo CA, Krawczyk CM, Divangahi M, Jones RG (2015) The energy sensor AMPK regulates T cell metabolic adaptation and effector responses in vivo. *Immunity*. 42:41–54

Blouin CC, Pagé EL, Soucy GM, Richard DE (2004) Hypoxic gene activation by lipopolysaccharide in macrophages: implication of hypoxia-inducible factor 1α. Blood 103:1124–1130

Bolster DR, Crozier SJ, Kimball SR, Jefferson LS (2002) AMP-activated protein kinase suppresses protein synthesis in rat skeletal muscle through down-regulated mammalian target of rapamycin (mTOR) signaling. J Biol Chem 277:23977–23980

Bonfils G, Jaquenoud M, Bontron S, Ostrowicz C, Ungermann C, De Virgilio C (2012) Leucyl-tRNA synthetase controls TORC1 via the EGO complex. Mol Cell 46:105–110

Brach MA, Gruss HJ, Herrmann F (1992) Prolongation of survival of human polymorphonuclear neutrophils by granulocyte-macrophage colony-stimulating factor is caused by inhibition of programmed cell death. Blood 80:2920–2924

Brinkmann V, Reichard U, Goosmann C, Fauler B, Uhlemann Y, Weiss DS, Weinrauch Y, Zychlinsky A (2004) Neutrophil extracellular traps kill bacteria. Science 303:1532–1535

Bronietzki AW, Schuster M, Schmitz I (2015) Autophagy in T-cell development, activation and differentiation. Immunol Cell Biol 93:25–34

Brown EJ, Albers MW, Shin TB, Ichikawa K, Keith CT, Lane WS, Schreiber SL (1994) A mammalian protein targeted by G1-arresting rapamycin–receptor complex. Nature 369:756

Brubaker SW, Bonham KS, Zanoni I, Kagan JC (2015) Innate immune pattern recognition: a cell biological perspective. Annu Rev Immunol 33:257–290

Brugarolas J, Lei K, Hurley RL, Manning BD, Reiling JH, Hafen E, Witters LA, Ellisen LW, Kaelin WG (2004) Regulation of mTOR function in response to hypoxia by REDD1 and the TSC1/TSC2 tumor suppressor complex. Genes Dev 18:2893–2904

Brunet A, Datta SR, Greenberg ME (2001) Transcription-dependent and-independent control of neuronal survival by the PI3K–Akt signaling pathway. Curr Opin Neurobiol 11:297–305

Buck MD, O'Sullivan D, Pearce EL (2015) T cell metabolism drives immunity. J Exp Med 212:1345–1360
Buckley AF, Kuo CT, Leiden JM (2001) Transcription factor LKLF is sufficient to program T cell quiescence via a c-Myc-dependent pathway. Nat Immunol 2:698–704
Buescher JM, Antoniewicz MR, Boros LG, Burgess SC, Brunengraber H, Clish CB, DeBerardinis RJ, Feron O, Frezza C, Ghesquiere B, Gottlieb E (2015) A roadmap for interpreting 13C metabolite labeling patterns from cells. Curr Opin Biotechnol 34:189–201
Byersdorfer CA, Tkachev V, Opipari AW, Goodell S, Swanson J, Sandquist S, Glick GD, Ferrara JL (2013) Effector T cells require fatty acid metabolism during murine graft-versus-host disease. Blood 122:3230–3237
Byrne CS, Chambers ES, Morrison DJ, Frost G (2015) The role of short chain fatty acids in appetite regulation and energy homeostasis. Int J Obesity 39:1331–1338
Cai Q, Lin T, Kamarajugadda S, Lu J (2013) Regulation of glycolysis and the Warburg effect by estrogen-related receptors. Oncogene 32:2079–2086
Cairns RA, Harris IS, Mak TW (2011) Regulation of cancer cell metabolism. Nat Rev Cancer 11:85–95
Cantorna MT, Nashold FE, Hayes CE (1995) Vitamin A deficiency results in a priming environment conducive for Th1 cell development. Eur J Immunol 25:1673–1679
Cao Y, Li H, Liu H, Zheng C, Ji H, Liu X (2010) The serine/threonine kinase LKB1 controls thymocyte survival through regulation of AMPK activation and Bcl-XL expression. *Cell Res* 20:99–108
Cao Y, Li H, Liu H, Zhang M, Hua Z, Ji H, Liu X (2011) LKB1 regulates TCR-mediated PLCγ1 activation and thymocyte positive selection. *EMBO J.* 30:2083–2093
Cao Y, Rathmell JC, Macintyre AN (2014) Metabolic reprogramming towards aerobic glycolysis correlates with greater proliferative ability and resistance to metabolic inhibition in CD8 versus CD4 T cells. PLoS One 9(8):e104104
Carr EL, Kelman A, Wu GS, Gopaul R, Senkevitch E, Aghvanyan A, Turay AM, Frauwirth KA (2010) Glutamine uptake and metabolism are coordinately regulated by ERK/MAPK during T lymphocyte activation. J Immunol 185:1037–1044
Cassatella MA (1999) Neutrophil-derived proteins: selling cytokines by the pound. Adv Immunol 73:369–509
Castrillo A, Joseph SB, Vaidya SA, Haberland M, Fogelman AM, Cheng G, Tontonoz P (2003) Crosstalk between LXR and toll-like receptor signaling mediates bacterial and viral antagonism of cholesterol metabolism. Mol Cell 12:805–816
Chabes A, Thelander L (2000) Controlled protein degradation regulates ribonucleotide reductase activity in proliferating mammalian cells during the normal cell cycle and in response to DNA damage and replication blocks. J Biol Chem 275:17747–17753
Chakravarty N (1968) Further observations on the inhibition of histamine release by 2-deoxyglucose. Acta Physiologica 72(4):425–432
Chakravarthy MV, Lodhi IJ, Yin L, Malapaka RR, Xu HE, Turk J, Semenkovich CF (2009) Identification of a physiologically relevant endogenous ligand for PPARα in liver. Cell 138:476–488
Cham CM, Gajewski TF (2005) Glucose availability regulates IFN-γ production and p70S6 kinase activation in $CD8^+$ effector T cells. J Immunol 174:4670–4677
Chan KL, Pillon NJ, Sivaloganathan DM, Costford SR, Liu Z, Théret M, Chazaud B, Klip A (2015) Palmitoleate reverses high fat-induced proinflammatory macrophage polarization via AMP-activated protein kinase (AMPK). J Biol Chem 290:16979–16988
Chandra RK (1990) Micronutrients and immune functions. Ann NY Acad Sci 587:9–16
Chandra RK (1997) Nutrition and the immune system: an introduction. Am J Clin Nutr 66:460S–463S
Chang SH, Chung Y, Dong C (2010) Vitamin D suppresses Th17 cytokine production by inducing C/EBP homologous protein (CHOP) expression. J Biol Chem 285:38751–38755

Chang CH, Curtis JD, Maggi LB, Faubert B, Villarino AV, O'Sullivan D, Huang SC, van der Windt GJ, Blagih J, Qiu J, Weber JD (2013) Posttranscriptional control of T cell effector function by aerobic glycolysis. Cell 153:1239–1251

Chang CH, Qiu J, O'Sullivan D, Buck MD, Noguchi T, Curtis JD, Chen Q, Gindin M, Gubin MM, van der Windt GJ, Tonc E (2015) Metabolic competition in the tumor microenvironment is a driver of cancer progression. Cell 162:1229–1241

Chantranupong L, Wolfson RL, Orozco JM, Saxton RA, Scaria SM, Bar-Peled L, Spooner E, Isasa M, Gygi SP, Sabatini DM (2014) The Sestrins interact with GATOR2 to negatively regulate the amino-acid-sensing pathway upstream of mTORC1. Cell Rep 9:1–8

Chauhan P, Shukla D, Chattopadhyay D, Saha B (2017) Redundant and regulatory roles for Toll-like receptors in *Leishmania* infection. Clin Exp Immunol 190:167–186. https://doi.org/10.1111/cei.13014

Chawla A, Boisvert WA, Lee CH, Laffitte BA, Barak Y, Joseph SB, Liao D, Nagy L, Edwards PA, Curtiss LK, Evans RM (2001) A PPARγ-LXR-ABCA1 pathway in macrophages is involved in cholesterol efflux and atherogenesis. Mol Cell 7:161–171

Chen Q, Ross AC (2007) Retinoic acid promotes mouse splenic B cell surface IgG expression and maturation stimulated by CD40 and IL-4. Cell Immunol 249:37–45

Chen C, Pore N, Behrooz A, Ismail-Beigi F, Maity A (2001) Regulation of glut1 mRNA by hypoxia-inducible factor-1 Interaction between H-ras and hypoxia. J Biol Chem 276:9519–9525

Chen S, Sims GP, Chen XX, Gu YY, Chen S, Lipsky PE (2007) Modulatory effects of 1,25-dihydroxyvitamin D3 on human B cell differentiation. J Immunol 179:1634–1647

Chi H (2012) Regulation and function of mTOR signalling in T cell fate decisions. Nat Rev Immunol 12:325–338

Chow CW, Rincón M, Davis RJ (1999) Requirement for transcription factor NFAT in interleukin-2 expression. Mol Cell Biol 19:2300–2307

Clambey ET, McNamee EN, Westrich JA, Glover LE, Campbell EL, Jedlicka P, de Zoeten EF, Cambier JC, Stenmark KR, Colgan SP, Eltzschig HK (2012) Hypoxia-inducible factor-1 alpha-dependent induction of FoxP3 drives regulatory T-cell abundance and function during inflammatory hypoxia of the mucosa. Proc Natl Acad Sci USA 109:E2784–E2793

Clark RB, Bishop-Bailey D, Estrada-Hernandez T, Hla T, Puddington L, Padula SJ (2000) The nuclear receptor PPARγ and immunoregulation: PPARγ mediates inhibition of helper T cell responses. J Immunol 164:1364–1371

Cohen HY, Miller C, Bitterman KJ, Wall NR, Hekking B, Kessler B, Howitz KT, Gorospe M, de Cabo R, Sinclair DA (2004) Calorie restriction promotes mammalian cell survival by inducing the SIRT1 deacetylase. Science 305:390–392

Contreras AV, Torres N, Tovar AR (2013) PPAR-α as a key nutritional and environmental sensor for metabolic adaptation. Adv Nutr 4:439–452

Coussens LM, Werb Z (2002) Inflammation and cancer. Nature 420:860–867

Covarrubias AJ, Aksoylar HI, Yu J, Snyder NW, Worth AJ, Iyer SS, Wang J, Ben-Sahra I, Byles V, Polynne-Stapornkul T, Espinosa EC (2016) Akt-mTORC1 signaling regulates Acly to integrate metabolic input to control of macrophage activation. Elife 5:e11612

Croft M (2009) Control of immunity by the TNFR-related molecule OX40 (CD134). Annu Rev Immunol 28:57–78

Crompton JG, Sukumar M, Roychoudhuri R, Clever D, Gros A, Eil RL, Tran E, Hanada KI, Yu Z, Palmer DC, Kerkar SP (2015) Akt inhibition enhances expansion of potent tumor-specific lymphocytes with memory cell characteristics. Cancer Res 75:296–305

Cunard R, Ricote M, DiCampli D, Archer DC, Kahn DA, Glass CK, Kelly CJ (2002) Regulation of cytokine expression by ligands of peroxisome proliferator activated receptors. J Immunol 168:2795–2802

Curiel R, Akin EA, Beaulieu G, DePalma L, Hashefi M (2011) PET/CT imaging in systemic lupus erythematosus. Ann NY Acad Sci 1228:71–80

Daitoku H, Sakamaki JI, Fukamizu A (2011) Regulation of FoxO transcription factors by acetylation and protein–protein interactions. BBA Mol Cell Res 1813:1954–1960

Dang CV (2013) MYC, metabolism, cell growth, and tumorigenesis. CSH Persp Med 3:a014217
Dang EV, Barbi J, Yang HY, Jinasena D, Yu H, Zheng Y, Bordman Z, Fu J, Kim Y, Yen HR, Luo W (2011) Control of TH17/Treg balance by hypoxia-inducible factor 1. Cell 146:772–784
Daniëls VW, Smans K, Royaux I, Chypre M, Swinnen JV, Zaidi N (2014) Cancer cells differentially activate and thrive on de novo lipid synthesis pathways in a low-lipid environment. PLoS One 9:e106913
Day CL, Kaufmann DE, Kiepiela P, Brown JA, Moodley ES, Reddy S, Mackey EW, Miller JD, Leslie AJ, DePierres C, Mncube Z (2006) PD-1 expression on HIV-specific T cells is associated with T-cell exhaustion and disease progression. Nature 443:350–354
De Pergola G, Silvestris F (2013) Obesity as a major risk factor for cancer. J Obesity:291546. https://doi.org/10.1155/2013/291546
DeBose-Boyd RA (2008) Feedback regulation of cholesterol synthesis: sterol-accelerated ubiquitination and degradation of HMG CoA reductase. Cell Res 18:609–621
Delgoffe GM, Kole TP, Zheng Y, Zarek PE, Matthews KL, Xiao B, Worley PF, Kozma SC, Powell JD (2009) The mTOR kinase differentially regulates effector and regulatory T cell lineage commitment. Immunity 30:832–844
Delgoffe GM, Pollizzi KN, Waickman AT, Heikamp E, Meyers DJ, Horton MR, Xiao B, Worley PF, Powell JD (2011) The kinase mTOR regulates the differentiation of helper T cells through the selective activation of signaling by mTORC1 and mTORC2. Nat Immunol 12:295–303
den Besten G, van Eunen K, Groen AK, Venema K, Reijngoud DJ, Bakker BM (2013) The role of short-chain fatty acids in the interplay between diet, gut microbiota, and host energy metabolism. J Lipid Res 54:2325–2340
Desvergne B, Wahli W (1999) Peroxisome proliferator-activated receptors: nuclear control of metabolism. Endocrine Rev 20:649–688
Devchand PR, Keller H, Peters JM, Vazquez M, Gonzalez FJ, Wahli W (1996) The PPARα–leukotriene B4 pathway to inflammation control. Nature 384:39
Dijkers PF, Medema RH, Pals C, Banerji L, Thomas NS, Lam EW, Burgering BM, Raaijmakers JA, Lammers JW, Koenderman L, Coffer PJ (2000) Forkhead transcription factor FKHR-L1 modulates cytokine-dependent transcriptional regulation of p27KIP1. Mol Cell Biol 20:9138–9148
Doedens AL, Phan AT, Stradner MH, Fujimoto JK, Nguyen JV, Yang E, Johnson RS, Goldrath AW (2013) Hypoxia-inducible factors enhance the effector responses of $CD8^+$ T cells to persistent antigen. Nat Immunol 14:1173–1182
Dong XY, Tang SQ, Chen JD (2012) Dual functions of Insig proteins in cholesterol homeostasis. Lipids Health Dis 11:173
Doyle AG, Herbein G, Montaner LJ, Minty AJ, Caput D, Ferrara P, Gordon S (1994) Interleukin-13 alters the activation state of murine macrophages in vitro: comparison with interleukin-4 and interferon-γ. Eur J Immunol 24:1441–1445
Duewell P, Kono H, Rayner KJ, Sirois CM, Vladimer G, Bauernfeind FG, Abela GS, Franchi L, Nunez G, Schnurr M, Espevik T (2010) NLRP3 inflammasomes are required for atherogenesis and activated by cholesterol crystals. Nature 464:1357–1361
Durham NM, Nirschl CJ, Jackson CM, Elias J, Kochel CM, Anders RA, Drake CG (2014) Lymphocyte Activation Gene 3 (LAG-3) modulates the ability of CD4 T-cells to be suppressed in vivo. PLoS One 9:e109080
Düvel K, Yecies JL, Menon S, Raman P, Lipovsky AI, Souza AL, Triantafellow E, Ma Q, Gorski R, Cleaver S, Vander Heiden MG (2010) Activation of a metabolic gene regulatory network downstream of mTOR complex 1. Mol Cell 39:171–183
Eckel RH (1997) Obesity and heart disease. Circulation 96:3248–3250
Egan DF, Shackelford DB, Mihaylova MM, Gelino S, Kohnz RA, Mair W, Vasquez DS, Joshi A, Gwinn DM, Taylor R, Asara JM (2011) Phosphorylation of ULK1 (hATG1) by AMP-activated protein kinase connects energy sensing to mitophagy. Science 331:456–461

Elstrom RL, Bauer DE, Buzzai M, Karnauskas R, Harris MH, Plas DR, Zhuang H, Cinalli RM, Alavi A, Rudin CM, Thompson CB (2004) Akt stimulates aerobic glycolysis in cancer cells. Cancer Res 64:3892–3899

Emanuelle S, Hossain MI, Moller IE, Pedersen HL, Meene AM, Doblin MS, Koay A, Oakhill JS, Scott JW, Willats WG, Kemp BE (2015) SnRK1 from *Arabidopsis thaliana* is an atypical AMPK. Plant J 82:183–192

Eriksson S, Gräslund A, Skog S, Thelander L, Tribukait B (1984) Cell cycle-dependent regulation of mammalian ribonucleotide reductase. The S phase-correlated increase in subunit M2 is regulated by de novo protein synthesis. J Biol Chem 259:11695–11700

Everts B, Amiel E, van der Windt GJ, Freitas TC, Chott R, Yarasheski KE, Pearce EL, Pearce EJ (2012) Commitment to glycolysis sustains survival of NO-producing inflammatory dendritic cells. Blood 120:1422–1431

Everts B, Amiel E, Huang SC, Smith AM, Chang CH, Lam WY, Redmann V, Freitas TC, Blagih J, Van Der Windt GJ, Artyomov MN (2014) TLR-driven early glycolytic reprogramming via the kinases TBK1-IKKε supports the anabolic demands of dendritic cell activation. Nat Immunol 15:323–332

Fajardo AM, Piazza GA, Tinsley HN (2014) The role of cyclic nucleotide signaling pathways in cancer: targets for prevention and treatment. Cancers 6:436–458

Fajas L, Egler V, Reiter R, Miard S, Lefebvre AM, Auwerx J (2003) PPARγ controls cell proliferation and apoptosis in an RB-dependent manner. Oncogene 22:4186–4193

Fang X, Yu SX, Lu Y, Bast RC, Woodgett JR, Mills GB (2000) Phosphorylation and inactivation of glycogen synthase kinase 3 by protein kinase A. Proc Natl Acad Sci USA 97:11960–11965

Faubert B, Boily G, Izreig S, Griss T, Samborska B, Dong Z, Dupuy F, Chambers C, Fuerth BJ, Viollet B, Mamer OA (2013) AMPK is a negative regulator of the Warburg effect and suppresses tumor growth in vivo. Cell Metab 17:113–124

Fernández-Riejos P, Najib S, Santos-Alvarez J, Martín-Romero C, Pérez-Pérez A, González-Yanes C, Sánchez-Margalet V (2010) Role of leptin in the activation of immune cells. Mediators Inflamm. https://doi.org/10.1155/2010/568343

Ferrannini E, DeFronzo RA (2015) Impact of glucose-lowering drugs on cardiovascular disease in type 2 diabetes. Eur Heart J 36:2288–2296

Fingar DC, Richardson CJ, Tee AR, Cheatham L, Tsou C, Blenis J (2004) mTOR controls cell cycle progression through its cell growth effectors S6K1 and 4E-BP1/eukaryotic translation initiation factor 4E. Mol Cell Biol 24:200–216

Fischer K, Hoffmann P, Voelkl S, Meidenbauer N, Ammer J, Edinger M, Gottfried E, Schwarz S, Rothe G, Hoves S, Renner K (2007) Inhibitory effect of tumor cell-derived lactic acid on human T cells. Blood 109:3812–3819

Flannagan RS, Heit B, Heinrichs DE (2016) Intracellular replication of *Staphylococcus aureus* in mature phagolysosomes in macrophages precedes host cell death, and bacterial escape and dissemination. Cell Microbiol 18:514–535

Fletcher M, Ramirez ME, Sierra RA, Raber P, Thevenot P, Al-Khami AA, Sanchez-Pino D, Hernandez C, Wyczechowska DD, Ochoa AC, Rodriguez PC (2015) L-Arginine depletion blunts antitumor T-cell responses by inducing myeloid-derived suppressor cells. Cancer Res 75:275–283

Forsythe JA, Jiang BH, Iyer NV, Agani F, Leung SW, Koos RD, Semenza GL (1996) Activation of vascular endothelial growth factor gene transcription by hypoxia-inducible factor 1. Mol Cell Biol 16:4604–4613

Fox CJ, Hammerman PS, Thompson CB (2005) Fuel feeds function: energy metabolism and the T-cell response. Nat Rev Immunol 5:844–852

Frauwirth KA, Riley JL, Harris MH, Parry RV, Rathmell JC, Plas DR, Elstrom RL, June CH, Thompson CB (2002) The CD28 signaling pathway regulates glucose metabolism. Immunity 16:769–777

Freedman SJ, Sun ZY, Poy F, Kung AL, Livingston DM, Wagner G, Eck MJ (2002) Structural basis for recruitment of CBP/p300 by hypoxia-inducible factor-1α. Proc Natl Acad Sci USA 99:5367–5372

Freemerman AJ, Johnson AR, Sacks GN, Milner JJ, Kirk EL, Troester MA, Macintyre AN, Goraksha-Hicks P, Rathmell JC, Makowski L (2014) Metabolic reprogramming of macrophages glucose transporter 1 (GLUT1)-mediated glucose metabolism drives a proinflammatory phenotype. J Biol Chem 289:7884–7896

Frieden TR (2015) The future of public health. N Engl J Med 373:1748–1754

Fujishima S, Hoffman AR, Vu T, Kim KJ, Zheng H, Daniel D, Kim Y, Wallace EF, Larrick JW, Raffin TA (1993) Regulation of neutrophil interleukin 8 gene expression and protein secretion by LPS, TNF-α, and IL-1β. J Cell Physiol 154:478–485

Fukuzumi M, Shinomiya H, Shimizu Y, Ohishi K, Utsumi S (1996) Endotoxin-induced enhancement of glucose influx into murine peritoneal macrophages via GLUT1. Infect Immun 64:108–112

Furusawa Y, Obata Y, Fukuda S, Endo TA, Nakato G, Takahashi D, Nakanishi Y, Uetake C, Kato K, Kato T, Takahashi M (2013) Commensal microbe-derived butyrate induces the differentiation of colonic regulatory T cells. Nature 504:446–450

Gajewski TF, Schreiber H, Fu YX (2013) Innate and adaptive immune cells in the tumor microenvironment. Nat Immunol 14:1014–1022

Galli SJ, Kalesnikoff J, Grimbaldeston MA, Piliponsky AM, Williams CM, Tsai M (2005) Mast cells as “tunable” effector and immunoregulatory cells: recent advances. Annu Rev Immunol 23:749–786

Galván-Peña S, O’Neill LA (2014) Metabolic reprograming in macrophage polarization. Front Immunol 5:420

Ganeshan K, Chawla A (2014) Metabolic regulation of immune responses. Annu Rev Immunol 32:609–634

Gao P, Tchernyshyov I, Chang TC, Lee YS, Kita K, Ochi T, Zeller KI, De Marzo AM, Van Eyk JE, Mendell JT, Dang CV (2009) c-Myc suppression of miR-23a/b enhances mitochondrial glutaminase expression and glutamine metabolism. Nature 458:762–765

Garçon F, Patton DT, Emery JL, Hirsch E, Rottapel R, Sasaki T, Okkenhaug K (2008) CD28 provides T-cell costimulation and enhances PI3K activity at the immune synapse independently of its capacity to interact with the p85/p110 heterodimer. Blood 111:1464–1471

Garedew A, Henderson SO, Moncada S (2010) Activated macrophages utilize glycolytic ATP to maintain mitochondrial membrane potential and prevent apoptotic cell death. Cell Death Diff 17:1540–1550

Gatza E, Wahl DR, Opipari AW, Sundberg TB, Reddy P, Liu C, Glick GD, Ferrara JL (2011) Manipulating the bioenergetics of alloreactive T cells causes their selective apoptosis and arrests graft-versus-host disease. Sci Transl Med 3:67ra8

Gazzerro P, Proto MC, Gangemi G, Malfitano AM, Kigali E, Pisanti S, Santoro A, Laezza C, Bifulco M (2012) Pharmacological actions of statins: a critical appraisal in the management of cancer. Pharmacol Rev 64:102–146

Geissmann F, Manz MG, Jung S, Sieweke MH, Merad M, Ley K (2010) Development of monocytes, macrophages, and dendritic cells. Science 327:656–661

Gerriets VA, Rathmell JC (2012) Metabolic pathways in T cell fate and function. Trends Immunol 33:168–173

Geyeregger R, Shehata M, Zeyda M, Kiefer FW, Stuhlmeier KM, Porpaczy E, Zlabinger GJ, Jäger U, Stulnig TM (2009) Liver X receptors interfere with cytokine-induced proliferation and cell survival in normal and leukemic lymphocytes. J Leukoc Biol 86:1039–1048

Ghaemi-Oskouie F, Shi Y (2011) The role of uric acid as an endogenous danger signal in immunity and inflammation. Curr Rheumatol Rep 13:160–166

Ghesquiere B, Wong BW, Kuchnio A, Carmeliet P (2014) Metabolism of stromal and immune cells in health and disease. Nature 511:167–176

Gigoux M, Shang J, Pak Y, Xu M, Choe J, Mak TW, Suh WK (2009) Inducible costimulator promotes helper T-cell differentiation through phosphoinositide 3-kinase. Proc Natl Acad Sci USA 106:20371–20376

Gingras AC, Raught B, Gygi SP, Niedzwiecka A, Miron M, Burley SK, Polakiewicz RD, Wyslouch-Cieszynska A, Aebersold R, Sonenberg N (2001) Hierarchical phosphorylation of the translation inhibitor 4E-BP1. Genes Dev 15:2852–2864

Ginhoux F, Greter M, Leboeuf M, Nandi S, See P, Gokhan S, Mehler MF, Conway SJ, Ng LG, Stanley ER, Samokhvalov IM (2010) Fate mapping analysis reveals that adult microglia derive from primitive macrophages. Science 330:841–845

Gordon S (2003) Alternative activation of macrophages. Nat Rev Immunol 3:23–35

Gordon S, Martinez FO (2010) Alternative activation of macrophages: mechanism and functions. Immunity 32:593–604

Goverse G, Labao-Almeida C, Ferreira M, Molenaar R, Wahlen S, Konijn T, Koning J, Veiga-Fernandes H, Mebius RE (2016) Vitamin A controls the presence of RORγ^+ innate lymphoid cells and lymphoid tissue in the small intestine. J Immunol 196:5148–5155

Gracias DT, Boesteanu AC, Fraietta JA, Hope JL, Carey AJ, Mueller YM, Kawalekar OU, Fike AJ, June CH, Katsikis PD (2016) Phosphatidylinositol 3-kinase p110δ isoform regulates CD8$^+$ T cell responses during acute viral and intracellular bacterial infections. J Immunol 196:1186–1198

Greene ME, Blumberg B, McBride OW, Yi HF, Kronquist K, Kwan K, Hsieh L, Greene G, Nimer SD (1995) Isolation of the human peroxisome proliferator activated receptor gamma cDNA: expression in hematopoietic cells and chromosomal mapping. Gene Exp 4:281–299

Greer EL, Dowlatshahi D, Banko MR, Villen J, Hoang K, Blanchard D, Gygi SP, Brunet A (2007) An AMPK-FOXO pathway mediates longevity induced by a novel method of dietary restriction in *C. elegans*. Curr Biol 17:1646–1656

Gubser PM, Bantug GR, Razik L, Fischer M, Dimeloe S, Hoenger G, Durovic B, Jauch A, Hess C (2013) Rapid effector function of memory CD8+ T cells requires an immediate-early glycolytic switch. Nat Immunol 14:1064–1072

Guertin DA, Sabatini DM (2007) Defining the role of mTOR in cancer. Cancer Cell 12:9–22

Guimarães-Costa AB, Nascimento MT, Froment GS, Soares RP, Morgado FN, Conceição-Silva F, Saraiva EM (2009) *Leishmania amazonensis* promastigotes induce and are killed by neutrophil extracellular traps. Proc Natl Acad Sci USA 106:6748–6753

Guo Y, Zhang Y, Hong K, Luo F, Gu Q, Lu N, Bai A (2014) AMPK inhibition blocks ROS-NFκB signaling and attenuates endotoxemia-induced liver injury. PLoS One 9:e86881

Guppy M, Greiner E, Brand K (1993) The role of the Crabtree effect and an endogenous fuel in the energy metabolism of resting and proliferating thymocytes. FEBS J 212:95–99

Gwinn DM, Shackelford DB, Egan DF, Mihaylova MM, Mery A, Vasquez DS, Turk BE, Shaw RJ (2008) AMPK phosphorylation of raptor mediates a metabolic checkpoint. Mol Cell 30:214–226

Hahn-Windgassen A, Nogueira V, Chen CC, Skeen JE, Sonenberg N, Hay N (2005) Akt activates the mammalian target of rapamycin by regulating cellular ATP level and AMPK activity. J Biol Chem 280:32081–32089

Hamilton SR, O'Donnell JB, Hammet A, Stapleton D, Habinowski SA, Means AR, Kemp BE, Witters LA (2002) AMP-activated protein kinase kinase: detection with recombinant AMPK α1 subunit. Biochem Biophy Res Commun 293:892–898

Hampton MB, Kettle AJ, Winterbourn CC (1998) Inside the neutrophil phagosome: oxidants, myeloperoxidase, and bacterial killing. Blood 92:3007–3017

Han JM, Jeong SJ, Park MC, Kim G, Kwon NH, Kim HK, Ha SH, Ryu SH, Kim S (2012) Leucyl-tRNA synthetase is an intracellular leucine sensor for the mTORC1-signaling pathway. Cell 149:410–424

Hancock T, Takigawa I, Mamitsuka H (2010) Mining metabolic pathways through gene expression. Bioinformatics 26:2128–2135

Hand TW, Cui W, Jung YW, Sefik E, Joshi NS, Chandele A, Liu Y, Kaech SM (2010) Differential effects of STAT5 and PI3K/AKT signaling on effector and memory CD8 T-cell survival. Proc Natl Acad Sci USA 107:16601–16606

Harada H, Itasaka S, Kizaka-Kondoh S, Shibuya K, Morinibu A, Shinomiya K, Hiraoka M (2009) The Akt/mTOR pathway assures the synthesis of HIF-1α protein in a glucose-and reoxygenation-dependent manner in irradiated tumors. J Biol Chem 284:5332–5342

Hardie DG (2002) A homologue of AMP-activated protein kinase in *Drosophila melanogaster* is sensitive to AMP and is activated by ATP depletion. Biochem J 367:179–186

Hardie DG, Hawley SA (2001) AMP-activated protein kinase: the energy charge hypothesis revisited. BioEssays 23:1112–1119

Hardie DG, Scott JW, Pan DA, Hudson ER (2003) Management of cellular energy by the AMP-activated protein kinase system. FEBS Lett 546:113–120

Harding FA, JG MA, Gross JA, Raulet DH, Allison JP (1992) CD28-mediated signalling co-stimulates murine T cells and prevents induction of anergy in T-cell clones. Nature 356:607–609

Harding HP, Novoa I, Zhang Y, Zeng H, Wek R, Schapira M, Ron D (2000) Regulated translation initiation controls stress-induced gene expression in mammalian cells. Mol Cell 6:1099–1108

Harding HP, Zhang Y, Zeng H, Novoa I, Lu PD, Calfon M, Sadri N, Yun C, Popko B, Paules R, Stojdl DF (2003) An integrated stress response regulates amino acid metabolism and resistance to oxidative stress. Mol Cell 11:619–633

Harris SG, Phipps RP (2001) The nuclear receptor PPAR gamma is expressed by mouse T lymphocytes and PPAR gamma agonists induce apoptosis. Eur J Immunol 31:1098–1105

Haschemi A, Kosma P, Gille L, Evans CR, Burant CF, Starkl P, Knapp B, Haas R, Schmid JA, Jandl C, Amir S (2012) The sedoheptulose kinase CARKL directs macrophage polarization through control of glucose metabolism. Cell Metab 15:813–826

Hawley SA, Davison M, Woods A, Davies SP, Beri RK, Carling D, Hardie DG (1996) Characterization of the AMP-activated protein kinase kinase from rat liver and identification of threonine 172 as the major site at which it phosphorylates AMP-activated protein kinase. J Biol Chem 271:27879–27887

Hawley SA, Boudeau J, Reid JL, Mustard KJ, Udd L, Mäkelä TP, Alessi DR, Hardie DG (2003) Complexes between the LKB1 tumor suppressor, STRADα/β and MO25α/β are upstream kinases in the AMP-activated protein kinase cascade. J Biol 2:28

Haxhinasto S, Mathis D, Benoist C (2008) The AKT–mTOR axis regulates de novo differentiation of $CD4^+$ $Foxp3^+$ cells. J Exp Med 205:565–574

Hay N, Sonenberg N (2004) Upstream and downstream of mTOR. Genes Dev 18:1926–1945

Hayashi K, Jutabha P, Endou H, Sagara H, Anzai N (2013) LAT1 is a critical transporter of essential amino acids for immune reactions in activated human T cells. J Immunol 191:4080–4085

Hedbacker K, Carlson M (2008) SNF1/AMPK pathways in yeast. Front Biosci: J Virtual Lib 13:2408

Heine G, Anton K, Henz BM, Worm M (2002) 1α, 25-dihydroxyvitamin D3 inhibits anti-CD40 plus IL-4-mediated IgE production in vitro. Eur J Immunol 32:3395–3404

Hemmi H, Akira S (2005) TLR signalling and the function of dendritic cells. Mech Epithelial Def 86:120–135

Henao-Mejia J, Elinav E, Jin C, Hao L, Mehal WZ, Strowig T, Thaiss CA, Kau AL, Eisenbarth SC, Jurczak MJ, Camporez JP (2012) Inflammasome-mediated dysbiosis regulates progression of NAFLD and obesity. Nature 482:179–185

Henin N, Vincent MF, Gruber HE, Van den Berghe G (1995) Inhibition of fatty acid and cholesterol synthesis by stimulation of AMP-activated protein kinase. FASEB J 9:541–546

Hershey JW (1991) Translational control in mammalian cells. Annu Rev Biochem 60:717–755

Hinnebusch AG (2005) Translational regulation of GCN4 and the general amino acid control of yeast. Annu Rev Microbiol 59:407–450

Ho PC, Bihuniak JD, Macintyre AN, Staron M, Liu X, Amezquita R, Tsui YC, Cui G, Micevic G, Perales JC, Kleinstein SH (2015) Phosphoenolpyruvate is a metabolic checkpoint of anti-tumor T cell responses. Cell 162:1217–1228
Hoag KA, Nashold FE, Goverman J, Hayes CE (2002) Retinoic acid enhances the T helper 2 cell development that is essential for robust antibody responses through its action on antigen-presenting cells. J Nutr 132:3736–3739
Hoebe K, Janssen E, Beutler B (2004) The interface between innate and adaptive immunity. Nat Immunol 5:971–974
Hoeffel G, Wang Y, Greter M, See P, Teo P, Malleret B, Leboeuf M, Low D, Oller G, Almeida F, Choy SH (2012) Adult Langerhans cells derive predominantly from embryonic fetal liver monocytes with a minor contribution of yolk sac-derived macrophages. J Exp Med 209:1167–1181
Hong SP, Leiper FC, Woods A, Carling D, Carlson M (2003) Activation of yeast Snf1 and mammalian AMP-activated protein kinase by upstream kinases. Proc Natl Acad Sci USA 100:8839–8843
Horton JD, Goldstein JL, Brown MS (2002) SREBPs: activators of the complete program of cholesterol and fatty acid synthesis in the liver. J Clin Invest 109:1125–1131
Hotamisligil GS (2006) Inflammation and metabolic disorders. Nature 444:860–867
Høyer-Hansen M, Bastholm L, Szyniarowski P, Campanella M, Szabadkai G, Farkas T, Bianchi K, Fehrenbacher N, Elling F, Rizzuto R, Mathiasen IS (2007) Control of macroautophagy by calcium, calmodulin-dependent kinase kinase-β, and Bcl-2. Mol Cell 25:193–205
Hu CJ, Wang LY, Chodosh LA, Keith B, Simon MC (2003) Differential roles of hypoxia-inducible factor 1α (HIF-1α) and HIF-2α in hypoxic gene regulation. Mol Cell Biol 23:9361–9374
Hua X, Thompson CB (2001) Quiescent T cells: actively maintaining inactivity. Nat Immunol 2:1097–1098
Huang BP, Lin CH, Chen HM, Lin JT, Cheng YF, Kao SH (2015) AMPK activation inhibits expression of proinflammatory mediators through downregulation of PI3K/p38 MAPK and NF-κB signaling in murine macrophages. DNA Cell Biol 34:133–141
Hubbard VM, Valdor R, Patel B, Singh R, Cuervo AM, Macian F (2010) Macroautophagy regulates energy metabolism during effector T-cell activation. J Immunol 185:7349–7357
Hudson CC, Liu M, Chiang GG, Otterness DM, Loomis DC, Kaper F, Giaccia AJ, Abraham RT (2002) Regulation of hypoxia-inducible factor 1α expression and function by the mammalian target of rapamycin. Mol Cell Biol 22:7004–7014
Inoki K, Li Y, Zhu T, Wu J, Guan KL (2002) TSC2 is phosphorylated and inhibited by Akt and suppresses mTOR signalling. Nat Cell Biol 4:648–657
Inoki K, Ouyang H, Zhu T, Lindvall C, Wang Y, Zhang X, Yang Q, Bennett C, Harada Y, Stankunas K, Wang CY (2006) TSC2 integrates Wnt and energy signals via a coordinated phosphorylation by AMPK and GSK3 to regulate cell growth. Cell 126:955–968
Inoki K, Kim J, Guan KL (2012) AMPK and mTOR in cellular energy homeostasis and drug targets. Annu Rev Pharmacol Toxicol 52:381–400
Iyer SS, Chatraw JH, Tan WG, Wherry EJ, Becker TC, Ahmed R, Kapasi ZF (2012) Protein energy malnutrition impairs homeostatic proliferation of memory CD8 T cells. J Immunol 188:77–84
Jablonski KA, Amici SA, Webb LM, de Dios Ruiz-Rosado J, Popovich PG, Partida-Sanchez S, Guerau-de-Arellano M (2015) Novel markers to delineate murine M1 and M2 macrophages. PLoS One 10:e0145342
Jacinto E, Loewith R, Schmidt A, Lin S, Rüegg MA, Hall A, Hall MN (2004) Mammalian TOR complex 2 controls the actin cytoskeleton and is rapamycin insensitive. Nat Cell Biol 6:1122–1128
Jacobs SR, Herman CE, MacIver NJ, Wofford JA, Wieman HL, Hammen JJ, Rathmell JC (2008) Glucose uptake is limiting in T cell activation and requires CD28-mediated Akt-dependent and independent pathways. J Immunol 180:4476–4486
Jelkmann W (2011) Regulation of erythropoietin production. J Physiol 589:1251–1258

Jeon SH, Lee MY, Kim SJ, Joe SG, Kim GB, Kim IS, Kim NS, Hong CU, Kim SZ, Kim JS, Kang HS (2007) Taurine increases cell proliferation and generates an increase in [Mg2+] I accompanied by ERK 1/2 activation in human osteoblast cells. FEBS Lett 581:5929–5934

Jewell JL, Kim YC, Russell RC, Yu FX, Park HW, Plouffe SW, Tagliabracci VS, Guan KL (2015) Differential regulation of mTORC1 by leucine and glutamine. Science 347:194–198

Jhun BS, Oh YT, Lee JY, Kong Y, Yoon KS, Kim SS, Baik HH, Ha J, Kang I (2005) AICAR suppresses IL-2 expression through inhibition of GSK-3 phosphorylation and NF-AT activation in Jurkat T cells. Biochem Biophys Res Commun 332:339–346

Jing E, Gesta S, Kahn CR (2007) SIRT2 regulates adipocyte differentiation through FOXO1 acetylation/deacetylation. Cell Metab 6:105–114

Johansen T (1979) Adenosine triphosphate levels during anaphylactic histamine release in rat mast cells in vitro. Effects of glycolytic and respiratory inhibitors. Eur J Pharmacol 58:107–115

Johnson AR, Justin Milner J, Makowski L (2012) The inflammation highway: metabolism accelerates inflammatory traffic in obesity. Immunol Rev 249:218–238

Jones RG, Thompson CB (2007) Revving the engine: signal transduction fuels T cell activation. Immunity 27:173–178

Jones DC, Manning BM, Daynes RA (2002) A role for the peroxisome proliferator-activated receptor α in T-cell physiology and ageing immunobiology. Proc Nutr Soc 61:363–369

Joshi MB, Lad A, Prasad B, Alevoor S, Balakrishnan A, Ramachandra L, Satyamoorthy K (2013) High glucose modulates IL-6 mediated immune homeostasis through impeding neutrophil extracellular trap formation. FEBS Lett 587:2241–2246

Jump DB (2004) Fatty acid regulation of gene transcription. Crit Rev Clin Lab Sci 41:41–78

Jung CH, Ro SH, Cao J, Otto NM, Kim DH (2010) mTOR regulation of autophagy. FEBS Lett 584:1287–1295

Kaesler S, Sobiesiak M, Kneilling M, Volz T, Kempf WE, Lang PA, Lang KS, Wieder T, Heller-Stilb B, Warskulat U, Häussinger D (2012) Effective T-cell recall responses require the taurine transporter Taut. Eur J Immunol 42:831–841

Kamphorst JJ, Cross JR, Fan J, de Stanchina E, Mathew R, White EP, Thompson CB, Rabinowitz JD (2013) Hypoxic and Ras-transformed cells support growth by scavenging unsaturated fatty acids from lysophospholipids. Proc Natl Acad Sci USA 110:8882–8887

Kane LP, Weiss A (2003) The PI-3 kinase/Akt pathway and T cell activation: pleiotropic pathways downstream of PIP3. Immunol Rev 192:7–20

Kawai T, Akira S (2009) The roles of TLRs, RLRs and NLRs in pathogen recognition. Int Immunol 21:317–337

Kawai T, Akira S (2011) Toll-like receptors and their crosstalk with other innate receptors in infection and immunity. Immunity 34:637–650

Keating SE, Zaiatz-Bittencourt V, Loftus RM, Keane C, Brennan K, Finlay DK, Gardiner CM (2016) Metabolic reprogramming supports IFN-γ production by CD56bright NK cells. J Immunol 196:2552–2560

Kelly B, O'Neill LA (2015) Metabolic reprogramming in macrophages and dendritic cells in innate immunity. Cell Res 25:771–784

Keppel MP, Saucier N, Mah AY, Vogel TP, Cooper MA (2015) Activation-specific metabolic requirements for NK cell IFN-γ production. J Immunol 194:1954–1962

Kerdiles YM, Beisner DR, Tinoco R, Dejean AS, Castrillon DH, DePinho RA, Hedrick SM (2009) Foxo1 links homing and survival of naive T cells by regulating L-selectin, CCR7 and interleukin 7 receptor. Nat Immunol 10:176–184

Kersten S (2014) Integrated physiology and systems biology of PPARα. Mol Metabol 3:354–371

Kidani Y, Elsaesser H, Hock MB, Vergnes L, Williams KJ, Argus JP, Marbois BN, Komisopoulou E, Wilson EB, Osborne TF, Graeber TG (2013) Sterol regulatory element-binding proteins are essential for the metabolic programming of effector T cells and adaptive immunity. Nat Immunol 14:489–499

Kim JW, Tchernyshyov I, Semenza GL, Dang CV (2006a) HIF-1-mediated expression of pyruvate dehydrogenase kinase: a metabolic switch required for cellular adaptation to hypoxia. Cell Metab 3:177–185

Kim KY, Kim JK, Han SH, Lim JS, Kim KI, Cho DH, Lee MS, Lee JH, Yoon DY, Yoon SR, Chung JW (2006b) Adiponectin is a negative regulator of NK cell cytotoxicity. J Immunol 176:5958–5964

Kim J, Kundu M, Viollet B, Guan KL (2011) AMPK and mTOR regulate autophagy through direct phosphorylation of Ulk1. Nat Cell Biol 13:132–141

Kim MV, Ouyang W, Liao W, Zhang MQ, Li MO (2013a) The transcription factor Foxo1 controls central-memory $CD8^+$ T cell responses to infection. Immunity 39:286–297

Kim SG, Buel GR, Blenis J (2013b) Nutrient regulation of the mTOR complex 1 signaling pathway. Mol Cells 35:463–473

Kim MH, Taparowsky EJ, Kim CH (2015) Retinoic acid differentially regulates the migration of innate lymphoid cell subsets to the gut. Immunity 43:107–119

Kliewer SA, Umesono K, Noonan DJ, Heyman RA, Evans RM (1992) Convergence of 9-cis retinoic acid and peroxisome proliferator signalling pathways through heterodimer formation of their receptors. Nature 358:771–774

Kliewer SA, Forman BM, Blumberg B, Ong ES, Borgmeyer U, Mangelsdorf DJ, Umesono K, Evans RM (1994) Differential expression and activation of a family of murine peroxisome proliferator-activated receptors. Proc Natl Acad Sci USA 91:7355–7359

Klimova T, Chandel NS (2008) Mitochondrial complex III regulates hypoxic activation of HIF. Cell Death Differ 15:660–666

Klotz L, Dani I, Edenhofer F, Nolden L, Evert B, Paul B, Kolanus W, Klockgether T, Knolle P, Diehl L (2007) Peroxisome proliferator-activated receptor γ control of dendritic cell function contributes to development of $CD4^+$ T cell anergy. J Immunol 178:2122–2131

Kojima H, Sitkovsky MV, Cascalho M (2003) HIF-1α deficiency perturbs T and B cell functions. Curr Pharm Des 9:1827–1832

Kola B (2008) Role of AMP-activated protein kinase in the control of appetite. J Neuroendocrinol 20:942–951

Kolev M, Dimeloe S, Le Friec G, Navarini A, Arbore G, Povoleri GA, Fischer M, Belle R, Loeliger J, Develioglu L, Bantug GR (2015) Complement regulates nutrient influx and metabolic reprogramming during Th1 cell responses. Immunity 42:1033–1047

Kong S, Yeung P, Fang D (2013) The class III histone deacetylase sirtuin 1 in immune suppression and its therapeutic potential in rheumatoid arthritis. J Genet Genomics 40:347–354

Kopf H, Gonzalo M, Howard OZ, Chen X (2007) Rapamycin inhibits differentiation of Th17 cells and promotes generation of FoxP3+ T regulatory cells. Int Pharmacol 7:1819–1824

Kovacic S, Soltys CL, Barr AJ, Shiojima I, Walsh K, Dyck JR (2003) Akt activity negatively regulates phosphorylation of AMP-activated protein kinase in the heart. J Biol Chem 278:39422–39427

Kramer PA, Ravi S, Chacko B, Johnson MS, Darley-Usmar VM (2014) A review of the mitochondrial and glycolytic metabolism in human platelets and leukocytes: implications for their use as bioenergetic biomarkers. Redox Biol 2:206–210

Krawczyk CM, Holowka T, Sun J, Blagih J, Amiel E, DeBerardinis RJ, Cross JR, Jung E, Thompson CB, Jones RG, Pearce EJ (2010) Toll-like receptor-induced changes in glycolytic metabolism regulate dendritic cell activation. Blood 115:4742–4749

Kuschak TI, Kuschak BC, Taylor CL, Wright JA, Wiener F, Mai S (2002) c-Myc initiates illegitimate replication of the ribonucleotide reductase R2 gene. Oncogene 21:909

Laplante M, Sabatini DM (2009) mTOR signaling at a glance. J Cell Sci 122(20):3589–3594

Larsson C (2006) Protein kinase C and the regulation of the actin cytoskeleton. Cell Signal 18:276–284

Lawrence T, Natoli G (2011) Transcriptional regulation of macrophage polarization: enabling diversity with identity. Nat Rev Immunol 11:750–761

LeBleu VS, O'Connell JT, Herrera KN, Wikman H, Pantel K, Haigis MC, De Carvalho FM, Damascena A, Chinen LT, Rocha RM, Asara JM (2014) PGC-1α mediates mitochondrial biogenesis and oxidative phosphorylation in cancer cells to promote metastasis. Nat Cell Biol 16:992–1003

Lee WJ, Kim M, Park HS, Kim HS, Jeon MJ, Oh KS, Koh EH, Won JC, Kim MS, Oh GT, Yoon M (2006) AMPK activation increases fatty acid oxidation in skeletal muscle by activating PPARα and PGC-1. Biochem Biophys Res Commun 340:291–295

Lee H, Cho JS, Lambacher N, Lee J, Lee SJ, Lee TH, Gartner A, Koo HS (2008) The *Caenorhabditis elegans* AMP-activated protein kinase AAK-2 is phosphorylated by LKB1 and is required for resistance to oxidative stress and for normal motility and foraging behavior. J Biol Chem 283(22):14988–14993

Lee K, Gudapati P, Dragovic S, Spencer C, Joyce S, Killeen N, Magnuson MA, Boothby M (2010) Mammalian target of rapamycin protein complex 2 regulates differentiation of Th1 and Th2 cell subsets via distinct signaling pathways. Immunity 32:743–753

Lee JS, Cella M, McDonald KG, Garlanda C, Kennedy GD, Nukaya M, Mantovani A, Kopan R, Bradfield CA, Newberry RD, Colonna M (2012) AHR drives the development of gut ILC22 cells and postnatal lymphoid tissues via pathways dependent on and independent of Notch. Nat Immunol 13:144–151

Lee JV, Carrer A, Shah S, Snyder NW, Wei S, Venneti S, Worth AJ, Yuan ZF, Lim HW, Liu S, Jackson E (2014) Akt-dependent metabolic reprogramming regulates tumor cell histone acetylation. Cell Metab 20:306–319

Lee JH, Elly C, Park Y, Liu YC (2015) E3 ubiquitin ligase VHL regulates hypoxia-inducible factor-1α to maintain regulatory T cell stability and suppressive capacity. Immunity 42:1062–1074

Lee JH, Cho US, Karin M (2016) Sestrin regulation of TORC1: is Sestrin a leucine sensor? *Sci Signal* 9:re5

Levring TB, Hansen AK, Nielsen BL, Kongsbak M, von Essen MR, Woetmann A, Ødum N, Bonefeld CM, Geisler C (2012) Activated human $CD4^+$ T cells express transporters for both cysteine and cystine. Sci Rep 2:266

L'homme L, Esser N, Riva L, Scheen A, Paquot N, Piette J, Legrand-Poels S (2013) Unsaturated fatty acids prevent activation of NLRP3 inflammasome in human monocytes/macrophages. J Lipid Res 54:2998–3008

Li Y, Innocentin S, Withers DR, Roberts NA, Gallagher AR, Grigorieva EF, Wilhelm C, Veldhoen M (2011a) Exogenous stimuli maintain intraepithelial lymphocytes via aryl hydrocarbon receptor activation. Cell 147:629–640

Li Y, Xu S, Mihaylova MM, Zheng B, Hou X, Jiang B, Park O, Luo Z, Lefai E, Shyy JY, Gao B (2011b) AMPK phosphorylates and inhibits SREBP activity to attenuate hepatic steatosis and atherosclerosis in diet-induced insulin-resistant mice. Cell Metab 13:376–388

Li S, Sullivan NL, Rouphael N, Yu T, Banton S, Maddur MS, McCausland M, Chiu C, Canniff J, Dubey S, Liu K (2017) Metabolic phenotypes of response to vaccination in humans. Cell 169:862–877

Lin Y, Tang Y, Wang F (2016) The protective effect of HIF-1α in t lymphocytes on cardiac damage in diabetic mice. Ann Clin Lab Sci 46:32–43

Liu L, Cash TP, Jones RG, Keith B, Thompson CB, Simon MC (2006) Hypoxia-induced energy stress regulates mRNA translation and cell growth. Mol Cell 21:521–531

Liu J, Divoux A, Sun J, Zhang J, Clément K, Glickman JN, Sukhova GK, Wolters PJ, Du J, Gorgun CZ, Doria A (2009) Genetic deficiency and pharmacological stabilization of mast cells reduce diet-induced obesity and diabetes in mice. Nat Med 15:940–945

Liu T, Li J, Liu Y, Xiao N, Suo H, Xie K, Yang C, Wu C (2012) Short-chain fatty acids suppress lipopolysaccharide-induced production of nitric oxide and proinflammatory cytokines through inhibition of NF-κB pathway in RAW264.7 cells. Inflammation 35:1676–1684

Liu G, Bi Y, Xue L, Zhang Y, Yang H, Chen X, Lu Y, Zhang Z, Liu H, Wang X, Wang R (2015) Dendritic cell SIRT1–HIF1α axis programs the differentiation of $CD4^+$ T cells through IL-12 and TGF-β1. Proc Natl Acad Sci USA 112:E957–E965

Lizcano JM, Göransson O, Toth R, Deak M, Morrice NA, Boudeau J, Hawley SA, Udd L, Mäkelä TP, Hardie DG, Alessi DR (2004) LKB1 is a master kinase that activates 13 kinases of the AMPK subfamily, including MARK/PAR-1. EMBO J 23:833–843

Lochner M, Berod L, Sparwasser T (2015) Fatty acid metabolism in the regulation of T cell function. Trends Immunol 36:81–91

Long X, Lin Y, Ortiz-Vega S, Yonezawa K, Avruch J (2005) Rheb binds and regulates the mTOR kinase. Curr Biol 15:702–713

Lord GM, Matarese G, Howard JK, Baker RJ, Bloom SR, Lechler RI (1998) Leptin modulates the T-cell immune response and reverses starvation-induced immunosuppression. Nature 394:897–901

Lucas CL, Kuehn HS, Zhao F, Niemela JE, Deenick EK, Palendira U, Avery DT, Moens L, Cannons JL, Biancalana M, Stoddard J (2014) Dominant-activating germline mutations in the gene encoding the PI(3)K catalytic subunit p110δ result in T cell senescence and human immunodeficiency. Nat Immunol 15:88–97

Luo J, Manning BD, Cantley LC (2003) Targeting the PI3K-Akt pathway in human cancer: rationale and promise. Cancer Cell 4:257–262

MacDonald PE, Joseph JW, Rorsman P (2005) Glucose-sensing mechanisms in pancreatic β-cells. Philos Trans R Soc Lond B Biol Sci 360:2211–2225

Macintyre AN, Finlay D, Preston G, Sinclair LV, Waugh CM, Tamas P, Feijoo C, Okkenhaug K, Cantrell DA (2011) Protein kinase B controls transcriptional programs that direct cytotoxic T cell fate but is dispensable for T cell metabolism. Immunity 34:224–236

Macintyre AN, Gerriets VA, Nichols AG, Michalek RD, Rudolph MC, Deoliveira D, Anderson SM, Abel ED, Chen BJ, Hale LP, Rathmell JC (2014) The glucose transporter Glut1 is selectively essential for CD4 T cell activation and effector function. Cell Metab 20:61–72

MacIver NJ, Blagih J, Saucillo DC, Tonelli L, Griss T, Rathmell JC, Jones RG (2011) The liver kinase B1 is a central regulator of T cell development, activation, and metabolism. J Immunol 187:4187–4198

MacIver NJ, Michalek RD, Rathmell JC (2013) Metabolic regulation of T lymphocytes. Annu Rev Immunol 31:259–283

MacMicking J, Xie QW, Nathan C (1997) Nitric oxide and macrophage function. Annu Rev Immunol 15:323–350

Maekawa Y, Ishifune C, Tsukumo SI, Hozumi K, Yagita H, Yasutomo K (2015) Notch controls the survival of memory CD4+ T cells by regulating glucose uptake. Nat Med 21:55–61

Manning BD, Cantley LC (2007) AKT/PKB signaling: navigating downstream. Cell 129:1261–1274

Mantovani A, Allavena P, Sica A, Balkwill F (2008) Cancer-related inflammation. Nature 454:436–444

Marçais A, Viel S, Grau M, Henry T, Marvel J, Walzer T (2013) Regulation of mouse NK cell development and function by cytokines. Front Immunol 4:450

Marçais A, Cherfils-Vicini J, Viant C, Degouve S, Viel S, Fenis A, Rabilloud J, Mayol K, Tavares A, Bienvenu J, Gangloff YG (2014) The metabolic checkpoint kinase mTOR is essential for IL-15 signaling during the development and activation of NK cells. Nat Immunol 15:749–757

Mariathasan S, Newton K, Monack DM, Vucic D, French DM, Lee WP, Roose-Girma M, Erickson S, Dixit VM (2004) Differential activation of the inflammasome by caspase-1 adaptors ASC and Ipaf. Nature 430:213–218

Mariathasan S, Weiss DS, Newton K, McBride J, O'rourke K, Roose-Girma M, Lee WP, Weinrauch Y, Monack DM, Dixit VM (2006) Cryopyrin activates the inflammasome in response to toxins and ATP. Nature 440:228–232

Marsin AS, Bouzin C, Bertrand L, Hue L (2002) The stimulation of glycolysis by hypoxia in activated monocytes is mediated by AMP-activated protein kinase and inducible 6-phosphofructo-2-kinase. J Biol Chem 277:30778–30783

Martinez FO, Gordon S (2014) The M1 and M2 paradigm of macrophage activation: time for reassessment. F1000Prime Rep 6:13
Martinon F, Burns K, Tschopp J (2002) The inflammasome: a molecular platform triggering activation of inflammatory caspases and processing of proIL-β. Mol Cell 10:417–426
Maslowski KM, Mackay CR (2011) Diet, gut microbiota and immune responses. Nat Immunol 12:5–9
Masters SL, Dunne A, Subramanian SL, Hull RL, Tannahill GM, Sharp FA, Becker C, Franchi L, Yoshihara E, Chen Z, Mullooly N (2010) Activation of the NLRP3 inflammasome by islet amyloid polypeptide provides a mechanism for enhanced IL-1β in type 2 diabetes. Nat Immunol 11:897–904
Masui R, Sasaki M, Funaki Y, Ogasawara N, Mizuno M, Iida A, Izawa S, Kondo Y, Ito Y, Tamura Y, Yanamoto K (2013) G protein-coupled receptor 43 moderates gut inflammation through cytokine regulation from mononuclear cells. Inflamm Bowel Dis 19:2848–2856
Maxwell PH, Wiesener MS, Chang GW, Clifford SC, Vaux EC, Cockman ME, Wykoff CC, Pugh CW, Maher ER, Ratcliffe PJ (1999) The tumour suppressor protein VHL targets hypoxia-inducible factors for oxygen-dependent proteolysis. Nature 399:271–275
Mayne CG, Williams CB (2013) Induced and natural regulatory T cells in the development of inflammatory bowel disease. Inflamm Bowel Dis 19:1772–1788
McCarthy JJ, Esser KA (2010) Anabolic and catabolic pathways regulating skeletal muscle mass. Curr Opin Clin Nutr Metab Care 13:230
McNamee EN, Johnson DK, Homann D, Clambey ET (2013) Hypoxia and hypoxia-inducible factors as regulators of T cell development, differentiation, and function. Immunol Res 55:58–70
Medzhitov R, Janeway C (2000) Innate immune recognition: mechanisms and pathways. Immunol Rev 173:89–97
Mendler AN, Hu B, Prinz PU, Kreutz M, Gottfried E, Noessner E (2012) Tumor lactic acidosis suppresses CTL function by inhibition of p38 and JNK/c-Jun activation. Int J Cancer 131:633–640
Meydani SN, Barklund MP, Liu S, Meydani M, Miller RA, Cannon JG, Morrow FD, Rocklin R, Blumberg JB (1990) Vitamin E supplementation enhances cell-mediated immunity in healthy elderly subjects. Am J Clin Nutr 52:557–563
Mezrich JD, Fechner JH, Zhang X, Johnson BP, Burlingham WJ, Bradfield CA (2010) An interaction between kynurenine and the aryl hydrocarbon receptor can generate regulatory T cells. J Immunol 185:3190–3198
Michalek RD, Rathmell JC (2010) The metabolic life and times of a T-cell. Immunol Rev 236:190–202
Michalek RD, Gerriets VA, Jacobs SR, Macintyre AN, MacIver NJ, Mason EF, Sullivan SA, Nichols AG, Rathmell JC (2011a) Cutting edge: distinct glycolytic and lipid oxidative metabolic programs are essential for effector and regulatory $CD4^+$ T cell subsets. J Immunol 186:3299–3303
Michalek RD, Gerriets VA, Nichols AG, Inoue M, Kazmin D, Chang CY, Dwyer MA, Nelson ER, Pollizzi KN, Ilkayeva O, Giguere V (2011b) Estrogen-related receptor-α is a metabolic regulator of effector T-cell activation and differentiation. Proc Natl Acad Sci USA 108:18348–18353
Michelini RH, Doedens AL, Goldrath AW, Hedrick SM (2013) Differentiation of CD8 memory T cells depends on Foxo1. J Exp Med 210(6):1189–1200. https://doi.org/10.1084/jem.20130392
Mills CD, Kincaid K, Alt JM, Heilman MJ, Hill AM (2000) M-1/M-2 macrophages and the Th1/Th2 paradigm. J Immunol 164:6166–6173
Mills EL, Kelly B, Logan A, Costa AS, Varma M, Bryant CE, Tourlomousis P, Däbritz JH, Gottlieb E, Latorre I, Corr SC (2016) Succinate dehydrogenase supports metabolic repurposing of mitochondria to drive inflammatory macrophages. Cell 167:457–470
Misawa T, Takahama M, Kozaki T, Lee H, Zou J, Saitoh T, Akira S (2013) Microtubule-driven spatial arrangement of mitochondria promotes activation of the NLRP3 inflammasome. Nat Immunol 14:454–460

Mogensen TH (2009) Pathogen recognition and inflammatory signaling in innate immune defenses. Clin Microbiol Rev 22:240–273
Mok CC, Lau CS (2003) Pathogenesis of systemic lupus erythematosus. J Clin Pathol 56:481–490
Momcilovic M, Hong SP, Carlson M (2006) Mammalian TAK1 activates Snf1 protein kinase in yeast and phosphorylates AMP-activated protein kinase in vitro. J Biol Chem 281:25336–25343
Mora JR, Iwata M, Von Andrian UH (2008) Vitamin effects on the immune system: vitamins A and D take centre stage. Nat Rev Immunol 8:685–698
Moretta L, Moretta A (2004) Unravelling natural killer cell function: triggering and inhibitory human NK receptors. EMBO J 23(2):255–259
Morris SM (2007) Arginine metabolism: boundaries of our knowledge. J Nutr 137:1602S–1609S
Mortha A, Chudnovskiy A, Hashimoto D, Bogunovic M, Spencer SP, Belkaid Y, Merad M (2014) Microbiota-dependent crosstalk between macrophages and ILC3 promotes intestinal homeostasis. Science 343:1249288
Mosmann TR, Coffman RL (1989) TH1 and TH2 cells: different patterns of lymphokine secretion lead to different functional properties. Annu Rev Immunol 7:145–173
Mosmann TR, Cherwinski H, Bond MW, Giedlin MA, Coffman RL (1986) Two types of murine helper T cell clone. I. Definition according to profiles of lymphokine activities and secreted proteins. J Immunol 136:2348–2357
Mosser DM, Edwards JP (2008) Exploring the full spectrum of macrophage activation. Nat Rev Immunol 8:958–969
Mottaghi A, Ebrahimof S, Angoorani P, Saboor-Yaraghi AA (2014) Vitamin A supplementation reduces IL-17 and RORc gene expression in atherosclerotic patients. Scand J Immunol 80:151–157
Moussaieff A, Rouleau M, Kitsberg D, Cohen M, Levy G, Barasch D, Nemirovski A, Shen-Orr S, Laevsky I, Amit M, Bomze D (2015) Glycolysis-mediated changes in acetyl-CoA and histone acetylation control the early differentiation of embryonic stem cells. Cell Metab 21:392–402
Mucida D, Park Y, Kim G, Turovskaya O, Scott I, Kronenberg M, Cheroutre H (2007) Reciprocal T_H17 and regulatory T cell differentiation mediated by retinoic acid. Science 317:256–260
Mukhopadhyay D, Mukherjee S, Roy S, Dalton JE, Kundu S, Sarkar A, Das NK, Kaye PM, Chatterjee M (2015) M2 polarization of monocytes-macrophages is a hallmark of Indian post kala-azar dermal leishmaniasis. PLoS Negl Trop Dis 9:e0004145
Munn DH, Sharma MD, Lee JR, Jhaver KG, Johnson TS, Keskin DB, Marshall B, Chandler P, Antonia SJ, Burgess R, Slingluff CL (2002) Potential regulatory function of human dendritic cells expressing indoleamine 2, 3-dioxygenase. Science 297:1867–1870
Munn DH, Sharma MD, Baban B, Harding HP, Zhang Y, Ron D, Mellor AL (2005) GCN2 kinase in T cells mediates proliferative arrest and anergy induction in response to indoleamine 2, -3-dioxygenase. Immunity 22:633–642
Murphy MP, Siegel RM (2013) Mitochondrial ROS fire up T cell activation. Immunity 38:201–202
Murray PJ, Allen JE, Biswas SK, Fisher EA, Gilroy DW, Goerdt S, Gordon S, Hamilton JA, Ivashkiv LB, Lawrence T, Locati M (2014) Macrophage activation and polarization: nomenclature and experimental guidelines. Immunity 41:14–20
Nakahama T, Hanieh H, Nguyen NT, Chinen I, Ripley B, Millrine D, Lee S, Nyati KK, Dubey PK, Chowdhury K, Kawahara Y (2013) Aryl hydrocarbon receptor-mediated induction of the microRNA-132/212 cluster promotes interleukin-17-producing T-helper cell differentiation. Proc Natl Acad Sci USA 110:11964–11969
Nakaya M, Xiao Y, Zhou X, Chang JH, Chang M, Cheng X, Blonska M, Lin X, Sun SC (2014) Inflammatory T cell responses rely on amino acid transporter ASCT2 facilitation of glutamine uptake and mTORC1 kinase activation. Immunity 40:692–705
Navarro F, Bacurau AV, Vanzelli A, Meneguello-Coutinho M, Uchida MC, Moraes MR, Almeida SS, Wasinski F, Barros CC, Würtele M, Araújo RC (2010) Changes in glucose and glutamine lymphocyte metabolisms induced by type I interferon α. Mediators Inflamm. https://doi.org/10.1155/2010/364290

Netea MG, Simon A, van de Veerdonk F, Kullberg BJ, Van der Meer JW, Joosten LA (2010) IL-1β processing in host defense: beyond the inflammasomes. PLoS Pathog 6:e1000661
Newsholme P, Newsholme EA (1989) Rates of utilization of glucose, glutamine and oleate and formation of end-products by mouse peritoneal macrophages in culture. Biochem J 261:211–218
Newsholme P, Curi R, Gordon S, Newsholme EA (1986) Metabolism of glucose, glutamine, long-chain fatty acids and ketone bodies by murine macrophages. Biochem J 239:121–125
Niccoli T, Partridge L (2012) Ageing as a risk factor for disease. Curr Biol 22:R741–R752
Nicklin P, Bergman P, Zhang B, Triantafellow E, Wang H, Nyfeler B, Yang H, Hild M, Kung C, Wilson C, Myer VE (2009) Bidirectional transport of amino acids regulates mTOR and autophagy. Cell 136:521–534
Nunn AV, Bell J, Barter P (2007) The integration of lipid-sensing and anti-inflammatory effects: how the PPARs play a role in metabolic balance. Nucl Recept 5:1
O'Brien TF, Gorentla BK, Xie D, Srivatsan S, McLeod IX, He YW, Zhong XP (2011) Regulation of T-cell survival and mitochondrial homeostasis by TSC1. Eur J Immunol 41:3361–3370
O'Sullivan D, van der Windt GJ, Huang SC, Curtis JD, Chang CH, Buck MD, Qiu J, Smith AM, Lam WY, DiPlato LM, Hsu FF (2014) Memory CD8+ T cells use cell-intrinsic lipolysis to support the metabolic programming necessary for development. Immunity 41:75–88
Odegaard JI, Chawla A (2011) Alternative macrophage activation and metabolism. Annu Rev Pathol Mech Dis 6:275–297
Ohh M, Park CW, Ivan M, Hoffman MA, Kim TY, Huang LE, Pavletich N, Chau V, Kaelin WG (2000) Ubiquitination of hypoxia-inducible factor requires direct binding to the β-domain of the von Hippel–Lindau protein. Nat Cell Biol 2:423–427
Okkenhaug K, Ali K, Vanhaesebroeck B (2007) Antigen receptor signalling: a distinctive role for the p110δ isoform of PI3K. Trends Immunol 28:80–87
O'Neill LA (2014) Glycolytic reprogramming by TLRs in dendritic cells. Nat Immunol 15:314–315
O'Neill LA, Hardie DG (2013) Metabolism of inflammation limited by AMPK and pseudo-starvation. Nature 493:346–355
Opitz CA, Litzenburger UM, Sahm F, Ott M, Tritschler I, Trump S, Schumacher T, Jestaedt L, Schrenk D, Weller M, Jugold M (2011) An endogenous tumour-promoting ligand of the human aryl hydrocarbon receptor. Nature 478:197–203
O'Reilly KE, Rojo F, She QB, Solit D, Mills GB, Smith D, Lane H, Hofmann F, Hicklin DJ, Ludwig DL, Baselga J (2006) mTOR inhibition induces upstream receptor tyrosine kinase signaling and activates Akt. Cancer Res 66:1500–1508
Oren R, Farnham AE, Saito K, Milofsky E, Karnovsky ML (1963) Metabolic patterns in three types of phagocytizing cells. J Cell Biol 17:487–501
O'Sullivan D, Pearce EL (2015) Targeting T cell metabolism for therapy. Trends Immunol 36:71–80
Ouyang W, Beckett O, Flavell RA, Li MO (2009) An essential role of the Forkhead-box transcription factor Foxo1 in control of T cell homeostasis and tolerance. Immunity 30:358–371
Pai CC, Kearsey SE (2017) A critical balance: dNTPs and the maintenance of genome stability. Genes 8:57
Palazon A, Goldrath AW, Nizet V, Johnson RS (2014) HIF transcription factors, inflammation, and immunity. Immunity 41:518–528
Paludan C, Schmid D, Landthaler M, Vockerodt M, Kube D, Tuschl T, Münz C (2005) Endogenous MHC class II processing of a viral nuclear antigen after autophagy. Science 307:593–596
Park J, Kim M, Kang SG, Jannasch AH, Cooper B, Patterson J, Kim CH (2015) Short-chain fatty acids induce both effector and regulatory T cells by suppression of histone deacetylases and regulation of the mTOR–S6K pathway. Mucosal Immunol 8:80–93
Parry RV, Chemnitz JM, Frauwirth KA, Lanfranco AR, Braunstein I, Kobayashi SV, Linsley PS, Thompson CB, Riley JL (2005) CTLA-4 and PD-1 receptors inhibit T-cell activation by distinct mechanisms. Mol Cell Biol 25:9543–9553

Patsoukis N, Brown J, Petkova V, Liu F, Li L, Boussiotis VA (2012) Selective effects of PD-1 on Akt and Ras pathways regulate molecular components of the cell cycle and inhibit T cell proliferation. Sci Signal 5:ra46

Patsoukis N, Bardhan K, Chatterjee P, Sari D, Liu B, Bell LN, Karoly ED, Freeman GJ, Petkova V, Seth P, Li L (2015) PD-1 alters T-cell metabolic reprogramming by inhibiting glycolysis and promoting lipolysis and fatty acid oxidation. Nat Commun 6:6692

Pauly M, Daussin F, Burelle Y, Li T, Godin R, Fauconnier J, Koechlin-Ramonatxo C, Hugon G, Lacampagne A, Coisy-Quivy M, Liang F (2012) AMPK activation stimulates autophagy and ameliorates muscular dystrophy in the mdx mouse diaphragm. Am J Pathol 181:583–592

Pearce EJ, Everts B (2015) Dendritic cell metabolism. Nat Rev Immunol 15:18–29

Pearce EL, Pearce EJ (2013) Metabolic pathways in immune cell activation and quiescence. Immunity 38:633–643

Pearce EL, Walsh MC, Cejas PJ, Harms GM, Shen H, Wang LS, Jones RG, Choi Y (2009) Enhancing CD8 T-cell memory by modulating fatty acid metabolism. Nature 460:103–107

Pearce EL, Poffenberger MC, Chang CH, Jones RG (2013) Fueling immunity: insights into metabolism and lymphocyte function. Science 342:1242454

Peng YJ, Nanduri J, Khan SA, Yuan G, Wang N, Kinsman B, Vaddi DR, Kumar GK, Garcia JA, Semenza GL, Prabhakar NR (2011) Hypoxia-inducible factor 2α (HIF-2α) heterozygous-null mice exhibit exaggerated carotid body sensitivity to hypoxia, breathing instability, and hypertension. Proc Natl Acad Sci USA 108:3065–3070

Peters NC, Egen JG, Secundino N, Debrabant A, Kimblin N, Kamhawi S, Lawyer P, Fay MP, Germain RN, Sacks D (2008) In vivo imaging reveals an essential role for neutrophils in leishmaniasis transmitted by sand flies. Science 321:970–974

Peterson TR, Sengupta SS, Harris TE, Carmack AE, Kang SA, Balderas E, Guertin DA, Madden KL, Carpenter AE, Finck BN, Sabatini DM (2011) mTOR complex 1 regulates lipin 1 localization to control the SREBP pathway. Cell 146:408–420

Peyssonnaux C, Datta V, Cramer T, Doedens A, Theodorakis EA, Gallo RL, Hurtado-Ziola N, Nizet V, Johnson RS (2005) HIF-1α expression regulates the bactericidal capacity of phagocytes. J Clin Invest 115:1806–1815

Phong B, Avery L, Menk AV, Delgoffe GM, Kane LP (2017) Cutting edge: murine mast cells rapidly modulate metabolic pathways essential for distinct effector functions. J Immunol 198:640–644

Piemonti L, Monti P, Sironi M, Fraticelli P, Leone BE, Dal Cin E, Allavena P, Di Carlo V (2000) Vitamin D3 affects differentiation, maturation, and function of human monocyte-derived dendritic cells. J Immunol 164:4443–4451

Porstmann T, Santos CR, Griffiths B, Cully M, Wu M, Leevers S, Griffiths JR, Chung YL, Schulze A (2008) SREBP activity is regulated by mTORC1 and contributes to Akt-dependent cell growth. Cell Metab 8:224–236

Pot Kreis C (2012) Aryl hydrocarbon receptor controls regulatory $CD4^+$ T cell function. Swiss Med Weekly 142:w13592

Powell JD, Delgoffe GM (2010) The mammalian target of rapamycin: linking T cell differentiation, function, and metabolism. Immunity 33:301–311

Powell JD, Lerner CG, Schwartz RH (1999) Inhibition of cell cycle progression by rapamycin induces T cell clonal anergy even in the presence of costimulation. J Immunol 162:2775–2784

Pua HH, Dzhagalov I, Chuck M, Mizushima N, He YW (2007) A critical role for the autophagy gene Atg5 in T cell survival and proliferation. J Exp Med 204:25–31

Puigserver P, Rhee J, Donovan J, Walkey CJ, Yoon JC, Oriente F, Kitamura Y, Altomonte J, Dong H, Accili D, Spiegelman BM (2003) Insulin-regulated hepatic gluconeogenesis through FOXO1–PGC-1α interaction. Nature 423:550–555

Quintana FJ, Basso AS, Iglesias AH, Korn T, Farez MF, Bettelli E, Caccamo M, Oukka M, Weiner HL (2008) Control of Treg and TH17 cell differentiation by the aryl hydrocarbon receptor. Nature 453:65–71

Rajamäki K, Lappalainen J, Öörni K, Välimäki E, Matikainen S, Kovanen PT, Eklund KK (2010) Cholesterol crystals activate the NLRP3 inflammasome in human macrophages: a novel link between cholesterol metabolism and inflammation. PLoS One 5:e11765

Ramsay G, Cantrell D (2015) Environmental and metabolic sensors that control T cell biology. Front Immunol 6:99

Rathmell JC, Elstrom RL, Cinalli RM, Thompson CB (2003) Activated Akt promotes increased resting T cell size, CD28-independent T cell growth, and development of autoimmunity and lymphoma. Eur J Immunol 33:2223–2232

Raught B, Gingras AC, Sonenberg N (2001) The target of rapamycin (TOR) proteins. Proc Natl Acad Sci USA 98:7037–7044

Ravnskjaer K, Frigerio F, Boergesen M, Nielsen T, Maechler P, Mandrup S (2010) PPARδ is a fatty acid sensor that enhances mitochondrial oxidation in insulin-secreting cells and protects against fatty acid-induced dysfunction. J Lipid Res 51:1370–1379

Reiling JH, Hafen E (2004) The hypoxia-induced paralogs Scylla and Charybdis inhibit growth by down-regulating S6K activity upstream of TSC in *Drosophila*. Genes Dev 18:2879–2892

Reimer T, Brcic M, Schweizer M, Jungi TW (2008) Poly (I: C) and LPS induce distinct IRF3 and NF-kappaB signaling during type-I IFN and TNF responses in human macrophages. J Leukoc Biol 83:1249–1257

Riese MJ, Grewal J, Das J, Zou T, Patil V, Chakraborty AK, Koretzky GA (2011) Decreased diacylglycerol metabolism enhances ERK activation and augments $CD8^+$ T cell functional responses. J Biol Chem 286:5254–5265

Rinella ME, Sanyal AJ (2016) Management of NAFLD: a stage-based approach. Nat Rev Gastroenterol Hepatol 13:196–205

Rius J, Guma M, Schachtrup C, Akassoglou K, Zinkernagel AS, Nizet V, Johnson RS, Haddad GG, Karin M (2008) NF-κB links innate immunity to the hypoxic response through transcriptional regulation of HIF-1α. Nature 453:807–811

Roos D, Loos JA (1973a) Effect of phytohaemagglutinin on the carbohydrate metabolism of human blood lymphocytes after inhibition of the oxidative phosphorylation. Exp Cell Res 77:121–126

Roos D, Loos JA (1973b) Changes in the carbohydrate metabolism of mitogenically stimulated human peripheral lymphocytes: II. Relative importance of glycolysis and oxidative phosphorylation on phytohaemagglutinin stimulation. Exp Cell Res 77:127–135

Rutter GA, da Silva XG, Leclerc I (2003) Roles of 5′-AMP-activated protein kinase (AMPK) in mammalian glucose homoeostasis. Biochem J 375:1–6

Ryan HE, Lo J, Johnson RS (1998) HIF-1α is required for solid tumor formation and embryonic vascularization. EMBO J 17:3005–3015

Sakai J, Duncan EA, Rawson RB, Hua X, Brown MS, Goldstein JL (1996) Sterol-regulated release of SREBP-2 from cell membranes requires two sequential cleavages, one within a transmembrane segment. Cell 85:1037–1046

Sakamoto T, Seiki M (2009) Mint3 enhances the activity of hypoxia-inducible factor-1 (HIF-1) in macrophages by suppressing the activity of factor inhibiting HIF-1. J Biol Chem 284:30350–30359

Sallusto F, Lenig D, Förster R, Lipp M, Lanzavecchia A (1999) Two subsets of memory T lymphocytes with distinct homing potentials and effector functions. Nature 401:708–712

Salminen A, Hyttinen JM, Kaarniranta K (2011) AMP-activated protein kinase inhibits NF-κB signaling and inflammation: impact on healthspan and lifespan. J Mol Med 89:667–676

Sancak Y, Peterson TR, Shaul YD, Lindquist RA, Thoreen CC, Bar-Peled L, Sabatini DM (2008) The Rag GTPases bind raptor and mediate amino acid signaling to mTORC1. Science 320:1496–1501

Sancak Y, Bar-Peled L, Zoncu R, Markhard AL, Nada S, Sabatini DM (2010) Ragulator-Rag complex targets mTORC1 to the lysosomal surface and is necessary for its activation by amino acids. Cell 141:290–303

Satoh-Takayama N, Vosshenrich CA, Lesjean-Pottier S, Sawa S, Lochner M, Rattis F, Mention JJ, Thiam K, Cerf-Bensussan N, Mandelboim O, Eberl G (2008) Microbial flora drives interleukin

22 production in intestinal NKp46+ cells that provide innate mucosal immune defense. Immunity 29:958–970
Sauer S, Bruno L, Hertweck A, Finlay D, Leleu M, Spivakov M, Knight ZA, Cobb BS, Cantrell D, O'Connor E, Shokat KM (2008) T cell receptor signaling controls Foxp3 expression via PI3K, Akt, and mTOR. Proc Natl Acad Sci USA 105:7797–7802
Sayed BA, Christy A, Quirion MR, Brown MA (2008) The master switch: the role of masT cells in autoimmunity and tolerance. Annu Rev Immunol 26:705–739
Scharping NE, Menk AV, Moreci RS, Whetstone RD, Dadey RE, Watkins SC, Ferris RL, Delgoffe GM (2016) The tumor microenvironment represses T cell mitochondrial biogenesis to drive intratumoral T cell metabolic insufficiency and dysfunction. Immunity 45:374–388
Scheu S, Stetson DB, Reinhardt RL, Leber JH, Mohrs M, Locksley RM (2006) Activation of the integrated stress response during T helper cell differentiation. Nat Immunol 7:644–651
Schmelzle T, Hall MN (2000) TOR, a central controller of cell growth. Cell 103:253–262
Schuhmacher M, Kohlhuber F, Hölzel M, Kaiser C, Burtscher H, Jarsch M, Bornkamm GW, Laux G, Polack A, Weidle UH, Eick D (2001) The transcriptional program of a human B cell line in response to Myc. Nucleic Acids Res 29:397–406
Schwartz RH (2003) T cell anergy. Annu Rev Immunol 21:305–334
Seidel UJ, Schlegel P, Lang P (2013) Natural killer cell mediated antibody-dependent cellular cytotoxicity in tumor immunotherapy with therapeutic antibodies. Front Immunol 4:76
Semenza GL, Wang GL (1992) A nuclear factor induced by hypoxia via de novo protein synthesis binds to the human erythropoietin gene enhancer at a site required for transcriptional activation. Mol Cell Biol 12:5447–5454
Sena LA, Li S, Jairaman A, Prakriya M, Ezponda T, Hildeman DA, Wang CR, Schumacker PT, Licht JD, Perlman H, Bryce PJ (2013) Mitochondria are required for antigen-specific T cell activation through reactive oxygen species signaling. Immunity 38:225–236
Sengupta S, Peterson TR, Sabatini DM (2010) Regulation of the mTOR complex 1 pathway by nutrients, growth factors, and stress. Mol Cell 40:310–322
Shang L, Chen S, Du F, Li S, Zhao L, Wang X (2011) Nutrient starvation elicits an acute autophagic response mediated by Ulk1 dephosphorylation and its subsequent dissociation from AMPK. Proc Natl Acad Sci USA 108:4788–4793
Shaw RJ, Cantley LC (2006) Ras, PI(3)K and mTOR signalling controls tumour cell growth. Nature 441:424–430
Shaw RJ, Kosmatka M, Bardeesy N, Hurley RL, Witters LA, DePinho RA, Cantley LC (2004) The tumor suppressor LKB1 kinase directly activates AMP-activated kinase and regulates apoptosis in response to energy stress. Proc Natl Acad Sci USA 101:3329–3335
Shaw RJ, Lamia KA, Vasquez D, Koo SH, Bardeesy N, DePinho RA, Montminy M, Cantley LC (2005) The kinase LKB1 mediates glucose homeostasis in liver and therapeutic effects of metformin. Science 310:1642–1646
Shi LZ, Wang R, Huang G, Vogel P, Neale G, Green DR, Chi H (2011) HIF1α-dependent glycolytic pathway orchestrates a metabolic checkpoint for the differentiation of TH17 and Treg cells. J Exp Med 208:1367–1376
Shimobayashi M, Hall MN (2016) Multiple amino acid sensing inputs to mTORC1. Cell Res 26:7–20
Shirai T, Nazarewicz RR, Wallis BB, Yanes RE, Watanabe R, Hilhorst M, Tian L, Harrison DG, Giacomini JC, Assimes TL, Goronzy JJ (2016) The glycolytic enzyme PKM2 bridges metabolic and inflammatory dysfunction in coronary artery disease. J Exp Med 213:337–354
Sica A, Schioppa T, Mantovani A, Allavena P (2006) Tumour-associated macrophages are a distinct M2 polarised population promoting tumour progression: potential targets of anti-cancer therapy. Eur J Cancer 42:717–727
Sims JE, Smith DE (2010) The IL-1 family: regulators of immunity. Nat Rev Immunol 10:89–102
Sinclair LV, Rolf J, Emslie E, Shi YB, Taylor PM, Cantrell DA (2013) Control of amino-acid transport by antigen receptors coordinates the metabolic reprogramming essential for T cell differentiation. Nat Immunol 14:500–508

Singh NP, Singh UP, Singh B, Price RL, Nagarkatti M, Nagarkatti PS (2011) Activation of aryl hydrocarbon receptor (AhR) leads to reciprocal epigenetic regulation of FoxP3 and IL-17 expression and amelioration of experimental colitis. PLoS One 6:e23522

Smith EM, Finn SG, Tee AR, Browne GJ, Proud CG (2005) The tuberous sclerosis protein TSC2 is not required for the regulation of the mammalian target of rapamycin by amino acids and certain cellular stresses. J Biol Chem 280:18717–18727

Smith PM, Howitt MR, Panikov N, Michaud M, Gallini CA, Bohlooly-y M, Glickman JN, Garrett WS (2013) The microbial metabolites, short-chain fatty acids, regulate colonic Treg cell homeostasis. Science 341:569–573

Smith-Garvin JE, Koretzky GA, Jordan MS (2009) T cell activation. Annu Rev Immunol 27:591–619

Song MY, Wang J, Lee Y, Lee J, Kwon KS, Bae EJ, Park BH (2016) Enhanced M2 macrophage polarization in high n-3 polyunsaturated fatty acid transgenic mice fed a high-fat diet. Mol Nutr Food Res 60:2481–2492

Sood R, Porter AC, Olsen D, Cavener DR, Wek RC (2000) A mammalian homologue of GCN2 protein kinase important for translational control by phosphorylation of eukaryotic initiation factor-2α. Genetics 154:787–801

Stone KD, Prussin C, Metcalfe DD (2010) IgE, mast cells, basophils, and eosinophils. J Allergy Clin Immunol 125:S73–S80

Straub RH, Cutolo M, Buttgereit F, Pongratz G (2010) Energy regulation and neuroendocrine–immune control in chronic inflammatory diseases. J Internal Med 267:543–560

Sukumar M, Roychoudhuri R, Restifo NP (2015) Nutrient competition: a new axis of tumor immunosuppression. Cell 162:1206–1208

Sun X, Kanwar JR, Leung E, Lehnert K, Wang D, Krissansen GW (2001) Gene transfer of antisense hypoxia inducible factor-1 α enhances the therapeutic efficacy of cancer immunotherapy. Gene Therapy 8:638

Sun Y, Connors KE, Yang DQ (2007) AICAR induces phosphorylation of AMPK in an ATM-dependent, LKB1-independent manner. Mol Cell Biochem 306:239–245

Sun K, Kusminski CM, Scherer PE (2011a) Adipose tissue remodeling and obesity. J Clin Invest 121:2094–2101

Sun Q, Chen X, Ma J, Peng H, Wang F, Zha X, Wang Y, Jing Y, Yang H, Chen R, Chang L (2011b) Mammalian target of rapamycin up-regulation of pyruvate kinase isoenzyme type M2 is critical for aerobic glycolysis and tumor growth. Proc Natl Acad Sci USA 108:4129–4134

Sundrud MS, Koralov SB, Feuerer M, Calado DP, Kozhaya AE, Rhule-Smith A, Lefebvre RE, Unutmaz D, Mazitschek R, Waldner H, Whitman M (2009) Halofuginone inhibits TH17 cell differentiation by activating the amino acid starvation response. Science 324:1334–1338

Széles L, Töröcsik D, Nagy L (2007) PPARγ in immunity and inflammation: cell types and diseases. BBA Mol Cell Biol Lipids 1771:1014–1030

Tabor CW, Tabor H (1984) Polyamines. Annu Rev Biochem 53:749–790

Taildeman J, Pérez-Novo CA, Rottiers I, Ferdinande L, Waeytens A, De Colvenaer V, Bachert C, Demetter P, Waelput W, Braet K, Cuvelier CA (2009) Human mast cells express leptin and leptin receptors. Histochem Cell Biol 131:703–711

Tamás P, Hawley SA, Clarke RG, Mustard KJ, Green K, Hardie DG, Cantrell DA (2006) Regulation of the energy sensor AMP-activated protein kinase by antigen receptor and $Ca2^+$ in T lymphocytes. J Exp Med 203:1665–1670

Tamás P, Macintyre A, Finlay D, Clarke R, Feijoo-Carnero C, Ashworth A, Cantrell D (2010) LKB1 is essential for the proliferation of T-cell progenitors and mature peripheral T cells. Eur J Immunol 40:242–253

Tannahill GM, Curtis AM, Adamik J, Palsson-McDermott EM, McGettrick AF, Goel G, Frezza C, Bernard NJ, Kelly B, Foley NH, Zheng L (2013) Succinate is an inflammatory signal that induces IL-1β through HIF-1α. Nature 496:238–242

Tennessen JM, Baker KD, Lam G, Evans J, Thummel CS (2011) The *Drosophila* estrogen-related receptor directs a metabolic switch that supports developmental growth. Cell Metab 13:139–148

Tontonoz P, Nagy L, Alvarez JG, Thomazy VA, Evans RM (1998) PPARγ promotes monocyte/macrophage differentiation and uptake of oxidized LDL. Cell 93:241–252

Topham NJ, Hewitt EW (2009) Natural killer cell cytotoxicity: how do they pull the trigger? Immunology 128:7–15

Traut TW (1994) Physiological concentrations of purines and pyrimidines. Mol Cell Biochem 140 (1):1–22

Tzameli I, Fang H, Ollero M, Shi H, Hamm JK, Kievit P, Hollenberg AN, Flier JS (2004) Regulated production of a peroxisome proliferator-activated receptor-γ ligand during an early phase of adipocyte differentiation in 3T3-L1 adipocytes. J Biol Chem 279:36093–36102

Uematsu S, Akira S (2007) Toll-like receptors and type I interferons. J Biol Chem 282:15319–15323

Uyttenhove C, Pilotte L, Théate I, Stroobant V, Colau D, Parmentier N, Boon T, Van den Eynde BJ (2003) Evidence for a tumoral immune resistance mechanism based on tryptophan degradation by indoleamine 2, 3-dioxygenase. Nat Med 9:1269–1274

Valledor AF, Hsu LC, Ogawa S, Sawka-Verhelle D, Karin M, Glass CK (2004) Activation of liver X receptors and retinoid X receptors prevents bacterial-induced macrophage apoptosis. Proc Natl Acad Sci USA 101:17813–17818

van der Windt GJ, Everts B, Chang CH, Curtis JD, Freitas TC, Amiel E, Pearce EJ, Pearce EL (2012) Mitochondrial respiratory capacity is a critical regulator of $CD8^+$ T cell memory development. Immunity 36:68–78

Van Dyken SJ, Locksley RM (2013) Interleukin-4- and interleukin-13-mediated alternatively activated macrophages: roles in homeostasis and disease. Annu Rev Immunol 31:317–343

Van Kessel KP, Bestebroer J, van Strijp JA (2014) Neutrophil-mediated phagocytosis of *Staphylococcus aureus*. Front Immunol 5:467

van Loosdregt J, Fleskens V, Tiemessen MM, Mokry M, van Boxtel R, Meerding J, Pals CE, Kurek D, Baert MR, Delemarre EM, Gröne A (2013) Canonical Wnt signaling negatively modulates regulatory T cell function. Immunity 39:298–310

Van Raam BJ, Sluiter W, De Wit E, Roos D, Verhoeven AJ, Kuijpers TW (2008) Mitochondrial membrane potential in human neutrophils is maintained by complex III activity in the absence of supercomplex organisation. PLoS One 3:e2013

Van YH, Lee WH, Ortiz S, Lee MH, Qin HJ, Liu CP (2009) All-trans retinoic acid inhibits type 1 diabetes by T regulatory (Treg)-dependent suppression of interferon-γ-producing T-cells without affecting Th17 cells. Diabetes 58:146–155

Vander Heiden MG, Cantley LC, Thompson CB (2009) Understanding the Warburg effect: the metabolic requirements of cell proliferation. Science 324:1029–1033

Vander Heiden MG, Lunt SY, Dayton TL, Fiske BP, Israelsen WJ, Mattaini KR, Vokes NI, Stephanopoulos G, Cantley LC, Metallo CM, Locasale JW (2011) Metabolic pathway alterations that support cell proliferation. Cold Spring Harbor Symp Quant Biol 76:325–334

Vattem KM, Wek RC (2004) Reinitiation involving upstream ORFs regulates ATF4 mRNA translation in mammalian cells. Proc Natl Acad Sci USA 101:11269–11274

Veldhoen M, Hirota K, Christensen J, O'Garra A, Stockinger B (2009) Natural agonists for aryl hydrocarbon receptor in culture medium are essential for optimal differentiation of Th17 T cells. J Exp Med 206:43–49

Viel S, Marçais A, Guimaraes FS, Loftus R, Rabilloud J, Grau M, Degouve S, Djebali S, Sanlaville A, Charrier E, Bienvenu J (2016) TGF-β inhibits the activation and functions of NK cells by repressing the mTOR pathway. Sci Signal 9:ra19

Vinolo MA, Rodrigues HG, Nachbar RT, Curi R (2011) Regulation of inflammation by short chain fatty acids. Nutrients 3:858–876

Vita AJ, Terry RB, Hubert HB, Fries JF (1998) Aging, health risks, and cumulative disability. N Engl J Med 338:1035–1041

Walmsley SR, Farahi N, Peyssonnaux C, Johnson RS, Cramer T, Sobolewski A, Condliffe AM, Cowburn AS, Johnson N, Chilvers ER (2005) Hypoxia-induced neutrophil survival is mediated by HIF-1α-dependent NF-κB activity. J Exp Med 201:105–115

Walsh CM, Fruman DA (2014) Too much of a good thing: immunodeficiency due to hyperactive PI3K signaling. J Clin Invest 124:3688–3690

Wang R, Green DR (2012) Metabolic checkpoints in activated T cells. Nat Immunol 13:907–915

Wang X, Proud CG (2006) The mTOR pathway in the control of protein synthesis. Physiology 21:362–369

Wang X, Sato R, Brown MS, Hua X, Goldstein JL (1994) SREBP-1, a membrane-bound transcription factor released by sterol-regulated proteolysis. Cell 77:53–62

Wang GL, Jiang BH, Rue EA, Semenza GL (1995) Hypoxia-inducible factor 1 is a basic-helix-loop-helix-PAS heterodimer regulated by cellular O2 tension. Proc Natl Acad Sci USA 92:5510–5514

Wang R, Dillon CP, Shi LZ, Milasta S, Carter R, Finkelstein D, McCormick LL, Fitzgerald P, Chi H, Munger J, Green DR (2011) The transcription factor Myc controls metabolic reprogramming upon T lymphocyte activation. Immunity 35:871–882

Wang S, Tsun ZY, Wolfson RL, Shen K, Wyant GA, Plovanich ME, Yuan ED, Jones TD, Chantranupong L, Comb W, Wang T (2015) Lysosomal amino acid transporter SLC38A9 signals arginine sufficiency to mTORC1. Science 347:188–194

Ward PS, Thompson CB (2012) Metabolic reprogramming: a cancer hallmark even Warburg did not anticipate. Cancer Cell 21:297–308

Warden SM, Richardson C, O'Donnell J, Stapleton D, Witters LA (2001) Post-translational modifications of the β-1 subunit of AMP-activated protein kinase affect enzyme activity and cellular localization. Biochem J 354:275–283

Watson E, Yilmaz LS, Walhout AJ (2015) Understanding metabolic regulation at a systems level: metabolite sensing, mathematical predictions, and model organisms. Annu Rev Genet 49:553–575

Waugh C, Sinclair L, Finlay D, Bayascas JR, Cantrell D (2009) Phosphoinositide (3,4,5)-triphosphate binding to phosphoinositide-dependent kinase 1 regulates a protein kinase B/Akt signaling threshold that dictates T-cell migration, not proliferation. Mol Cell Biol 29:5952–5962

Wek RC, Jiang HY, Anthony TG (2006) Coping with stress: eIF2 kinases and translational control. Biochem Soc Trans 34:7–11

Wen H, Gris D, Lei Y, Jha S, Zhang L, Huang MT, Brickey WJ, Ting JP (2011) Fatty acid-induced NLRP3-ASC inflammasome activation interferes with insulin signaling. Nat Immunol 12:408–415

Wen H, Ting JP, O'Neill LA (2012) A role for the NLRP3 inflammasome in metabolic diseases—did Warburg miss inflammation? Nat Immunol 13:352–357

Wenes M, Shang M, Di Matteo M, Goveia J, Martín-Pérez R, Serneels J, Prenen H, Ghesquière B, Carmeliet P, Mazzone M (2016) Macrophage metabolism controls tumor blood vessel morphogenesis and metastasis. Cell Metab 24:701–715

Weng NP, Akbar AN, Goronzy J (2009) CD28− T cells: their role in the age-associated decline of immune function. Trends Immunol 30:306–312

Wensveen FM, Jelenčić V, Valentić S, Šestan M, Wensveen TT, Theurich S, Glasner A, Mendrila D, Štimac D, Wunderlich FT, Brüning JC (2015) NK cells link obesity-induced adipose stress to inflammation and insulin resistance. Nat Immunol 16:376–385

West AP, Brodsky IE, Rahner C, Woo DK, Erdjument-Bromage H, Tempst P, Walsh MC, Choi Y, Shadel GS, Ghosh S (2011) TLR signalling augments macrophage bactericidal activity through mitochondrial ROS. Nature 472:476–480

Wheaton WW, Weinberg SE, Hamanaka RB, Soberanes S, Sullivan LB, Anso E, Glasauer A, Dufour E, Mutlu GM, Budigner GS, Chandel NS (2014) Metformin inhibits mitochondrial complex I of cancer cells to reduce tumorigenesis. elife 3

Wherry EJ (2011) T cell exhaustion. Nat Immunol 12:492–499

Wherry JC, Schreiber RD, Unanue ER (1991) Regulation of gamma interferon production by natural killer cells in SCID mice: roles of tumor necrosis factor and bacterial stimuli. Infect Immun 59:1709–1715

Wieman HL, Wofford JA, Rathmell JC (2007) Cytokine stimulation promotes glucose uptake via phosphatidylinositol-3 kinase/Akt regulation of Glut1 activity and trafficking. Mol Biol Cell 18:1437–1446
Wise DR, DeBerardinis RJ, Mancuso A, Sayed N, Zhang XY, Pfeiffer HK, Nissim I, Daikhin E, Yudkoff M, McMahon SB, Thompson CB (2008) Myc regulates a transcriptional program that stimulates mitochondrial glutaminolysis and leads to glutamine addiction. Proc Natl Acad Sci USA 105:18782–18787
Wofford JA, Wieman HL, Jacobs SR, Zhao Y, Rathmell JC (2008) IL-7 promotes Glut1 trafficking and glucose uptake via STAT5-mediated activation of Akt to support T-cell survival. Blood 111:2101–2111
Wojtaszewski JF, Jørgensen SB, Hellsten Y, Hardie DG, Richter EA (2002) Glycogen-dependent effects of 5-aminoimidazole-4-carboxamide (AICA)-riboside on AMP-activated protein kinase and glycogen synthase activities in rat skeletal muscle. Diabetes 51:284–292
Woo SR, Corrales L, Gajewski TF (2015) Innate immune recognition of cancer. Annu Rev Immunol 33:445–474
Woods A, Vertommen D, Neumann D, Türk R, Bayliss J, Schlattner U, Wallimann T, Carling D, Rider MH (2003) Identification of phosphorylation sites in AMP-activated protein kinase (AMPK) for upstream AMPK kinases and study of their roles by site-directed mutagenesis. J Biol Chem 278:28434–28442
Wu X, Huang M (2008) Dif1 controls subcellular localization of ribonucleotide reductase by mediating nuclear import of the R2 subunit. Mol Cell Biol 28:7156–7167
Wu Q, Liu Y, Chen C, Ikenoue T, Qiao Y, Li CS, Li W, Guan KL, Liu Y, Zheng P (2011) The tuberous sclerosis complex–mammalian target of rapamycin pathway maintains the quiescence and survival of naive T cells. J Immunol 187:1106–1112
Wullschleger S, Loewith R, Hall MN (2006) TOR signaling in growth and metabolism. Cell 124:471–484
Wumesh KC, Satpathy AT, Rapaport AS, Briseño CG, Wu X, Albring JC, Russler-Germain EV, Kretzer NM, Durai V, Persaud SP, Edelson BT (2014) L-Myc expression by dendritic cells is required for optimal T-cell priming. Nature 507:243–247
Xie X, Zhang D, Zhao B, Lu MK, You M, Condorelli G, Wang CY, Guan KL (2011) IκB kinase ε and TANK-binding kinase 1 activate AKT by direct phosphorylation. Proc Natl Acad Sci USA 108:6474–6479
Xiao B, Sanders MJ, Underwood E, Heath R, Mayer FV, Carmena D, Jing C, Walker PA, Eccleston JF, Haire LF, Saiu P (2011) Structure of mammalian AMPK and its regulation by ADP. Nature 472:230
Xu J, Wagoner G, Douglas JC, Drew PD (2009) Liver X receptor agonist regulation of Th17 lymphocyte function in autoimmunity. J Leukoc Biol 86:401–409
Yang XY, Wang LH, Chen T, Hodge DR, Resau JH, DaSilva L, Farrar WL (2000) Activation of human T lymphocytes is inhibited by peroxisome proliferator-activated receptor γ (PPARγ) agonists PPARγ co-association with transcription factor NFAT. J Biol Chem 275:4541–4544
Yang K, Neale G, Green DR, He W, Chi H (2011) The tumor suppressor Tsc1 enforces quiescence of naive T cells to promote immune homeostasis and function. Nat Immunol 12:888–897
Yao R, Zhang Z, An X, Bucci B, Perlstein DL, Stubbe J, Huang M (2003) Subcellular localization of yeast ribonucleotide reductase regulated by the DNA replication and damage checkpoint pathways. Proc Natl Acad Sci USA 100:6628–6633
Zhao E, Maj T, Kryczek I, Li W, Wu K, Zhao L, Wei S, Crespo J, Wan S, Vatan L, Szeliga W (2015) Cancer mediates effector T cell dysfunction by targeting microRNAs and EZH2 via glycolysis restriction. Nat Immunol 17:95–103
Zheng Y, Collins SL, Lutz MA, Allen AN, Kole TP, Zarek PE, Powell JD (2007) A role for mammalian target of rapamycin in regulating T cell activation versus anergy. J Immunol 178:2163–2170
Zheng Y, Delgoffe GM, Meyer CF, Chan W, Powell JD (2009) Anergic T cells are metabolically anergic. J Immunol 183:6095–6101

Zhou G, Myers R, Li Y, Chen Y, Shen X, Fenyk-Melody J, Wu M, Ventre J, Doebber T, Fujii N, Musi N (2001) Role of AMP-activated protein kinase in mechanism of metformin action. J Clin Invest 108:1167–1174

Zhou X, Clister TL, Lowry PR, Seldin MM, Wong GW, Zhang J (2015a) Dynamic visualization of mTORC1 activity in living cells. Cell Rep 10:1767–1777

Zhou Y, Yu X, Chen H, Sjöberg S, Roux J, Zhang L, Ivoulsou AH, Bensaid F, Liu CL, Liu J, Tordjman J (2015b) Leptin deficiency shifts mast cells toward anti-inflammatory actions and protects mice from obesity and diabetes by polarizing M2 macrophages. Cell Metab 22:1045–1058

Zhu J, Yamane H, Paul WE (2009) Differentiation of effector CD4 T cell populations. Annu Rev Immunol 28:445–489

Zhu L, Yang T, Li L, Sun L, Hou Y, Hu X, Zhang L, Tian H, Zhao Q, Peng J, Zhang H (2014) TSC1 controls macrophage polarization to prevent inflammatory disease. Nat Commun 5:4696

Zindy F, Eischen CM, Randle DH, Kamijo T, Cleveland JL, Sherr CJ, Roussel MF (1998) Myc signaling via the ARF tumor suppressor regulates p53-dependent apoptosis and immortalization. Genes Dev 12:2424–2433

Zoncu R, Efeyan A, Sabatini DM (2011) mTOR: from growth signal integration to cancer, diabetes and ageing. Nat Rev Mol Cell Biol 12:21–35

Chapter 4
Alterations on Cellular Redox States upon Infection and Implications for Host Cell Homeostasis

Inês Mesquita, Baptiste Vergnes, and Ricardo Silvestre

Abstract The cofactors nicotinamide adenine dinucleotide (NAD^+) and its phosphate form, $NADP^+$, are crucial molecules present in all living cells. The delicate balance between the oxidized and reduced forms of these molecules is tightly regulated by intracellular metabolism assuring the maintenance of homeostatic conditions, which are essential for cell survival and proliferation. A recent cluster of data has highlighted the importance of the intracellular NAD^+/NADH and $NADP^+$/NADPH ratios during host–pathogen interactions, as fluctuations in the levels of these cofactors and in precursors' bioavailability may condition host response and, therefore, pathogen persistence or elimination. Furthermore, an increasing interest has been given towards how pathogens are capable of hijacking host cell proteins in their own advantage and, consequently, alter cellular redox states and immune function. Here, we review the basic principles behind biosynthesis and subcellular compartmentalization of NAD^+ and $NADP^+$, as well as the importance of these cofactors during infection, with a special emphasis on pathogen-driven modulation of host NAD^+/$NADP^+$ levels and contribution to the associated immune response.

Keywords Nicotinamide adenine dinucleotide (NAD^+) · Host–pathogen interaction · NAD^+/NADH ratio · NADPH · Sirtuins · L-Tryptophan

I. Mesquita · R. Silvestre (✉)
Life and Health Sciences Research Institute (ICVS), School of Health Sciences, University of Minho, Braga, Portugal

ICVS/3B's-PT Government Associate Laboratory, Braga/Guimarães, Portugal
e-mail: ricardosilvestre@med.uminho.pt

B. Vergnes
MIVEGEC (IRD 224-CNRS 5290-Université Montpellier), Institut de Recherche pour le Développement (IRD), Montpellier, France

R. Silvestre, E. Torrado (eds.), *Metabolic Interaction in Infection*, Experientia Supplementum 109, https://doi.org/10.1007/978-3-319-74932-7_4

4.1 Introduction

The maintenance of an adequate cellular redox state, characterized by a tight regulation of the intracellular levels of nicotinamide adenine dinucleotide (NAD^+) and its phosphate form, $NADP^+$, is a vital premise for cellular homeostasis. NAD^+ is reduced to NADH during catabolic reactions, which yields a high quantity of mitochondrial respiration-driven energy to fulfil cellular needs. NADPH is regenerated from $NADP^+$ through a variety of redox reactions that support reductive biosynthesis contributing to maintain the cellular pool of reduced glutathione (GSH), essential for the development of detoxification mechanisms. Consistent with these important roles in cell biology, the intracellular NAD^+/NADH and $NADP^+$/NADPH ratios have arisen as central players in metabolic homeostasis, as well as cell survival and proliferation. A growing body of evidences has begun to unravel the importance of NAD^+ and $NADP^+$ bioavailability and fluctuations during host-pathogen interactions. In this chapter, we will revisit the alterations on cellular redox states and further implications for host cell homeostasis and defence against distinct infectious agents and argue for NAD^+ and $NADP^+$ metabolism as potential therapeutic targets.

4.2 The Biosynthetic Machinery Behind NAD^+ and $NADP^+$ Production

In eukaryotic cells, NAD^+ is originated either through de novo synthesis or by salvage pathways. The de novo pathway relies on the uptake and metabolism of dietary L-tryptophan in N-formylkynurenine, a rate-limiting step catalysed by indoleamine 2,3-dioxygenase (IDO) or tryptophan 2,3-dioxygenase (TDO). Following subsequent enzymatic reactions, quinolinic acid is generated and converted in nicotinic acid mononucleotide (NaMN), which then originates nicotinic acid adenine dinucleotide (NaAD), through the action of nicotinamide mononucleotide adenylyltransferases (NMNATs 1–3). The final step in de novo synthesis is the conversion of NaAD in NAD^+ by the glutamine-dependent NAD^+ synthase (Magni 2008). However, NAD^+ pools can also be replenished through salvage pathways, which rely on the uptake of biosynthetic precursors, namely, nicotinic acid (NA), nicotinamide (NAM) or nicotinamide riboside (NR). Both these biosynthetic pathways may converge by activation of the Preiss–Handler pathway, in which NA is converted to NaMN by the NA phosphoribosyltransferase (NAPT). Additionally, NAM can originate NAD^+ via NAM mononucleotide (NMN) intermediate, through the combined action of the rate-limiting NAM phosphoribosyltransferase (NAMPT) and NMNAT (Revollo et al. 2007). Finally, NR can also be phosphorylated in NMN by nicotinamide ribose kinase (NRK) and then adenylated to NAD^+ by NMNAT (Fig. 4.1).

The conversion of NAD^+ to $NADP^+$ allows the cells to sustain a high demand in reducing equivalents, such as in conditions where high levels of nucleotide, protein

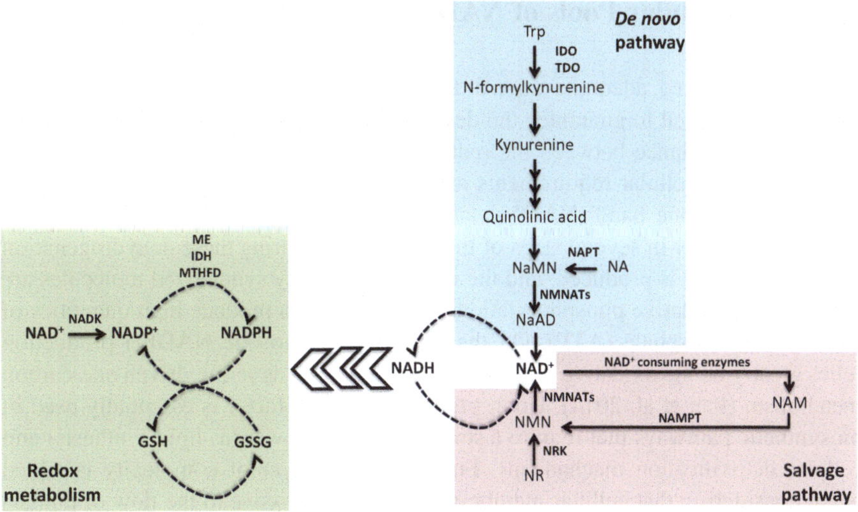

Fig. 4.1 Intracellular machinery for NAD^+/$NADP^+$ synthesis in mammals. De novo (blue) and salvage (red) pathways for NAD^+ production, as well as $NADP^+$ production and utilization during redox metabolism (green) are depicted. *GSSG* oxidized glutathione, *GSH* reduced glutathione, *IDH* isocitrate dehydrogenase, *IDO* indoleamine 2,3-dioxygenase, *ME* malic enzyme, *MTHFD* methylene tetrahydrofolate dehydrogenase, *NAD^+*, nicotinamide adenine dinucleotide, *NaAD* nicotinic acid dinucleotide, *NADP* nicotinamide adenine dinucleotide phosphate, *NADK* NAD^+ kinase, *NAM* nicotinamide, *NaMN* nicotinic acid mononucleotide, *NAMPT* nicotinamide phosphoribosyltransferase, *NAPT* nicotinic acid phosphoribosyltransferase, *NMNAT* nicotinamide mononucleotide adenylyltransferase, *NMN* nicotinamide mononucleotide, *NR* nicotinamide riboside, *NRK* nicotinamide riboside kinase, *TDO* tryptophan 2,3-dioxygenase, *Trp* tryptophan

and lipid synthesis, or detoxification mechanisms are required. The only known enzyme responsible for this replenishment through NAD^+ phosphorylation is NAD^+ kinase (NADK). NADPH regeneration may be achieved in the cytosol and/or mitochondria through the activity of several enzymes: glucose-6-phosphate dehydrogenase (G6PDH) and 6-phosphogluconate dehydrogenase in the oxidative branch of pentose phosphate pathway (PPP), methylenetetrahydrofolate dehydrogenase (MTHFD) and aldehyde dehydrogenases (ALDHs) during folate metabolism and isocitrate dehydrogenases (IDHs) and malic enzyme (ME) in the tricarboxylic acid (TCA) cycle.

The use of ^{13}C-labelled tracers has helped to address the metabolic activity during cellular events. However, using this technology to address NADPH biosynthesis may originate misleading data, as the same carbon source may originate both NADH or NADPH, depending on the isoenzyme involved: some enzymes that catalyse NADPH synthesis, as MTHFD, ALDHs and IDHs may also generate NADH (Liu et al. 2016; Tibbetts and Appling 2010; Wise et al. 2011).

4.3 Intracellular Pools of NAD^+ and $NADP^+$

As abovementioned, adequate ratios of these cofactors within the intracellular milieu must be maintained to guarantee the development of adequate metabolic functions. The exquisite balance between the reduced and oxidized forms of these cofactors depends on the cellular requirements and the chosen metabolic pathways that will fulfil them. On one hand, NAD^+ catalyses catabolic reactions, as it is used as a hydrogen acceptor in several steps of the TCA cycle. During these dehydrogenation reactions, NADH is produced, and the majority of newly synthetized molecules are redirected to oxidative phosphorylation, in an attempt to produce high quantities of adenosine triphosphate (ATP). On the other hand, cytosolic NADPH production relies greatly on the oxidative branch of the PPP, as well as serine-driven one-carbon metabolism (Fan et al. 2014). Upon production, this cofactor is essentially used in biosynthetic pathways that require a strong reductive power (as lipid synthesis) and cellular detoxification mechanisms. Furthermore, a layer of complexity is added when considering that cellular membranes are not permissive to the flow of numerous molecules and that these cofactors are not only essential in the cytosol but are also key elements within distinct organelles. As so, cells are capable of preserving compartmentalized and independent pools, as in the case of NAD^+ (Dölle et al. 2010). Interestingly, intracellular $NADP^+$ content is somewhat limiting, when considering the vast number of metabolic pathways that rely on the bioavailability of this cofactor (proliferation, lipid synthesis and defence against stressful conditions). Similar to what happens with NAD^+, $NADP^+$ is also compartmentalized in different organelles, and electron transport across membranes is only achieved through shuttles. Accordingly, in eukaryotic cells, metabolic reactions can take place in different organelles that will condition and impact the further utilization of specific metabolites and precursors. Consequently, a cluster of studies has started to highlight the importance of certain pathways in the replenishment of organelle-specific pools of NAD^+ and $NADP^+$.

4.3.1 Subcellular NAD^+ Compartmentalization

The main cellular pools of NAD^+ are the cytosolic and mitochondrial ones (VanLinden et al. 2015), with the cytosolic being around 100 times superior than the mitochondrial counterpart. It is estimated that while the cytosolic pool displays NAD^+/NADH ratios between 60 and 700, the mitochondrial one ranges from 7 to 8 (Williamson et al. 1967). Considering the need for distinct intermediates and cofactors during different metabolic processes, it is well-defined the importance of NAD^+ compartmentalization when considering the impermeability of cellular membranes. One of the main enzymes responsible for this subcellular specificity and control of NAD^+ pool distribution is the biosynthetic enzyme NMNAT, which has three distinctly localized isoforms: NMNAT1 is a nuclear enzyme, NMNAT2 has a cytosolic localization and

NMNAT3 is present at the mitochondria (Mori et al. 2014). It is known that the isoform 1 is the most efficient enzyme, responsible for adenylyl transference or pyrophosphorylysis and for the maintenance of the nuclear NAD^+ pool. However, whether NMNAT1-driven NAD^+ can shuttle to the cytoplasm, by exchange through nuclear pores, is still questionable, as cytosolic NAD^+ may be replenished through NMNAT2 activity. The mitochondrial NAD^+ pool is maintained by NMNAT3 (Lau et al. 2010; Di Stefano and Conforti 2013), which is the enzyme with lowest selectivity for purine nucleotides, and it appears to be independent of other enzymes in the NAD^+ biosynthetic pathway as NAMPT inhibition by FK866 does not affect NAD^+ intramitochondrial levels (Pittelli et al. 2010). The maintenance of individually compartmentalized regions is essential for preventing short-term cell death in the event of massive NAD^+ depletion. This protective capacity, mainly mediated by the activation of a SIRT3/SIRT4/NAMPT-mediated axis, is known as the 'mitochondrial oasis effect', during which cell viability is maintained for a short period of time after cytosolic NAD^+ depletion, due to the preservation of mitochondrial function (Yang et al. 2007).

4.3.2 Subcellular $NADP^+$ Compartmentalization

One of the most important characteristics associated with eukaryotic cell metabolism is the compartmentalization of metabolic reactions within organelles. Similar to what happens with NAD^+, multistep shuttles are used to transfer NADPH-reducing power between the different cellular compartments to ensure correct organellar metabolic processes. The use of isotope tracing experiments has provided new insights regarding cytosolic and mitochondrial production of NADPH, namely, the contribution of the PPP for cytosolic pools and serine/glycine metabolism for the mitochondrial one. The major source of NADPH production is the oxidative branch of PPP in the presence of a high demand of nucleotide synthesis. As abovementioned, other potential sources of cytosolic NADPH include reactions catalysed by IDH, malic enzyme (ME), ALDH and MTHFD (Pollak et al. 2007; Tibbetts and Appling 2010). However, some isoforms of these enzymes are involved in mitochondrial reactions, which may explain the transference of reducing equivalents between cytosol and mitochondria. For instances, IDH2 catalyses the reductive carboxylation of alpha-ketoglutarate (αKG) to isocitrate in the mitochondria, at the expense of one molecule of mitochondrial NADPH. Next, citrate/isocitrate may be shuttled to the cytosolic compartment, where it is oxidized by IDH1, with a concomitant production of cytosolic NADPH (Wise et al. 2011). ME is a $NADP^+$-dependent enzyme, which through malate decarboxylation generates NADPH to fuel fatty acid synthesis, the most NADPH-demanding pathway in eukaryotes. The three isoforms (1, 2 and 3) are distinctively compartmentalized (ME1 is cytosolic, while ME2 and ME3 are mitochondrial), thus contributing for separate pools of NADPH. Lewis et al. (2014) have recently developed a reporter system that allows the tracing of compartmentalized sources of NADPH. In this work, they took advantage of a neomorphic mutant IDH, which catalyses the conversion of αKG into (*D*)2-hydroxyglutarate (2HG), through transference of a hydride from NADPH. The production of 2HG serves as an end-product readout, as this metabolite is not further metabolized. By supplementing the cells with [3-^{2}H]-glucose and assessing the enrichment of ^{2}H in the 2HG pool, the authors were able to discriminate between mitochondrial versus cytosolic NADPH production. Furthermore and considering that serine and glycine interconversion generates NADPH, Lewis et al. sought to evaluate the contribution of this metabolic pathway in NADPH compartmentalization, thus showing that serine metabolism is capable of replenishing mitochondrial NADPH. Mitochondrial NADPH may also be produced through activity of a nicotinamide nucleotide transhydrogenase (NNT), which transfers reducing equivalents from NADH to NADPH (Gameiro et al. 2013). This purely cofactor-modulating enzyme has been shown to modulate central carbon metabolism, namely, through coordination of glucose and glutamine metabolism in the TCA cycle, as NNT knockdown affected glucose and glutamine consumption in cancer cell lines.

4.4 The Role of NAD^+ and $NADP^+$ During Host–Pathogen Interactions

As abovementioned, the maintenance of homeostatic levels of both NAD^+ and $NADP^+$ is essential for a correct cellular function that supports survival and proliferation. This is particularly important when acknowledging that, upon infection, the metabolic requirements of host cells are altered, in order to privilege certain pathways to fulfil their highly demanding cellular needs. This suggests that the bioavailability of certain precursors within the intracellular milieu is essential in dictating if a host cell is capable of arresting or sustaining the infectious process, which will ultimately allow for the pathogen to be cleared or to thrive and persist within the host. When considering the importance of adequate NAD^+/NADH and $NADP^+$/NADPH ratios in the establishment of an appropriate metabolic environment, it seems clear that fluctuations in these factors or in the bioavailability of their building blocks impact host defence and pathogen success. In fact, it has already been demonstrated the importance of NAD^+ and $NADP^+$ levels during infection and how these ratios may be altered upon pathogen proliferation (Mesquita et al. 2016).

4.4.1 Pathogen-Driven Modulation of NAD^+ Levels

Several reports have shown how the modulation of host NAD^+ levels is imperious for achieving effective pathogen colonization and survival. However, additional layers of complexity arise when considering that distinct cell types and pathogenic agents may drive different alterations in host cell energy status. For instance, upon erythrocyte infection, the parasitic agent *Plasmodium falciparum* increases the intracellular levels of NAD^+, which is due to increased activity of NAMPT and NAPT (Zerez et al. 1990). The observed elevation of NAD^+ levels is expected to be essential for supporting a high glycolytic rate in the infected erythrocyte, which displays a concomitant upregulation of lactate production (Olszewski et al. 2009). Conversely, human immunodeficiency virus (HIV) infection of human peripheral blood lymphocytes (PBMCs) leads to a decrease in intracellular NAD^+ levels (Murray et al. 1995). Having in consideration that advanced states of HIV infection display common features with pellagra (a clinical state that is associated with niacin deficiency), it was also shown that exogenous administration of NAM inhibits HIV replication in PBMCs (Murray and Srinivasan 1995). Consistent with these findings, it has been recently suggested that a disruption of host metabolic machinery by HIV (and also other viruses) may contribute to persistence of the HIV reservoir and, consequently, disease and pathogenesis progression (Palmer et al. 2016). Infection of macrophages with the protozoan agent *Leishmania infantum* was shown to induce a transient increase in NAD^+ levels, 18–24 h postinfection, which is consistent with a metabolic switch towards mitochondrial respiration at later stages of infection (Moreira et al. 2015).

Group A streptococci (GAS) has a noteworthy capacity of modulating NAD^+ levels in infected cells. Among other virulence factors, GAS produce a NAD^+ glycohydrolase that cleaves NAD^+ to produce NAM, adenosine diphosphate (ADP)-ribose and cyclic ADP-ribose (Tatsuno et al. 2010). Along with streptolysin O, these virulence factors promote intracellular survival within host cells, mainly by preventing phagolysosome acidification and thus evading the immune system (Bastiat-Sempe et al. 2014; Sharma et al. 2016). In a mechanistic perspective, it was shown that NAD^+ glycohydrolase depletes intracellular NAD^+ and ATP levels (Michos et al. 2006), which disrupts several host defence mechanisms and contributes to dissemination and chronic persistence. Moreover, NAD^+ glycohydrolase expression also protects against xenophagic killing, thus allowing for increased bacterial survival in pharyngeal keratinocytes (O'Seaghdha and Wessels 2013). In an experimental model of pneumococcal meningitis, it was demonstrated that an activation of the kynurenine pathway and consequent accumulation of 3-hydroxykynurenine in the hippocampus were correlated with extended apoptotic damage (Bellac et al. 2006). Consistently, it was also shown that pharmacological inhibition of 3-hydroxylase and kynureninase leads to a decreased level of NAD^+, thus originating energy failure and consequent cell death. These data seem to suggest that the replenishment of cellular NAD^+ pools during infection may serve as a neuroprotective mechanism (Bellac et al. 2010).

Recent evidences have also suggested the link between liver X receptor (LXR) pathway and NAD^+ intracellular metabolism during macrophage infection with invasive bacteria. Matalonga et al. (2017) proposed that LXR activation originates a reduction of intracellular NAD^+ levels, through a CD38-dependent mechanism. This activation decreased macrophage bacterial burden, due to alterations in cytoskeleton dynamics. Additionally, exogenous administration of NAD^+ rescued the infective ability of the bacteria. Furthermore, using an in vivo model of infection with *Salmonella* Typhimurium, an amelioration of the clinical symptoms and protection against bacteria dissemination upon treatment with LXR agonists was also observed, which emphasizes the relevance of LXR agonists or other inducers of CD38 NADase activity as potential therapies against infection.

Despite all these recent examples showing that infection with intracellular pathogens can directly increase or decrease host NAD^+ levels, further studies are required to determine the druggability of this metabolism for therapeutic intervention.

4.4.2 The Impact of Enzymes Involved in NAD^+ Metabolism in Host Immune Response Towards Invading Pathogens

4.4.2.1 Indoleamine 2,3-Dioxygenase

The most well-studied enzyme in NAD^+ de novo synthesis is IDO, which catalyses the rate-limiting step in tryptophan catabolism and, consequently, contributes to tryptophan depletion. More than a metabolic hub of amino acid metabolism, this enzyme has arisen as an important regulator of immune plasticity and homeostasis, particularly during autoimmunity, infection and neoplasia. It has been shown, using different models, that IDO has an immunosuppressive role, characterized by an impairment of effector T cell proliferation and induction of regulatory T cells (Prendergast et al. 2014). During tryptophan starvation, naïve $CD4^+$ T cells become tolerogenic, and autoimmune-preventive T regulatory cells are generated, through a general control nonderepressible 2 (GCN2)-dependent mechanism (Fallarino et al. 2006). In opposition, IDO activity may also be controlled by interferons, which are major players in the development of an immune response. This crosstalk between tryptophan catabolism and immune modulation indicates IDO as a major metabolic rheostat, which suggests that it may be involved in host immune response towards invading pathogens.

It has been shown in distinct infectious models that genetic or pharmacological modulation of IDO alters infection outcome (Munn and Mellor 2013; Schmidt and Schultze 2014). Furthermore, considering that several important human pathogens are tryptophan auxotrophs, it is expected that IDO-mediated depletion of this metabolite may further influence microorganism survival (Brown et al. 2008). IDO has already been demonstrated to be essential for gut homeostasis and development of a protective response against bacteria and other pathogens. Consistently, in a context of experimental colitis, IDO inhibition resulted in higher mortality, when compared with the placebo-treated animals, as well as increased immunopathology and elevated colonic pro-inflammatory cytokine expression (Gurtner et al. 2003). Furthermore, using a *Clostridium difficile* infection model, IDO knockout mice displayed increased mucosal destruction and caecal haemorrhage, along with increased levels of interferon (IFN)-γ-producing neutrophils (El-Zaatari et al. 2014). This phenotype appears to indicate that tryptophan catabolism may be a central mechanism associated with the regulation of tissue pathology and control of bacterial burden.

Several research groups addressed the role of IDO during HIV infection. This chronic infection is characterized by a progressive and severe immune dysfunction, associated with increased tryptophan metabolism by IDO (denominated as *tryptophan starvation*). The induction of this enzyme is driven by host- and virus-derived factors, such as cytotoxic T lymphocyte-associated protein 4 (CTLA-4)/B7 engagement; IFN-α and IFN-γ production by the innate and adaptive immune system, respectively; and HIV-associated proteins (Boasso et al. 2007). The immunosuppressive role of IDO is mainly correlated with its capacity to impair CD4 and CD8 T cells (Boasso et al. 2007, 2008), which results in the gradual establishment of a state of T cell anergy, characteristic of HIV infection. Additionally, Potula and colleagues showed that IDO inhibition by the competitive inhibitor 1-methyl-tryptophan (1-MT) increases cytotoxic CD8 T cells and decreases HIV-infected macrophages in a murine model of HIV encephalitis (Potula et al. 2005). In line with the already described importance of IDO during HIV infection, Vujkovic-Cvijin et al. showed that during both acute and chronic simian immunodeficiency virus (SIV) infection of rhesus macaques, a specific depletion of gut-resident *Lactobacillus* is correlated with increased levels of IDO1 and loss of T helper (Th)17 cell abundance. The observed phenotype suggests that IDO1 inhibition by *Lactobacillus* species may serve as a protective mechanism that prevents Th17 loss (Vujkovic-Cvijin et al. 2015). Consequently, the maintenance of gut homeostasis may prevent microbial translocation into the bloodstream, which is a hallmark of chronic immune activation and disease progression in HIV patients.

Other viral infections have been associated with induction of IDO expression in immune cells. A lipooligosaccharide expressed by the pathogenic agent of chancroid, *Haemophilus ducreyi*, is responsible for IDO induction in dendritic cells, via type I interferon- and tumour necrosis factor (TNF)-α-dependent mechanisms, as well as modulation of the mitogen-associated protein kinase (MAPK), nuclear factor kappa B (NF-κB) and Janus kinase/signal transducers and activators of transcription (JAK-STAT) pathways (Li et al. 2011). In a similar fashion, IDO expression was also increased during macrophage infection with Epstein-Barr virus (EBV), which is commonly associated with infectious mononucleosis and human malignancies. Furthermore, EBV-induced IDO expression was linked to an impairment of the cytotoxic activity of CD8 T cells, which are vital for the control of viral particles, as well as cancer progression (Liu et al. 2014). West Nile virus induces IDO expression on infected macrophages through a NF-κB-dependent manner. Although IDO is not required for the control of flavivirus replication, its expression prior to macrophage exposure to West Nile virus prevents infection (Yeung et al. 2012). These results suggest that, although IDO is commonly associated with immune suppression and dysfunction, its impact during infection may be time-dependent, as well as pathogen-dependent.

Although IDO has a very noteworthy role during host-pathogen interactions, other enzymes involved in NAD^+ anabolism and catabolism may be associated with host susceptibility versus resistance during infection.

4.4.2.2 Nicotinamide Phosphoribosyltransferase

NAMPT, a biosynthetic enzyme from the salvage pathway, inhibits HIV replication in monocytes by preventing the integration of proviral DNA in host genome (Van den Bergh et al. 2010). However, a multifunctional transactivator encoded by HIV, the Tat protein, is capable of inhibiting NAMPT activity and consequently depleting intracellular NAD^+ pools. Consistently, an imbalance in NAD^+ content in the intracellular milieu further impacts the activity of NAD^+-consuming protein, as sirtuin (SIRT)1 (Zhang et al. 2010). The inhibition of SIRT1 activation during HIV infection has been associated to Tat-induced p53 activation and decreased deacetylation of SIRT1 (Thakur et al. 2012). Pretreatment of HeLa-CD4-long terminal repeat (LTR)-β-gal (MAGI) cells transfected with Tat plasmid with resveratrol, a natural product that activates SIRT1, resulted in increased NAD^+ levels and attenuates Tat-induced HIV transactivation (Zhang et al. 2010). Furthermore, pretreatment with SIRT1 inhibitor, nicotinamide, results in the opposite phenotype, which indicates that SIRT1 inhibition by Tat is crucial for an adequate viral proliferation (Zhang et al. 2009). Another natural compound, tanshinone II A, has been shown to inhibit Tat-regulated HIV transactivation. After treatment of TZM-bl cells with tanshinone II A, SIRT1 and NAMPT activity is rescued, as well as intracellular NAD^+ levels. Furthermore, the increase in NAMPT expression appears to result from activation of adenosine monophosphate-activated protein kinase (AMPK) signalling pathway (Zhang et al. 2014).

This cluster of data insinuates that this important link between the regulation of NAD^+ levels and viral replication can be used in the development of new therapeutic strategies against HIV infection, with a particular emphasis towards the NAMPT/SIRT1 axis. It is important to emphasize that alterations in NAD^+/NADH levels, driven by changes in anabolic versus catabolic mechanisms, are also associated with metabolic disorders and altered cellular functions, as seen, for instance, during sepsis. Sepsis is potentially life-threatening complication, derived from a generalized bacterial or fungal infection, which is associated to an acute inflammatory response and eventually a late immunosuppressive state. It has been recently shown that cellular bioenergetics is essential in dictating the functional fate of immune cells, which are rendered tolerant upon septic shock. Using an in vitro model of immunotolerance, Chen et al. demonstrated that immunotolerant monocytes, when restimulated with lipopolysaccharide (LPS), showed a defect in glucose fermentation, characterized by decreased production of lactate and NAD^+ levels, when compared with non-tolerant ones. Furthermore, LPS-stimulated PBMCs from immunotolerant patients with sepsis displayed a similar defective phenotype, which was only restored upon patient recovery from septic shock (Cheng et al. 2016). The shift from an acute to a late inflammatory stage during sepsis is associated with a switch from a glycolytic phenotype, highly dependent on NAD^+ availability, to an oxidative metabolism, characterized by fatty acid oxidation. It was demonstrated that during Toll-like receptor (TLR) 4 stimulation, this switch is coordinated by a crosstalk between NAMPT, SIRT1 and SIRT6: while NAMPT drives NAD^+ production to allow a metabolic coupling between the acute and late stage on inflammation, NAD^+-dependent SIRT1 and SIRT6 coordinate the increase in fatty acid oxidation and decrease in glucose fermentation, thus integrating the metabolic reprogramming with the immune response during sepsis (Liu et al. 2012). Furthermore, during the phase of cellular adaptation to sepsis, it was shown that the sequential activity of NAD^+-dependent SIRT1 and mitochondrial SIRT3 is essential in driving a functional mitochondrial biogenesis (Liu et al. 2015). This suggests that NAD^+ availability is essential in dictating the outcome after septic shock, as well as the metabolic reprogramming during endotoxin tolerance, such as the activity of several proteins, like sirtuins, are highly dependent on intracellular NAD^+ pools.

4.4.2.3 NAD^+-Consuming Proteins

NAD^+ is a vital cofactor for three distinct classes of proteins: sirtuins, poly(ADP-ribose) polymerases (PARPs) and CD38/157, where it serves as a donor of ADP-ribose and, therefore, controls the levels of extracellular and intracellular NAD^+. These nucleotide-metabolizing proteins are *metabolic sensors*, responsible for the tight regulation of energy metabolism, cell survival and proliferation.

CD38/157 Membrane Proteins

The class of cADP-ribose synthases, also known as the lymphocyte antigens CD38 and CD157, are involved in the production of secondary messengers that play important roles in distinct signalling pathways. These proteins are the main regulators of NAD^+ levels, as around 100 molecules of NAD^+ require processing to yield a single molecule of cADP-ribose (Dousa et al. 1996; de Toledo et al. 2000). As membrane proteins, these molecules possess an extracellular domain, and its rapid activity is responsible for maintaining low levels of extracellular NAD^+ (Seman et al. 2004). Consistently, during homeostasis, low quantities of NAD^+ are found in mice serum (Kim et al. 1993). However, upon infection or tissue damage, NAD^+ levels can quickly rise and be secreted. This extracellular NAD^+ activates $P2Y_{11}$ purinoceptors at the surface of human granulocytes, which contributes for their functional activation (Moreschi et al. 2006). Following CD38 activation, Ca^{2+} levels are dramatically increased, which promotes the migration of innate immune cells (as monocytes, neutrophils and dendritic cells) to danger sites and to secondary lymphoid organs (Partidá-Sánchez et al. 2007). The role of CD38 in the regulation of immune responses has already started to be elucidated in vivo, by taking advantage of $CD38^{-/-}$ mice. Partida-Sánchez et al. (2001) showed that in the absence of these proteins, mice become more susceptible to *Streptococcus pneumonia*, as neutrophils appear to be incapable of accumulating in the lung of infected animals and are irresponsive to the bacterial chemoattractant fMLP. Likewise, using an experimental infection with *Listeria monocytogenes*, Lischke et al. (2013) showed that the absence of CD38 renders the mice more susceptible 3 days postinfection, which suggests a role of this molecule in the innate immune response. $CD38^{-/-}$ mice also display higher susceptibility to *Mycobacterium avium* infection, as Th1 differentiation and polarization seem to be impaired. Furthermore, alterations in the granulomatous barrier were also observed in these animals, which suggest that the absence of CD38 promotes bacterial dissemination and growth (Viegas et al. 2007). Likewise, $CD38^{-/-}$ mice display a reduced hepatic elimination of *Entamoeba histolytica*, which is associated with a limited inflammatory response (Estrada-Figueroa et al. 2011).

This cluster of data suggests that a deeper understanding of the role of CD38 and extracellular NAD^+ levels is essential for achieving a specific modulation of the inflammatory responses, with the aim of better controlling the infectious process.

PARPs

Mono- or poly(ADP-ribosyl)ation (PARylation) reactions, which are mediated by ADP-ribose transferases (ARTs) or PARPs, are also dependent on NAD^+ availability. PARP-1 is the most abundantly expressed protein of the PARP family, and it is involved in DNA damage response, cell death by apoptosis and epigenetic alterations in mammalian cells (Schreiber et al. 2006). This protein induces the translocation of ADP-ribose molecules from NAD^+ to acceptor proteins or to a previously formed poly(ADP-ribose) chain. In situations of PARP-1 overactivity, which may be driven by metabolic insults, intracellular NAD^+ pools are depleted, and the cell enters in a state of bioenergetic failure, which culminates with the initiation of the apoptotic pathway (Sodhi et al. 2010; Yu et al. 2002). During infection, some pathogens may hijack host PARP-1 activity in their own advantage. Recently, it was demonstrated that PARP-1 is overexpressed in *Helicobacter pylori*-infected gastric mucosa and that antibiotic treatment reduces gastric inflammation, through decreasing PARP-1 expression and NF-kB activation (Lee et al. 2016). In EBV-infected B cells, pharmacological inhibition, as well as genetic ablation of PARP-1 suppressed the expression of latent membrane protein (LMP) 1-activated genes, which are essential for driving EBV latency and tumorigenesis (Martin et al. 2016). Consistently with this report, a recent study highlighted the role of PARP-1 in the inhibition of EBV lytic reactivation, through binding at the *BZLF1* lytic switch promoter. Furthermore, EBV reactivation has a negative impact on PARP-1 activity, which reinforces the restrictive function of this molecule towards EBV reactivation (Lupey-Green et al. 2017). In a very similar fashion, it was demonstrated that Kaposi's sarcoma-associated herpesvirus (KSHV) also downregulates PARP-1 upon reactivation. This is due to the expression of viral processivity factor of KSHV (PF-8), which directly interacts with PARP-1 and triggers proteasome-dependent degradation (Cheong et al. 2015). In line with these data, Navarro et al. developed a mouse model that exhibited PARP-1-deficiency with a *Cd4*-promoter-driven conditional deletion of PARP-2 in T cells. Mice bearing a double deficiency in PARP-1 and -2 displayed an accumulation of DNA damage, which conditioned the replenishment of the memory pool in response to vaccinia virus infection (Navarro et al. 2017). The role of PARP-1 during HIV infection has been vastly studied; however, it has also suffered immense debate: several contradictory reports suggest that PARP-1 activation is essential for HIV DNA integration (Ha et al. 2001; Kameoka et al. 2005), while other state that it is not only dispensable (Siva and Bushman 2002), but also associated with an impairment of the viral process (Gutierrez et al. 2016; Rom et al. 2015). This suggests that further studies are required to understand the exact impact of PARP-1 activity during viral infections.

The activity of PARP-1 has also been addressed during bacterial infections, although at lower extent. Following *Mycobacteria* infection of alveolar macrophages, necrosis appears to be induced as a possible evasion strategy that allows infectious bacteria to escape and disseminate to new host cells. Wu and colleagues showed that, upon Bacillus Calmette-Guerin (BCG) infection of RAW264.7 macrophages, NAD^+ levels are decreased, which is consistent with an increase in PARP-1 expression. Furthermore, they demonstrated that activation of Wnt/β-catenin signalling prevents BCG-induced macrophage necrosis, which appears to be correlated with increased levels of intracellular glutathione and downregulation of PARP-1 expression (Wu et al. 2015). Similarly, PARP-1 activation was found in the brainstem of patients that died with *Plasmodium falciparum* malaria (Medana et al. 2001). PARP-1 inhibition with 3-aminobenzamide was shown to attenuate NAD^+ depletion and pneumococci-induced cytotoxicity, in a context of experimental pneumococcal meningitis (Koedel et al. 2002). Using a lipopolysaccharide-induced model of acute lung injury, it was demonstrated that PARP-1 inhibitor, 3, 4-Dihydro-5[4-(1-piperindinyl) butoxy]-1(2H)-isoquinoline, reduced lung inflammation and vascular leakage, as well as neutrophil infiltration and the establishment of an inflammatory environment, thus promoting a protective phenotype (Wang et al. 2013).

These reports suggest that the modulation of PARP-1 activity may be explored in the context of potential therapies against infections. However, it is necessary to have in consideration that alterations in PARP-1 activity may be cell- and context-specific, so each individual case should be properly addressed.

Sirtuins

As abovementioned, sirtuins are a class of NAD^+-dependent proteins, and this family of proteins includes seven members (SIRT1-7), which display distinct subcellular localizations (Michan and Sinclair 2007). Sirtuins are activated in situations of energy deficit and, therefore, prompt metabolic shifts to increase ATP production, namely, through utilization of noncarbohydrate energy sources, as fatty acids (Houtkooper et al. 2012).

It was recently shown that SIRT1 is downregulated during in vitro *Mycobacterium tuberculosis* (Mtb) infection of monocytes/macrophages and the rescue of SIRT1 activation during the course of infection leads to dampened inflammatory responses, through deacetylation of RelA/p65 and impaired binding to the promoter of inflammatory genes. Furthermore, in *Mtb*-infected mice, SIRT1 activators improved lung pathology, decreased inflammatory responses and augmented the efficacy of anti-TB drugs (Cheng et al. 2017). This suggests that SIRT1 activators may disclose a potential utilization during host-directed therapies for the treatment of tuberculosis. Similarly, treatment of *Trypanosoma cruzi*-infected mice with SIRT1 agonists, resveratrol or SRT1720, decreased cardiac pathology and improved heart function in chagasic mice (Wan et al. 2016). In line with this, it was shown that SIRT1 knockdown or inhibition by NAD^+ precursor NAM or sirtinol in KSHV-infected cells originated an increase in the quantity of infectious virions (He and Gao 2014; Li et al. 2014). Ganesan and colleagues showed that macrophage infection with virulent *Salmonella enterica* serovar Typhimurium is characterized by a quick drop in ATP and NAD^+ levels, which originated a transient activation of AMPK. This metabolic sensor appears to be regulated by the complex liver kinase B1 (LKB1)/SIRT1, which is degraded in lysosomes, culminating in the disruption of AMPK-derived modulation of mammalian target of rapamycin (mTOR). This suggests that *S. Typhimurium* degrades host cell proteins as an evasion strategy to avoid xenophagy and consequent bacterial elimination (Ganesan et al. 2017). Interestingly, during macrophage infection with the protozoan parasite *Leishmania infantum*, Moreira et al. observed a metabolic switch towards a respiratory phenotype, which was consistent with AMPK activation 10–14 h postinfection. Furthermore, the increase in NAD^+ levels of 18 h postinfection was concomitant with the observed decrease in SIRT1 expression, which suggests that *L. infantum* is capable of modulating host SIRT1 activity by altering the redox status of the host cell (Moreira et al. 2015). Likewise, the parasite *Cryptosporidium parvum* is also capable of altering SIRT1 expression, as infected human biliary epithelial cells display higher levels of SIRT1, which regulates NF-kB-driven innate immune response (Xie et al. 2014). SIRT1 expression was also found to be increased during hepatitis B virus (HBV) infection, with its inhibition by sirtinol being associated with a suppression of viral replication (Ren et al. 2014).

Although SIRT1 is the most studied member of the sirtuin family, some reports regarding the role of the other members during infections have started to appear, namely, SIRT2. Upon *L. monocytogenes* infection, SIRT2 is translocated to the nucleus, where it deacetylates histone H3 on lysine 18 (H3K18) and possibly reprogrammes host cells to allow bacterial proliferation. Consistently, experimental infection of $SIRT2^{-/-}$ mice resulted in an impairment in bacterial proliferation, which suggests that SIRT2-mediated H3K18 deacetylation has an essential role during *L. monocytogenes* infection (Eskandarian et al. 2013). However, the role of this protein during infection is most likely pathogen-specific, as the modulation of SIRT2 during *Mtb* infection did not alter the outcome of chronic infection (Cardoso et al. 2015).

Although sirtuins have recently arisen as very promising drug targets for metabolic, neurodegenerative, cardiovascular and neoplastic diseases, it is wise to further study their implication during infectious processes, especially when considering their possible modulation by invading pathogens. The major mechanisms of action of NAD^+-consuming proteins and metabolic consequences during host–pathogen interaction is synthetized on Fig. 4.2.

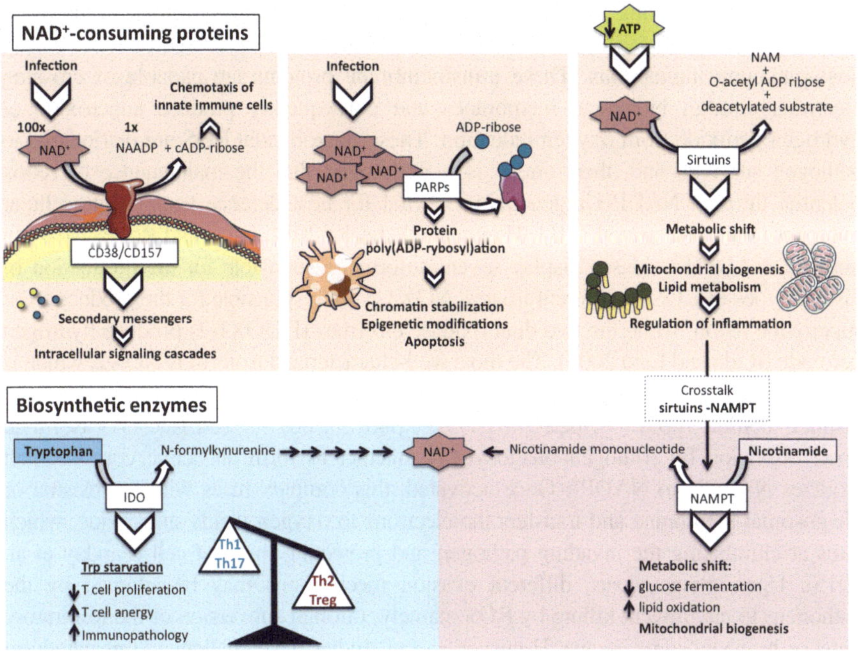

Fig. 4.2 Mechanism of action of NAD⁺-consuming proteins and metabolic consequences during host–pathogen interaction. Three major classes of NAD⁺-consuming proteins are involved in NAD⁺ catabolism upon infection: CD38/157, PARPs and sirtuins. The ectoenzymes CD38/157 are responsible for the breakdown of 100 molecules of NAD⁺, with concomitant production of one molecule of NAADP and cADP-ribose, which are involved in the activation of intracellular signalling pathways and in the chemotactic process of innate immune cells, such as neutrophils, dendritic cells and monocytes, to the sites of inflammation. PARPs respond to DNA damage with a simultaneous poly (ADP-rybosyl)ation of target protein, which may be at the genesis of epigenetic alterations. The consequent NAD⁺ depletion may reduce intracellular energy reservoirs and induce apoptotic cell death. Sirtuins (SIRT1-7) may be modulated by energy stress associated with the infectious process. Upon activation, sirtuins degrade NAD⁺ and originate a metabolic shift that aims at restoring energetic homeostasis, with a possible function in regulating inflammation. Biosynthetic enzymes, such as IDO and NAMPT, may be targeted by pathogens and consequently alter intracellular NAD⁺ levels. Tryptophan starvation, originated by hyperactivation of IDO, originated an unbalance in T cell response, characterized by anergy and a decrease in pro-inflammatory subsets, such as Th1 and Th17. Activation of NAMPT during infection may also originate a metabolic switch towards an oxidative phenotype, which is most likely connected with sirtuin activity and function. *ADP* adenosine diphosphate, *ATP* adenosine triphosphate, *IDO* indoleamine 2,3-dioxygenase, *NAD⁺* nicotinamide adenine dinucleotide, *NAADP* nicotinic acid dinucleotide phosphate, *NAMPT* nicotinamide phosphoribosyltransferase, *PARPs* poly(ADP-ribose) polymerases, *Th* T helper, *Trp* tryptophan

4.5 Pathogen-Driven Modulation of NADP⁺ Levels

NADPH can provide reducing equivalents for anabolic processes, as fatty acid synthesis, and to protect cells against oxidative burst-induced damage. It acts as a cofactor for NADPH-dependent glutathione reductases that yield reduced glutathione, essential for detoxifying reactive oxygen species (ROS), mainly formed in the mitochondria. The NOX family of NADPH oxidase is the major source of ROS during host–pathogen interactions. These transmembrane proteins are capable of crossing electrons through biological membranes and consequently produce superoxide or hydrogen peroxide from oxygen reduction. These microbicidal ROS are detrimental to pathogen survival, and, therefore, it is hypothesized that the maintenance of redox potential through NADPH regeneration is vital for host defence against intracellular microbes (Van Assche et al. 2011; Hogan and Wheeler 2014; Paiva and Bozza 2014). In humans, NADPH oxidases display seven isoforms, responsible for the production of distinct types of ROS in different tissues. NOX1-5 are responsible for the production of superoxide anion, while the two dual oxidase enzymes (DUOX1-2) produce hydrogen peroxide (Rada and Leto 2008). The most well-characterized isoform is NOX2, which is found in macrophage phagosomes and in neutrophil membranes. NOX2 possesses six distinct subunits (gp91^{phox}, p22^{phox}, p47^{phox}, p40^{phox}, p67^{phox} and Rac GTPase) that, upon activation by pathogen phagocytosis, interact to form an active complex that oxidizes NADPH to NADP^{+}. Once activated, this complex fuses with the plasma or phagosomal membrane and transfers the electrons to oxygen yields superoxide, which aims at eliminating the invading pathogen and protecting the host cell (Panday et al. 2015). Upon phagocytosis, different evasion mechanisms may be adopted by the pathogens to circumvent killing by ROS, namely, through subversion of the respiratory burst or by phagosome escape. However, recent studies have challenged this paradigm by demonstrating that some antioxidants can be used as therapeutic agents against certain pathogens (Mittal et al. 2014). Furthermore, exacerbated ROS levels may induce tissue damage and increased pathological inflammation, which may be detrimental for the host.

DUOX1 and 2 are mainly expressed in mucosa-associated epithelium, as the surface of salivary glands, airways and the gastrointestinal tract. The role of these enzymes was initially described using a *Drosophila* model, where a knockdown of DUOX rendered the flies more sensitive to gut infections (Kim and Lee 2014). This susceptible phenotype was partially attributed to alterations in gut homeostasis as well as decreased oxidative burst and, consequently, increased pathogen proliferation (Ha 2005). DUOX2 expression may be regulated by NOX1 activity during intestinal infection with *Citrobacter rodentium*. Accordingly, Cyba (p22^{phox}) deficiency, which inactivates NOX 1–4, was shown to protect against gut infections. This was mediated by activation of a compensatory mechanism in which gut commensals rescue impaired hydrogen peroxide production, which negatively regulates *C. rodentium* virulence factors (Pircalabioru et al. 2016).

One of the most studied mechanisms associated with host defence is the development of an adequate respiratory burst, during which several types of ROS contribute to pathogen clearance. Consistently, an impairment in the establishment of a hostile microenvironment may contribute to pathogen thriving. Chronic granulomatous disease (CGD) is a primary immunodeficiency caused by mutations in any of the five genes (*CYBB*, *CYBA*, *NCF1*, *NCF2* and *NCF4*) that encode for components of the phagocyte NADPH oxidase complex. It is essentially characterized by an impairment of intracellular pathogens killing, due to a lack of NADPH oxidase-derived ROS. Due to this defect, CGD patients develop aggressive forms of bacterial and fungal infections, as well as dysregulated inflammatory responses

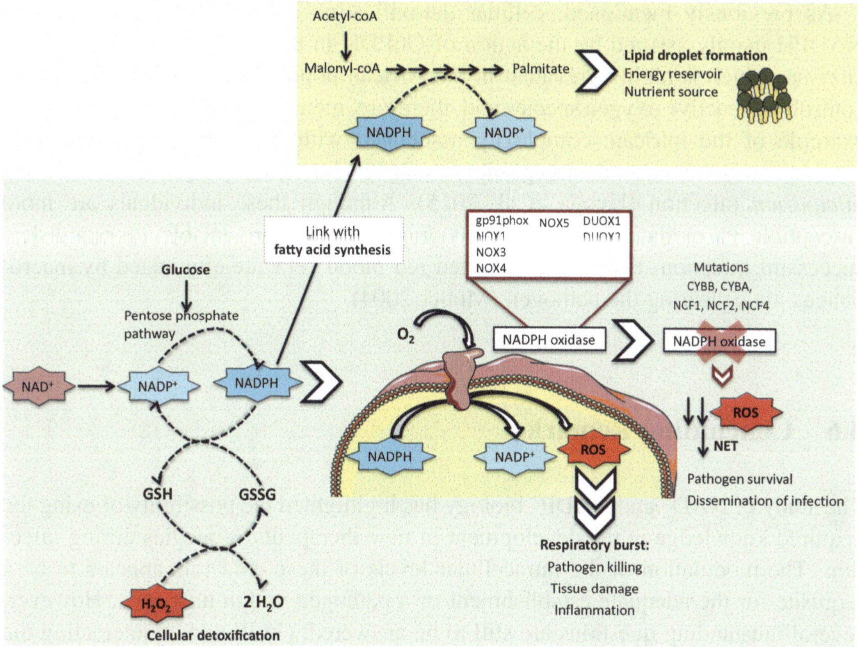

Fig. 4.3 NADP+/NADPH metabolism and function in host cell metabolism, homeostasis and defence mechanisms. NADP+ may be obtained through phosphorylation of NAD+. The major contributor for intracellular NADP+ pools is the pentose phosphate pathway. NADPH may be spent to regenerate GSH, which is used for cellular detoxification of ROS. NADPH oxidases (gp91phox, NOX1, NOX3, NOX4, NOX5, DUOX1, DUOX2) are responsible for the production of microbicidal ROS (respiratory burst), at the expense of NADPH molecules. Mutations in different subunits of the NOX family of NADPH oxidases (*CYBB*, *CYBA*, *NCF1*, *NCF2*, *NCF4* genes) originate decreased levels of ROS, which downregulates host cell microbicidal mechanisms and allow pathogen survival and dissemination. NADPH may equally be used during fatty acid synthesis, thus contributing for an increased pool of intracellular lipid droplets. *GSSG* oxidized glutathione, *GSH* reduced glutathione, *NAD+* nicotinamide adenine dinucleotide, *NADP+* nicotinamide adenine dinucleotide phosphate, *NET* neutrophil extracellular trap, *ROS* reactive oxygen species

being particularly susceptible to mycobacteriosis and aspergillosis (Conti et al. 2016; Deffert et al. 2014; Segal and Romani 2009). The available treatment for the comorbidities linked to this disease poses a challenge, particularly when having in consideration that the majority of immunomodulators can exponentiate the degree of immune suppression already observed in these patients. Recently, the utilization of gene therapy as a curative treatment of this primary immunodeficiency has been suggested, namely, for the treatment of fungal infections (Grez et al. 2011). During CGD, the defects in NADPH oxidase originate decreased pathogen killing, since there is a downregulation in the production of neutrophil extracellular traps (NETs), which is highly dependent of ROS production. Consequently, *Aspergillus* hyphae can infect and disseminate. The restoration of NETs formation by gene therapy results in increased ability to control aspergillosis (Bianchi et al. 2009), in a process driven by calprotectin released by NETs, which prevented spreading of fungal components (Bianchi et al. 2011; Fig. 4.3).

As previously mentioned, cellular detoxification depends mainly on reduced NADPH mainly assured by the action of G6PDH in the PPP. A deficiency in this enzyme, which is highly prevalent in the African population, results in a defect in controlling reactive oxygen species and, therefore, increased oxidative stress. A clear example of the intricate complexity associated with NADPH in host–pathogen interactions is the reduced susceptibility of G6PDH-deficient people to *Plasmodium falciparum* infection (Uyoga et al. 2015). Although these individuals are more susceptible to oxidative stress, *Plasmodium* parasites are unable to establish a successful infection. Eventually, infected red blood cells are eliminated by macrophages, thus clearing the pathogen (Müller 2004).

4.6 Concluding Remarks

The study of NAD^+ and $NADP^+$ biology has highlighted the possibility of using the acquired knowledge in the development of new therapeutic strategies during infection. The modulation of the intracellular levels of these cofactors appears to be a requisite for the adequate establishment of a pathogen within host cells. However, several outstanding questions are still to be answered: (1) Would counteracting the observed modulation of $NAD(P)^+$/NAD(P)H be an effective strategy to tackle infection? (2) After modulation of biosynthetic pathways, are pathogens capable to bypass the altered intracellular milieu and thrive within the host cell? (3) Can pathogen auxotrophy be the driving force behind evasion mechanisms? (4) Is it possible to target $NAD(P)^+$/NAD(P)H metabolism in infected cells, without compromising the homeostasis of both uninfected and bystander cells? Although it seems clear that $NAD(P)^+$/NAD(P)H metabolism is a central hub during infectious diseases, further studies are mandatory to understand how we can effectively modulate these pathways. Additionally, it is important to acknowledge that the majority of the studies where the role of these cofactors during host–pathogen interactions was elucidated lack the intrinsic complexity of living beings, and notably the fluctuations in NAD^+ biosynthesis related to the circadian clock (Peek et al. 2013). Targeting $NAD(P)^+$/NAD(P)H biology during an experimental in vitro infection is vital for understanding the underlying cellular and molecular mechanisms associated with resistance versus susceptibility. However, transposing this knowledge for clinical therapy may be a challenging and a long-lasting protocol. Nonetheless, important steps towards the utilization of $NAD(P)^+$/NAD(P)H biology to fight infectious diseases have been made, and the following years will be crucial for developing new strategies.

Funding Statement This work was supported by the Northern Portugal Regional Operational Programme (NORTE 2020), under the Portugal 2020 Partnership Agreement, through the European Regional Development Fund (FEDER) (NORTE-01-0145-FEDER-000013) and the Fundação para a Ciência e Tecnologia (FCT) (contracts SFRH/BD/120127/2016 to IM and IF/00021/2014 to RS), and Infect-Era (project INLEISH).

References

Bastiat-Sempe B, Love JF, Lomayesva N, Wessels MR (2014) Streptolysin O and NAD-glycohydrolase prevent phagolysosome acidification and promote group A Streptococcus survival in macrophages. MBio 5:e01690-14

Bellac CL, Coimbra RS, Christen S, Leib SL (2006) Pneumococcal meningitis causes accumulation of neurotoxic kynurenine metabolites in brain regions prone to injury. Neurobiol Dis 24:395–402

Bellac CL, Coimbra RS, Christen S, Leib SL (2010) Inhibition of the Kynurenine-NAD$^+$ pathway leads to energy failure and exacerbates apoptosis in pneumococcal meningitis. J Neuropathol Exp Neurol 69:1096–1104

Bianchi M, Hakkim A, Brinkmann V, Siler U, Seger RA, Zychlinsky A, Reichenbach J (2009) Restoration of NET formation by gene therapy in CGD controls aspergillosis. Blood 114:2619–2622

Bianchi M, Niemiec MJ, Siler U, Urban CF, Reichenbach J (2011) Restoration of anti-Aspergillus defense by neutrophil extracellular traps in human chronic granulomatous disease after gene therapy is calprotectin-dependent. J Allergy Clin Immunol 127:1243–1252

Boasso A, Herbeuval JP, Hardy AW, Anderson SA, Dolan MJ, Fuchs D, Shearer GM (2007) HIV inhibits CD4$^+$ T-cell proliferation by inducing indoleamine 2,3-dioxygenase in plasmacytoid dendritic cells. Blood 109:3351–3359

Boasso A, Hardy AW, Anderson SA, Dolan MJ, Shearer GM (2008) HIV-induced type I interferon and tryptophan catabolism drive T cell dysfunction despite phenotypic activation. PLoS One 3: e2961

Brown SA, Palmer KL, Whiteley M (2008) Revisiting the host as a growth medium. Nat Rev Microbiol 6:657–666

Cardoso F, Castro F, Moreira-Teixeira L, Sousa J, Torrado E, Silvestre R, Castro AG, Saraiva M, Pais TF (2015) Myeloid sirtuin 2 expression does not impact long-term *Mycobacterium tuberculosis* control. PLoS One 10:e0131904

Cheng SC, Scicluna BP, Arts RJW, Gresnigt MS, Lachmandas E, Giamarellos-Bourboulis EJ, Kox M, Manjeri GR, Wagenaars JAL, Cremer OL et al (2016) Broad defects in the energy metabolism of leukocytes underlie immunoparalysis in sepsis. Nat Immunol 17:406–413

Cheng CY, Gutierrez NM, Marzuki MB, Lu X, Foreman TW, Paleja B, Lee B, Balachander A, Chen J, Tsenova L et al (2017) Host sirtuin 1 regulates mycobacterial immunopathogenesis and represents a therapeutic target against tuberculosis. Sci Immunol 2:eaaj1789

Cheong W-C, Park J-H, Kang H-R, Song MJ (2015) Down-regulation of poly (ADP-ribose) polymerase-1 by a viral processivity factor facilitates gammaherpesvirus lytic replication. J Virol 89:9676–9682

Conti F, Lugo-Reyes SO, Blancas Galicia L, He J, Aksu G, Borges de Oliveira E, Deswarte C, Hubeau M, Karaca N, de Suremain M et al (2016) Mycobacterial disease in patients with chronic granulomatous disease: a retrospective analysis of 71 cases. J Allergy Clin Immunol 138:241–248.e3

de Toledo FG, Cheng J, Liang M, Chini EN, Dousa TP (2000) ADP-Ribosyl cyclase in rat vascular smooth muscle cells: properties and regulation. Circ Res 86:1153–1159

Deffert C, Cachat J, Krause K-H (2014) Phagocyte NADPH oxidase, chronic granulomatous disease and mycobacterial infections. Cell Microbiol 16:1168–1178

Di Stefano M, Conforti L (2013) Diversification of NAD biological role: the importance of location. FEBS J 280:4711–4728

Dölle C, Niere M, Lohndal E, Ziegler M (2010) Visualization of subcellular NAD pools and intra-organellar protein localization by poly-ADP-ribose formation. Cell Mol Life Sci 67:433–443

Dousa TP, Chini EN, Beers KW (1996) Adenine nucleotide diphosphates: emerging second messengers acting via intracellular Ca2$^+$ release. Am J Physiol 271:C1007–C1024

El-Zaatari M, Chang Y-M, Zhang M, Franz M, Shreiner A, McDermott AJ, van der Sluijs KF, Lutter R, Grasberger H, Kamada N et al (2014) Tryptophan catabolism restricts IFN-γ–expressing neutrophils and *Clostridium difficile* immunopathology. J Immunol 193:807–816

Eskandarian HA, Impens F, Nahori M-A, Soubigou G, Coppee J-Y, Cossart P, Hamon MA (2013) A role for SIRT2-dependent histone H3K18 deacetylation in bacterial infection. Science 341:1238858

Estrada-Figueroa LA, Ramírez-Jiménez Y, Osorio-Trujillo C, Shibayama M, Navarro-García F, García-Tovar C, Talamás-Rohana P (2011) Absence of CD38 delays arrival of neutrophils to the liver and innate immune response development during hepatic amoebiasis by *Entamoeba histolytica*. Parasite Immunol 33:661–668

Fallarino F, Grohmann U, You S, McGrath BC, Cavener DR, Vacca C, Orabona C, Bianchi R, Belladonna ML, Volpi C et al (2006) The combined effects of tryptophan starvation and tryptophan catabolites down-regulate T Cell receptor-chain and induce a regulatory phenotype in naive T cells. J Immunol 176:6752–6761

Fan J, Ye J, Kamphorst JJ, Shlomi T, Thompson CB, Rabinowitz JD (2014) Quantitative flux analysis reveals folate-dependent NADPH production. Nature 510:298–302

Gameiro PA, Laviolette LA, Kelleher JK, Iliopoulos O, Stephanopoulos G (2013) Cofactor balance by nicotinamide nucleotide transhydrogenase (NNT) coordinates reductive carboxylation and glucose catabolism in the tricarboxylic acid (TCA) cycle. J Biol Chem 288:12967–12977

Ganesan R, Hos NJ, Gutierrez S, Fischer J, Stepek JM, Daglidu E, Krönke M, Robinson N (2017) *Salmonella* Typhimurium disrupts Sirt1/AMPK checkpoint control of mTOR to impair autophagy. PLoS Pathog 13:e1006227

Grez M, Reichenbach J, Schwäble J, Seger R, Dinauer MC, Thrasher AJ (2011) Gene therapy of chronic granulomatous disease: the engraftment dilemma. Mol Ther 19:28–35

Gurtner GJ, Newberry RD, Schloemann SR, McDonald KG, Stenson WF (2003) Inhibition of indoleamine 2,3-dioxygenase augments trinitrobenzene sulfonic acid colitis in mice. Gastroenterology 125:1762–1773

Gutierrez DA, Valdes L, Serguera C, Llano M (2016) Poly(ADP-ribose) polymerase-1 silences retroviruses independently of viral DNA integration or heterochromatin formation. J Gen Virol 97:1686–1692

Ha E-M (2005) A direct role for dual oxidase in *Drosophila* gut immunity. Science 310:847–850

Ha HC, Juluri K, Zhou Y, Leung S, Hermankova M, Snyder SH (2001) Poly(ADP-ribose) polymerase-1 is required for efficient HIV-1 integration. Proc Natl Acad Sci U S A 98:3364–3368

He M, Gao SJ (2014) A novel role of SIRT1 in gammaherpesvirus latency and replication. Cell Cycle 13:3328–3330

Hogan D, Wheeler RT (2014) The complex roles of NADPH oxidases in fungal infection. Cell Microbiol 16:1156–1167

Houtkooper RH, Pirinen E, Auwerx J (2012) Sirtuins as regulators of metabolism and healthspan. Nat Rev Mol Cell Biol 13:225–238

Kameoka M, Nukuzuma S, Itaya A, Tanaka Y, Ota K, Inada Y, Ikuta K, Yoshihara K (2005) Poly (ADP-ribose)polymerase-1 is required for integration of the human immunodeficiency virus type 1 genome near centromeric alphoid DNA in human and murine cells. Biochem Biophys Res Commun 334:412–417

Kim S-H, Lee W-J (2014) Role of DUOX in gut inflammation: lessons from *Drosophila* model of gut-microbiota interactions. Front Cell Infect Microbiol 3:116

Kim UH, Kim MK, Kim JS, Han MK, Park BH, Kim HR (1993) Purification and characterization of NAD glycohydrolase from rabbit erythrocytes. Arch Biochem Biophys 305:147–152

Koedel U, Winkler F, Angele B, Fontana A, Pfister HW (2002) Meningitis-associated central nervous system complications are mediated by the activation of poly(ADP-ribose) polymerase. J Cereb Blood Flow Metab 22:39–49

Lau C, Dölle C, Gossmann TI, Agledal L, Niere M, Ziegler M (2010) Isoform-specific targeting and interaction domains in human nicotinamide mononucleotide adenylyltransferases. J Biol Chem 285:18868–18876

Lee WP, Hou MC, Lan KH, Li CP, Chao Y, Lin HC, Lee SD (2016) *Helicobacter pylori*-induced chronic inflammation causes telomere shortening of gastric mucosa by promoting PARP-1-mediated non-homologous end joining of DNA. Arch Biochem Biophys 606:90–98

Lewis CA, Parker SJ, Fiske BP, McCloskey D, Gui DY, Green CR, Vokes NI, Feist AM, Vander Heiden MG, Metallo CM (2014) Tracing compartmentalized NADPH metabolism in the cytosol and mitochondria of mammalian cells. Mol Cell 55:253–263

Li W, Katz BP, Spinola SM (2011) *Haemophilus ducreyi* lipooligosaccharides induce expression of the immunosuppressive enzyme indoleamine 2,3-dioxygenase via type I interferons and tumor necrosis factor alpha in human dendritic cells. Infect Immun 79:3338–3347

Li Q, He M, Zhou F, Ye F, Gao S-J (2014) Activation of Kaposi's sarcoma-associated herpesvirus (KSHV) by inhibitors of class III histone deacetylases: identification of sirtuin 1 as a regulator of the KSHV life cycle. J Virol 88:6355–6367

Lischke T, Heesch K, Schumacher V, Schneider M, Haag F, Koch-Nolte F, Mittrücker HW (2013) CD38 controls the innate immune response against listeria monocytogenes. Infect Immun 81:4091–4099

Liu TF, Vachharajani VT, Yoza BK, McCall CE (2012) NAD^+-dependent sirtuin 1 and 6 proteins coordinate a switch from glucose to fatty acid oxidation during the acute inflammatory response. J Biol Chem 287:25758–25769

Liu W, Lin Y, Xiao H, Xing S, Chen H, Chi P, Zhang G (2014) Epstein-Barr virus infection induces indoleamine 2,3-dioxygenase expression in human monocyte-derived macrophages through p38/mitogen-activated protein kinase and NF-κB pathways: impairment in T cell functions. J Virol 88:6660–6671

Liu TF, Vachharajani V, Millet P, Bharadwaj MS, Molina AJ, McCall CE (2015) Sequential actions of SIRT1-RELB-SIRT3 coordinate nuclear-mitochondrial communication during immunometabolic adaptation to acute inflammation and sepsis. J Biol Chem 290:396–408

Liu L, Shah S, Fan J, Park JO, Wellen KE, Rabinowitz JD (2016) Malic enzyme tracers reveal hypoxia-induced switch in adipocyte NADPH pathway usage. Nat Chem Biol 12:345–352

Lupey-Green LN, Moquin SA, Martin KA, McDevitt SM, Hulse M, Caruso LB, Pomerantz RT, Miranda JL, Tempera I (2017) PARP1 restricts Epstein Barr Virus lytic reactivation by binding the BZLF1 promoter. Virology 507:220–230

Magni G (2008) Enzymology of mammalian NAD metabolism in health and disease. Front Biosci 3:6135

Martin KA, Lupey LN, Tempera I (2016) Epstein-Barr Virus oncoprotein LMP1 mediates epigenetic changes in host gene expression through PARP1. J Virol 90:8520–8530

Matalonga J, Glaria E, Bresque M, Escande C, Carbó JM, Kiefer K, Vicente R, León TE, Beceiro S, Pascual-García M et al (2017) The nuclear receptor LXR limits bacterial infection of host macrophages through a mechanism that impacts cellular NAD metabolism. Cell Rep 18:1241–1255

Medana IM, Mai NTH, Day NPJ, Hien TT, Bethell D, Phu NH, Farrar J, White NJ, Turner GDH (2001) Cellular stress and injury responses in the brains of adult Vietnamese patients with fatal *Plasmodium falciparum* malaria. Neuropathol Appl Neurobiol 27:421–433

Mesquita I, Varela P, Belinha A, Gaifem J, Laforge M, Vergnes B, Estaquier J, Silvestre R (2016) Exploring NAD^+ metabolism in host-pathogen interactions. Cell Mol Life Sci 73:1225–1236

Michan S, Sinclair D (2007) Sirtuins in mammals: insights into their biological function. Biochem J 404:1–13

Michos A, Gryllos I, Håkansson A, Srivastava A, Kokkotou E, Wessels MR (2006) Enhancement of streptolysin O activity and intrinsic cytotoxic effects of the group A streptococcal toxin, NAD-glycohydrolase. J Biol Chem 281:8216–8223

Mittal M, Siddiqui MR, Tran K, Reddy SP, Malik AB (2014) Reactive oxygen species in inflammation and tissue injury. Antioxid Redox Signal 20:1126–1167

Moreira D, Rodrigues V, Abengozar M, Rivas L, Rial E, Laforge M, Li X, Foretz M, Viollet B, Estaquier J et al (2015) Leishmania infantum modulates host macrophage mitochondrial metabolism by hijacking the SIRT1-AMPK axis. PLoS Pathog 11:1–24

Moreschi I, Bruzzone S, Nicholas RA, Fruscione F, Sturla L, Benvenuto F, Usai C, Meis S, Kassack MU, Zocchi E et al (2006) Extracellular NAD^+ is an agonist of the human P2Y 11 purinergic receptor in human granulocytes. J Biol Chem 281:31419–31429

Mori V, Amici A, Mazzola F, Di Stefano M, Conforti L, Magni G, Ruggieri S, Raffaelli N, Orsomando G (2014) Metabolic profiling of alternative NAD biosynthetic routes in mouse tissues. PLoS One 9:e113939

Müller S (2004) Redox and antioxidant systems of the malaria parasite *Plasmodium falciparum*. Mol Microbiol 53:1291–1305

Munn DH, Mellor AL (2013) Indoleamine 2,3 dioxygenase and metabolic control of immune responses. Trends Immunol 34:137–143

Murray MF, Srinivasan A (1995) Nicotinamide inhibits HIV-1 in both acute and chronic in vitro infection. Biochem Biophys Res Commun 210:954–959

Murray MF, Nghiem M, Srinivasan A (1995) HIV infection decreases intracellular nicotinamide adenine dinucleotide [NAD]. Biochem Biophys Res Commun 212:126–131

Navarro J, Gozalbo-López B, Méndez AC, Dantzer F, Schreiber V, Martínez C, Arana DM, Farrés J, Revilla-Nuin B, Bueno MF et al (2017) PARP-1/PARP-2 double deficiency in mouse T cells results in faulty immune responses and T lymphomas. Sci Rep 7:41962

O'Seaghdha M, Wessels MR (2013) Streptolysin O and its co-toxin NAD-glycohydrolase protect group A Streptococcus from xenophagic killing. PLoS Pathog 9:e1003394

Olszewski KL, Morrisey JM, Wilinski D, Burns JM, Vaidya AB, Rabinowitz JD, Llinás M (2009) Host-parasite interactions revealed by *Plasmodium falciparum* metabolomics. Cell Host Microbe 5:191–199

Paiva CN, Bozza MT (2014) Are reactive oxygen species always detrimental to pathogens? Antioxid Redox Signal 20:1000–1037

Palmer CS, Cherry CL, Sada-Ovalle I, Singh A, Crowe SM (2016) Glucose metabolism in T cells and monocytes: new perspectives in HIV pathogenesis. EBioMedicine 6:31–41

Panday A, Sahoo MK, Osorio D, Batra S (2015) NADPH oxidases: an overview from structure to innate immunity-associated pathologies. Cell Mol Immunol 12:5–23

Partida-Sánchez S, Cockayne DA, Monard S, Jacobson EL, Oppenheimer N, Garvy B, Kusser K, Goodrich S, Howard M, Harmsen A et al (2001) Cyclic ADP-ribose production by CD38 regulates intracellular calcium release, extracellular calcium influx and chemotaxis in neutrophils and is required for bacterial clearance in vivo. Nat Med 7:1209–1216

Partidá-Sánchez S, Rivero-Nava L, Shi G, Lund FE (2007) CD38: an ecto-enzyme at the crossroads of innate and adaptive immune responses. Adv Exp Med Biol 590:171–183

Peek CB, Affinati AH, Ramsey KM, Kuo H-Y, Yu W, Sena LA, Ilkayeva O, Marcheva B, Kobayashi Y, Omura C et al (2013) Circadian clock NAD^+ cycle drives mitochondrial oxidative metabolism in mice. Science 342:1243417–1243417

Pircalabioru G, Aviello G, Kubica M, Zhdanov A, Paclet MH, Brennan L, Hertzberger R, Papkovsky D, Bourke B, Knaus UG (2016) Defensive mutualism rescues NADPH oxidase inactivation in gut infection. Cell Host Microbe 19:651–663

Pittelli M, Formentini L, Faraco G, Lapucci A, Rapizzi E, Cialdai F, Romano G, Moneti G, Moroni F, Chiarugi A (2010) Inhibition of nicotinamide phosphoribosyltransferase: cellular bioenergetics reveals a mitochondrial insensitive NAD pool. J Biol Chem 285:34106–34114

Pollak N, Dölle C, Ziegler M (2007) The power to reduce: pyridine nucleotides – small molecules with a multitude of functions. Biochem J 402:205–218

Potula R, Poluektova L, Knipe B, Chrastil J, Heilman D, Dou H, Takikawa O, Munn DH, Gendelman HE, Persidsky Y (2005) Inhibition of indoleamine 2,3-dioxygenase (IDO) enhances elimination of virus-infected macrophages in an animal model of HIV-1 encephalitis. Blood 106:2382–2390

Prendergast GC, Smith C, Thomas S, Mandik-Nayak L, Laury-Kleintop L, Metz R, Muller AJ (2014) Indoleamine 2,3-dioxygenase pathways of pathogenic inflammation and immune escape in cancer. Cancer Immunol Immunother 63:721–735

Rada B, Leto T (2008) Oxidative innate immune defenses by Nox/Duox family NADPH oxidases. Contrib Microbiol 15:164–187

Ren J-H, Tao Y, Zhang Z-Z, Chen W-X, Cai X-F, Chen K, Ko BCB, Song C-L, Ran L-K, Li W-Y et al (2014) Sirtuin 1 regulates hepatitis B virus transcription and replication by targeting transcription factor AP-1. J Virol 88:2442–2451

Revollo JR, Körner A, Mills KF, Satoh A, Wang T, Garten A, Dasgupta B, Sasaki Y, Wolberger C, Townsend RR et al (2007) Nampt/PBEF/Visfatin regulates insulin secretion in beta cells as a systemic NAD biosynthetic enzyme. Cell Metab 6:363–375

Rom S, Reichenbach NL, Dykstra H, Persidsky Y (2015) The dual action of poly(ADP-ribose) polymerase 1 (PARP 1) inhibition in HIV-1 infection: HIV-1 ltr inhibition and diminution in Rho GTPase activity. Front Microbiol 6:878

Schmidt SV, Schultze JL (2014) New insights into IDO biology in bacterial and viral infections. Front Immunol 5:384

Schreiber V, Dantzer F, Amé JC, De Murcia G (2006) Poly(ADP-ribose): novel functions for an old molecule. Nat Rev Mol Cell Biol 7:517–528

Segal BH, Romani LR (2009) Invasive aspergillosis in chronic granulomatous disease. Med Mycol 47(Suppl 1):S282–S290

Seman M, Adriouch S, Haag F, Koch-Nolte F (2004) Ecto-ADP-ribosyltransferases (ARTs): emerging actors in cell communication and signaling. Curr Med Chem 11:857–872

Sharma O, O'Seaghdha M, Velarde JJ, Wessels MR (2016) NAD^+-glycohydrolase promotes intracellular survival of group A Streptococcus. PLoS Pathog 12:e1005468

Siva AC, Bushman F (2002) Poly(ADP-ribose) polymerase 1 is not strictly required for infection of murine cells by retroviruses. J Virol 76:11904–11910

Sodhi RK, Singh N, Jaggi AS (2010) Poly(ADP-ribose) polymerase-1 (PARP-1) and its therapeutic implications. Vascul Pharmacol 53:77–87

Tatsuno I, Isaka M, Minami M, Hasegawa T (2010) NADase as a target molecule of in vivo suppression of the toxicity in the invasive M-1 group A Streptococcal isolates. BMC Microbiol 10:144

Thakur BK, Chandra A, Dittrich T, Welte K, Chandra P (2012) Inhibition of SIRT1 by HIV-1 viral protein Tat results in activation of p53 pathway. Biochem Biophys Res Commun 424:245–250

Tibbetts AS, Appling DR (2010) Compartmentalization of mammalian folate-mediated one-carbon metabolism. Annu Rev Nutr 30:57–81

Uyoga S, Ndila CM, Macharia AW, Nyutu G, Shah S, Peshu N, Clarke GM, Kwiatkowski DP, Rockett KA, Williams TN (2015) Glucose-6-phosphate dehydrogenase deficiency and the risk of malaria and other diseases in children in Kenya: a case-control and a cohort study. Lancet Haematol 2:e437–e444

Van Assche T, Deschacht M, Da Luz RAI, Maes L, Cos P (2011) Leishmania-macrophage interactions: insights into the redox biology. Free Radic Biol Med 51:337–351

Van den Bergh R, Florence E, Vlieghe E, Boonefaes T, Grooten J, Houthuys E, Tran H, Gali Y, De Baetselier P, Vanham G et al (2010) Transcriptome analysis of monocyte-HIV interactions. Retrovirology 7:53

VanLinden MR, Dölle C, Pettersen IK, Kulikova VA, Niere M, Agrimi G, Dyrstad SE, Palmieri F, Nikiforov AA, Tronstad KJ, Ziegler M (2015) Subcellular distribution of NAD+ between cytosol and mitochondria determines the metabolic profile of human cells. J Biol Chem 290 (46):27644–27659. https://doi.org/10.1074/jbc.M115.654129

Viegas MS, do Carmo A, Silva T, Seco F, Serra V, Lacerda M, Martins TC (2007) CD38 plays a role in effective containment of mycobacteria within granulomata and polarization of Th1 immune responses against *Mycobacterium avium*. Microbes Infect 9:847–854

Vujkovic-Cvijin I, Swainson LA, Chu SN, Ortiz AM, Santee CA, Petriello A, Dunham RM, Fadrosh DW, Lin DL, Faruqi AA et al (2015) Gut-resident *Lactobacillus* abundance associates with IDO1 inhibition and Th17 dynamics in SIV-infected macaques. Cell Rep 13:1589–1597

Wan X, Wen JJ, Koo SJ, Liang LY, Garg NJ (2016) SIRT1-PGC1α-NFκB pathway of oxidative and inflammatory stress during *Trypanosoma cruzi* infection: benefits of SIRT1-targeted therapy in improving heart function in Chagas disease. PLoS Pathog. 12:e1005954

Wang G, Huang X, Li Y, Guo K, Ning P, Zhang Y (2013) PARP-1 inhibitor, DPQ, attenuates LPS-induced acute lung injury through inhibiting NF-κB-mediated inflammatory response. PLoS One 8:e79757

Williamson D, Lund P, Krebs H (1967) The redox state of free nicotinamide-adenine dinucleotide in the cytoplasm and mitochondria of rat liver. Biochem J 103:514–527

Wise DR, Ward PS, Shay JES, Cross JR, Gruber JJ, Sachdeva UM, Platt JM, DeMatteo RG, Simon MC, Thompson CB (2011) Hypoxia promotes isocitrate dehydrogenase-dependent carboxylation of -ketoglutarate to citrate to support cell growth and viability. Proc Natl Acad Sci 108:19611–19616

Wu X, Deng G, Li M, Li Y, Ma C, Wang Y, Liu X (2015) Wnt/β-Catenin signaling reduces Bacillus Calmette-Guerin-induced macrophage necrosis through a ROS-mediated PARP/AIF-dependent pathway. BMC Immunol 16:16

Xie H, Lei N, Gong AY, Chen XM, Hu G (2014) *Cryptosporidium parvum* induces SIRT1 expression in host epithelial cells through downregulating let-7i. Hum Immunol 75:760–765

Yang H, Yang T, Baur JA, Perez E, Matsui T, Carmona JJ, Lamming DWW, Souza-Pinto NC, Bohr VA, Rosenzweig A et al (2007) Nutrient-sensitive mitochondrial NAD^+ levels dictate cell survival. Cell 130:1095–1107

Yeung AWS, Wu W, Freewan M, Stocker R, King NJC, Thomas SR (2012) Flavivirus infection induces indoleamine 2,3-dioxygenase in human monocyte-derived macrophages via tumor necrosis factor and NF-κB. J Leukoc Biol 91:657–666

Yu SW, Wang H, Poitras MF, Coombs C, Bowers WJ, Federoff HJ, Poirier GG, Dawson TM, Dawson VL (2002) Mediation of poly(ADP-ribose) polymerase-1 – dependent cell death by apoptosis-inducing factor. Science 297:259–263

Zerez CR, Roth EF, Schulman S, Tanaka KR (1990) Increased nicotinamide adenine dinucleotide content and synthesis in *Plasmodium falciparum*-infected human erythrocytes. Blood 75:1705–1710

Zhang HS, Zhou Y, Wu MR, Zhou HS, Xu F (2009) Resveratrol inhibited Tat-induced HIV-1 LTR transactivation via NAD^+-dependent SIRT1 activity. Life Sci 85:484–489

Zhang H-S, Sang W-W, Wang Y-O, Liu W (2010) Nicotinamide phosphoribosyltransferase/sirtuin 1 pathway is involved in human immunodeficiency virus type 1 Tat-mediated long terminal repeat transactivation. J Cell Biochem 110:1464–1470

Zhang HS, Chen XY, Wu TC, Zhang FJ (2014) Tanshinone II A inhibits tat-induced HIV-1 transactivation through redox-regulated AMPK/Nampt pathway. J Cell Physiol 229:1193–1201

Chapter 5
Mitochondrial Bioenergetics and Dynamics During Infection

Cynthia Soultawi, Yasmina Fortier, Calaiselvy Soundaramourty, Jérôme Estaquier, and Mireille Laforge

Abstract Microbes have developed a series of strategies to overcome the defense mechanisms of the infected host. During pathogen–host coevolution, they develop strategy to manipulate cellular machinery particularly in subverting mitochondrion function. Mitochondria are highly dynamic organelles that constantly remodel their structure. In particular, shaping and cellular distribution of the mitochondrial network is maintained in large part by the conserved activities of mitochondrial division, fusion, motility, and tethering. Mitochondria have been long recognized for their role in providing energy production, calcium metabolism, and apoptosis. More recently, mitochondria have been also shown to serve as a platform for innate immune response. In this context, mitochondrial dynamics and shaping is not only essential to maintain cristae structure and bioenergetic to fuel cellular demands but contribute to regulate cellular function such as innate immune response and mitochondrial permeabilization. Due to their key role in cell survival, mitochondria represent attractive targets for pathogens. Therefore, microbes by manipulating mitochondrial dynamics may escape to host cellular control. Herein, we describe how mitochondrial bioenergetics, dynamics, and shaping are impacted during microbe infections and how this interplay benefits to pathogens contributing to the diseases.

Keywords Mitochondrial dynamics · Bioenergetics · Pathogens

C. Soultawi · Y. Fortier · C. Soundaramourty · M. Laforge (✉)
CNRS FR3636, Faculty of Medecine des Saint-Pères, Paris Descartes University, Paris, France

J. Estaquier (✉)
CNRS FR3636, Faculty of Medecine des Saint-Pères, Paris Descartes University, Paris, France

Centre Hospitalier Universitaire (CHU) de Québec Research Center, Faculty of Medicine, Laval University, Québec, QC, Canada

R. Silvestre, E. Torrado (eds.), *Metabolic Interaction in Infection*, Experientia Supplementum 109, https://doi.org/10.1007/978-3-319-74932-7_5

5.1 Mitochondria

Mitochondria, which are found in nearly all eukaryotes, including plants, animals, fungi, and protists, are large enough to be observed with a light microscope and were first discovered in the 1800s. Albert von Kölliker described in 1857 what he called "granules" in the cells of muscles. Thus, the name of the organelles was coined to reflect the way they looked to the first scientists to observe them, stemming from the Greek words for "thread" and "granule."

Mitochondria are considered the power of the cell, converting oxygen and nutrients into adenosine triphosphate (ATP). Otto Heinrich Warburg described for the first time in 1912 that an enzyme within cells enabled the processing of oxygen. Thereafter, in 1923, David Keilin demonstrates how the oxidation state within cytochromes (hemoproteins responsible for electron transport) was changed during respiration. He later identified the cytochrome c, part of the inner membrane of mitochondria, which was described thereafter in 1996 as central hub in regulating apoptosis (Liu et al. 1996). In 1950, Eugene Kennedy and Albert Lehninger demonstrate mitochondrial oxidation, while in 1978, Peter D. Mitchell described the diffusion of hydrogen ions across membranes, and its relation to ATP during respiration in eukaryotic cells. Briefly, in aerobic respiration, oxygen is essential for ATP production. In the presence of oxygen, glycolysis is the first stage of aerobic cellular respiration. Thus, theoretically this process yields 38 ATP molecules that can be made per oxidized glucose molecule during cellular respiration. Thus, 2 come from the glycolysis, 2 from the Krebs cycle (or the tricarboxylic acid cycle), and about 34 from the electron transport system (oxidative phosphorylation which occurs in the mitochondrial cristae). Thus, pyruvic acid generated by glycolysis crosses the mitochondrial membrane and used by the acetyl coenzyme A (acetyl CoA) in the initial step of the TCA cycle that takes place in the matrix of cell mitochondria. Without oxygen, pyruvate is not metabolized by cellular respiration. Thus, the pyruvate is not transported into the mitochondrion but remains in the cytoplasm, where it is converted into lactic acid. This process of fermentation is less efficient since only two ATP are produced per glucose. Therefore, mitochondrion is of crucial importance for providing and responding for energy demands to fuel the cells. Thus, mitochondrial mass increases with the energy demands.

5.2 Mitochondrial Dynamics and Impact on Bioenergetics

Maintenance of mitochondrial function and quality control are intimately related to morphology and fusion, fission, and turnover. Mammalian cells maintain their mitochondrial network by balancing the opposing processes of mitochondrial fusion and fission (Chan 2012). The physiological relevance of mitochondrial dynamics came to light with the discovery that mutations or alterations in mitochondrial dynamics proteins have been associated with several human pathologies; these included neuropathies, such as Charcot–Marie–Tooth Syndrome and Autosomal Dominant Hereditary Optic Atrophy (ADOA) (Zuchner et al. 2004; Alexander et al. 2000; Delettre et al. 2000); neurodegenerative diseases, such as Parkinson's disease (PD) (McCoy and Cookson 2012); atherosclerosis (Chen et al. 2004); as well as metabolic syndromes, such as obesity and type 2 diabetes mellitus (T2DM) (Sebastian et al. 2012; Schneeberger et al. 2013).

Mitochondrial fission generates discrete and fragmented mitochondria that can increase reactive oxygen species (ROS) production (Yu et al. 2006), facilitate mitophagy (Frank et al. 2012; Toyama et al. 2016), mediate apoptosis (Youle and Karbowski 2005; Manor et al. 2015; Lee et al. 2016), and accelerate cell proliferation (Taguchi et al. 2007). Fission and fusion involve distinct sets of proteins. Dynamin-related protein 1 (Drp1) is a cytosolic protein that translocates to the outer mitochondrial membrane (OMM) where it not only oligomerizes into ring-like structures at future sites of division (Youle and Karbowski 2005) but also contributes to regulate mitochondrial outer membrane permeabilization (Estaquier and Arnoult 2007). Mitochondria in cells lacking DRP1 adopt a hypertubular configuration in yeast, worms, and mammals. Several DRP1 receptors and recruitment factors on the OM have been identified, such as FIS1, DDP/TIMM8a, MFF, MiD49, and MiD51, which exert partially overlapping functions (Labbe et al. 2014; Arnoult et al. 2005b).

In mammals, Drp1 contributes to mitochondrial permeabilization by colocalizing with Bax (Montessuit et al. 2010). The posttranslational modifications of DRP1 that affect its mitochondrial recruitment and function include phosphorylation, ubiquitylation, and SUMOylation (van der Bliek et al. 2013).

Mitochondrial fusion generates linear or tubular networks facilitating the exchange of mitochondrial contents, prevents mutations in mtDNA (Santel et al. 2003), maximizes OXPHOS activity and ATP production (Manor et al. 2015; Cogliati et al. 2013), and enhances ER interactions important for Ca^{2+} flux (de Brito and Scorrano 2008). Conversely to fission, mitochondria fusion involves the mammalian orthologs mitofusin 1 and mitofusin 2 (Mfn1 and Mfn2) that orchestrate OM fusion depending on GTP hydrolysis. Mfn synthesis is regulated by transcriptional and posttranscriptional mechanisms, while proteasome-mediated degradation is regulated by ubiquitylation and phosphorylation. Deletion of either Mfn1 or Mfn2 or abrogation of GTPase activity prevents mitochondrial fusion, causing mitochondrial network fragmentation (Mishra and Chan 2014). Optic atrophy 1 (OPA1) is a dynamin-like GTPase anchored to the IM by an N-terminal transmembrane domain (Olichon et al. 2002). The bulk of the protein, including the GTP-binding and GTPase effector domains, is exposed to the IM space (IMS). Multiple isoforms of OPA1 are found in mammalian cells and tissues, which are the result of alternative splicing and proteolytic cleavage. Genetic ablation of Opa1 blocks fusion and causes mitochondria to fragment and is associated with several diseases (Alexander et al. 2000) (Delettre et al. 2000). Thus, OPA1 level is lower in muscles from elderly subjects and was associated to a decline in mitochondrial function (Joseph et al. 2012). Importantly, most of skeletal muscle biopsies from autosomal dominant optic atrophy (DOA) patients show mtDNA defects and COX-negative muscle fibers (Yu-Wai-Man et al. 2010). Recently, a role of the noncanonical NF-kB pathway has been proposed to regulate mitochondrial dynamics and OPA1 expression suggesting novel avenues in the regulation of mitochondrial function (Laforge et al. 2015).

OPA1 maintains the cristae architecture and is essential for the stability of the OXPHOS machinery (Duvezin-Caubet et al. 2006). Indeed, complexes that form the OXPHOS machinery involve macro-complexes in which the electrons are transferred. This electron transfer process is accompanied by the pumping of protons out of the matrix by complexes I, III, and IV, leading to the generation of mitochondrial trans-membrane potential ($\Delta\Psi m$). Complex V (F_0F_1-ATP synthase) uses this potential to convert ADP into ATP (Boatright et al. 2003; Vafai and Mootha 2012). Therefore, it has been shown that the absence of OPA1 destabilizes this process. In particular, it has been shown that the absence of OPA1 favors the release of cytochrome C induced by apoptotic stimuli. Furthermore, OPA1 is released during apoptosis (Arnoult et al. 2005a; Olichon et al. 2003; Cipolat et al. 2006; Ishihara et al. 2006; Varanita et al. 2015; Cogliati et al. 2013).

Disruption of the electron transport chain can also lead to an increase of reactive oxygen species (ROS) production (Figueira et al. 2013). The accumulation or persistence of ROS can result in damage to biomolecules located in the vicinity of the ETC and mitochondrial dysfunction and disease (Wallace 2005; Baker and Haynes 2011). Similar to OPA1, it has been also reported that depletion of Mfn2 in mammalian cells induces mitochondrial dysfunction, characterized by a decrease in oxygen consumption coupled to ATP synthesis, increased proton leak, reduced mitochondrial membrane potential, increased production of ROS, and reduced Coenzyme Q levels, thereby leading to reduced glucose, pyruvate, and fatty acid oxidation (Sebastian et al. 2012; Bach et al. 2003; Chen et al. 2005; Mourier et al. 2015). These alterations appear to be specific to MFN2, because ablation of Mfn1 in MEF cells or mouse liver has been reported to lead to the opposite phenotype, characterized by an increase in ATP-coupled and maximal respiration (Kulkarni et al. 2016). Furthermore, repression of Drp1 or the expression of its dominant negative mutant form reduces oxygen consumption and induces an increase in proton leak (Twig et al. 2008; Zepeda et al. 2014; Ikeda et al. 2015). In cardiomyocytes, Drp1 was reported to inhibit mitochondrial respiration, and this effect appeared to be independent of mitochondrial fission (Zhang et al. 2017). Altogether these results indicate the critical impact of mitochondrial dynamics and shaping on mitochondrial functions.

5.3 Pathogens, Mitochondrial Dynamics, and Bioenergetics

Mitochondria can be directly targeted by pathogenic proteins or influenced by the physiological alterations to cellular environment during pathogenesis like oxidative stress, hypoxia, deregulated calcium homeostasis, endoplasmic reticulum stress, and apoptosis. Those physiological alterations can directly affect the mitochondrial dynamics and bioenergetics. We discuss, below, some of pathogens, which have been characterized for their effect on mitochondrial dynamics and bioenergetics highlighting the significance of these alterations in the infectious process.

5.3.1 Viral Infections

Viruses interfere with the mitochondrial pathways and disrupt mitochondrial dynamics to facilitate their proliferation. In particular, several groups have reported a critical role of Drp1 as a target for viral subversion. Thus, hepatitis B virus (HBV) and hepatitis C virus (HCV) utilize the alterations in mitochondrial dynamics for their persistence. HCV and HBV disrupt mitochondrial dynamics by promoting mitochondrial fission by stimulating the expression and inducing the phosphorylation of DRP1 on serine residue (Ser616) (Chang and Blackstone 2007a, b). Thus, altered mitochondrial dynamics and induction of mitophagy which protects virus-infected hepatocytes from apoptotic cell death facilitate the persistence of HBV-infected hepatocytes. Furthermore, DRP1 silencing suppressed HCV secretion, which is associated with a decrease in cellular glycolysis and ATP levels and an increase in interferon synthesis. Silencing Drp1 caused significant increase in apoptotic signaling in HCV-infected cells. These results suggest that HCV induced mitochondrial fission to attenuate apoptosis contributing to persistent infection (Kim et al. 2013, 2014). The classical swine fever virus (CSFV), which causes typical clinical characteristics in piglets, including hemorrhagic syndrome and immunosuppression, induced mitochondrial fragmentation and Drp1 translocation into mitochondria. Silencing of Drp1 in CSFV-infected cells diminished viral replication. Thus, CSFV induced mitochondrial fission to enhance cell survival and viral persistence (Gou et al. 2017). Porcine reproductive and respiratory syndrome virus (PRRSV), which causes acute mitochondrial dysfunction, also induced mitochondrial fission to attenuate apoptosis in Marc145 cells. Similarly to HCV and HBV infections, PRRSV infection induced the expression of Drp1, enhanced its phosphorylation at Ser616, and its translocation to mitochondria (Li et al. 2016).

Epstein–Barr virus (EBV) is an oncogenic virus belonging to the herpes virus family and is implicated in various lymphoid and epithelial cancers (Young and Rickinson 2004). The viral latent membrane protein 2A (LMP2A) has been reported to cause elevated mitochondrial fission in gastric and breast cancer cells. The authors observed higher level of Drp1 expression in LMP2A-transfected cells. The authors also establish that altered mitochondrial dynamics plays a critical role in cell migration and metastasis in epithelial cancers (Pal et al. 2014). Although the role of Drp1 was not addressed, it has been reported that the pseudorabies virus (PRV), an alphaherpesvirus, localized to mitochondria affects mitochondrial function. Thus, PRV and herpes simplex virus type 1 (HSV-1) infection of neurons disrupts mitochondrial motility and morphology, which has been linked to intracellular Ca^{2+} increases (Kramer and Enquist 2012).

It is well known that neurotoxicity of human immunodeficiency virus-1 (HIV) includes neuronal apoptosis in which the envelope glycoprotein and tat have proposed to play a crucial role (Mattson et al. 2005). In this context, it has been reported that in a model transgenic mouse expressing the envelope glycoprotein, the gp120, in neurons, mitochondria are damaged and elongated. The authors reported a slight decrease in the levels of the phosphorylated form of DRP1 on Ser616 and an increase of MFN1 (Fields et al. 2016). Interestingly, in HIV individuals, associated with neurocognitive disorders (HAND), the authors observed a decrease in DRP1 in the frontal cortex tissues, along with enlarged and elongated mitochondria localized to the soma of damaged neurons (Fields et al. 2016).

Interestingly, Dengue virus (DENV) which interacts with cellular organelles, including mitochondria, to successfully replicate in cells, also decreased, like HIV infection, the levels of Drp1, as well as the phosphorylated form of Drp1 on serine 616. The authors also observed that inhibiting Drp1 with a mutant of serine inhibits viral replication (Barbier et al. 2017). Furthermore, the DENV protease NS2B3 has been reported to cleave MFN1 and MFN2 impairing mitochondrial fusion infection (Yu et al. 2015). It is interestingly to note that two different functions of MFN1 and MFN2 in infected cells have been proposed. MFN1 overexpression induced higher levels of IFNβ mRNA and blocked the apoptotic cascade induced by DENV. Silencing of MFN1 attenuated mitochondrial antiviral signaling protein (MAVS)-triggering IRF3 phosphorylation and IFNβ induction. In contrary, MFN2 overexpression did not affect IRF3 phosphorylation, but mitochondrial membrane permeabilization. Thus, MFNs participate in host defense against DENV infection by promoting antiviral response and cell survival (Yu et al. 2015). Although the role of MFN2 has received less attention than Drp1, CSFV-infected cells were associated with MFN2 ubiquitination and degradation (Gou et al. 2017). Altogether, these different results highlight how viruses subvert Drp1 and MFN1/2 in controlling innate immune response and cell survival that may have an impact on viral persistence.

5.3.2 Bacteria Infection

Several bacteria have also been reported to target mitochondria interfering with the host cell apoptotic machinery. However, only few reports are available on the effects of intracellular pathogens on mitochondrial dynamics. *Listeria monocytogenes* alters mitochondrial morphology, inducing mitochondrial network fragmentation without inducing cell death (Stavru et al. 2011; Stavru and Cossart 2011). The authors identified the secreted pore-forming toxin listeriolysin O as the bacterial factor responsible for mitochondrial network disruption at early time points of infection. Strikingly, mitochondrial fragmentation does not require Drp1 as fragmentation proceeds efficiently in cells with impaired Drp1 function. Furthermore, mitochondrial fragmentation is independent of OPA1. However, the fragmentation sites were marked by the endoplasmic reticulum (ER), suggesting that although the initial constriction machinery is shared, molecules other than Drp1 can terminate mitochondrial abscission downstream of the ER (Stavru et al. 2013; Lebreton et al. 2015).

The pathogenic bacteria *Helicobacter pylori* target mitochondria to modulate the host's apoptotic machinery. The vacuolating cytotoxin A (VacA), which induces MOMP, (Willhite and Blanke 2004) also induces mitochondrial network fragmentation through the recruitment of Drp1. Interestingly, the authors report that Bax is not required for Drp1-dependent mitochondrial fission, but Cyt *c* release was significantly inhibited in the presence of the Drp1 inhibitor mdivi-1 within VacA-intoxicated cells. These results indicate that enhanced mitochondrial fission contributes to MOMP (Jain et al. 2011).

The pathogenic bacteria *Legionella pneumophila* which encodes a type IV secretion system (T4SS) also induces mitochondrial network fragmentation. Thus, effector proteins that inject into macrophages are essential for bacterial multiplication within vacuoles (LCV). These vacuoles interact thereafter with mitochondria. Interestingly, both T4SS-independent and T4SS-dependent mechanisms contribute to mitochondrial dysfunction. The authors have identified among the partner of T4SS, a Ran GTPase activator, MitF, which is required for fission and associated with the accumulation of Drp1 at the mitochondrial membrane. But the contributing role of both T4SS mechanisms seems to be more complex with contrasting effect on mitochondrial respiration and glycolysis. This latter seems to be essential for *s* creating the optimal environment for bacteria multiplication (Escoll et al. 2017).

Chlamydia trachomatis had a marked effect on the mitochondrial network. *Chlamydia* requires mitochondrial ATP for normal development and preserves mitochondrial integrity through a miR-30c-5p-dependent inhibition of Drp1-mediated mitochondrial fission (Chowdhury et al. 2017). Another model of bacteria infection that causes mitochondrial dysfunction is Shigella. Shigella infection increases lactate dehydrogenase (LDH) release and induces cell death accompanied by mitochondrial fragmentation linked to DRP1 activation. Another study had demonstrated that Drp1 interacts with septins to enhance mitochondrial fission (Sirianni et al. 2016; Lum and Morona 2014). Thus, like for viral infections, DRP1 seems to be a key actor in regulating mitochondrial dynamics in the context of host's bacteria interaction.

5.3.3 Parasite Infections

The pathogenesis of infection with the helminth parasite *Schistosoma mansoni* in mice has been reported to involve a T helper Th1 to Th2 cytokine switch, in which IL-10 play a critical role in regulating T cell apoptosis of Th1 cells. This infection is associated with a pathogenic granulomatous response in response to parasite egg deposit in the liver in which the occurrence of apoptotic cells in this tissue coincides with the pathogenicity (Estaquier et al. 1997). Recent findings indicate that *S. mansoni* is associated with mitochondrial network mitochondrial fragmentation in the liver of infected mice. This is associated with lower levels of ATP. However, several genes associated with mitochondrial biogenesis were upregulated that could be a mechanism to compensate lower ATP production (Chen et al. 2015).

Recently, it has been reported that infection of macrophages of *Leishmania infantum* is associated with AMPK activation and ATP lowering, which induced a compensatory mechanism associated with higher expression of genes related to mitochondrial biogenesis and glycolysis. Thus, macrophages switch from an early glycolytic metabolism to an oxidative phosphorylation, and this metabolic process involves SIRT1 and LKB1. Thus, these observations indicate the importance of SIRT1 and AMPK energetic sensors in regulating mitochondrial function favoring intracellular survival and proliferation of *Leishmania* (Moreira et al. 2015). In a recent study, infection of macrophages with *Leishmania donovani* is associated with mitochondrial depolarization and reduces mitochondrial dynamics in infected host cells. The infection is associated with a reduction in the levels of MFN2, which is essential for mitochondria–ER tethering (Ishihara et al. 2004; de Brito and Scorrano 2008). The authors propose that *L. donovani* infection by impacting on mitochondrial dynamics represents a mechanism to limit inflammatory cytokine expression that can be detrimental for the parasite (Chakrabarty and Bhattacharyya 2017). However, this contrasts with earlier report indicating that *L donovani* infection induces gene expression for inflammatory cytokines from bone marrow-derived macrophages and to prevent apoptosis (Moore and Matlashewski 1994). Furthermore, it was thereafter shown that infection with *Leishmania* major prevents mitochondrial depolarization and cytochrome c release induced by apoptotic insults both in neutrophils and macrophages (Aga et al. 2002; Akarid et al. 2004; Donovan et al. 2009). Therefore, it is of importance to determine the contributing role of mitochondrial dynamics in the context of *Leishmania* species.

5.3.4 Non Conventional Pathogens

Prion disease has been previously reported with decreased mitochondrial activity. In a model of transgenic mice overexpressing prion protein (Tg7) infected with the prion strain 263K, a severe deficit in mitochondrial oxygen consumption was reported. Proteome analysis reveals that mitochondrial profile is modified in this rodent model involving changes in the relative abundance of proteins involved in electron transport, mitochondrial dynamics, and mitochondrial protein synthesis (Faris et al. 2017). Another study indicates significant decreases of Drp1 in the brain of scrapie-infected rodents at the terminal stage of the infection (Yang et al. 2017) and considered by the author as a read out of mitochondrial dynamics dysregulation.

5.4 Concluding Remarks

In this book chapter, we have summarized recent exciting findings on how mitochondrial dynamics and bioenergetics are affected during pathogen infection. In recent years, a growing number of studies have highlighted that pathogens can alter mitochondrial morphology and dynamics and therefore bioenergetics turning this defect on their own profit for proliferation and persistence. Further investigation of the effect of mitochondrial dynamics and bioenergetics during infections will enhance our understanding of pathogen–host interactions and their role in pathogenesis. Such an understanding will be useful in the therapeutic design of new defense strategies. Currently, our understanding of mitochondrial dynamics during infections is still in its infancy, but certainly represents an emerging avenue of molecular investigations in infection diseases.

Funding Statement This work was supported by grants to ML and JE from the Agence Nationale de Recherches sur le Sida et les HÕpatites Virales (ANRS). CS is supported by a fellowship from Lebanon and ANRS, and YF is supported by fellowships from ANRS. The work was supported by Infect-ERA (Project INLEISH).

References

Aga E, Katschinski DM, Van Zandbergen G, Laufs H, Hansen B, Muller K, Solbach W, Laskay T (2002) Inhibition of the spontaneous apoptosis of neutrophil granulocytes by the intracellular parasite *Leishmania* major. J Immunol 169:898–905

Akarid K, Arnoult D, Micic-Polianski J, Sif J, Estaquier J, Ameisen JC (2004) *Leishmania* major-mediated prevention of programmed cell death induction in infected macrophages is associated with the repression of mitochondrial release of cytochrome c. J Leukoc Biol 76:95–103

Alexander C, Votruba M, Pesch UE, Thiselton DL, Mayer S, Moore A, Rodriguez M, Kellner U, Leo-Kottler B, Auburger G, Bhattacharya SS, Wissinger B (2000) OPA1, encoding a dynamin-related GTPase, is mutated in autosomal dominant optic atrophy linked to chromosome 3q28. Nat Genet 26:211–215

Arnoult D, Grodet A, Lee YJ, Estaquier J, Blackstone C (2005a) Release of OPA1 during apoptosis participates in the rapid and complete release of cytochrome c and subsequent mitochondrial fragmentation. J Biol Chem 280:35742–35750

Arnoult D, Rismanchi N, Grodet A, Roberts RG, Seeburg DP, Estaquier J, Sheng M, Blackstone C (2005b) Bax/Bak-dependent release of DDP/TIMM8a promotes Drp1-mediated mitochondrial fission and mitoptosis during programmed cell death. Curr Biol 15:2112–2118

Bach D, Pich S, Soriano FX, Vega N, Baumgartner B, Oriola J, Daugaard JR, Lloberas J, Camps M, Zierath JR, Rabasa-Lhoret R, Wallberg-Henriksson H, Laville M, Palacin M, Vidal H, Rivera F, Brand M, Zorzano A (2003) Mitofusin-2 determines mitochondrial network architecture and mitochondrial metabolism. A novel regulatory mechanism altered in obesity. J Biol Chem 278:17190–17197

Baker BM, Haynes CM (2011) Mitochondrial protein quality control during biogenesis and aging. Trends Biochem Sci 36:254–261

Barbier V, Lang D, Valois S, Rothman AL, Medin CL (2017) Dengue virus induces mitochondrial elongation through impairment of Drp1-triggered mitochondrial fission. Virology 500:149–160

Boatright KM, Renatus M, Scott FL, Sperandio S, Shin H, Pedersen IM, Ricci JE, Edris WA, Sutherlin DP, Green DR, Salvesen GS (2003) A unified model for apical caspase activation. Mol Cell 11:529–541

Chakrabarty Y, Bhattacharyya SN (2017) *Leishmania donovani* restricts mitochondrial dynamics to enhance miRNP stability and target RNA repression in host macrophages. Mol Biol Cell 28:2091–2105

Chan DC (2012) Fusion and fission: interlinked processes critical for mitochondrial health. Annu Rev Genet 46:265–287

Chang CR, Blackstone C (2007a) Cyclic AMP-dependent protein kinase phosphorylation of Drp1 regulates its GTPase activity and mitochondrial morphology. J Biol Chem 282:21583–21587

Chang CR, Blackstone C (2007b) Drp1 phosphorylation and mitochondrial regulation. EMBO Rep 8:1088–1089. author reply 1089–90

Chen KH, Guo X, Ma D, Guo Y, Li Q, Yang D, Li P, Qiu X, Wen S, Xiao RP, Tang J (2004) Dysregulation of HSG triggers vascular proliferative disorders. Nat Cell Biol 6:872–883

Chen H, Chomyn A, Chan DC (2005) Disruption of fusion results in mitochondrial heterogeneity and dysfunction. J Biol Chem 280:26185–26192

Chen TT, Wu LS, Hsu PW, Pang CY, Lee KM, Cheng PC, Peng SY (2015) Mitochondrial dynamics in the mouse liver infected by *Schistosoma mansoni*. Acta Trop 148:13–23

Chowdhury SR, Reimer A, Sharan M, Kozjak-Pavlovic V, Eulalio A, Prusty BK, Fraunholz M, Karunakaran K, Rudel T (2017) Chlamydia preserves the mitochondrial network necessary for replication via microRNA-dependent inhibition of fission. J Cell Biol 216:1071–1089

Cipolat S, Rudka T, Hartmann D, Costa V, Serneels L, Craessaerts K, Metzger K, Frezza C, Annaert W, D'Adamio L, Derks C, Dejaegere T, Pellegrini L, D'Hooge R, Scorrano L, De Strooper B (2006) Mitochondrial rhomboid PARL regulates cytochrome c release during apoptosis via OPA1-dependent cristae remodeling. Cell 126:163–175

Cogliati S, Frezza C, Soriano ME, Varanita T, Quintana-Cabrera R, Corrado M, Cipolat S, Costa V, Casarin A, Gomes LC, Perales-Clemente E, Salviati L, Fernandez-Silva P, Enriquez JA, Scorrano L (2013) Mitochondrial cristae shape determines respiratory chain supercomplexes assembly and respiratory efficiency. Cell 155:160–171

de Brito OM, Scorrano L (2008) Mitofusin 2 tethers endoplasmic reticulum to mitochondria. Nature 456:605–610

Delettre C, Lenaers G, Griffoin JM, Gigarel N, Lorenzo C, Belenguer P, Pelloquin L, Grosgeorge J, Turc-Carel C, Perret E, Astarie-Dequeker C, Lasquellec L, Arnaud B, Ducommun B, Kaplan J, Hamel CP (2000) Nuclear gene OPA1, encoding a mitochondrial dynamin-related protein, is mutated in dominant optic atrophy. Nat Genet 26:207–210

Donovan MJ, Maciuba BZ, Mahan CE, McDowell MA (2009) *Leishmania* infection inhibits cycloheximide-induced macrophage apoptosis in a strain-dependent manner. Exp Parasitol 123:58–64

Duvezin-Caubet S, Jagasia R, Wagener J, Hofmann S, Trifunovic A, Hansson A, Chomyn A, Bauer MF, Attardi G, Larsson NG, Neupert W, Reichert AS (2006) Proteolytic processing of OPA1 links mitochondrial dysfunction to alterations in mitochondrial morphology. J Biol Chem 281:37972–37979

Escoll P, Song OR, Viana F, Steiner B, Lagache T, Olivo-Marin JC, Impens F, Brodin P, Hilbi H, Buchrieser C (2017) *Legionella pneumophila* modulates mitochondrial dynamics to trigger metabolic repurposing of infected macrophages. Cell Host Microbe 22:302–316 e7

Estaquier J, Arnoult D (2007) Inhibiting Drp1-mediated mitochondrial fission selectively prevents the release of cytochrome c during apoptosis. Cell Death Differ 14:1086–1094

Estaquier J, Marguerite M, Sahuc F, Bessis N, Auriault C, Ameisen JC (1997) Interleukin-10-mediated T cell apoptosis during the T helper type 2 cytokine response in murine *Schistosoma mansoni* parasite infection. Eur Cytokine Netw 8:153–160

Faris R, Moore RA, Ward A, Sturdevant DE, Priola SA (2017) Mitochondrial respiration is impaired during late-stage hamster prion infection. J Virol 91:e00524-17

Fields JA, Serger E, Campos S, Divakaruni AS, Kim C, Smith K, Trejo M, Adame A, Spencer B, Rockenstein E, Murphy AN, Ellis RJ, Letendre S, Grant I, Masliah E (2016) HIV alters neuronal mitochondrial fission/fusion in the brain during HIV-associated neurocognitive disorders. Neurobiol Dis 86:154–169

Figueira TR, Barros MH, Camargo AA, Castilho RF, Ferreira JC, Kowaltowski AJ, Sluse FE, Souza-Pinto NC, Vercesi AE (2013) Mitochondria as a source of reactive oxygen and nitrogen species: from molecular mechanisms to human health. Antioxid Redox Signal 18:2029–2074

Frank M, Duvezin-Caubet S, Koob S, Occhipinti A, Jagasia R, Petcherski A, Ruonala MO, Priault M, Salin B, Reichert AS (2012) Mitophagy is triggered by mild oxidative stress in a mitochondrial fission dependent manner. Biochim Biophys Acta 1823:2297–2310

Gou H, Zhao M, Xu H, Yuan J, He W, Zhu M, Ding H, Yi L, Chen J (2017) CSFV induced mitochondrial fission and mitophagy to inhibit apoptosis. Oncotarget 8:39382–39400

Ikeda Y, Shirakabe A, Maejima Y, Zhai P, Sciarretta S, Toli J, Nomura M, Mihara K, Egashira K, Ohishi M, Abdellatif M, Sadoshima J (2015) Endogenous Drp1 mediates mitochondrial autophagy and protects the heart against energy stress. Circ Res 116:264–278

Ishihara N, Eura Y, Mihara K (2004) Mitofusin 1 and 2 play distinct roles in mitochondrial fusion reactions via GTPase activity. J Cell Sci 117:6535–6546

Ishihara N, Fujita Y, Oka T, Mihara K (2006) Regulation of mitochondrial morphology through proteolytic cleavage of OPA1. Embo J 25:2966–2977

Jain P, Luo ZQ, Blanke SR (2011) *Helicobacter pylori* vacuolating cytotoxin A (VacA) engages the mitochondrial fission machinery to induce host cell death. Proc Natl Acad Sci U S A 108:16032–16037

Joseph AM, Adhihetty PJ, Buford TW, Wohlgemuth SE, Lees HA, Nguyen LM, Aranda JM, Sandesara BD, Pahor M, Manini TM, Marzetti E, Leeuwenburgh C (2012) The impact of aging on mitochondrial function and biogenesis pathways in skeletal muscle of sedentary high- and low-functioning elderly individuals. Aging Cell 11:801–809

Kim SJ, Khan M, Quan J, Till A, Subramani S, Siddiqui A (2013) Hepatitis B virus disrupts mitochondrial dynamics: induces fission and mitophagy to attenuate apoptosis. PLoS Pathog 9: e1003722

Kim SJ, Syed GH, Khan M, Chiu WW, Sohail MA, Gish RG, Siddiqui A (2014) Hepatitis C virus triggers mitochondrial fission and attenuates apoptosis to promote viral persistence. Proc Natl Acad Sci U S A 111:6413–6418

Kramer T, Enquist LW (2012) Alphaherpesvirus infection disrupts mitochondrial transport in neurons. Cell Host Microbe 11:504–514

Kulkarni SS, Joffraud M, Boutant M, Ratajczak J, Gao AW, Maclachlan C, Hernandez-Alvarez MI, Raymond F, Metairon S, Descombes P, Houtkooper RH, Zorzano A, Canto C (2016) Mfn1 deficiency in the liver protects against diet-induced insulin resistance and enhances the hypoglycemic effect of metformin. Diabetes 65:3552–3560

Labbe K, Murley A, Nunnari J (2014) Determinants and functions of mitochondrial behavior. Annu Rev Cell Dev Biol 30:357–391

Laforge M, Rodrigues V, Silvestre R, Gautier C, Weil R, Corti O, Estaquier J (2015) NF-kappaB pathway controls mitochondrial dynamics. Cell Death Differ 23:89–98

Lebreton A, Stavru F, Cossart P (2015) Organelle targeting during bacterial infection: insights from Listeria. Trends Cell Biol 25:330–338

Lee JE, Westrate LM, Wu H, Page C, Voeltz GK (2016) Multiple dynamin family members collaborate to drive mitochondrial division. Nature 540:139–143

Li S, Wang J, Zhou A, Khan FA, Hu L, Zhang S (2016) Porcine reproductive and respiratory syndrome virus triggers mitochondrial fission and mitophagy to attenuate apoptosis. Oncotarget 7:56002–56012

Liu X, Kim CN, Yang J, Jemmerson R, Wang X (1996) Induction of apoptotic program in cell-free extracts: requirement for dATP and cytochrome c. Cell 86:147–157

Lum M, Morona R (2014) Dynamin-related protein Drp1 and mitochondria are important for Shigella flexneri infection. Int J Med Microbiol 304:530–541

Manor U, Bartholomew S, Golani G, Christenson E, Kozlov M, Higgs H, Spudich J, Lippincott-Schwartz J (2015) A mitochondria-anchored isoform of the actin-nucleating spire protein regulates mitochondrial division. Elife 4:e08828

Mattson MP, Haughey NJ, Nath A (2005) Cell death in HIV dementia. Cell Death Differ 12(Suppl 1):893–904

McCoy MK, Cookson MR (2012) Mitochondrial quality control and dynamics in Parkinson's disease. Antioxid Redox Signal 16:869–882

Mishra P, Chan DC (2014) Mitochondrial dynamics and inheritance during cell division, development and disease. Nat Rev Mol Cell Biol 15:634–646

Montessuit S, Somasekharan SP, Terrones O, Lucken-Ardjomande S, Herzig S, Schwarzenbacher R, Manstein DJ, Bossy-Wetzel E, Basanez G, Meda P, Martinou JC (2010) Membrane remodeling induced by the dynamin-related protein Drp1 stimulates Bax oligomerization. Cell 142:889–901

Moore KJ, Matlashewski G (1994) Intracellular infection by *Leishmania donovani* inhibits macrophage apoptosis. J Immunol 152:2930–2937

Moreira D, Rodrigues V, Abengozar M, Rivas L, Rial E, Laforge M, Li X, Foretz M, Viollet B, Estaquier J, Cordeiro Da Silva A, Silvestre R (2015) *Leishmania infantum* modulates host macrophage mitochondrial metabolism by hijacking the SIRT1-AMPK axis. PLoS Pathog 11: e1004684

Mourier A, Motori E, Brandt T, Lagouge M, Atanassov I, Galinier A, Rappl G, Brodesser S, Hultenby K, Dieterich C, Larsson NG (2015) Mitofusin 2 is required to maintain mitochondrial coenzyme Q levels. J Cell Biol 208:429–442

Olichon A, Emorine LJ, Descoins E, Pelloquin L, Brichese L, Gas N, Guillou E, Delettre C, Valette A, Hamel CP, Ducommun B, Lenaers G, Belenguer P (2002) The human dynamin-related protein OPA1 is anchored to the mitochondrial inner membrane facing the inter-membrane space. FEBS Lett 523:171–176

Olichon A, Baricault L, Gas N, Guillou E, Valette A, Belenguer P, Lenaers G (2003) Loss of OPA1 perturbates the mitochondrial inner membrane structure and integrity, leading to cytochrome c release and apoptosis. J Biol Chem 278:7743–7746

Pal AD, Basak NP, Banerjee AS, Banerjee S (2014) Epstein-Barr virus latent membrane protein-2A alters mitochondrial dynamics promoting cellular migration mediated by Notch signaling pathway. Carcinogenesis 35:1592–1601

Santel A, Frank S, Gaume B, Herrler M, Youle RJ, Fuller MT (2003) Mitofusin-1 protein is a generally expressed mediator of mitochondrial fusion in mammalian cells. J Cell Sci 116:2763–2774

Schneeberger M, Dietrich MO, Sebastian D, Imbernon M, Castano C, Garcia A, Esteban Y, Gonzalez-Franquesa A, Rodriguez IC, Bortolozzi A, Garcia-Roves PM, Gomis R, Nogueiras R, Horvath TL, Zorzano A, Claret M (2013) Mitofusin 2 in POMC neurons connects ER stress with leptin resistance and energy imbalance. Cell 155:172–187

Sebastian D, Hernandez-Alvarez MI, Segales J, Sorianello E, Munoz JP, Sala D, Waget A, Liesa M, Paz JC, Gopalacharyulu P, Oresic M, Pich S, Burcelin R, Palacin M, Zorzano A (2012) Mitofusin 2 (Mfn2) links mitochondrial and endoplasmic reticulum function with insulin signaling and is essential for normal glucose homeostasis. Proc Natl Acad Sci U S A 109:5523–5528

Sirianni A, Krokowski S, Lobato-Marquez D, Buranyi S, Pfanzelter J, Galea D, Willis A, Culley S, Henriques R, Larrouy-Maumus G, Hollinshead M, Sancho-Shimizu V, Way M, Mostowy S (2016) Mitochondria mediate septin cage assembly to promote autophagy of Shigella. EMBO Rep 17:1029–1043

Stavru F, Cossart P (2011) Listeria infection modulates mitochondrial dynamics. Commun Integr Biol 4:364–366

Stavru F, Bouillaud F, Sartori A, Ricquier D, Cossart P (2011) Listeria monocytogenes transiently alters mitochondrial dynamics during infection. Proc Natl Acad Sci U S A 108:3612–3617

Stavru F, Palmer AE, Wang C, Youle RJ, Cossart P (2013) Atypical mitochondrial fission upon bacterial infection. Proc Natl Acad Sci U S A 110:16003–16008

Taguchi N, Ishihara N, Jofuku A, Oka T, Mihara K (2007) Mitotic phosphorylation of dynamin-related GTPase Drp1 participates in mitochondrial fission. J Biol Chem 282:11521–11529

Toyama EQ, Herzig S, Courchet J, Lewis TL Jr, Loson OC, Hellberg K, Young NP, Chen H, Polleux F, Chan DC, Shaw RJ (2016) Metabolism. AMP-activated protein kinase mediates mitochondrial fission in response to energy stress. Science 351:275–281

Twig G, Elorza A, Molina AJ, Mohamed H, Wikstrom JD, Walzer G, Stiles L, Haigh SE, Katz S, Las G, Alroy J, Wu M, Py BF, Yuan J, Deeney JT, Corkey BE, Shirihai OS (2008) Fission and selective fusion govern mitochondrial segregation and elimination by autophagy. EMBO J 27:433–446

Vafai SB, Mootha VK (2012) Mitochondrial disorders as windows into an ancient organelle. Nature 491:374–383

van der Bliek AM, Shen Q, Kawajiri S (2013) Mechanisms of mitochondrial fission and fusion. Cold Spring Harb Perspect Biol 5:a011072

Varanita T, Soriano ME, Romanello V, Zaglia T, Quintana-Cabrera R, Semenzato M, Menabo R, Costa V, Civiletto G, Pesce P, Viscomi C, Zeviani M, Di Lisa F, Mongillo M, Sandri M,

Scorrano L (2015) The OPA1-dependent mitochondrial cristae remodeling pathway controls atrophic, apoptotic, and ischemic tissue damage. Cell Metab 21:834–844

Wallace DC (2005) A mitochondrial paradigm of metabolic and degenerative diseases, aging, and cancer: a dawn for evolutionary medicine. Annu Rev Genet 39:359–407

Willhite DC, Blanke SR (2004) *Helicobacter pylori* vacuolating cytotoxin enters cells, localizes to the mitochondria, and induces mitochondrial membrane permeability changes correlated to toxin channel activity. Cell Microbiol 6:143–154

Yang X, Shi Q, Sun J, Lv Y, Ma Y, Chen C, Xiao K, Zhou W, Dong XP (2017) Aberrant alterations of mitochondrial factors Drp1 and Opa1 in the brains of scrapie experiment rodents. J Mol Neurosci 61:368–378

Youle RJ, Karbowski M (2005) Mitochondrial fission in apoptosis. Nat Rev Mol Cell Biol 6:657–663

Young LS, Rickinson AB (2004) Epstein-Barr virus: 40 years on. Nat Rev Cancer 4:757–768

Yu T, Robotham JL, Yoon Y (2006) Increased production of reactive oxygen species in hyperglycemic conditions requires dynamic change of mitochondrial morphology. Proc Natl Acad Sci U S A 103 (2658):2653

Yu CY, Liang JJ, Li JK, Lee YL, Chang BL, Su CI, Huang WJ, Lai MM, Lin YL (2015) Dengue virus impairs mitochondrial fusion by cleaving mitofusins. PLoS Pathog 11:e1005350

Yu-Wai-Man P, Griffiths PG, Gorman GS, Lourenco CM, Wright AF, Auer-Grumbach M, Toscano A, Musumeci O, Valentino ML, Caporali L, Lamperti C, Tallaksen CM, Duffey P, Miller J, Whittaker RG, Baker MR, Jackson MJ, Clarke MP, Dhillon B, Czermin B, Stewart JD, Hudson G, Reynier P, Bonneau D, Marques W Jr, Lenaers G, McFarland R, Taylor RW, Turnbull DM, Votruba M, Zeviani M, Carelli V, Bindoff LA, Horvath R, Amati-Bonneau P, Chinnery PF (2010) Multi-system neurological disease is common in patients with OPA1 mutations. Brain 133:771–786

Zepeda R, Kuzmicic J, Parra V, Troncoso R, Pennanen C, Riquelme JA, Pedrozo Z, Chiong M, Sanchez G, Lavandero S (2014) Drp1 loss-of-function reduces cardiomyocyte oxygen dependence protecting the heart from ischemia-reperfusion injury. J Cardiovasc Pharmacol 63:477–487

Zhang H, Wang P, Bisetto S, Yoon Y, Chen Q, Sheu SS, Wang W (2017) A novel fission-independent role of dynamin-related protein 1 in cardiac mitochondrial respiration. Cardiovasc Res 113:160–170

Zuchner S, Mersiyanova IV, Muglia M, Bissar-Tadmouri N, Rochelle J, Dadali EL, Zappia M, Nelis E, Patitucci A, Senderek J, Parman Y, Evgrafov O, Jonghe PD, Takahashi Y, Tsuji S, Pericak-Vance MA, Quattrone A, Battaloglu E, Polyakov AV, Timmerman V, Schroder JM, Vance JM (2004) Mutations in the mitochondrial GTPase mitofusin 2 cause Charcot-Marie-Tooth neuropathy type 2A. Nat Genet 36:449–451

Chapter 6
Computational Systems Biology of Metabolism in Infection

Müberra Fatma Cesur, Ecehan Abdik, Ünzile Güven-Gülhan, Saliha Durmuş, and Tunahan Çakır

Abstract A systems approach to elucidate the effect of infection on cell metabolism provides several opportunities from a better understanding of molecular mechanisms to the identification of potential biomarkers and drug targets. This is obvious from the fact that we have witnessed the accelerated use of computational systems biology in the last five years to study metabolic changes in pathogen and/or host cells in response to infection. In this chapter, we aim to present a comprehensive review of the recent research by focusing on genome-scale metabolic network models of pathogen-host systems and genome-wide metabolomics and fluxomics analysis of infected cells.

Keywords Genome-scale metabolic network · Flux balance analysis · Drug target · Metabolomics · Fluxomics · Pathogen-host interaction

6.1 Introduction

Microorganisms can communicate with their hosts via hundreds of interactions that may be neutral, beneficial, or deleterious. In pathogen-host interactions, pathogens try to exploit the metabolites of the host cells by manipulating the host metabolism. They can rewire the cellular networks of the host with the major objective of sustained survival. Regarding the host side, the organisms try to prevent the disruption caused by pathogens in their cellular networks. Thus, several metabolic changes are induced in both pathogen and host sides in the course of infection. Understanding these alterations is crucial to develop novel diagnostics and treatment strategies for infectious diseases.

Computational systems biology, the integrated computational analysis of cellular interaction networks and biological data at genome-scale, has contributed to several

M. F. Cesur · E. Abdik · Ü. Güven-Gülhan · S. Durmuş · T. Çakır (✉)
Computational Systems Biology Group, Department of Bioengineering, Gebze Technical University, Gebze, Kocaeli, Turkey
e-mail: tcakir@gtu.edu.tr

R. Silvestre, E. Torrado (eds.), *Metabolic Interaction in Infection*, Experientia Supplementum 109, https://doi.org/10.1007/978-3-319-74932-7_6

scientific developments (Kitano 2002; Francke et al. 2005; Cassman et al. 2007). Since around 2010, we have witnessed the application of the approach in the infection research in terms of the effect on metabolism (Durmus et al. 2015). On the one hand, genome-scale metabolic networks (GEMs) of pathogens were reconstructed from genome sequences and used as scaffolds to predict growth rates of the pathogens as well as pathways activated/suppressed during infection. This modeling framework also enabled in silico identification of candidate biomarkers and drug targets. On the other hand, genome-wide data collection using metabolomics and fluxomics approaches enabled data-driven identification of several metabolites, fluxes, and pathways, which were used to develop new hypotheses to investigate molecular mechanisms of infection. In this chapter, we first review GEMs reconstructed for infection research. In the next section, the structure-based and sequence-based identification and analysis of drug-drug target candidates derived from metabolic network analyses are presented. The last section reviews metabolome and fluxome studies for infection mechanisms.

6.2 Genome-Scale Metabolic Network Models for Infection Research

Genome-derived reconstruction of organism-specific metabolic networks has been on the rise since it enables the simulation of several molecular details of condition-specific behavior of the organism. The most common modeling approach which is based on the use of GEM structure as a constraint is known as constraint-based modeling approach. Constraint-based modeling methods are widely used to simulate metabolic fluxes in many pathogens and host cells at the system level (Fig. 6.1). To generate metabolism-specific holistic pictures regarding the complex infection processes and to identify effective drug targets for the pathogen, constraint-based modeling of the pathogens has been increasingly applied since 2009 (Raghunathan et al. 2009). In silico metabolic modeling via the constraint-based approach is useful to address some core questions in infection research: Are there any possible links between the pathogenicity and pathogen metabolism? Which metabolic processes and metabolites support the adaptation of pathogens to the host cell? What are the impacts of genetic differences on the cellular metabolisms? Are the metabolic network models sufficient to elucidate crucial metabolites and enzymes as putative drug targets or biomarkers for infectious diseases?

The most common constraint-based modeling approach is called flux balance analysis (FBA). FBA approach is based on the information about the stoichiometry of a metabolic network. It focuses on the optimization of an objective function (mostly maximization of growth) that is crucial to predict internal flux distribution at steady state (Kauffman et al. 2003). Using some constraints (e.g., mass balance, flux limitations, and reaction reversibility) to reduce the solution space, optimization is often implemented, and the outcomes are mostly utilized to evaluate gene deletion

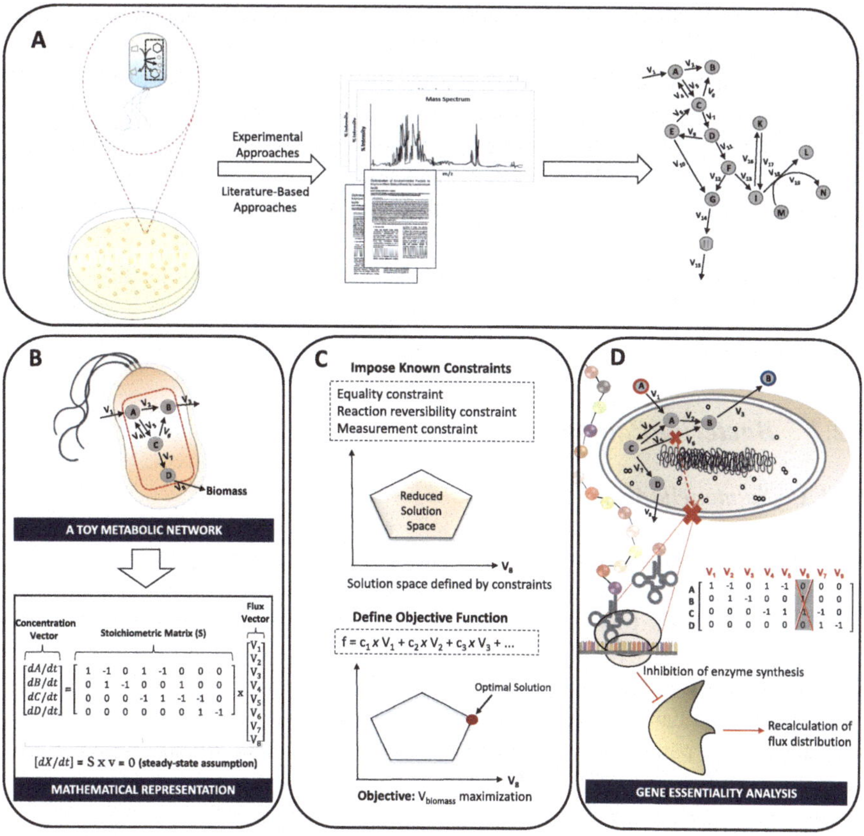

Fig. 6.1 Reconstruction and analysis of a genome-scale metabolic network (GEM). (**a**) A genome-scale metabolic network for the pathogen of interest is reconstructed based on genome data as well as experimental results from literature. (**b**) A reconstructed GEM can be used as a constraint to flux balance analysis (FBA) method. The method is based on mass balance equations around metabolites in the GEM. (**c**) Use of other constraints shrinks the solution space considerably. However, still, an optimization method is required to solve for a solution – the metabolic flux values. This is based on the definition of an objective function in accordance with the physiology of the organism. (**d**) Gene essentiality analysis is a computational approach that is based on FBA. It is helpful to identify potential metabolic drug targets

results, growth profile, and metabolic behavior of a cell (Kauffman et al. 2003; Huthmacher et al. 2010; Kim and Lun 2014) (Fig. 6.1). Although FBA is highly beneficial to analyze metabolic networks in large-scale, some other approaches integrating regulatory mechanisms to the metabolic network models have also been frequently used. In these modeling approaches, experimental transcriptome or proteome data is incorporated to identify active pathways to improve phenotype predictions (Lewis et al. 2012). Gene inactivity moderated by metabolism and expression (GIMME) (Becker and Palsson 2008) and metabolic adjustment by differential expression (MADE) (Jensen and Papin 2011) are among the computational approaches used for context-specific network extraction by incorporating transcriptome/proteome data.

In this section, we review the recent related literature for major pathogenic genera from the aspect of the use of constraint-based modeling for genome-scale pathogen and/or host metabolic networks. For each genus, we present the major findings of the systems-based approach to the metabolism (Table 6.1). Here, we are concerned with the studies using constraint-based modeling for the elucidation of the metabolic alterations that accompany infection progression and for the detection of the metabolites and enzymes acting as potential biomarkers, drug targets, or positive regulators of antibiotic efficacy.

6.2.1 Bacterial Pathogens

6.2.1.1 *Salmonella* Genus

The genus *Salmonella* is a member of the family *Enterobacteriaceae*, and it is related phylogenetically to *Escherichia coli*, prokaryotic workhorse of molecular genetics. It includes two species, *Salmonella enterica* and *Salmonella bongori*. *S. enterica* contains over 2600 serovars with different antigenic characteristics. *S. enterica* serovar typhimurium (hereafter referred to as *S. typhimurium*) is the primary intracellular pathogen causing human gastroenteritis, and this multidrug-resistant enteric pathogen is one of the global public health threats (Bäumler and Fang 2013; Gal-mor et al. 2014). Computational modeling of biological systems can provide a comprehensive understanding of the bacterial virulence and open up new avenues of drug target identification. A GEM model of *S. typhimurium* including 1083 genes and 1087 reactions was developed via modifying a preexisting *E. coli* model by Raghunathan and colleagues. Using a condition mimicking the host nutrient environment to identify essential genes in survival and in virulence phenotypes of the bacteria, they demonstrated that in silico virulence predictions were predominantly in accordance with the experimental virulence characterization. Furthermore, the use of gene expression data provided the identification of active enzymes and pathways over the course of infection (Raghunathan et al. 2009). Regulation of *S. typhimurium* metabolism is essential for adaptation into the host environment, and this process is mainly managed by metabolic coupling between pathogen and host. To elucidate the alterations in *S. typhimurium* metabolism for virulence, metabolite profiles of the pathogen were investigated in LB (a rich medium) and LPM (a virulence-inducing medium). The analyses of omics data using the omics-constrained models showed that there is a possible molecular coupling between pathogen and host to obtain the nutrients and metabolites required for energy demand satisfaction and adaptation to hostile pathogenic environment. Moreover, the turnover fluxes for metabolites related to fatty acid and proteoglycan synthesis and utilization of amino acids were altered to induce the adaptation and survival of the pathogen in the phagosome environment. Collectively, the pathogen

Table 6.1 An overview of pathogenic GEMs highlighted in the chapter

Pathogen name	Pathogen type	Number of metabolites	Number of genes	Number of reactions	References
E. coli K-12	Gram-negative bacterium	625	904	931	Reed et al. (2003)
H. pylori	Gram-negative bacterium	485	341	554	Thiele et al. (2005)
S. aureus	Gram-positive bacterium	571	619	640	Becker and Palsson (2005)
M. tuberculosis	Atypical bacterium	740	661	939	Jamshidi and Palsson (2007)
P. aeruginosa	Gram-negative bacterium	760	1056	883	Oberhardt et al. (2008)
S. typhimurium LT2	Gram-negative bacterium	744	1083	1087	Raghunathan et al. (2009)
S. aureus	Gram-positive bacterium	1399–1437	522–547	1444–1497	Lee et al. (2009)
P. aeruginosa	Gram-negative bacterium	760	1056	883	Oberhardt et al. (2010)
A. baumannii	Gram-negative bacterium	778	650	891	Kim et al. (2010)
M. tuberculosis	Atypical bacterium	3400	2071	4489	Bordbar et al. (2010)
P. falciparum	Protozoan parasite	616	366	1001	Plata et al. (2010)
P. falciparum	Protozoan parasite	1622	579	998	Huthmacher et al. (2010)
V. vulnificus CMCP6	Gram-negative bacterium	765	673	943	Kim et al. (2011)
K. pneumoniae MGH 78578	Gram-negative bacterium	not stated	1188	1970	Liao et al. (2011)
P. falciparum	Protozoan parasite	1622	579	1394	Bazzani et al. (2012)
M. pneumonia	Atypical bacterium	216	145	306	Wodke et al. (2013)
S. typhimurium LT2	Gram-negative bacterium	1088	824	1097	Hartman et al. (2014)
C. difficile	Gram-positive bacterium	703	806	1031	Larocque et al. (2014)
E. faecalis V583	Gram-positive bacterium	642	668	706	Veith et al. (2015)
M. tuberculosis	Atypical bacterium	not stated	810	938	Ma et al. (2015)
T. gondii	Protozoan parasite	1019	527	1089	Tymoshenko et al. (2015)

(continued)

Table 6.1 (continued)

Pathogen name	Pathogen type	Number of metabolites	Number of genes	Number of reactions	References
P. aeruginosa	Gram-negative bacterium	760	1056	883	Opperman and Shachar-hill (2016)
A. fumigatus	Saprophytic fungi	204	102	162	Kaltdorf et al. (2016)
M. tuberculosis	Gram-negative and gram-positive bacterium	826	661	1025	Gaurav and Hasija (2017)
S. aureus		not stated	619	640	
S. typhimurium		1802	1271	2545	
H. pylori		485	341	554	
L. donovani	Protozoan parasite	1135	604	1159	Sharma et al. (2017)
M. hyopneumoniae strain 11	Gram-positive bacterium	298	133	284	Kamminga et al. (2017)
P. aeruginosa	Gram-negative bacterium	1286 1284	1129 1146	1495 1493	Bartell et al. (2017)

The table is given in chronological order

can achieve the adaptation by rewiring some cellular pathways and by changing its own metabolism (Kim et al. 2013). In recent years, researchers have focused on the identification of lethal genes of *Salmonella* (Hartman et al. 2014). To this end, a new GEM model with 824 genes, 1097 reactions, and 1088 metabolites was developed de novo via ScrumPy software package unlike the model of Raghunathan et al., which was derived from a previous *E. coli* model by neglecting the impact of the genetic differences on the metabolism. This study concentrated on the detection of a subset of the reactions indispensable for ATP synthesis (denoted as catabolic core) as putative drug targets to, at least, reduce virulence. Using linear programming, 34 reactions (e.g., glycolysis, the tricarboxylic acid (TCA) cycle, and the Entner-Doudoroff (ED) pathway) were defined as the catabolic core (Hartman et al. 2014). Using the catabolic core, the minimal cut set algorithm of Klamt et al. determined 11 minimal cut sets whose inactivation causes lethality via a failure in ATP production (Klamt and Gilles 2004; Hartman et al. 2014). Among these, seven cut sets with the reactions related to the amino acid biosynthesis/degradation and the ED pathway were also experimentally unraveled to be significant for the bacterial growth (Hartman et al. 2014).

6.2.1.2 *Mycobacterium* Genus

Mycobacterium tuberculosis is a member of the genus *Mycobacterium* causing one of the most devastating diseases, tuberculosis. It is associated with more than a million deaths worldwide each year owing to the growing antibiotic resistance crisis

(Gengenbacher and Kaufmann 2012). The infection process of *M. tuberculosis* begins with the inhalation of contaminated aerosols. Then, the bacteria encounter alveolar macrophages in the lower respiratory tract. Once recognized, the pathogens are engulfed by the macrophages, but vitality of the pathogens is still maintained. Upon the activation of innate immune system, various chemokines and cytokines are produced, and other immune cells (e.g., dendritic cells, neutrophils, and lymphocytes) as well as more macrophages are recruited to the sites of infection to support removal of the pathogens (Silva Miranda et al. 2012; Ehlers and Schaible 2013). This cascade of intracellular events induces the formation of a collection of the recruited immune cells known as granuloma (i.e., organized aggregates of multinucleated giant cells, epithelioid cells, T cells, infected and uninfected macrophages) to cope with the bacteria (Silva Miranda et al. 2012; Pienaar et al. 2016). However, *M. tuberculosis* can exploit even granuloma as a safety shelter against other immune cells (Silva Miranda et al. 2012). Therefore, a deep understanding of the molecular interactions between the pathogen and its host is crucial to shed light on its infection mechanism.

To provide more insight into the interactions between *M. tuberculosis* and its host (i.e., human alveolar macrophage), an integrated metabolic network model including 2071 genes and 4489 reactions was developed. Besides, gene expression data from distinct *M. tuberculosis* infectious states (latent, pulmonary, and meningeal) were mapped onto the integrated model, and the differential metabolism of human alveolar macrophages based on the infection types was highlighted (Bordbar et al. 2010). Elucidation of regulatory relationships among transcription factors (TFs), metabolic target genes, and environmental conditions can open new avenues for the identification of novel drug targets. In this context, Ma and colleagues (Ma et al. 2015) constituted an improved model for *M. tuberculosis* to examine the effects of 104 TF knockout and upregulation events on the bacterial growth by using the Probabilistic Regulation of Metabolism (PROM) framework (Chandrasekaran and Price 2010; Simeonidis et al. 2013). It was revealed that this expanded model has a stronger ability to predict the metabolic consequences of the TF perturbations in comparison with the original *M. tuberculosis* model. More recently, Pienaar et al. investigated *M. tuberculosis* metabolism within granulomas to identify new drug targets (Pienaar et al. 2016). They developed a novel multi-scale computational model (GranSim-CBM) combining an agent-based model of granuloma generation and a dynamic constraint-based model of the pathogen metabolism and growth. The combined model reflects comprehensive rules and biological background of the host immune system, hostile environment, and *M. tuberculosis* metabolism within the granuloma (e.g., *M. tuberculosis* adaptation to hypoxia, lipid inclusion formation, nutrient limitation, oxygen levels, and so on). The GranSim-CBM represents dynamic profiles of *M. tuberculosis* residing within the granuloma, covering the abilities of the bacteria for actively dividing and responding to the changes in the dynamic granuloma microenvironments (e.g., changing oxygen levels and immune pressure). Thus, the microenvironment is the key factor affecting bacterial growth and adaptation since oxygen-related bacterial properties were found to be significant factors determining infection outcome. Furthermore, they showed that the bypass pathways (e.g., cytochrome *bd* complex compensating the absence of cytochrome bc1 protein complex within the cytochrome *c* system) were essential for metabolic enzyme knockout mutants to compensate the effect of defective bacterial growth. This result highlights the importance of the use of multiple drugs with a synergistic effect. Lastly, the authors suggested that this model can provide a solid basis for the investigation of antituberculosis drugs (Pienaar et al. 2016).

6.2.1.3 *Mycoplasma* Genus

The genus *Mycoplasma* belongs to the class *Mollicutes*, and they are among the smallest microorganisms in terms of genome size (Fadiel et al. 2007). Thus, they are devoid of some metabolic pathways and structural cell components. These *cell wall*-less microorganisms possess only trilaminar plasma membrane, and this property renders them resistant against the antibiotics (e.g., penicillin) that act by interfering with bacterial cell wall formation. *M. pneumoniae*, *M. hominis*, and *M. genitalium* are some of the most known pathogenic *Mycoplasma* species infecting human (Fadiel et al. 2007; Harvey et al. 2007). *M. pneumoniae* is the principal cause of atypical pneumonia. Since it has a limited metabolic capability, it must obtain its needs from the host organism (Harvey et al. 2007; Atkinson et al. 2008). This pathogen has some deficiencies in the central carbon metabolism due to the lack of TCA cycle that provides various intermediates essential for the biosynthesis of many amino acids (e.g., glutamate synthesis from α-glutamate) and due to the lack of a complete electron transport chain. Therefore, it uses glycolysis as the energy-yielding system via substrate-level phosphorylation managed by phosphoglyceric acid kinase, pyruvate kinase, and acetate kinase activities (Atkinson et al. 2008; Dutow et al. 2010). Pathogens must alter their metabolisms to adapt to the dynamic and hostile environment and to effectively utilize the sources of the host over the course of the infection. For instance, the induction of pentose phosphate metabolism in the case of infection was revealed for various pathogens (Richardson et al. 2015).

To examine the energetic metabolism of this intracellular pathogen in the course of the adaptation, a genome-wide metabolic model including 145 genes, 306 reactions, and 216 metabolites was reconstructed and analyzed via experimental validation (Wodke et al. 2013). The pentose phosphate pathway (PPP) and nucleotide metabolism were found to be responsible for the adaptation events. The authors suggested that the nonessential genes may be important to preserve the network robustness by contributing to the adaptation to environmental changes. Of note, they highlighted that *M. pneumoniae* invests most of its energy to maintain the cell homeostasis instead of proliferation, as another part of the adaptation process (Wodke et al. 2013).

Another pneumonia-causing pathogen, *M. hyopneumoniae*, is capable of infecting pigs, and concurrent infection with some viral pathogens causes more severe form of the mycoplasmal pneumonia. This pathogen can be directly transmitted between pigs, and it leads to important economic losses for swine industry worldwide (Hwang et al. 2010; Kamminga et al. 2017).

To tackle the infection, antibiotic administration and vaccination are frequently implemented. Although *M. hyopneumoniae* is sensitive to various antibiotics (e.g., fluoroquinolones, tetracyclines, spiramycin, and tiamulin) that target translation machinery, replication, and transcription enzymes, growing antibiotic resistance of this pathogen poses an obstacle in the eradication of the *Mycoplasma* infections (Le Carrou et al. 2006; Hwang et al. 2010). Therefore, vaccination including inactivated whole-cell bacteria is another strategy for protection from the mycoplasmal pneumonia. Even if this approach is a beneficial option, the cultivation of *M. hyopneumoniae* is a tedious and time-consuming task owing to long-lasting incubation and a fastidious medium requirement (Hwang et al. 2010; Kamminga et al. 2017). Therefore, the optimization of the culture medium is indispensable to improve insufficient biomass yield of this pathogen likely due to low protein-coding capacity. Recently, a constraint-based GEM model of *M. hyopneumoniae* strain 11 including 284 reactions and 298 metabolites was constructed in order to unravel the cellular energy balances and determine essential components for optimal bacterial growth (Kamminga et al. 2017). The researchers aimed to improve the production of a *M. hyopneumoniae* vaccine by maximizing the biomass production. The model revealed that this pathogen spends most of its energy (84%) for non-growth-associated maintenance as in the case of *M. pneumoniae.* The model-driven experiments demonstrated that the ratio of the cellular energy devoted to the bacterial growth increased when the medium was supplemented with pyruvate. Thus, the model provided crucial insights into improving fermentation process of *M. hyopneumoniae* and hence vaccine development in larger scale (Kamminga et al. 2017).

6.2.1.4 *Staphylococcus* Genus

Staphylococcus aureus along with coagulase negative *Staphylococci* leads to 30% of hospital-acquired infections with average treatment costs of $2 billion each year (Conlan et al. 2012). The life-threatening diseases including bacteremia, severe sepsis, toxic shock syndrome, endocarditis, and meningitis are elicited in humans upon *S. aureus* infection (Lee et al. 2009; Otto 2010). In clinical settings, antibiotic resistance in many *Staphylococci* poses a major obstacle in the treatment of the diseases caused by these pathogens. Especially methicillin-resistant *S. aureus* (MRSA) strains are prevalent in hospitals (Otto 2010). Moreover, *S. aureus* strains can inhabit diverse ecological niches in human from the skin to respiratory tract, and they can overcome or evade the immune response by means of several immune resistance factors, toxins, and various evasion factors (Bosi et al. 2016). Differences in the genotypes of these strains result in phenotypic diversity, and there are some techniques to investigate the impact of the genomic diversity among strains within a species on virulence phenotypes (Conlan et al. 2012; Bosi et al. 2016). An emerging approach to characterize the strains is the "pan-genome" analysis. The concept of "pan-genome" refers to a collection of strain-specific genes or unique genes found in some strains (dispensable or accessory genome) and a subset of the genes shared by all the given strains (core genome) within a species (Conlan et al. 2012; Chaves-Moreno et al. 2015; Guimarães et al. 2015; Bosi et al. 2016). The core genome likely includes the genes related to basic molecular mechanisms, while the dispensable genome provides selective advantages (e.g., antibiotic resistance and adaptation ability) to the microorganisms under certain environmental conditions (Bosi et al. 2016). Genomics is an indispensably useful approach to compare different strains within the same species. However, the studies focusing on the metabolic networks are also important in terms of paving the way for the detection of metabolic determinants of pathogenicity as drug targets in different strains. Identification of common drug targets for different *S. aureus* strains can enable a way to cope with this pathogen more efficiently by the development of broad therapeutics. To this end, metabolic networks of 13 multidrug-resistant *S. aureus* strains including 522 to 547 enzymes, 1444 to 1497 reactions (cellular, transport and exchange), and 1399 to 1437 metabolites were reconstructed for the discovery of common drug targets (Lee et al. 2009). The researchers used ERGO bioinformatics suite (Overbeek et al. 2003) and the KEGG ligand/reaction database (Goto et al. 2002). To determine the essential enzymes in all 13 *S. aureus* strains, they applied FBA-based in silico gene deletion studies. They reported some common enzymes that are potential drug targets due to their roles in the cellular processes such as amino acid and fatty acid metabolic pathways. A subsequent study assembled pan-genomic data of *S. aureus* species, constituting an open pan-genome of 7411 genes and a core genome of 1441 genes (Bosi et al. 2016). They reconstructed GEMs of 64 strains of *S. aureus* based on the pan-genome in order to determine the diversity in metabolic profiles of these strains in different conditions. Despite the conserved nature of the metabolism in the core genome of the strains, some differences were also detected in the amino acid and nucleotide biosynthesis pathways. All strains modeled were predicted as niacin (vitamin B3) and thiamin (vitamin B1) auxotrophs. The model also predicted the strain-specific auxotrophies for some compounds such as riboflavin, guanine, and leucine. Further, strains were classified based on the metabolic capabilities and virulence factors (VFs) through the model to explore new drug targets via determining strain-specific biomarkers and metabolic determinants of virulence (Bosi et al. 2016).

6.2.1.5 *Pseudomonas* Genus

The genus *Pseudomonas* within the family of *Pseudomonadaceae* includes many species that can inhabit distinct niches due to their diverse metabolic characters and genetic differences. Nevertheless, different species can exhibit similarities in terms of their basic central metabolism that connects the catabolism of the carbon sources to the synthesis of amino acids and nucleotides (Kahlon 2016). TCA cycle is found in all species of *Pseudomonas* (Goldman and Green 2015). Due to the lack of phosphofructokinase enzyme, the conversion of glucose to pyruvate is facilitated by the archetypal glycolysis (Embden-Meyerhof-Parnas (EMP) pathway) or another alternative route so-called ED pathway, utilized by many *Pseudomonas* species (e.g., *P. putida*) (Goldman and Green 2015; Nikel et al. 2015). PPP and glyoxylate shunt (GS) are among the other important metabolic pathways in *Pseudomonas*. GS enables the utilization of simple carbon compounds such as acetate and fatty acids under conditions of glucose hunger. A vital role was reported for GS in regulating pathogenesis, the hypoxic conditions, and oxidative stress evoked by some bactericidal antibiotics or immune response (Ahn et al. 2016; Bergkessel et al. 2016). For instance, anoxic and carbon-limiting conditions lead to a tremendous drop in the fluxes through the TCA cycle and electron transport chain in *P. aeruginosa*, and GS is introduced for the conversion of the acetyl-CoA to some useful TCA cycle precursors in order to replenish the TCA cycle or to enter the gluconeogenic pathway (Bergkessel et al. 2016). This opportunistic pathogen can infect a wide host range (e.g., humans, animals, fungi, and plants), and it is the causal agent of many diseases including pneumonia, otitis media, meningitis, and keratitis (Sousa and Pereira 2014; Opperman and Shachar-Hill 2016). *P. aeruginosa* is also the leading underlying cause of deaths and morbidity for people with cystic fibrosis (CF) (Sousa and Pereira 2014; Bhagirath et al. 2016). CF leads to respiratory, digestive, and reproductive system dysfunctions as well as secretory gland damages (Sousa and Pereira 2014). The large genome with almost 6000 genes allows for the survival and adaptation of the pathogen in a specific set of conditions. A substantial part of the differentially expressed bacterial genes during chronic infection is associated with metabolic and physiological activities. Thus, the examination of the pathogen metabolism is crucial to deal with this severe pathogenic threat.

Oberhardt et al. presented the first network-driven analysis of changing *Pseudomonas* metabolism by using transcriptome data at distinct time points in the course of a CF lung infection, at distinct time points during 44 months (Oberhardt et al. 2010). They used the transcriptome data as a constraint of a GEM model with 1056 genes, 760 metabolites, and 883 reactions. This study resulted in the identification of the genes that are vital for the pathogen in the infection process. More recently, ^{13}C-based metabolic flux analysis (^{13}C-MFA) in addition to FBA analysis was also implemented in order to understand the metabolic profiles of one reference and five *Pseudomonas* isolates from the patients with CF (Opperman and Shachar-Hill 2016). A strong relationship of oxidative PPP and ED pathway with the growth of the pathogen was discovered. Moreover, ^{13}C-MFA revealed a correlation between the fluxes through the GS and intrinsic growth rates across strains.

Even if blocking the bacterial growth by disturbing growth-essential functions is a widely used approach, this growth-based selection elicits development of antimicrobial resistance at rapid pace, and this situation leads to substantial cost increases for treatment as well as increasing morbidity and mortality rates. On the other hand, the inhibition of VF synthesis and virulence-linked genes may preclude this unsatisfactory outcome. An effort has been undertaken to explore the complex relationship between pathogen metabolism and virulence-linked pathways (e.g., synthesis pathways of VFs) (Bartell et al. 2017). In the study, a newly reconstructed GEMs was presented for *P. aeruginosa* strain PA14 (mPA14) with 1129 genes, 1495 reactions, and 1286 metabolites as well as an updated GEM for *P. aeruginosa* PAO1 strain (mPAO1) with 1146 genes, 1493 reactions, and 1284 metabolites. Many new virulence-linked genes (112 genes for mPAO1 and 108 genes for mPA14) were included in these models in the expert curation step. The models were validated through comparative approaches using a carbon source utilization dataset and a published gene essentiality dataset. Using in silico gene knockouts to investigate the relationship between growth and virulence, forty-six genes were found to be essential for both biomass production and for the synthesis of up to 10 VFs out of a total of 17 VFs. Thus, this study represented a novel insight for the discovery of drug targets using metabolic models linking biomass and virulence of the pathogens (Bartell et al. 2017).

6.2.1.6 *Vibrio* Genus

Vibrio vulnificus is a halophilic bacterium that inhabits estuarine and coastal waters (Kim et al. 2011). To prevent or inhibit the infection, diverse strategies are implemented on the side of the host. Complement cascade activity is an early immune response to *V. vulnificus* entering into bloodstream. Another defense mechanism implemented by the host is highly acidic gastric environment. However, the pathogen can withstand low pH by neutralizing the environment through breakdown of lysine to form cadaverine using lysine decarboxylase. As well as acid neutralization, cadaverine also provides an oxidative stress tolerance (Jones and Oliver 2009). Thus, this opportunistic human pathogen can lead to gastroenteritis or severe wound infections (Cañigral et al. 2010). Even if antibiotic therapy with third-generation cephalosporin, doxycycline, and fluoroquinolones is usually effective to overcome *V. vulnificus* isolates, the development of the resistance by a considerable number of the isolates threatens the continued effectiveness of many of these antimicrobials. Rapid technological advances have provided unprecedented opportunities for large-scale sequencing. Therefore, an increase in the number of studies trying to close the gaps in linking the genomic data to the development of more effective drugs is not surprising. In such a study, Kim and colleagues reconstructed a GEM (VvuMBEL consisting of 943 reactions, 765 metabolites, and 673 genes) for *V. vulnificus* CMCP6 from the re-sequenced and re-annotated genome using MetaFluxNet (Kim et al. 2011). The validated GEM was used to identify potential antimicrobial targets. Putative targets including five survival-related metabolites (i.e., 2-amino-4-hydroxy-6-hydroxymethyl-7,8-dihydropteridine (AHHMP), D-glutamate (DGLU), 2,3-dihydrodipicolinate (DHDP), 1-deoxy-D-xylulose 5-phosphate (DX5P), and 4-aminobenzoate (PABA)) were selected out of 193 essential metabolites involved in the metabolisms of cofactors, vitamins, lipids, and amino acids, and they were experimentally validated. The structural analogs of these metabolites were found through a chemical library. Of these analogs, compound 24837 (an analog of PABA) was found to be useful in terms of antimicrobial capacity (Kim et al. 2011).

6.2.1.7 *Enterococcus* Genus

The genus *Enterococcus* is a part of the normal flora in the human gastrointestinal tract, and it is capable of tolerating extreme temperature and growth conditions such as high salinity and alkaline (Frank et al. 2014; Zhang et al. 2015; Großeholz et al. 2016). This inherent adaptation ability makes the bacteria superior among other bacterial populations in their niches (Frank et al. 2014). *Enterococcus* comprises over 30 species significantly affecting human population by causing urinary tract and surgical wound infections, hepatobiliary sepsis, endocarditis, and bacteraemia (Kajihara et al. 2015; Zhang et al. 2015). Of these species, *E. faecalis* accounts for a considerable portion of the enterococcal infections (Frank et al. 2014). In addition to the high prevalence of *E. faecalis* infection, increasing multidrug resistance of the pathogen prompts the scientists to investigate the adaptation capability of this pathogen under the extreme conditions (e.g., low pH of the habitat). The regulation of vacuolar-type ATPase activity is a prominent strategy to maintain cellular pH homeostasis in *E. faecalis* (Großeholz et al. 2016). Such enzymes are responsible for the translocation of excessive protons (H^+) from cytosol to the extracellular or intraluminal space in an irreversible manner. Thus, both a membrane potential and a pH gradient are formed during the proton pumping (Sachs and Keeling 2001). *E. faecalis* regulates and maintains the transmembrane pH gradient through the enhanced activity of the H^+-pumping ATPase enzyme.

To examine the cellular changes over the course of the adaptation of *E. faecalis* V583 to decreasing pH, Großeholz et al. integrated quantitative whole-cell proteome data into the reconstructed GEM, which included additional pH-dependent processes (Großeholz et al. 2016). The proteome data facilitated the reorganization of the model based on a physiological state through the assignment of the significant changes in the proteins as additional constraints on flux bounds. The turnover rates of some metabolites were also included in the model to compute flux distributions with the aim of comparison of different metabolic states. Using flux variability analysis method, acid-adaptive phenomena were identified, as observed through the enhanced proton pump activity of the V-type ATPase, the tendency for homolactic fermentation, and the increased glucose, L-arginine, and L-serine consumptions. Thus, the reduction of the proton production in the central metabolism and decreased membrane permeability were found as important adaptations against lowering pH. The authors highlighted that the approach of integrating proteome data with GEM models holds promise for reproducing a more realistic picture of the metabolism in certain conditions (Großeholz et al. 2016).

Großeholz and colleagues used a modified GEM in the study, originally published by Veith et al. (2015). The original model with 642 metabolites and 706 reactions was reconstructed using the software PySCeS CBMPy (http://cbmpy.sourceforge.net) to identify vulnerable points in central *E. faecalis* V583 metabolism for an effective control of this multiple antibiotic-resistant pathogen (Veith et al. 2015). All metabolic reactions necessary for the cell growth were included into the model including central energy metabolism, nucleotide biosynthesis, amino acid metabolism, the synthesis of vitamins and cofactors, the biosynthesis of peptidoglycan, and fatty acid biosynthesis reactions. Several databases were used in the reconstruction of the GEM, including Uniprot (The Uniprot Consortium 2017), KEGG (Kanehisa et al. 2008), EnteroCyc (http://enterocyc.broadinstitute.org), TransportDB (Ren et al. 2007), and TCDB (Saier et al. 2016). Experimentally identified auxotrophies for seven amino acids were considered to validate the metabolic network. The model could predict all amino acid auxotrophies properly. In addition, double leave-out simulations were performed, and an impaired growth was observed when L-glutamine and L-glutamic acid were simultaneously omitted, in agreement with the experimental findings. The model was also used to understand the altered metabolic fluxes in a glutamine synthetase deletion mutant to shed light on glutamate/glutamine metabolic hubs (Veith et al. 2015).

6.2.1.8 *Chromobacterium* Genus

Chromobacterium violaceum belongs to the genus *Chromobacterium*, and it can cause lethal septicemia with many liver and lung abscesses by rarely infecting especially healthy and young individuals (Yang and Li 2011). Versatile energy metabolism of the bacteria provides the ability to utilize different energy resources. Hence, it can catabolize the energy sources via Embden-Meyerhof, TCA, and GS in aerobic conditions, while it can survive in anaerobic conditions via the breakdown of glucose to acetic and formic acid (Creczynski-pasa and Antônio 2004). Moreover, one of the most prominent properties of this pathogen is the ability to synthesize a well-characterized blue-violet pigment so-called violacein upon the conversion of tryptophan under aerobic conditions (Creczynski-pasa and Antônio 2004; Banerjee et al. 2017). Production of the pigment is under the control of quorum-sensing mechanisms. Some potential therapeutic activities (e.g., antitumoral, antileishmanial, antioxidant, and antibiotic activities) were attributed to this pigment in different studies (Creczynski-pasa and Antônio 2004). The antimicrobial activity of violacein on gram-positive bacterial strains gains special attention to tackle especially multidrug-resistant pathogens like *S. aureus* (Choi et al. 2015).

A recent study simulated and comprehensively analyzed the metabolic profile of the *C. violaceum* strain ATCC 12472T in order to provide inferences about the metabolic basis of antibiotic efficacy (Banerjee et al. 2017). In the study, the chloramphenicol and streptomycin-resistant *C. violaceum* strain ATCC 12472T (ChlR and StrpR) populations were developed under laboratory conditions. When exposed to these antibiotics in high doses, the resistant populations exhibited improved growth rate and biomass yield in comparison with the wild type (WT). The researchers studied the molecular alterations in the resistant pathogen via the analysis of genome sequence, protein structures, and metabolic profiles of the bacteria. Growth and viability testing detected that four exogenous metabolites (pyruvate, maleate, succinate, and oxoadipate) increased the susceptibility of the resistant populations against chloramphenicol and streptomycin antibiotics. They proposed these metabolites as candidates for antibiotic therapy. Moreover, the metabolomics based on liquid chromatography-high-resolution mass spectrometry with tandem mass spectrometry (LCHRMS with MS/MS) revealed that the abundance levels of PABA, succinate, leucine, hypoxanthine, and violacein were significantly changed between the resistant and WT populations. Thus, the violacein was suggested as a putative biomarker indicating the resistance. Constraint-based flux balance modeling was also used to investigate the contribution of the benign metabolites to antibiotic efficacy. The disruption of redox homeostasis was found to be an important result of the antibiotic activity. Lastly, it was highlighted that the resistant pathogen compensated the effect of the impaired NAD/NADH ratios by rewiring its metabolic network (Banerjee et al. 2017).

6.2.1.9 *Clostridium* Genus

Clostridium difficile is a pathogenic bacterium which alters the composition of gut microbiota by colonizing in the mammalian intestine after antibiotic treatment, affecting roughly 600,000 people in the USA each year. The infection causes diarrhea and intestinal inflammation, with a potential to lead to fatal colitis (Steinway et al. 2015). A study reconstructed the metabolic network of *C. difficile* strain 630 at genome-scale, with 806 genes and 1031 reactions (Larocque et al. 2014). FBA and synthetic accessibility (Wunderlich and Mirny 2006) were used to predict essential genes via in silico gene deletion analyses. Seventy-six essential genes and 39 essential gene pairs were predicted to be specific to the strain. Some of these essential genes are aspartate-semialdehyde dehydrogenase, diaminopimelate epimerase, isoprenoid synthase, UDP-N-acetylenolpyruvoylglucosamine reductase, and NH_3-dependent NAD(+) synthetase that play role in essential amino acid metabolisms. One hundred twenty-five molecules (most of them are in experimental phase) that bind to 29 protein homologs of potential *C. difficile* targets were identified using DrugBank database (Wishart et al. 2008; Knox et al. 2011). Twenty-two of these inhibitor molecules were found to be approved drugs which can be evaluated for drug-repositioning approach. The others may be potential inhibitor drugs that may lead to development of new antibiotics (Larocque et al. 2014).

6.2.1.10 *Acinetobacter* Genus

Acinetobacter baumannii is effective in immunocompromised patients in the hospital environment, causing severe infections in the lungs, blood, brain, and respiratory and urinary tracts. A GEM analysis on a drug-resistant *A. baumannii AYE* strain was evaluated by constraint-based flux analysis in silico (Kim et al. 2010). Essentiality analysis, when compared with experimental data, revealed nine essential genes involved in sulfur metabolism, histidine metabolism, lysine, phenylalanine, tyrosine and tryptophan biosynthesis, urea cycle, and metabolism of amino groups. Filtering metabolites for druggability potential revealed 97 essential metabolites that play a role in cell growth. Out of these 97 metabolites, 9 of them were found unique to the pathogen, not interfering with human homologous enzymes. These metabolites can be used as drug target candidates. They are AHHMP, DGLU, DHDP, 2-amino-4-hydroxy-6-(D-erythro-1,2,3-trihydroxypropyl)-7,8-dihydropteridine, 3-dehydroshikimate, DX5P, 3-dehydroquinate, 2-dehydro-3-deoxy-D-octanoate, and PABA (Kim et al. 2010). These metabolites may serve as broad-spectrum drug targets since they were also reported to be essential for other pathogens such as *E. coli* (Reed et al. 2003), *Helicobacter pylori* (Thiele et al. 2005), *M. tuberculosis* (Jamshidi and Palsson 2007), *P. aeruginosa* (Oberhardt et al. 2008), and *S. aureus* (Becker and Palsson 2005).

6.2.1.11 *Klebsiella* Genus

Klebsiella pneumoniae is another pathogen that causes hospital-acquired infections, and new drugs are necessary due to the resistance of this gram-negative bacterium to existing antibiotics. To study its metabolic functions at systems-level, a GEM model was reconstructed in silico for *K. pneumoniae MGH 78578* strain and then experimentally verified (Liao et al. 2011). The network included 1228 genes controlling 1970 reactions catalyzed by 1188 enzymes. Applying FBA to the GEM, they validated the reconstructed model by comparing experimental and in silico growth rates on different carbon sources. In addition, in silico gene deletion analysis was performed, and 118 essential genes were predicted computationally. Eight of them were unique to *K. pneumoniae MGH 78578* when compared to the computational predictions by *E. coli* and *S. typhimurium* models, which are from the same family *Enterobacteriaceae*. The unique genes were involved in lipopolysaccharide biosynthesis, capsule synthesis, and lipid biosynthesis pathways.

6.2.2 Protozoan Pathogens

6.2.2.1 *Plasmodium* Genus

Plasmodium falciparum is the most virulent protozoan parasite within the known *Plasmodium* species infecting human. It is the major etiological agent of a mosquito-borne disease, malaria, in especially young children and pregnant women (Plata et al. 2010; Josling and Llinás 2015). These parasites invade the circulating red blood cells (RBCs) by leaving the liver, and they release many enzymes including kinases, lipases, and proteases in order to utilize host nutrients via the manipulation of the RBC membrane permeability (Tilley et al. 2011). In other words, the pathogen acquires the essential metabolites to proliferate in a host-dependent manner. The lack of an approved vaccine against malaria and increasing antimalarial drug resistance prompted the researchers to establish a comprehensive metabolic network of this pathogen and investigate the novel drug targets (Huthmacher et al. 2010; Plata et al. 2010; Bazzani et al. 2012; Abdelrazig et al. 2017).

Plata and colleagues established a GEM of *P. falciparum* consisting of 1001 reactions and 616 metabolites by using various resources (i.e., PlasmoDB (Aurrecoechea et al. 2009), PlasmoCyc (Yeh et al. 2004), and KEGG (Kanehisa et al. 2008)). They suggested 40 potential enzymatic drug targets representing low sequence similarity with human proteins by using FBA. Of these targets, nicotinate mononucleotide adenylyltransferase is important in NAD^+ metabolism. Moreover, the authors validated the model to a great extent by using gene expression data (Plata et al. 2010).

Investigation of the metabolism in terms of only pathogen cannot ensure to generate a holistic picture of infection process, so the metabolic networks of pathogens can be embedded into the host models. In the study of Huthmacher et al., a metabolic network of *P. falciparum* was compiled from online resources including Malaria Parasite Metabolic Pathways (MPMP) (Ginsburg 2006), BioCyc (Karp et al. 2005), KEGG (Kanehisa et al. 2008), Reactome (Matthews et al. 2009), BRENDA (Chang et al. 2009), and TCDB (Saier et al. 2016). They examined the stage-specific metabolism of the parasite development by mapping the transcriptomics data sets onto the combined *Plasmodium*-human erythrocyte metabolic network. The authors also presumed some antimalarial drug targets via knock-out simulations (Huthmacher et al. 2010).

A crucial step in the drug target detection is selectivity analysis in an effort to determine targets that are essential for only pathogen but not for the host. One such study was performed by Bazzani and colleagues who combined and pruned a set of literature-based enzyme lists of *P. falciparum* and evaluated them using an updated GEM with 1394 reactions controlled by 579 genes and the modified liver metabolic network HepatoNet1 that includes 2539 reactions, 704 genes, and 1149 metabolites. Using in silico gene deletions, they reported 24 drug targets out of the 48 enzymes that were highly selective (Bazzani et al. 2012).

6.2.2.2 *Leishmania* Genus

Leishmania donovani is another insect vector-transmitted parasite, and this pathogen belonging to the genus *Leishmania* is the causal agent of visceral leishmaniasis (VL) in vertebrates (Dostálová and Volf 2012). The World Health Organization has recently reported that parasites of *Leishmania* cause over 20,000 deaths per year. Lack of the effective vaccines to combat these pathogens and emerging resistance of the parasites render them one of the major health threats in the developing world. The computational approaches enable systems-level analyses that hold great promise for the understanding of *Leishmania* metabolism and improving an efficient treatment of VL. In a recent study (Sharma et al. 2017), a genome-scale, well-curated metabolic model of *L. donovani* containing 1159 reactions, 1135 metabolites, 604 genes, and 8 subcellular compartments was generated based on both a laborious search in the literature and Web databases (LeishCyc (Doyle et al. 2009), TCDB (Saier et al. 2016), MetRxn 1.0 (Kumar et al. 2012), MetaCyc (Caspi et al. 2016), BRENDA (Chang et al. 2009), KEGG (Kanehisa et al. 2008), TransportDB (Ren et al. 2007)). Eighteen formerly uncharacterized or mis-annotated genes were re-annotated during model refinement. Moreover, in silico gene knockout simulations culminated in the detection of 28 human nonhomologous genes that are mainly essential in the lipid metabolism of *L. donovani*. It is important to note that the incorporation of gene regulatory information into the metabolic network model via GIMME algorithm (Becker and Palsson 2008) provides concordant results with the experimental findings.

6.2.2.3 *Toxoplasma* Genus

Toxoplasma gondii is a zoonotic protozoan parasite belonging to the phylum of Apicomplexa similar to *Plasmodium* species. It causes one of the most common parasitic human diseases known as toxoplasmosis (Tymoshenko et al. 2015; Wilking et al. 2016). It is presumed that this pathogen affects up to one third of the world's population. The infection results in mostly flu-like symptoms in humans, but some severe symptoms (e.g., fetal death, hydrocephalus, calcifications, and blindness) can also be encountered (Reid et al. 2012; Wilking et al. 2016). To simulate the growth of *T. gondii*, a GEM model was reconstructed by Tymoshenko and colleagues (Tymoshenko et al. 2015). The model includes 527 genes controlling 867 metabolic reactions. Some metabolic functions were assumed for the non-annotated genes in the study. FBA was applied to the GEM, and a minimal set of at least 53 enzyme-coding genes and 76 reactions were found to be essential for *T. gondii* replication in the human cells.

6.2.3 *Comparative Analysis of Different Pathogens*

Microbes rarely exist alone in nature. Instead, they frequently share a wide range of environment changing from plants to teeth surface (Liu et al. 2017). One of the most remarkable hallmarks regarding the microbial communities is the relationship of the microbiome with health and disease. Much attention has been paid to metabolism-oriented studies recently in order to treat or prevent infection by comparatively analyzing metabolic changes in the microbial cells. In this section, we highlight a few of the recent findings that employ GEM reconstructions of the involved pathogens for the documentation of metabolic alterations at genome-scale.

Biofilm refers to an organized mixed aggregate of microorganisms surrounded by a self-produced extracellular matrix, known as extracellular polymeric substance, including exopolysaccharides, nucleic acids, and proteins (Bendouah et al. 2006). Biofilm formation is a prerequisite for more efficient protection of a microbial community against the host defense system and antibiotics (Bendouah et al. 2006; Phalak et al. 2016). Therefore, this association plays an important role in chronic illnesses such as CF, periodontitis, endocarditis, and chronic wounds (Bendouah et al. 2006; Zhao et al. 2013). The chronic wounds offer an ideal condition for bacterial colonization since the necrotic debris mediates bacterial adherence to host tissue (Zhao et al. 2013). Polymicrobial flora colonized in this environment leads to impediment to the healing of chronic wounds and higher mortality rates in comparison with monocultures (Phalak et al. 2016). In this context, understanding of the medical importance of biofilms and the role of quorum sensing in regulation of biofilm formation has turned the development and dynamics of the biofilm formation into a very hot topic (Xavier 2016). Furthermore, the cells found within biofilms also attract an increasing attention due to their differentiated metabolic rates.

The biofilm-oriented studies have predominantly focused on two prominent microorganisms, aerobe *P. aeruginosa* and facultative anaerobe *S. aureus* (Bendouah et al. 2006; Zhao et al. 2013; Phalak et al. 2016). By combining the genome-scale metabolic reconstructions with partial differential equations reflecting transport equations (i.e., diffusion and metabolism of nutrients in the biofilm), Phalak et al. developed spatiotemporal metabolic models of the biofilm consortia including *P. aeruginosa* and *S. aureus* either alone or both as the parts of a chronic wound biofilm consortium. They aimed to compare metabolic behaviors of each pathogen and examine spatiotemporal metabolism of the multispecies biofilm. Model simulations have revealed that *S. aureus* supported much thicker biofilms in comparison to *P. aeruginosa*, but this value slightly decreased for *S. aureus* when grown in the mixed biofilm. The authors highlighted a possible inhibitory (antagonistic) metabolic effect of *P. aeruginosa* on *S. aureus*. Another superiority of *S. aureus* is higher anaerobic growth rate. This likely supported the production of much thicker biofilms and more efficient lactate production as a putative cross-fed metabolite. The computational analysis correctly predicted the two important mechanisms contributing to the competitiveness of *P. aeruginosa* in aerobic regions: the lysis of *S. aureus* mediated by *P. aeruginosa*-secreted pyocyanin and the aerotaxis mechanism that triggers the localization of *P. aeruginosa* toward upper portion of the biofilm (whereas *S. aureus* being dominated in the lower anaerobic region). Taken together, this study provided an overview about the metabolic profiles of two common elements (pathogens) of the biofilm consortium in the chronic wounds (Phalak et al. 2016).

The study of essential genes and reactions that are vital components of the biological networks paves the way for the elucidation of novel drug targets as well as the identification of minimal requirements for the cells (Gaurav and Hasija 2017). Another useful approach, so-called synthetic lethal (SL) screening, provides the identification of a pair of nonessential genes whose simultaneous deletion results in the loss of viability (Li et al. 2014; Gaurav and Hasija 2017). Recently, this approach has come into prominence as a highly selective anticancer strategy upon drug-induced inhibition of the SL partner of a cancer-specific mutation. However, this intriguing system has some insufficiencies such as the scarcity of known SL gene pairs in human, the challenge to identify SL genes, and the heterogeneity of cancer cells (Li et al. 2014). This approach can be used for different aims as well as gene/reaction essentiality studies. Such a study has been recently published with a focus on the comparative analysis of metabolic networks belonging to *M. tuberculosis*, *H. pylori*, *S. aureus*, and *S. typhimurium*. The genome-scale networks of *M. tuberculosis*, *H. pylori*, *S. aureus*, and *S. typhimurium* were used (Table 6.1) . By performing FBA with a single objective (maximization of growth) and gene deletions, similarities and differences of growth rates were evaluated for all the four metabolic network models of the pathogens in this comparative study. The ability of the bacteria to grow both aerobically and anaerobically was analyzed. The differences among fluxes were attributed to important metabolic pathways (glycolysis, oxidative phosphorylation, and amino acid biosynthesis). A greater number of essential gene predictions were associated with *S. typhimurium* than with the other three pathogens. This situation was found to be valid for SL genes. Considering connectivity of the metabolites, very few metabolites were identified as highly connected (in analogy to "hubs" in protein-protein interaction (PPI) networks) suggesting these molecules as "global players" (Gaurav and Hasija 2017).

As mentioned above, respiratory disorders have a severe economic impact in pig-breeding worldwide. The porcine respiratory tract infections are initiated via colonization of some bacteria including *M. flocculare*, *M. hyopneumoniae*, and *M. hyorhinis* (Ferrarini et al. 2016). Even if these three bacteria are closely related according to 16S rRNA sequence comparison, they differ in terms of pathogenicity characters. To date, a disease has not been elicited related to *M. flocculare*, while *M. hyopneumoniae* is associated with porcine respiratory disease complex and mycoplasmal pneumonia and *M. hyorhinis* with polyserositis and arthritis (Siqueira et al. 2013; Ferrarini et al. 2016). To induce pneumonia, *M. hyopneumoniae* adheres to the cilia by causing epithelial damage, whereas the adherence of *M. flocculare* does not induce any damage, likely due to the use of different adhesins (Siqueira et al. 2013). Both genome and transcriptome profiles of these three mycoplasmas have been evaluated based on comparative analyses (Siqueira et al. 2013, 2014). However, the mechanisms of virulence were not completely unsolved. Therefore, a comparative characterization of the metabolic profiles of the bacteria based on the reconstructed GEM models has been recently reported by Ferrarini and colleagues (Ferrarini et al. 2016). Using FBA simulations and some experimental setups, various distinctions (e.g., differences in the carbohydrate uptake ability and the capacity to metabolize myoinositol, amino sugar, carbohydrates, glycerol) were highlighted among the bacteria in the study. These results were suggested to determine the metabolic behavior and virulence of the bacteria. It is important to note that the authors suggested a possible relationship between the high virulence of *M. hyopneumoniae* and myoinositol catabolism. The metabolism of glycerol was also attributed to the pathogenicity differences between the mycoplasma species considering the conversion of the glycerol to toxic hydrogen peroxide. Taken together, this work opened new avenues to explain the diversity in the pathogenic characters of these closely related mycoplasma species (Ferrarini et al. 2016).

The human intestinal tract harbors a microbiome consisting of numerous microbes (bacteria, archaea, viruses, and eukaryotic microbes) that has tremendous importance in health and disease. The gut microbiome interacts with the host physiology by contributing to the metabolic functions and protection against pathogens (Shreiner et al. 2015). Moreover, it can be disrupted in terms of the structure and function through the infection status of the host and the antibiotic treatment

(Shreiner et al. 2015; Bäumler and Sperandio 2016). As mentioned above, *C. difficile* causes antibiotic-associated diarrhea through disturbing gut microbiota. To unravel the impact of the antibiotic administration and *C. difficile* infection on the gut microbiota and to explore a probiotic against *C. difficile* infection, Steinway et al. introduced the first Boolean dynamic model based on time series meta-genomic sequence information and integrated the model with GEM reconstructions (Steinway et al. 2015). Ten most abundant genera in the gut microbiota were considered in the modelling approach. GEMs of all the ten genera were reconstructed at the genus level. For this, species-level reconstructions for each genus were obtained by feeding RAST server annotations of the species genomes to model SEED (Henry et al. 2010), and, afterward, all species-level reconstructions were united to generate genus-level GEMs. The comparative analysis of GEMs enabled the investigation of metabolic basis of the microbial interactions. The modeling results suggested that the overabundance of *Barnesiella*, a member of gut flora, can inhibit the growth of *C. difficile*. This was validated by further experimental work. Comparative GEM analyses predicted that *Barnesiella intestinihominis* likely suppresses *C. difficile* growth via some non-metabolic mechanisms. Based on the follow-up experimental results supporting computational predictions, *B. intestinihominis* was suggested as a probiotic to reduce *C. difficile* growth (Steinway et al. 2015).

Bacillus anthracis, *Francisella tularensis*, and *Yersinia pestis* are the bioterrorism agents that cause anthrax, tularemia, and bubonic plague, respectively. Metabolic network analysis on the whole genome of these agents was performed using FBA analysis to identify common antimicrobial drug targets (Ahn et al. 2014). For each agent, 35–40 essential metabolic enzymes were identified. Of those, nine enzymes were found to be common between the three pathogens, and they were involved in coenzyme A, folate, phosphatidyl-ethanolamine, and nucleic acid pathways. Among those enzymes, dihydrofolate synthase/tetrahydrofolate synthase enzyme in folate metabolism was found to be targeted by existing antimicrobials (trimethoprim and Rab1), while others may be potential novel targets for broad-spectrum antibiotics against bioterrorism agents. The study provides a computational approach for combinatorial therapeutics.

6.3 Identification of Anti-infection Therapeutics Combining Metabolic Pathway Analysis with Structure-Based and Sequence-Based Analyses

Compared to the traditional drug design approach that targets single enzymes, a system-wide approach constitutes a better approach for the discovery of broad-spectrum drugs and a better solution for multiple drug resistance. Analysis of metabolic networks at genome-scale enables systems-level understanding of infectious diseases, as documented in the previous section, and aids in the efficient drug design. In this approach, metabolic pathways of the pathogen of interest are shortlisted based on some criteria such as selectivity and network centrality with respect to host metabolic pathways, and the resulting drug target candidates are further structurally analyzed by computation by considering 3D structure. Sequence-based homology search using BLASTP can also be performed to filter candidate proteins. This section reviews the recent literature that combines metabolic pathway analysis with computational structure-based and/or sequence-based analysis to identify drug targets.

Using the genome-wide metabolic network analysis, metabolic pathways that are important for pathogen but not involved in host system can be extracted. Then, with the structure-based methods, essential proteins in the pathogen-specific metabolic pathways are further evaluated, and novel disease-related genes are predicted that leads to drug target discovery. This structure-based drug design approach enables the discovery of more valuable drug targets. Regarding this phenomenon, a recent study was performed in order to discover novel drug targets for the treatment of MRSA infection (Uddin et al. 2015). MRSA has been known to exhibit extreme resistance to conventional antibiotics, and metabolic network analysis approach aims to discover novel antibiotics that will not lead to resistance. Ninety-three and 264 metabolic pathways for MRSA and human host were retrieved from KEGG database (Kanehisa et al. 2008), and 22 pathogen-specific pathways for MRSA were identified and further explored in the study. Out of these 22 pathways, 12 enzymes essential for pathogen survival were found to be potential novel drug targets. Among these, further computational structural analyses were performed for glutamate synthase (involved in nitrogen and glutamate metabolism) using the protein sequence. The 3D structure prediction, homology modeling, binding site detection, and molecular docking were the methods used to characterize the enzyme (Uddin et al. 2015). Glutamate synthase has also been proposed as drug target for *M. tuberculosis* (Cheung and Tanner 2011).

Another study proposed common antibiotic targets for *E. coli* and *S. aureus* by the analysis of essential reactions in their metabolic networks and the determination of essential enzymes for infection (Shen et al. 2010). Then, they performed virtual screening of potential inhibitors on these enzymes and experimentally approved the in vitro effect. Using FBA approach, they predicted 38 essential metabolic reactions and the enzymes catalyzing these reactions. Seven indispensable reactions that are related to the metabolic pathways playing role in bacterial cell wall, amino acid, and porphyrin biosyntheses were found to be common for *E. coli* MG1655 and 13 *S. aureus* strains. Type-II bacterial fatty acid biosynthesis (FAS II) pathway that is indispensable for 13 *S. aureus* strains was further analyzed for the structural prediction of the essential enzymes. Crystal structure information and homology models of FAS II enzymes (FabD, FabH, FabB/F, FabG, FabA/Z, and FabI) were used for the virtual screening of inhibitors and molecular docking analysis in active sites of the enzymes. Forty-one small molecules that potentially inhibit FAS II enzymes were identified, and their inhibitory effects were experimentally tested in vitro. Eight of the molecules were found to be highly active, and nine were found to be weakly active (Shen et al. 2010).

Another comparative metabolic pathway analysis between *Streptococcus pyogenes* pathogen and its human host revealed 105 enzymes important in metabolic reactions among 65 metabolic pathways (Singh et al. 2016). Out of 105 enzymes, 53 enzymes that are not homologous in human were selected for further analysis, and 28 enzymes were found as essential enzymes for the pathogenicity of *S. pyogenes* SF370. Subcellular localizations of these enzymes were found via CELLO (Yu et al. 2006). Eighteen enzymes were predicted to be located in cytoplasm, and the remaining ten enzymes were located in the membrane. Since membrane proteins are more accessible from the extracellular surface, they are more considered as important drug targets (Rajendran et al. 2016). Ten membrane-located enzymes were found to be involved in several metabolic pathways including carbohydrate metabolism, amino acid metabolism, lipid metabolism, energy metabolism, vitamin and cofactor biosynthesis, and nucleotide metabolism. Of these enzymes, dihydropteroate synthase was considered as a novel potential drug target, and its crystal structure was obtained via homology modeling (Singh et al. 2016). Further investigations on discovering its inhibitors by virtual screening and interaction studies via molecular docking need to be considered during drug development process.

Campylobacter jejuni is a food-borne pathogen with global health burden. A recent study revealed novel drug targets effective on this pathogen via the metabolomics guided drug discovery approach (Mehla and Ramana 2015). Upon the comparative metabolic pathway analysis between the pathogen and human host, 16 metabolic pathways were identified as specific for *C. jejuni*. Four hundred forty-six proteins were involved in the identified pathways. Comparative homology analysis via BLASTP revealed that 115 proteins out of 446 were essential in pathogenesis, and 66 of them are nonhomologous in human. According to the druggability potential analysis using DrugBank database (Wishart et al. 2008; Knox et al. 2011), 34 proteins were found to play crucial roles in bacterial growth and survival, serving as potential therapeutic targets.

As discussed in the previous section, increasing resistance to existing drugs is an emerging obstacle in the treatment of tuberculosis. In order to overcome this limitation, drug resistance pathways of *M. tuberculosis* were identified by metabolic pathway analysis, and putative drug targets unique to *M. tuberculosis* were discovered (Cloete et al. 2016). Eighteen drug-resistant gene products were mapped to 15 metabolic pathways (retrieved from KEGG (Kanehisa et al. 2008)) that were identified using available microarray data. Three of the pathways (including 105 genes) were reported to be essential for mycobacterial growth and latent tuberculosis, which are pyrimidine (42 genes), oxidative phosphorylation (47 genes), and nicotinate-nicotinamide metabolisms (16 genes) (Gerdes et al. 2002; Sassetti et al. 2003; Murphy and Brown, 2007; Samant et al. 2008). Upon the screening of these 105 gene products, 9 essential and nonhuman homolog genes were found, and 1 putative novel target (Rv1712, cytidylate kinase) was selected for structural analysis and functional characterization. The crystal structure of Rv1712 protein was generated via molecular dynamics simulation. According to the computational docking studies on this protein, ten compounds with higher affinities were selected by virtual screening as potential Rv1712 inhibitors. Experimental validation is needed for further analysis of this potential drug (Cloete et al. 2016).

M. ulcerans is the third most common disease-causing agent after tuberculosis and leprosy, and there is an increasing need for better therapeutics against emerging resistance. The increasing availability of omics data leads the way for the identification of novel therapeutic candidates via metabolic pathway analysis. In a study, 424 essential genes were predicted from the *M. ulcerans* genome using homology searching (Butt et al. 2012). Metabolic pathway analysis revealed that most essential genes are involved in carbohydrate and amino acid metabolisms. Of these genes, 236 proteins were found to be nonhomologous to human host; thus they can be potential drug candidates that eliminate side effects in human systems. Experimental and computational 3D structures were retrieved from the Protein Data Bank (PDB) (Sussman et al. 1998) and ModBase (Pieper et al. 2011) databases, respectively. Biological significance and subcellular localization of therapeutic targets were predicted via a computational prediction method called MycobacSVM (Lin et al. 2010), and the druggability potential of these targets was assessed through the DrugBank database (Wishart et al. 2008; Knox et al. 2011). Based on these analyses, essential enzymes involved in cell wall synthesis, thiamine biosynthesis, protein biosynthesis, and histidine biosynthesis were reported as potential drug targets against *M. ulcerans* (Butt et al. 2012).

Brucella melitensis is a gram-negative bioterrorism agent that causes zoonotic diseases (Brucellosis). According to a metabolic pathway analysis study, there were 87 pathways in *B. melitensis*, and, when compared to human host, 27 of them were unique to the pathogen (Gupta et al. 2017). In addition, 142 nonhomologous proteins were identified through BLASTP analysis. The sequences of nonhomologous proteins were queried in the database of essential genes, DEG, (Zhang et al. 2004) to identify essential genes in *B. melitensis*. A further computational choke point analysis for the nonhomologous and essential genes filtered the amount to 18 genes, which were identified as potential drug targets. PPIs for these target proteins were analyzed via STRING database (Szklarczyk et al. 2017). Phosphoribosyl-AMP cyclohydrolase (HisI) was selected for further structural characterization since it had the highest number of degree in the network. HisI was found to be involved in histidine metabolism, secondary metabolites, and amino acid biosynthesis pathways, which were reported to be targeted in the antimicrobial therapy of different pathogens (Lee et al. 2009; Henriksen et al. 2010; Lunardi et al. 2013). The homology model of HisI was constructed via MODELLER (Webb et al. 2017). Molecular docking analysis through DockBlaster tool (Irwin et al. 2009) revealed ZINC04880153 compound with the lowest binding energy among 500 molecules. Molecular dynamics simulations were performed for the complex HisI-ZINC04880153 in order to understand the behavior of protein-ligand interactions. Since HisI showed stability to the ligand, ZINC04880153 can be a good drug candidate that needs to be further studied in wet-lab experiments (Gupta et al. 2017).

Enterobacteriaceae pathogens are causative agents of gastrointestinal infections and urinary and respiratory tract infections. A recent study focusing on one of the important pathogens within this group (*Shigella flexneri*) revealed potential drug targets overcoming the emerging resistance problem (Hadizadeh et al. 2017). Of 115 metabolic pathways for *S. flexneri* that were derived from KEGG database (Kanehisa et al. 2008), 35 were specific to the pathogen with no common human pathways. 1495 proteins that were identified to be nonhomologous to human proteome based on BLASTP search were further analyzed for essentiality using DEG database (Zhang et al. 2004), and 874 essential proteins were identified for *S. flexneri*. Another homology search using BLASTP against the whole proteome of gut microbiota covering more than 80 species led to 125 essential proteins with less than ten hits. One hundred twenty-five proteins were further filtered by the similarity analysis among nine other *Enterobacteriaceae* species. Thirty-five proteins with 50% identity and more were selected as drug target candidates. The candidates were subjected to function prediction, subcellular localization, broad-spectrum condition, and druggability analyses using several Web-based tools or databases. Such analyses are important to evaluate suitability of drug target candidates. These 35 proteins were involved in some vital pathways such as metabolism, biosynthesis of macromolecule, and cell division. Thirty of them were found to be potential novel drug targets, and 13 of them were considered as broad-spectrum targets for *Enterobacteriaceae* pathogens (Hadizadeh et al. 2017).

GEM models of the pathogen *S. typhimurium*, which leads to food-borne illnesses, were already discussed in the previous section. Very recently, a metabolic network analysis was performed in order to find common pathways including potential drug targets, which were further subjected to computational structural analysis (Mehla and Ramana 2017). They used a previously reconstructed GEM (Raghunathan et al. 2009) in their work as a constraint to FBA and minimization of metabolic adjustment (MOMA) analyses (Segrè et al. 2002). Seventy-three genes involved in important metabolic pathways including peptidoglycan biosynthesis, tyvelose, folate, fatty acid, and riboflavin biosynthesis pathways were identified to be critical through in silico gene knockout studies of *S. typhimurium LT2* genome. Essentiality analysis focusing on the experimentally predicted essential genes of *Salmonella* species revealed 54 essential unique proteins that were further subjected to structural and druggability analyses. The druggability analysis via BLASTP against DrugBank (Wishart et al. 2008; Knox et al. 2011) revealed that 47 proteins have higher similarity to known drug targets. Fourteen of them were targeted by FDA-approved drugs, while others were in experimental phase. Of 54 proteins, the crystal structures were available for 8 proteins (yadF, ribH, fabD, aroD, rfbG, rfbD, ppnK, and rib) in PDB (Sussman et al. 1998), and 37 had computationally determined structure via homology modeling, available in ModBase database (Pieper et al. 2011). Nine proteins (mraY, fabI, ribA, pgsA, folB, rfaL, dfp, cdsA, and psd) did not have any available structural data. The final list of drug target candidate proteins was studied for their interactions with each other using STRING database (Szklarczyk et al. 2017). The interacting proteins were mapped to peptidoglycan, folate, and L-rhamnose synthesis pathways. These proteins and the metabolites involved in these pathways can be candidate drug targets for therapeutic approaches upon experimental validations in order to combat the multidrug resistance among *S. typhimurium* infections (Mehla and Ramana 2017).

Fungal pathogens are also important infectious agents that affect human life. *Aspergillus fumigatus* is one of them leading to aspergillosis especially in immunocompromised patients, and, therefore, effective therapeutics needs to be developed. In this regard, a study focusing on discovering novel drug targets for *A. fumigatus* infections compiled major metabolic pathways of the pathogen, leading to 162 unique metabolic reactions with 102 enzymes and 204 metabolites that play a role in the central carbohydrate and amino acid metabolisms, lipid synthesis and degradation, and vitamin synthesis (Kaltdorf et al. 2016). In order to identify a list of drug target candidates, they combined the results of three different bioinformatics pipelines: (1) computational analysis of the metabolic network structure, (2) analysis of transcriptome datasets to identify highly expressed genes encoding essential enzymes, and (3) analysis of PPIs and identification of pathogen-specific enzymes using BLASTP. The resulting set included 64 protein targets. Some of these targets (UaZ, AFUA_2G06230; AroC, AFUA_5G13130; FasA, AFUA_3G04210; FasB, AFUA_3G04220) have already been confirmed to be good drug targets with experimental methods (Sasse et al. 2016). Chorismate synthase (ARO2, AFUA_1G06940) and chorismate mutase (AroC, AFUA_5G13130) were also experimentally validated as antimycotics, pointing to the power of the combined approach to predict metabolic drug targets (Kaltdorf et al. 2016).

6.4 Metabolism-Focused Genome-Wide Data for Infection Research: Metabolome and Fluxome

6.4.1 Metabolomics for Infection Research

Metabolomics is the field of the identification and quantification of numerous small molecules (metabolites) which are biochemical fingerprints of the metabolism in biological samples. Metabolites are chemical phenotypes of organisms resulting from activities on transcriptome and proteome levels. In this way, metabolomics provides an integrated global profile of the condition of an organism (Zhou et al. 2015). This omics approach does not require any a priori hypothesis, so it can provide a comprehensive characterization of all the metabolites of the sample. After obtaining metabolomics data, multivariate statistical analyses can be applied to get the information of interest from huge and complex data. Bioinformatics tools allow the discrimination and classification between different groups of data, such as patients with different clinical conditions (Bozzetto et al. 2017). This is powerful to identify differentiation of noninfected and infected samples and to reveal pathogen-host metabolic interactions and progression in infection (Mortensen et al. 2016). Metabolomics can elicit new or unexpected metabolites and makes possible to characterize unknown metabolic pathways and identify new biomarkers or therapeutic targets (Bozzetto et al. 2017). As a part of an integrated systems biology approach, outputs of metabolomics provide a unique molecular fingerprint for an individual. In addition, metabolome-wide association studies, which promote integrated models, can be carried out through new high-throughput technologies for personalized medicine by correlating specific metabolites with a specific disease (Cribbs et al. 2016).

6.4.1.1 Basic Computational Tools in Metabolomics

Multivariate data analysis methods, which include principal component analysis (PCA) and partial least square-discriminant analysis (PLS-DA), are applied to investigate and visualize possible metabolic changes between study groups (Abdelrazig et al. 2017). An alternative open name for PLS-DA is projection to latent structures-discriminant analysis. PCA is used to identify data groups and outliers and to investigate main differences between the groups. PLS-DA is performed to examine variations on the discriminate variables. Then, it creates a predictive classification model of the differential metabolites. The quality of the model is confirmed by some methods such as analysis of variance (ANOVA) (Zhou et al. 2016; Banoei et al. 2017). Another method, area under the ROC (receiver operating characteristic) curve (AUC), is also used to assess specificity and selectivity of prediction models (Zhou et al. 2015). Significantly changed metabolomic features can be searched for the exact molecular mass on online databases such as HMDB (Wishart et al. 2013), METLIN (Smith et al. 2005), and KEGG (Kanehisa et al. 2008) to identify the corresponding metabolites (Zhou et al. 2015). Web-based MetaboAnalyst tool (Xia et al. 2015) can also be used to carry out all statistical analyses and pathway annotations.

6.4.1.2 Infection Metabolomics and Microbiota

Infection conditions may cause alterations on the microbial composition in the host, and these microbes may affect the host via their metabolites. Therefore, investigation of the relationship between the responsible microbiota and metabolic profiles of the host can provide functional information on microbe-host interactions. In a recent study, 16S ribosomal RNA (rRNA) gene sequencing was used to identify bacterial species of the lung's bacterial microbiota, and LCHRMS with MS/MS was used for the metabolome analysis of bronchoalveolar lavage fluid of HIV-infected and HIV-uninfected individuals (Cribbs et al. 2016). Thirty-nine HIV-infected subjects and 20 HIV-uninfected controls were used for the investigation. Twelve of the 5930 mass-to-charge ratio (m/z) features were found to be significantly different between the infected and uninfected individuals, whereas 79 additional m/z features were identified by correlation-based network analysis. Significant overrepresentation of cystine, two complex carbohydrates, and 3,5-dibromo-L-tyrosine was observed. Then, sparse partial least squares regression was applied for the integration of metabolomic features with the microbial species. All 5930 m/z features were integrated with the 153 microbial taxa. *Caulobacteraceae*, *Staphylococcaceae*, and

Nocardioidaceae families and *Streptococcus* genus demonstrated the highest association with the m/z features in HIV infection. Metabolite annotation and pathway enrichment analysis were performed for the three bacteria families, and lineolate, glycerophospholipid, and fatty acid metabolism were found as correlated pathways within these bacteria. These correlations and associations play a role in the pathogenesis of pneumonia in HIV-infected individuals, and they propose that interactions between the lung's bacterial microbiota and the host may have functional metabolic effect in the lung (Cribbs et al. 2016).

In a different approach (Seekatz et al. 2017), the authors used fecal microbiota transplantation (FMT) as a treatment method for recurrent *C. difficile* infection. They examined associations between microbial community members and metabolites of the patients after FMT over time to identify bacteria groups which have critical restorative functions. Fecal microbiota and metabolome states were identified in pre-FMT and post-FMT stages. It is found that the level of the secondary bile acid deoxycholic acid and the short-chain fatty acids butyrate, acetate, and propionate increased in post-FMT stage. By applying a generalized estimating equation approach to model metabolite concentrations, this alteration was found to be associated with members classified within the *Lachnospiraceae*, *Ruminococcaceae*, and unclassified *Clostridiales* families (Seekatz et al. 2017).

6.4.1.3 Infection Metabolomics in Hepatitis

Comprehensive analysis of metabolic pathways leads to the discovery of novel biomarkers for infectious diseases. Such an example is the discovery of new small molecules that can be used as biomarkers for early diagnosis of hepatitis B virus (HBV) (Zhang et al. 2013a) and hepatitis C virus (HCV) (Zhang et al. 2013b). The authors mapped the metabolome analysis results on metabolic pathways using MetaboAnalyst tool (Xia et al. 2015). The urine metabolic profile was obtained by high-throughput MS techniques, and 11 and 20 differential metabolites were found for HBV and HCV, respectively. The dysregulated metabolic pathways for HBV were metabolism of xenobiotics by cytochrome P450, phenylalanine metabolism, and amino sugar and nucleotide sugar metabolism, whereas the differential pathways for HCV played a role in tyrosine metabolism, glutathione metabolism, phenylalanine metabolism, ascorbate and aldarate metabolism, and arginine and proline metabolisms. Out of these 11 and 20 metabolites, potential metabolite markers were identified. The metabolite markers for HBV included biotin sulfone, 5-oxoheneicosanoic acid, D-glucosaminide, and 2-methylhippuric acid (Zhang et al. 2013a). HCV metabolite markers were succinic acid, carnitine, ethyl citrate, indolebutyric acid, valine, taurine, and L-tyrosine (Zhang et al. 2013b). These potential metabolite markers can be further evaluated for diagnostic use in the infected individuals.

6.4.1.4 Infection Metabolomics in HCMV Infection

Human cytomegalovirus (HCMV), a member of the *Herpesviridae* family, has linear DNA genome which encodes viral and cellular gene products (Varnum et al. 2004; Scarborough et al. 2017). HCMV infection is the most common viral infection in pregnancy. In a complex study, the metabolomic profile in amniotic fluid samples of HCMV-infected and HCMV-uninfected fetuses was investigated to define changes in metabolic pathways during congenital HCMV infection and to discover new potential diagnostic and/or prognostic biomarkers (Fattuoni et al. 2016). They carried out a retrospective cohort study which included 63 pregnant women: 20 of them had primary HCMV infection during pregnancy and had transmitted the virus to the fetus (transmitters); 20 of them had the infection without transmitting the virus to the fetus (non-transmitters); and 23 women who underwent amniocentesis for cytogenetic-based diagnosis were assumed as controls. Gas chromatography-mass spectrometry (GC-MS) and a multivariate statistical approach were used for metabolomics analysis. They created PLS-DA models using MetaboAnalyst 3.0 (Xia et al. 2015) for network mapping. They performed multivariate statistical analysis and pathway analysis also in MetaboAnalyst tool (Xia et al. 2015), which combines pathway enrichment analysis with pathway topology analysis. The latter is based on the identification of the most important nodes of the network by betweenness centrality (the number of shortest paths passing through a certain node). Metabolic pathway models from KEGG (Kanehisa et al. 2008) were also used to complete the analysis and produce an interactive system for data exploration. When they compared with the controls, significantly increased levels of glutamine, glycine, serine, pyruvic acid, threonine, threonic acid, and cystine were observed in HCMV transmitters in addition to the decreased levels of unknowns U1437, U1715 and U1804, glutamic acid, fructose, sugar-like A203003 and A203005, and tyrosine. In non-transmitters, increased levels of glutamine, serine, glycine, threonic acid, threonine, 1-monostearin, urea, and cystine were observed, as sorbitol, unknowns U1751 and U1804, sugar-like A203003, xylitol, leucine, and fructose were decreased. When transmitters and non-transmitters were compared, they could not build a statistically significant model. The activation of glutamine-glutamate and pyrimidine metabolic pathways was found common in HCMV-infected subjects (transmitters and non-transmitters). An important result of the study is the separation of asymptomatic versus symptomatic HCMV cases due to the alteration of fatty acids biosynthesis. These findings can contribute to the improvement of the accuracy of early diagnosis and to the administration of HCMV pregnancy-related infections (Fattuoni et al. 2016).

6.4.1.5 Infection Metabolomics with Longitudinal Data

To understand the progression mechanism of a disease, it is crucial to examine different stages of the disease over time. Understanding the molecular changes during disease progression provides new perspectives to improve therapeutic and diagnostic tests. *T. gondii*, as mentioned in a previous section, is the etiological agent of the common parasitic disease known as toxoplasmosis. In a study, the dynamic changes in the metabolomic profile of mouse sera during *T. gondii* infection were investigated using LC-MS with electrospray ionization (ESI) source (Zhou et al. 2016). A global metabolome analysis was applied to the serum samples of infected and control mice at 7, 14, and 21 days post-infection (DPI). The most differentially altered metabolomic profile was observed at 7 DPI. When the infection time increased, the number of differentially changed metabolites decreased. This behavior indicated that dysregulation of metabolic state was restored in a time-dependent manner. Seventy-nine differentially expressed metabolites in ESI+ mode and 74 in ESI− mode were detected in the sera of *T. gondii*-infected mice by multivariate statistical analysis. Furthermore, 19 dysregulated metabolites (5 in ESI+ mode and 14 in ESI− mode) that are related to the amino acids metabolism and energy metabolism were identified by PCA and PLS-DA. The KEGG pathway database (Kanehisa et al. 2008) was used to list the related pathways of metabolites. The potential utility of these metabolites as diagnostic biomarkers was validated by ROC curve analysis in MetaboAnalyst 3.0 (Xia et al. 2015; Zhou et al. 2016).

In another study, dengue (an acute febrile illness) caused by dengue virus (DENV) was examined with the same approach of following disease progression over time (Cui et al. 2017). Human CD34+ fetal liver cells were transferred into NOD-*scid Il2rg-/-* mice to construct a humanized mice (humice) model that develops significant levels of human platelets, monocytes/macrophages, and hepatocytes. Serum metabolomic profiling of the humice was performed and analyzed at 0, 3, 7, 14, and 28 DPI by LC-MS analyses. An initial PCA analysis showed proper grouping of the data. HMDB (Wishart et al. 2013), METLIN (Smith et al. 2005), LIPID MAPS (Sud et al. 2012), and MassBank (Horai et al. 2010) were used for exact mass searching. The identified metabolomic features were analyzed with one-way ANOVA (p-value < 0.05) with Benjamini-Hochberg multiple testing correction and further filtered with the fold change threshold criteria of 1.5. In the results, 38 differential metabolites belonging to fatty acids, purines and pyrimidines, acylcarnitines, acylglycines, phospholipids, sphingolipids, amino acid and derivatives, free fatty acids, and bile acid were identified in the humice serum. These metabolites reversibly changed over time, exhibiting significant perturbation at 3 or 7 DPI and returning back to the control levels at 14 or 28 DPI. They can be considered as prognostic markers of the disease. Purine and pyrimidine metabolism, fatty acid β-oxidation, phospholipid catabolism, arachidonic acid and linoleic acid metabolism, sphingolipid metabolism, tryptophan metabolism, phenylalanine metabolism, lysine biosynthesis and degradation, and bile acid biosynthesis (related to liver impairment) were found as the major differential metabolic pathways by pathway analysis tool in MetaboAnalyst (Xia et al. 2015). Most of their findings overlapped with the findings in a longitudinal cohort of adult human dengue patients across different infection stages. They demonstrated that the humice could be a useful small animal model for the study of dengue pathogenesis and the development of dengue therapeutics (Cui et al. 2017).

Helicobacter pylori, a gram-negative microaerophilic bacterium, chronically colonizes the gastric epithelium of more than half of the world's population. In another recent study with the similar approach of time-dependent identification of metabolome profiles, *H. pylori* infection was investigated at 1, 3, or 6 months by analyzing the metabolites in the stomach tissue of mice that had been infected with *H. pylori* SS1 (Nishiumi et al. 2017). The analysis was performed using LC-MS and GC-MS. The F-test was applied to compare the variances of each group, and then the student's *t*-test (p-value < 0.05) was used to measure the significance of differentiation between each group. At 1 month after the infection, upregulation was observed in the glycolytic pathway, the TCA cycle, and the choline pathway. The urea cycle was found to be downregulated after 6 months of the infection. High levels of some amino acids were detected at 1 month, while low levels of many amino acids were identified at 3 and 6 months. According to the results, they suggested that *H. pylori* infection causes various metabolic alterations at lesional sites, and this can be the cause behind the transition of disease conditions (Nishiumi et al. 2017).

6.4.1.6 Infection Metabolomics in Coinfection

DENV and chikungunya virus (CHIKV) are included in the genus *Flavivirus* and the genus *Togavirus*, respectively. Dengue and CHIKV are included among the most important vector-borne diseases, and they affect approximately 40 million people globally. In a study, CHIKV and dengue mono-/coinfections were investigated by identifying the global metabolome of patient sera infected with CHIKV and dengue using nuclear magnetic resonance (NMR) spectroscopy (Shrinet et al. 2016). MetaboAnalyst Web tool (Xia et al. 2015) was used for all of the statistical analyses covering PCA and PLS-DA and pathway annotation analyses. One-way ANOVA, fold change analysis with threshold value 2, and *t*-test with a cutoff p-value of 0.05 were applied to identify the metabolites changed between the groups. Many metabolites were found as significantly changed in both CHIKV mono-infection and CHIKV-dengue coinfection. Glycine, serine, threonine, galactose, and pyrimidine metabolisms were detected as the most perturbed pathways in both mono- and coinfection conditions. These pathways are related to the clinical symptoms such as fever, inflammation, energy deprivation, and joint pain during the infections. (Shrinet et al. 2016).

In another coinfection study, metabolic alteration mechanism of intestinal helminth infection, with the increased bacterial pathogenicity, was investigated in coinfection conditions with the murine helminth *Heligmosomoides polygyrus* and the bacteria *S. typhimurium*. Intestinal contents of the infected mice were analyzed by LC-MS. Welch t-test with unequal variances was carried out for the statistical analysis. MetaboAnalyst 3.0 (Xia et al. 2015) was used for multivariate and clustering analyses. Out of 4593 metabolite features, 362 (41 upregulated and 321 downregulated) were significantly changed during *H. polygyrus* infection. Helminth colonization also caused alteration in intestinal microbiota and enhanced the production of microbiota-derived short-chain fatty acids, which enhanced the expression of virulence genes in *S. typhimurium*. These results for the first time demonstrated that helminth infection enhanced colonization by *Salmonella* bacteria (Reynolds et al. 2017).

6.4.1.7 Other Studies

Urine metabolome was collected with NMR spectroscopy in *C. difficile* infection (CDI), which is one of the most common hospital-acquired infections (Kao et al. 2016). Urine samples were collected from 31 adults who were at various stages of recurrent CDI. Multivariate statistics were carried out using MetaboAnalyst 3.0 (Xia et al. 2015). A model of PLS-DA was created using 53 metabolites, which clearly discriminated healthy controls and CDI patients. Choline, which significantly increased in CDI patients, was detected as the most important urine metabolite for the separation of healthy controls from CDI patients. Later, meta-genomic analysis of stool samples was collected from six patients, with >3 episodes of CDI, and compared to seven controls. Decreased abundance of choline-metabolizing bacteria was observed in microbiota from recurrent CDI patients. The study is important since it shows the importance of urine metabolomics in discriminating CDI patients (Kao et al. 2016).

H1N1 influenza infection is a major health subject that can be fatal, especially among the elderly and patients with comorbid diseases. It is also the predominant viral cause of community-acquired pneumonia, with relatively high mortality in adults. Metabolite profiles of 42 patients with H1N1 pneumonia, 31 ventilated control subjects in the intensive care unit, and 30 culture-positive plasma samples from patients with bacterial community-acquired pneumonia were analyzed by ^{1}H nuclear NMR and GC-MS within 24 h of hospital admission (Banoei et al. 2017). Multivariate data analysis was used to find significantly altered metabolites. Then, prediction models of H1N1 diagnosis and mortality were built with orthogonal PLS-DA. Cross-validation ANOVA and AUC methods were used to test the model. Pathway analysis, univariate analysis, and student t-test were carried out using MetaboAnalyst (Xia et al. 2015). They identified many metabolites which discriminate H1N1 pneumonia from bacterial pneumonia and nonsurvivors from survivors of H1N1 pneumonia. Increased concentrations of dimethylamine, β-alanine, formate, and quinic acid and a decreased concentration of alanine were observed in H1N1 pneumonia when compared with the two other cohorts. The prognosis of mortality in H1N1 pneumonia was presented by a PCA model which separated the survivor and nonsurvivor cohort samples on the basis of 63 metabolites (Banoei et al. 2017).

Recurrent respiratory infections (RRI) are included in the most common complaints in the pediatric groups and can cause severe morbidity. Urine samples of 13 children with RRI and 15 healthy children were analyzed and compared before and after the pidotimod treatment using LC-MS and multivariate statistical analysis (Bozzetto et al. 2017). Pidotimod is an immunostimulant, which was shown to be protective against respiratory infections. METLIN (Smith et al. 2005) and HMDB

(Wishart et al. 2013) online databases were used for the identification of the m/z values of the compounds. The PCA, PLS-DA, and univariate data analyses were performed in R platform. One hundred thirty-eight metabolites were found as significantly changed between RRI and healthy groups, and 35 of them were still observed after the treatment. The 35 metabolites that cannot be restored by pidotimod treatment were further investigated and found to be related to the composition of gut microbiota. Based on these results, the authors suggested that the combined use of immunostimulants and probiotics has potential synergic effect to prevent children from RRI, which needs further experimental support (Bozzetto et al. 2017).

P. falciparum is an intracellular parasite that is the most common cause of malaria, as discussed in a previous section. In a recent study, 21 adults with infection and 25 controls were used to analyze urinary samples to find urinary biomarkers in falciparum malaria (Abdelrazig et al. 2017). Urinary samples were collected before any treatment and after the fourth week of antimalarial medication. They used LC-MS to screen small molecules in urinary samples. Then, they used PCA and PLS-DA for the identification of possible metabolic changes between all classes. After statistically significant malaria biomarkers were predicted, they used the Urine Metabolome Database (Bouatra et al. 2013) for tentative identification and tandem MS for the verification of the chemical identity of these biomarkers. As a result, differences between the urinary metabolomic profiles of patients with *P. falciparum* infection and healthy controls were detected. After treatment with antimalarial medication, the metabolomic profile of patients showed similarity with that of healthy controls. Twenty-nine urinary metabolites were identified as significantly altered. High levels of urinary pipecolic acid, taurine, N-acetylspermidine, N-acetylputrescine, and 1, 3-diacetylpropane imply that these metabolites can be used as potential biomarkers of *P. falciparum* malaria (Abdelrazig et al. 2017).

Dengue fever (DF) and dengue hemorrhagic fever/dengue shock syndrome (DHF/DSS) are two different outcomes of dengue disease. They were studied with metabolomics approach to characterize the serum metabolome of patients (Voge et al. 2016). Serum samples from dengue patients from Nicaragua and Mexico were analyzed by LC-MS to identify metabolites that were statistically different between DHF/DSS, DF, and non-dengue (ND) diagnosis groups. METLIN (Smith et al. 2005), HMDB (Wishart et al. 2013), and the Omics discovery pipeline (Riley et al. 2010) were used to identify metabolites putatively. Samples of dengue disease diagnosis groups were clustered by PCA. In the Nicaraguan samples, they found 191 metabolites between DF and ND outcomes and 83 metabolites between DHF/DSS and DF outcomes as statistically different. In the Mexican samples, 306 metabolites between DF and ND and 37 between DHF/DSS and DF outcomes were also found to be statistically different. The structural identities of 13 metabolites that are statistically different between DHF/DSS, DF, and ND disease outcomes in at least one of the pairwise comparisons of the diagnosis groups were verified by MS/MS. Statistical analysis of serum samples from patients who progressed to DHF/DSS from DF yielded 65 metabolites that differentiated dengue disease outcomes. Differential perturbation of the serum metabolome was observed during infection with different DENV serotypes and during primary and secondary DENV infections. These results demonstrated that a serum-based metabolomics approach can be used to identify metabolites or small molecule biomarkers that are responsible for different DENV infections and disease outcomes. Metabolic pathways and pathogenic and immunologic mechanisms related to dengue disease severity can also be clarified by investigating differential metabolites (Voge et al. 2016).

6.4.2 ^{13}C-Based Fluxomics in Infection Research

Metabolic research focuses on understanding and altering the metabolic mechanism of organism through some purposes such as restoring metabolic functions in undesirable conditions like diseases, blocking the metabolism of pathogens, and optimizing metabolic production rates of microorganisms used in biotechnological areas (Ng et al. 2016). An important step in understanding metabolic function of a particular cell is the reconstruction of the metabolic network topology, which provides a static picture about the cell's metabolic capacity (see Sect. 6.2 for details). On the other hand, metabolic flux provides a dynamic profile as the rate of flow of metabolite atoms through the metabolic network. It directly measures the activity level of a metabolic pathway. However, only a small number of fluxes such as nutrient uptake, product secretion, and biomass production are measurable by direct measurements. Since the intracellular fluxes, which have important roles on metabolic pathways, are not directly measurable, different modeling strategies have been developed. ^{13}C-based flux analyses are based on the analysis of labeled amino acids in proteins. ^{13}C-MFA and ^{13}C-metabolic flux ratio analyses are the most common ^{13}C-based approaches (Ng et al. 2016). The general strategy in these approaches is that the cells are fed by ^{13}C-labeled substrates and the carbon backbone of the substrates is distributed to different metabolic pathways. Then, the isotopic composition of intermediates and products can be measured by spectroscopic methods like NMR or GC-MS. A mathematical modeling framework is later used based on ^{13}C-labeling measurements to infer the fluxes operating in the metabolic network (Ng et al. 2016).

Enteroinvasive *E. coli* (EIEC) and *S. typhimurium* are typical intracellular pathogens that cause, respectively, bloody diarrhea and gastroenteritis in human. In a study, the intracellular carbon metabolism of EIEC HN280, EIEC 4608-58, and *S. typhimurium* strain 14028 replicating in epithelial colorectal adenocarcinoma cells (Caco-2) was analyzed by ^{13}C-isotopologue profiling (Götz et al. 2010). [U-^{13}C6] glucose was used to feed Caco-2 cells infected with the bacterial strains. Then, MS was used to determine ^{13}C-isotopologue patterns of protein-derived amino acids from the bacteria and the host cells. According to the data, both EIEC HN280 and *S. typhimurium* strain 14028 utilized glucose. However, they did not metabolize glucose 6-phosphate, other phosphorylated carbohydrates, gluconate, or fatty acids as major carbon substrates. EIEC 4608-58, on the other hand, used C3-compound(s) as carbon source besides glucose. The fluxome results revealed strain-specific differences in the intracellular metabolism of the studied pathogens (Götz et al. 2010).

In another study, the diet and metabolism of intracellular *M. tuberculosis* in interaction with its host were examined by a systems-based approach using ^{13}C-flux spectral analysis (^{13}C-FSA) (Beste et al. 2013). They used the THP-1 cell line as model system to investigate the metabolic interactions between *M. tuberculosis* and macrophages. Classical ^{13}C-MFA is only applicable for the metabolic systems at steady state. Therefore, they developed ^{13}C-FSA to examine the non-steady-state metabolism of intracellular *M. tuberculosis*. The isotopomer composition of macrophage amino acids (obtained from the measured labeling patterns); the estimated isotopomer compositions of macrophage glucose, acetate, and glycerol pools (also obtained from the macrophage data and knowledge of the input ^{13}C6 glucose); and the *M. tuberculosis* labeling data are inputs of ^{13}C-FSA. Then, the automated simulations were used to test various hypotheses for nutrient uptake by *M. tuberculosis* for the purpose of identifying substrates and flux distributions. A range of potential substrates (glucose, glycerol, acetate, or single amino acids) and their combinations, which *M. tuberculosis* may obtain from the macrophage, was scanned by ^{13}C-FSA. As a result, they demonstrated that intracellular *M. tuberculosis* can obtain and metabolize amino acids, C1, and C2 substrates in the host. C2 compound, which is highly likely to be acetate- or acetyl-CoA, was defined as the major carbon source that feeds the central metabolism. They indicated CO2 as the carbon source of the C1 substrate. Contribution of glycolytic C3 substrates to the intracellular diet of *M. tuberculosis* was also demonstrated using a mutant strain. Accessibility of *M. tuberculosis* to a diverse diet within the host cell has significant implications for designing drugs against this and other intracellular pathogens (Beste et al. 2013).

P. aeruginosa is a human pathogen which causes numerous acute and opportunistic infections, commonly urinary tract and catheter-associated urinary tract infections. In a study, quantitative analysis of metabolic fluxes in the model strain *P. aeruginosa* PAO1 and 17 clinically isolates from patients with urinary tract infections was carried out by ^{13}C-MFA (Berger et al. 2014). A large-scale model of the central metabolism of *P. aeruginosa* was constructed by the help of the GEM model of *P. aeruginosa* PAO1, KEGG database (Kanehisa et al. 2008), and the *Pseudomonas* genome database (Winsor et al. 2011). Then, the OpenFLUX software (Quek et al. 2009) was used to compute metabolic fluxes through the network by using the carbon-labeling data. The results demonstrated that all *P. aeruginosa* strains catabolize glucose through the ED pathway with fully respiratory metabolism and without overflow. This high-flux pathway supplied much more NADPH than required by anabolism. By this way, the pathogen can resist oxidative stress enforced by the host. *P. aeruginosa* also recruited the PPP especially for biosynthesis. While glycolytic metabolism was conserved among all the isolates, the flux through pyruvate metabolism, the TCA cycle, and the GS showed high variability as a result of adaptive processes in individual strains during infection. This was also mostly due to the isolation of strains from different environments: the urinary tract and the surface of the catheters infections. Glucose uptake rate and growth efficiency of the isolates also indicated strong differences. Projection into the theoretical flux space showed that *P. aeruginosa* metabolism is optimized for efficient growth and indicated significant potential for increasing NADPH supply to drive oxidative stress response (Berger et al. 2014).

Leishmania strains are the cause of the human disease leishmaniasis, as discussed in Sect. 6.2. In a recent study, in vivo fractional fluxes into two or more metabolite nodes were calculated simultaneously by a new ^{13}C-based metabolic flux approach, based on the isotopologue analysis of free metabolite pools in *L. mexicana* (Ng et al. 2016). This is the first study that calculates quantitative in vivo fractional fluxes of central carbon metabolism in a human parasite. GS-MS was used as the spectrometric method. A simple model network including lower glycolysis, TCA cycle, glycosomal succinate fermentation pathway, and several anaplerotic reactions was created by the help of the previously published data for *L. mexicana*. A total of six flux ratios into the oxaloacetate and malate pools were successfully calculated in the rapidly dividing *L. mexicana* by using the new approach. The study is especially valuable for demonstrating the calculation of experimental metabolic fluxes based on the labeling of free metabolite pools rather than the use of labeled amino acids in proteins (Ng et al. 2016).

6.5 Concluding Remarks

Interactions between the host and pathogens play an active role in rewiring the metabolic networks during infection. Concomitant alterations can result in serious metabolic disorders. Therefore, understanding these changes in both pathogen and host sides can provide an insight into the molecular background of the infection. In this regard, the systems biology approach opens up new opportunities to unravel the alterations via metabolome-/fluxome-based or model-driven analyses of microbial infection processes.

The reconstructed GEMs are often combined with constraint-based modeling techniques to improve the understanding of the molecular mechanisms behind the pathological process. Although many current GEMs describe the metabolism per se via isolating it from other cellular systems by ignoring the connection of the metabolism with other cellular processes (e.g., transcriptional regulation), studies which integrate metabolic networks with transcriptomics/proteomics data sets are also available (Huthmacher et al. 2010; Jerby and Ruppin 2012). It is important to note that the use of metabolic models including different types of pathophenotype data can provide significant clues regarding the condition-specific changes during the infection process, and it can identify the relationship between diseases and transcriptome/proteome and metabolome data. Thus, it can enable the prediction of novel drug targets and biomarkers. The biomarkers are key molecules in especially "personalized medicine." They allow the detection of proper drug dose, prediction of prognosis, therapeutic efficacy, and adverse drug reactions (Landeck et al. 2016). Thus, personalized medicine is linked to the systems medicine via biomarkers.

An increasing global issue is the emerging antibiotic resistance crisis, and it can be generally attributed to horizontal gene transfer or genetic mutations causing enhanced drug efflux, drug target modification, and production of the enzymes responsible for antibiotic inhibition. Analysis of the metabolic pathways at genome-scale is an intriguing way to address this challenge in the therapeutic process. Characterization of the metabolic changes and unexplored portion of the genomes of the resistant pathogens can shed light on the adaptation mechanisms and identify new drug targets. For instance, a novel systematic approach, developed by Zampieri et al., elucidated some weaknesses and superiorities of the antibiotic-resistant *E. coli* lineages (Zampieri et al. 2017). To combat the antibiotic-resistant pathogens, the efforts to characterize the metabolic changes in the infection process can be extended by the structure-based and/or the sequenced-based approaches. Thus, both essential metabolites in the pathogen-specific pathways can be analyzed more efficiently by reducing the adverse effect on the host organism, and the drug candidates can be proposed by structural modeling. In addition, drug targets in common metabolic pathways for different pathogens can be detected in order to tackle a broad range of pathogens.

Alternative methods have been also implemented for system-level studies of metabolism. The concept of "metabolomics" includes identification and quantification of the endogenous and exogenous metabolites to represent the condition-specific profile of a biological system (Zhou et al. 2015). This information serves as a mirror to reflect the small variations between two different states (e.g., infected vs. healthy). This approach is based on the collection of analytical data and multivariate statistical analysis by some approaches (e.g., unsupervised and/or supervised approaches) so as to classify different metabolites/samples in the general sense (Noto et al. 2014). Thus, metabolomics is also a powerful tool for the personalized medicine considering the ability of this method to associate the specific metabolites with specific diseases (Cribbs et al. 2016). Other approaches, ^{13}C-MFA and ^{13}C-metabolic flux ratio analysis, comprise experimental quantification of the intracellular fluxes. Such methods are based on the distribution of the isotope labels through the metabolic network and measurement of isotopic composition of the metabolites by MS or NMR (Kim and Lun 2014; Ng et al. 2016). These methods can provide system-wide quantitative insights into the intracellular metabolic fluxes for the determination of the metabolic profiles of the pathogens. On the other hand, they have some drawbacks such as a limited capacity to measure the fluxes and the need for extensive instrumentation (Kim and Lun 2014). Therefore, novel instrumentation and flux analysis strategies must be introduced to quantify the fluxes experimentally.

Briefly, distinct approaches have been developed in order to elucidate dynamics of the metabolic networks so far. Since all the given approaches have some pros and cons, further research strategies must be employed for the following issues: (1) understanding the impact of the genetic differences/diverse conditions on the pathogen metabolism and resistance/adaptation mechanism, (2) elucidation of the molecular background of the infection strategies, and (3) identification of novel drug targets and prediction of the biomarkers. Further consideration of the biological properties of the pathogens in reconstructing a metabolic model at genome scale and the spread of the comprehensive models taking both the host metabolism and the pathogen metabolism into account will be useful to improve current drug-based searches. Such considerations will contribute to the development of more selective drugs with low side effects in the foreseeable future.

Acknowledgments This work was financially supported by the Turkish Academy of Sciences—Outstanding Young Scientists Award Program (TUBA-GEBIP), and by TUBITAK, The Scientific and Technological Research Council of Turkey (Project Code: 316S005).

References

Abdelrazig S, Ortori CA, Davey G, Deressa W, Mulleta D, Barrett DA, Amberbir A, Fogarty AW (2017) A metabolomic analytical approach permits identification of urinary biomarkers for *Plasmodium falciparum* infection: a case-control study. Malar J 16(1):229. https://doi.org/10.1186/s12936-017-1875-z

Ahn S, Jung J, Jang IA, Madsen EL, Park W (2016) Role of glyoxylate shunt in oxidative stress response. J Biol Chem 291(22):11928–11938. https://doi.org/10.1074/jbc.M115.708149

Ahn YY, Lee DS, Burd H, Blank W, Kapatral V (2014) Metabolic network analysis-based ıdentification of antimicrobial drug targets in category a bioterrorism agents. PLoS One 9(1): e85195. https://doi.org/10.1371/journal.pone.0085195

Atkinson TP, Balish MF, Waites KB (2008) Epidemiology, clinical manifestations, pathogenesis and laboratory detection of *Mycoplasma pneumoniae* infections. FEMS Microbiol Rev 32 (6):956–973. https://doi.org/10.1111/j.1574-6976.2008.00129.x

Aurrecoechea C, Brestelli J, Brunk BP, Dommer J, Fischer S, Gajria B, Gao X, Gingle A, Grant G, Harb OS, Heiges M, Innamorato F, Iodice J, Kissinger JC, Kraemer E, Li W, Miller JA, Nayak V, Pennington C, Pinney DF, Roos DS, Ross C, Stoeckert CJ Jr, Treatman C, Wang H (2009) PlasmoDB: a functional genomic database for malaria parasites. Nucleic Acids Res 37:539–543. https://doi.org/10.1093/nar/gkn814

Banerjee D, Parmar D, Bhattacharya N, Ghanate AD, Panchagnula V, Raghunathan A (2017) A scalable metabolite supplementation strategy against antibiotic resistant pathogen *Chromobacterium violaceum* induced by $NAD^+/NADH^+$ imbalance. BMC Syst Biol 11 (1):51. https://doi.org/10.1186/s12918-017-0427-z

Banoei MM, Vogel HJ, Weljie AM, Kumar A, Yende S, Angus DC, Winston BW, Canadian Critical Care Translational Biology Group (CCCTBG) (2017) Plasma metabolomics for the diagnosis and prognosis of H1N1 influenza pneumonia. Crit Care 21(1):97. https://doi.org/10.1186/s13054-017-1672-7

Bartell JA, Blazier AS, Yen P, Thøgersen JC, Jelsbak L, Goldberg JB, Papin JA (2017) Reconstruction of the metabolic network of *Pseudomonas aeruginosa* to interrogate virulence factor synthesis. Nat Commun 8:14631. https://doi.org/10.1038/ncomms14631

Bäumler A, Fang FC (2013) Host specificity of bacterial pathogens. Cold Spring Harb Perspect Med 3(12):a010041

Bäumler A, Sperandio V (2016) Interactions between the microbiota and pathogenic bacteria in the gut. Nature 535(7610):85–93. https://doi.org/10.1038/nature18849

Bazzani S, Hoppe A, Holzhütter H (2012) Network-based assessment of the selectivity of metabolic drug targets in *Plasmodium falciparum* with respect to human liver metabolism. BMC Syst Biol 6:118

Becker SA, Palsson BO (2008) Context-specific metabolic networks are consistent with experiments. PLoS Comput Biol 4(5):e1000082. https://doi.org/10.1371/journal.pcbi.1000082

Becker SA, Palsson BØ (2005) Genome-scale reconstruction of the metabolic network in *Staphylococcus aureus* N315: an initial draft to the two-dimensional annotation. BMC Microbiol 5:8. https://doi.org/10.1186/1471-2180-5-8

Bendouah Z, Barbeau J, Hamad WA, Desrosiers M (2006) Biofilm formation by *Staphylococcus aureus* and *Pseudomonas aeruginosa* is associated with an unfavorable evolution after surgery for chronic sinusitis and nasal polyposis. Otolaryngol Head Neck Surg 134 (6):991–996. https://doi.org/10.1016/j.otohns.2006.03.001

Berger A, Dohnt K, Tielen P, Jahn D, Becker J, Wittmann C (2014) Robustness and plasticity of metabolic pathway flux among uropathogenic isolates of *Pseudomonas aeruginosa*. PLoS One 9(4):e88368. https://doi.org/10.1371/journal.pone.0088368

Bergkessel M, Basta DW, Newman DK (2016) The physiology of growth arrest: uniting molecular and environmental microbiology. Nat Rev Microbiol 14(9):549–562. https://doi.org/10.1038/nrmicro.2016.107

Beste D, Nöh K, Niedenführ S, Mendum T, Hawkins N, Ward J, Beale M, Wiechert W, McFadden J (2013) ^{13}C-flux spectral analysis of host-pathogen metabolism reveals a mixed diet for intracellular *Mycobacterium tuberculosis*. Chem Biol 20(8):1012–1021

Bhagirath AY, Li Y, Somayajula D, Dadashi M, Badr S, Duan K (2016) Cystic fibrosis lung environment and *Pseudomonas aeruginosa* infection. BMC Pulm Med 16(1):174. https://doi.org/10.1186/s12890-016-0339-5

Bordbar A, Lewis NE, Schellenberger J, Palsson BØ, Jamshidi N (2010) Insight into human alveolar macrophage and *M. tuberculosis* interactions via metabolic reconstructions. Mol Syst Biol 6:422. https://doi.org/10.1038/msb.2010.68

Bosi E, Monk JM, Aziz RK, Fondi M, Nizet V, Palsson BØ (2016) Comparative genome-scale modelling of *Staphylococcus aureus* strains identifies strain-specificmetabolic capabilities linked to pathogenicity. Proc Natl Acad Sci USA 113(26):3801–3809. https://doi.org/10.1073/pnas.1523199113

Bouatra S, Aziat F, Mandal R, Guo AC, Wilson MR, Knox C, Bjorndahl TC, Krishnamurthy R, Saleem F, Liu P, Dame ZT, Poelzer J, Huynh J, Yallou FS, Psychogios N, Dong E, Bogumil R, Roehring C, Wishart DS (2013) The human urine metabolome. PLoS One 8(9). https://doi.org/10.1371/journal.pone.0073076

Bozzetto S, Pirillo P, Carraro S, Berardi M, Cesca L, Stocchero M, Giordano G, Zanconato S, Baraldi E (2017) Metabolomic profile of children with recurrent respiratory infections. Pharmacol Res 115:162–167

Butt AM, Nasrullah I, Tahir S, Tong Y (2012) Comparative genomics analysis of *Mycobacterium ulcerans* for the identification of putative essential genes and therapeutic candidates. PLoS One 7(8). https://doi.org/10.1371/journal.pone.0043080

Cañigral I, Moreno Y, Alonso J, González A, Ferrús M (2010) Detection of *Vibrio vulnificus* in seafood, seawater and wastewater samples from a Mediterranean coastal area. Microbiol Res 165(8):657–664. https://doi.org/10.1016/j.micres.2009.11.012

Caspi R, Billington R, Ferrer L, Foerster H, Fulcher CA, Keseler IM, Kothari A, Krummenacker M, Latendresse M, Mueller A, Ong Q, Paley S, Subhraveti P, Weaver DS, Karp D (2016) The MetaCyc database of metabolic pathways and enzymes and the BioCyc collection of pathway/genome databases. Nucleic Acids Res 44(1):471–480. https://doi.org/10.1093/nar/gkv1164

Cassman M, Arkin A, Doyle F, Katagiri F, Lauffenburger D, Stokes C (2007) Systems biology: international research and development. Springer Science & Business Media, Dordrecht

Chandrasekaran S, Price ND (2010) Probabilistic integrative modeling of genome-scale metabolic and regulatory networks in *Escherichia coli* and *Mycobacterium tuberculosis*. Proc Natl Acad Sci USA 107(41):17845–17850. https://doi.org/10.1073/pnas.1005139107

Chang A, Scheer M, Grote A, Schomburg I (2009) BRENDA, AMENDA and FRENDA the enzyme information system: new content and tools in 2009. Nucleic Acids Res 37:588–592. https://doi.org/10.1093/nar/gkn820

Chaves-Moreno D, Wos-oxley ML, Jáuregui R, Medina E, Oxley AP (2015) Application of a novel "pan-genome"-based strategy for assigning rnaseq transcript reads to *Staphylococcus aureus* strains. PLoS One 10(12):e0145861. https://doi.org/10.1371/journal.pone.0145861

Cheung YW, Tanner JA (2011) Targeting glutamate synthase for tuberculosis drug development. Hong Kong Med J 17(Suppl 2):32–34

Choi SY, Yoon K, Lee JI, Mitchell RJ (2015) Violacein: properties and production of a versatile bacterial pigment. Biomed Res Int 2015:465056. https://doi.org/10.1155/2015/465056

Cloete R, Oppon E, Murungi E, Schubert W, Christoffels A (2016) Resistance related metabolic pathways for drug target identification in *Mycobacterium tuberculosis*. BMC Bioinformatics 17 (75). https://doi.org/10.1186/s12859-016-0898-8

Conlan S, Mijares LA, Comparative N, Program S, Becker J, Blakesley RW, Bouffard GG, Brooks S, Coleman H, Gupta J, Gurson N, Park M, Schmidt B, Thomas PJ, Otto M, Kong HH, Murray PR, Segre JA (2012) *Staphylococcus epidermidis* pan-genome sequence analysis reveals diversity of skin commensaland hospital infection-associated isolates. Genome Biol 13 (7):R64. https://doi.org/10.1186/gb-2012-13-7-r64

Creczynski-pasa TB, Antônio RV (2004) Energetic metabolism of *Chromobacterium violaceum*. Genet Mol Res 3(1):162–166

Cribbs SK, Uppal K, Li S, Jones DP, Huang L, Tipton L, Fitch A, Greenblatt RM, Kingsley L, Guidot DM, Ghedin E, Morris A (2016) Correlation of the lung microbiota with metabolic profiles in bronchoalveolar lavage fluid in HIV infection. Microbiome 4:3. https://doi.org/10.1186/s40168-016-0147-4

Cui L, Hou J, Fang J, Lee Y, Costa V, Won L, Chen Q, Ooi E, Tannenbaum S, Chen J, Ong C (2017) Serum metabolomics ınvestigation of humanized mouse model of dengue virus infection. J Virol 91(14):e00386-1. https://doi.org/10.1128/JVI.00386-17

Dostálová A, Volf P (2012) Leishmania development in sand flies: parasite-vector interactions overview. Parasit Vectors 5:276

Doyle MA, Macrae JI, De Souza DP, Saunders EC, Mcconville MJ, Likić VA (2009) LeishCyc: a biochemical pathways database for Leishmania major. BMC Syst Biol 3:57. https://doi.org/10.1186/1752-0509-3-57

Durmus S, Çakir T, Özgür A, Guthke R (2015) A review on computational systems biology of pathogen-host interactions. Front Microbiol 6:235. https://doi.org/10.3389/fmicb.2015.00235

Dutow P, Schmidl SR, Ridderbusch M, Stülke J (2010) Interactions between glycolytic enzymes of *Mycoplasma pneumoniae*. J Mol Microbiol Biotechnol 19(3):134–139. https://doi.org/10.1159/000321499

Ehlers S, Schaible UE (2013) The granuloma in tuberculosis: dynamics of a host-pathogen collusion. Front Immunol 3:411. https://doi.org/10.3389/fimmu.2012.00411

Fadiel A, Eichenbaum KD, El Semary N, Epperson B (2007) Mycoplasma genomics: tailoring the genome for minimal life requirements through reductive evolution. Front Biosci 12:2020–2028. https://doi.org/10.2741/2207

Fattuoni C, Palmas F, Noto A, Barberini L, Mussap M, Grapov D, Dessì A, Casu M, Casanova A, Furione M, Arossa A, Spinillo A, Baldanti F, Fanos V, Zavattoni M (2016) Primary HCMV infection in pregnancy from classic data towards metabolomics: an exploratory analysis. Clin Chim Acta 460:23–32

Ferrarini MG, Siqueira FM, Mucha SG, Palama TL, Jobard É, Elena-herrmann B, Vasconcelos ATR, Tardy F, Schrank IS, Zaha A, Sagot MF (2016) Insights on the virulence of swine respiratory tract mycoplasmas through genome-scale metabolic modeling. BMC Genomics 17:353. https://doi.org/10.1186/s12864-016-2644-z

Francke C, Siezen RJ, Teusink B (2005) Reconstructing the metabolic network of a bacterium from its genome. Trends Microbiol 13(11):550–558. https://doi.org/10.1016/j.tim.2005.09.001

Frank KL, Colomer-winter C, Grindle SM, Lemos JA, Schlievert PM, Dunny GM (2014) Transcriptome analysis of *Enterococcus faecalis* during mammalian ınfection shows cells undergo adaptation and exist in a stringent response state. PLoS One 9(12):e115839. https://doi.org/10.1371/journal.pone.0115839

Gal-mor O, Boyle EC, Grassl GA (2014) Same species, different diseases: how and why typhoidal and non-typhoidal *Salmonella enterica* serovars differ. Front Microbiol 5:391. https://doi.org/10.3389/fmicb.2014.00391

Gaurav K, Hasija Y (2017) Comparative analysis of metabolic network of pathogens. Front Biol 12 (2):139–150. https://doi.org/10.1007/s11515-017-1440-8

Gengenbacher M, Kaufmann S (2012) *Mycobacterium tuberculosis*: success through dormancy. FEMS Microbiol Rev 36(3):514–532. https://doi.org/10.1111/j.1574-6976.2012.00331.x. Mycobacterium

Gerdes SY, Scholle MD, Souza MD, Bernal A, Baev MV, Farrell M, Kurnasov OV, Daugherty MD, Mseeh F, Polanuyer BM, Campbell JW, Anantha S, Shatalin KY, Chowdhury SAK, Fonstein MY, Osterman AL (2002) From genetic footprinting to antimicrobial drug targets: examples in cofactor biosynthetic pathways. J Bacteriol 184(16):4555–4572. https://doi.org/10.1128/JB.184.16.4555

Ginsburg H (2006) Progress in in silico functional genomics: the malaria metabolic pathways database. Trends Parasitol 22(6):238–240

Goldman E, Green L (2015) Practical handbook of microbiology. CRC Press, Boca Raton, FL

Goto S, Okuno Y, Hattori M, Nishioka T, Kanehisa M (2002) LIGAND: database of chemical compounds and reactions in biological pathways. Nucleic Acids Res 30(1):402–404. https://doi.org/10.1093/nar/30.1.402

Götz A, Eylert E, Eisenreich W, Goebel W (2010) Carbon metabolism of *Enterobacterial* human pathogens growing in epithelial colorectal adenocarcinoma (Caco-2) cells. PLoS One 5(5): e10586. https://doi.org/10.1371/journal.pone.0010586

Großeholz R, Koh C, Veith N, Fiedler T, Strauss M, Olivier B, Collins BC, Schubert OT, Bergmann F, Kreikemeyer B, Aebersold R, Kummer U (2016) Integrating highly quantitative proteomics and genome-scale metabolic modeling to study pH adaptation in the human pathogen *Enterococcus faecalis*. NPJ Syst Biol Appl 2:16017. https://doi.org/10.1038/npjsba.2016.17

Großhennig S, Schmidl SR, Schmeisky G, Busse J, Stülke J (2013) Implication of glycerol and phospholipid transporters in mycoplasma pneumoniae growth and virulence. Infect Immun 81 (3):896–904. https://doi.org/10.1128/IAI.01212-12

Guimarães LC, Florczak-Wyspianska J, de Jesus LB, Viana MV, Silva A, Ramos RT, de Carvalho Azevedode Castro Soares S, de Carvalho Azevedo VA (2015) Inside the pan-genome - methods and software overview. Curr Genomics 16(4):245–252

Gupta M, Prasad Y, Sharma SK, Kumar C (2017) Identification of phosphoribosyl-amp cyclohydrolase, as drug target and its inhibitors in *Brucella melitensis* bv. 1 16M using metabolic pathway analysis. J Biomol Struct Dyn 35(2):287–299. https://doi.org/10.1080/07391102.2015.1137229

Hadizadeh M, Tabatabaiepour SN, Tabatabaiepour SZ, Hosseini Nave H, Mohammadi M, Sohrabi S (2017) Genome-wide Identification of potential drug target in *Enterobacteriaceae* family: a homology-based method. Microb Drug Resist. https://doi.org/10.1089/mdr.2016.0259

Hartman HB, Fell DA, Rossell S, Jensen PR, Woodward MJ, Thorndahl L, Jelsbak L, Olsen JE, Raghunathan A, Daefler S, Poolman MG (2014) Identification of potential drug targets in *Salmonella enterica* sv. Typhimurium using metabolic modelling and experimental validation. Microbiology 160(Pt 6):1252–1266. https://doi.org/10.1099/mic.0.076091-0

Harvey R, Champe P, Fisher B (2007) Lippincotts illustrated reviews: microbiology. Lippincott Williams & Wilkins, Philadelphia, PA

Henriksen S, Liu J, Estiu G, Oltvai Z, Wiest O (2010) Identification of novel bacterial histidine biosynthesis inhibitors using docking, ensemble rescoring, and whole-cell assays. Bioorg Med Chem 18(14):5148–5156. https://doi.org/10.1016/j.bmc.2010.05.060.Identification

Henry CS, Dejongh M, Best AA, Frybarger PM, Linsay B, Stevens RL (2010) High-throughput generation, optimization and analysis of genome-scale metabolic models. Nat Biotechnol 28 (9):977–982. https://doi.org/10.1038/nbt.1672

Horai H, Arita M, Kanaya S, Nihei Y, Ikeda T, Suwa K, Ojima Y, Tanaka K, Tanaka S, Aoshima K, Oda Y, Kakazu Y, Kusano M, Tohge T, Matsuda F, Sawada Y, Hirai M, Nakanishi H, Ikeda K, Akimoto N, Maoka T, Takahash H, Ara T, Sakurai N, Suzuki H, Shibata D, Neumann S, Iida T, Tanaka K, Funatsu K, Matsuura F, Soga T, Taguchi R, Saito K, Nishioka T (2010) MassBank: a public repository for sharing mass spectral data for life sciences. J Mass Spectrom 45 (7):703–714. https://doi.org/10.1002/jms.1777

Huthmacher C, Hoppe A, Bulik S, Holzhütter H (2010) Antimalarial drug targets in *Plasmodium falciparum* predicted by stage-specific metabolic network analysis. BMC Syst Biol 4:120

Hwang M, Damte D, Cho M, Kim Y, Park S (2010) Optimization of culture media of pathogenic *Mycoplasma hyopneumoniae* by a response surface methodology. J Vet Sci 11(4):327–332. https://doi.org/10.4142/jvs.2010.11.4.327

Irwin JJ, Shoichet BK, Mysinger MM, Huang N, Colizzi F, Wassam P, Cao Y (2009) Automated docking screens: a feasibility study. J Med Chem 52(18):5712–5720. https://doi.org/10.1021/jm9006966

Jamshidi N, Palsson BØ (2007) Investigating the metabolic capabilities of *Mycobacterium tuberculosis* H37Rv using the in silico strain iNJ661 and proposing alternative drug targets. BMC Syst Biol 1(26). https://doi.org/10.1186/1752-0509-1-26

Jensen PA, Papin JA (2011) Functional integration of a metabolic network model and expression data without arbitrary thresholding. Bioinformatics 27(4):541–547. https://doi.org/10.1093/bioinformatics/btq702

Jerby L, Ruppin E (2012) Predicting drug targets and biomarkers of cancer via genome-scale metabolic modeling. Clin Cancer Res 18(20):5572–5584. https://doi.org/10.1158/1078-0432.CCR-12-1856

Jones MK, Oliver JD (2009) *Vibrio vulnificus*: disease and pathogenesis. Infect Immun 77 (5):1723–1733. https://doi.org/10.1128/IAI.01046-08

Josling GA, Llinás M (2015) Sexual development in Plasmodium parasites: knowing when its time to commit. Nat Rev Microbiol 13(9):573–587. https://doi.org/10.1038/nrmicro3519

Kahlon R (2016) Pseudomonas: molecular and applied biology. Springer, Switzerland, pp 82–83

Kajihara T, Nakamura S, Iwanaga N, Oshima K, Takazono T, Miyazaki T, Izumikawa K, Yanagihara K, Kohno N, Kohno S (2015) Clinical characteristics and risk factors of enterococcal infections in Nagasaki, Japan: a retrospective study. BMC Infect Dis 15:426. https://doi.org/10.1186/s12879-015-1175-6

Kaltdorf M, Srivastava M, Gupta SK, Liang C, Krappmann S, Dandekar T (2016) Systematic identification of anti-fungal drug targets by a metabolic network approach. Front Mol Biosci 3:22. https://doi.org/10.3389/fmolb.2016.00022

Kamminga T, Slagman S, Bijlsma J, Martins Dos Santos V, Suarez-Diez M, Schaap P (2017) Metabolic modeling of energy balances in *Mycoplasma hyopneumoniae* shows that pyruvate addition increases growth rate. Biotechnol Bioeng 114(10):2339–2347. https://doi.org/10.1002/bit.26347

Kanehisa M, Araki M, Goto S, Hattori M, Hirakawa M, Itoh M, Katayama T, Kawashima S, Okuda S, Tokimatsu T, Yamanishi Y (2008) KEGG for linking genomes to life and the environment. Nucleic Acids Res 36:480–484. https://doi.org/10.1093/nar/gkm882

Kao D, Ismond KP, Tso V, Millan B, Hotte N, Fedorak RN (2016) Urine-based metabolomic analysis of patients with *Clostridium difficile* infection: a pilot study. Metabolomics 12:135. https://doi.org/10.1007/s11306-016-1080-z

Karp PD, Ouzounis CA, Moore-Kochlacs C, Goldovsky L, Kaipa P, Ahrén D, Tsoka S, Darzentas N, Kunin V, López-Bigas N (2005) Expansion of the BioCyc collection of pathway/genome databases to 160 genomes. Nucleic Acids Res 33(19):6083–6089. https://doi.org/10.1093/nar/gki892

Kauffman KJ, Prakash P, Edwards JS (2003) Advances in flux balance analysis. Curr Opin Biotechnol 14(5):491–496. https://doi.org/10.1016/j.copbio.2003.08.001

Kim HU, Kim T, Lee YS (2010) Genome-scale metabolic network analysis and drug targeting of multi-drug resistant pathogen *Acinetobacter baumannii* AYE. Mol Biosyst 6(2):339–348. https://doi.org/10.1039/b916446d

Kim HU, Kim SY, Jeong H, Kim TY, Kim JJ, Choy HE, Yi KY, Rhee JH, Lee SY (2011) Integrative genome-scale metabolic analysis of *Vibrio vulnificus* for drug targeting and discovery. Mol Syst Biol 7:460. https://doi.org/10.1038/msb.2010.115

Kim M, Lun D (2014) Methods for integration of transcriptomic data in genome-scale metabolic models. Comput Struct Biotechnol J 11(18):59–65. https://doi.org/10.1016/j.csbj.2014.08.009

Kim Y, Schmidt B, Kidwai A, Jones M, Deatherage Kaise B, Brewer H, Mitchell H, Palsson B, McDermott J, Heffron F, Smith R, Peterson S, Ansong C, Hyduk D, Metz T, Adkins J (2013) *Salmonella* modulates metabolism during growth under conditions that induce expression of virulence genes. Mol Biosyst 9(6):1522–1534. https://doi.org/10.1039/c3mb25598k

King ZA, Lu J, Miller P, Federowicz S, Lerman JA, Ebrahim A, Palsson BO, Lewis NE (2016) BiGG models: a platform for integrating, standardizing and sharing genome-scale models. Nucleic Acids Res 44:515–522. https://doi.org/10.1093/nar/gkv1049

Kitano H (2002) Computational systems biology. Nature 420(6912):206–210

Klamt S, Gilles ED (2004) Minimal cut sets in biochemical reaction networks. Bioinformatics 20 (2):226–234. https://doi.org/10.1093/bioinformatics/btg395

Knox C, Law V, Jewison T, Liu P, Ly S, Frolkis A, Pon A, Banco K, Mak C, Neveu V, Djoumbou Y, Eisner R, Guo AC, Wishart DS (2011) DrugBank 3.0: a comprehensive resource for 'omics' research on drugs. Nucleic Acids Res 39:1035–1041. https://doi.org/10.1093/nar/gkq1126

Kumar A, Suthers PF, Maranas CD (2012) MetRxn: a knowledgebase of metabolites and reactions spanning metabolic models and databases. BMC Bioinformatics 13:6. https://doi.org/10.1186/1471-2105-13-6

Landeck L, Kneip C, Reischl J, Asadullah K (2016) Biomarkers and personalized medicine: current status and further perspectives with special focus on dermatology. Exp Dermatol 25 (5):333–339. https://doi.org/10.1111/exd.12948

Larocque M, Chénard T, Najmanovich R (2014) A curated *C. difficile* strain 630 metabolic network: prediction of essential targets and inhibitors. BMC Syst Biol 8:117

Le Carrou J, Laurentie M, Kobisch M, Gautier-Bouchardon AV (2006) Persistence of *Mycoplasma hyopneumoniae* in experimentally infected pigs after marbofloxacin treatment and detection of mutations in the *parC* gene. Antimicrob Agents Chemother 50(6):1959–1966. https://doi.org/10.1128/AAC.01527-05

Le Novère N, Bornstein B, Broicher A, Courtot M, Donizelli M, Dharuri H, Li L, Sauro H, Schilstra M, Shapiro B, Snoep J, Hucka M (2006) BioModels database: a free, centralized database of curated, published, quantitative kinetic models of biochemical and cellular systems. Nucleic Acids Res 34(Database issue):D689–D691. https://doi.org/10.1093/nar/gkj092

Lee DS, Burd H, Liu J, Almaas E, Wiest O, Barabási AL, Oltvai ZN, Kapatral V (2009) Comparative genome-scale metabolic reconstruction and flux balance analysis of multiple *Staphylococcus aureus* genomes identify novel antimicrobial drug targets. J Bacteriol 191 (12):4015–4024. https://doi.org/10.1128/JB.01743-08

Lewis NE, Nagarajan H, Palsson BO (2012) Constraining the metabolic genotype-phenotype relationship using a phylogeny of in silico methods. Nat Rev Microbiol 10(4):291–305. https://doi.org/10.1038/nrmicro2737

Li X, Mishra SK, Wu M, Zhang F, Zheng J (2014) Syn-lethality: an integrative knowledge base of synthetic lethality towards discovery of selective anticancer therapies. Biomed Res Int 2014:19603. https://doi.org/10.1155/2014/196034

Liao Y, Huang T, Chen F, Charusanti P, Hong JSJ, Chang H, Tsai S, Palsson BO, Hsiung CA (2011) An experimentally validated genome-scale metabolic reconstruction of *Klebsiella pneumoniae* MGH 78578, iYL1228. J Bacteriol 193(7):1710–1717. https://doi.org/10.1128/JB.01218-10

Lin H, Ding H, Guo FB, Huang J (2010) Prediction of subcellular location of mycobacterial protein using feature selection techniques. Mol Divers 14(4):667–671. https://doi.org/10.1007/s11030-009-9205-1

Liu A, Archer AM, Biggs MB, Papin JA (2017) Growth-altering microbial interactions are responsive to chemical context. PLoS One 12(3):e0164919

Lunardi J, Nunes J, Bizarro C, Basso L, Santos D, Machado P (2013) Targeting the histidine pathway in *Mycobacterium tuberculosis*. Curr Top Med Chem 13(22):2866–2884. https://doi.org/10.2174/15680266113136660203

Ma S, Minch KJ, Rustad TR, Hobbs S, Zhou SL, Sherman DR, Price ND (2015) Integrated modeling of gene regulatory and metabolic networks in *Mycobacterium tuberculosis*. PLoS Comput Biol 11(11):e1004543. https://doi.org/10.1371/journal.pcbi.1004543

Matthews L, Gopinath G, Gillespie M, Caudy M, Croft D, De Bono B, Garapati P, Hemish J, Hermjakob H, Jassal B, Kanapin A, Lewis S, Mahajan S, May B, Schmidt E, Vastrik I, Wu G, Birney E, Stein L, D'Eustachio P (2009) Reactome knowledgebase of human biological pathways and processes. Nucleic Acids Res 37:619–622. https://doi.org/10.1093/nar/gkn863

Mehla K, Ramana J (2015) Novel drug targets for food-borne pathogen *Campylobacter jejuni*: an Integrated subtractive genomics and comparative metabolic pathway study. OMICS 19 (7):393–406

Mehla K, Ramana J (2017) Tapping into *Salmonella typhimurium* LT2 genome in a quest to explore its therapeutic arsenal: a metabolic network modeling approach. Biomed Pharmacother 86:57–66. https://doi.org/10.1016/j.biopha.2016.11.129

Mortensen N, Mercier K, Mcritchie S, Cavallo T, Pathmasiri W, Stewart D, Sumner SJ (2016) Microfluidics meets metabolomics to reveal the impact of *Campylobacter jejuni* infection on biochemical pathways. Biomed Microdevices 18(3):51. https://doi.org/10.1007/s10544-016-0076-9

Murphy DJ, Brown JR (2007) Identification of gene targets against dormant phase *Mycobacterium tuberculosis* infections. BMC Infect Dis 7:84. https://doi.org/10.1186/1471-2334-7-84

Ng M, Saunders E, Olshansky M, McConville M, Likic V (2016) ^{13}C metabolic flux ratio analysis by direct measurement of free metabolic intermediates in *L. mexicana* using gas chromatography-mass spectrometry. Bio Rxiv:73676. https://doi.org/10.1101/073676

Nikel PI, Chavarría M, Fuhrer T, Sauer U, De Lorenzo V (2015) *Pseudomonas putida* KT2440 strain metabolizes glucose through a cycle formed by enzymes of the entner- doudoroff, embden-meyerhof-parnas, and pentose phosphate pathways. J Biol Chem 290 (43):25920–25932. https://doi.org/10.1074/jbc.M115.687749

Nishiumi S, Yoshida M, Azuma T (2017) Alterations in metabolic pathways in stomach of mice infected with *Helicobacter pylori*. Microb Pathog 109:78–85

Noto A, Dessi A, Puddu M, Mussap M, Fanos V (2014) Metabolomics technology and their application to the study of the viral infection. J Matern Fetal Neonatal Med 27(Suppl 2):53–57. https://doi.org/10.3109/14767058.2014.955963

Oberhardt M, Puchałka J, Fryer K, Martins dos Santos V, Papin J (2008) Genome-scale metabolic network analysis of the opportunistic pathogen *Pseudomonas aeruginosa* PAO1. J Bacteriol 190(8):2790–2803. https://doi.org/10.1128/JB.01583-07

Oberhardt M, Goldberg J, Hogardt M, Papin J (2010) Metabolic network analysis of *Pseudomonas aeruginosa* during chronic cystic fibrosis lung ınfection. J Bacterio 192(20):5534–5548. https://doi.org/10.1128/JB.00900-10

Opperman M, Shachar-Hill Y (2016) Metabolic flux analyses of *Pseudomonas aeruginosa* cystic fibrosis isolates. Metab Eng 38:251–263. https://doi.org/10.1016/j.ymben.2016.09.002

Otto M (2010) *Staphylococcus* colonization of the skin and antimicrobial peptides. Expert Rev Dermatol 5(2):183–195. https://doi.org/10.1586/edm.10.6.Staphylococcus

Overbeek R, Larsen N, Walunas T, Souza M, Pusch G, Selkov E Jr, Liolios K, Joukov V, Kaznadzey D, Anderson I, Bhattacharyya A, Burd H, Gardner W, Hanke P, Kapatral V, Mikhailova N, Vasieva O, Osterman A, Vonstein V, Fonstein M, Ivanova N, Kyrpides N

(2003) The ERGOTM genome analysis and discovery system. Nucleic Acids Res 31 (1):164–171. https://doi.org/10.1093/nar/gkg148

Phalak P, Chen J, Carlson R, Henson M (2016) Metabolic modeling of a chronic wound biofilm consortium predicts spatial partitioning of bacterial species. BMC Syst Biol 10(1):90. https://doi.org/10.1186/s12918-016-0334-8

Pienaar E, Matern WM, Linderman JJ, Bader JS, Kirschner DE (2016) Multiscale model of *Mycobacterium tuberculosis* infection maps metabolite and gene perturbations to granuloma sterilization. Infect Immun 84(5):1650–1669. https://doi.org/10.1128/IAI.01438-15.Editor

Pieper U, Webb BM, Barkan DT, Schneidman-duhovny D, Schlessinger A, Braberg H, Yang Z, Meng EC, Pettersen EF, Huang CC, Datta RS, Sampathkumar P, Ferrin TE, Burley SK, Madhusudhan MS, Sjölander K, Ferrin TE, Burley SK, Sali A (2011) ModBase, a database of annotated comparative protein structure models, and associated resources. Nucleic Acids Res 39 (Database issue):D465–D474. https://doi.org/10.1093/nar/gkq1091

Plata G, Hsiao T-L, Olszewski KL, Llinás M, Vitkup D (2010) Reconstruction and flux-balance analysis of the *Plasmodium falciparum* metabolic network. Mol Syst Biol 6:408. https://doi.org/10.1038/msb.2010.60

Quek L-E, Wittmann C, Nielsen LK, Krömer JO (2009) OpenFLUX: efficient modelling software for ^{13}C-based metabolic flux analysis. Microb Cell Fact 8:25. https://doi.org/10.1186/1475-2859-8-25

Raghunathan A, Reed J, Shin S, Palsson B, Daefler S (2009) Constraint-based analysis of metabolic capacity of *Salmonella typhimurium* during host-pathogen interaction. BMC Syst Biol 3:38. https://doi.org/10.1186/1752-0509-3-38

Rajendran R, May A, Sherry L, Kean R, Williams C, Jones BL, Burgess KV, Heringa J, Abeln S, Brandt BW, Munro CA, Ramage G (2016) Integrating *Candida albicans* metabolism with biofilm heterogeneity by transcriptome mapping. Sci Rep 6:35436. https://doi.org/10.1038/srep35436

Reed JL, Vo TD, Schilling CH, Palsson BO (2003) An expanded genome-scale model of *Escherichia coli* K-12 (iJR904 GSM/GPR). Genome Biol 4(9):R54

Reid AJ, Vermont SJ, Cotton JA, Harris D, Hill-Cawthorne GA, Latham SM, Mourier T, Norton R, Quail MA, Sanders M, Shanmugam D, Sohal A, Wasmuth JD, Brunk B, Grigg ME, Howard JC, Parkinson J, Roos DS, Trees AJ, Berriman M, Pain A, Wastling JM (2012) Comparative genomics of the apicomplexan parasites *Toxoplasma gondii* and *Neospora caninum*: Coccidia differing in host range and transmission strategy. PLoS Pathog 8(3):e1002567. https://doi.org/10.1371/journal.ppat.1002567

Ren Q, Chen K, Paulsen IT (2007) TransportDB: a comprehensive database resource for cytoplasmic membrane transport systems and outer membrane channels. Nucleic Acids Res 35(Database issue):D274–D279. https://doi.org/10.1093/nar/gkl925

Reynolds L, Redpath S, Yurist-Doutsch S, Gill N, Brown E, Van Der Heijden J, Brosschot TP, Han J, Marshall N, Woodward S, Valdez Y, Borchers C, Perona-Wright G, Finlay B (2017) Enteric helminths promote *Salmonella* coinfection by altering the intestinal metabolome. J Infect Dis 215(8):1245–1254

Richardson A, Somerville G, Sonenshein A (2015) Regulating the ıntersection of metabolism and pathogenesis in Gram-positive bacteria. Microbiol Spectr 3(3). https://doi.org/10.1128/microbiolspec.MBP-0004-2014.Regulating

Riley CP, Gough ES, He J, Jandhyala SS, Kennedy B, Orcun S, Ouzzani M, Buck C, Roumani AM, Zhang X (2010) The proteome discovery pipeline – a data analysis pipeline for mass spectrometry-based differential proteomics discovery. Open Proteomics J 3:8–19

Sachs G, Keeling D (2001) Ion motive ATPases: V- and P-type ATPases. eLS, Wiley.

Saier MH, Reddy VS, Tsu BV, Ahmed MS, Li C, Moreno-hagelsieb G (2016) The Transporter Classification Database (TCDB): recent advances. Nucleic Acids Res 44(Database issue): D372–D379. https://doi.org/10.1093/nar/gkv1103

Samant S, Lee H, Ghassemi M, Chen J, Cook JL, Mankin AS, Neyfakh AA (2008) Nucleotide biosynthesis is critical for growth of bacteria in human blood. PLoS Pathog 4(2):e37. https://doi.org/10.1371/journal.ppat.0040037

Sasse A, Hamer S, Amich J, Binder J, Krappmann S (2016) Mutant characterization and in vivo conditional repression identify aromatic amino acid biosynthesis to be essential for *Aspergillus fumigatus* virulence. Virulence 7(1):56–62. https://doi.org/10.1080/21505594.2015.1109766

Sassetti CM, Boyd DH, Rubin EJ (2003) Genes required for mycobacterial growth defined by high density mutagenesis. Mol Microbiol 48(1):77–84

Scarborough JA, Paul J, Spencer J (2017) Evolution of the ability to modulate host chemokine networks via gene duplication in human cytomegalovirus (HCMV). Infect Genet Evol 51:46–53. https://doi.org/10.1016/j.meegid.2017.03.013

Seekatz A, Casey M, Krishna R, Yu-Ming C, Alison E, John Y, Vincent B, Freeman E, Kao JY, Young VB (2017) Inter-individual recovery of the microbiota and metabolome over time following fecal microbiota transplantation ın patients with recurrent *Clostridium difficile* infection. bioRxiv. https://doi.org/10.1101/141846

Segrè D, Vitkup D, Church GM (2002) Analysis of optimality in natural and perturbed. Proc Natl Acad Sci USA 99(23):15112–15117

Sharma M, Shaikh N, Yadav S, Singh S, Garg P (2017) A systematic reconstruction and constraint-based analysis of *Leishmania donovani* metabolic network: identification of potential antileishmanial drug targets. Mol Biosyst 13(5):955–969. https://doi.org/10.1039/C6MB00823B

Shen Y, Liu J, Estiu G, Isin B, Ahn Y, Lee D, Barabási A, Kapatral V, Wiest O, Oltvai Z (2010) Blueprint for antimicrobial hit discovery targeting metabolic networks. Proc Natl Acad Sci USA 107(3):1082–1087. https://doi.org/10.1073/pnas.0909181107

Shreiner A, Kao J, Young V (2015) The gut microbiome in health and in disease. Curr Opin Gastroenterol 31(1):69–75. https://doi.org/10.1097/MOG.0000000000000139.The

Shrinet J, Shastri J, Gaind R, Bhavesh N, Sunil S (2016) Serum metabolomics analysis of patients with chikungunya and dengue mono/co-infections reveals distinct metabolite signatures in the three disease conditions. Sci Rep 6:36833. https://doi.org/10.1038/srep36833

Silva Miranda M, Breiman A, Allain S, Deknuydt F, Altare F (2012) The tuberculous granuloma: an unsuccessful host defence mechanism providing a safety shelter for the bacteria? Clin Dev Immunol 2012:13912. https://doi.org/10.1155/2012/139127

Simeonidis E, Chandrasekaran S, Price N (2013) A guide to integrating transcriptional regulatory and metabolic networks using PROM (probabilistic regulation of metabolism). Methods Mol Biol:103–112. https://doi.org/10.1007/978-1-62703-299-5

Singh S, Singh D, Singh A, Gautam B, Ram G, Dwivedi S, Ramteke P (2016) An approach for identification of novel drug targets in Streptococcus pyogenes SF370 through pathway analysis. Interdiscip Sci 8(4):388–394

Siqueira F, Thompson C, Virginio V, Gonchoroski T, Reolo L, Almeida L, da Fonsêca M, de Souza R, Prosdocimi F, Schrank I, Ferreira H, de Vasconcelos A, Zaha A (2013) New insights on the biology of swine respiratory tract mycoplasmas from a comparative genome analysis. BMC Genomics 14:175

Siqueira F, Gerber A, Guedes R, Almeida L, Schrank I, Vasconcelos A, Zaha A (2014) Unravelling the transcriptome profile of the Swine respiratory tract mycoplasmas. PLoS One 9(10):e110327. https://doi.org/10.1371/journal.pone.0110327

Smith C, OMaille G, Want E, Qin C, Trauger S, Brandon T, Custodio D, Abagyan R, Siuzdak G (2005) METLIN: a metabolite mass spectral database. Ther Drug Monit 27(6):747–751

Sousa A, Pereira M (2014) *Pseudomonas aeruginosa* diversification during ınfection development in cystic fibrosis lungs – a review. Pathogens 3(3):680–703. https://doi.org/10.3390/pathogens3030680

Steinway S, Biggs M, Loughran TP Jr, Papin JA, Albert R (2015) Inference of network dynamics and metabolic interactions in the gut microbiome. PLoS Comput Bio 11(5):e1004338. https://doi.org/10.1371/journal.pcbi.1004338

Sud M, Fahy E, Cotter D, Dennis E, Subramaniam S (2012) LIPID MAPS-Nature lipidomics gateway: an online resource for students and educators interested in lipids. J Chem Educ 89 (2):291–292. https://doi.org/10.1021/ed200088u.LIPID

Sussman JL, Lin D, Jiang J, Manning NO, Prilusky J (1998) Protein Data Bank (PDB): database of three-dimensional structural information of biological macromolecules. Acta Crystallogr D Biol Crystallogr 54(Pt 6 Pt 1):1078–1084. https://doi.org/10.1107/S0907444998009378

Szklarczyk D, Morris J, Cook H, Kuhn M, Wyder S, Simonovic M, Santos A, Doncheva N, Roth A, Bork P, Jensen L, Mering C (2017) The STRING database in 2017: quality-controlled protein-protein association networks, made broadly accessible. Nucleic Acids Res 45(D1):D362–D368. https://doi.org/10.1093/nar/gkw937

The Uniprot Consortium (2017) UniProt: the universal protein knowledgebase. Nucleic Acids Res 45(D1):D158–D169. https://doi.org/10.1093/nar/gkw1099

Thiele I, Vo TD, Price ND, Palsson BØ (2005) Expanded metabolic reconstruction of *Helicobacter pylori* (iIT341 GSM/GPR): an *in silico* genome-scale characterization of single- and double-deletion mutants. J Bacteriol 187(16):5818–5830. https://doi.org/10.1128/JB.187.16.5818

Tilley L, Dixon MW, Kirk K (2011) The *Plasmodium falciparum*-infected red blood cell. Int J Biochem Cell Biol 43(6):839–842. https://doi.org/10.1016/j.biocel.2011.03.012

Tymoshenko S, Oppenheim R, Agren R, Nielsen J, Soldati-Favre D, Hatzimanikatis V (2015) Metabolic needs and capabilities of *Toxoplasma gondii* through combined computational and experimental analysis. PLoS Comput Biol 11(5):e1004261. https://doi.org/10.1371/journal.pcbi.1004261

Uddin R, Saeed K, Khan W, Azam S, Wadood A (2015) Metabolic pathway analysis approach: identification of novel therapeutic target against methicillin resistant *Staphylococcus aureus*. Gene 556(2):213–226

Varnum S, Streblow D, Monroe M, Smith P, Auberry K, Pasa-Tolic L, Wang D, Camp D 2nd, Rodland K, Wiley S, Britt W, Shenk T, Smith R, Nelson JA (2004) Identification of proteins in human cytomegalovirus (HCMV) particles: the HCMV proteome. J Virol 78(20):10960–10966. https://doi.org/10.1128/JVI.78.20.10960

Veith N, Solheim M, Van Grinsven KWA, Olivier BG, Levering J, Grosseholz R, Hugenholtz J, Holo H, Nes I, Teusink B (2015) Using a genome-scale metabolic model of *Enterococcus faecalis* V583 to assess amino acid uptake and its impact on central metabolism. Appl Environ Microbiol 81(5):1622–1633. https://doi.org/10.1128/AEM.03279-14

Voge N, Perera R, Mahapatra S, Gresh L, Balmaseda A, Loro MA, Hopf-jannasch A, Belisle J, Harris E, Blair C, Beaty B (2016) Metabolomics-based discovery of small molecule biomarkers in serum associated with dengue virus infections and disease outcomes. PLoS Negl Trop Dis 10 (2):e0004449. https://doi.org/10.1371/journal.pntd.0004449

Webb B, Sali A, Francisco S (2017) Comparative protein structure modeling using MODELLER. Curr Protoc Bioinformatics 54. https://doi.org/10.1002/cpbi.3.Comparative

Wilking H, Thamm M, Stark K, Aebischer T, Seeber F (2016) Prevalence, incidence estimations, and risk factors of *Toxoplasma gondii* infection in Germany: a serological study. Sci Rep 6 (22551). https://doi.org/10.1038/srep22551

Winsor G, Lam D, Fleming L, Lo R, Whiteside M, Yu N, Hancock R, Brinkman F (2011) *Pseudomonas* genome database: improved comparative analysis and population genomics capability for *Pseudomonas* genomes. Nucleic Acids Res 39(Database issue):D596–D600. https://doi.org/10.1093/nar/gkq869

Wishart DS, Knox C, Guo AC, Cheng D, Shrivastava S, Tzur D, Gautam B, Hassanali M, Tg C (2008) DrugBank: a knowledgebase for drugs, drug actions and drug targets. Nucleic Acids Res 36(Database issue):D901–D906. https://doi.org/10.1093/nar/gkm958

Wishart DS, Jewison T, Guo AC, Wilson M, Knox C, Liu Y, Djoumbou Y, Mandal R, Aziat F, Dong E, Bouatra S, Sinelnikov I, Arndt D, Xia J, Liu P, Yallou F, Bjorndahl T, Perez-pineiro R, Eisner R, Allen F, Neveu V, Greiner R, Scalbert A (2013) HMDB 3.0 — the human metabolome database in 2013. Nucleic Acids Res 41(Database issue):D801–D807. https://doi.org/10.1093/nar/gks1065

Wodke J, Puchałka J, Lluch-Senar M, Marcos J, Yus E, Godinho M, Gutiérrez-Gallego R, dos Santos V, Serrano L, Klipp E, Maier T (2013) Dissecting the energy metabolism in *Mycoplasma pneumoniae* through genome-scale metabolic modeling. Mol Syst Biol 9:653. https://doi.org/10.1038/msb.2013.6

Wunderlich Z, Mirny L (2006) Using the topology of metabolic networks to predict viability of mutant strains. Biophys J 91(6):2304–2311. https://doi.org/10.1529/biophysj.105.080572

Xavier J (2016) Sociomicrobiology and pathogenic bacteria. Microbiol Spectr 4(3). https://doi.org/10.1128/microbiolspec.VMBF-0019-2015.Sociomicrobiology

Xia J, Sinelnikov IV, Han B, Wishart DS (2015) MetaboAnalyst 3.0 — making metabolomics more meaningful. Nucleic Acids Res 43(W1):W251–W257. https://doi.org/10.1093/nar/gkv380

Yang C, Li Y (2011) *Chromobacterium violaceum* infection: a clinical review of an important but neglected infection. J Chin Med Assoc 74(10):435–441. https://doi.org/10.1016/j.jcma.2011.08.013

Yeh I, Hanekamp T, Tsoka S, Karp P, Altman R (2004) Computational analysis of *Plasmodium falciparum* metabolism: organizing genomic information to facilitate drug discovery. Genome Res 14(5):917–924. https://doi.org/10.1101/gr.2050304

Yu C, Chen Y, Lu C, Hwang J (2006) Prediction of protein subcellular localization. Proteins 64 (3):643–651. https://doi.org/10.1002/prot

Zampieri M, Enke T, Chubukov V, Ricci V, Piddock L, Sauer U (2017) Metabolic constraints on the evolution of antibiotic resistance. Mol Syst Biol 13(3):917. https://doi.org/10.15252/msb.20167028

Zhang A, Sun H, Han Y, Yan G, Wang X (2013a) Urinary metabolic biomarker and pathway study of hepatitis B virus infected patients based on UPLC-MS system. PLoS One 8(5):e64381. https://doi.org/10.1371/journal.pone.0064381

Zhang A, Sun H, Han Y, Yan G, Yuan Y, Song G, Yuan X, Xie N, Wang X (2013b) Ultraperformance liquid chromatography–mass spectrometry based comprehensive metabolomics combined with pattern recognition and network analysis methods for characterization of metabolites and metabolic pathways from biological data sets. Anal Chem 85 (15):7606–7612. https://doi.org/10.1021/ac401793d

Zhang R, Ou H, Zhang C (2004) DEG: a database of essential genes. Nucleic Acids Res 32 (Database issue):D271–D272. https://doi.org/10.1093/nar/gkh024

Zhang S, Wang D, Wang Y, Hasman H, Aarestrup F, Alwathnani H, Zhu Y, Rensing C (2015) Genome sequences of copper resistant and sensitive *Enterococcus faecalis* strains isolated from copper-fed pigs in Denmark. Stand Genomic Sci 10:35. https://doi.org/10.1186/s40793-015-0021-1

Zhao G, Usui ML, Lippman SI, James GA, Stewart PS, Fleckman P, Olerud JE (2013) Biofilms and inflammation in chronic wounds. Adv Wound Care 2(7):389–399. https://doi.org/10.1089/wound.2012.0381

Zhou C, Zhou D, Elsheikha H, Liu G, Suo X (2015) Global metabolomic profiling of mice brains following experimental infection with the cyst-forming *Toxoplasma gondii*. PLoS One 10(10): e0139635. https://doi.org/10.1371/journal.pone.0139635

Zhou C, Zhou D, Elsheikha H, Zhao Y, Suo X (2016) Metabolomic profiling of mice serum during toxoplasmosis progression using liquid spectrometry. Sci Rep 6:19557. https://doi.org/10.1038/srep19557

Chapter 7
Metabolomic-Based Methods in Diagnosis and Monitoring Infection Progression

Miguel Fernández-García, David Rojo, Fernanda Rey-Stolle, Antonia García, and Coral Barbas

Abstract A robust biomarker screening and validation is crucial for overcoming the current limits in the clinical management of infectious diseases. In this chapter, a general workflow for metabolomics is summarized. Subsequently, an overview of the major contributions of this omics science to the field of biomarkers of infectious diseases is discussed. Different approaches using a variety of analytical platforms can be distinguished to unveil the key metabolites for the diagnosis, prognosis, response to treatment and susceptibility for infectious diseases. To allow the implementation of such biomarkers into the clinics, the performance of large-scale studies employing solid validation criteria becomes essential. Focusing on the etiological agents and after an extensive review of the field, we present a comprehensive revision of the main metabolic biomarkers of viral, bacterial, fungal, and parasitic diseases. Finally, we discussed several articles which show the strongest validation criteria. Following these research avenues, precious clinical resources will be revealed, allowing for reduced misdiagnosis, more efficient therapies, and affordable costs, ultimately leading to a better patient management.

Keywords Metabolomics · Infectious diseases · Diagnostics · Biomarkers · Biomarker discovery

7.1 Introduction

The new challenge in the study of the biological processes is to understand the interactions among all biological agents (genes, transcripts, proteins, and metabolites). The *metabolome*, defined as the set of all metabolites (Oliver et al. 1998), represents the closest level to the phenotype which can be directly correlated with the status of the organism. In other words, changes in endogenous and environmental

M. Fernández-García (✉) · D. Rojo · F. Rey-Stolle · A. García · C. Barbas
Center for Metabolomics and Bioanalysis (CEMBIO), CEU San Pablo University, Boadilla del Monte, Spain
e-mail: mig.fernandez.ce@ceindo.ceu.es

R. Silvestre, E. Torrado (eds.), *Metabolic Interaction in Infection*, Experientia Supplementum 109, https://doi.org/10.1007/978-3-319-74932-7_7

factors are likely to impact on the metabolome, which acts as a major chemical bridge connecting the environment with the different levels of a biological system (Fig. 7.1).

The interactions among organisms are uncountable. However, only a few of them produce a dysregulation of homeostatic processes, eventually leading to disease. Infectious diseases are one of the leading causes of mortality (WHO 2017a), and a better management of patients is undoubtedly necessary. Unfortunately, some diagnostic methods can be slow and tedious and lack adequate specificity and sensitivity (Brand et al. 2010; Jain 2010). In this context, metabolomics analyses can be performed to screen for metabolic biomarkers, which display novel potential advantageous features such as more accurate diagnosis, dynamic disease evaluation, non-invasive sampling, or personalized treatment assessment (Mayeux 2004). A biomarker can be defined as the biological molecule which is able to create a predictive model that can be used to classify a new sample or person into a specific group (e.g., healthy vs. diseased) with enough sensitivity and specificity (Xia et al. 2013). This task is technically challenging (Xia et al. 2013). Note that a PubMed search from 1990 to 2009 for the term "biomarker" revealed a total of 441,510

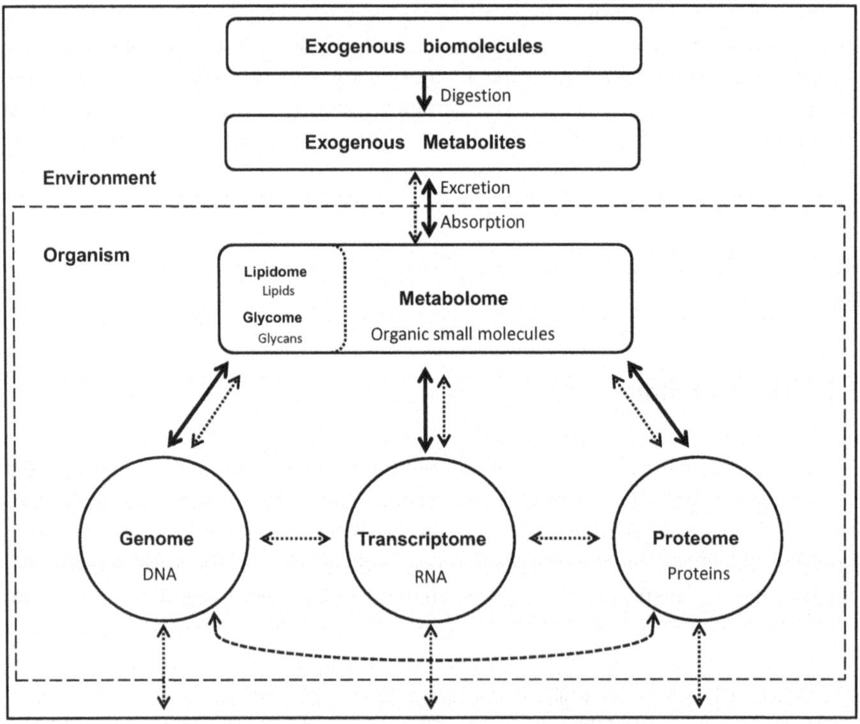

Fig. 7.1 General scheme showing the major mass fluxes (normal arrows) and molecular interactions (dashed arrows) between the different systems of an organism and its environment

Table 7.1 Overview of the metabolome coverage of the main analytical techniques

Analytical technique	Metabolome coverage	Compound requirements	Polarity	Examples
GC-MS	Low	Thermostable compounds with acidic H or volatiles	High–mid	Amino acids, small organic acids, carbohydrates, fatty acids, terpenes, cholesterol derivatives
LC-MS	High	Soluble compounds	Depending on column	Amino acids, oligopeptides, small organic acids, carbohydrates, phosphate compounds nucleosides, lipids, nucleotides, carnitines
CE-MS	Medium	Charged compounds	Only polar	Amino acids, oligopeptides, small organic acids, phosphate compounds nucleosides, nucleotides, carnitines
^{1}H-NMR	Low	Abundant compounds with non-exchangeable protons	–	Small organic acids, carbohydrates, amino acids
^{31}P-NMR	Low	Abundant phosphorus-containing compounds	–	Nucleotides, nucleosides, polyphosphate, phosphate enzyme cofactors, sugar phosphates

published articles, an enormous academic investment not translated into the expected improvements for patient care (Ptolemy and Rifai 2010).

Herein, we present a robust analytical workflow for biomarker discovery and validation with special attention to experimental design, sample treatment, analytical platforms, and data analysis. Subsequently, we will focus on the contributions of metabolomics to the identification of biomarkers in the context of infectious diseases. Table 7.1 depicts the strenghts, weaknesses and typical applications in metabolomics of the main analytical techniques.

7.2 Metabolomics Workflow for Biomarker Discovery

Sensu stricto, metabolomics corresponds with a non-target analysis—also named as global profiling or fingerprinting (Fiehn 2002)—in which the total metabolome is measured and the data are treated following differential analysis methodologies. Metabolomics is ideal for biomarker discovery, since no previous hypotheses are needed. Nevertheless, before they can be routinely used in clinics, biomarkers need to be validated which, from the analytical point of view, requires the performance of a classic target analysis.

The fingerprinting workflow for biomarkers screening (Fig. 7.2) encompasses several steps such as experimental design, sample treatment, analytical measurement and data processing, which will be below described in detail. Note that each step

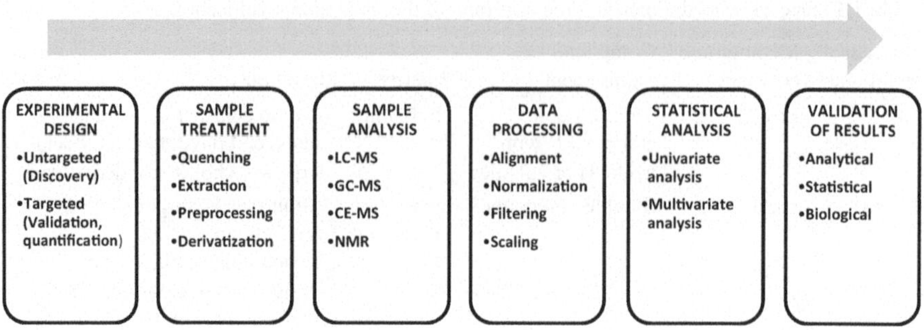

Fig. 7.2 General workflow in a metabolomics experiment for biomarker discovery and validation

represents a potential source of error which may be accumulated. Special attention should be paid to the biospecimen selection and data meta-analysis, the two major bottlenecks in biomarker screening (Voskuil 2015).

7.2.1 *Experimental Design*

A certain amount of variability is inherent to biology. However, ideally, the characteristic under study should be the only source of differences among the sample groups. Therefore, group selection becomes critical for a successful differential analysis. In order to minimize this problem, the size of the study cohort can be increased to mathematically dilute the interindividual variability. Nevertheless, a long analytical sequence can cause a problem in the signal stability. Although there are several normalization strategies (Godzien et al. 2013), it is highly recommended to periodically analyze quality controls (QC) through the analytical batch to monitor any potential bias. Ideally QCs should be a pool of aliquots obtained from each biological sample. However, when this is not possible, other alternatives are acceptable (Dunn et al. 2012; Godzien et al. 2015).

7.2.2 *Sample Treatment*

All the samples should be obtained following a strict protocol. What is more, their transport and storage are crucial to avoid metabolite degradation and subsequent potential bias between batches. Specific standard operating procedures (SOP) should be adapted to each type of matrix, although samples should be generally preserved at −80 °C unless otherwise recommended. Another turning point for sample stability is a fast and efficient metabolic quenching. Enzymes can be inactivated by the addition of organic solvents, or abrupt temperature, or pH variations (Mushtaq et al. 2014). Typical metabolic quenching strategies comprise the use of liquid nitrogen or cold methanol/water mixtures (Dettmer et al. 2007). However, caution must be taken as aggressive quenching methods can damage plasma membranes, causing potential metabolite leakage (Winder et al. 2008). To overcome this, a softer quenching methodology based on a fast vacuum filtration of cells and subsequent addition of a quenching solution has been proposed (da Luz et al. 2014), although its relatively longer quenching time can alter the metabolome.

The next step along sample preparation is metabolite extraction. If the sample is a solid (e.g., tissue), matrix disaggregation is previously needed. In a fingerprinting approach, the selected extraction solvent should solubilize as many metabolites as possible. Double extraction with both polar and non-polar solvents is always recommended to enhance the metabolite coverage (note that independent series of extractions with such solvents are always more efficient than a single extraction with the mix). On the other hand, in a target analysis, the solvent of interest is the one that maximizes the analyte recovery. Liquid–liquid and solid-phase extractions are commonly used methods for liquid samples such as biofluids and cell culture supernatants. Furthermore, samples can be derivatized to modify the chemical properties of analytes. Derivatization is usually required for gas chromatography (GC) analysis which needs thermally stable and volatile metabolites.

7.2.3 Analytical Measurement

In metabolomics fingerprinting, the main analytical platforms are nuclear magnetic resonance (NMR) and mass spectrometry (MS). NMR spectroscopy is a nondestructive technique which allows in vivo evaluation of the metabolism. However, the low sensitivity limits its applicability. On the other hand, MS is usually coupled to a separation technique as far as it is not possible to perform the direct analysis of the metabolome due to its chemical complexity. A mass spectrometer is composed by a source (where the molecules are ionized), an analyzer (in which the metabolites are separated by their *m/z* ratio), and a detector (which register the number of counts per hit). Different types of analyzers are available for metabolomics, being the time of flight (TOF) one of the most used, considering that it renders mass accuracy with only several part per million (ppm) of error. The quadrupole time of flight (QTOF) allows ion separation and subsequent collision-induced dissociation for MS/MS identification analysis. Finally, the triple quadrupole (QqQ) is ideal for target quantification by multiple reaction monitoring (MRM) mode.

GC separation of metabolites is based on their different vapor pressure and their affinity for the stationary phase of the column. Generally, low bleeding capillary columns are used in GC-MS, being the wall-coated open-tubular columns the most popular option (Mirnaghi and Caudy 2014). In metabolomics, the most frequently used stationary phases are DB-5 and DB-50 ((5%-phenyl)-methylpolysiloxane and (50%-phenyl)-methylpolysiloxane, respectively) (Rojo 2014). The main types of ion sources in GC-MS are electron ionization (EI) and chemical ionization (CI). GC-EI-MS renders well-separated efficient peaks, with high reproducibility in the MS spectra. These advantageous features have promoted the development of GC-MS spectral libraries (e.g., FiehnLib, NIST 14), which are extremely useful for metabolite identification. GC-MS metabolite coverage is limited to thermally stable and volatile compounds such as esters, terpenes, certain alcohols, aldehydes, thiols, or aromatic compounds (Rowan 2011). To overcome this limitation, chemical derivatization procedures are routinely applied (e.g., silylation with N,O-Bis(trimethylsilyl) trifluoroacetamide and trimethylchlorosilane for sugars, amino acids, amines, amides, fatty acids, etc.).

In liquid chromatography (LC), MS coupled systems samples are injected into a packed column containing a stationary phase. Metabolites are separated due to their different partition coefficients between the stationary phase and the eluotropic mobile phase (either isocratic or in a gradient). Based on the stationary phase, it can be distinguished among reverse-phase LC (e.g., C18, C8), hydrophilic interaction liquid chromatography (HILIC), and normal-phase LC. Reverse-phase LC is the gold standard technique for the analysis of hydrophobic and medium polarity compounds (e.g., lipids), whereas HILIC is preferred for polar metabolites although its reproducibility is challenging. Normal phase is not commonly used in metabolomics since highly hydrophobic mobile phases are required (e.g., hexane). The subsequent ionization of the metabolites is generally driven by an electrospray source either in positive or negative mode. The use of a QTOF analyzer enables metabolite annotation by comparing their accurate masses with public databases (e.g., HMDB, METLIN, and LIPID MAPS) and eventually their structural elucidation by MS/MS analysis.

Capillary electrophoresis (CE) comprises a wide variety of electrokinetic separation methods, from which capillary zone electrophoresis is the most commonly used in metabolomics (Zhang et al. 2017). Such technique is based on the application of high voltages between two electrodes, which are connected by a buffer solution that flows through a capillary. Net charges are produced in the capillary inner wall, which interacts with the oppositely charged ions contained in the buffer solution. When voltages are applied, ions migrate toward the oppositely charged electrode, carrying solvent molecules. Consequently, a plug-like flow defined as electroosmotic flow is produced. Electrophoretic mobility is an additional force that appears only in charged compounds and produces their displacement toward the electrode with opposite charge. CE-MS instruments can operate in two different modes depending on the charge of metabolites and MS detector polarity. Cation separation is mainly performed in bare fused silica capillaries by application of high voltage with positive MS ionization. It is ideal for metabolites such as nucleotides, nucleosides, amino acids, acylcarnitines, and other amines and amides. On the contrary, electrophoresis with reversed polarity required electroosmotic flow suppression or inversion by neutral or positively coated (e.g., polybrene-dextran sulfate-polybrene) capillaries (Ramautar et al. 2011). Although it is highly interesting for the analysis of short-chain organic acids or sugar phosphates, reversed polarity CE-MS capabilities have been limited by certain technical difficulties that occur whenever positive potential is applied to the stainless steel needle (e.g., a special ESI platinum needle is required under ESI in negative ionization mode).

In NMR, radio-frequency pulses are applied to the liquid samples which are placed in a magnetic field. Such pulses are capable of altering nuclei with non-zero magnetic momentum as ^{1}H or ^{13}C. Since the specific chemical environment of a nucleus determines its spin momentum dynamic evolution and coupling with other nuclei, precise acquisition information about the molecular structure information is acquired. Hence, accurate metabolite identification and quantification can be performed. Advantages of NMR are highly reproducible spectra, fast speed acquisition rates, minimum sample preparation, and relatively low cost of analysis. NMR can also be used for the analysis of phosphorus-containing metabolites, rendering relatively simplified spectra. These metabolites are either indicative of various cellular processes, energy metabolism (e.g., ATP, GMP, NADPH), central carbon metabolism processes (e.g., sugar phosphates, phosphoenol-3-pyruvate), or structural and signaling functions (e.g., phospholipids). To increase the sensitivity and metabolome coverage of NMR, chemical derivatization with ^{13}C and ^{15}N isotopic tags has been described (Gowda et al. 2012).

7.2.4 Data Processing

Data processing workflow is technique-dependent. Generally speaking, the first task once the analysis is finished is to check whether all profiles with the corresponding internal standard peaks and QC samples show a good repeatability and if the pressure or current curves are adequate. To deliver high-quality results, the reproducibility and suitability of every sample must be verified. Traceability of the analytical batch through an internal standard or based on QCs are the two main strategies to ensure a stable equipment performance.

A MS chromatogram/electropherogram contains three main pieces of information: MS spectrum, time, and abundance (ion current). Based on them, deconvolution algorithms reprocess the signal, and, according to the isotopic pattern distribution across the time, they create the metabolic features which ideally correspond to a chromatographic peak characterized by a defined area, elution time, and *m/z*. A particular case of deconvolution is done in GC-MS, in which the identification of the metabolic features is done by comparing the retention times and spectra from empirical data to reference libraries. Subsequently, features are aligned and a matrix is created. Alignment should be performed with a certain tolerance in retention time and mass windows to correct small fluctuations.

In some cases, normalization may be required to minimize alterations in the stability of analytical platforms. Data filtering is necessary to eliminate spurious signals produced after deconvolution (Godzien et al. 2015). First, a filtration by presence in QCs is performed. Secondly, filtration by presence in at least a minimum number of samples of one of the groups is also performed. As only stable signals pass the filter, random artifacts are reduced. As a result, only the metabolites that are strongly represented in the sample set are considered for statistical analysis. Data scaling is desirable to adjust the importance assigned to the elements of the data in fitting the model. By scaling, the weight of each variable is adjusted with a scaling factor estimated by either a dispersion criterion or a size measure.

Differential analysis statistical methods are used in fingerprinting analysis for biomarker discovery, considering each metabolite as a variable. Both univariate and multivariate statistical tests can be applied. Univariate models such as student's *t*-test contemplate each metabolite as an independent variable, whereas multivariate can measure the influence of each metabolite on others. Regarding to multivariate tests, both unsupervised and supervised analysis can be performed. Principal component analysis (PCA) is the most common unsupervised model based on which any potential sample outlier can be detected. What is more, an adequate reproducibility of the analyses can be easily assessed by observing tight clustering of QCs in PCA plots. On the contrary, supervised analyses aim to generate models with a priori assigned groups. The most common supervised multivariate models are partial least squares regression discriminant analysis (PLS-DA) and orthogonal PLS-DA (OPLS-DA) analyses. To overcome the statistical bias induced by model overfitting and finding spurious correlations, different statistical methodologies such as permutation tests, cross-validation, or jackknifing have been used. Note that a detailed explanation of these algorithms is out of the scope of this chapter.

7.3 Metabolomics Workflow for Biomarker Validation

In a broad sense, validation is the assessment of an action, instrument, or technique to establish that it is delivering the intended outcome. Considering the multidisciplinary nature of the metabolomics experiment, different validation steps must be performed across the whole workflow. From the analytical point of view, validity is assessed as reporting several parameters for a certain method, such as the accuracy, precision, linearity, and limit of quantification. With respect to a mathematical model, validation ensures that overfitting is absent, which can be calculated by applying several tests mentioned above. Regarding the biological interpretation, validations aim to confirm the initial hypothesis using several strategies like detecting the same group of metabolites with a different technique in an independent set of samples, integrating multi-omics information, or developing ad hoc experiments.

Focusing on biomarker discovery, there is a relatively little consistency in the way of validating candidate biomarkers (Xia et al. 2013). Besides their identification by structural elucidation or standard spiking, to validate a biomarker means to (1) demonstrate that it is still significant in a second sample cohort, bigger than the previous one and with a broader interindividual variability, and (2) to assess on it the discriminant capability of the candidate by reporting its sensitivity and specificity which should be calculated based on accurate concentration values. In other words, a strong biomarker with a potential use in clinics' daily routine should have a small ratio of false negatives (misdiagnosed diseased patients) and false positives (misdiagnosed healthy patients). A way to visualize both is performing the calculation of the receiver operator characteristic (ROC) curve. In this type of diagrams, specificity is plotted against selectivity for a given set of concentration values in order to determine the threshold between disease and healthy status. The area under the curve (AUC) of the ROC curve defines the strength of the biomarker. Generally, a strong biomarker displays AUC values bigger than 0.8, whereas fair biomarkers account for values between 0.8 and 0.6. An AUC below 0.6 does not satisfy the quality criteria for a good biomarker.

Any analytical platform is potentially suitable to measure accurate concentration values by using the classical quantification methods of standard curve, standard addition, or internal standard. Nevertheless, due to its superior capabilities in terms of sensitivity, the gold standard technique for quantification is a triple quadrupole (QqQ-MS), coupled to LC or GC and operating in MRM which reaches limits of detection of part per billion (μg/L). In brief, MRM mode is based on a selective fragmentation of the molecular ion of the MS spectra of the candidate, respectively, using its two most abundant fragments as quantification and qualification transitions. Thanks to MRM it is possible to avoid background MS noises.

7.4 Biomarkers for Infectious Diseases: Current Landscape

The analysis of biofluids or tissues from infected patients represents the most accurate methodology to describe the truly metabolic changes associated with disease. However, since sample availability is generally compromised in acute, low prevalent, or hazardous diseases, other experimental designs have been conceived, being most of these studies performed in human disease-mimicking models (Fig. 7.3). Although substantial differences between the metabolic profiles of disease models and human patients have been reported (Himmelreich et al. 2005; Salek et al. 2007), models show several advantages such as an enhanced evaluation of the disease prognosis. Model-based experiments comprise studies wherein animals are challenged with human or humanlike pathogens and in vitro studies wherein infection is performed under host cell lines. Other studies, while not containing truly present host–pathogen interactions, are capable of elucidating considerable aspects of the infectious process; these include, among others, experiments wherein a host–pathogen interaction has been previously established (e.g., analysis of cultures from clinical isolates) and studies wherein host–pathogen interactions are mimicked by means of challenging either host or pathogen cells with chemical conditions characteristics of disease (e.g., hypoxia (Eoh and Rhee 2013), addition of bacterial lipopolysaccharide (Laiakis et al. 2012)). Although a clear distinction can be observed between biomarker identification and unveiling disease's mechanism, it should be highlighted that metabolomics studies which aim to understand the pathogenesis of infectious diseases can also reveal potential biomarkers. Due to the large number of the studies in this area, we focused on selected series of examples which illustrate an overview of the state of the art in the field.

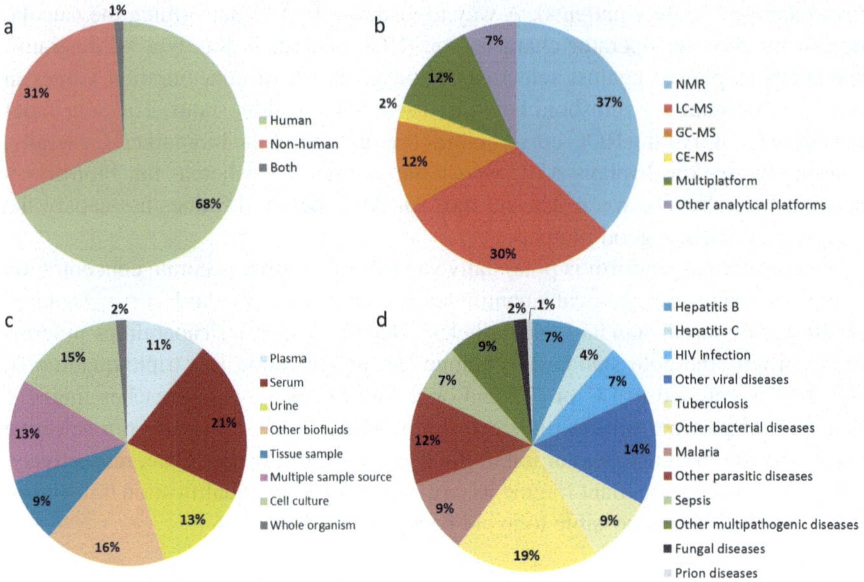

Fig. 7.3 Classification of metabolomics experiments in infectious diseases attending to (**a**) host nature, (**b**) analytical technique, (**c**) sample type, and (**d**) disease under study. Items in pie charts are displayed clockwise

7.4.1 Biomarker Discovery

7.4.1.1 Metabolomics of Viral Diseases

The area of metabolomics of viral diseases has been subject of extensive study and review (Sanchez and Lagunoff 2015). In this book chapter, we will describe the major applications of metabolomics in three relevant viral diseases [human immunodeficiency virus infection (HIV), hepatitis C, and dengue fever]. Since the beginning of AIDS epidemics, HIV has infected more than 70 million people, from which 36.7 million live with HIV/AIDS worldwide (WHO 2017d). Although diverse therapeutic treatments have been developed, more than 1 million deaths per year are still caused by AIDS-related diseases (WHO 2017d). The first metabolomics studies regarding to HIV aimed to find metabolic biomarkers for HIV infection/response to antiretroviral (ART) treatment. In this context, Hewer et al. (2006) studied the serum metabolic profiles from ART-treated/ART-naïve HIV+ patients and HIV—controls by ^{1}H-NMR. Changes in the levels of glucose, lipids, amino acids, glycerol, choline, and lipoproteins were observed, suggesting potentially diagnostic capabilities of this technique, alternative to the current gold standard (Western blot) (Syed et al. 2005). Other studies have been conducted to find a fast and non-invasive HIV diagnostic test; Ghannoum et al. (2013) compared the GC-MS and LC-MS metabolic profiles of oral lavage from healthy subjects and ART-naïve and ART-experienced patients. Interestingly, an increased phenylalanine/tyrosine ratio was found in ART-naïve patients, opening doors for new oral biomarkers. In contrast, a notable number of studies regarding the study of HIV-related conditions have been reported; since secondary lung infections are common among HIV+ patients, Cribbs et al. (2014) used LC-MS to study the differences between metabolic profiles of bronchoalveolar lavage fluid (BALF) of HIV+ patients and controls. Groups notably clustered in OPLS-DA models, and important altered levels of phospholipids, tripeptides, and pyochelin (a siderophore related with the opportunistic pathogen *Pseudomonas aeruginosa*) were observed. Additionally, HIV-related neurocognitive impairment has also been an object of study by Cassol et al. (2014). They performed a multiplatform, combined GC-MS and LC-MS analysis of the cerebrospinal fluid (CSF) of HIV+ individuals and controls, finding correlations between the levels of ketone bodies (1,2-propanediol, β-hydroxybutyrate), neurotransmitters (glutamate, N-acetylaspartate), and myoinositol with neurocognitive worsening.

Hepatitis C is a highly prevalent disease, affecting more than 71 million people worldwide (WHO 2017b). This liver disease is caused by the hepatitis C virus (HCV), a flavivirus comprehending six different genotypes. HCV infection usually displays mild symptoms in the acute phase of the disease and considerably high rates of chronification (85% of infected patients) (Pawlotsky 2004). Chronic HCV infection generates a progressive fibrosis of the liver, which is classified in four stages by the METAVIR system (F0 for absence of fibrosis, F1–2 for mild fibrosis, F3 for advanced fibrosis, and F4 for fully developed cirrhosis) (Bedossa and Poynard 1996). Such deterioration can ultimately lead to severe conditions, such as cirrhosis and hepatocellular carcinoma (HCC) (Pawlotsky 2004). Due to the considerable prevalence and such potentially fatal outcomes, HCV infection has been considerably studied by metabolomics (du Preez and Sithebe 2013). Additionally, given the invasiveness of the current gold standard for assessing the progression of HCV-related liver fibrosis (liver biopsy), many studies have focused on the development of alternative methods to assess disease progression. In this context, Zhang et al. (2006) conducted a targeted method using LC to quantify amino acids in plasma samples of HCV+ patients, generating models based on amino acid ratios from which a combined model ((Phe)/(Val) + (Thr + Met + Orn)/(Pro + Gly)) was capable of discriminating F0–2 from F3–F4 patients (ROC AUC = 0.92, 95% CI: 0.84–1.00) and F0–F3 from F4 patients (ROC AUC = 0.99, 95% CI: 0.96–1.01). Additionally, Cheng et al. (2012) performed a multiplatform LC-MS and GC-MS analysis of plasma metabolites comparing advanced fibrotic patients with hepatitis C patients not showing this condition. Notable alterations in primary bile acids (glycocholate, taurocholate), long-chain fatty acids (stearidonate, palmitoleate), and cysteine were observed. In another study conducted by Embade et al. (2016), a ^{1}H-NMR analysis on serum samples from fibrotic and non-fibrotic patients was performed, finding significant alterations in amino acids and lipoproteins, among others. From such alterations, a multivariate model based on choline, acetoacetate, and LDL1 levels which effectively discriminated between F0 and F4 patients (ROC AUC = 0.922, 95% CI: 0.85–0.97) was generated. On the other hand, potential biomarkers to predict the treatment outcome of hepatitis C have also been studied. Before direct-acting antivirals (DAAs)—the current first-line treatments—were developed, predictive models of the outdated treatment comprising alpha-interferon plus ribavirin were generated (Saito et al. 2013). Following a similar line, Ceccotti et al. (2016) performed a longitudinal study wherein ^{1}H-NMR was applied to serum samples from HCV+ patients before and after successful DAA therapy, finding differences in the levels of formate, acetate, and methionine between baseline and 12th week post-treatment samples from patients who achieved a sustained virological response.

Dengue fever (DF) is a widespread disease with an incidence of approximately 390 million people per year worldwide (Bhatt et al. 2013). DF is caused by dengue virus (DENV), from which four different serotypes have been described—DENV-1 to DENV-4. This mosquito-borne disease is usually asymptomatic (80% of cases) or manifested through mild symptoms (e.g., fever, rash, joint pain) (Martina et al. 2009). However, approximately 5% of DF cases progress to more severe forms, namely, dengue hemorrhagic fever (DHF) and dengue shock syndrome (DSS), which seriously affect the cardiovascular system of patients, potentially leading to death (Martina et al. 2009). Given the heterogeneity of possible disease outcomes and the lack of an effective biomarker for prediction of a severe prognosis (John et al. 2015), different metabolomics studies have been performed to elucidate the mechanism of this disease and to find possible diagnostic and prognostic biomarkers. Preliminary studies of DENV-infected cell cultures have been described, such as the performance of a combined ^{1}H-NMR and direct infusion MS analysis of DENV-infected endothelial cell lines (EA.hy926) by Birungi et al. (2010). These analyses revealed metabolic alterations in the levels of amino acids, fatty acids, and metabolites of the tricarboxylic acid cycle, showing a full separation among DENV serotypes in PCA plots, suggesting differences in the disease mechanism for each serotype. In another study from Fontaine et al. (2015), a combined LC-MS and GC-MS analysis of infected human foreskin fibroblasts unveiled alterations in the host central carbon metabolism (particularly in glycolytic intermediates). Different studies analyzing the serum, plasma, and urine of DF patients have been performed. For example, Shahfiza et al. (2015) identified by ^{1}H-NMR a series of relevant gender-dependent alterations of urinary metabolites during DF (e.g., acetaminophen, glycine, betaine, creatinine, creatine phosphate, N-methylhydantoin, and β-hydroxybutyrate), while El-Bacha et al. (2016), using the same analytical technique, found plasma evidences for a DF-induced liver dysfunction (e.g., branched-chain amino acids, LDL, and VLDL). LC-MS and subsequent MS/MS ex vivo analyses have also revealed important contributions to DENV diagnostics. In this context, Voge et al. (2016) identified a differential metabolic profile between patients infected with DENV-1 and DENV-2 (octadecatrienal, octadecadienal, octadecadienoic acid, octadecenamide, sphingosine, 25-hydroxy-hexadehydrovitamin D_3, 1,25-dihydroxycholesterol, and 3-deoxy-25-hydroxyvitamin D_3, among others). Another LC-MS study, performed by Cui et al. (2016), suggested candidate biomarkers for the prediction of the progression of mild DF to DHF/DSS (e.g., kynurenine, serotonin, bilirubin, oleamide, 13-E-docosenamide, deoxyinosine, and several glycerophospholipids) from which particularly serotonin, in combination with γ-interferon, was able to build a robust predictive model (ROC AUC = 0.92).

7.4.1.2 Metabolomics of Bacterial Diseases

Prokaryotic cells encompass the biggest set of living organisms, comprising thousands of eubacterial and archaeal species. Interestingly, from the human health point of view, some of these organisms can potentially establish a symbiotic or a pathogenic relationship. In some cases, such relationship can be tracked trough different specific non-eukaryotic metabolites which unfortunately are not always easy to measure. We overview this vast area focusing on selected types of bacterial diseases.

Urinary tract infection (UTI) is a common disease, which is potentially capable of leading to severe excretory system complications. Although the etiology of UTI can be diverse, the major causative pathogen is uropathogenic *Escherichia coli* (UPEC) (up to 80% of cases) (Ronald 2003). Conventional bacterial identification tests display several limitations, suggesting the necessity of a better diagnostic tool (Pearle 2004). Candidate biomarkers for the potential diagnosis of UPEC-produced UTI were suggested by Lv et al. (2011), which compared the LC-MS urinary metabolic profile between both female UTI patients and healthy individuals. PLS-DA analysis notably discriminated these two sample groups, from which the metabolites most contributing to the group separation were identified as diphosphomevalonate, N-acetylneuraminic acid, tetrahydroaldosterone-3-glucuronide, 3-carboxy-1-hydroxypropylthiamine diphosphate, 6-ketoprostaglandin F1α, cortolone-3-glucuronide and 21-hydroxypregnenolone, desmosine, and pregnanediol-3-glucuronide. Interestingly, some UTI diagnostic biomarkers have been validated (see Sect. 7.4.2).

Clostridium difficile is a gram-positive, frequently nosocomial pathogen, which affects approximately 0.5–3 million people per year (Rojo et al. 2015). *Clostridium difficile* infection (CDI) affects the gastrointestinal tract and constitutes a potentially fatal disease which shows high rates of recurrence (20% after an initial infection, 40% after a second episode, and 60–80% in subsequent episodes) (Kao et al. 2016). Given the shortcomings derived from the current tests for the differential diagnosis or prediction of recurrent CDI (Burnham and Carroll 2013), several metabolomics studies have been focused on the identification of such diagnostic and prognostic biomarkers. For example, Kao et al. (2016) performed a combined non-targeted and targeted ^{1}H-NMR analysis of the urine of patients with symptomatic CDI and healthy controls. Patients were grouped attending to the recurrence of CDI (one, two, three, or more than three episodes). When comparing patients and controls, choline was identified as the metabolite most contributing to the discrimination of both groups, while other relevant metabolites (e.g., hippurate, 3-methylhistidine, dimethylamine, trimethylamine, ethanolamine, leucine, valine, pyroglutamate, asparagine, threonine, and glutamine) were also identified. Differences between patients showing primary CDI and recurrent CDI infections were also clearly observed in PCA plots: recurrent patients showing more than two CDI episodes were effectively discriminated from the rest of CDI patients. Metabolites most contributing to this separation were identified as trans-aconitic acid and histidine, although other metabolites such as creatine, sucrose, O-acetylcarnitine, ethanol, tartrate, acetate, asparagine, 3-hydroxyvaleric acid, and carnitine were also found to be altered. In the same framework, Allegretti et al. (2016) performed a target LC-MS analysis directed toward the evaluation of bile salts on stool samples from patients with primary CDI and recurrent CDI and healthy controls. Statistically significant differences in the bile acid profile were found across the different groups under study; significantly decreased levels of litocholate and deoxycholate were found in all CDI patients when compared with controls. Similarly, the levels of deoxycholate were also found to be increased in primary CDI when compared to recurrent CDI patients. Recurrent CDI patients also showed higher abundance of primary bile salts and lower abundance of secondary bile salts than other groups, as well as higher levels of cholate and chenodeoxycholate compared with controls. Interestingly, the ratio of deoxycholate to glycoursodeoxycholate plus deoxycholate was identified as a potential biomarker combination model for the differential diagnosis between first and recurrent CDI. Regarding the search for potential biomarkers of CDI susceptibility, Dann et al. (2015) performed a longitudinal study wherein untargeted metabolomics were applied to stool samples of either patients at risk of developing CDI or patients being tested for CDI, finding increased concentrations of γ-aminobutyric acid in patients who developed CDI.

Enteric fever is a disease caused by the gram-negative bacteria *Salmonella enterica*, serotype typhi (typhoid fever), and *Salmonella enterica*, serotype paratyphi (paratyphoid fever). Enteric fever worldwide causes more than 25 million new infections per year (Buckle et al. 2012). Unfortunately, the current standard diagnostic tests for enteric fever lack of sensitivity and specificity, thus requesting the development of a potentially useful diagnostic biomarker (Parry et al. 2011). Furthermore, as both pathogens may display similar clinical symptoms and at the same time different antimicrobial resistance (Zaki and Karande 2011), biomarkers for the differential diagnosis may be potentially useful for a correct clinical practice. Näsström et al. (2014) compared the plasma metabolic profile of *Salmonella* typhi-/paratyphi-infected patients and healthy volunteers through GC-MS analysis. Although partial overlap in PCA scores was observed when comparing all groups, OPLS-DA analyses revealed correct group clustering, suggesting the potential discriminatory capacity of the analyzed metabolites. Subsequent selection of the most statistically significant metabolites identified a combination of six metabolites (gluconic acid, phenylalanine, ethanolamine, pipecolic acid, and two carbohydrates), which could discriminate between typhi and paratyphi patients (ROC AUC = 0.796), as well as between these patients and controls (ROC AUC = 0.923 and 0.948 for typhi and paratyphi patients, respectively).

The potential role of metabolomics in the diagnosis of lethal bacterial diseases can be exemplified by the studies of melioidosis. This disease, caused by the gram-negative bacteria *Burkholderia pseudomallei*, can have high mortality rates (up to 19%) (Currie et al. 2000). Melioidosis symptoms range from local abscesses to pneumonia or sepsis. Given the sensitivity lack of the current gold standard for the diagnosis of this disease (observation in cell cultures) (Limmathurotsakul et al. 2010), Lau et al. (2016) performed an untargeted LC-MS analysis of the plasma from patients with either melioidosis or bacteremia (distinct from *B. pseudomallei*) and healthy volunteers. PCA scores plot revealed notable clustering among the disease groups. About melioidosis patients, the most significant metabolites were found to be acylcarnitines, sphingomyelins, lysophosphatidylethanolamines, and phosphatidylcholines when compared to other disease groups/controls. Interestingly, sphingomyelin (d18:2/16:0) was found to be the best discriminator of melioidosis when compared with controls or other bacteremic patients.

Tuberculosis (TB) is another potentially lethal disease, which is primarily caused by *Mycobacterium tuberculosis*. Despite being treatable and curable, this widespread mycobacterial disease constitutes the first mortality cause among all infectious diseases, with estimated incidence and mortality rates of 10.4 and 1.4 million people per year worldwide, respectively (WHO 2017d). Given that infection with *M. tuberculosis* is usually acquired through bacilli inhalation, TB is most likely primarily developed in the lungs, causing pulmonary disease (active pulmonary TB). However, most of the infections remain subclinical, causing a latent infection (latent TB) that can progress to active pulmonary TB (10% of latent TB cases) (Fatima et al. 2017). In addition, *M. tuberculosis* can affect a wide range of extrapulmonary organs and tissues, such as the CNS, the genitourinary tract, and the lymphatic system (lymph node TB or LNTB) (Fatima et al. 2017). For instance, CNS invasion by *M. tuberculosis* can cause a varied range of serious conditions associated with high mortality and disability rates, such as tubercular meningitis (TBM), which is more common in children. In contrast, genital tuberculosis (GTB) specially constitutes a notable factor for women infertility. Active TB must be treated for at least 6 months, being the first-line treatment a combination of isoniazid, rifampicin, ethambutol, and pyrazinamide. Unfortunately, multidrug-resistant tuberculosis (MDR-TB) and extensively drug-resistant tuberculosis (XDR-TB) constitute a major concern for public health, mainly in developing countries (Günther 2014). Given that current standard diagnostic methods (such as identification of the bacteria in cell cultures) show important drawbacks (time-consuming, low specificity, and sensitivity), most studies have focused on the development of biomarkers for the diagnosis of different forms of TB. To find such diagnostic biomarkers for active TB diagnosis, different experiments have compared the metabolic profile of active TB patients with controls. For instance, du Preez and Loots (2013) performed a non-targeted GC-MS analysis of sputum samples from patients suspected of having TB. After outlier removal, clustering of active TB patients and healthy controls was observed in PCA score plots. Altered levels of carbohydrates (glucosamine, N-acetylglucosamine, 2-deoxy-erythro-pentitol, glucopyranose, mannopyranose, gluconic acid δ-lactone), fatty acids (nonadecanoic acid, oleic acid, sebacic acid, C17:1ω7c), and other compounds (citramalic acid, glutaric acid, ethane, butanal, γ-aminobutyric acid, 3,4-dihydroxybutanoic acid, and normetanephrine) were found in TB-positive patients. In the same line, Frediani et al. (2014a) performed a LC-MS profiling of plasma. Interestingly, increased metabolites were found in the TB patient group from this study, such as lipids, choline derivatives, and glutamate. A similar experiment was designed by Zhong et al. (2016), who performed a non-targeted LC-MS analysis of serum from TB patients. The correspondent PCA scores plot revealed discrimination of TB-positive patients and controls. Interestingly, 99 significant metabolites were found to be statistically significant between patients and controls (e.g., fatty acids, phosphatidylcholines, and phosphatidylethanolamines). In the same framework, Luier and Loots (2016) performed a GC-MS analysis of urine samples from active TB and healthy controls, revealing significantly altered metabolites in the urine of TB patients (e.g., 5-hydroxyhexanoate, phenylacetate, 2-octenoate, 2-C-methylglycerol, 5-hydroxyhydantoin, oxalic acid, rhamnulose, quinolinic acid, ribitol, indole-3-carboxylate, kynurenate, and glycerol monostearate).

Other studies have also explored diagnostic biomarkers for latent TB and the differential diagnosis between active and latent TB. The typical experimental design of these studies comprises the grouping of individuals in active TB patients, latent TB patients, and controls. An example of this is the study published by Weiner et al. (2012), which evaluated the serum profiles of TB patients by means of a combined GC-MS and LC-MS analysis. Metabolites capable of discriminating between active TB patients and the rest of groups under study were identified as kynurenine, pyroglutamine, phenylalanine, histidine, cysteine, glutamine, citrulline, creatine, and tryptophan, among others. In contrast, 3-carboxy-4-methyl-5-propyl-2-furanpropanoic acid, N-acetylneuraminate, xanthine, hypoxanthine, and inosine levels were reported as statistically significant between active and latent TB patients. Even more interestingly, some compounds differed between latent TB and healthy controls (e.g., glycylvaline, 5-oxoproline, and inosine), suggesting the presence of potential biomarkers for latent TB diagnosis. Another potential application of metabolomics regarding the TB study is the suggestion of biomarkers that differentiate TB from other diseases that may show similar clinical features. These studies generally include groups of patients suffering from other diseases in their analysis cohort. For instance, Feng et al. (2015) compared the serum profiles obtained by LC-MS analysis of patients showing active TB, lung cancer, pneumonia, bronchiectasis, and chronic obstructive pulmonary disease. OPLS-DA models showed a full discrimination of TB patients when compared with other diseases, except for the pneumonia group. Interestingly, a combination of potential biomarkers (behenic acid, lysophosphatidylcholine (18:0), threoninyl-γ-glutamate, and presqualene diphosphate) displayed remarkable discriminatory capacity between TB patients and controls (ROC AUC = 0.991, 95% CI: 0.982–1.000). Only lysophosphatidylcholine (16:0) was able to statistically discriminate between TB patients and the rest of disease patients. Another approximation for the differential diagnosis of TB was conducted by Das et al. (2015) who compared the GC-MS urinary profile of symptomatic individuals suspected of TB infection. Urine analysis revealed notable clustering in PLS-DA analyses from which 44 statistically significant metabolites (e.g., lactic acid, norepinephrine, malic acid, hydroquinone, glucose, and 4-hydroxybenzoic acid) were used to build a model capable of differentiating TB patients (ROC AUC = 0.855, 95% CI: 0.72–0.96). Another potential capacity of metabolomics regarding TB is to unravel biomarkers capable of predicting early treatment response. In this context, Mahapatra et al. (2014) compared the LC-MS profiles of the urine from pulmonary TB at different times of treatment (baseline, month 1, month 2, and month 6). A comparison of metabolic profiles was further performed, ultimately leading to the selection of six molecular features that showed the highest statistical significance between baseline and early (1 month) treatment samples, from which four features were tentatively annotated as p-aminobenzoic acid, formimino-L-glutamic acid, Leu-Ala, and L-alpha-aspartyl-L-hydroxyproline. From another point of view, in the line of correct treatment management, few preliminary studies have been documented to evaluate possible biomarkers for distinguishing drug susceptible TB (DS-TB) from drug-resistant TB. Based on the results of an untargeted LC-MS analysis performed in MDR and DS-TB patients, Frediani et al. (2014a, b) suggested a relationship between the profile of D-resolvins, phosphatidylinositol, and trehalose-6-mycolate. In this context, the same group performed another LC-MS analysis comparing both MDR-TB and DS-TB patients. Statistical analysis revealed 66 significant features (e.g., mannose, isopentenyl diphosphate, etc.) (Frediani et al. 2015).

The aforementioned studies principally evaluated either latent or active pulmonary tuberculosis. Apart from this, other complications of TB have also been studied. With regard to the lymphatic system, Singh et al. (2014) performed a high-resolution magic-angle spinning (NMR technique used to perform experiments in solid state) analysis of biopsies from lymph node tissue of tuberculous cervical lymphadenitis and non-specific cervical lymphadenitis patients in order to find potential biomarkers for differential diagnosis. Successful separation between both patient groups was observed in PCA score plots. Statistically significant metabolites comprised monounsaturated and polyunsaturated fatty acids, amino acids (e.g., Ala, Lys, Gly, and Tyr), glucose, phosphocholine, and lactate, among others. Diagnostic biomarkers have also been suggested for tubercular meningitis in children. By ^{1}H-NMR, Mason et al. (2016) compared the urinary profiles among TBM patients, non-TBM patients manifesting neurological disease, and healthy controls. In PCA analysis, the TBM group clustered and separated totally from controls, although partially from non-TBM patients. Further statistical analysis revealed a series of metabolites that widely contributed to the separation of groups. These included mostly small organic acids (4-hydroxyhippuric acid, quinolinic acid, 2-ketoglutaric acid, 3-hydroxyisovaleric acid, 2-hydroxybutyric acid, 3-hydroxybutyric acid, succinic acid, methylcitric acid) and amino acids (tryptophan, glutamic acid). Further evaluation of these metabolites revealed a combined model of four metabolites (4-hydroxyhippuric, 2-ketoglutaric, quinolinic, and methylcitric acids), which could differentiate TBM patients at early onset of disease (ROC AUC = 0.974). In order to find biomarkers for genital tuberculosis, Subramani et al. (2016) performed a ^{1}H-NMR analysis of the serum from women with dormant GTB and controls (unexplained infertile women, women undergoing recurrent spontaneous miscarriage (RSM), and healthy proven fertile women undergoing voluntary sterilization). OPLS-DA revealed several metabolites that separate GTB women from the rest of the groups. These included mostly small organic acids (3-hydroxybutyrate, citrate, acetate, succinate) and amino acids (glutamine, glutamic acid, threonine), suggesting potential biomarkers for the differential diagnosis of this disease.

7.4.1.3 Metabolomics of Parasitic Diseases

Parasites are defined as a broad biological group encompassing different types of organisms ranging from protozoa to helminths. In this context, certain apicomplexan protozoa cause relevant human diseases, such as malaria (*Plasmodium* spp.), babesiosis (*Babesia* spp.), toxoplasmosis (*Toxoplasma gondii*), and cryptosporidiosis (*Cryptosporidium* spp.). There is a wide variety of metabolomics experiments of *Plasmodium* spp., which have been extensively reviewed (Lakshmanan et al. 2011; Olszewski and Llinás 2011; Sonawat and Sharma 2012). In contrast, few studies regarding *Toxoplasma gondii* and *Cryptosporidium parvum* have been described (Bisanz et al. 2006; Ng et al. 2012; Ramakrishnan et al. 2012). For these reasons, we focus on the major contributions of metabolomics to the study of malaria in truly infected patients.

With more than 212 million infected individuals and 429,000 deaths per year worldwide reported (2015) (WHO 2017d), malaria is a mosquito-borne disease shows high incidence and considerable mortality rates. Etiological agents of malaria comprehend different obligate intracellular parasites of the genus *Plasmodium* (*P. falciparum*, *P. malariae*, *P. vivax*, *P. ovale*, and *P. knowlesi*). These species parasitize erythrocytes and hepatocytes, causing human malaria with different disease severity and persistence of infection profiles. Although symptoms of mild malaria are often unspecific, severe malaria, usually caused by *P. falciparum*, can cause severe problems such as respiratory distress, renal failure, coagulopathy, shock, encephalopathy, and multiple organ dysfunction (Autino et al. 2012). The gold standard for malaria diagnosis implies direct observation of the parasites in blood smears. However, shortcomings are derived from this method (Hänscheid 2003), which could be potentially overcome by other diagnostic biomarkers. In this context, Lakshmanan et al. (2012) performed a LC-MS analysis of plasma samples from both *P. falciparum*-infected and *P. falciparum*-non-infected individuals, finding excellent group clustering in PLS-DA analyses. Traumatin, a plantlike metabolite was found to be considerably altered between both groups, as well as other statistically significant metabolites such as amino acids, amino acid derived compounds, lipids, and metabolites of the central carbon metabolism. Other studies, apart from suggesting potential diagnostic biomarkers, have been focused on the identification of biomarkers for the prognosis and differential diagnosis of malaria. Severe cases of malaria with CNS affectation—cerebral malaria (CM) or cerebral malaria with multiple organ dysfunction (CMMOD)—are an important source of mortality. With regard to this, Surowiec et al. (2015) performed a GC-MS analysis to compare the plasma profile of pediatric patients suffering from mild and severe malaria and healthy controls. Different metabolites were found to be altered between binary comparisons of disease groups that contributed to the generation of OPLS-DA models with calculated ROC curves. Mild malaria and severe malaria patients could be distinguished from the control groups, as well as mild malaria from severe malaria patients (ROC AUC values of 0.8442, 0.9165, and 0.7637, respectively). The most significantly altered metabolites found in this study included urea, glucuronic acid, histidine, β-hydroxybutyric acid, cysteine, tryptophan, palmitoleic acid, and octadecatrienoic acid. In the same line, Sengupta et al. (2016) performed a ^{1}H-NMR experiment which compared the plasma profiles of differently affected (mild malaria, severe non-cerebral malaria, CM, and CMMOD) *P. falciparum*-infected patients, as well as patients presenting sepsis and viral encephalitis. Partial segregation between all patient groups (all forms of malaria, sepsis, viral encephalitis, and controls) was observed in PCA and OPLS-DA score plots. Important metabolites differentiating malaria patients and controls were lipoproteins (LDL/VLDL), lactate, and glycoproteins. Metabolites discriminating between *P. falciparum*-infected patients and patients suffering from sepsis/encephalitis included LDL/VLDL, lactic acid, and isoleucine. It is worthwhile to comment that a decrease in the glycoprotein levels was observed in CM and CMMOD patients when compared to other malaria patients, suggesting their potential applicability as CM diagnostic biomarkers. Pappa et al. (2015) studied the association between brain volume (a strong predictor of the outcome of cerebral malaria) and the plasma metabolic profile of children with CM. Based on a GC-MS and LC-MS multiplatform analysis, different metabolites were found to be correlated with brain volume. These included, among others, metabolites from the phospholipase A_2 pathway (Arachidonic acid, 1-eicosatrienoylglycerophosphoethanolamine, 1-oleoylglycerophosphoetanolamine), mannitol, urobilin, fatty acids (pentadecanoic acid) and fatty acid-related metabolites (5-hydroxyhexanoic acid). Diagnostic biomarkers of vivax malaria have been proposed by Sengupta et al. (2011), who compared the ^{1}H-NMR urinary profiles of *P. vivax*-infected patients, non-malarial fever patients, and healthy controls. Increased (valerylglycine, N-acetylornithine, salicylurate, phenylpyruvic acid, pipecolic acid, biopterin-3-hydroxybutyrate) and decreased (tyrosine, glucose, guanidinoacetate, alanine, creatine/phosphocreatine, N-acetylglutamate) metabolites were suggested as potential biomarkers for differentiating between malaria patients and controls. In contrast increased (glucose, glutamine, alanine, ornithine, hippurate, phenylalanine) and decreased (N-butyrate, acetate) metabolites were suggested as biomarkers for the differential diagnosis of fever due to either malaria or viral CNS affectation. Lastly, a ^{1}H-NMR longitudinal study performed by the same group searched for diagnostic biomarkers and biomarkers for response to treatment in urine samples from *P. vivax*-infected patients and patients suffering from viral encephalitis (Sengupta et al. 2015). Urine samples were collected from day 1 (admission day) to day 5 and at day 30 after treatment, being this last sample set used as a self-negative control group. Day 30 samples were compared by PCA analyses with samples from day 1 to day 5, revealing progressive loss of observable clustering, which completely disappeared at day 5. Altered metabolites comprised, among others, alanine, glycine, 3-methylhistidine, hippuric acid, taurine, guanidinoacetate, and citric acid. Predictive models were generated, which were able to identify *P. vivax* malaria patients with up to a value of 85% success rate in a second sample cohort. Discriminatory metabolites were identified when comparing the urine of viral fever patients and *P. vivax*-infected patients from day 1 (particularly taurine and citrate, with ROC AUC values of 0.66 and 0.69, respectively).

From the clinical point of view, kinetoplastids are another group of relevant protozoal organisms in which various etiological agents of important human diseases such as human African trypanosomiasis (HAT) and Chagas disease are included. More than 65 million people are at risk of developing HAT, which is caused by *Trypanosoma brucei gambiense* and *Trypanosoma brucei rhodesiense* (WHO 2017c). Two stages of HAT can be clearly distinguished. In the first stage, parasites live and replicate extracellularly in the circulatory and lymphatic systems, whereas in the second stage (sleeping sickness) CNS invasion occurs, causing neurological disorders and, ultimately, coma and death if untreated (Kennedy 2004). Since important toxicity is derived from the current therapies for HAT advanced stages, a sensitive and specific test for disease staging is required (Kennedy 2004). However, the current surrogate standard for HAT staging lacks sensitivity and is invasive (requiring a lumbar puncture), since it implies the direct observation of either the parasite or an increase in the white blood cells count in the CSF (Vincent et al. 2016). To potentially overcome these limitations and study the global metabolic alterations of HAT, Vincent et al. (2016) performed a LC-MS analysis which compared the plasma, urine, and CSF metabolic profiles of *T. brucei*-infected patients of first and second HAT stages with controls. In CSF analysis, neopterin could discriminate between disease stages with 100% sensitivity and specificity. Furthermore, the levels of 5-hydroxytryptophan, a metabolite associated with somnolence, and other relevant metabolites, such as kynurenine, O-acetylcarnitine, N-acetylhistidine, tryptophan, N-acetyl-aspartylglutamate, linoleamide, and oleamide, were also found to be altered. Altered plasma metabolites were used to construct a model with two unidentified features that effectively discriminated both disease stages, (ROC AUC = 0.92) suggesting a potentially useful, less invasive method for the staging of this potentially fatal disease.

Chagas disease, caused by *Trypanosoma cruzi*, affects 8 million people worldwide, causing 15,000 deaths per year (Gironès et al. 2014). In contrast with *T. brucei* life cycle, *T. cruzi* requires to invade host cells to replicate as amastigotes, ultimately causing cellular lysis. Given that cardiac muscle cells are particularly invaded by this parasite, serious cardiovascular alterations are generally presented in *T. cruzi*-infected patients (Teixeira et al. 2011). Chagas disease can be manifested in an acute or chronic form and is usually asymptomatic (Teixeira et al. 2011). Current treatment efficacy is poorly assessed, due to the absence of a surrogate indicator of therapeutic success (Gironès et al. 2014). A preliminary study performed by Gironès et al. (2014) aimed to study the general metabolic alterations found in heart tissue and plasma of *T. brucei*-infected mice by untargeted GC-MS and LC-MS analysis. Metabolites such as p-cresol sulfate (a toxic metabolite from microbial origin), kynurenine, and allantoin were suggested as potential biomarkers of this disease, due to their considerable concentration changes found in serum.

Focusing on helminthiasis, several illustrative examples of how metabolomics has helped to identify biomarkers can be found in the study of onchocerciasis. This disease, commonly known as river blindness, affects more than 37 million people worldwide (Globisch et al. 2013). Onchocerciasis is a blackfly-borne (*Simulium* spp.) disease caused by filariae (*Onchocerca volvulus*), which typically affects the eyes and skin, causing inflammation and blindness (Globisch et al. 2013). Denery et al. (2010) performed a combined non-targeted and targeted analysis to evaluate the metabolic profile from serum and plasma of *O. volvulus* positive and negative individuals from different regions (African countries, Guatemala, USA, India) and compared such profile with other possible comorbidities (Chagas disease, leishmaniasis, HAT). After data filtering, 14 features were selected for LC-MS/MS and Fourier transform ion cyclotron resonance mass spectrometry analysis. These included hexacosenoic and pentacosenoic acids, a phosphorylated sphingolipid, a phosphorylated sphingolipid, sterol lipids, and fatty acids. Considering such candidate biomarkers, PCA analysis revealed discrimination of *O. volvulus* positive patients for the African samples, although group overlapping was observed (mainly in Guatemalan samples and samples from patients affected by either Chagas disease or leishmaniasis). Subsequent models considering these 14 features were generated by machine learning algorithms. The highest discriminatory capacity for the entire sample set was achieved using a random forests algorithm (88.6% of sensitivity and 88.3% of specificity), whereas for the African sample set, a functional trees algorithm achieved a 100% sensitivity and specificity. In another study of the same group, Globisch et al. (2013) performed a non-targeted LC-MS analysis in urine samples, identifying a highly enriched feature in *O. volvulus* positive African patients. Further validation of this compound will be discussed (see Sect. 7.4.2).

7.4.1.4 Metabolomics of Fungal Diseases

To the best of our knowledge, few metabolomics studies of fungal infections have been performed, being these particularly focused on the study of invasive fungal infections (IFIs) (de Francisco et al. 2015; Dessì et al. 2014; Koo et al. 2014; Liew et al. 2016). IFIs are generally caused by fungi from the genera *Aspergillus* (invasive aspergillosis or IA), *Cryptococcus* (cryptococcosis), and *Candida* (invasive candidiasis). Because of the inhalation of conidia, initial infection by *Aspergillus* spp. and *Cryptococcus* spp. is usually localized in the lungs. However, as in infections caused by *Candida* spp., fungi can spread to other regions such as the blood or CNS, developing potentially fatal diseases (Badiee and Hashemizadeh 2014). For this reason, IFIs constitute a notable source of mortality among immunocompromised patients (Enoch et al. 2017). The gold standard for IFI diagnosis is cell culture, which is usually slow (>72 h) and lacks sensitivity (Enoch et al. 2017). Koo et al. (2014) evaluated the volatile organic compounds (VOCs) from the breath of pulmonary IA suspected patients by GC-MS, finding a metabolic signature in IA patients (increased levels of β-trans-bergamotene, α-trans-bergamotene, β-vatirenene, and trans-geranylacetone). Focusing on cryptococcosis, a study published by Liew et al. (2016) simulated a primary *C. neoformans* infection using lung epithelial cell cultures. Two initial doses of yeast cells were used for the study, which consisted in the incubation and analysis of yeast positive and negative controls at 6, 12, and 18 h. GC-MS analysis revealed altered levels of metabolites in the supernatants of *C. neoformans*-infected cultures. Independently of the inoculation dose, cysteine was identified as a discriminative metabolite at 6 h of incubation. In contrast lactic acid, tyrosine, fructose, fumaric acid, and pantothenic acid were identified at 12 and 18 h of incubation in low inoculated cell cultures, whereas malic acid, 3-hydroxyvaleric acid, 3-phenyllactic acid, and adonitol were significantly altered in highly inoculated cell cultures. Another form of IFI, invasive candidiasis of the bloodstream, displays high mortality rates in newborns (up to 30%) (Dessì et al. 2014). A preliminary study performed by Dessì et al. (2014) compared the GC-MS urinary metabolic profiles between healthy controls and a newborn presenting fungal sepsis, obtained at different time points. Groups clustered separately in PCA analysis and the most relevant metabolites altered in the fungal septic patient were identified as amino acids (glycine, serine, threonine), carbohydrates (glucose, maltose), and both organic and fatty acids (citrate, hexadecanoate, octadecanoate), suggesting potential biomarkers for the diagnosis of this potentially fatal disease.

7.4.1.5 Metabolomics of Diseases with Diverse Etiological Agents

In clinics, one of the main potential goals of biomarkers is to discriminate between different etiological agents that can cause a disease with the same symptoms, contributing to a better patient management and a more sensitive use of pharmacological resources. In this line, Slupsky et al. (2009) performed a study which aimed to identify potential biomarkers for community-acquired pneumonia (CAP). CAP is a serious condition, generally caused by *Streptococcus pneumoniae*, wherein patients show a high mortality rate (up to 25%) (MMWR Recommendations and reports 1996). Currently, the standard diagnosis for CAP is based on pathogen isolation and culture from biofluids. This process has serious disadvantages: *S. pneumoniae* culture is time-consuming (typically more than 36 h) (Marston et al. 1997) and displays notable false positive rates (up to 65% in children) (Faden et al. 2002), since this microorganism can be found in the commensal human microbiota. In order to perform a preliminary approach for adequate biomarker identification, ^{1}H-NMR analysis was performed in urine samples, wherein PCA analysis revealed notable discrimination of *S. pneumoniae*-infected and control groups. To further assess selectivity of these biomarkers, the *S. pneumoniae*-infected group was compared with other patients with lung infections that shared common clinical features with CAP (caused by *Legionella pneumophila*, *Mycobacterium tuberculosis*, *Staphylococcus aureus*, *Coxiella burnetii*, *Haemophilus influenzae*, *Mycoplasma pneumoniae*, *Escherichia coli*, *Enterococcus faecalis*, *Moraxella catarrhalis*, *Streptococcus viridans*, *Streptococcus anginosus*, influenza A virus, picornavirus, respiratory syncytial virus, parainfluenza viruses, coronaviruses, human metapneumovirus, or hantavirus). PCA analysis revealed notable discrimination between pneumococcal pneumonia and other bacterial and viral pulmonary infections except for *L. pneumophila*, evidencing the notable differential diagnostic capacity achieved by this study.

Other studies are specifically focused on the identification of biomarkers for differential diagnosis, such as the one performed by Li et al. (2017), who explored the possibility of discriminating tubercular meningitis from viral meningitis. Meningitis is a serious condition wherein inflammation of the meninges is produced, caused by a multitude of virus, bacteria (including mycobacteria), fungi, or parasites. Clinical symptoms (e.g., fever, neck stiffness, headache, photophobia, and vomiting) are common among diverse etiological agent, and therefore a fast patient management becomes crucial in order to reduce meningitis mortality. For this reason, given the wide variety of agents causing meningitis, pathogen-specific diagnostic methods have been developed (Bahr and Boulware 2011). In the same framework, this study reported a ^{1}H-NMR analysis of the CSF from TBM and viral meningitis patients. While PCA analysis showed partial clustering of sample groups, OPLS-DA analysis allowed their complete discrimination. Metabolites most contributing to such discrimination were amino acids (glycine, tyrosine, glutamine, threonine, valine, alanine, aspartic acid, asparagine), carbohydrates (glucose, sucrose, fructose), small organic acids (acetate, citrate, malonic acid, pyruvic acid, malonate), lipoproteins, and others (betaine, glycerine, putrescine, choline), thus confirming a different metabolic signature among TBM and viral meningitis.

Another differential analysis example was proposed by Soga et al. (2011). In this study, the capacity of a CE-MS analysis of serum to discriminate among different forms of liver disease was evaluated. Patients suffering from infectious diseases at different disease stages, such as asymptomatic hepatitis B virus infection (AHB), chronic hepatitis B (CHB), chronic hepatitis C (CHC) hepatitis C with persistently normal alanine transaminase (CNALT), cirrhosis type C (CIR), hepatocellular carcinoma, non-alcoholic steatohepatitis (NASH), simple steatosis (SS), and drug-induced liver injury (DI), were included. Analysis revealed, among others, 49 statistically significant metabolites, mainly corresponding to γ-glutamyl dipeptides which were generally found to be increased in liver disease. The discriminatory capacity of models comprising the levels of γ-glutamyl dipeptides alone or in combination with biochemical parameters (aspartate aminotransferase (AST) and alanine aminotransferase (ALT), methionine sulfoxide) was tested in a first training and a second validation cohort. ROC AUC for both cohorts were able to effectively each liver disease with the rest of liver diseases and healthy controls (values of the ROC AUCs from the validation cohorts were 0.895, 0.707, 0.993, 0.803, 0.967, 0.849, 0.763, and 0.762 for CNALT (γ-Glu-Taurine, ALT), CHC (γ-Glu-Lys), CIR (γ-Glu-Ala, γ-Glu-Leu, γ-Glu-Ser, γ-Glu-Taurine), HCC (γ-Glu-Ala, γ-Glu-Citrulline, γ-Glu-Thr, γ-Glu-Phe), C (γ-Glu-Phe), DI (γ-Glu-Citrulline, ALT), AHB (Methionine sulfoxide, AST), and CHB (γ-Glu-Thr), respectively), brilliantly evidencing the potential of metabolomics in differential diagnosis.

Independently of the disease-causing pathogen, metabolic biomarkers for the diagnosis of multipathogenic diseases can be potentially identified. In this line, Su et al. (2014) identified several diagnostic and prognostic biomarkers of sepsis by LC-MS/MS analysis. Sepsis is a serious condition caused by multiple possible pathogens that spread from a local infection focus through the blood, generating a systemic infection, ultimately leading to septic shock and multiple organ failure. Even with the correct treatment, patients with sepsis show considerably high mortality rates (Martin 2012). The study was performed on controls, septic patients, and patients showing non-septic systemic inflammatory response syndrome (SIRS group), a sepsis-like condition. Further subdivision of the sepsis patients' group was performed attending to different clinical criteria, classifying these in sepsis, severe sepsis, septic shock, and death patients (death occurred within 48 h after the analysis). Serum samples were obtained within 24 h after emergency room admission except for the death group, from which samples were collected within 48 h before death. Several metabolites were found to be altered between the different disease groups. Different statistical analyses were performed; PCA analysis revealed differentiation of controls from the rest of group patients, while OPLS-DA allowed the observation of discrimination between SIRS and sepsis groups. Further PLS-DA analysis differentiated mild sepsis from severe sepsis and septic shock. Additionally, PLS-DA analysis also differentiated patients of the death subgroup from the rest of septic patients. Eventually, the ROC curves for 12 selected significant metabolites among the different groups were calculated. Regarding sepsis diagnosis, lactitol dehydrate, N-nonanoyl-glycine, S-phenyl-cysteine, and S-(3-methylbutanoyl)-dihydrolipoamide-E were identified as potential biomarkers, from which this last obtained the best ROC values (AUC = 0.900; 95% CI: 0.799–1.001). Severity of sepsis could be determined by N,N-dimethyllysine, glycerylphosphorylethanolamine, cysteine, and 2-phenylacetamide, accounting this last metabolite for the highest ROC value between these potential biomarkers (AUC = 0.804; 95% CI: 0.647–0.962). Predictive biomarkers of death were identified as phosphatidylglycerol (22:2/0:0), S-(3-methylbutanoyl)-dihydrolipoamide-E, S-succinyl glutathione, and glycerophosphocholine), being phosphatidylglycerol (22:2/0:0) the best predictor of death within these groups (ROC AUC = 0.96; 95% CI: 0.863–1.057). In conclusion, this study was able to clearly assess disease diagnosis, differential diagnosis, and prognosis of sepsis between the different patients.

7.4.2 Biomarker Validation

To our knowledge, most of infectious disease metabolomics experiments have been focused on a discovery phase combining, in some cases, non-targeted and targeted analysis on the same samples. Fully validation following the criteria mentioned above is not so common.

One of the most successful examples is the validation of a diagnostic method for urinary tract infection (see Sect. 7.4.1.2) by Gupta et al. (2012). In this target experiment, the selection of the analytes was done based on the findings of previous fingerprinting studies (Gupta et al. 2009; Lenz et al. 2003, 2004; Sweatman et al. 1993), which suggested an increase of certain end products of bacterial metabolism in contaminated urine samples (lactate, acetate, succinate, formate, ethanol, citrate, trimethylamine, urea, glycine, hippurate, creatinine, trimethylamine, trimethylamine-N-oxide). For the biomarker validation, different experiments were designed: first, an ex vivo study was performed over a cohort of 682 suspected urine patients and

50 healthy controls to test the diagnostic capabilities of the method; secondly, an in vitro study was carried out in pathogen inoculated culture media in which samples were generated by urine sterilization supplementation and uropathogen inoculation. The results showed significant statistical variations in the levels of the measured from which acetate, formate, lactate, and succinate urine concentrations were selected to generate a model that was able to remarkably discriminate UTI patients from healthy controls (ROC AUC = 0.996; 95% CI: 0.992–1.001). Moreover, the levels of these metabolites were able to differentiate urine samples infected by either gram-positive cocci or gram-negative bacilli (ROC AUC = 0.975; 95% CI: 0.957–0.993), indicating the potential differential diagnostic capabilities of this method; finally, on the third place, a longitudinal study was performed, wherein the urine 10 UTI patients were sampled before and after 7, 14, and 21 days of antibiotic treatment. Further validation of the method was confirmed by the observation of a control-like urine profile in successfully antibiotic treatment patients, indicating eradication of the infection. Therefore, the remarkable ability of this large-scale study to identify and discriminate the causal agent of UTI was stated.

In relation with onchocerciasis, as mentioned above (see Sect. 7.4.1.3), Globisch et al. (2013) identified a highly increased feature in African *O. volvulus* positive urine samples (*m/z* 356.1340). In order to decipher the molecular structure of such feature, the fragmentation patterns of a subsequent LC-MS/MS analysis revealed losses of glucuronic acid, an aliphatic N-acetyl moiety, water, and ethylene. As a result, N-acetyltyramine-O,β-glucuronide (NATOG) was proposed as candidate structure. Structure assignment was confirmed by LC purification of NATOG from samples and subsequent comparative analysis of the *m/z* and retention times of both purified and de novo synthesized NATOG. However, when analysis of NATOG was performed in Guatemalan samples, lower levels of NATOG were detected, suggesting phylogenetic differences between African and American *O. volvulus*. NATOG capacities as prognostic biomarker were further tested in a comparative analysis wherein NATOG levels were measured in urine samples from *O. volvulus* positive patients receiving either doxycycline or placebo at 4–6 weeks posttreatment. Diminished levels of NATOG were observed in successfully treated patients, thus proving its potential use in disease progression evaluation.

7.5 Conclusions and Future Trends

The examples discussed in previous sections just illustrate selected representative possibilities of metabolomics which are far from being fully summarized. We have observed few biomarkers that have been integrated in clinics. Important obstacles to this are the lack of a consensus on validation criteria and standardization of metabolomics experiments. There is a need for a refinement of both analytical (e.g., pathogen isolation methods) and computational techniques (e.g., refinement of computational models of host–pathogen interactions) to ease the conclusions inferred from analysis. Given the current landscape of the actual development of metabolic biomarkers in this area, it can be concluded that it is time to take advantage of the evidence supported by previous knowledge found in preliminary studies. This would ultimately permit wider integration of metabolomics in the context of a more translational research, allowing this yet relatively recent omics science to reach one of its paramount objectives: to become a standard, well-stablished tool for biomarker identification which eventually will need to be competitive in cost/benefit with the classical clinical test.

Funding Statement The co-authors would like to acknowledge funding from the Spanish Ministry of Economy and Competitiveness (CTQ2014-55279-R). Author M. F.-G. acknowledges Fundación Universitaria San Pablo CEU for his PhD fellowship.

References

Allegretti JR, Kearney S, Li N, Bogart E, Bullock K, Gerber GK, Bry L, Clish CB, Alm E, Korzenik JR (2016) Recurrent *Clostridium difficile* infection associates with distinct bile acid and microbiome profiles. Aliment Pharmacol Ther 43(11):1142–1153. https://doi.org/10.1111/apt.13616

Autino B, Corbett Y, Castelli F, Taramelli D (2012) Pathogenesis of malaria in tissues and blood. Mediterr J Hematol Infect Dis. https://doi.org/10.4084/mjhid.2012.061

Badiee P, Hashemizadeh Z (2014) Opportunistic invasive fungal infections: diagnosis & clinical management. Indian J Med Res 139(2):195–204

Bahr NC, Boulware DR (2014) Methods of rapid diagnosis for the etiology of meningitis in adults. Biomark Med 8(9):1085–1103. https://doi.org/10.2217/bmm.14.67

Bedossa P, Poynard T (1996) An algorithm for the grading of activity in chronic hepatitis C. Hepatology 24(2):289–293. https://doi.org/10.1002/hep.510240201

Bhatt S, Gething PW, Brady OJ, Messina JP, Farlow AW, Moyes CL, Drake JM, Brownstein JS, Hoen AG, Sankoh O, Myers MF, George DB, Jaenisch T, Wint GRW, Simmons CP, Scott TW, Farrar JJ, Hay SI (2013) The global distribution and burden of dengue. Nature 496(7446):504–507. https://doi.org/10.1038/nature12060

Birungi G, Chen SM, Loy BP, Ng ML, Li SFY (2010) Metabolomics approach for investigation of effects of dengue virus infection using the EA hy926 cell line. J Proteome Res 9(12):6523–6534. https://doi.org/10.1021/pr100727m

Bisanz C, Bastien O, Grando D, Jouhet J, Maréchal E, Cesbron-Delauw MF (2006) *Toxoplasma gondii* acyl-lipid metabolism: de novo synthesis from apicoplast-generated fatty acids versus scavenging of host cell precursors. Biochem J 394:197–205. https://doi.org/10.1042/bj20050609

Brand HK, Hermans PW, de Groot R (2010) Host biomarkers and paediatric infectious diseases: from molecular profiles to clinical application. In: Finn A, Curtis N, Pollard AJ (eds) Hot topics in infection and immunity in children VI, 1st edn. Springer-Verlag, New York, pp 19–31

Buckle GC, Walker CL, Black RE (2012) Typhoid fever and paratyphoid fever: systematic review to estimate global morbidity and mortality for 2010. J Glob Health 2(1):010401. https://doi.org/10.7189/jogh.02.010401

Burnham CA, Carroll KC (2013) Diagnosis of *Clostridium difficile* infection: an ongoing conundrum for clinicians and for clinical laboratories. Clin Microbiol Rev 26(3):604–630. https://doi.org/10.1128/cmr.00016-13

Cassol E, Misra V, Dutta A, Morgello S, Gabuzda D (2014) Cerebrospinal fluid metabolomics reveals altered waste clearance and accelerated aging in HIV patients with neurocognitive impairment. AIDS 28(11):1579–1591. https://doi.org/10.1097/qad.0000000000000303

Ceccotti G, Meoni G, Tenori L, Gragnani L, Fognani E, Gianni E, Luchinat C, Zignego AL (2016) A serum metabolomic analysis of HCV-infected patients successfully treated with IFN-free DAA regimens. Hepatology 64:378A–378A

Cheng J, Joyce A, Sterling RK, Sanyal AJ (2012) Metabolomic profiling identifies fibrosis-associated differences in hepatitis C virus infection. Gastroenterology 142(5):S968

Cribbs SK, Park Y, Guidot DM, Martin GS, Brown LA, Lennox J, Jones DP (2014) Metabolomics of bronchoalveolar lavage differentiate healthy HIV-1-infected subjects from controls. AIDS Res Hum Retroviruses 30(6):579–585. https://doi.org/10.1089/aid.2013.0198

Cui L, Lee YH, Thein TL, Fang J, Pang J, Ooi EE, Leo YS, Ong CN, Tannenbaum SR (2016) Serum metabolomics reveals serotonin as a predictor of severe dengue in the early phase of dengue fever. Plos Negl Trop Dis. https://doi.org/10.1371/journal.pntd.0004607

Currie BJ, Fisher DA, Howard DM, Burrow JN, Lo D, Selva-Nayagam S, Anstey NM, Huffam SE, Snelling PL, Marks PJ, Stephens DP, Lum GD, Jacups SP, Krause VL (2000) Endemic melioidosis in tropical northern Australia: a 10-year prospective study and review of the literature. Clin Infect Dis 31(4):981–986. https://doi.org/10.1086/318116

da Luz JA, Hans E, Zeng A-P (2014) Automated fast filtration and on-filter quenching improve the intracellular metabolite analysis of microorganisms. Eng Life Sci 14(2):135–142. https://doi.org/10.1002/elsc.201300099

Dann S, Aitken S, Ross C, Tessier MEM, Loeffelholz M, Koussoulas K, Bornstein JC, Feng H, Versalovic J, Pothoulakis C, Garey KW, Savidge T (2015) Zolpidem confers disease susceptibility to *Clostridium difficile* infection. Gastroenterology 148(4):S727–S728

Das MK, Bishwal SC, Das A, Dabral D, Badireddy VK, Pandit B, Varghese GM, Nanda RK (2015) Deregulated tyrosine-phenylalanine metabolism in pulmonary tuberculosis patients. J Proteome Res 14(4):1947–1956. https://doi.org/10.1021/acs.jproteome.5b00016

de Francisco TM, Zaramella IF, Gasparetto JC, Cerqueira LB, Piantavini MS, Pontarolo R, Campos FR (2015) Rapid detection of aspergillosis in immunocompromised patients using DIMS and chemometric analysis. Anal Methods 7(15):6346–6351. https://doi.org/10.1039/c5ay01146a

Denery JR, Nunes AAK, Hixon MS, Dickerson TJ, Janda KD (2010) Metabolomics-based discovery of diagnostic biomarkers for onchocerciasis. PloS Negl Trop Dis. https://doi.org/10.1371/journal.pntd.0000834

Dessì A, Liori B, Caboni P, Corsello G, Giuffrè M, Noto A, Serraino F, Stronati M, Zaffanello M, Fanos V (2014) Monitoring neonatal fungal infection with metabolomics. J Matern Fetal Neonatal Med 27:34–38. https://doi.org/10.3109/14767058.2014.954787

Dettmer K, Aronov PA, Hammock BD (2007) Mass spectrometry-based metabolomics. Mass Spectrom Rev 26(1):51–78. https://doi.org/10.1002/mas.20108

du Preez I, Loots DT (2013) New sputum metabolite markers implicating adaptations of the host to *Mycobacterium tuberculosis*, and vice versa. Tuberculosis (Edinb) 93(3):330–337. https://doi.org/10.1016/j.tube.2013.02.008

du Preez I, Sithebe NP (2013) The use of metabolomics as a tool to investigate hepatitis C. Metabolomics 9(2):497–505. https://doi.org/10.1007/s11306-012-0467-8

Dunn WB, Wilson ID, Nicholls AW, Broadhurst D (2012) The importance of experimental design and QC samples in large-scale and MS-driven untargeted metabolomic studies of humans. Bioanalysis 4(18):2249–2264. https://doi.org/10.4155/bio.12.204

El-Bacha T, Struchiner CJ, Cordeiro MT, Almeida FCL, Marques ET Jr, Da Poian AT (2016) ^{1}H nuclear magnetic resonance metabolomics of plasma unveils liver dysfunction in dengue patients. J Virol 90(16):7429–7443. https://doi.org/10.1128/jvi.00187-16

Embade N, Marino Z, Diercks T, Cano A, Lens S, Cabrera D, Navasa M, Falcón-Pérez JM, Caballería J, Castro A, Bosch J, Mato JM, Millet O (2016) Metabolic characterization of advanced liver fibrosis in HCV patients as studied by serum ^{1}H-NMR spectroscopy. PLoS One 11(5). https://doi.org/10.1371/journal.pone.0155094

Enoch DA, Yang H, Aliyu SH, Micallef C (2017) The changing epidemiology of invasive fungal infections. Methods Mol Biol 1508:17–65. https://doi.org/10.1007/978-1-4939-6515-1_2

Eoh H, Rhee KY (2013) Multifunctional essentiality of succinate metabolism in adaptation to hypoxia in *Mycobacterium tuberculosis*. Proc Natl Acad Sci U S A 110(16):6554–6559. https://doi.org/10.1073/pnas.1219375110

Faden H, Heimerl M, Varma C, Goodman G, Winkelstein P (2002) Urinary excretion of pneumococcal cell wall polysaccharide in children. Pediatr Infect Dis J 21(8):791–793. https://doi.org/10.1097/01.inf.0000022727.27339.3e

Fatima Z, Hameed S, Saibabu V, Sharma S, Hans S (2017) Tuberculosis: propagation beyond lungs. In: Bhargava S (ed) Diagnosis & management of tuberculosis, 1st edn. Open Access eBooks, Wilmington, pp 1–21

Feng S, Du Y-Q, Zhang L, Zhang L, Feng R-R, Liu S-Y (2015) Analysis of serum metabolic profile by ultra-performance liquid chromatography-mass spectrometry for biomarkers discovery: application in a pilot study to discriminate patients with tuberculosis. Chin Med J (Engl) 128(2):159–168. https://doi.org/10.4103/0366-6999.149188

Fiehn O (2002) Metabolomics – the link between genotypes and phenotypes. Plant Mol Biol 48(1–2):155–171. https://doi.org/10.1023/a:1013713905833

Fontaine KA, Sanchez EL, Camarda R, Lagunoff M (2015) Dengue virus induces and requires glycolysis for optimal replication. J Virol 89(4):2358–2366. https://doi.org/10.1128/jvi.02309-14

Frediani JK, Jones D, Tukvadze N, Sanikidze E, Kipiani M, Uppal K, Kurani S, Hebbar G, Colas R, Dalli J, Serhan C, Tangpricha V, Blumberg H, Ziegler T (2014a) Plasma high-resolution metabolomic profiling reveals upregulation of specific resolvins in patients with pulmonary tuberculosis. FASEB J 28(1)

Frediani JK, Jones DP, Tukvadze N, Uppal K, Sanikidze E, Kipiani M, Tran VT, Hebbar G, Walker DI, Kempker RR, Kurani SS, Colas RA, Dalli J, Tangpricha V, Serhan CN, Blumberg HM, Ziegler TR (2014b) Plasma metabolomics in human pulmonary tuberculosis disease: a pilot study. PLoS ONE 9. https://doi.org/10.1371/journal.pone.0108854

Frediani JK, Chong E, Jones D, Yu T, Zhang L, Tukvadze N, Sanikidze E, Kipiani M, Hebbar G, Tran V, Kempker R, Tangpricha V, Blumberg H, Ziegler T (2015) Comparative plasma high-resolution metabolomic profiling in patients with drug-susceptible and multi-drug resistant pulmonary tuberculosis. FASEB J 29

Ghannoum MA, Mukherjee PK, Jurevic RJ, Retuerto M, Brown RE, Sikaroodi M, Webster-Cyriaque J, Gillevet PM (2013) Metabolomics reveals differential levels of oral metabolites in HIV-infected patients: toward novel diagnostic targets. OMICS 17(1):5–15. https://doi.org/10.1089/omi.2011.0035

Gironès N, Carbajosa S, Guerrero NA, Poveda C, Chillón-Marinas C, Fresno M (2014) Global metabolomic profiling of acute myocarditis caused by *Trypanosoma cruzi* infection. PloS Negl Trop Dis. https://doi.org/10.1371/journal.pntd.0003337

Globisch D, Moreno AY, Hixon MS, Nunes AAK, Denery JR, Specht S, Hoerauf A, Janda KD (2013) Onchocerca volvulus-neurotransmitter tyramine is a biomarker for river blindness. Proc Natl Acad Sci U S A 110(11):4218–4223. https://doi.org/10.1073/pnas.1221969110

Godzien J, Ciborowski M, Angulo S, Barbas C (2013) From numbers to a biological sense: how the strategy chosen for metabolomics data treatment may affect final results. A practical example based on urine fingerprints obtained by LC-MS. Electrophoresis 34(19):2812–2826. https://doi.org/10.1002/elps.201300053

Godzien J, Alonso-Herranz V, Barbas C, Armitage EG (2015) Controlling the quality of metabolomics data: new strategies to get the best out of the QC sample. Metabolomics 11(3):518–528. https://doi.org/10.1007/s11306-014-0712-4

Gowda GA, Shanaiah N, Raftery D (2012) Isotope enhanced approaches in metabolomics. Adv Exp Med Biol 992:147–164. https://doi.org/10.1007/978-94-007-4954-2_8

Günther G (2014) Multidrug-resistant and extensively drug-resistant tuberculosis: a review of current concepts and future challenges. Clin Med (Lond) 14(3):279–285

Gupta A, Dwivedi M, Mahdi AA, Gowda GA, Khetrapal CL, Bhandari M (2009) ^{1}H-nuclear magnetic resonance spectroscopy for identifying and quantifying common uropathogens: a metabolic approach to the urinary tract infection. BJU Int 104(2):236–244. https://doi.org/10.1111/j.1464-410X.2009.08448.x

Gupta A, Dwivedi M, Mahdi AA, Khetrapal CL, Bhandari M (2012) Broad identification of bacterial type in urinary tract infection using ^{1}H NMR spectroscopy. J Proteome Res 11(3):1844–1854. https://doi.org/10.1021/pr2010692

Hänscheid T (2003) Current strategies to avoid misdiagnosis of malaria. Clin Microbiol Infect 9(6):497–504. https://doi.org/10.1046/j.1469-0691.2003.00640.x

Hewer R, Vorster J, Steffens FE, Meyer D (2006) Applying biofluid 1H NMR-based metabonomic techniques to distinguish between HIV-1 positive/AIDS patients on antiretroviral treatment and HIV-1 negative individuals. J Pharm Biomed Anal 41(4):1442–1446. https://doi.org/10.1016/j.jpba.2006.03.006

Himmelreich U, Accurso R, Malik R, Dolenko B, Somorjai PR, Gupta RK, Gomes L, Mountford CE (2005) Identification of *Staphylococcus aureus* brain abscesses: rat and human studies with ^{1}H MR spectroscopy. Radiology 236(1):261–270. https://doi.org/10.1148/radiol.2361040869

Jain KK (2010) Role of Biomarkers in Health Care. In: Jain KK (ed) The Handbook of Biomarkers, 1st edn. Humana Press, New York, pp 134–149

John DV, Lin Y-S, Perng GC (2015) Biomarkers of severe dengue disease – a review. J Biomed Sci 22:83. https://doi.org/10.1186/s12929-015-0191-6

Kao D, Ismond KP, Tso V, Millan B, Hotte N, Fedorak RN (2016) Urine-based metabolomic analysis of patients with *Clostridium difficile* infection: a pilot study. Metabolomics 12(8). https://doi.org/10.1007/s11306-016-1080-z

Kennedy PG (2004) Human African trypanosomiasis of the CNS: current issues and challenges. J Clin Invest 113(4):496–504. https://doi.org/10.1172/jci200421052

Koo S, Thomas HR, Daniels SD, Lynch RC, Fortier SM, Shea MM, Rearden P, Comolli JC, Baden LR, Marty FM (2014) A breath fungal secondary metabolite signature to diagnose invasive aspergillosis. Clin Infect Dis 59(12):1733–1740. https://doi.org/10.1093/cid/ciu725

Laiakis EC, Hyduke DR, Fornace AJ Jr (2012) Comparison of mouse urinary metabolic profiles after exposure to the inflammatory stressors gamma radiation and lipopolysaccharide. Radiat Res 177(2):187–199. https://doi.org/10.1667/rr2771.1

Lakshmanan V, Rhee KY, Daily JP (2011) Metabolomics and malaria biology. Mol Biochem Parasitol 175(2):104–111. https://doi.org/10.1016/j.molbiopara.2010.09.008

Lakshmanan V, Rhee KY, Wang W, Yu Y, Khafizov K, Fiser A, Wu P, Ndir O, Mboup S, Ndiaye D, Daily JP (2012) Metabolomic analysis of patient plasma yields evidence of plant-like alpha-linolenic acid metabolism in *Plasmodium falciparum*. J Infect Dis 206(2):238–248. https://doi.org/10.1093/infdis/jis339

Lau SK, Lee KC, Lo GC, Ding VS, Chow WN, Ke TY, Curreem SO, To KK, Ho DT, Sridhar S, Wong SC, Chan JF, Hung IF, Sze KH, Lam CW, Yuen KY, Woo PC (2016) Metabolomic profiling of plasma from melioidosis patients using UHPLC-QTOF MS reveals novel biomarkers for diagnosis. Int J Mol Sci 17(3). https://doi.org/10.3390/ijms17030307

Lenz EM, Bright J, Wilson ID, Morgan SR, Nash AF (2003) A ^{1}H NMR-based metabonomic study of urine and plasma samples obtained from healthy human subjects. J Pharm Biomed Anal 33(5):1103–1115. https://doi.org/10.1016/s0731-7085(03)00410-2

Lenz EM, Bright J, Wilson ID, Hughes A, Morrisson J, Lindberg H, Lockton A (2004) Metabonomics, dietary influences and cultural differences: a ^{1}H NMR-based study of urine samples obtained from healthy British and Swedish subjects. J Pharm Biomed Anal 36(4):841–849. https://doi.org/10.1016/j.jpba.2004.08.002

Li Z, Du B, Li J, Zhang J, Zheng X, Jia H, Xing A, Sun Q, Liu F, Zhang Z (2017) Cerebrospinal fluid metabolomic profiling in tuberculous and viral meningitis: screening potential markers for differential diagnosis. Clin Chim Acta 466:38–45. https://doi.org/10.1016/j.cca.2017.01.002

Liew KL, Jee JM, Yap I, Yong PVC (2016) In vitro analysis of metabolites secreted during infection of lung epithelial cells by *Cryptococcus neoformans*. PLoS ONE. https://doi.org/10.1371/journal.pone.0153356

Limmathurotsakul D, Jamsen K, Arayawichanont A, Simpson JA, White LJ, Lee SJ, Wuthiekanun V, Chantratita N, Cheng A, Day NPJ, Verzilli C, Peacock SJ (2010) Defining the true sensitivity of culture for the diagnosis of melioidosis using Bayesian latent class models. PLoS One 5(8). https://doi.org/10.1371/journal.pone.0012485

Luier L, Loots DT (2016) Tuberculosis metabolomics reveals adaptations of man and microbe in order to outcompete and survive. Metabolomics 12(3). https://doi.org/10.1007/s11306-016-0969-x

Lv H, Hung CS, Chaturvedi KS, Hooton TM, Henderson JP (2011) Development of an integrated metabolomic profiling approach for infectious diseases research. Analyst 136(22):4752–4763. https://doi.org/10.1039/c1an15590c

Mahapatra S, Hess AM, Johnson JL, Eisenach KD, DeGroote MA, Gitta P, Joloba ML, Kaplan G, Walzl G, Boom WH, Belisle JT (2014) A metabolic biosignature of early response to anti-tuberculosis treatment. BMC Infect Dis 14:53. https://doi.org/10.1186/1471-2334-14-53

Marston BJ, Plouffe JF, File TM, Hackman BA, Salstrom SJ, Lipman HB, Kolczak MS, Breiman RF (1997) Incidence of community-acquired pneumonia requiring hospitalization – results of a population-based active surveillance study in Ohio. Arch Intern Med 157(15):1709–1718

Martin GS (2012) Sepsis, severe sepsis and septic shock: changes in incidence, pathogens and outcomes. Expert Rev Anti Infect Ther 10(6):701–706. https://doi.org/10.1586/eri.12.50

Martina BE, Koraka P, Osterhaus AD (2009) Dengue virus pathogenesis: an integrated view. Clin Microbiol Rev 22(4):564–581. https://doi.org/10.1128/cmr.00035-09

Mason S, van Furth AMT, Solomons R, Wevers RA, van Reenen M, Reinecke CJ (2016) A putative urinary biosignature for diagnosis and follow-up of tuberculous meningitis in children: outcome of a metabolomics study disclosing host-pathogen responses. Metabolomics 12(7):110. https://doi.org/10.1007/s11306-016-1053-2

Mayeux R (2004) Biomarkers: potential uses and limitations. NeuroRx 1(2):182–188. https://doi.org/10.1602/neurorx.1.2.182

Mirnaghi FS, Caudy AA (2014) Challenges of analyzing different classes of metabolites by a single analytical method. Bioanalysis 6(24):3393–3416. https://doi.org/10.4155/bio.14.236

MMWR Recommendations and reports (1996) Defining the public health impact of drug-resistant *Streptococcus pneumoniae*: report of a working group. https://www.cdc.gov/mmwr/preview/mmwrhtml/00040449.htm. Accessed 16 May 2017

Mushtaq MY, Choi YH, Verpoorte R, Wilson EG (2014) Extraction for metabolomics: access to the metabolome. Phytochem Anal 25(4):291–306. https://doi.org/10.1002/pca.2505

Näsström E, Thieu NT, Dongol S, Karkey A, Voong Vinh P, Ha Thanh T, Johansson A, Amit A, Thwaites G, Dolecek C, Basnyat B, Baker S, Antti H (2014) *Salmonella* typhi and *Salmonella* paratyphi A elaborate distinct systemic metabolite signatures during enteric fever. eLife. https://doi.org/10.7554/eLife.03100

Ng JS, Ryan U, Trengove RD, Maker GL (2012) Development of an untargeted metabolomics method for the analysis of human faecal samples using *Cryptosporidium*-infected samples. Mol Biochem Parasitol 185(2):145–150. https://doi.org/10.1016/j.molbiopara.2012.08.006

Oliver SG, Winson MK, Kell DB, Baganz F (1998) Systematic functional analysis of the yeast genome. Trends Biotechnol 16(9):373–378. https://doi.org/10.1016/s0167-7799(98)01214-1

Olszewski KL, Llinás M (2011) Central carbon metabolism of *Plasmodium* parasites. Mol Biochem Parasitol 175(2):95–103. https://doi.org/10.1016/j.molbiopara.2010.09.001

Pappa V, Seydel K, Gupta S, Feintuch CM, Potchen MJ, Kampondeni S, Goldman-Yassen A, Veenstra M, Lopez L, Kim RS, Berman JW, Taylor T, Daily JP (2015) Lipid metabolites of the phospholipase A_2 pathway and inflammatory cytokines are associated with brain volume in paediatric cerebral malaria. Malar J 14:513. https://doi.org/10.1186/s12936-015-1036-1

Parry CM, Wijedoru L, Arjyal A, Baker S (2011) The utility of diagnostic tests for enteric fever in endemic locations. Expert Rev Anti Infect Ther 9(6):711–725. https://doi.org/10.1586/eri.11.47

Pawlotsky JM (2004) Pathophysiology of hepatitis C virus infection and related liver disease. Trends Microbiol 12(2):96–102. https://doi.org/10.1016/j.tim.2003.12.005

Pearle MS (2004) Does the midstream urine culture and sensitivity test reliably predict infected urine near a ureteral obstruction? Nat Clin Pract Urol 1(1):16–17. https://doi.org/10.1038/ncpuro0004

Ptolemy AS, Rifai N (2010) What is a biomarker? Research investments and lack of clinical integration necessitate a review of biomarker terminology and validation schema. Scand J Clin Lab Invest Suppl 242:6–14. https://doi.org/10.3109/00365513.2010.493354

Ramakrishnan S, Docampo MD, MacRae JI, Pujol FM, Brooks CF, van Dooren GG, Hiltunen JK, Kastaniotis AJ, McConville MJ, Striepen B (2012) Apicoplast and endoplasmic reticulum cooperate in fatty acid biosynthesis in apicomplexan parasite *Toxoplasma gondii*. J Biol Chem 287(7):4957–4971. https://doi.org/10.1074/jbc.M111.310144

Ramautar R, Nevedomskaya E, Mayboroda OA, Deelder AM, Wilson ID, Gika HG, Theodoridis GA, Somsen GW, de Jong GJ (2011) Metabolic profiling of human urine by CE-MS using a positively charged capillary coating and comparison with UPLC-MS. Mol Biosyst 7(1):194–199. https://doi.org/10.1039/c0mb00032a

Rojo D (2014) Metabolómica e integración multiómica en organismos unicelulares. Hacia la comprensión de sistemas biológicos. Dissertation, CEU San Pablo University

Rojo D, Gosalbes MJ, Ferrari R, Pérez-Cobas AE, Hernández E, Oltra R, Buesa J, Latorre A, Barbas C, Ferrer M, Moya A (2015) *Clostridium difficile* heterogeneously impacts intestinal community architecture but drives stable metabolome responses. ISME J 9(10):2206–2220. https://doi.org/10.1038/ismej.2015.32

Ronald A (2003) The etiology of urinary tract infection: traditional and emerging pathogens. Dis Mon 49(2):71–82. https://doi.org/10.1067/mda.2003.8

Rowan DD (2011) Volatile metabolites. Metabolites 1(1):41–63. https://doi.org/10.3390/metabo1010041

Saito T, Sugimoto M, Igarashi K, Saito K, Shao L, Katsumi T, Tomita K, Sato C, Okumoto K, Nishise Y, Watanabe H, Tomita M, Ueno Y, Soga T (2013) Dynamics of serum metabolites in patients with chronic hepatitis C receiving pegylated interferon plus ribavirin: a metabolomics analysis. Metabolism 62(11):1577–1586. https://doi.org/10.1016/j.metabol.2013.07.002

Salek RM, Maguire ML, Bentley E, Rubtsov DV, Hough T, Cheeseman M, Nunez D, Sweatman BC, Haselden JN, Cox RD, Connor SC, Griffin JL (2007) A metabolomic comparison of urinary changes in type 2 diabetes in mouse, rat, and human. Physiol Genomics 29(2):99–108. https://doi.org/10.1152/physiolgenomics.00194.2006

Sanchez EL, Lagunoff M (2015) Viral activation of cellular metabolism. Virology 479–480:609–618. https://doi.org/10.1016/j.virol.2015.02.038

Sengupta A, Ghosh S, Basant A, Malusare S, Johri P, Pathak S, Sharma S, Sonawat HM (2011) Global host metabolic response to *Plasmodium vivax* infection: a ^{1}H NMR based urinary metabonomic study. Malar J 10:384. https://doi.org/10.1186/1475-2875-10-384

Sengupta A, Ghosh S, Pathak S, Gogtay N, Thatte U, Doshi M, Sharma S, Sonawat HM (2015) Metabolomic analysis of urine samples of vivax malaria in-patients for biomarker identification. Metabolomics 11(5):1351–1362. https://doi.org/10.1007/s11306-015-0791-x

Sengupta A, Ghosh S, Das BK, Panda A, Tripathy R, Pied S, Ravindran B, Pathak S, Sharma S, Sonawat HM (2016) Host metabolic responses to *Plasmodium falciparum* infections evaluated by ^{1}H NMR metabolomics. Mol Biosyst 12(11):3324–3332. https://doi.org/10.1039/c6mb00362a

Shahfiza N, Osman H, Hock TT, Shaari K, Abdel-Hamid AH (2015) Metabolomics for characterization of gender differences in patients infected with dengue virus. Asian Pac J Trop Med 8(6):452–457. https://doi.org/10.1016/s1995-7645(14)60359-x

Singh S, Srivastava S, Roy R, Gaurav K, Kumar S, Sonkar AA, Goel MM, Garg R (2014) Metabolic profiling of cervical tubercular lymphadenitis tissues by proton HR-MAS NMR spectroscopy. Metabolomics 10(5):975–985. https://doi.org/10.1007/s11306-014-0626-1

Slupsky CM, Rankin KN, Fu H, Chang D, Rowe BH, Charles PG, McGeer A, Low D, Long R, Kunimoto D, Sawyer MB, Fedorak RN, Adamko DJ, Saude EJ, Shah SL, Marrie TJ (2009) Pneumococcal pneumonia: potential for diagnosis through a urinary metabolic profile. J Proteome Res 8(12):5550–5558. https://doi.org/10.1021/pr9006427

Soga T, Sugimoto M, Honma M, Mori M, Igarashi K, Kashikura K, Ikeda S, Hirayama A, Yamamoto T, Yoshida H, Otsuka M, Tsuji S, Yatomi Y, Sakuragawa T, Watanabe H, Nihei K, Saito T, Kawata S, Suzuki H, Tomita M, Suematsu M (2011) Serum metabolomics reveals gamma-glutamyl dipeptides as biomarkers for discrimination among different forms of liver disease. J Hepatol 55(4):896–905. https://doi.org/10.1016/j.jhep.2011.01.031

Sonawat HM, Sharma S (2012) Host responses in malaria disease evaluated through nuclear magnetic resonance-based metabonomics. Clin Lab Med 32(2):129–142. https://doi.org/10.1016/j.cll.2012.04.005

Su L, Huang Y, Zhu Y, Xia L, Wang R, Xiao K, Wang H, Yan P, Wen B, Cao L, Meng N, Luan H, Liu C, Li X, Xie L (2014) Discrimination of sepsis stage metabolic profiles with an LC/MS-MS-based metabolomics approach. BMJ Open Respir Res. https://doi.org/10.1136/bmjresp-2014-000056

Subramani E, Dutta M, Jothiramajayam M, Joshi M, Srivastava S, Mukherjee A, Chakravarty B, Chaudhury K (2016) Identification of serum metabolic markers for diagnosis of women with dormant genital tuberculosis. Metabolomics 12(6):99. https://doi.org/10.1007/s11306-016-1042-5

Surowiec I, Orikiiriza J, Karlsson E, Nelson M, Bonde M, Kyamanwa P, Karenzi B, Bergström S, Trygg J, Normark J (2015) Metabolic signature profiling as a diagnostic and prognostic tool in pediatric *Plasmodium falciparum* malaria. Open Forum Infect Dis 2(2):ofv062. https://doi.org/10.1093/ofid/ofv062

Sweatman BC, Farrant RD, Lindon JC (1993) NMR of biofluids: detection of ^{2}H-acetate and ^{2}H-formate in urine as an indicator of microbiological contamination. J Pharm Biomed Anal 11(2):169–172. https://doi.org/10.1016/0731-7085(93)80138-q

Syed IH, Balakrishnan P, Solomon SS, Murugavel KG, Kumarasamy N, Vidya S, Martin SP, Thyagarajan SP, Mayer KH, Solomon S (2005) HIV-1 western blot assay: what determines an indeterminate status? Indian J Med Sci 59(10):443–450

Teixeira AR, Hecht MM, Guimaro MC, Sousa AO, Nitz N (2011) Pathogenesis of Chagas' disease: parasite persistence and autoimmunity. Clin Microbiol Rev 24(3):592–630. https://doi.org/10.1128/cmr.00063-10

Vincent IM, Daly R, Courtioux B, Cattanach AM, Bieler S, Ndung'u JM, Bisser S, Barrett MP (2016) Metabolomics identifies multiple candidate biomarkers to diagnose and stage human African trypanosomiasis. PloS Negl Trop Dis. https://doi.org/10.1371/journal.pntd.0005140

Voge NV, Perera R, Mahapatra S, Gresh L, Balmaseda A, Loroño-Pino MA, Hopf-Jannasch AS, Belisle JT, Harris E, Blair CD, Beaty BJ (2016) Metabolomics-based discovery of small molecule biomarkers in serum associated with dengue virus infections and disease outcomes. PLoS Negl Trop Dis. https://doi.org/10.1371/journal.pntd.0004449

Voskuil J (2015) How difficult is the validation of clinical biomarkers? F1000Res 4:101. https://doi.org/10.12688/f1000research.6395.1

Weiner J 3rd, Parida SK, Maertzdorf J, Black GF, Repsilber D, Telaar A, Mohney RP, Arndt-Sullivan C, Ganoza CA, Faé KC, Walzl G, Kaufmann SH (2012) Biomarkers of inflammation, immunosuppression and stress with active disease are revealed by metabolomic profiling of tuberculosis patients. PLoS ONE 7(7). https://doi.org/10.1371/journal.pone.0040221

WHO (2017a) The top 10 causes of death. http://www.who.int/mediacentre/factsheets/fs310/en/. Accessed 29 May 2017

WHO (2017b) Global hepatitis report 2017. World Health Organization, Geneva

WHO (2017c) Trypanosomiasis, human African (sleeping sickness). http://www.who.int/mediacentre/factsheets/fs259/en/. Accessed 28 May 2017

WHO (2017d) World health statistics 2017: monitoring health for the SDGs, sustainable development goals. World Health Organization, Geneva

Winder CL, Dunn WB, Schuler S, Broadhurst D, Jarvis R, Stephens GM, Goodacre R (2008) Global metabolic profiling of *Escherichia coli* cultures: an evaluation of methods for quenching and extraction of intracellular metabolites. Anal Chem 80(8):2939–2948. https://doi.org/10.1021/ac7023409

Xia J, Broadhurst DI, Wilson M, Wishart DS (2013) Translational biomarker discovery in clinical metabolomics: an introductory tutorial. Metabolomics 9(2):280–299. https://doi.org/10.1007/s11306-012-0482-9

Zaki SA, Karande S (2011) Multidrug-resistant typhoid fever: a review. J Infect Dev Ctries 5(5):324–337

Zhang QW, Takahashi M, Noguchi Y, Sugimoto T, Kimura T, Okumura A, Ishikawa T, Kakumu S (2006) Plasma amino acid profiles applied for diagnosis of advanced liver fibrosis in patients with chronic hepatitis C infection. Hepatol Res 34(3):170–177. https://doi.org/10.1016/j.hepres.2005.12.006

Zhang W, Hankemeier T, Ramautar R (2017) Next-generation capillary electrophoresis-mass spectrometry approaches in metabolomics. Curr Opin Biotechnol 43:1–7. https://doi.org/10.1016/j.copbio.2016.07.002

Zhong L, Zhou J, Chen X, Yin Y (2016) Serum metabolomic study for the detection of candidate biomarkers of tuberculosis. Int J of Clin Exp Pathol 9(3):3256–3266

Part II
Microbial Metabolic Adaptation and Pathogen Rewiring of Host Cell Metabolism During Infection

Chapter 8
Metabolic Host Response to Intracellular Infections

Catarina M. Ferreira, Ana Margarida Barbosa, Inês M. Pereira, and Egídio Torrado

Abstract The interaction between intracellular bacterial pathogens with the host immune response can result in multiple outcomes that range from asymptomatic clearance to the establishment of infection. At its core, these interactions result in multiple metabolic adaptations of both the pathogen and its host cell. There is growing evidence that the host metabolic response plays a key role in the development of immune responses against the invading pathogen. However, successful intracellular pathogens have developed multiple mechanisms to circumvent the host response to thrive in the intracellular compartment. Here, we provide a brief overview on the crucial role of fundamental metabolic host responses in the generation of protective immunity to intracellular bacterial pathogens and discuss some of the mechanisms used by these pathogens to exploit the host metabolic response to their own advantage. This understanding will further our knowledge in host-pathogen interactions and may provide new insights for the development of novel therapies.

Keywords Host-pathogen interactions · Intracellular bacteria · Infection · Immunity · Metabolism

8.1 Introduction

The interaction between the eukaryotic host with prokaryotic organisms, in general, and intracellular bacteria, in particular, may lead to metabolic adaptations of both interacting organisms resulting into beneficial, neutral, or detrimental outcomes for either organism (Peleg et al. 2008). An example of beneficial metabolic adaptations between eukaryotic and prokaryotic organisms occurs in mammalians and their

---

C. M. Ferreira · A. M. Barbosa · I. M. Pereira · E. Torrado (✉)
Life and Health Sciences Research Institute (ICVS), School of Medicine, University of Minho, Braga, Portugal

ICVS/3B's - PT Government Associate Laboratory, University of Minho, Braga, Portugal
e-mail: egidiotorrado@med.uminho.pt

R. Silvestre, E. Torrado (eds.), *Metabolic Interaction in Infection*, Experientia Supplementum 109, https://doi.org/10.1007/978-3-319-74932-7_8

microbiota (Nicholson et al. 2012). These metabolic adaptations are critical to maintain a symbiotic relationship and, in many cases, to prevent disease development. However, these interactions are extremely complex and unstable as they rely on external factors, such as diet and immune status, thus making it challenging to assign specific metabolic pathways to individual organisms and to dissect their impact in health and disease. On the other hand, infecting bacterial agents establish more stable relationships with their host and therefore provide a more suitable model to address the impact of metabolic alterations in the outcome of infection. In this regard, intracellular bacterial pathogens are of particular interest and have been widely used as model agents to analyze metabolic host responses. Indeed, intracellular bacteria usually have one cell type as their target host cells. Therefore, the infected host cell represents a well-defined metabolic entity, providing important clues on the host metabolic adaptations triggered by the infectious process and how these adaptations impact control or progression of the infection (Munoz-Elias and McKinney 2006). Indeed, the intracellular pathogen lifestyle represents a well-adapted condition to bacterial growth and survival within the host, as it protects the pathogen from humoral and complement-mediated immunity, while the infecting agent interferes with host metabolic pathways, gaining access to a wide range of nutrients and metabolites (Ribet and Cossart 2015).

Emerging data now show that intracellular bacterial pathogens interfere with the metabolic response of their host cell during their first interaction, through the engagement by the host cell of highly conserved structural components of pathogens, but also throughout infection by translocating effector molecules into the host cell. These interactions modulate the host response at various levels, such as regulating the expression of nutrient sensors and transporters that control the uptake of substrates such as glucose, amino acids, oligopeptides, acetyl-CoA, vitamins, and fatty acids (Eisenreich et al. 2013). Additionally, the intracellular pathways activated by the engagement of pathogen recognition receptors (PRRs) by pathogen-associated molecular patterns (PAMPs) trigger inflammatory responses and the release of soluble mediators, such as cytokines, that also modulate the metabolic response of host cells (Meylan et al. 2006; Kumar et al. 2011). Although these responses may be directed at controling infection, well-adapted intracellular pathogens have evolved and adapted to take advantage of these responses to thrive in the intracellular compartment. For this reason, it is not always clear whether the host metabolic adaptation is directed at controlling infection or is being modulated by the invading pathogen to take advantage of the host's nutrients and metabolites to satisfy its own bioenergetic and biosynthetic requirements.

In this chapter, we will discuss some of the interactions between the host and intracellular bacterial pathogens from a metabolic point of view. We begin by presenting a brief overview of the major metabolic pathways controlling the metabolism of mammalian cells and discuss their importance in the differentiation process of immune cells with critical roles in the control of infections by intracellular pathogens. As these processes are potential targets for the interaction with bacterial pathogens, we will then discuss common and pathogen-specific metabolic adaptations triggered by intracellular bacterial pathogens. Understanding the importance of the metabolic host response during intracellular bacterial infections and how these pathogens modulate these responses to their advantage may provide novel and attractive therapeutic targets for antibacterial therapies.

8.2 Metabolic Host Cell Adaptations in the Response to Intracellular Bacterial Pathogens

Cellular metabolism is one of the most important processes to maintain homeostatic status in living organisms. This phenomenon comprises different biochemical pathways that produce or consume energy and are critical for living cells to proliferate and adapt to environmental changes. In broad terms, cellular metabolism is usually divided in two major categories: anabolism and catabolism. During catabolic reactions complex molecules are broken down to release energy and small molecules, such as nucleic acids and amino acids. The energy and small molecules produced during catabolism are then used in anabolic processes to generate complex macromolecules required for cell survival and proliferation (DeBerardinis and Thompson 2012). Therefore, although these processes are different, and occur in different locations within the cell, the different metabolic pathways are interconnected by interchangeable intermediates that serve as precursors in different pathways (for more detail see Chap. 1).

Phagocytic cells such as macrophages and dendritic cells represent the first line of defense to invading pathogens. These innate immune cells rapidly sense the presence of invading microorganisms through the engagement of their PRRs by PAMPs (Nathan and Shiloh 2000; Odegaard and Chawla 2011). These interactions trigger important core host metabolic responses that are crucial to control intracellular bacterial pathogens, such as the generation of reactive oxygen species (ROS) by NADPH oxidase, reactive nitrogen species by the inducible nitric oxide synthase (iNOS), and autophagy.

Autophagy is a highly conserved and regulated catabolic process controlled by autophagy-related genes (Atg) which relies on the formation of autophagosomes that engulfs intracellular components, including phagocytosed bacteria, and delivers them to degradation in the lysosome (Mizushima et al. 2011). Therefore, autophagy has emerged as a key process both in the control intracellular pathogen growth and also in the development of acquired immune responses (Desai et al. 2015; Levine et al. 2011). Indeed, proteins and peptides degraded in the autophagosome can then be loaded in major histocompatibility complex (MHC) molecules, thereby inducing T cell activation (Mantegazza et al. 2013; Fraga et al. 2017). The importance of autophagy in limiting intracellular bacterial growth has been demonstrated using mice deficient in different components of the autophagy process (Casanova 2017; Jo et al. 2013; Pareja and Colombo 2013). It is therefore not surprising that well-adapted intracellular pathogens have evolved different mechanisms to interfere with the autophagic process of its host cell. For example, one of the first

Mycobacterium tuberculosis virulence traits identified and studied was the active inhibition of the fusion of the mycobacteria-containing phagosomes with acidified lysosomes (Armstrong and Hart 1971). In this regard, several factors produced by *M. tuberculosis* have been identified to interfere with the autophagic process, affecting not only the ability of macrophages to control infection but also the development of protective T cell responses (Fraga et al. 2017; Deretic et al. 2009; Castillo et al. 2012). Although *M. tuberculosis* is probably one of the most well-adapted intracellular pathogens, interfering with the endocytic pathway and autophagy is not a unique feature of this pathogen. Indeed, *Legionella pneumophila* has been recently shown to inhibit autophagy by secreting the protein effector RavZ that directly uncouple Atg8 proteins attached to phosphatidylethanolamine on autophagosome membranes (Choy et al. 2012). Other intracellular pathogens such as *Listeria monocytogenes* have evolved a mechanism to rapidly escape the phagosome or autophagosome of its macrophage host cell, enabling replication in the cytosol and cell-to-cell spread by the bacteria through a bacterial-driven actin-based motility mechanism (Kuehl et al. 2015).

One of the most relevant transcription factors involved in the metabolic adaptation to infection is nuclear factor-kappa B (NF-kB), a transcription factor activated by multiple stimuli including the interaction of PRRs of phagocytic cells with PAMPs (Baker et al. 2011; Lutay et al. 2014; Kim et al. 2016). In addition to playing a key role in the stress response, differentiation and proliferation processes, and cell death, NF-kB also mediates pro-inflammatory responses required for host control of many microbial pathogens. For instance, NF-kB is responsible for the expression of several pro-inflammatory cytokines produced by pro-inflammatory macrophages (M1) such as IL-6, IL-1, and TNF-α that assist in the activation of T lymphocytes (Baker et al. 2011). This role of NF-kB in immune cell differentiation can also be linked with its ability to upregulate the expression of enzymes that catalyze important metabolic reactions. Among NF-kB-induced metabolic enzymes is indolamine-2,3-dioxygenase (IDO) that catalyzes the first step in the kynurenine pathway thus playing a critical role in tryptophan degradation. IDO has a critical role in T cell proliferation and in the induction of apoptosis in activated T cells (Tas et al. 2007). The importance of NF-kB in the control of intracellular bacteria is evidenced in studies showing that the genetic disruption of the p50 subunit of NF-kB renders mice more susceptible to *M. tuberculosis* infection (Yamada et al. 2001). Additionally, pharmacological inhibition of NF-kB activation results in a decrease in viable intracellular *M. tuberculosis* in human macrophages (Bai et al. 2013). On the other hand, NF-kB-mediated pro-inflammatory mechanisms may also be a target for bacterial pathogenesis. In this regard, the modulation of IDO has been shown to be implicated in immune modulation through its ability to limit T cell function (Munn and Mellor 2013). These properties of IDO have been shown during infections by mycobacteria (Plain et al. 2011; Blumenthal et al. 2012). Additionally, *L. monocytogenes* cell surface protein InIB, phospholipases, and listeriolysin O that mediate phagosome escape and cell-to-cell spread also induce NF-kB activation, thus driving the inflammatory response (Mansell et al. 2000). Therefore, the outcome of infections is determined by a close interplay between immune effectors directed at damaging the pathogen and the countermeasures deployed by the invader.

Although this framework is helpful to understand and describe host-pathogen interactions through a metabolic point of view, it generally overlooks the obligatory and wide-ranging connections between the metabolism of the host and the microorganism. For example, at the site of infection, the host and pathogen share substrates and generate metabolic products that can be used by both interacting organisms. This metabolic interplay is most obvious when the pathogen is intracellular as it replicates within host cells. The fact that the majority of the metabolites used in prokaryotic and eukaryotic signaling are similar suggests important connections between the physiology of the infecting pathogen and its host cell. Indeed, slight alterations to the metabolic crosstalk between host and pathogen can have a significant impact in the outcome of infection. On the one hand, the host provides the pathogen with nutrients, such as carbon sources, amino acids, and phosphorus, as well as electron acceptors that can be used for the metabolism of the pathogen. On the other hand, the products that result from the metabolism of the bacteria also have impact in the inflammatory state of the host, thus changing the dynamics of the metabolic interactions between the two interacting partners. For instance, pathogen respiratory pathways produce carbon dioxide and water that may impact the host inflammatory response. Furthermore, a fermentative state of the bacteria may lead to the production of organic acids that subvert host metabolic homeostatic regulation. For example, acetate can directly activate host inflammatory pathways, while citrate can lead to the production of ROS. Additionally, succinate derived from bacterial fermentation can promote host inflammation by activating hypoxia-inducible factor 1-alpha (HIF-1α). The activation of this transcription factor is involved in the production of glucose transporters, especially glucose transporter 1 (GLUT1) and lactate dehydrogenase A (LDHA), having a direct impact in the fermentative capacity of host cells (Masoud and Li 2015). Furthermore, butyrate produced by bacterial fermentation can induce host histone acetylation resulting in the expression of pro-inflammatory genes. The induction of inflammatory pathways by the host is critical for limiting the availability of electron acceptors, macronutrients, and micronutrients for the pathogen, thus influencing its physiological state. For this reason, the interaction between microbial pathogens and host immunity is a metabolic competition wherein there is an interplay between immune effectors and pathogen countermeasures. Therefore, chief among the challenges in the study of metabolism during infections is differentiating a host-induced metabolic response from virulence-adapted metabolic pathways.

8.3 Metabolic Pathways Drive Immune Cell Differentiation Critical to Control Intracellular Bacterial Infections

Due to the fundamental role of metabolism in controlling cell fate, recent research has brought to light the key role of metabolism in the differentiation and effector function of immune cells. Indeed, the intrinsic relationship between the cellular metabolic status of immune cells with their phenotype and activation state is now well recognized (van der Windt and Pearce 2012; Buck et al. 2015). In this regard, immune cells may remain quiescent for long periods. In this quiescent form, these cells metabolize glucose to pyruvate or they use fatty acid oxidation to generate ATP (Pearce and Pearce 2013; van der Windt and Pearce 2012). Pyruvate is then used in the tricarboxylic acid (TCA) cycle to generate large amounts of ATP via oxidative phosphorylation. Interestingly, upon activation, immune cells switch from using oxidative phosphorylation to fuel their energetic flux and rely mostly on glycolysis and subsequent lactic acid fermentation (Michalek and Rathmell 2010; Wahl et al. 2012; Pearce et al. 2013; Krawczyk et al. 2010), despite the fact that glycolysis is a less efficient process to generate ATP than oxidative phosphorylation, per molecule of glucose. This metabolic switch has been interpreted as a means to rapidly support the metabolic requirements of ATP for proliferation (Buck et al. 2015). This is particularly important in T cells that begin clonal expansion upon recognizing their cognate antigen in the presence of costimulation. Indeed, T cell-specific deletion of GLUT1 impairs CD4 T cell activation, clonal expansion, and survival (Macintyre et al. 2014). However, upon PRR activation both macrophages and dendritic cells also switch from oxidative phosphorylation to glycolysis, without increasing their proliferation (Kelly and O'Neill 2015). These data suggest that, in addition to the energy needed for proliferation, glycolysis may also provide other advantages during immune cell activation and differentiation. This question is of paramount importance during infections by intracellular pathogens as their control is dependent upon the development of T cells with the T helper (Th)1 phenotype, characterized by the production of interferon-gamma (IFN-γ), a critical cytokine in the activation of microbicidal mechanisms of macrophages. The role of these responses in the control of intracellular bacterial pathogens has been demonstrated by the heightened susceptibility of both mice and humans deficient in the IL-12-IFN-γ axis to intracellular infections, including *M. tuberculosis* (Cooper et al. 1997; Azad et al. 2012; Ramirez-Alejo and Santos-Argumedo 2014), *L. monocytogenes* (Way et al. 2007; Wagner et al. 1994), *Salmonella enterica* serovar Typhimurium (Price et al. 2007; Jouanguy et al. 1999), and *L. pneumophila* (Matsunaga et al. 2001).

With the key role of IFN-γ-producing T cells in the control of intracellular bacterial infections, recent studies have addressed the role of glycolysis in the development of this cell population (Chang et al. 2013; Peng et al. 2016). These studies show that, rather than being required for T cell proliferation and survival, glycolysis is essential for the differentiation process of naïve T cells into the Th1 phenotype (Chang et al. 2013; Peng et al. 2016). Indeed, while activated T cells can fuel their proliferation needs either through glycolysis or oxidative phosphorylation interchangeably, their ability to produce IFN-γ was markedly compromised when glycolysis, but not when oxidative phosphorylation, was inhibited (Chang et al. 2013). In this study, activated T cells were cultured with galactose instead of glucose as a carbon source, to prevent ATP generation by fermentation (Chang et al. 2013). Using this approach the authors showed that in the absence of glycolysis, the enzyme glyceraldehyde-3-phosphate dehydrogenase (GAPDH) binds to AU-rich elements within the 3' untranslated region of the IFN-γ mRNA, inhibiting its translation

(Chang et al. 2013). These data show that glycolysis permits cytokine production by modulating cytokine translation mechanisms (Chang et al. 2013) (Fig. 8.1).

The importance of glycolysis in the development of IFN-γ-producing T cells has been recently confirmed (Peng et al. 2016). In this study, the authors used a genetic

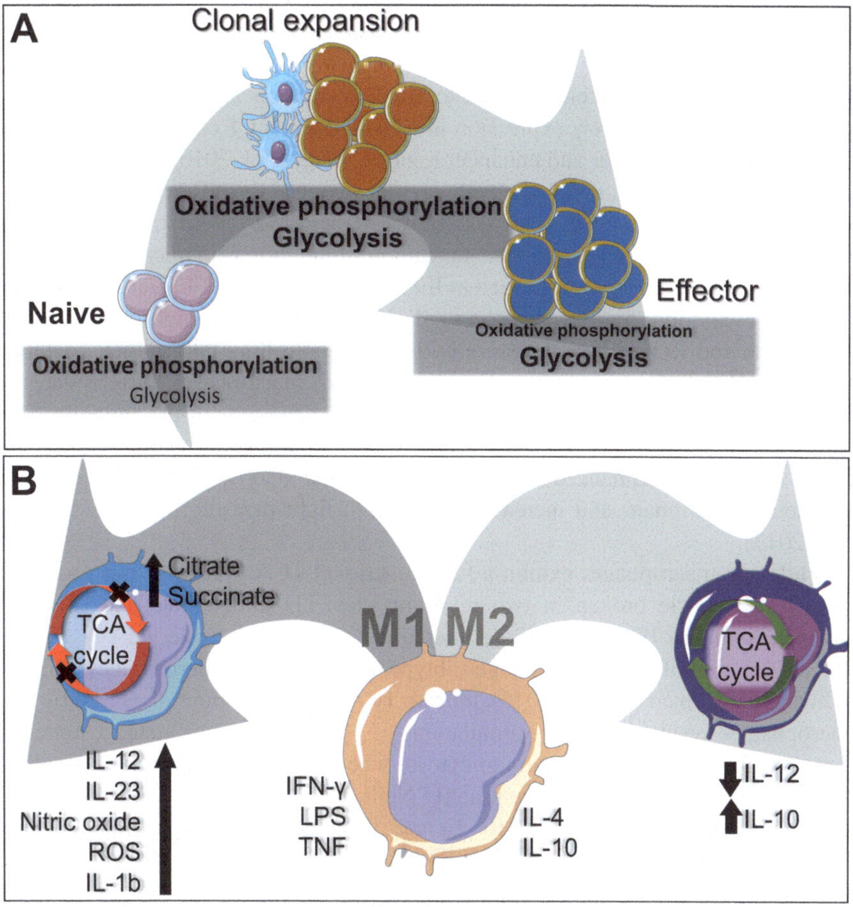

Fig. 8.1 Metabolism drives immune cell differentiation. (**a**) Naïve T cells generate ATP by oxidative phosphorylation and glycolysis. Upon stimulation in secondary lymphoid organs, naïve T cells begin clonal expansion. While ATP generated by oxidative phosphorylation or glycolysis can be used to support proliferation, glycolysis is essential for their differentiation into the Th1 phenotype essential to control intracellular bacteria. (**b**) Pro-inflammatory stimuli promote the differentiation of M1 macrophages (classical activation). These cells produce IL-12 and IL-23, nitric oxide, and ROS and are typically associated with control of intracellular pathogens. On the other hand, IL-10 and IL-4 trigger the differentiation of M2 macrophages (alternative activation) that produce low amounts of pro-inflammatory cytokines but high amounts of IL-10. These macrophages are associated with tissue repair and control of parasitic infections. M2 macrophages display a fully functional TCA cycle, while M1 macrophages exhibit a broken TCA cycle that results in the accumulation of citrate and succinate

model wherein LDHA, the enzyme that defines the biochemical reaction of lactic acid fermentation that converts pyruvate to lactate with the regeneration of nicotinamide adenine dinucleotide (NAD+), was specifically deleted in T cells (Peng et al. 2016). Instead of a translational mechanism of glycolysis in IFN-γ production, this study showed that glycolysis might primarily relieve the burden of the mitochondria to generate ATP resulting in more citrate exported out of the mitochondria to generate acetyl-CoA, which is required for histone acetylation of the *Ifng* locus (Peng et al. 2016). As acetyl-CoA can be generated by acetyl-CoA synthetase from acetate independently of citrate, acetate supplementation augmented acetyl-CoA production, corrected IFN-γ expression in LDHA-deficient T cells, and normalized H3K9Ac in *Ifng* promoter and enhancer regions (Peng et al. 2016) (Fig. 8.1).

In addition to T cells, metabolic pathways also control the differentiation process of macrophages into the M1 (classical activated) and M2 (alternatively activated) phenotypes. M1 macrophages differentiate upon stimulation with microbial products and pro-inflammatory cytokines, such as IFN-γ and/or LPS or TNF, and display high antigen presentation, high production of IL-12 and IL-23, and high production of nitric oxide and ROS, being therefore critical for controlling intracellular bacterial pathogens (Verreck et al. 2004). By contrast, M2 macrophages differentiate in response to IL-4, IL-10, and IL-13; produce reduced amounts of pro-inflammatory cytokines, including IL-12, but high levels of anti-inflammatory cytokines, such as IL-10; and display enhanced scavenging of cellular debris, promotion of tissue remodeling and repair, and increased capacity to fight parasitic infection (Alfano et al. 2013).

While M2 macrophages exhibit a fully functional TCA cycle, M1 macrophages have a TCA cycle broken in two different sites: (1) after citrate and (2) after succinate synthase (Fig. 8.1). These alterations in the TCA cycle are key for the effector function of each macrophage subtype, as several key intermediates of other pathways are formed (O'Neill 2015). Indeed, TCA cycle alterations displayed by M1 macrophages result in the accumulation of citrate and succinate. Citrate is transported from the mitochondria to cytosol where it is used for fatty acid production required for membrane biogenesis (O'Neill 2015; O'Neill et al. 2016). On the other hand, succinate stabilizes HIF-1α and the production of pro-inflammatory cytokines, such as IL-1β, one of the hallmarks of inflammation (Kelly and O'Neill 2015; Tannahill et al. 2013) (Fig. 8.1).

The development of M2 macrophages, and their anti-inflammatory effect mediated by IL-10, has been recently shown to be regulated at the metabolic level (Ip et al. 2017; Minton 2017). Specifically, IL-10 was shown to have inhibitory effects on the glycolytic flux by regulating GLUT1 translocation from intracellular vesicles to the plasma membrane after lipopolysaccharide stimulation. This effect of IL-10 was mediated by a reduction in the accumulation of dysfunctional mitochondria and ROS production from the accumulated mitochondria through induction of autophagy and inhibition of the mammalian target of rapamycin (mTOR), which resulted in reduced caspase-1-dependent inflammasome activation via induction of DNA-damage-inducible transcript 4 and reduced IL-1β production (Ip et al. 2017) (Fig. 8.1).

Taken together, these data show that fundamental metabolic response plays critical roles in the differentiation and effector function of T cells and macrophages (Fig. 8.1).

8.4 Metabolic Host Response to Specific Intracellular Bacterial Pathogens

8.4.1 Chlamydia trachomatis *and* C. pneumoniae

Chlamydia species are Gram-negative obligate intracellular pathogens that affect millions of people worldwide. Although most of the *Chlamydia* species are symbionts of different organisms (Bachmann et al. 2014), *C. trachomatis* and *C. pneumonia* are two important human pathogens responsible for a wide range of diseases (Elwell et al. 2016). Specifically, infections by *C. trachomatis* are the leading cause of severe ocular, respiratory, and urogenital infections and cause trachoma and sexually transmitted diseases (Byrne 2010). Untreated chlamydial infection can cause epididymitis and proctitis in men, whereas in women it can result in salpingitis, endometritis, pelvic inflammatory disease, ectopic pregnancy, and tubal factor infertility (Malhotra et al. 2013). On the other hand, *C. pneumoniae* typically infect the upper respiratory tract causing community-acquired pneumonia and chronic bronchitis (Kalman et al. 1999). Infections by *C. pneumoniae* have also been associated with atherosclerosis and neurodegenerative diseases (Boman and Hammerschlag 2002; Stratton and Sriram 2003).

The developmental life cycle of *Chlamydia* species is biphasic, where the organism alternates between two morphologically and functionally distinct forms: noninfectious cell type known as the infectious elementary body and an intracellular replicative form called replicative reticulate body (Vromman et al. 2014). The infectious elementary body possesses a rigid cell wall and densely packed DNA (Vromman et al. 2014). Although the elementary body has been described as metabolically dormant, recent data show that it actually displays high levels of metabolic and biosynthetic activities although different from those of the replicative reticulate body (Omsland et al. 2012). Upon entering the host cell (typically an epithelial cell), the infectious elementary body is maintained in a specialized vacuole of the cytosol called inclusion. Eight to 12 h after phagocytosis, the pathogen differentiates into the replicative reticulate body. This way, chlamydiae replicate in the inclusion giving rise to progeny that undergoes a secondary differentiation process leading again to the infectious elementary body form (Kading et al. 2014; Hatch et al. 1984).

Chlamydia species are unable to produce several essential metabolites, including nucleotides (purines and pyrimidines), lipids (such as ceramide), and amino acids (such as tryptophan) (Omsland et al. 2012; Tipples and McClarty 1993; Hatch 1975; Derre et al. 2011; Wylie et al. 1997; Cocchiaro et al. 2008; Scidmore et al. 1996;

Carabeo et al. 2003; Elwell et al. 2011). Indeed, *Chlamydia* species exhibit an incomplete primary carbon metabolism as they lack most anabolic pathways, hence their reduced genome (1.2 Mb or less) (Eisenreich et al. 2015). Therefore, these pathogens have developed strategies to manipulate the host cell metabolism and acquire metabolites and energy from the host cell (Brothwell et al. 2016). In this regard, amino acids and lipids appear to be the major carbon and energy sources required for *Chlamydia* to fulfill its impaired metabolic and replicative needs.

Among host-derived lipids, sphingomyelin and cholesterol are present in purified chlamydiae, suggesting that these host-derived lipids traverse the inclusion and are incorporated by the bacterium (van Ooij et al. 2000; Robertson et al. 2009). Indeed, *C. trachomatis* has been shown to interrupt an exocytic pathway to acquire host sphingolipids from Golgi-derived vesicles in transit to the plasma membrane (Hackstadt et al. 1996). The critical role of sphingomyelin was recently demonstrated in *C. trachomatis*-infected epithelial cell lines through the pharmacological inhibition of sphingomyelin biosynthesis and trafficking (Robertson et al. 2009). This study showed that inhibition of sphingomyelin biosynthesis in epithelial cell lines infected with *C. trachomatis* resulted in a marked loss of inclusion membrane integrity, disruption of the inclusion, and early release of intracellular bacteria (Robertson et al. 2009). Additionally, using in vitro models of IFN-γ or penicillin-induced persistence wherein sphingomyelin biosynthesis was inhibited during the recovery phase resulted in disruption in inclusion membrane integrity and failure of persistent forms to completely reactivate to infectious elementary body (Robertson et al. 2009).

To acquire amino acids, an important carbon and energy sources for chlamydiae, both *C. trachomatis* and *C. pneumoniae* rely on the external protein degradation in the host cell lysosomal compartment (Ouellette et al. 2011). Indeed, pharmacological inhibition of vacuolar H(+)/ATPase that blocks lysosomal acidification and functions impairs the growth of both *C. trachomatis* and *C. pneumoniae* (Ouellette et al. 2011). In this condition, chlamydial growth inhibition was associated with the accumulation of undigested material within the lysosomal lumen. Accordingly, inhibition of protein synthesis inhibited the ability of lysosomal inhibitors to block chlamydial growth (Ouellette et al. 2011). In addition to lysosomal amino acids, these bacteria also secrete to the cytosol of their host cell the proteases CPAF and the chlamydial high-temperature requirement protein A (cHtrA) (Haferkamp 2017; Chen et al. 2012). These proteins have been proposed to play key roles in the proteolysis strategy of these bacteria to acquire amino acids and manipulate host cell signaling pathways. Although it is clear that these proteases are important chlamydial virulence factors, there are still no definitive data on their role (Haferkamp 2017; Chen et al. 2012).

In addition to lipids and amino acids, a recent study showed that C. *trachomatis* infection downregulates the tumor suppressor p53 through an activation of the PI3K/Akt pathway resulting in E3 ubiquitin-protein ligase HDM2 phosphorylation of Ser166 and p53 proteasomal degradation (Siegl et al. 2014). The reduced P53 in infected cells resulted in increased chlamydial progeny formation, whereas pharmacological stabilization of p53 impaired inclusion formation. Importantly, live bacteria were required to induce downregulation of P53 as heat inactivation of *Chlamydia* prior to infection prevented the reduction of p53 levels, suggesting that this mechanism is independent on PAMP sensing by host cells (Siegl et al. 2014). While p53 plays a key role in the downregulation of glucose transport and glycolysis (Schwartzenberg-Bar-Yoseph et al. 2004), the data suggested that it was the p53-mediated inhibition of glucose-6-P-dehydrogenase and subsequent inhibition of the pentose phosphate pathway that impaired chlamydial growth. Accordingly, overexpression of glucose-6-P-dehydrogenase overcomes the inhibitory activity of stabilized p53 and rescues chlamydial development (Schwartzenberg-Bar-Yoseph et al. 2004). Thus, live *Chlamydia*-induced downregulation of p53 prevented repression of the pentose phosphate pathway, which is responsible for DNA damage repair; provides important metabolites for bacterial proliferation, such as nucleotide precursors; and protects bacteria against oxidative stress (Siegl et al. 2014). Modulation of P53 is not unique during *Chlamydia* infection, but many other bacterial pathogens have been shown to inhibit p53 using various mechanisms (Zaika et al. 2015).

Since there are no established in vivo models, *Chlamydia*-infected host cell lines and tissues are the best approach to gene expression profiling studies. In this regard, a study where gene expression analysis was performed in swab samples of patients with active trachoma and healthy controls showed that among the most upregulated genes in *Chlamydia*-infected samples were associated with nutrient transporters (Natividad et al. 2010; Eisenreich et al. 2013). These data suggest that the increased expressing of host nutrient transporter may be an important metabolic response of the host cell to *Chlamydia* infection. As discussed above, chlamydiae require most of the nutrients transported by the proteins encoded by these genes. Therefore, it is tempting to speculate that this host metabolic response to infection may favor bacterial growth. However, subsequent transcriptomic or proteomic approaches failed to detect upregulation of these nutrient transporters during productive or persistent *Chlamydia* infection in both HeLa and HL cells (Mannonen et al. 2011). Therefore, further data and new models of study are required to define the relevance of these metabolic alterations for the outcome of infection. The development of these models will provide important answers on the relevance of metabolic adaptation of host cells in response to *Chlamydia* infection and how these adaptations impact the outcome of infection.

Altogether, these data show that metabolic host response to *Chlamydia* infections is characterized by an increased glucose uptake and flux, through glycolysis. Since chlamydial infections promote host anabolic pathways during the proliferative phase (such as protein and lipid synthesis pathways), the pentose phosphate pathway is also enhanced through a P53-dependent mechanism. The notable upregulation of genes that encode nutrient transporters in human samples with active *Chlamydia* infection may play key roles in the progression of the infection. Future research directed at identifying chlamydial factors that modulate the host metabolic response, including nutrient uptake by infected cells and their impact in chlamydial growth, may provide important targets to treat these diseases.

8.4.2 Coxiella burnetii

Coxiella burnetii is a Gram-negative intracellular bacterial pathogen and the etiological agent of acute and chronic human Q (query) fever, a zoonosis normally presented as an atypical pneumonia (van Schaik et al. 2013). As for *Chlamydia* spp., *C. burnetii* is an obligate intracellular pathogen. However, *C. burnetii* replicates in the hostile environment of the phagosomes and autophagosome characterized by low pH, high proteolytic, and oxidative activities (Moffatt et al. 2015). Indeed, after internalization by macrophages, *C. burnetii* phagosome is directed down the canonical pathway while also engaging autophagy to form a unique lysosome-derived vacuole niche called *Coxiella*-containing vacuole (Kohler and Roy 2015). While the lumen of the mature *Coxiella*-containing vacuole is acidic and contains active lysosomal proteases and the typical lysosomal proteins and lysosomal properties, this organelle displays distinct features, including homotypic fusion and a cholesterol-enriched membrane (Kohler and Roy 2015). To form this unique niche, the bacteria utilize a type IV secretion system responsible for translocating over 130 effector proteins into the host cell (Moffatt et al. 2015). The use of *Coxiella* mutants in some of these effector proteins revealed significant intracellular growth defects suggesting a key role of the type IV secretion system and the effector proteins translocated to the host cell in the formation of the *Coxiella*-containing vacuole and therefore in the intracellular growth of *C. burnetii* (Moffatt et al. 2015).

One of the most common approaches used to study the host metabolic response to *C. burnetii* infection is the human macrophage-like cell line THP-1. Metabolic genes upregulated in *C. burnetii*-infected THP-1 cells include (1) *SLC2A6*, which encodes a glucose transporter; (2) *TDO2*, encoding tryptophan 2,3-dioxygenase that catalyzes the first step of tryptophan degradation; (3) *CYBB* which is functionally linked to the phagosomal NADPH oxidase; (4) *FTH1*, encoding the ferritin heavy chain that oxidizes Fe (II) to Fe (III); (5) *RENB*, encoding N-acylglucosamine 2-epimerase important on the amino and nucleotide sugar metabolism; (6) *PIR*, encoding the iron-binding protein and transcriptional cofactor pirin; and (7) *ATP6V1H*, encoding the subunit of the V-ATPase, a vacuolar proton pump (Ren et al. 2003).

The upregulation of the *FTH1* gene is not unique to THP-1 cells, but it has been also observed in the murine macrophage-like J774A.1 cell line (Howe and Mallavia 1999). Specifically in this cell line, infection with viable *C. burnetii*, but not inactivated, induces the expression of transferrin receptors which results in a 2.5-fold increase in total intracellular iron (Howe and Mallavia 1999). That the increase in cellular iron is critical for *C. burnetii* replication was demonstrated by reducing the intracellular free iron pool using the cell-permeant iron chelator (Howe and Mallavia 1999). These data suggest that the upregulation of transferrin receptors may be essential to supply the iron needs of *C. burnetii* during intracellular growth, thus representing an important virulence mechanism (Howe and Mallavia 1999). As only viable bacteria induced the upregulation of transferrin receptors, these data also suggest that the *C. burnetii* type IV secretion system may play a key role in this mechanism. Future research will be important to clarify this issue.

As discussed above, one of the distinctive features of the *Coxiella*-containing vacuole is a cholesterol-enriched membrane (Kohler and Roy 2015). Indeed, cholesterol trafficking increases significantly after *C. burnetii* infection (>73%), and the pharmacological inhibition of cholesterol biosynthesis and uptake disrupts the *Coxiella*-containing vacuole which inhibits *C. burnetii* replication (Howe and Heinzen 2006). These data show that the pathogenicity of *C. burnetii* is highly dependent on cholesterol for its biogenesis (Howe and Heinzen 2006). In this regard, *C. burnetii* infection increased host cell cholesterol content by 1.75-fold with a coincident upregulation of host genes in both cholesterol uptake and biosynthesis (Howe and Heinzen 2006). The upregulated genes include (1) the low-density lipoprotein receptor (*LDLR*), responsible for uptake of exogenous cholesterol; (2) 3-hydroxy-3-methylglutaryl-CoA reductase (*HMGCR*), which catalyzes the rate-limiting step in the cholesterol biosynthetic pathway; (3) lanosterol synthase (*LSS*), which catalyzes the conversion of squalene to lanosterol; and (4) -24-dehydrocholesterol reductase (*DHCR24*), which catalyzes the final step in cholesterol biosynthesis (Howe and Heinzen 2006). These data show that *C. burnetii* actively modulates host metabolism to growth in the intracellular compartment. Therefore, targeting cholesterol metabolism may represent an important target to control infection. In this regard, a recent study performed a screening of FDA-approved drugs that disrupt cholesterol homeostasis of the host cell and determine their impact in intracellular *C. burnetii* growth (Mulye et al. 2017). This study showed that some of these drugs did prevent intracellular *C. burnetii* growth (Mulye et al. 2017). Moreover, the impact of these drugs appears to be more pronounced during *C. burnetii* infection than during *Legionella pneumophila*, *Rickettsia conorii*, or *Brucella abortus*, showing that *C. burnetii* is uniquely sensitive to this metabolic host response (Czyz et al. 2014).

An interesting hallmark of the host metabolic response to *C. burnetii* infection is the reduced neutrophil NADPH oxidase activity (Siemsen et al. 2009; Nguyen et al. 2017). Indeed, when compared to other bacterial infection (such as *Staphylococcus aureus* or *Escherichia coli*), the phagocytosis of *C. burnetii* results in a reduced aggregation of the oxidase into the phagosomal membrane and a significant reduction on ROS production (Siemsen et al. 2009). The inhibition of this process appears to be dependent on the secretion of an acid phosphatase by the *Coxiella* type IV secretion system (Hill and Samuel 2011). Therefore, these data show that the *Coxiella* type IV secretion system and the effectors translocated to the host cell are involved in modulating multiple host metabolic responses.

Taken together, the data discussed above show that iron uptake, modulation of NADPH oxidase activity, and increased cholesterol uptake and biosynthesis are important host metabolic responses that may benefit *C. burnetii* intracellular replication. Among these responses, targeting cholesterol metabolism may represent a novel approach to control *C. burnetii* infection. On the other hand, not much is known on primary carbon metabolism during *C. burnetii* infection. Future experiments using primary cells will be critical to address this issue. This data will provide a broader picture of the altered host metabolic pathways and their impact in the outcome of infection.

8.4.3 Legionella pneumophila

L. pneumophila is a Gram-negative bacterial pathogen that grows ubiquitously in aquatic environments in the form of biofilms, but it relies on the intracellular environment of protozoa, such as amoebae, to replicate (Cianciotto 2001; Escoll et al. 2013). Of the multiple species of *Legionella*, *L. pneumophila* is the most common cause of human infections with a clinical spectrum ranging from the self-resolving to the fatal pneumonic form, the latter known as Legionnaires' disease (Harrison et al. 2007; Fields et al. 2002).

Infections by *L. pneumophila* begin with the inhalation of contaminated aerosols that are phagocytozed by alveolar macrophages (Allegra et al. 2016). After internalization, *L. pneumophila* relies on a type IV secretion system to translocate over 300 protein effectors across the bacterial membrane into the host cell cytosol (Ham et al. 2011; Hubber and Roy 2010; Manske and Hilbi 2014). These protein effectors inhibit the maturation of the *Legionella*-containing vacuole protecting the bacteria from the acidic and proteolytic environment of the phagolysosome, which is critical for the differentiation of *L. pneumophila* into the replicative form (Ninio and Roy 2007; Ham et al. 2011). In this regard, the differentiation of *L. pneumophila* from the replicative to the transmissive form appears to be linked with its metabolic environment and nutrient availability (Byrne and Swanson 1998; Dalebroux et al. 2009; Edwards et al. 2009; Hovel-Miner et al. 2010). For example, reduced amounts of amino acids as well as the accumulation of the short-chain fatty acids formate, acetate, propionate, and butyrate appear to trigger the switch from the replicative to the transmissive *L. pneumophila* form (Edwards et al. 2009; Fonseca and Swanson 2014). These data suggest that amino acids are an important carbon source for *L. pneumophila*. In support of this hypothesis is the observation that, when *L. pneumophila*-infected monocytes were cultured in defined medium lacking L-cysteine, L-glutamine, or L-serine, bacterial replication was significantly reduced (Wieland et al. 2005). This study further showed that these monocytes upregulate the amino acid transporter SLC1A5 upon infection and pharmacological or genetic inhibition of these transporter functions resulted in reduced growth of *L. pneumophila* (Wieland et al. 2005). In addition to the upregulation of SLC1A5, both amoeba and human proteasomal degradation have been also shown to be critical to generate amino acids to support *L. pneumophila* intracellular growth (Price et al. 2011).

In the majority of immunocompetent mice and humans, *L. pneumophila* infection is controlled by the innate immune response within the first 5–7 days of infection. Despite this control, *L. pneumophila* inhibits the transcriptional response of infected macrophages through a mechanism that is dependent on effectors translocated by the type IV secretion system (Hempstead and Isberg 2015). Specifically, these effectors have been shown to interfere with protein translation, triggering a stress response by the host that is characterized by the expression of NF-kB-regulated genes, including *Il1a*, *Il1b*, *Tnfa*, *Il23a*, *Csf1*, and *Csf2*, that are not translated into proteins (Asrat et al. 2014). As some of these cytokines are essential to control *L. pneumophila*

(Barry et al. 2013; Friedman et al. 2002), these data suggest that the inhibition of the transcriptional response may be an important *L. pneumophila*-immune evasion mechanism. However, a subset of inflammatory genes bypasses this inhibition, being translated in the presence of *L. pneumophila* effectors (Asrat et al. 2014). These data show that despite the inhibitory action of *L. pneumophila*, the host cell is capable of producing inflammatory cytokines and mounts an inflammatory response. Whether the balance between cytokines that are inhibited and cytokines that are produced is important in defining the outcome of infection has not been determined.

The importance of *L. pneumophila* effectors translocated to the host cytosol is also important to modulate the host metabolic response. In this regard, a recent study showed that type IV secreted effectors play critical roles in the mitochondrial dynamics and macrophage metabolic response during infection. Specifically, *L. pneumophila* infection induced mitochondria fragmentation at a very early time point after infection, through a mechanism that is independent on the type IV translocated effectors, resulting in the inhibition of oxidative phosphorylation (Escoll et al. 2017). Therefore, the host cell resorts to glycolysis to generate ATP. Importantly, while inhibition of oxidative phosphorylation did not impact *L. pneumophila* growth, the reduction of cellular glycolysis significantly impaired its replication (Escoll et al. 2017).

These data show that the host metabolic response plays a crucial role in supporting intracellular *L. pneumophila* growth. Clearly, *L. pneumophila* is well-adapted to take advantage of host pathways (proteasomal degradation and the glycolytic pathway) for its intracellular growth. With over 300 effectors translocated by type IV secretion system, future research will be essential to identify novel effectors used by *L. pneumophila* to modulate the metabolism of its host cell.

8.4.4 Listeria monocytogenes

Listeria monocytogenes is the etiological agent of listeriosis and the only human pathogenic species of the genus *Listeria*. Clinical manifestations of listeriosis are variable, ranging from febrile gastroenteritis and sepsis to abortion. One of the most distinctive features of this facultative intracellular pathogen is its ability to escape from vacuoles of infected cells and to subsequently replicate in the cytosol and spread from cell to cell. Several proteins including listeriolysin O, a pore-forming cholesterol-dependent cytolysin, a phosphatidylinositol-specific phospholipase C, and a broad-range phospholipase mediate this process (Vazquez-Boland et al. 2001; Portnoy et al. 2002). These proteins are therefore critical for the pathogenicity of *L. monocytogenes*.

To define the host metabolic response to *L. monocytogenes* infection, a recent study compared the metabolic response of bone marrow-derived macrophages and J744A.1 cell line infected with *L. monocytogenes* (Gillmaier et al. 2012). This study showed that primary cells respond to *L. monocytogenes* infection with increased glucose uptake and glycolysis with no alterations in TCA cycle and glutaminolysis

(Gillmaier et al. 2012). As expected, these effects were not observed in *L. monocytogenes*-infected J744A.1 macrophages, because of the enhanced glycolysis of this macrophage cell line even without infection (Gillmaier et al. 2012). Importantly, the increased glucose uptake and glycolysis was associated with enhanced bacterial growth, specifically in J744A.1 macrophages where the number of infected cells increased markedly in the first hours of infection (Gillmaier et al. 2012). In accordance with these data are the in vivo experiments wherein the transcription profile of mice intestinal epithelia infected with *L. monocytogenes* displays an increase in glycolytic pathway-associated genes (Lecuit et al. 2007). Specifically upregulated was the gene encoding for hexokinase II which phosphorylates glucose to produce glucose-6-phosphate, one of the intermediates of glycolytic pathway (Eisenreich et al. 2013). In this regard, it has been previously shown that *L. monocytogenes* exploits hexose phosphates from the host cell as a source of carbon and energy to fuel fast intracellular growth, in a mechanism that is dependent on the virulence factor Hpt permease (Chico-Calero et al. 2002). These data suggest that *L. monocytogenes* can take advantage of the glycolytic state of the host cell to grow in the cytosol and spread to new cells.

In addition to the increased glycolytic state, *L. monocytogenes*-infected cells have also been shown to downregulate several genes of cytochrome P450 families involved in the biosynthesis of oxysterols and bile acids (Pikuleva 2006) while upregulate *Ch25h*, a gene that encodes the cholesterol 25-hydroxylase responsible for the formation of 25-hydroxycholesterol, in a type I IFN-dependent way (Zou et al. 2011). Knockdown or overexpression of this gene in ZBM2 macrophage cell line infected with *L. monocytogenes* showed significant differences in the growth of the bacteria that correlated with *Ch25h* expression levels. Specifically, cells that had higher levels of *Ch25h* had more bacteria, while cell lines with reduced *Ch25h* had fewer bacteria (Zou et al. 2011). Mechanistically, 25-hydroxycholesterol serves as an endogenous ligand for LXR nuclear receptor, which promotes macrophage survival and an environment for *L. monocytogenes* growth (Joseph et al. 2004).

In addition to its effects in macrophage survival, 25-hydroxycholesterol has been associated with IL-1β production (Rosklint et al. 2002). In this regard, IL-1 receptor blockade exacerbates listeriosis in mice (Havell et al. 1992). Furthermore, IL-1 receptor antagonist and IL-1 receptor-deficient mice are less and more susceptible, respectively, to listeriosis (Hirsch et al. 1996). Taken together, these data disclose a direct link between metabolic and inflammatory pathways with impact in the control of *L. monocytogenes* infection.

Using *Drosophila melanogaster* as model of *L. monocytogenes* infection, a recent study showed a decrease in triglycerides and glycogen levels upon *L. monocytogenes* infection (Chambers et al. 2012). In line with this, infected flies displayed a reduction in the concentration of substrates for mitochondrial beta-oxidation and glycolysis. However, whether this reduction was a result of consumption of host resources by the bacteria or a host defense mechanism was not clearly determined. Furthermore, *L. monocytogenes*-infected flies displayed a reduction in the levels of uric acid, a potent antioxidant (Kand'ar et al. 2006). While it is plausible that the decrease in antioxidant levels could improve the effectiveness of the immune effector response, through ROS production, stabilization of uric acid levels had no effect on *L. monocytogenes* burdens (Kand'ar et al. 2006). These data suggest that the decreased antioxidant potential does not impact control of infection. However, changes in uric acid metabolism may have important implications in metabolism and immunity still undisclosed. Further research is required to determine impact of uric acid metabolism in *L. monocytogenes* infection.

Taken together, these data show that host cells undergo important glycolytic alterations upon infection with *L. monocytogenes*. However, these alterations appear to favor bacterial growth rather than the host antibacterial response. Indeed, *L. monocytogenes* infection promotes glucose uptake and steroid and carbon metabolism of its host cell, leading to the production of glucose-6-phosphate, serine, and glycerol, which can be utilized by the bacteria for intracellular growth.

8.4.5 Mycobacterium tuberculosis

The intracellular pathogen *M. tuberculosis*, the etiological agent of tuberculosis, belongs to a very restrict group of bacterial pathogens that have humans as its only known host. One of the key features of the success of *M. tuberculosis* as a human pathogen is its ability to modulate and circumvent the activation and effector functions of its host cells, typically macrophages (Fraga et al. 2017).

Upon gaining access to the lower airways, *M. tuberculosis* is phagocytozed by lung-resident macrophages. Although macrophages are the primary host cell for *M. tuberculosis*, these cells are also the effector cell of the immune response. Upon recognition and phagocytosis, the *M. tuberculosis*-containing phagosome is directed through the canonical endocytic pathway, fusing with lysosomes while also engaging autophagy (Fraga et al. 2017). In parallel, the engagement of *M. tuberculosis* PAMPs and the cytokine microenvironment of the infection focus promote the production of ROS and nitric oxide by infected macrophages. Although both ROS and nitric oxide may play important roles in the control intracellular *M. tuberculosis*, mice deficient in iNOS are more susceptible to *M. tuberculosis* infection than PHOX-deficient mice (Adams et al. 1997; Ehrt et al. 2001). These data show that iNOS activity is critical to control *M. tuberculosis* infection. The key role of nitric oxide during *M. tuberculosis* infection is now recognized to be mediated both by direct effects on the intracellular bacteria and by regulating the inflammatory response (Mishra et al. 2017; Mishra et al. 2013). On the other hand, *M. tuberculosis* resistance in ROS-mediated killing mechanisms is likely mediated by the expression of the enzymes catalase and peroxidase properties encoded by KatG (Ng et al. 2004; Manca et al. 1999) and the upregulation of the Mn-dependent superoxide dismutase that acts as an antioxidant by quenching ROS (Bresciani et al. 2015; Liao et al. 2013; Ragno et al. 2001). In addition to nitric oxide and ROS, *M. tuberculosis*-infected macrophages also upregulate the production of heme oxygenase-1 (Sinnis and Ernst 2008). This enzyme produces carbon monoxide that, together with nitric oxide, induces the transcription of the *M. tuberculosis* dormancy regulon DosR (Shiloh et al. 2008). This genetic program is induced in conditions that inhibit aerobic respiration and prevent *M. tuberculosis* replication by shifting metabolism away from aerobic respiration and maintaining energy levels and redox balance during anaerobiosis (Leistikow et al. 2010). Furthermore, DosR is also critical for the rapid resumption of growth once *M. tuberculosis* exits an anaerobic or nitric oxide-induced nonrespiring state (Leistikow et al. 2010). This mechanism highlights the extraordinary adaptability of *M. tuberculosis* to the hostile environment of the activated macrophage.

Recent studies using RNA-seq technology and immunohistochemistry have shown the occurrence of the Warburg effect in *M. tuberculosis*-infected cells during early stages of granuloma formation in mice (Shi et al. 2015). This finding is supported by the upregulation of genes encoding GLUT1, 3, and 6, hexokinases 2 and 3, members of phosphofructokinase family 1 and 2, GAPDH, phosphoglycerate kinase 1, LDHA, and monocarboxylate transporter 4 (Shi et al. 2015). This gene expression profile was also validated by increased protein expression (Shi et al. 2015) and enhanced glycolytic flux with increased accumulation of lactate in the lungs of *M. tuberculosis*-infected mice (Shin et al. 2011). Given that the Warburg effect in mice occurs in parallel with the expression of host adaptive immunity (Liberti and Locasale 2016; Shi et al. 2003, 2015), these data suggest that the Warburg effect may be essential for the bioenergetic requirement of the host to mount an efficient host responses against *M. tuberculosis*. This hypothesis is supported by zebra fish data showing that pharmacological or genetical stabilization of HIF-1α, one of the main regulators of the Warburg effect, at early stages of infection reduced *M. marinum* burdens of infected larvae through a mechanism that was dependent on iNOS (Elks et al. 2013). Furthermore, the relevance of HIF-1α is also clear by the impaired resistance to *M. avium* infection and premature development of necrotic granulomata in mice lacking HIF-1α (Cardoso et al. 2015). Noteworthy is the differential expression of Warburg effect-associated markers and HIF-1α within mouse lung granulomata. Indeed, infected macrophages within the center of the granuloma display a reduced Warburg effect accompanied by an increased glucose metabolism, contrasting to peripheral infected macrophages that presented reduced glucose metabolism (Shi et al. 2015). In this regard, mycolic acids in the cell wall of *M. tuberculosis* have been shown to divert host cells glycolysis toward ketone body biosynthesis by promoting glucose uptake (Singh et al. 2015). This process actively disrupts macrophage metabolic homeostasis and promotes the accumulation of lipid bodies resulting in the development of foamy macrophages, an important reservoir for *M. tuberculosis* that supports the transition of the bacteria into the nonreplicating state (Peyron et al. 2008; Singh et al. 2012). In this regard, a recent study suggested that the expression of the early secretory antigen target 6 Kd (ESAT-6) induced the formation of foamy macrophages by interacting directly with the host cell glycolytic enzymes enolase 1 and phosphoglycerate kinase 1 (Singh et al. 2015). These data show that *M. tuberculosis* has evolved several mechanisms to induce metabolic flux perturbations with important implications in the control of infection.

A recent study exploring mycobacterial genes critical for survival of *M. tuberculosis* within an immunocompetent host reinforced the role of metabolism in host-pathogen dynamics during infection (Marrero et al. 2010). This study showed that the most fundamental genes upregulated by the bacteria during intracellular growth are associated with gluconeogenesis (Marrero et al. 2010). The finding that gluconeogenesis plays a relevant role in bacterial pathogenesis is supported by other studies showing that *M. tuberculosis* is able to dysregulate host lipid metabolism toward an increased production of carbon sources such as cholesterol, triacylglycerides, and glycosphingolipids that can be used for gluconeogenesis by the pathogen, specifically in caseating lesions (Kim et al. 2010). Many of the upregulated genes responsible for the accumulation of lipids within granulomata can be induced by TNF-α, suggesting that this response is a consequence of sustained inflammation within the *M. tuberculosis* infection foci. Furthermore, mycobacterial genes important for a functional tryptophan biosynthetic pathway are also critical for *M. tuberculosis* pathogenesis and survival, as they provide protection from the protective acquired immune response (Zhang et al. 2013). In fact, immune-mediated IDO upregulation limits tryptophan availability to the pathogen. Nevertheless, this antimicrobial response may not be effective due to the ability of *M. tuberculosis* to synthesize tryptophan de novo. For this reason, the use of targets for *M. tuberculosis* tryptophan biosynthetic machinery, such as anthranilate synthase, has been recently shown to be bactericidal in vitro and to limit bacterial growth in vivo (Zhang et al. 2013).

Several studies have also addressed the role of *M. tuberculosis* in protein turnover (Rao and Li 2009; Macallan et al. 1998). This catabolic process results in the release of amino acids followed by their oxidation, contributing to energy storage and nitrogen excretion. Under normal conditions, this process provides a rapid supply of substrate for acute-phase protein synthesis and immune cell proliferation and activation. However, excessive amounts of protein turnover may also deplete protein, resulting adverse effects on the function of several tissues, particularly skeletal muscle. In this regard, there is an increase in the amount of several amino acids in lung tissues and in the serum of *M. tuberculosis*-infected mice and guinea pigs, suggesting enhanced proteolysis and/or catabolism of ingested amino acids (Shin et al. 2011; Somashekar et al. 2011). Future research will be crucial to clarify this issue and determine its impact in control of infection.

Altogether, the data discussed above show that *M. tuberculosis* interfere with metabolic and host defense mechanisms, by promoting lipid metabolism, glucose consumption, and proteolysis. On the one hand, these mechanisms are critical for the effectiveness of the immune response as cellular proliferation and biosynthetic and secretory activities of immune cells require high-energy consumption. Accordingly, glucose availability has been shown to play crucial roles in lymphocyte activation and expansion, lipid turnover, and synthesis, which are critical to the microbicidal capacity of macrophages (Calder 2013). However, it is also clear that these metabolic adaptations also benefit the bacteria, thus highlighting the success of *M. tuberculosis* as a human pathogen.

8.4.6 Salmonella enterica

Salmonella enterica is Gram-negative rod-shaped bacteria of the *Enterobacteriaceae* family responsible for approximately 93.8 million cases of infections per year worldwide, resulting in thousands of fatal cases (Ao et al. 2015; Feasey et al. 2012). Like other *Gammaproteobacteria*, *S. enterica* is present in various environments, including sewage, soils, and water systems, and in the gut flora (Dandekar et al. 2014). In humans, different *S. enterica* forms drive distinct physiologic outcomes. Specifically, the non-typhoidal serovars (*S. enterica* serovar Typhimurium) are the main cause of severe bloodstream infections in developing countries, while typhoidal forms (*S. enterica* serovars Typhi and Paratyphi) cause severe systemic infections, including typhoid fever (Gal-Mor et al. 2014; Feasey et al. 2012). Infections by *S. enterica* serovar Typhimurium are leading causes of food-borne illness and diarrheal disease worldwide (Anderson and Kendall 2017). In mice, however, infections with *S. enterica* serovar Typhimurium or Typhi result in similar outcomes. These pathogens are very well-adapted organisms that have evolved multiple mechanisms to circumvent the antimicrobial properties of the host phagocytic cells (Olive and Sassetti 2016). Indeed, upon infection, the pathogen survives in the intracellular compartment in nutritionally deprived membrane compartment termed *Salmonella*-containing vacuole. It is only after reaching the mesenteric lymph nodes that the infection can spread to other organs, especially the liver and spleen (Dandekar et al. 2014).

Animal models of *Salmonella* infection have provided important advances in the understanding of the host metabolic responses to these pathogens. In this regard, a recent study reported the transcriptional profile of the mouse intestinal mucosa responses in an early (8 h) and later stage (4 days) of infection analyzing 28.000 genes (Liu et al. 2010). Overall, 856 genes were differentially expressed at 8 h post-infection and 7558 genes after 4 days post-infection. The most downregulated genes during the early stage of infection were associated with the oxidative phosphorylation pathway (Liu et al. 2010). On the other hand, at later time points post-infection, there was a downregulation of genes belonging to many metabolic pathways, whereas genes associated with the immune and inflammatory response, proliferation, and apoptosis were upregulated (Liu et al. 2010). In accordance with these data, a different study determined metabolite concentration in the intestine and liver of *S. enterica* serovar Typhimurium-infected mice and compared their concentrations with uninfected mice (Antunes et al. 2011). This study showed a significant decrease in metabolites derived from glucose metabolism, in the liver but not in the intestine of infected mice (Antunes et al. 2011). Additionally, increased metabolites derived from the metabolism of arachidonic acid, the main precursor of the eicosanoid pathway, were also determined (Antunes et al. 2011). This increased concentration was through transcriptional regulation of the enzymes involved in eicosanoid synthesis (Antunes et al. 2011). Taken together, these data show that *Salmonella* infection has a pronounced impact in glucose and arachidonic acid metabolism in the liver but not in the intestine. Whether this is a result of the high-energy required for the immune control of the infection remains to be clarified.

The invasion processes of *Salmonella* in warm-blooded animals have been shown to depend on *Salmonella*-pathogenicity islands (SPI)-1 and SPI-2, which are encoded by two different type III secretion systems. As described above, *Salmonella* reside in very restrict environment of the *Salmonella*-containing vacuole, and therefore, SPI-1 and SPI-2 are induced in distinct conditions (Jennings et al. 2017; Cascales 2017). SPI-1 effectors are induced in a nutrient rich environment and potentiate the internalization of *Salmonella* by nonprofessional phagocytes, whereas SPI-2 effectors are induced in a nutrient-poor environment and are essential for intracellular replication. As these type III secretion systems play a key role in the pathogenicity of *Salmonella*, it is not surprising that some of the effectors secreted by these systems also modulate host metabolism. For example, AvrA can reduce oxidative phosphorylation and activate mTOR, NF-kB, and p53, in *Salmonella*-infected intestinal tissue (Lin et al. 2016), thus impacting carbon metabolism. Specifically, p53 activation by AvrA shuts off glycolysis, while mTOR activation stimulates amino acid uptake and catabolism (Santos et al. 2009). SopB is another important effector predominately involved in promoting *Salmonella* invasion. After host cell-pathogen contact, SopB activates Akt, which in its turn inhibits apoptosis of infected host cells promoting *Salmonella* invasion of epithelial cells and macrophages (Vonaesch et al. 2014; Scanu et al. 2015). Nevertheless, activated Akt also leads to Mdm2 phosphorylation and p53 ubiquitination followed by its degradation, which in turn may promote glycolysis. SopE, on the other hand, induces iNOS in the host intestinal cells (Cherayil et al. 2000). Although, iNOS activity is a crucial antimicrobial defense mechanism against intracellular bacterial infections, nitric oxide produced by iNOS activity can be converted into nitrate, which in turn can be used by bacteria as electron acceptor in anaerobic respiration (Lopez et al. 2012). Under these conditions, *Salmonella* can use ethanolamine present in the intestine and have a growth advantage over intestinal anaerobes that are unable to use ethanolamine as carbon source (Thiennimitr et al. 2011). Finally, SpiC activates the ERK1/2 signal transduction pathway to enhance COX-2 expression and PGE2 synthesis in infected macrophages (Uchiya and Nikai 2004). *Salmonella*-induced PGE_2 activates the protein kinase A pathway and upregulates IL-10 production by macrophages, promoting an immunosuppressive phenotype and impaired killing ability. However, it appears that there is a temporal role of COX-2 expression in vivo (Bowman and Bost 2004). Indeed, at early stages of acute infection, COX-2 inhibition results in increased bacterial burdens in the mesenteric lymph nodes, whereas at later stages of infection, COX-2 inhibition enhances host survival (Bowman and Bost 2004). Therefore, PGE2 is essential for acute *Salmonella* control, but it may promote chronic disease.

Taken together, these data show that specific effector molecules produced by *Salmonella* interfere with key regulators of the host metabolic response. Specifically, the activation of mTOR, NF-kB, and p53 may have important implication in carbon metabolism of host cell. However, more data is required to determine the impact of these changes in the control of infection.

8.5 Conclusion

The outcome of infections caused by intracellular bacteria depends on complex interactions between the pathogen and the host immune system. Throughout this chapter, we have provided a brief review on the metabolic adaptations of both the host cell and the pathogen and discussed their impact in the outcome of infection. On the one hand, it is clear that the host metabolic responses play crucial roles in the differentiation and function of immune cells that are required to control infection. On the other hand, successful intracellular pathogens have evolved several strategies to take advantage of the host metabolic adaptation or modulate these metabolic adaptations to their own benefit. The identification of some of the pathways by which intracellular pathogens modulate the host metabolic response has already proven viable as targets for antibacterial therapy. With the increased resistance of bacterial pathogens to antibiotics, it is of crucial importance that we identify novel pathways that may be targeted to improve infection control. Only by determining these pathways will we be able to effectively intervene.

Acknowledgments Our work is funded by the project NORTE-01-0145-FEDER-000013, supported by the Northern Portugal Regional Operational Programme (NORTE 2020), under the Portugal 2020 Partnership Agreement, through the European Regional Development Fund (FEDER), and the Fundação para a Ciência e Tecnologia (FCT) through the FCT investigator grant IF/01390/2014 to E.T. and the PhD fellowship SFRH/BD/120371/2016 to A.M.B.

References

Adams LB, Dinauer MC, Morgenstern DE, Krahenbuhl JL (1997) Comparison of the roles of reactive oxygen and nitrogen intermediates in the host response to *Mycobacterium tuberculosis* using transgenic mice. Tuber Lung Dis 78(5–6):237–246. https://doi.org/10.1016/S0962-8479(97)90004-6

Alfano M, Graziano F, Genovese L, Poli G (2013) Macrophage polarization at the crossroad between HIV-1 infection and cancer development. Arterioscler Thromb Vasc Biol 33(6):1145–1152. https://doi.org/10.1161/ATVBAHA.112.300171

Allegra S, Leclerc L, Massard PA, Girardot F, Riffard S, Pourchez J (2016) Characterization of aerosols containing *Legionella* generated upon nebulization. Sci Rep 6:33998. https://doi.org/10.1038/srep33998

Anderson CJ, Kendall MM (2017) *Salmonella enterica* Serovar Typhimurium strategies for host adaptation. Front Microbiol 8:1983. https://doi.org/10.3389/fmicb.2017.01983

Antunes LC, Arena ET, Menendez A, Han J, Ferreira RB, Buckner MM, Lolic P, Madilao LL, Bohlmann J, Borchers CH, Finlay BB (2011) Impact of *salmonella* infection on host hormone metabolism revealed by metabolomics. Infect Immun 79(4):1759–1769. https://doi.org/10.1128/IAI.01373-10

Ao TT, Feasey NA, Gordon MA, Keddy KH, Angulo FJ, Crump JA (2015) Global burden of invasive nontyphoidal *Salmonella* disease, 2010(1). Emerg Infect Dis 21(6):941. https://doi.org/10.3201/eid2106.140999

Armstrong JA, Hart PD (1971) Response of cultured macrophages to *Mycobacterium tuberculosis*, with observations on fusion of lysosomes with phagosomes. J Exp Med 134(3 Pt 1):713–740. https://doi.org/10.1084/jem.134.3.713

Asrat S, Dugan AS, Isberg RR (2014) The frustrated host response to *Legionella pneumophila* is bypassed by MyD88-dependent translation of pro-inflammatory cytokines. PLoS Pathog 10(7): e1004229. https://doi.org/10.1371/journal.ppat.1004229

Azad AK, Sadee W, Schlesinger LS (2012) Innate immune gene polymorphisms in tuberculosis. Infect Immun 80(10):3343–3359. https://doi.org/10.1128/IAI.00443-12

Bachmann NL, Polkinghorne A, Timms P (2014) *Chlamydia* genomics: providing novel insights into chlamydial biology. Trends Microbiol 22(8):464–472. https://doi.org/10.1016/j.tim.2014.04.013

Bai X, Feldman NE, Chmura K, Ovrutsky AR, Su WL, Griffin L, Pyeon D, McGibney MT, Strand MJ, Numata M, Murakami S, Gaido L, Honda JR, Kinney WH, Oberley-Deegan RE, Voelker DR, Ordway DJ, Chan ED (2013) Inhibition of nuclear factor-kappa B activation decreases survival of *Mycobacterium tuberculosis* in human macrophages. PLoS One 8(4):e61925. https://doi.org/10.1371/journal.pone.0061925

Baker RG, Hayden MS, Ghosh S (2011) NF-kappaB, inflammation, and metabolic disease. Cell Metab 13(1):11–22. https://doi.org/10.1016/j.cmet.2010.12.008

Barry KC, Fontana MF, Portman JL, Dugan AS, Vance RE (2013) IL-1alpha signaling initiates the inflammatory response to virulent *Legionella pneumophila* in vivo. J Immunol 190 (12):6329–6339. https://doi.org/10.4049/jimmunol.1300100

Blumenthal A, Nagalingam G, Huch JH, Walker L, Guillemin GJ, Smythe GA, Ehrt S, Britton WJ, Saunders BM (2012) *M. tuberculosis* induces potent activation of IDO-1, but this is not essential for the immunological control of infection. PLoS One 7(5):e37314. https://doi.org/10.1371/journal.pone.0037314

Boman J, Hammerschlag MR (2002) *Chlamydia pneumoniae* and atherosclerosis: critical assessment of diagnostic methods and relevance to treatment studies. Clin Microbiol Rev 15(1):1–20

Bowman CC, Bost KL (2004) Cyclooxygenase-2-mediated prostaglandin E2 production in mesenteric lymph nodes and in cultured macrophages and dendritic cells after infection with *Salmonella*. J Immunol 172(4):2469–2475

Bresciani G, da Cruz IB, Gonzalez-Gallego J (2015) Manganese superoxide dismutase and oxidative stress modulation. Adv Clin Chem 68:87–130. https://doi.org/10.1016/bs.acc.2014.11.001

Brothwell JA, Muramatsu MK, Toh E, Rockey DD, Putman TE, Barta ML, Hefty PS, Suchland RJ, Nelson DE (2016) Interrogating genes that mediate *Chlamydia trachomatis* survival in cell culture using conditional mutants and recombination. J Bacteriol 198(15):2131–2139. https://doi.org/10.1128/JB.00161-16

Buck MD, O'Sullivan D, Pearce EL (2015) T cell metabolism drives immunity. J Exp Med 212 (9):1345–1360. https://doi.org/10.1084/jem.20151159

Byrne B, Swanson MS (1998) Expression of *Legionella pneumophila* virulence traits in response to growth conditions. Infect Immun 66(7):3029–3034

Byrne GI (2010) *Chlamydia trachomatis* strains and virulence: rethinking links to infection prevalence and disease severity. J Infect Dis 201(Suppl 2):S126–S133. https://doi.org/10.1086/652398

Calder PC (2013) Feeding the immune system. Proc Nutr Soc 72(3):299–309. https://doi.org/10.1017/S0029665113001286

Carabeo RA, Mead DJ, Hackstadt T (2003) Golgi-dependent transport of cholesterol to the *Chlamydia trachomatis* inclusion. Proc Natl Acad Sci USA 100(11):6771–6776. https://doi.org/10.1073/pnas.1131289100

Cardoso MS, Silva TM, Resende M, Appelberg R, Borges M (2015) Lack of the transcription factor Hypoxia-Inducible Factor 1alpha (HIF-1alpha) in macrophages accelerates the necrosis of *Mycobacterium avium*-induced granulomas. Infect Immun 83(9):3534–3544. https://doi.org/10.1128/IAI.00144-15

Casanova JE (2017) Bacterial autophagy: offense and defense at the host-pathogen interface. Cell Mol Gastroenterol Hepatol 4(2):237–243. https://doi.org/10.1016/j.jcmgh.2017.05.002

Cascales E (2017) Inside the chamber of secrets of the Type III secretion system. Cell 168 (6):949–951. https://doi.org/10.1016/j.cell.2017.02.028

Castillo EF, Dekonenko A, Arko-Mensah J, Mandell MA, Dupont N, Jiang S, Delgado-Vargas M, Timmins GS, Bhattacharya D, Yang H, Hutt J, Lyons CR, Dobos KM, Deretic V (2012) Autophagy protects against active tuberculosis by suppressing bacterial burden and inflammation. Proc Natl Acad Sci USA 109(46):E3168–E3176. https://doi.org/10.1073/pnas.1210500109

Chambers MC, Song KH, Schneider DS (2012) *Listeria monocytogenes* infection causes metabolic shifts in *Drosophila melanogaster*. PLoS One 7(12):e50679. https://doi.org/10.1371/journal.pone.0050679

Chang CH, Curtis JD, Maggi LB Jr, Faubert B, Villarino AV, O'Sullivan D, Huang SC, van der Windt GJ, Blagih J, Qiu J, Weber JD, Pearce EJ, Jones RG, Pearce EL (2013) Posttranscriptional control of T cell effector function by aerobic glycolysis. Cell 153(6):1239–1251. https://doi.org/10.1016/j.cell.2013.05.016

Chen AL, Johnson KA, Lee JK, Sutterlin C, Tan M (2012) CPAF: a Chlamydial protease in search of an authentic substrate. PLoS Pathog 8(8):e1002842. https://doi.org/10.1371/journal.ppat.1002842

Cherayil BJ, McCormick BA, Bosley J (2000) Salmonella enterica serovar typhimurium-dependent regulation of inducible nitric oxide synthase expression in macrophages by invasins SipB, SipC, and SipD and effector SopE2. Infect Immun 68(10):5567–5574

Chico-Calero I, Suarez M, Gonzalez-Zorn B, Scortti M, Slaghuis J, Goebel W, Vazquez-Boland JA, European Listeria Genome C (2002) Hpt, a bacterial homolog of the microsomal glucose-6-phosphate translocase, mediates rapid intracellular proliferation in *Listeria*. Proc Natl Acad Sci USA 99(1):431–436. https://doi.org/10.1073/pnas.012363899

Choy A, Dancourt J, Mugo B, O'Connor TJ, Isberg RR, Melia TJ, Roy CR (2012) The *Legionella* effector RavZ inhibits host autophagy through irreversible Atg8 deconjugation. Science 338 (6110):1072–1076. https://doi.org/10.1126/science.1227026

Cianciotto NP (2001) Pathogenicity of *Legionella pneumophila*. Int J Med Microbiol 291 (5):331–343. https://doi.org/10.1078/1438-4221-00139

Cocchiaro JL, Kumar Y, Fischer ER, Hackstadt T, Valdivia RH (2008) Cytoplasmic lipid droplets are translocated into the lumen of the *Chlamydia trachomatis* parasitophorous vacuole. Proc Natl Acad Sci USA 105(27):9379–9384. https://doi.org/10.1073/pnas.0712241105

Cooper AM, Magram J, Ferrante J, Orme IM (1997) Interleukin 12 (IL-12) is crucial to the development of protective immunity in mice intravenously infected with *Mycobacterium tuberculosis*. J Exp Med 186(1):39–45. https://doi.org/10.1084/jem.186.1.39

Czyz DM, Potluri LP, Jain-Gupta N, Riley SP, Martinez JJ, Steck TL, Crosson S, Shuman HA, Gabay JE (2014) Host-directed antimicrobial drugs with broad-spectrum efficacy against intracellular bacterial pathogens. MBio 5(4):e01534-01514. https://doi.org/10.1128/mBio.01534-14

Dalebroux ZD, Edwards RL, Swanson MS (2009) SpoT governs *Legionella pneumophila* differentiation in host macrophages. Mol Microbiol 71(3):640–658. https://doi.org/10.1111/j.1365-2958.2008.06555.x

Dandekar T, Fieselmann A, Fischer E, Popp J, Hensel M, Noster J (2014) *Salmonella*-how a metabolic generalist adopts an intracellular lifestyle during infection. Front Cell Infect Microbiol 4:191. https://doi.org/10.3389/fcimb.2014.00191

DeBerardinis RJ, Thompson CB (2012) Cellular metabolism and disease: what do metabolic outliers teach us? Cell 148(6):1132–1144. https://doi.org/10.1016/j.cell.2012.02.032

Deretic V, Delgado M, Vergne I, Master S, De Haro S, Ponpuak M, Singh S (2009) Autophagy in immunity against *Mycobacterium tuberculosis*: a model system to dissect immunological roles of autophagy. Curr Top Microbiol Immunol 335:169–188. https://doi.org/10.1007/978-3-642-00302-8_8

Derre I, Swiss R, Agaisse H (2011) The lipid transfer protein CERT interacts with the *Chlamydia inclusion* protein IncD and participates to ER-*Chlamydia* inclusion membrane contact sites. PLoS Pathog 7(6):e1002092. https://doi.org/10.1371/journal.ppat.1002092

Desai M, Fang R, Sun J (2015) The role of autophagy in microbial infection and immunity. Immunotargets Ther 4:13–26. https://doi.org/10.2147/ITT.S76720

Edwards RL, Dalebroux ZD, Swanson MS (2009) *Legionella pneumophila* couples fatty acid flux to microbial differentiation and virulence. Mol Microbiol 71(5):1190–1204. https://doi.org/10.1111/j.1365-2958.2009.06593.x

Ehrt S, Schnappinger D, Bekiranov S, Drenkow J, Shi S, Gingeras TR, Gaasterland T, Schoolnik G, Nathan C (2001) Reprogramming of the macrophage transcriptome in response to interferon-gamma and *Mycobacterium tuberculosis*: signaling roles of nitric oxide synthase-2 and phagocyte oxidase. J Exp Med 194(8):1123–1140

Eisenreich W, Heesemann J, Rudel T, Goebel W (2013) Metabolic host responses to infection by intracellular bacterial pathogens. Front Cell Infect Microbiol 3:24. https://doi.org/10.3389/fcimb.2013.00024

Eisenreich W, Heesemann J, Rudel T, Goebel W (2015) Metabolic adaptations of intracellullar bacterial pathogens and their mammalian host cells during infection ("Pathometabolism"). Microbiol Spectr 3(3). https://doi.org/10.1128/microbiolspec.MBP-0002-2014

Elks PM, Brizee S, van der Vaart M, Walmsley SR, van Eeden FJ, Renshaw SA, Meijer AH (2013) Hypoxia inducible factor signaling modulates susceptibility to mycobacterial infection via a nitric oxide dependent mechanism. PLoS Pathog 9(12):e1003789. https://doi.org/10.1371/journal.ppat.1003789

Elwell C, Mirrashidi K, Engel J (2016) *Chlamydia* cell biology and pathogenesis. Nat Rev Microbiol 14(6):385–400. https://doi.org/10.1038/nrmicro.2016.30

Elwell CA, Jiang S, Kim JH, Lee A, Wittmann T, Hanada K, Melancon P, Engel JN (2011) *Chlamydia trachomatis* co-opts GBF1 and CERT to acquire host sphingomyelin for distinct roles during intracellular development. PLoS Pathog 7(9):e1002198. https://doi.org/10.1371/journal.ppat.1002198

Escoll P, Rolando M, Gomez-Valero L, Buchrieser C (2013) From amoeba to macrophages: exploring the molecular mechanisms of *Legionella pneumophila* infection in both hosts. Curr Top Microbiol Immunol 376:1–34. https://doi.org/10.1007/82_2013_351

Escoll P, Song OR, Viana F, Steiner B, Lagache T, Olivo-Marin JC, Impens F, Brodin P, Hilbi H, Buchrieser C (2017) *Legionella pneumophila* modulates mitochondrial dynamics to trigger metabolic repurposing of infected macrophages. Cell Host Microbe 22(3):302–316. https://doi.org/10.1016/j.chom.2017.07.020

Feasey NA, Dougan G, Kingsley RA, Heyderman RS, Gordon MA (2012) Invasive non-typhoidal *salmonella* disease: an emerging and neglected tropical disease in Africa. Lancet 379(9835):2489–2499. https://doi.org/10.1016/S0140-6736(11)61752-2

Fields BS, Benson RF, Besser RE (2002) *Legionella* and Legionnaires' disease: 25 years of investigation. Clin Microbiol Rev 15(3):506–526

Fonseca MV, Swanson MS (2014) Nutrient salvaging and metabolism by the intracellular pathogen *Legionella pneumophila*. Front Cell Infect Microbiol 4:12. https://doi.org/10.3389/fcimb.2014.00012

Fraga AG, Barbosa AM, Ferreira CM, Fevereiro J, Pedrosa J, Torrado E (2017) Immune-evasion strategies of mycobacteria and their implications for the protective immune response. Curr Issues Mol Biol 25:169–198. https://doi.org/10.21775/cimb.025.169

Friedman H, Yamamoto Y, Klein TW (2002) *Legionella pneumophila* pathogenesis and immunity. Semin Pediatr Infect Dis 13(4):273–279. https://doi.org/10.1053/spid.2002.127206

Gal-Mor O, Boyle EC, Grassl GA (2014) Same species, different diseases: how and why typhoidal and non-typhoidal *Salmonella enterica* serovars differ. Front Microbiol 5:391. https://doi.org/10.3389/fmicb.2014.00391

Gillmaier N, Gotz A, Schulz A, Eisenreich W, Goebel W (2012) Metabolic responses of primary and transformed cells to intracellular *Listeria monocytogenes*. PLoS One 7(12):e52378. https://doi.org/10.1371/journal.pone.0052378

Hackstadt T, Rockey DD, Heinzen RA, Scidmore MA (1996) *Chlamydia trachomatis* interrupts an exocytic pathway to acquire endogenously synthesized sphingomyelin in transit from the Golgi apparatus to the plasma membrane. EMBO J 15(5):964–977

Haferkamp I (2017) Crossing the border – solute entry into the chlamydial inclusion. Int J Med Microbiol. https://doi.org/10.1016/j.ijmm.2017.08.006

Ham H, Sreelatha A, Orth K (2011) Manipulation of host membranes by bacterial effectors. Nat Rev Microbiol 9(9):635–646. https://doi.org/10.1038/nrmicro2602

Harrison TG, Doshi N, Fry NK, Joseph CA (2007) Comparison of clinical and environmental isolates of *Legionella pneumophila* obtained in the UK over 19 years. Clin Microbiol Infect 13(1):78–85. https://doi.org/10.1111/j.1469-0691.2006.01558.x

Hatch TP (1975) Competition between *Chlamydia psittaci* and L cells for host isoleucine pools: a limiting factor in chlamydial multiplication. Infect Immun 12(1):211–220

Hatch TP, Allan I, Pearce JH (1984) Structural and polypeptide differences between envelopes of infective and reproductive life cycle forms of *Chlamydia* spp. J Bacteriol 157(1):13–20

Havell EA, Moldawer LL, Helfgott D, Kilian PL, Sehgal PB (1992) Type I IL-1 receptor blockade exacerbates murine listeriosis. J Immunol 148(5):1486–1492

Hempstead AD, Isberg RR (2015) Inhibition of host cell translation elongation by *Legionella pneumophila* blocks the host cell unfolded protein response. Proc Natl Acad Sci USA 112(49):E6790–E6797. https://doi.org/10.1073/pnas.1508716112

Hill J, Samuel JE (2011) *Coxiella burnetii* acid phosphatase inhibits the release of reactive oxygen intermediates in polymorphonuclear leukocytes. Infect Immun 79(1):414–420. https://doi.org/10.1128/IAI.01011-10

Hirsch E, Irikura VM, Paul SM, Hirsh D (1996) Functions of interleukin 1 receptor antagonist in gene knockout and overproducing mice. Proc Natl Acad Sci USA 93(20):11008–11013

Hovel-Miner G, Faucher SP, Charpentier X, Shuman HA (2010) ArgR-regulated genes are derepressed in the *Legionella*-containing vacuole. J Bacteriol 192(17):4504–4516. https://doi.org/10.1128/JB.00465-10

Howe D, Heinzen RA (2006) *Coxiella burnetii* inhabits a cholesterol-rich vacuole and influences cellular cholesterol metabolism. Cell Microbiol 8(3):496–507. https://doi.org/10.1111/j.1462-5822.2005.00641.x

Howe D, Mallavia LP (1999) *Coxiella burnetii* infection increases transferrin receptors on J774A. 1 cells. Infect Immun 67(7):3236–3241

Hubber A, Roy CR (2010) Modulation of host cell function by *Legionella pneumophila* type IV effectors. Annu Rev Cell Dev Biol 26:261–283. https://doi.org/10.1146/annurev-cellbio-100109-104034

Ip WKE, Hoshi N, Shouval DS, Snapper S, Medzhitov R (2017) Anti-inflammatory effect of IL-10 mediated by metabolic reprogramming of macrophages. Science 356(6337):513–519. https://doi.org/10.1126/science.aal3535

Jennings E, Thurston TLM, Holden DW (2017) *Salmonella* SPI-2 Type III secretion system effectors: molecular mechanisms and physiological consequences. Cell Host Microbe 22(2):217–231. https://doi.org/10.1016/j.chom.2017.07.009

Jo EK, Yuk JM, Shin DM, Sasakawa C (2013) Roles of autophagy in elimination of intracellular bacterial pathogens. Front Immunol 4:97. https://doi.org/10.3389/fimmu.2013.00097

Joseph SB, Bradley MN, Castrillo A, Bruhn KW, Mak PA, Pei L, Hogenesch J, O'Connell RM, Cheng G, Saez E, Miller JF, Tontonoz P (2004) LXR-dependent gene expression is important for macrophage survival and the innate immune response. Cell 119(2):299–309. https://doi.org/10.1016/j.cell.2004.09.032

Jouanguy E, Doffinger R, Dupuis S, Pallier A, Altare F, Casanova JL (1999) IL-12 and IFN-gamma in host defense against mycobacteria and *salmonella* in mice and men. Curr Opin Immunol 11(3):346–351

Kading N, Szaszak M, Rupp J (2014) Imaging of *Chlamydia* and host cell metabolism. Future Microbiol 9(4):509–521. https://doi.org/10.2217/fmb.14.13

Kalman S, Mitchell W, Marathe R, Lammel C, Fan J, Hyman RW, Olinger L, Grimwood J, Davis RW, Stephens RS (1999) Comparative genomes of *Chlamydia pneumoniae* and *C. trachomatis*. Nat Genet 21(4):385–389. https://doi.org/10.1038/7716

Kand'ar R, Zakova P, Muzakova V (2006) Monitoring of antioxidant properties of uric acid in humans for a consideration measuring of levels of allantoin in plasma by liquid chromatography. Clin Chim Acta 365(1–2):249–256. https://doi.org/10.1016/j.cca.2005.09.002

Kelly B, O'Neill LA (2015) Metabolic reprogramming in macrophages and dendritic cells in innate immunity. Cell Res 25(7):771–784. https://doi.org/10.1038/cr.2015.68

Kim MJ, Wainwright HC, Locketz M, Bekker LG, Walther GB, Dittrich C, Visser A, Wang W, Hsu FF, Wiehart U, Tsenova L, Kaplan G, Russell DG (2010) Caseation of human tuberculosis granulomas correlates with elevated host lipid metabolism. EMBO Mol Med 2(7):258–274. https://doi.org/10.1002/emmm.201000079

Kim YK, Shin JS, Nahm MH (2016) NOD-like receptors in infection, immunity, and diseases. Yonsei Med J 57(1):5–14. https://doi.org/10.3349/ymj.2016.57.1.5

Kohler LJ, Roy CR (2015) Biogenesis of the lysosome-derived vacuole containing *Coxiella burnetii*. Microbes Infect 17(11–12):766–771. https://doi.org/10.1016/j.micinf.2015.08.006

Krawczyk CM, Holowka T, Sun J, Blagih J, Amiel E, DeBerardinis RJ, Cross JR, Jung E, Thompson CB, Jones RG, Pearce EJ (2010) Toll-like receptor-induced changes in glycolytic metabolism regulate dendritic cell activation. Blood 115(23):4742–4749. https://doi.org/10.1182/blood-2009-10-249540

Kuehl CJ, Dragoi AM, Talman A, Agaisse H (2015) Bacterial spread from cell to cell: beyond actin-based motility. Trends Microbiol 23(9):558–566. https://doi.org/10.1016/j.tim.2015.04.010

Kumar H, Kawai T, Akira S (2011) Pathogen recognition by the innate immune system. Int Rev Immunol 30(1):16–34. https://doi.org/10.3109/08830185.2010.529976

Lecuit M, Sonnenburg JL, Cossart P, Gordon JI (2007) Functional genomic studies of the intestinal response to a foodborne enteropathogen in a humanized gnotobiotic mouse model. J Biol Chem 282(20):15065–15072. https://doi.org/10.1074/jbc.M610926200

Leistikow RL, Morton RA, Bartek IL, Frimpong I, Wagner K, Voskuil MI (2010) The *Mycobacterium tuberculosis* DosR regulon assists in metabolic homeostasis and enables rapid recovery from nonrespiring dormancy. J Bacteriol 192(6):1662–1670. https://doi.org/10.1128/JB.00926-09

Levine B, Mizushima N, Virgin HW (2011) Autophagy in immunity and inflammation. Nature 469 (7330):323–335. https://doi.org/10.1038/nature09782

Liao D, Fan Q, Bao L (2013) The role of superoxide dismutase in the survival of *Mycobacterium tuberculosis* in macrophages. Jpn J Infect Dis 66(6):480–488

Liberti MV, Locasale JW (2016) The Warburg effect: how does it benefit cancer cells? Trends Biochem Sci 41(3):211–218. https://doi.org/10.1016/j.tibs.2015.12.001

Lin Z, Zhang YG, Xia Y, Xu X, Jiao X, Sun J (2016) *Salmonella* enteritidis effector AvrA stabilizes intestinal tight junctions via the JNK pathway. J Biol Chem 291(52):26837–26849. https://doi.org/10.1074/jbc.M116.757393

Liu X, Lu R, Xia Y, Sun J (2010) Global analysis of the eukaryotic pathways and networks regulated by *Salmonella* typhimurium in mouse intestinal infection in vivo. BMC Genomics 11:722. https://doi.org/10.1186/1471-2164-11-722

Lopez CA, Winter SE, Rivera-Chavez F, Xavier MN, Poon V, Nuccio SP, Tsolis RM, Baumler AJ (2012) Phage-mediated acquisition of a type III secreted effector protein boosts growth of *salmonella* by nitrate respiration. MBio 3(3). https://doi.org/10.1128/mBio.00143-12

Lutay N, Hakansson G, Alaridah N, Hallgren O, Westergren-Thorsson G, Godaly G (2014) Mycobacteria bypass mucosal NF-kB signalling to induce an epithelial anti-inflammatory IL-22 and IL-10 response. PLoS One 9(1):e86466. https://doi.org/10.1371/journal.pone.0086466

Macallan DC, McNurlan MA, Kurpad AV, de Souza G, Shetty PS, Calder AG, Griffin GE (1998) Whole body protein metabolism in human pulmonary tuberculosis and undernutrition: evidence for anabolic block in tuberculosis. Clin Sci (Lond) 94(3):321–331

Macintyre AN, Gerriets VA, Nichols AG, Michalek RD, Rudolph MC, Deoliveira D, Anderson SM, Abel ED, Chen BJ, Hale LP, Rathmell JC (2014) The glucose transporter Glut1 is selectively essential for CD4 T cell activation and effector function. Cell Metab 20(1):61–72. https://doi.org/10.1016/j.cmet.2014.05.004

Malhotra M, Sood S, Mukherjee A, Muralidhar S, Bala M (2013) Genital *Chlamydia trachomatis*: an update. Indian J Med Res 138(3):303–316

Manca C, Paul S, Barry CE 3rd, Freedman VH, Kaplan G (1999) *Mycobacterium tuberculosis* catalase and peroxidase activities and resistance to oxidative killing in human monocytes in vitro. Infect Immun 67(1):74–79

Mannonen L, Markkula E, Puolakkainen M (2011) Analysis of *Chlamydia pneumoniae* infection in mononuclear cells by reverse transcription-PCR targeted to chlamydial gene transcripts. Med Microbiol Immunol 200(3):143–154. https://doi.org/10.1007/s00430-011-0184-3

Mansell A, Braun L, Cossart P, O'Neill LA (2000) A novel function of InIB from *Listeria monocytogenes*: activation of NF-kappaB in J774 macrophages. Cell Microbiol 2(2):127–136

Manske C, Hilbi H (2014) Metabolism of the vacuolar pathogen *Legionella* and implications for virulence. Front Cell Infect Microbiol 4:125. https://doi.org/10.3389/fcimb.2014.00125

Mantegazza AR, Magalhaes JG, Amigorena S, Marks MS (2013) Presentation of phagocytosed antigens by MHC class I and II. Traffic 14(2):135–152. https://doi.org/10.1111/tra.12026

Marrero J, Rhee KY, Schnappinger D, Pethe K, Ehrt S (2010) Gluconeogenic carbon flow of tricarboxylic acid cycle intermediates is critical for *Mycobacterium tuberculosis* to establish and maintain infection. Proc Natl Acad Sci USA 107(21):9819–9824. https://doi.org/10.1073/pnas.1000715107

Masoud GN, Li W (2015) HIF-1alpha pathway: role, regulation and intervention for cancer therapy. Acta Pharm Sin B 5(5):378–389. https://doi.org/10.1016/j.apsb.2015.05.007

Matsunaga K, Klein TW, Newton C, Friedman H, Yamamoto Y (2001) *Legionella pneumophila* suppresses interleukin-12 production by macrophages. Infect Immun 69(3):1929–1933. https://doi.org/10.1128/IAI.69.3.1929-1933.2001

Meylan E, Tschopp J, Karin M (2006) Intracellular pattern recognition receptors in the host response. Nature 442(7098):39–44. https://doi.org/10.1038/nature04946

Michalek RD, Rathmell JC (2010) The metabolic life and times of a T-cell. Immunol Rev 236:190–202. https://doi.org/10.1111/j.1600-065X.2010.00911.x

Minton K (2017) Immune regulation: IL-10 targets macrophage metabolism. Nat Rev Immunol 17(6):345. https://doi.org/10.1038/nri.2017.57

Mishra BB, Rathinam VA, Martens GW, Martinot AJ, Kornfeld H, Fitzgerald KA, Sassetti CM (2013) Nitric oxide controls the immunopathology of tuberculosis by inhibiting NLRP3 inflammasome-dependent processing of IL-1beta. Nat Immunol 14(1):52–60. https://doi.org/10.1038/ni.2474

Mishra BB, Lovewell RR, Olive AJ, Zhang G, Wang W, Eugenin E, Smith CM, Phuah JY, Long JE, Dubuke ML, Palace SG, Goguen JD, Baker RE, Nambi S, Mishra R, Booty MG, Baer CE, Shaffer SA, Dartois V, McCormick BA, Chen X, Sassetti CM (2017) Nitric oxide prevents a pathogen-permissive granulocytic inflammation during tuberculosis. Nat Microbiol 2:17072. https://doi.org/10.1038/nmicrobiol.2017.72

Mizushima N, Yoshimori T, Ohsumi Y (2011) The role of Atg proteins in autophagosome formation. Annu Rev Cell Dev Biol 27:107–132. https://doi.org/10.1146/annurev-cellbio-092910-154005

Moffatt JH, Newton P, Newton HJ (2015) *Coxiella burnetii*: turning hostility into a home. Cell Microbiol 17(5):621–631. https://doi.org/10.1111/cmi.12432

Mulye M, Samanta D, Winfree S, Heinzen RA, Gilk SD (2017) Elevated cholesterol in the *Coxiella burnetii* intracellular niche is bacteriolytic. MBio 8(1). https://doi.org/10.1128/mBio.02313-16

Munn DH, Mellor AL (2013) Indoleamine 2,3 dioxygenase and metabolic control of immune responses. Trends Immunol 34(3):137–143. https://doi.org/10.1016/j.it.2012.10.001

Munoz-Elias EJ, McKinney JD (2006) Carbon metabolism of intracellular bacteria. Cell Microbiol 8(1):10–22. https://doi.org/10.1111/j.1462-5822.2005.00648.x

Nathan C, Shiloh MU (2000) Reactive oxygen and nitrogen intermediates in the relationship between mammalian hosts and microbial pathogens. Proc Natl Acad Sci USA 97(16):8841–8848

Natividad A, Freeman TC, Jeffries D, Burton MJ, Mabey DC, Bailey RL, Holland MJ (2010) Human conjunctival transcriptome analysis reveals the prominence of innate defense in *Chlamydia trachomatis* infection. Infect Immun 78(11):4895–4911. https://doi.org/10.1128/IAI.00844-10

Ng VH, Cox JS, Sousa AO, MacMicking JD, McKinney JD (2004) Role of KatG catalase-peroxidase in mycobacterial pathogenesis: countering the phagocyte oxidative burst. Mol Microbiol 52(5):1291–1302. https://doi.org/10.1111/j.1365-2958.2004.04078.x

Nguyen GT, Green ER, Mecsas J (2017) Neutrophils to the ROScue: mechanisms of NADPH oxidase activation and bacterial resistance. Front Cell Infect Microbiol 7:373. https://doi.org/10.3389/fcimb.2017.00373

Nicholson JK, Holmes E, Kinross J, Burcelin R, Gibson G, Jia W, Pettersson S (2012) Host-gut microbiota metabolic interactions. Science 336(6086):1262–1267. https://doi.org/10.1126/science.1223813

Ninio S, Roy CR (2007) Effector proteins translocated by *Legionella pneumophila*: strength in numbers. Trends Microbiol 15(8):372–380. https://doi.org/10.1016/j.tim.2007.06.006

O'Neill LA (2015) A broken krebs cycle in macrophages. Immunity 42(3):393–394. https://doi.org/10.1016/j.immuni.2015.02.017

O'Neill LA, Kishton RJ, Rathmell J (2016) A guide to immunometabolism for immunologists. Nat Rev Immunol 16(9):553–565. https://doi.org/10.1038/nri.2016.70

Odegaard JI, Chawla A (2011) Alternative macrophage activation and metabolism. Annu Rev Pathol 6:275–297. https://doi.org/10.1146/annurev-pathol-011110-130138

Olive AJ, Sassetti CM (2016) Metabolic crosstalk between host and pathogen: sensing, adapting and competing. Nat Rev Microbiol 14(4):221–234. https://doi.org/10.1038/nrmicro.2016.12

Omsland A, Sager J, Nair V, Sturdevant DE, Hackstadt T (2012) Developmental stage-specific metabolic and transcriptional activity of *Chlamydia trachomatis* in an axenic medium. Proc Natl Acad Sci USA 109(48):19781–19785. https://doi.org/10.1073/pnas.1212831109

Ouellette SP, Dorsey FC, Moshiach S, Cleveland JL, Carabeo RA (2011) *Chlamydia* species-dependent differences in the growth requirement for lysosomes. PLoS One 6(3):e16783. https://doi.org/10.1371/journal.pone.0016783

Pareja ME, Colombo MI (2013) Autophagic clearance of bacterial pathogens: molecular recognition of intracellular microorganisms. Front Cell Infect Microbiol 3:54. https://doi.org/10.3389/fcimb.2013.00054

Pearce EL, Pearce EJ (2013) Metabolic pathways in immune cell activation and quiescence. Immunity 38(4):633–643. https://doi.org/10.1016/j.immuni.2013.04.005

Pearce EL, Poffenberger MC, Chang CH, Jones RG (2013) Fueling immunity: insights into metabolism and lymphocyte function. Science 342(6155):1242454. https://doi.org/10.1126/science.1242454

Peleg AY, Tampakakis E, Fuchs BB, Eliopoulos GM, Moellering RC Jr, Mylonakis E (2008) Prokaryote-eukaryote interactions identified by using *Caenorhabditis elegans*. Proc Natl Acad Sci USA 105(38):14585–14590. https://doi.org/10.1073/pnas.0805048105

Peng M, Yin N, Chhangawala S, Xu K, Leslie CS, Li MO (2016) Aerobic glycolysis promotes T helper 1 cell differentiation through an epigenetic mechanism. Science 354(6311):481–484. https://doi.org/10.1126/science.aaf6284

Peyron P, Vaubourgeix J, Poquet Y, Levillain F, Botanch C, Bardou F, Daffe M, Emile JF, Marchou B, Cardona PJ, de Chastellier C, Altare F (2008) Foamy macrophages from tuberculous patients' granulomas constitute a nutrient-rich reservoir for *M. tuberculosis* persistence. PLoS Pathog 4(11):e1000204. https://doi.org/10.1371/journal.ppat.1000204

Pikuleva IA (2006) Cytochrome P450s and cholesterol homeostasis. Pharmacol Ther 112(3):761–773. https://doi.org/10.1016/j.pharmthera.2006.05.014

Plain KM, de Silva K, Earl J, Begg DJ, Purdie AC, Whittington RJ (2011) Indoleamine 2,3-dioxygenase, tryptophan catabolism, and *Mycobacterium avium* subsp. paratuberculosis: a model for chronic mycobacterial infections. Infect Immun 79(9):3821–3832. https://doi.org/10.1128/IAI.05204-11

Portnoy DA, Auerbuch V, Glomski IJ (2002) The cell biology of *Listeria monocytogenes* infection: the intersection of bacterial pathogenesis and cell-mediated immunity. J Cell Biol 158(3):409–414. https://doi.org/10.1083/jcb.200205009

Price CT, Al-Quadan T, Santic M, Rosenshine I, Abu Kwaik Y (2011) Host proteasomal degradation generates amino acids essential for intracellular bacterial growth. Science 334(6062):1553–1557. https://doi.org/10.1126/science.1212868

Price JD, Simpfendorfer KR, Mantena RR, Holden J, Heath WR, van Rooijen N, Strugnell RA, Wijburg OL (2007) Gamma interferon-independent effects of interleukin-12 on immunity to *Salmonella enterica* serovar Typhimurium. Infect Immun 75(12):5753–5762. https://doi.org/10.1128/IAI.00971-07

Ragno S, Romano M, Howell S, Pappin DJ, Jenner PJ, Colston MJ (2001) Changes in gene expression in macrophages infected with *Mycobacterium tuberculosis*: a combined transcriptomic and proteomic approach. Immunology 104(1):99–108

Ramirez-Alejo N, Santos-Argumedo L (2014) Innate defects of the IL-12/IFN-gamma axis in susceptibility to infections by mycobacteria and *salmonella*. J Interferon Cytokine Res 34(5):307–317. https://doi.org/10.1089/jir.2013.0050

Rao PK, Li Q (2009) Protein turnover in mycobacterial proteomics. Molecules 14(9):3237–3258. https://doi.org/10.3390/molecules14093237

Ren Q, Robertson SJ, Howe D, Barrows LF, Heinzen RA (2003) Comparative DNA microarray analysis of host cell transcriptional responses to infection by *Coxiella burnetii* or *Chlamydia trachomatis*. Ann N Y Acad Sci 990:701–713

Ribet D, Cossart P (2015) How bacterial pathogens colonize their hosts and invade deeper tissues. Microbes Infect 17(3):173–183. https://doi.org/10.1016/j.micinf.2015.01.004

Robertson DK, Gu L, Rowe RK, Beatty WL (2009) Inclusion biogenesis and reactivation of persistent *Chlamydia trachomatis* requires host cell sphingolipid biosynthesis. PLoS Pathog 5(11):e1000664. https://doi.org/10.1371/journal.ppat.1000664

Rosklint T, Ohlsson BG, Wiklund O, Noren K, Hulten LM (2002) Oxysterols induce interleukin-1beta production in human macrophages. Eur J Clin Invest 32(1):35–42

Santos RL, Raffatellu M, Bevins CL, Adams LG, Tukel C, Tsolis RM, Baumler AJ (2009) Life in the inflamed intestine, *Salmonella* style. Trends Microbiol 17(11):498–506. https://doi.org/10.1016/j.tim.2009.08.008

Scanu T, Spaapen RM, Bakker JM, Pratap CB, Wu LE, Hofland I, Broeks A, Shukla VK, Kumar M, Janssen H, Song JY, Neefjes-Borst EA, te Riele H, Holden DW, Nath G, Neefjes J (2015) *Salmonella* manipulation of host signaling pathways provokes cellular transformation associated with Gallbladder carcinoma. Cell Host Microbe 17(6):763–774. https://doi.org/10.1016/j.chom.2015.05.002

Schwartzenberg-Bar-Yoseph F, Armoni M, Karnieli E (2004) The tumor suppressor p53 down-regulates glucose transporters GLUT1 and GLUT4 gene expression. Cancer Res 64(7):2627–2633

Scidmore MA, Fischer ER, Hackstadt T (1996) Sphingolipids and glycoproteins are differentially trafficked to the *Chlamydia trachomatis* inclusion. J Cell Biol 134(2):363–374

Shi L, Jung YJ, Tyagi S, Gennaro ML, North RJ (2003) Expression of Th1-mediated immunity in mouse lungs induces a *Mycobacterium tuberculosis* transcription pattern characteristic of nonreplicating persistence. Proc Natl Acad Sci USA 100(1):241–246. https://doi.org/10.1073/pnas.0136863100

Shi L, Salamon H, Eugenin EA, Pine R, Cooper A, Gennaro ML (2015) Infection with *Mycobacterium tuberculosis* induces the Warburg effect in mouse lungs. Sci Rep 5:18176. https://doi.org/10.1038/srep18176

Shiloh MU, Manzanillo P, Cox JS (2008) *Mycobacterium tuberculosis* senses host-derived carbon monoxide during macrophage infection. Cell Host Microbe 3(5):323–330. https://doi.org/10.1016/j.chom.2008.03.007

Shin JH, Yang JY, Jeon BY, Yoon YJ, Cho SN, Kang YH, Ryu DH, Hwang GS (2011) (1)H - NMR-based metabolomic profiling in mice infected with *Mycobacterium tuberculosis*. J Proteome Res 10(5):2238–2247. https://doi.org/10.1021/pr101054m

Siegl C, Prusty BK, Karunakaran K, Wischhusen J, Rudel T (2014) Tumor suppressor p53 alters host cell metabolism to limit *Chlamydia trachomatis* infection. Cell Rep 9(3):918–929. https://doi.org/10.1016/j.celrep.2014.10.004

Siemsen DW, Kirpotina LN, Jutila MA, Quinn MT (2009) Inhibition of the human neutrophil NADPH oxidase by *Coxiella burnetii*. Microbes Infect 11(6-7):671–679. https://doi.org/10.1016/j.micinf.2009.04.005

Singh V, Jamwal S, Jain R, Verma P, Gokhale R, Rao KV (2012) *Mycobacterium tuberculosis*-driven targeted recalibration of macrophage lipid homeostasis promotes the foamy phenotype. Cell Host Microbe 12(5):669–681. https://doi.org/10.1016/j.chom.2012.09.012

Singh V, Kaur C, Chaudhary VK, Rao KV, Chatterjee S (2015) *M. tuberculosis* secretory protein ESAT-6 induces metabolic flux perturbations to drive foamy macrophage differentiation. Sci Rep 5:12906. https://doi.org/10.1038/srep12906

Sinnis P, Ernst JD (2008) CO-opting the host HO-1 pathway in tuberculosis and malaria. Cell Host Microbe 3(5):277–279. https://doi.org/10.1016/j.chom.2008.04.009

Somashekar BS, Amin AG, Rithner CD, Troudt J, Basaraba R, Izzo A, Crick DC, Chatterjee D (2011) Metabolic profiling of lung granuloma in *Mycobacterium tuberculosis* infected guinea pigs: ex vivo 1H magic angle spinning NMR studies. J Proteome Res 10(9):4186–4195. https://doi.org/10.1021/pr2003352

Stratton CW, Sriram S (2003) Association of *Chlamydia pneumoniae* with central nervous system disease. Microbes Infect 5(13):1249–1253

Tannahill GM, Curtis AM, Adamik J, Palsson-McDermott EM, McGettrick AF, Goel G, Frezza C, Bernard NJ, Kelly B, Foley NH, Zheng L, Gardet A, Tong Z, Jany SS, Corr SC, Haneklaus M, Caffrey BE, Pierce K, Walmsley S, Beasley FC, Cummins E, Nizet V, Whyte M, Taylor CT, Lin H, Masters SL, Gottlieb E, Kelly VP, Clish C, Auron PE, Xavier RJ, O'Neill LA (2013) Succinate is an inflammatory signal that induces IL-1beta through HIF-1alpha. Nature 496 (7444):238–242. https://doi.org/10.1038/nature11986

Tas SW, Vervoordeldonk MJ, Hajji N, Schuitemaker JH, van der Sluijs KF, May MJ, Ghosh S, Kapsenberg ML, Tak PP, de Jong EC (2007) Noncanonical NF-kappaB signaling in dendritic cells is required for indoleamine 2,3-dioxygenase (IDO) induction and immune regulation. Blood 110(5):1540–1549. https://doi.org/10.1182/blood-2006-11-056010

Thiennimitr P, Winter SE, Winter MG, Xavier MN, Tolstikov V, Huseby DL, Sterzenbach T, Tsolis RM, Roth JR, Baumler AJ (2011) Intestinal inflammation allows *Salmonella* to use ethanolamine to compete with the microbiota. Proc Natl Acad Sci USA 108(42):17480–17485. https://doi.org/10.1073/pnas.1107857108

Tipples G, McClarty G (1993) The obligate intracellular bacterium *Chlamydia trachomatis* is auxotrophic for three of the four ribonucleoside triphosphates. Mol Microbiol 8(6):1105–1114

Uchiya K, Nikai T (2004) *Salmonella enterica* serovar Typhimurium infection induces cyclooxygenase 2 expression in macrophages: involvement of *Salmonella* pathogenicity island 2. Infect Immun 72(12):6860–6869. https://doi.org/10.1128/IAI.72.12.6860-6869.2004

van der Windt GJ, Pearce EL (2012) Metabolic switching and fuel choice during T-cell differentiation and memory development. Immunol Rev 249(1):27–42. https://doi.org/10.1111/j.1600-065X.2012.01150.x

van Ooij C, Kalman L, van I, Nishijima M, Hanada K, Mostov K, Engel JN (2000) Host cell-derived sphingolipids are required for the intracellular growth of *Chlamydia trachomatis*. Cell Microbiol 2(6):627–637

van Schaik EJ, Chen C, Mertens K, Weber MM, Samuel JE (2013) Molecular pathogenesis of the obligate intracellular bacterium *Coxiella burnetii*. Nat Rev Microbiol 11(8):561–573. https://doi.org/10.1038/nrmicro3049

Vazquez-Boland JA, Kuhn M, Berche P, Chakraborty T, Dominguez-Bernal G, Goebel W, Gonzalez-Zorn B, Wehland J, Kreft J (2001) *Listeria* pathogenesis and molecular virulence determinants. Clin Microbiol Rev 14(3):584–640. https://doi.org/10.1128/CMR.14.3.584-640.2001

Verreck FA, de Boer T, Langenberg DM, Hoeve MA, Kramer M, Vaisberg E, Kastelein R, Kolk A, de Waal-Malefyt R, Ottenhoff TH (2004) Human IL-23-producing type 1 macrophages promote but IL-10-producing type 2 macrophages subvert immunity to (myco)bacteria. Proc Natl Acad Sci USA 101(13):4560–4565. https://doi.org/10.1073/pnas.0400983101

Vonaesch P, Sellin ME, Cardini S, Singh V, Barthel M, Hardt WD (2014) The *Salmonella* Typhimurium effector protein SopE transiently localizes to the early SCV and contributes to intracellular replication. Cell Microbiol 16(12):1723–1735. https://doi.org/10.1111/cmi.12333

Vromman F, Laverriere M, Perrinet S, Dufour A, Subtil A (2014) Quantitative monitoring of the *Chlamydia trachomatis* developmental cycle using GFP-expressing bacteria, microscopy and flow cytometry. PLoS One 9(6):e99197. https://doi.org/10.1371/journal.pone.0099197

Wagner RD, Steinberg H, Brown JF, Czuprynski CJ (1994) Recombinant interleukin-12 enhances resistance of mice to *Listeria monocytogenes* infection. Microb Pathog 17(3):175–186

Wahl DR, Byersdorfer CA, Ferrara JL, Opipari AW Jr, Glick GD (2012) Distinct metabolic programs in activated T cells: opportunities for selective immunomodulation. Immunol Rev 249(1):104–115. https://doi.org/10.1111/j.1600-065X.2012.01148.x

Way SS, Havenar-Daughton C, Kolumam GA, Orgun NN, Murali-Krishna K (2007) IL-12 and type-I IFN synergize for IFN-gamma production by CD4 T cells, whereas neither are required for IFN-gamma production by CD8 T cells after *Listeria monocytogenes* infection. J Immunol 178(7):4498–4505

Wieland H, Ullrich S, Lang F, Neumeister B (2005) Intracellular multiplication of Legionella pneumophila depends on host cell amino acid transporter SLC1A5. Mol Microbiol 55(5):1528–1537. https://doi.org/10.1111/j.1365-2958.2005.04490.x

Wylie JL, Hatch GM, McClarty G (1997) Host cell phospholipids are trafficked to and then modified by *Chlamydia trachomatis*. J Bacteriol 179(23):7233–7242

Yamada H, Mizuno S, Reza-Gholizadeh M, Sugawara I (2001) Relative importance of NF-kappaB p50 in mycobacterial infection. Infect Immun 69(11):7100–7105. https://doi.org/10.1128/IAI.69.11.7100-7105.2001

Zaika AI, Wei J, Noto JM, Peek RM (2015) Microbial regulation of p53 tumor suppressor. PLoS Pathog 11(9):e1005099. https://doi.org/10.1371/journal.ppat.1005099

Zhang YJ, Reddy MC, Ioerger TR, Rothchild AC, Dartois V, Schuster BM, Trauner A, Wallis D, Galaviz S, Huttenhower C, Sacchettini JC, Behar SM, Rubin EJ (2013) Tryptophan biosynthesis protects mycobacteria from CD4 T-cell-mediated killing. Cell 155(6):1296–1308. https://doi.org/10.1016/j.cell.2013.10.045

Zou T, Garifulin O, Berland R, Boyartchuk VL (2011) *Listeria monocytogenes* infection induces prosurvival metabolic signaling in macrophages. Infect Immun 79(4):1526–1535. https://doi.org/10.1128/IAI.01195-10

Chapter 9
Protozoan Parasite Auxotrophies and Metabolic Dependencies

Elodie Gazanion and Baptiste Vergnes

Abstract Diseases caused by protozoan parasites have a major impact on world health. These early branching eukaryotes cause significant morbidity and mortality in humans and livestock. During evolution, protozoan parasites have evolved toward complex life cycles in multiple host organisms with different nutritional resources. The conservation of functional metabolic pathways required for these successive environments is therefore a prerequisite for parasitic lifestyle. Nevertheless, parasitism drives genome evolution toward gene loss and metabolic dependencies (including strict auxotrophy), especially for obligatory intracellular parasites. In this chapter, we will compare and contrast how protozoan parasites have perfected this metabolic adaptation by focusing on specific auxotrophic pathways and scavenging strategies used by clinically relevant apicomplexan and trypanosomatid parasites to access host's nutritional resources. We will further see how these metabolic dependencies have in turn been exploited for therapeutic purposes against these human pathogens.

Keywords Auxotrophy · Purines · Polyamines · Vitamins · Folates · NAD^+ · Heme

9.1 Introduction

Parasitic protozoa are an evolutionarily divergent group of unicellular eukaryotes that are responsible of a wide range of human and veterinary diseases. The most notorious and clinically relevant protozoan parasites are the apicomplexans *Plasmodium* spp. and *Toxoplasma gondii* (the causative agents of malaria and toxoplasmosis, respectively) and the trypanosomatids *Trypanosoma brucei*, *Trypanosoma cruzi*, and *Leishmania* spp., which cause African sleeping sickness, Chagas' disease, and leishmaniasis, respectively. Malaria is one of the leading causes of death and

E. Gazanion · B. Vergnes (✉)
MIVEGEC, IRD, CNRS, University of Montpellier, Montpellier, France
e-mail: baptiste.vergnes@ird.fr

R. Silvestre, E. Torrado (eds.), *Metabolic Interaction in Infection*, Experientia Supplementum 109, https://doi.org/10.1007/978-3-319-74932-7_9

morbidity worldwide, especially in the developing countries. Despite recent success of malaria control programs in reducing incidence and mortality rates to more than 20%, there is an estimated 429,000 malaria deaths in 2015 that mostly occurred in the African region (World Malaria Report 2016). *T. gondii* parasites are found in around a third of human population and can be fatal in immunocompromised individuals, while the trypanosomatids can be classified as the second most deadly parasites after *Plasmodium*, causing approximately 150,000 deaths annually (Nussbaum et al. 2010). Resistance to the frontline antiparasitic drugs, together with the lack of an effective vaccine, poses today considerable risks to global health (Fairlamb et al. 2016). There is therefore an urgent need to identify new drug targets for therapeutic alternatives. With the exception of *T. brucei* that remains extracellular during all its life cycle, apicomplexan and trypanosomatid parasites are readily adapted to intracellular lifestyle. For instance, *P. falciparum* develops within the gut of female *Anopheles* mosquitoes and is transmitted to vertebrate hosts during the blood meal, where they first invade hepatocytes and then enter the bloodstream to infect red blood cells (RBC). In this nutritionally complex environment, *Plasmodium* relies on the host cell cytosol and hemoglobin digestion to fulfill its nutrient requirements (Krugliak et al. 2002). *Leishmania* parasites are transmitted to vertebrate hosts through sand fly bites and are internalized by macrophages where they differentiate into amastigotes inside the phagolysosome. The survival of *Leishmania* amastigotes in this acidic environment is consistent with their optimal metabolism at acid pH (Mukkada et al. 1985). In contrast, *T. gondii* and *T. cruzi* are both able to invade a wide range of nucleated mammalian cells.

Before genome sequencing, all our knowledge about protozoan parasite metabolism came essentially from biochemical studies carried out on specific life cycle stages that could be readily cultured in vitro. These stages are usually those found in insect vectors in which parasites are living extracellularly but with a metabolism that is not necessarily similar to that found in mammal host and inside host's cells. During the 2000s, the access to the genomes of most of the human protozoan parasites suddenly disclosed a global view of the parasite's metabolic capacities. Detailed metabolic maps were drawn from the genome-derived sequences, highlighting new putative targets for drug discovery. These include parasite-specific metabolic pathways but also those for which parasites are auxotrophic. In opposition to prototrophy, auxotrophy defines the inability of an organism to synthetize a particular organic compound required for its growth. Protozoan parasites are auxotrophic for a wide range of amino acids, purines, vitamins, lipids, and other metabolites that must be imported at sufficient levels to sustain parasite growth. For a therapeutic perspective, auxotrophic pathways are very attractive targets to starve and impair parasite proliferation in its vertebrate host, with potentially high specificity.

Through the course of evolution, parasitic organisms have gained or lost specific metabolic pathways to optimize their life cycles in respect to their host environment(s). This generally concerns the loss of unnecessary or redundant functional pathways and gain of salvage routes to import and assimilate metabolic intermediates. The situation is even more complex for dixenous parasites that alternate between an insect vector and a vertebrate host, offering different nutritional resources that obligate parasites to behave as metabolically and phenotypically different microorganisms. The conservation of metabolic pathways required in these successive host environments is therefore essential for protozoan parasite life cycle completion. In that context, it can be puzzling to consider that parasites evolved toward genome compaction and metabolic auxotrophy at the expense of their metabolic autonomy. Moreover, how can metabolic dependency cause a fitness benefit? For all these questions, auxotrophy should not be considered as a metabolic "laziness" but more likely as a consequence of a fine-tuned biology. Experimental acquisition of metabolic dependencies in protozoan parasites is lacking, but this phenomenon is well documented in bacteria (D'Souza et al. 2014). For example, amino acid auxotrophic genotypes can emerge in less than 2000 generations for *E. coli* bacteria populations evolving in amino acid-rich environments. Interestingly, these auxotrophic mutants gained a significant fitness benefit over the evolutionary ancestor during competition experiments (D'Souza and Kost 2016). In *Legionella pneumophila*, a parasite of amoebae (and an accidental pathogen of humans causing severe pneumonia), the continuous passages inside mouse macrophages for hundreds of generations have fixed mutations in multiple steps of the lysine biosynthesis pathway resulting in lysine auxotrophy (Ensminger et al. 2012). Conversely, macrophage-adapted bacteria show a reduced capacity to replicate in amoebae. These results demonstrate that host restriction can rapidly modify pathogen fitness and support therefore that host cycling drives the conservation of metabolic pathways that are deleterious for growth in only specific hosts.

For space limitation, this chapter will focus on selected auxotrophic pathways and metabolic dependencies that have been validated as true or promising therapeutic targets against apicomplexan and trypanosomatid parasites. A general feature that characterizes parasite auxotrophies is the existence of multiple routes of synthesis and interconversion, combined to scavenging possibilities at different metabolic steps. These sophisticated bypass systems are readily used by parasites to circumvent the inhibition of specific enzymes. Nevertheless, such alteration of basic metabolisms can have a major impact on parasite virulence in vivo and, as detailed below, few examples of success stories exist.

9.2 What Genome Sequencing Can Tell Us About Parasites and Parasitism?

The release of the complete genome sequences of several important protozoan parasites over a decade ago provides a rich resource for understanding the biology and evolution of these pathogens. Comparative genomic analysis raises central questions on gene repertoires, gain/loss of genes, and genome organization and provides a global view of the metabolic potential of these parasites, irrespective of the life cycle stage. The comparison of the TriTryp genomes (*T. brucei*/*T. cruzi*/ *Leishmania major*) has highlighted that both gene order and gene repertoire (synteny) are broadly conserved despite a divergence estimated between 200 and 500 million years ago, predating the divergence of mammals (Berriman et al. 2005;

El-Sayed et al. 2005a, b; Ivens et al. 2005). This conserved core genome of trypanosomatids comprises about 6200 genes ($\sim$70% of the genome) and reflects strong selective constraints on genome structure. Additional comparative genomic analysis from clinically relevant *Leishmania* species (*L. major*, *L. infantum*, and *L. braziliensis* causing human cutaneous, visceral, and mucocutaneous leishmaniasis, respectively) confirmed a greater extend of synteny with gene loss and pseudogene formation being the principal factors that shaped *Leishmania* species genomes through evolution (Peacock et al. 2007).

In addition to *Plasmodium* and *Toxoplasma*, the phylum Apicomplexa contains several pathogens of medical and veterinary importance including *Cryptosporidium*, *Eimeria*, *Theileria*, and *Babesia*. Genome content of these organisms revealed a massive gene loss in both nuclear and organellar genomes compared to most model eukaryotes. About 50% of the genes found in *Plasmodium* species have no orthologs in other apicomplexan, while only 85% of the genome content is conserved between *P. falciparum* and rodent malaria parasites (Kooij et al. 2006). The apicomplexan genome is highly dynamic with significant genome rearrangement between lineages. In contrast to trypanosomatids, syntenic regions are rare between lineages and appear to be totally absent across the phylum. It seems therefore that there are different criteria governing genome evolution within the Apicomplexa phylum relative to other unicellular eukaryotes and notably trypanosomatids (DeBarry and Kissinger 2011).

One of the most intriguing cellular features of the phylum Apicomplexa is the presence of a non-photosynthetic secondary plastid, termed the apicoplast. This organelle derived by secondary endosymbiosis of an alga and retention of the algal plastid. During their evolution toward parasitism, most members of Apicomplexa, including *Plasmodium* and *Toxoplasma*, lost the ability to photosynthesize while maintaining in the apicoplast the biosynthesis of fatty acids, isoprenoids, iron-sulfur clusters, and heme (Lim and McFadden 2010). One striking feature of the apicoplast is that it is essential for parasite growth and development and therefore a validated target for drug therapy. In *Plasmodium*, the sequencing of the small circular apicoplast genome (35 kb encoding less than 50 proteins) brought little insight into its function. Most of the apicoplast-encoded genes are ascribed to "house-keeping" functions such as transcription and translation (Wilson et al. 1996). To find out possible functions of the apicoplast, Waller et al. used publicly available *T. gondii* EST sequences and preliminary sequence data for *P. falciparum* to identify nuclear-encoded apicoplast proteins based on sequence similarity to proteins of known function in plant and algal systems (Waller et al. 1998). The identification of an apicoplast-targeting signal sequence has facilitated the prediction of proteins residing into apicoplast and the reconstruction of metabolic pathways found in this organelle (Zuegge et al. 2001; Foth et al. 2003).

All the parasite genome sequences and their functional annotation are available in the portal Eukaryotic Pathogen Database Resource (EuPathDB, http://eupathdb.org) (Aurrecoechea et al. 2017) that offers the access to large genomic-scale data set from diverse eukaryotic pathogens including *Plasmodium* (http://plasmodb.org), *Toxoplasma* (http://toxodb.org), and trypanosomatids (http://tritrypdb.org). Reconstruction of parasite metabolic network at genome-scale was first realized in an automated fashion and further improved by manual refinement and curation, some being implemented with metabolomics data sets. Among the most up-to-date metabolic resources are the "Malaria Parasite Metabolic Pathways" (MPMP), the "Library of Apicomplexan Metabolic Pathways" (LAMP) (Shanmugasundram et al. 2013), and the *T. brucei* TrypanoCyc database (Shameer et al. 2015).

With the increasing number of published complete genome sequences, including not only new parasite strains and subspecies but also early free living relatives, we are now able to delineate an accurate picture regarding how parasites have evolved from free-living organisms (Janouskovec and Keeling 2016). *Bodo saltans* is the closest known nonparasitic relative of trypanosomatids. It lives in water habitats and feeds on bacteria by phagocytosis. Comparing the genomes of *B. saltans* and parasitic trypanosomatids reveals that the transition from a free-living to a parasitic lifestyle has resulted in the loss of approximately 50% of protein-coding genes. Gene loss is notably marked in macromolecular digestion and ion transport, while parasite innovations (gene gains) mostly concern cell-surface gene families and membrane transporters (Jackson et al. 2016). *B. saltans* does not contain however any complete metabolic pathways that are absent in parasitic trypanosomatids (Opperdoes et al. 2016). Interestingly, some similar observations can be made from comparative phylogenomic analyses between apicomplexan parasites and their closest free-living and photosynthetic relatives called chrompodellids (Woo et al. 2015; Janouskovec et al. 2015). In this case, however, chrompodellids harbor a wider range of metabolic capacities than apicomplexan parasites, notably those involved in photosynthesis.

Overall, comparative phylogenomic analyses support that the emergence of parasitism may not be determined by acquisition of novel components, but rather by loss and modification of characteristics already present in their free-living ancestors (Janouskovec and Keeling 2016).

9.3 Protozoan Parasite Auxotrophies

9.3.1 Purines

The nucleotide metabolism is considered as promising therapeutic targets against protozoan parasites since almost four decades and has consequently been the focus of considerable scientific investigations. Pyrimidines and purines are essential for the biosynthesis of DNA, RNA, and sugar nucleotides. In addition, certain of these molecules and related derivatives are particularly important for energy production (ATP, GTP), vitamin-derived cofactors synthesis (NAD+, FAD, FMN, SAM, or acetyl-CoA), and cell signaling involving intracellular second messengers (cAMP, cGMP). The purine pathway has garnered extensive attention because, unlike their vertebrate hosts, all protozoan parasites lack the capacity to synthesize the purine ring de novo. Most of them synthetize however pyrimidine de novo, similarly to mammalian cells. In order to survive and proliferate, purine auxotrophic parasites developed diverse strategies to salvage purines through an array of interconnected and seemingly redundant pathways. Marked differences in the salvage pathways and the preferred substrates between parasites exist however, reflecting host metabolic specificities. This critical participation of purine/pyrimidine nucleotides in many aspects of parasite function and replication makes the nucleotide metabolism an "Achilles' heel" for drug intervention.

Trypanosomatids *Leishmania* and *Trypanosoma* lack nine of the ten genes required to make IMP from phosphoribosyl pyrophosphate (PRPP) and are thus unable to synthetize purines de novo. To salvage purines from the host, *Leishmania* possess a group of membrane and secreted nucleotidase/nuclease hydrolases (NH) that mediate the degradation of extracellular nucleosides, nucleotides, and nucleic acids with low substrate specificity. Several purine transporters are encoded in the *Leishmania* genome. The first parasite purine transporters identified are the nucleoside transporters (NTs) LdNT1 from *Leishmania donovani* (specific for adenosine and the pyrimidine nucleosides) (Vasudevan et al. 1998) and LdNT2 (for inosine, guanosine, xanthosine) (Carter et al. 2000b). Two other purine nucleobase transporters (NT3 and NT4) for hypoxanthine, xanthine, adenine, and guanine are also present in *Leishmania*, the latter having an optimal transport activity at acidic pH which correspond to the conditions found by parasites inside the parasitophorous vacuole (Ortiz et al. 2009). Once internalized, nucleosides and nucleobase are interconverted by a complex network of purine salvage enzymes (Fig. 9.1a). The genetic dissection of this pathway using reverse genetic tools revealed functionally redundant routes for purine acquisition and pinpointed the preferred purine source for *Leishmania* parasite. Metabolic flux and gene inactivation studies reported indeed that despite a superficial complexity, the majority of purine flux into nucleotides in *Leishmania* appeared limited to two main routes: HGPRT and XPRT (Boitz et al. 2012). Further studies showed that *Leishmania* is able to sense variations in extracellular purine concentration by upregulating the translation of purine transporters and salvage enzymes in a coordinated manner (Carter et al. 2010; Ortiz et al. 2010; Martin et al. 2014). Following purine removal from the culture media, *Leishmania* upregulate primarily purine transporters (NT1-3) and nucleases (NH), while other key purine salvage components, notably HGPRT and XPRT, are upregulated later and to a lesser extent (Martin et al. 2014). The purine salvage pathways of *Leishmania* and *Trypanosoma* are remarkably similar with one notable exception: while *Leishmania* has two distinct routes of adenine metabolism (AAH and APRT, see Fig. 9.1a) and the former being the preferred one, both *T. brucei* and *T. cruzi* lack AAH ortholog enzyme (Boitz and Ullman 2013). Adenosine, the preferred source of purine in *T. brucei*, is then successively cleaved to adenine by specific nuclease and phosphorylated to AMP by APRT (Fig. 9.1a). Remarkably, the adenosine/adenine transporter of *T. brucei* (TbAT1) is also the transporter of the arsenical melarsoprol and pentamidine, two important drugs currently used for the treatment of African sleeping sickness (Baker et al. 2013). In contrast to *Leishmania*, trypanosomes encode a complex family of at least 12 closely related nucleoside and nucleobase transporters resulting from a series

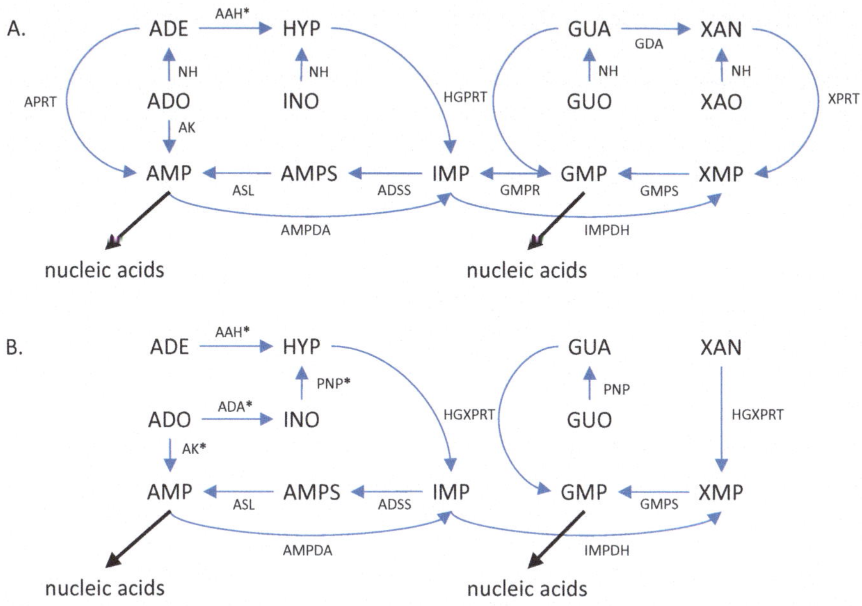

Fig. 9.1 Simplified illustration of the purine salvage pathway and interconversion enzymes in trypanosomatid (**a**) and apicomplexan (**b**) protozoan parasites. Asterisks correspond to enzymes found in specific parasite species (see text for details). Abbreviations: *ADE* adenine, *HYP* hypoxanthine, *GUA* guanine, *XAN* xanthine, *ADO* adenosine, *INO* inosine, *GUO* guanosine, *XAO* xanthosine, *AAH* adenine aminohydrolase, *GDA* guanine deaminase, *NH* nucleoside hydrolase, *AK* adenosine kinase, *APRT* adenine phosphoribosyltransferase, *HGPRT* hypoxanthine-guanine phosphoribosyltransferase, *XPRT* xanthine phosphoribosyltransferase, *HGXPRT* hypoxanthine-guanine-xanthine phosphoribosyltransferase, *ADSS* adenylosuccinate synthetase, *ASL* adenylosuccinate lyase, *AMPDA* AMP deaminase, *IMPDH* inosine monophosphate dehydrogenase, *GMPS* GMP synthase, *GMPR* GMP reductase, *NH* nucleoside hydrolase, *PNP* purine nucleoside phosphorylase

of gene duplication events (Landfear et al. 2004; de Koning et al. 2005). The reason for the expansion of this gene family in trypanosomes compared to *Leishmania* is still not clear.

Apicomplexans Marked differences in purine scavenging abilities can be noticed between *Toxoplasma* and *Plasmodium* parasites. Three transporters involved in the uptake of purine bases and nucleosides have been identified in *T. gondii*. TgAT1 is specific to adenosine and inosine; TgAT2 possesses broad substrate specificity, while TgNBT1 transports hypoxanthine, xanthine, and guanine. In the case of *Plasmodium*, purine nucleosides and nucleobases are transported across the plasma membrane by a unique transporter PfNT1 (Carter et al. 2000a). In contrast to trypanosomatids, apicomplexans share a single phosphoribosyltransferase enzyme (HGXPRT) to catalyze the synthesis of IMP, XMP, and GMP from imported hypoxanthine, xanthine, and guanine, respectively (Fig. 9.1b). *Toxoplasma* mostly produces AMP from adenosine using adenosine kinase (AK), which is the most active enzyme of this pathway (Krug et al. 1989). Surprisingly, this enzyme is absent in *P. falciparum*, and AMP generation in this parasite strictly depends on adenylosuccinate synthase and lyase activities from IMP, the common precursor for the synthesis of all purine nucleotides in *Plasmodium*. The *Plasmodium* purine salvage pathway is actually highly simplified and relies on the activity of the three enzymes HGXPRT, purine nucleoside phosphorylase (PNP), and adenosine deaminase (ADA) (Cassera et al. 2011). These enzymes ensure the sequential conversion of exogenous adenosine to inosine and hypoxanthine, which will be then phosphoribosylated to form IMP (Fig. 9.1b). It has been suggested however that hypoxanthine is the major purine precursor incorporated into parasite during the intraerythrocytic stage of *Plasmodium* (Gero and O'Sullivan 1990). This implies that HGXPRT should be considered as the most promising drug target of the purine pathway in *P. falciparum*.

Purine-Based Chemotherapy Different strategies have been developed to specifically target the purine synthesis in parasitic protozoans. Basically, the two main approaches consist of either inhibiting specific enzymes using substrates analogs generally derived from anticancer or antiviral drugs or using subversive substrates that will be activated to toxic products by a parasite-specific purine salvage enzyme. For comprehensive reviews of the best purine targets and the different class of purine inhibitors that have been tested against protozoan parasites, see references (Kouni 2003; Berg et al. 2010). Due to the high redundancy of the interconversion pathways in protozoan purine metabolism, it is thought that two or more inhibitors might be required for an effective chemotherapy against these parasites. Additional in vivo experiments are required to determine whether parasites can readily bypass the inhibition of a specific scavenging route of purine acquisition without loss of fitness and virulence.

9.3.2 Polyamines and Arginase

Polyamines defined small cationic molecules with more than two amine groups and are found in all living organisms. In eukaryotes, the most common polyamines are putrescine, spermidine, and spermine. Due to the wide range of cellular processes in which polyamines are involved, it is still difficult to delineate the precise functions of these metabolites that comprise gene regulation, cell proliferation, and stress response (Miller-Fleming et al. 2015). The polyamine biosynthetic pathway is ubiquitous and highly conserved among eukaryotes. Marked differences exist however in protozoan parasites (Birkholtz et al. 2011). For trypanosomatids, polyamine biosynthesis plays an essential role for the synthesis of the trypanosomatid-specific thiol trypanothione (TSH_2) that controls redox regulation and defense against oxidative stress damages. The essential role of polyamines in cell proliferation and their association with cancer has led to the development of drug compounds with antitumoral properties. While being not enough satisfying for human application, these polyamine-related molecules were shown to be highly effective against protozoan parasites, notably *T. brucei* (see below). The canonical polyamine pathway in eukaryotes consists of five enzymes: arginase (ARG), ornithine decarboxylase (ODC), S-adenosylmethionine decarboxylase (AdoMetDC), spermidine synthase (SpdS), and spermine synthase (SpmS). Trypanosomatids and *Plasmodium* parasites lack the final SpmS enzyme. ARG, the first step in polyamine biosynthesis, converts L-arginine (Arg) to L-ornithine (Orn) that is subsequently metabolized to the diamine putrescine by the catalytic action of ODC. Putrescine is then converted into spermidine (Spd) and spermine (Spm) by the addition of aminopropyl groups provided by decarboxylated S-adenosylmethionine (dcSAM). This addition is mediated by SpdS and SpmS, respectively, while dcSAM results from the activity of AdoMetDC on S-adenosylmethionine (SAM) (Fig. 9.2).

Trypanosomatids The redox system in trypanosomatids is essentially based on a specific dithiol called trypanothione (TSH_2) that is composed of two molecules of glutathione (GSH) joined by a spermidine linker (Krauth-Siegel and Comini 2008) (Fig. 9.2a). Given the role of spermidine as a major component of the trypanothione synthesis, polyamine inhibition is believed to be highly efficient against trypanosomatid parasites. Actually, the last approved drug to treat African trypanosomiasis caused by *T. b. gambiense* is α-difluoromethylornithine (DFMO; eflornithine), an inhibitor of ODC initially developed as an anticancer agent. This is

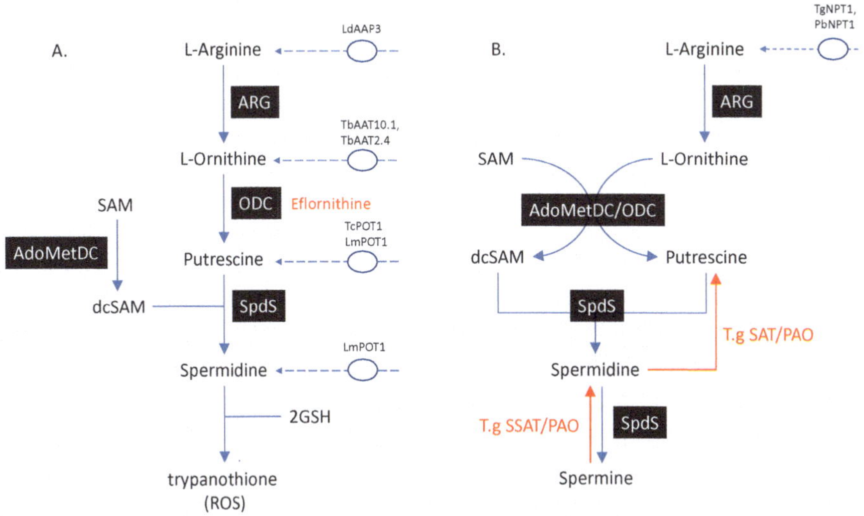

Fig. 9.2 Polyamine metabolism and transporters in trypanosomatid (**a**) and apicomplexan (**b**) parasites. Key parasite enzymes are boxed in black. The backward polyamine biosynthesis pathway of *T. gondii* is indicated by the presence of *T. gondii* spermine/spermidine N-acetyltransferase (SSAT), spermidine N-acetyltransferase (SAT), and polyamine oxidase (PAO) enzymes. Abbreviations: *ARG* arginase, *ODC* ornithine decarboxylase, *SpdS* spermidine synthase, *SAM*, S-adenosylmethionine, *dcSAM* decarboxylated SAM, *AdoMetDC* adenosylmethionine decarboxylase, *ROS* reactive oxygen species

the only antitrypanosomal drug whose molecular target is known. Concomitant with its action on ODC activity, eflornithine reduces the intracellular levels of putrescine, spermidine, and trypanothione in bloodstream form of *T. brucei* (Fairlamb et al. 1987). The reasons why DFMO is more efficient against *T. brucei* than the other trypanosomatids or even mammalian ODC are multiple. The main explanation is that *T. brucei* parasites are directly in contact to the drug in the bloodstream and are not able to import exogenous polyamines in enough quantity. Finally, TbODC has a relatively slow turnover as compared to the human ortholog. Recently, two high-affinity Orn transporters that belong to the amino acid transporter family (TbAAT10-1, TbAAT2-4) have been described in *T. brucei* (Macedo et al. 2017) (Fig. 9.2a). Downregulation of these transporters renders parasites hypersensitive to DFMO (taken up by a distinct TbAAT6 transporter). Surprisingly, *ODC* gene is missing in *T. cruzi* genome, making this parasite totally dependent on putrescine uptake to synthetize spermidine and trypanothione. In contrast to *T. brucei*, polyamine transport in *T. cruzi* is indeed highly efficient and essentially mediated by the TcPOT1 transporter (also named TcPAT12). This transporter has a high affinity for putrescine (but not for spermidine), and deletion of *TcPOT1* gene impairs the ability of *T. cruzi* parasites to sustain a robust infection in mammalian cells (Hasne et al. 2016). The *Leishmania* ortholog of TcPOT1 is LmPOT1, the first eukaryotic polyamine transporter cloned in eukaryotes. It is selective for putrescine and spermidine uptake but does not appear to be expressed in the amastigote stage in mammal host (Hasne and Ullman 2005). The *Leishmania* polyamine biosynthesis diverges from that of other trypanosomatids by the fact that *Leishmania* can use ARG for the de novo production of Orn from Arg. Interestingly, *Leishmania* parasites are able to sense changes in the external Arg concentration. Indeed, to fulfill its polyamine needs following macrophage infection, intracellular *Leishmania* parasites activate a coordinated "Arg deprivation response" (ADR) by rapidly increasing the expression of the high-affinity parasite Arg transporter (LdAAP3) (Goldman-Pinkovich et al. 2016). Double knockout of the *ARG* gene in several *Leishmania* species showed a significant reduction of infectivity in mice as compared to wild-type parasites. Recent data support however that inside the parasitophorous vacuole of macrophages, *ARG* is not essential for parasite survival. *L. donovani* amastigotes prefer indeed salvaging host Orn, and to a lesser extend host spermidine pools, while host putrescine levels appear too limited to support robust infection (Boitz et al. 2017).

Apicomplexans The most distinctive feature of polyamine biosynthesis in *Plasmodium* parasites is the presence of a bifunctional AdoMetDC/ODC enzyme where the active site of both ODC and AdoMetDC are located on a single polypeptide (Fig. 9.2b). This bifunctional AdoMetDC/ODC enzyme appears to be important for the regulation of the polyamine pools in the parasite (Birkholtz et al. 2004). Deletion of the *AdoMetDC/ODC* gene in *Plasmodium yoelii* significantly reduced blood-stage parasitemia in mice and completely abolished *Anopheles* transmission (Hart et al. 2016). Furthermore, a single SpdS enzyme ensures the synthesis of both spermidine and spermine (Fig. 9.2b). Arg uptake and transformation to polyamines by its own ARG and AdoMetDC/ODC enzymes also appear to be determinant for *Plasmodium* development in the liver (Meireles et al. 2017). The apicomplexan Arg transporter has been recently characterized in *T. gondii* (TgNPT1) and *P. berghei* (PbNPT1), and gene deletion experiments have confirmed their role in parasite's survival and virulence (Rajendran et al. 2017). Studies on polyamine biosynthesis in *T. gondii* are more limited. Surprisingly it appears that *T. gondii* is devoid of ODC and ARG enzymes and can only synthetize putrescine from host spermine in backward reactions through the combined action of *Toxoplasma* spermine/spermidine N-acetyltransferase (SSAT) and polyamine oxidase (PAO) (Cook et al. 2007) (Fig. 9.2b). Additionally, *T. gondii* parasites possess a high-affinity putrescine transporter and can also scavenge Orn from the culture medium (Cook et al. 2007), while none of these transporters have been identified up to now. Therefore, while polyamine supply in *Plasmodium* mainly depends on biosynthesis from arginine to spermidine, *Toxoplasma* can only rely on polyamines scavenging and interconversion.

Polyamine-Based Chemotherapy The success story of DFMO (eflornithine) in African trypanosomiasis treatment supports the legitimacy of developing inhibitors interfering with polyamine homeostasis in parasites. Unfortunately, DFMO efficiency against other protozoan parasites is far from being satisfying, mainly because of their capacity to bypass ODC enzymatic inhibition by increasing the uptake of preformed polyamines from the host. This dual source of polyamines (biosynthesis and scavenging) is certainly the key issue for future development of drugs against ODC or the different enzymes of the polyamine biosynthesis pathway (i.e., ARG, AdoMet, SpdS). Recent reviews on polyamine-based therapeutic strategies against protozoan parasites can be found in Birkholtz et al. (2011), Roberts and Ullman (2017), and Jagu et al. (2017).

9.3.3 Cofactors and Vitamins

Vitamins define organic compounds and vital nutrients that are not synthetized by an organism but that are required in limited amounts. They are used as precursors of cofactors for many enzymes involved in different metabolisms. Thirteen vitamins are universally recognized at present, but the evolution of vitamin-related pathways has resulted in profound differentiation of auxotrophy among species, including parasites. The most striking example is vitamin C (ascorbate), which is a vitamin for humans but not for most other organisms that can synthetize it de novo. By contrast, microorganisms such as protozoan parasites can synthesize certain vitamins de novo, either completely or partially. The complex pattern of vitamin auxotrophy across the eukaryotic tree of life is intimately connected with the interdependence between organisms (Helliwell et al. 2013). Genomic analysis revealed that trypanosomatids are auxotrophs for thiamine (B1), riboflavin (B2), nicotinic acid (B3), pantothenic acid (B5), pyridoxamine (B6), biotin (B7), and folic acid (B9) (Klein et al. 2013). Interestingly, nonpathogenic trypanosomatids harboring endosymbionts are only auxotrophs for thiamine, biotin, and nicotinic acid, most of the genes required for the synthesis of other vitamins being present in the symbiont genome. Apicomplexans possess the ability to synthetize several vitamins de novo including vitamins B1, B6, and B9. Some divergences in their biosynthesis pathway exist however between *Plasmodium* and *Toxoplasma* (Müller and Kappes 2007). The inability of human to synthetize vitamins implies therefore that the inhibition of parasite vitamin biosynthetic pathways might be a good therapeutic strategy to selectively interfere with parasite development. This is the case for the folate pathway that has proven to be a valuable target for drug intervention in *Plasmodium* (see below). Conversely, the lack of a specific de novo vitamin biosynthetic pathway in parasites also offers the possibility to block precursor assimilation and conversion to active cofactors. This is notably the case for vitamin B3, the unique precursor of NAD+ in protozoan parasites.

9.3.3.1 Folate Metabolism (Vitamin B9)

The term folates designs several structurally similar compounds that are used as cofactors in a variety of "one-carbon" (C1) transfer reactions. In their reduced tetrahydrofolate (THF) form, folates are notably involved in the biosynthesis of thymidine (DNA synthesis), the conversion of serine to glycine, methionine biosynthesis (DNA and protein methylation reactions), histidine catabolism, and purine biosynthesis. For these reasons, folate pathway has long been considered to be an attractive pathway to develop anti-parasite drugs. Folates are composed of a pterin ring linked to para-aminobenzoate (pABA) and L-glutamate (Fig. 9.3a). Inside the cell, they are further polyglutamylated to enhance cellular retention. *Plasmodium* and *Toxoplasma* parasites possess both biosynthetic and salvage pathways and are then able to synthetize pterin and pABA compounds and to perform folate assembly de novo. Pterin is synthesized from GTP (guanosine 5′-triphosphate), whereas pABA is obtained from chorismate. Instead, trypanosomatids are auxotrophs for both folates and pterins and must import them from exogenous sources.

Trypanosomatids Most of our knowledge about folate metabolism in *Leishmania* and other trypanosomatids came from studies using the anti-DHFR cancer drug methotrexate (MTX) (Ouellette et al. 2002; Vickers and Beverley 2011). *Leishmania* can acquire folates (and MTX) by at least three members of the folate-biopterin transporter (FBT) family named FT1, FT5, and BT1, the latter being a biopterin transporter that can also import folates but not MTX. This *Leishmania* FBT family comprises in total 14 members, while only 7 are present in *T. brucei*. A genome-wide RNA interference experiment has recently identified a tandem array of folate transporters (TbFT1–3) as major contributors of folates and antifolate drugs uptake in *T. brucei* (Dewar et al. 2016). The enzymatic reduction of folate to tetrahydrofolate (THF) in trypanosomatids is catalyzed by a bifunctional dihydrofolate reductase-thymidylate synthase (DHFR-TS). Interestingly, a similar reduction can also be carried out by an alternative enzyme called pteridine reductase 1 (PTR1)

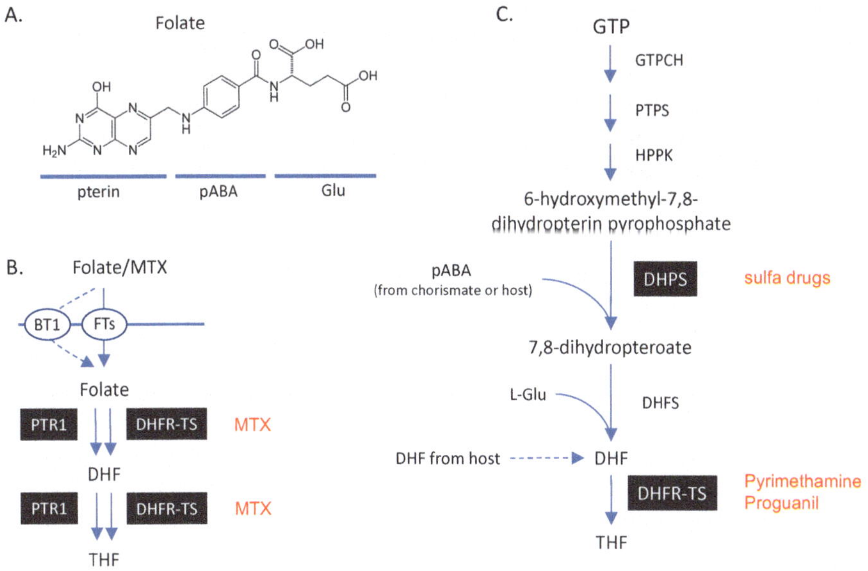

Fig. 9.3 Simplified illustration of folate metabolism in protozoan parasites. (**a**) Structure of folate composed of three building blocks: pterin, para-aminobenzoate (pABA), and glutamic acid (Glu) (from https://commons.wikimedia.org/wiki/File:Folic_acid_structure.svg). (**b**) Folate transport and reduction in trypanosomatids mediated by DHFR-TS (the target of the drug methotrexate (MTX)) and PTR1 enzymes. (**c**) De novo synthesis of folates in apicomplexans. Drugs targeting DHPS and DHFR-TS enzymes (boxed in black) are indicated in red. Abbreviations: *BT1* biopterin transporter 1, *FTs* folate transporters, *DHFR-TS* dihydrofolate reductase-thymidylate synthase, *PTR1* pterin reductase 1, *DHF* dihydrofolate, *THF* tetrahydrofolate, *GTPCH* GTP cyclohydrolase 1, *PTPS* 6-pyruvoyl-tetrahydropterin synthase, *HPPK* hydromethyldihydropteridine pyrophosphokinase, *DHPS* dihydropteroate synthase, *DHFS* dihydrofolate synthase

(Fig. 9.3b). PTR1 can reduce both unconjugated pteridines and folates. This gene was first discovered in MTX-resistant *Leishmania* parasites in which both *PTR1* and *BT1* genes were found amplified in a common extrachromosomal element (Ouellette et al. 2002). PTR1 was shown to contribute to 10% of the reduction of folates in wild-type parasite and is 2000-fold less susceptible to inhibition by MTX than DHFR-TS (Nare et al. 1997). This PTR1 bypass of DHFR-TS reduces the effectiveness of classical antifolate drugs in *Leishmania* parasites and supports therefore that both enzymes have to be targeted simultaneously. Despite numerous efforts, such an inhibitor has not been identified up to now.

Apicomplexans Apicomplexans can both synthetize folates de novo and salvage them from their host. Nevertheless, antifolates have a long history of successful application in malaria chemotherapy, and despite widespread parasite resistance, the antifolates still play an important role in malaria control (Müller and Hyde 2013). The biosynthesis pathway starts by converting GTP to 6-hydroxymethyl-7,8-dihydropterin pyrophosphate by three successive enzymatic steps (Fig. 9.3c). The next dihydropteroate synthase enzyme (DHPS) catalyzes the formation of 7,8-dihydropteroate from pABA and the product of the hydromethyldihydropteridine pyrophosphokinase (HPPK) reaction. DHPS is the target of sulfa drugs (sulfadoxine), the first class of *Plasmodium* antifolates drugs that are structural analogues of pABA. The de novo synthesis of folates is completed by the action of the dihydrofolate synthase (DHFS) that converts 7,8-dihydropteroate to dihydrofolate (DHF) by conjugating L-Glu residue to pABA. The final reduction of DHF to THF is mediated by a bifunctional DHFR-TS enzyme, which is the second target of antifolate drugs like pyrimethamine and proguanil. Since the plasmodial DHFR-TS is highly susceptible to these inhibitors, pyrimethamine in combination with sulfadoxine has long been used in the treatment of malaria until drug-resistant parasites with mutations in both *DHF* and *DHFR* genes have emerged. Two folate transporters (PfFT1 and PfFT2) have been also characterized in *Plasmodium* and can mediate folate as well as pABA uptake from the host (Salcedo-Sora et al. 2011). This ability of *Plasmodium* to salvage folates can directly alter the efficacy of antifolate drugs in pregnant woman taking folate supplementation treatment (Nzila et al. 2014). Currently, antifolate drugs such as pyrimethamine, proguanil, and sulfonamides are still part of the treatment of uncomplicated malaria in combination with other drugs such as artemisinin or amodiaquine (Müller and Hyde 2013). Nevertheless, several enzymes of the folate pathway are still considered as promising targets for drug development against *Plasmodium* parasites (Salcedo-Sora and Ward 2013).

9.3.3.2 Niacin (Vitamin B3) and NAD+

Niacin or nicotinate (Na), together with its amide form nicotinamide (Nam), defines the group of vitamin B3 complex. They are all precursors of the nicotinamide adenine dinucleotide (NAD+), an essential cofactor in all living cells that carries electrons from one reaction to another in redox reactions. By its redox function, NAD+ and its derived phosphorylated form (NADP+) are involved in hundreds of reactions in many metabolic pathways including glycolysis, TCA cycle, pentose phosphate pathway, or fatty acid biosynthesis. In the last decades, non-redox roles have been also attributed to NAD+. NAD+ is consumed by a number of important cellular enzymes that participate in various cellular functions such as epigenetic regulation, DNA repair, or signal transduction. In mammals, those enzymes include class III protein deacetylases (sirtuins), poly(ADP)-ribose polymerases (PARPs), or cADP-ribose synthases (CD38 or CD157). These NAD+-dependent signaling processes involve the release of Nam and thus require constant replenishment of cellular NAD+ pools. The role of NAD+ in host-pathogen interactions and its potential for new treatments of infectious diseases have been recently reviewed (Mesquita et al. 2016). The characterization of this metabolism in protozoan parasites is quite recent. Apicomplexan and trypanosomatid parasites cannot synthesize NAD+ de novo and must salvage NAD+ precursors from host or culture medium. The different enzymes involved in NAD+ salvage pathway have been notably characterized in *Leishmania* and *Plasmodium* parasites (Gazanion et al. 2011; O'Hara et al. 2014). In contrast, human can synthetize NAD+ either by a salvage pathway or by de

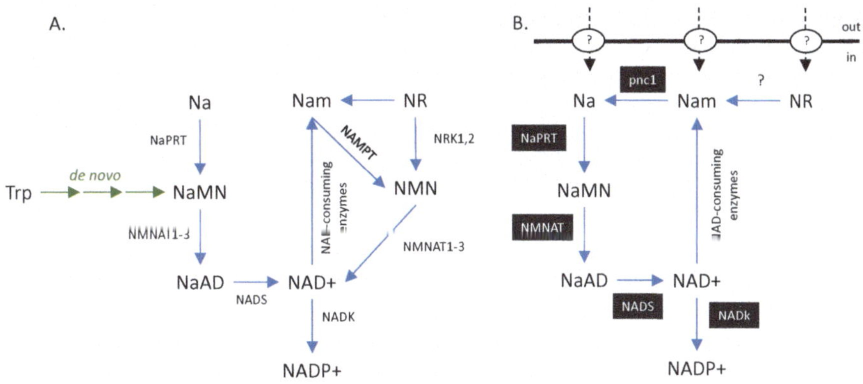

Fig. 9.4 NAD+ metabolism in human cells (**a**) and protozoan parasites (**b**). Abbreviations: *Na* nicotinic acid, *Nam* nicotinamide, *NR* nicotinamide riboside, *NaMN* nicotinic acid mononucleotide, *NaAD* nicotinic acid dinucleotide, *NMN* nicotinamide mononucleotide, *NAD+* nicotinamide adenine dinucleotide, *NADP+* nicotinamide adenine dinucleotide phosphate, *NaPRT* nicotinamide phosphoribosyltransferase, *NMNAT* nicotinamide mononucleotide adenylyltransferase, *NADS* NAD+ synthase, *NADK* NAD+ kinase, *NAMPT* nicotinamide phosphoribosyltransferase, *NRK* nicotinamide riboside kinase, *Trp* tryptophan, *pnc1* nicotinamidase

novo synthesis from tryptophan, both pathways converging on nicotinic acid phosphoribosyltransferase (NaMN) (Fig. 9.4a). Briefly, Na enters the Preiss-Handler pathway and is converted to NAD+ in three enzymatic steps by the action of nicotinate phosphoribosyltransferase (NaPRT), nicotinate mononucleotide adenylyltransferase (NMNAT), and NAD+ synthase (NADS). In lower eukaryotes including protozoan parasites, Nam is converted to Na by a highly conserved enzyme called nicotinamidase (pnc1) (Fig. 9.4b). This enzyme is absent in humans in which Nam is first converted to Nam mononucleotide (NMN) through the activity of a Nam phosphoribosyltransferase (NAMPT), and then adenylated to NAD+ by NMNAT (Fig. 9.4a).

Trypanosomatids *Leishmania* parasites are NAD+ auxotrophs and can import Na, Nam, or NR to ensure intracellular NAD+ synthesis (Gazanion et al. 2011). The *Leishmania* nicotinamidase has been recently biochemically and functionally characterized in *L. infantum*. Deletion of this gene led to obtain NAD+ depleted parasites that cannot sustain infection in mice model nor in sand fly vector (Gazanion et al. 2011; Gazanion et al. 2012). This enzyme is therefore a validated drug target against *Leishmania* and possibly other trypanosomatid parasites (Michels and Avilán 2011). Based on comparative genomic analysis, all *Trypanosoma* and *Leishmania* species possess a canonical NAD+ salvage pathway. Other enzymes of this pathway, notably NMNAT, have recently been characterized in *L. braziliensis* and *T. cruzi* (Contreras et al. 2015; Sánchez-Lancheros et al. 2016).

Apicomplexans Some members of apicomplexans including *P. falciparum* or *T. gondii* have a NAD+ salvage pathway similar to that of trypanosomatids, although with subtle differences. In contrast to the cytosolic localization of the *Leishmania* nicotinamidase, the *Plasmodium* nicotinamidase concentrates mainly in the nucleus (O'Hara et al. 2014). The same study showed that PfNMNAT can be a valuable drug target for drug development and lead compounds were proposed based on previously identified bacterial NMNAT inhibitors. Surprisingly, other apicomplexans such as *Cryptosporidium*, *Babesia*, and *Theileria* species are not able to synthesis NAD+ from Na and Nam. They all possess NAD+ kinase orthologs, suggesting that they can directly salvage NAD+ from host and convert it to NADP+, but no NAD+ transporter has been identified to date, and no homologs of validated vitamin B3 transporters can be found in apicomplexans or trypanosomatids genome databases. How NAD+ or vitamin B3 precursors are imported into protozoan parasites remains to be determined, and such transporters might constitute new promising drug targets for chemical inhibition.

9.3.3.3 Other Vitamins and Vitamin-Based Chemotherapy

In addition to the two examples cited above, biosynthesis of other vitamins in protozoan parasites has been well documented for their suitability to identify new anti-parasite drugs. Most of the vitamin biosynthesis pathways are indeed absent in humans and constitute therefore a valuable reservoir of drug targets. While folate metabolism has been largely exploited for drug design, other vitamins such as thiamine (B1), riboflavin (B2), pyridoxamine (B6), or pantothenate (B5) constitute promising alternatives (Müller and Kappes 2007; Müller et al. 2010; Kronenberger et al. 2013). Pantothenate (B5), for example, is required for the synthesis of coenzyme A (CoA), an essential cofactor for both energy metabolism and in fatty acid biosynthesis. *T. gondii* parasites possess all four genes required for de novo synthesis of pantothenate from valine while these genes are absent in *P. falciparum*. Once taken up by the parasites, pantothenate is phosphorylated by the pantothenate kinase (PanK), the first enzyme and the rate-limiting step in the biosynthesis of CoA. Even if this enzyme has four isoforms in humans, the unique *Plasmodium* PanK is highly sensitive to pantothenate analogues. Resistant lines to pantothenate analogues with mutations in *PfPanK*1 gene can be generated however in vitro (Tjhin et al. 2017).

Riboflavin (B2) is the vital precursor of both flavin mononucleotide (FMN) and flavin adenine dinucleotide (FAD) used as cofactors by flavoenzymes. The synthesis of FMN and FAD in trypanosomatids is effective but dependent on efficient riboflavin uptake. The first family of protist riboflavin transporters (RibJ) has recently been identified in trypanosomatids and might constitute new attractive trypanosomatid targets (Balcazar et al. 2017).

9.3.4 Heme Synthesis

Heme, or iron protoporphyrin IX, is a ubiquitous cofactor composed of four pyrrole rings interconnected to each other and to a Fe^{2+} ion (Fig. 9.5a). Heme iron can exist in either a reduced (Fe^{2+}) or oxidized state (Fe^{3+}), and this transitional property is exploited by heme-dependent proteins to carry out diverse redox conversions within the cell. For parasites, heme is notably required for the synthesis of hemoproteins such as cytochromes a–c or P450 and is therefore essential for cellular respiration via the electron transport chain, the desaturation of fatty acids, and sterol synthesis (Kořený et al. 2013). Heme is also a central source of iron for parasites, and, as a component of parasite peroxidases and catalases, it helps parasites to deal with reactive oxygen species (ROS) and oxidative burst produced by the host cell. However, due to its reactive nature, free heme is also a potentially toxic molecule that can generate ROS and induce lipid peroxidation. Protozoan parasites have evolved therefore robust pathways to synthesize heme de novo or scavenge it from their environment together with efficient heme detoxification pathways (Kořený et al. 2013; Sigala and Goldberg 2014). In most eukaryotes, heme biosynthesis involves eight enzymatic steps that begin with the formation of aminolevulinic acid (ALA) in the mitochondria by the condensation of glycine and succinyl-CoA by ALA synthase (ALAS). The next enzymatic steps continue in the cytosol (or the

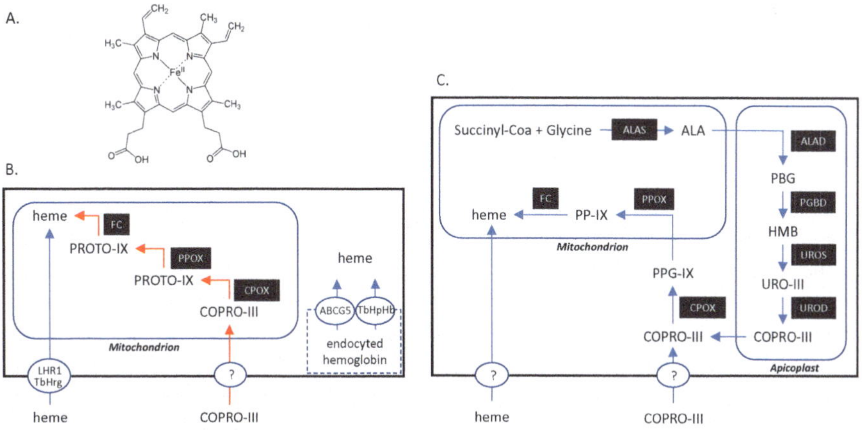

Fig. 9.5 Schematic representation of the heme metabolism in protozoan parasites. (**a**) Chemical structure of heme B (from https://commons.wikimedia.org/wiki/File:Heme_b.svg). (**b**) Heme auxotrophy in trypanosomatids parasites and the presence of the last three mitochondrial enzymes involved in heme synthesis for *Leishmania* spp. (boxed in black). (**c**) Heme biosynthesis in *P. falciparum* taking place into mitochondrion, apicoplast, and cytoplasm. Key enzymes are boxed in black. Abbreviations: *ALAS* aminolevulinic acid synthase, *ALA* 5-aminolevulinic acid, *ALAD* aminolevulinic acid dehydratase, *PBGD* porphobilinogen deaminase, *UROS* uroporphyrinogen synthase, *UROD* uroporphyrinogen decarboxylase, *CPOX* coproporphyrinogen oxidase, *PPOX* protoporphyrinogen oxidase, *FC* ferrochelatase

apicoplast for apicomplexans parasites) and finally return back to mitochondria to finalize heme synthesis by ferrochelatase enzyme (FC) (Fig. 9.5).

Trypanosomatids The evolution of the heme synthetic pathway in trypanosomatids is highly unusual. All trypanosomatid parasites are heme-auxotrophs and lost the ability to synthesize heme before acquiring a parasitic lifestyle since the closest nonparasitic relatives of trypanosomatids (bodonids) lack heme synthesis pathway (Kořený et al. 2010). To satisfy their heme requirements, *T. brucei* and *T. cruzi* take up heme directly from their hosts. In *Leishmania* spp., however, this pathway was partially rescued by the presence of three genes encoding enzymes for the last steps of heme synthesis (coproporphyrinogen oxidase (CPOX), protoporphyrinogen IX oxidase (PPOX), and FC). These genes were supposedly obtained from bacteria by lateral gene transfer (Kořený et al. 2010; Fig. 9.5b). The signification of this acquisition is not clear, but it may translate the ability of *Leishmania* parasites to use heme precursors from their hosts when the heme concentration is too low (Chang and Chang 1985; Zwerschke et al. 2014). In the bloodstream, haptoglobin-hemoglobin (HpHb) complexes are the major source of extracellular circulating heme. The bloodstream stage of *T. brucei* scavenges heme in hemoglobin via a specific HpHb receptor localized in the flagellar pocket (Dostert et al. 2008). In the tsetse fly however, the HpHb receptor is not expressed, and heme importation is mediated by the recently characterized TbHrg transporter (Horáková et al. 2017). In *Leishmania*, two independent pathways of heme uptake have been described. One involves a specific heme transporter named LHR1 (Huynh et al. 2012). The second one is mediated by hemoglobin endocytosis, lysosomal degradation, and heme salvage and involves the ATP-binding cassette protein LABCG5 (Campos-Salinas et al. 2011). *Leishmania* and some other trypanosomatids are currently the only known eukaryotes possessing a bacterial-type PPOX enzyme (Zwerschke et al. 2014). This enzyme, together with the characterized heme transporters, may constitute promising drug targets for trypanosomatid parasites.

Apicomplexans *Plasmodium* parasites require heme for growth and have access to abundant host-derived heme within infected red blood cells, the most heme-rich cells in the human body. Blood-stage *Plasmodium* parasites import up to 80% of the erythrocytes' hemoglobin into its digestive vacuole to fulfill its nutrient requirements. During hemoglobin digestion, malaria parasites scavenge amino acids and peptides but also generate millimolar concentration of host-derived heme. Nevertheless, *Plasmodium* also expresses a complete, although unusual, heme biosynthetic pathway that spans between the mitochondria, the plantlike non-photosynthetic apicoplast, and the cytosol (Fig. 9.5c). Heme is essential for malaria parasites as a precursor for cytochrome production in the mitochondrial electron transport chain. However, de novo heme synthesis pathway is dispensable during the blood stage of the parasite life cycle since the genes coding for ALAS and FC, the first and the last enzymes involved in heme biosynthesis, respectively, can be inactivated without effect on parasite growth (Nagaraj et al. 2013; Ke et al. 2014). These results support that *Plasmodium* can scavenge host-derived heme, either released from the digestive vacuole or derived from a partial heme synthesis pathway present in RBC (Sigala et al. 2015). No homologs of heme transporters identified in trypanosomatids and other eukaryotes can be found however in *Plasmodium* genome sequence. The de novo heme synthesis pathway is nonetheless essential for parasite development during mosquitoes and liver stages, justifying why this pathway is maintained in these parasites (Nagaraj et al. 2013; Ke et al. 2014). While being useful for many cellular processes, free heme is also a highly toxic molecule. Numerous heme detoxification mechanisms have been identified in blood feeding parasites including heme degradation by heme oxygenases (HO), heme containment by physical barriers, or heme conversion into inert crystalline structure called hemozoin (Toh et al. 2010). *Plasmodium* parasites sequester heme derived from hemoglobin digestion into crystals of insoluble hemozoin inside the parasite digestive vacuole. This heme detoxification process has proved to be an exploitable drug target against malaria parasites. Chloroquine (CQ) and other quinolone drugs are known to enter the food vacuole and interfere with the formation of hemozoin, resulting in increased free heme concentration and oxidative stress in the parasite. While the mode of action of CQ is not fully understood, parasite resistance is now widespread and generally involves the increase of drug efflux by mutations in several classes of efflux pumps (Fairlamb et al. 2016). Importantly, there is evidence that heme is also directly implicated in the activation of the most potent antimalarial drug artemisinin (Zhang et al. 2015).

Heme-Based Chemotherapy Pathways of heme biosynthesis, uptake, detoxification, and breakdown have provided several potential important targets for drug development in protozoan parasites. In *Plasmodium*, heme detoxification is intimately connected to the mode of action of several antimalarial drugs used in the clinic. Heme synthesis is only present in apicomplexans, and functional studies have shown that it is not essential for parasite growth during blood-stage infection. In the case of *Leishmania* parasites harboring a partial heme synthesis pathway, the last three enzymes, and notably the bacterial-type PPOX, might be attractive targets for drug inhibition. Interestingly, the enzymatic activity of PPOX can be exploited for photodynamic therapy (PDT), an alternative therapeutic approach to conventional drugs that uses light-activated molecules. PPOX ensures the formation of protoporphyrin IX (PP-IX) in mitochondria, which upon exposure to light generates cytotoxic ROS and triggers oxidative damage. This approach has been successfully tested against *Plasmodium* parasites (Sigala et al. 2015). The authors showed that infected RBC actively import exogenous ALA resulting in the accumulation of PP-IX through the concomitant action of parasites and vestigial human enzymes. Once photoactivated by nontoxic chemicals, PP-IX will selectively kill malaria parasites while leaving uninfected cells intact. The PDT strategy using light-excitable properties of porphyrins present clear advantages among which a low toxicity and the inability of treated pathogens to develop resistance mechanisms (Wainwright et al. 2017).

9.4 Concluding Remarks

Understanding the metabolic interactions of protozoan parasites with their hosts during their complex multistage life cycle has the potential to identify new key biological processes that could lead to new and improved antiparasitic drugs. It is generally admitted that the metabolic optimization of parasites culminates with the complete, or partial, loss of redundant metabolic pathways to reduce energy costs and increase fitness, two conditions required to be a good parasite. This loss of metabolic autonomy is usually compensated by the acquisition of a complex system of nutrients import and interconversion that have only been partially uncovered in protozoan parasites of medical importance. This chapter provides a non-exhaustive list of essential metabolic compounds for which protozoan parasites are totally, or partially, dependent to their host. Interestingly, it seems that while most of these parasites need to scavenge similar host-derived compounds, many differences exist in the way they are acquired and assimilated. For instance, all mentioned protozoan parasites are auxotrophs for purines, but marked differences exist within trypanosomatid and apicomplexan parasites in the routes of purine acquisition and the presence of enzymes that mediate purine interconversion. In the same way, polyamine biosynthesis in trypanosomatids is directly connected to their specific thiol metabolism, and if ODC enzyme is a validated drug target in *T. brucei*, this enzyme is absent in the related parasite *T. cruzi*. To fulfill their polyamine needs, *Plasmodium* parasites use a unique bifunctional AdoMetDC/ODC enzyme, whereas *Toxoplasma* only relies on backward reactions from host spermine, the final product of polyamine biosynthesis. Finally, if apicomplexans synthetize folates de novo and can further scavenge host folate intermediates, trypanosomatid parasites are auxotrophs for folates and pterins but can rely on a large family of folate transporters and can bypass the need of DHFR-TS by PTR1 enzymatic activity. With the exception of the NAD+ synthesis pathway that appears to be highly conserved between the main trypanosomatid and apicomplexan parasites, all other metabolic pathways presented here possess notable parasite-specific singularities that can be explained by the highly different environments in which these parasites evolve during their complex life cycle.

Until now, the polyamine biosynthesis pathway in *T. brucei*, the synthesis of folates, and the mechanism of heme detoxification in *Plasmodium* have shown to be genuine drug targets for the treatment of African sleeping sickness and malaria. Additional investigations are required to evaluate the druggability of alternative metabolic pathways for which protozoan parasites are also dependent to their host. The recent improvement of metabolomics analyses together with the new powerful reverse genomic tools available in these parasites (i.e., CRISPR/Cas9) will undoubtedly rapidly increase our comprehension of the different mechanisms used by protozoan parasites to hijack essential nutrients from their host, one of the most fascinating aspects of parasite's biology.

Funding Statement This work was supported by the IRD (Institut de Recherche pour le Développement) institutional funding. E.G. is a recipient of the FRM (Fondation pour la Recherche Médicale) postdoctoral fellowship (ARF2015093409).

References

Aurrecoechea C, Barreto A, Basenko EY et al (2017) EuPathDB: the eukaryotic pathogen genomics database resource. Nucleic Acids Res 45:D581–D591

Baker N, de Koning HP, Mäser P, Horn D (2013) Drug resistance in African trypanosomiasis: the melarsoprol and pentamidine story. Trends Parasitol 29:110–118

Balcazar DE, Vanrell MC, Romano PS et al (2017) The superfamily keeps growing: Identification in trypanosomatids of RibJ, the first riboflavin transporter family in protists. PLoS Negl Trop Dis 11:e0005513–e0005522

Berg M, Van der Veken P, Goeminne A et al (2010) Inhibitors of the purine salvage pathway: a valuable approach for antiprotozoal chemotherapy? Curr Med Chem 17:2456–2481

Berriman M, Ghedin E, Hertz-Fowler C et al (2005) The genome of the African trypanosome *Trypanosoma brucei*. Science 309:416–422

Birkholtz L-M, Wrenger C, Joubert F et al (2004) Parasite-specific inserts in the bifunctional S-adenosylmethionine decarboxylase/ornithine decarboxylase of *Plasmodium falciparum* modulate catalytic activities and domain interactions. Biochem J 377:439–448

Birkholtz L-M, Williams M, Niemand J et al (2011) Polyamine homoeostasis as a drug target in pathogenic protozoa: peculiarities and possibilities. Biochem J 438:229–244

Boitz JM, Ullman B (2013) Adenine and adenosine salvage in *Leishmania donovani*. Mol Biochem Parasitol 190:51–55

Boitz JM, Ullman B, Jardim A, Carter NS (2012) Purine salvage in *Leishmania*: complex or simple by design? Trends Parasitol:1–8

Boitz JM, Gilroy CA, Olenyik TD et al (2017) Arginase is essential for survival of *Leishmania donovani* promastigotes but not intracellular amastigotes. Infect Immun 85:e00554–16

Campos-Salinas J, Cabello-Donayre M, García-Hernández R et al (2011) A new ATP-binding cassette protein is involved in intracellular haem trafficking in *Leishmania*. Mol Microbiol 79:1430–1444

Carter NS, Ben Mamoun C, Liu W et al (2000a) Isolation and functional characterization of the PfNT1 nucleoside transporter gene from *Plasmodium falciparum*. J Biol Chem 275:10683–10691

Carter NS, Drew ME, Sanchez M et al (2000b) Cloning of a novel inosine-guanosine transporter gene from *Leishmania donovani* by functional rescue of a transport-deficient mutant. J Biol Chem 275:20935–20941

Carter NS, Yates PA, Gessford SK et al (2010) Adaptive responses to purine starvation in *Leishmania donovani*. Mol Microbiol 78:92–107

Cassera MB, Zhang Y, Hazleton KZ, Schramm VL (2011) Purine and pyrimidine pathways as targets in *Plasmodium falciparum*. Curr Top Med Chem 11:2103–2115

Chang CS, Chang KP (1985) Heme requirement and acquisition by extracellular and intracellular stages of *Leishmania mexicana* amazonensis. Mol Biochem Parasitol 16:267–276

Contreras LE, Neme R, Ramírez MH (2015) Identification and functional evaluation of *Leishmania braziliensis* nicotinamide mononucleotide adenylyltransferase. Protein Expr Purif 115:26–33

Cook T, Roos D, Morada M et al (2007) Divergent polyamine metabolism in the *Apicomplexa*. Microbiology 153:1123–1130

D'Souza G, Kost C (2016) Experimental evolution of metabolic dependency in bacteria. PLoS Genet 12:e1006364

D'Souza G, Waschina S, Pande S et al (2014) Less is more: selective advantages can explain the prevalent loss of biosynthetic genes in bacteria. Evolution 68:2559–2570

de Koning HP, Bridges DJ, Burchmore RJS (2005) Purine and pyrimidine transport in pathogenic protozoa: from biology to therapy. FEMS Microbiol Rev 29:987–1020
DeBarry JD, Kissinger JC (2011) Jumbled genomes: missing apicomplexan synteny. Mol Biol Evol 28:2855–2871
Dewar S, Sienkiewicz N, Ong HB et al (2016) The role of folate transport in antifolate drug action in *Trypanosoma brucei*. J Biol Chem 291:24768–24778
Dostert C, Pétrilli V, Van Bruggen R et al (2008) Innate immune activation through Nalp3 inflammasome sensing of asbestos and silica. Science 320:674–677
El-Sayed NMA, Myler PJ, Bartholomeu DC et al (2005a) The genome sequence of *Trypanosoma cruzi*, etiologic agent of Chagas disease. Science 309:409–415
El-Sayed NMA, Myler PJ, Blandin G et al (2005b) Comparative genomics of trypanosomatid parasitic protozoa. Science 309:404–409
Ensminger AW, Yassin Y, Miron A, Isberg RR (2012) Experimental evolution of *Legionella pneumophila* in mouse macrophages leads to strains with altered determinants of environmental survival. PLoS Pathog 8:e1002731–12
Fairlamb AH, Henderson GB, Bacchi CJ, Cerami A (1987) In vivo effects of difluoromethylornithine on trypanothione and polyamine levels in bloodstream forms of *Trypanosoma brucei*. Mol Biochem Parasitol 24:185–191
Fairlamb AH, Gow NAR, Matthews KR, Waters AP (2016) Drug resistance in eukaryotic microorganisms. Nat Microbiol 1:16092
Foth BJ, Ralph SA, Tonkin CJ et al (2003) Dissecting apicoplast targeting in the malaria parasite *Plasmodium falciparum*. Science 299:705–708
Gazanion E, Garcia D, Silvestre R et al (2011) The *Leishmania* nicotinamidase is essential for NAD (+) production and parasite proliferation. Mol Microbiol 82:21–38
Gazanion E, Seblova V, Votýpka J et al (2012) *Leishmania infantum* nicotinamidase is required for late-stage development in its natural sand fly vector, *Phlebotomus perniciosus*. Int J Parasitol 42:323–327
Gero AM, O'Sullivan WJ (1990) Purines and pyrimidines in malarial parasites. Blood Cells 16:467–484. discussion 485–98
Goldman-Pinkovich A, Balno C, Strasser R et al (2016) An arginine deprivation response pathway is induced in *Leishmania* during macrophage invasion. PLoS Pathog 12:e1005494–18
Hart RJ, Ghaffar A, Abdalal S et al (2016) *Plasmodium* AdoMetDC/ODC bifunctional enzyme is essential for male sexual stage development and mosquito transmission. Biology Open 5:1022–1029
Hasne M-P, Ullman B (2005) Identification and characterization of a polyamine permease from the protozoan parasite *Leishmania* major. J Biol Chem 280:15188–15194
Hasne M-P, Soysa R, Ullman B (2016) The *Trypanosoma cruzi* diamine transporter is essential for robust infection of mammalian cells. PLoS One 11:e0152715–17
Helliwell KE, Wheeler GL, Smith AG (2013) Widespread decay of vitamin-related pathways: coincidence or consequence? Trends Genet 29:469–478
Horáková E, Changmai P, Vancová M et al (2017) The *Trypanosoma brucei* TbHrg protein is a heme transporter involved in the regulation of stage-specific morphological transitions. J Biol Chem 292:6998–7010
Huynh C, Yuan X, Miguel DC et al (2012) Heme Uptake by *Leishmania* amazonensis is mediated by the transmembrane protein LHR1. PLoS Pathog 8:e1002795
Ivens AC, Peacock CS, Worthey EA et al (2005) The genome of the kinetoplastid parasite, *Leishmania* major. Science 309:436–442
Jackson AP, Otto TD, Aslett M et al (2016) Kinetoplastid phylogenomics reveals the evolutionary innovations associated with the origins of parasitism. Curr Biol 26:161–172
Jagu E, Pomel S, Pethe S et al (2017) Polyamine-based analogs and conjugates as antikinetoplastid agents. Eur J Med Chem 139:982–1015
Janouskovec J, Keeling PJ (2016) Evolution: causality and the origin of parasitism. Curr Biol 26: R174–R177

Janouskovec J, Tikhonenkov DV, Burki F et al (2015) Factors mediating plastid dependency and the origins of parasitism in apicomplexans and their close relatives. Proc Natl Acad Sci 112:10200–10207

Ke H, Sigala PA, Miura K et al (2014) The heme biosynthesis pathway is essential for *Plasmodium falciparum* development in mosquito stage but not in blood stages. J Biol Chem 289:34827–34837

Klein CC, Alves JMP, Serrano MG et al (2013) Biosynthesis of vitamins and cofactors in bacterium-harbouring trypanosomatids depends on the symbiotic association as revealed by genomic analyses. PLoS One 8:e79786

Kooij TWA, Janse CJ, Waters AP (2006) *Plasmodium* post-genomics: better the bug you know? Nat Rev Microbiol 4:344–357

Kořený L, Lukes J, Oborník M (2010) Evolution of the haem synthetic pathway in kinetoplastid flagellates: an essential pathway that is not essential after all? Int J Parasitol 40:149–156

Kořený L, Oborník M, Lukes J (2013) Make it, take it, or leave it: heme metabolism of parasites. PLoS Pathog 9:e1003088

Kouni el MH (2003) Potential chemotherapeutic targets in the purine metabolism of parasites. Pharmacol Ther 99:283–309

Krauth-Siegel RL, Comini MA (2008) Redox control in trypanosomatids, parasitic protozoa with trypanothione-based thiol metabolism. Biochim Biophys Acta 1780:1236–1248

Kronenberger T, Schettert I, Wrenger C (2013) Targeting the vitamin biosynthesis pathways for the treatment of malaria. Future Med Chem 5:769–779

Krug EC, Marr JJ, Berens RL (1989) Purine metabolism in *Toxoplasma gondii*. J Biol Chem 264:10601–10607

Krugliak M, Zhang J, Ginsburg H (2002) Intraerythrocytic *Plasmodium falciparum* utilizes only a fraction of the amino acids derived from the digestion of host cell cytosol for the biosynthesis of its proteins. Mol Biochem Parasitol 119:249–256

Landfear SM, Ullman B, Carter NS, Sanchez MA (2004) Nucleoside and nucleobase transporters in parasitic protozoa. Eukaryotic Cell 3:245–254

Lim L, McFadden GI (2010) The evolution, metabolism and functions of the apicoplast. Philos Trans R Soc Lond B Biol Sci 365:749–763

Macedo JP, Currier RB, Wirdnam C et al (2017) Ornithine uptake and the modulation of drug sensitivity in *Trypanosoma brucei*. FASEB J 31(10):4649–4660. fj.201700311R–17

Martin JL, Yates PA, Soysa R et al (2014) Metabolic reprogramming during purine stress in the protozoan pathogen *Leishmania donovani*. PLoS Pathog 10:e1003938

Meireles P, Mendes AM, Aroeira RI et al (2017) Uptake and metabolism of arginine impact *Plasmodium* development in the liver. Sci Rep 7:849–812

Mesquita I, Varela P, Belinha A et al (2016) Exploring NAD(+) metabolism in host-pathogen interactions. Cell Mol Life Sci 73:1225–1236

Michels PAM, Avilán L (2011) The NAD(+) metabolism of *Leishmania*, notably the enzyme nicotinamidase involved in NAD(+) salvage, offers prospects for development of anti-parasite chemotherapy. Mol Microbiol 82:4–8

Miller-Fleming L, Olín-Sandoval V, Campbell K, Ralser M (2015) Remaining mysteries of molecular biology: the role of polyamines in the cell. J Mol Biol 427:3389–3406

Mukkada AJ, Meade JC, Glaser TA, Bonventre PF (1985) Enhanced metabolism of *Leishmania donovani* amastigotes at acid pH: an adaptation for intracellular growth. Science 229:1099–1101

Müller IB, Hyde JE (2013) Folate metabolism in human malaria parasites—75 years on. Mol Biochem Parasitol 188:63–77

Müller S, Kappes B (2007) Vitamin and cofactor biosynthesis pathways in *Plasmodium* and other apicomplexan parasites. Trends Parasitol 23:112–121

Müller IB, Hyde JE, Wrenger C (2010) Vitamin B metabolism in *Plasmodium falciparum* as a source of drug targets. Trends Parasitol 26:35–43

Nagaraj VA, Sundaram B, Varadarajan NM et al (2013) Malaria parasite-synthesized heme is essential in the mosquito and liver stages and complements host heme in the blood stages of infection. PLoS Pathog 9:e1003522–e1003513
Nare B, Hardy LW, Beverley SM (1997) The roles of pteridine reductase 1 and dihydrofolate reductase-thymidylate synthase in pteridine metabolism in the protozoan parasite *Leishmania* major. J Biol Chem 272:13883–13891
Nussbaum K, Honek J, Cadmus CMCVC, Efferth T (2010) Trypanosomatid parasites causing neglected diseases. Curr Med Chem 17:1594–1617
Nzila A, Okombo J, Molloy AM (2014) Impact of folate supplementation on the efficacy of sulfadoxine/pyrimethamine in preventing malaria in pregnancy: the potential of 5-methyl-tetrahydrofolate. J Antimicrob Chemother 69:323–330
O'Hara JK, Kerwin LJ, Cobbold SA et al (2014) Targeting NAD+ metabolism in the human malaria parasite *Plasmodium falciparum*. PLoS One 9:e94061
Opperdoes FR, Butenko A, Flegontov P et al (2016) Comparative metabolism of free-living *Bodo saltans* and parasitic trypanosomatids. J Eukaryot Microbiol 63:657–678
Ortiz D, Sanchez MA, Koch HP et al (2009) An acid-activated nucleobase transporter from *Leishmania* major. J Biol Chem 284:16164–16169
Ortiz D, Valdes R, Sanchez MA et al (2010) Purine restriction induces pronounced translational upregulation of the NT1 adenosine/pyrimidine nucleoside transporter in *Leishmania* major. Mol Microbiol 78:108–118
Ouellette M, Drummelsmith J, El-Fadili A et al (2002) Pterin transport and metabolism in *Leishmania* and related trypanosomatid parasites. Int J Parasitol 32:385–398
Peacock CS, Seeger K, Harris D et al (2007) Comparative genomic analysis of three *Leishmania* species that cause diverse human disease. Nat Genet 39:839–847
Rajendran E, Hapuarachchi SV, Miller CM et al (2017) Cationic amino acid transporters play key roles in the survival and transmission of apicomplexan parasites. Nat Commun 8:1–13
Roberts S, Ullman B (2017) Parasite polyamines as pharmaceutical targets. Curr Pharm Des 23:1–17
Salcedo-Sora JE, Ward SA (2013) The folate metabolic network of falciparum malaria. Mol Biochem Parasitol 188:51–62
Salcedo-Sora JE, Ochong E, Beveridge S et al (2011) The molecular basis of folate salvage in *Plasmodium falciparum*: characterization of two folate transporters. J Biol Chem 286:44659–44668
Sánchez-Lancheros DM, Ospina-Giraldo LF, Ramírez-Hernández MH et al (2016) Nicotinamide mononucleotide adenylyltransferase of *Trypanosoma cruzi* (TcNMNAT): a cytosol protein target for serine kinases. Mem Inst Oswaldo Cruz 111:670–675
Shameer S, Logan-Klumpler FJ, Vinson F et al (2015) TrypanoCyc: a community-led biochemical pathways database for *Trypanosoma brucei*. Nucleic Acids Res 43:D637–D644
Shanmugasundram A, Gonzalez-Galarza FF, Wastling JM et al (2013) Library of apicomplexan metabolic pathways: a manually curated database for metabolic pathways of apicomplexan parasites. Nucleic Acids Res 41:D706–D713
Sigala PA, Goldberg DE (2014) The peculiarities and paradoxes of *Plasmodium* heme metabolism. Annu Rev Microbiol 68:259–278
Sigala PA, Crowley JR, Henderson JP, Goldberg DE (2015) Deconvoluting heme biosynthesis to target blood-stage malaria parasites. eLife 4:50
Tjhin ET, Spry C, Sewell AL et al (2017) Mutations in the pantothenate kinase of *Plasmodium falciparum* confer diverse sensitivity profiles to antiplasmodial pantothenate analogues. bioRxiv:137182
Toh SQ, Glanfield A, Gobert GN, Jones MK (2010) Heme and blood-feeding parasites: friends or foes? Parasit Vectors 3:108
Vasudevan G, Carter NS, Drew ME et al (1998) Cloning of *Leishmania* nucleoside transporter genes by rescue of a transport-deficient mutant. Proc Natl Acad Sci 95:9873–9878
Vickers TJ, Beverley SM (2011) Folate metabolic pathways in *Leishmania*. Essays Biochem 51:63–80

Wainwright M, Maisch T, Nonell S et al (2017) Photoantimicrobials-are we afraid of the light? Lancet Infect Dis 17:e49–e55

Waller RF, Keeling PJ, Donald RGK et al (1998) Nuclear-encoded proteins target to the plastid in *Toxoplasma gondii* and *Plasmodium falciparum*. Proc Natl Acad Sci 95:12352–12357

Wilson RJ, Denny PW, Preiser PR et al (1996) Complete gene map of the plastid-like DNA of the malaria parasite *Plasmodium falciparum*. J Mol Biol 261:155–172

Woo YH, Ansari H, Otto TD et al (2015) Chromerid genomes reveal the evolutionary path from photosynthetic algae to obligate intracellular parasites. eLife 4:441

World Malaria Report 2016: Summary. World Health Organization, Geneva; 2017 (WHO/HTM/GMP/2017.4). Licence: CC BY-NC-SA 3.0 IGO.

Zhang C-J, Chia WN, Loh CCY et al (2015) Haem-activated promiscuous targeting of artemisinin in *Plasmodium falciparum*. Nat Commun 6:1–11

Zuegge J, Ralph S, Schmuker M et al (2001) Deciphering apicoplast targeting signals – feature extraction from nuclear-encoded precursors of *Plasmodium falciparum* apicoplast proteins. Gene 280:19–26

Zwerschke D, Karrie S, Jahn D, Jahn M (2014) *Leishmania* major possesses a unique HemG-type protoporphyrinogen IX oxidase. Biosci Rep 34:391–400

Chapter 10
Viral Manipulation of the Host Metabolic Network

Inês Mesquita and Jérôme Estaquier

Abstract Viruses are intracellular parasites that rely on host machinery to replicate and achieve a successful infection. Viruses have evolved to retain a broad range of strategies to manipulate host cell metabolism and metabolic resources, channeling them toward the production of virion components leading to viral production. Although several viruses share similar strategies for manipulating host cell metabolism, these processes depend on several factors, namely, the viral life cycle and the metabolic and energetic status of the infected cell. Based on this knowledge, the development of new therapeutic approaches that circumvent viral spread through the target of altered metabolic pathways is an opportunity to tackle the infection. However, finding effective broad-spectrum strategies that aim at restoring to homeostasis the metabolic alterations induced upon virus infection is still a Holy Grail quest for antiviral therapies. Here, we review the strategies by which viruses manipulate host metabolism for their own benefit, with a particular emphasis on carbohydrate, glutamine, and lipid metabolism.

Keywords Host cell metabolism · Host–pathogen interaction · Viral infection · Metabolic manipulation

I. Mesquita
Life and Health Sciences Research Institute (ICVS), School of Medicine, University of Minho, Braga, Portugal

ICVS/3B's-PT Government Associate Laboratory, Braga, Guimarães, Portugal

J. Estaquier (✉)
Centre de Recherche du CHU de Québec, Université Laval, Québec, Canada

CNRS FR 3636, Université Paris Descartes, Paris, France

R. Silvestre, E. Torrado (eds.), *Metabolic Interaction in Infection*, Experientia Supplementum 109, https://doi.org/10.1007/978-3-319-74932-7_10

10.1 Introduction

The recent advances in the field of metabolomics have allowed a deeper study on how pathogenic agents are capable of hijacking cellular metabolism to fulfill their own energetic and biosynthetic needs. Several viral organisms induce an increase in the uptake and catabolism of nutrients in the host cell to support the production of new progeny. Indeed, the processes of genome replication and packaging are highly energy consuming requiring high quantities of adenosine triphosphate (ATP). Early studies from the mid-1950s to the early 1960s have begun to demonstrate how viruses are able to affect host cell metabolism. While these initial reports focused on general measures on carbohydrate and non-carbohydrate metabolism (Levy and Baron 1957; Physicochemical and Essex 1974; Singh et al. 1974), more recent studies are unveiling the molecular mechanisms by which viruses are capable of manipulating host cell metabolism. In addition to this, several efforts have been made in trying to understand how the modulation of host metabolism impacts the clinical outcome of infection and how current therapies decrease viral proliferation through metabolism-related mechanisms.

10.2 Carbohydrate Metabolism

Glucose oxidation is a primary carbon and energy source in mammalian cells during homeostasis. In the presence of oxygen, glucose is metabolized into pyruvate. Upon translocation to the mitochondria, pyruvate is catabolized in the tricarboxylic acid (TCA) cycle and drives the electron transport chain to produce a high amount of ATP. However, in hypoxic conditions or upon an event demanding a rapid requirement of energy, glucose may be shunted toward lactic acid fermentation. The first studies demonstrating an upregulation of the glycolytic machinery upon viral infection came from early studies in the mid-1950s. Eagle and Habel showed that poliovirus replication in HeLa cells was blocked in culture media composed solely of a balanced salt solution. However, upon glucose supplementation, viral proliferation and load were restored to normal levels (Eagle and Habel 1956). Several subsequent studies confirmed these findings, and recent works have highlighted the molecular mechanism associated with glucose metabolism and the impact during infection progression (Fig. 10.1).

Human immunodeficiency virus (HIV) is an enveloped virus that actively targets and depletes CD4 T cells. Once activated, T cells suffer a strong metabolic rewiring that potently stimulates cellular metabolism to support the elevated energetic and biosynthetic demands to allow growth, proliferation, and the development of adequate effector functions (Buck et al. 2015). Following T-cell receptor (TCR) engagement, CD28 co-stimulation is required to allow maximal glucose uptake by promoting upregulation and cell surface trafficking of the glucose transporter (GLUT) 1 (Jacobs et al. 2008). In metabolically quiescent T cells, reverse

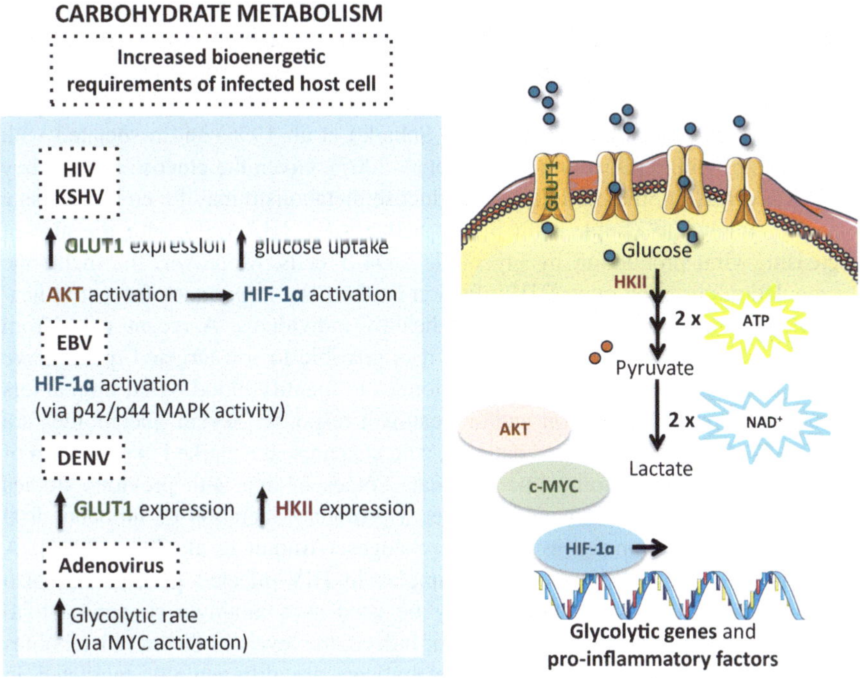

Fig. 10.1 Alterations on host carbohydrate metabolism upon virus infection. After infection, host cells upregulate the glycolytic machinery to fulfill the increased bioenergetic needs required to produce infectious virions. Increased glucose uptake, through GLUT1 expression, and fermentation into lactate originate a rapid production of ATP and NAD+. The activation of specific metabolic players, such as HIF-1α, culminates in the establishment of a highly glycolytic and pro-inflammatory phenotype. *ATP* adenosine triphosphate, *EBV* Epstein-Barr virus, *GLUT1* glucose transporter 1, *HIF-1α* hypoxia inducible factor 1 alpha, *HIV* human immunodeficiency virus, *HK* hexokinase, *KSHV* Kaposi's sarcoma-associated herpesvirus, *MAPK* mitogen-associated protein kinase, *NAD* nicotinamide adenosine dinucleotide

transcription is blocked, as well as integration of full-length HIV-1 DNA. Consequently, the HIV genome remains mostly in the linear and unintegrated form, which culminates in a dampening of viral replication (Bukrinsky et al. 1991; Chun et al. 1997). Upon T-cell activation, integration of viral extrachromosomal form occurs and infectious virions are produced (Bukrinsky et al. 1991; Stevenson et al. 1990). GLUT1 expression and concomitant glucose uptake appear to be essential for an adequate infection of CD4 T cells (Loisel-Meyer et al. 2012; Palmer et al. 2014). Th17 are characterized by the expression of C-C chemokine receptor (CCR) 6 and hypoxia-inducible factor 1 alpha (HIF-1α), a central regulator of glucose metabolism (Shi et al. 2011). It has been suggested that Th17 is a major reservoir for HIV (Gosselin et al. 2017; Stieh et al. 2016). Sun et al. (2015) showed that polarization toward a Th1/Th17 phenotype promotes HIV persistence during antiretroviral therapy (ART). Recent evidence showed that HIV-infected individuals treated with ART have decreased numbers of circulating GLUT1-expressing CD4 T cells (Palmer et al. 2014) that can be used as a measure of persistent immune activation associated with viral persistence. Additionally, groundbreaking studies using positron emission tomography imaging with intravenous 2-^{18}F-2-deoxyglucose have demonstrated that certain deep tissues (as the spleen and the mesenteric and ileocecal lymph nodes, which drain the inferior and superior portions of the intestine) are highly glycolytic sites (Santangelo et al. 2015; Scharko et al. 1996) and associated with major sites of viral replication (Cumont et al. 2007). Given the glucose dependency for HIV expression, the modulation of glucose metabolism may be envisaged as a synergic strategy to couple with highly active ART (HAART) with the aim of decreasing viral production by glycolytic CD4 T cells. Moreover, the metabolic dysregulation observed upon HIV infection is of crucial importance for a stratification of distinct groups of patients and healthy individuals. A recent work from Scarpelini et al. (2016) has shown that it is possible to use targeted quantitative tandem mass spectrometry (MS) metabolomics to identify blood-based biomarkers to track HIV natural evolution and/or treatment response. Several metabolites are statistically increased in infected patients, who also present a marked deregulation of acylcarnitine and sphingomyelin. This data comes in line with previous studies demonstrating that HIV infection induces a profound alteration of mitochondrial function, which culminates in systemic changes (Arnoult et al. 2003, 2004). A metabolomics analysis of oral wash samples in HIV-infected patients has also demonstrated that oral metabolites may be used as a noninvasive approach to determine immune status during infection. Indeed, the levels of distinct metabolites from amino acid and carbohydrate metabolisms are differentially regulated in HIV-infected individuals patients (Ghannoum et al. 2013). Cribbs et al. (2016) combined targeted sequencing of the 16S rRNA gene and high-resolution liquid chromatography LC-MS to study the relation between the respiratory microbiota and metabolic profiles in the bronchoalveolar lavage fluids of uninfected and HIV-infected outpatients. In this work, specific metabolites in the bronchoalveolar lavages of HIV-infected patients were correlated with bacterial organisms that have a role in pneumonia-related complications. LC-tandem MS (LC-MS/MS) technology has also been used to monitor specific metabolic alterations in HIV-infected cells (Hollenbaugh et al. 2011).

Macrophages are innate immune cells and professional phagocytes that are considered to be HIV reservoirs, in which HIV impacts on mitochondria function (Laforge et al. 2011). Similarly to CD4 T cells, HIV is capable of hijacking macrophage metabolism to allow its persistence. Using a stable-isotope labeling by amino acids in cell culture (SILAC) coupled with MS-based proteomics, Barrero et al. (2013) showed that the viral protein Vpr induces HIF-1α expression, as well as an upregulation of glycolytic enzymes, such as hexokinase (HK) and pyruvate kinase muscle type 2 (PKM2). The increased glycolytic flux allows for the establishment of a permissive niche, which is ideal for HIV replication and survival within macrophages. This is consistent with earlier reports indicating the requirement of active process for viral integration.

Several other viruses also hijack host glycolytic machinery for their own advantage. Kaposi's sarcoma-associated herpesvirus (KSHV), also known as human herpesvirus (HHV)-8, is a gammaherpesvirus, capable of infecting several cell types including monocytes, dendritic, oral epithelial, and endothelial cells. It was reported that KSHV shifts cell metabolism toward fermentation that is associated with the maintenance of cell latency. Through this process, KSHV limits host cell machinery to express a limited amount of viral genes, contributing to a successful infection (Delgado et al. 2010). It was recently demonstrated that a hyperglycemic environment is crucial for triggering KSHV reactivation, due to a sirtuin (SIRT) 1-induced epigenetic transactivation of KSHV lytic genes (Ye et al. 2016). Consistently, Gonnella et al. (2013) showed that KSHV infection of THP-1 cells resulted in increased levels of (i) AKT phosphorylation, a master regulator of glucose metabolism, and (ii) of GLUT1 membrane expression. The inhibition of the glycolytic flux with 2-deoxyglucose (2DG) rendered the infected cells more sensitive to the chemotherapeutic drug bortezomib, indicating a possible therapeutic approach by combining these inhibitors for KSHV-associated malignancies. Yogev and colleagues demonstrated that a KSHV viral-expressed miRNA cluster was capable of stabilizing HIF1α. The consequent activation of HIF1α culminates in a virus-induced metabolic reprogramming of the infected cell toward fermentation. Additionally, the expression of these miRNAs decreased mitochondrial mass in infected cells, which is associated with a survival advantage in hypoxic environments and with the persistence of cell latency (Yogev et al. 2014). HIF1α was also shown to be activated upon transfection of latent membrane protein 1 (LMP1), the principal Epstein-Barr virus (EBV) oncoprotein. The increase in HIF1α expression, controlled by p42/p44 mitogen-activated protein kinase (MAPK) activity, as well as hydrogen peroxide production, results in increased levels of vascular endothelial growth factor (VEGF) appearance in extracellular fluids, contributing to invasiveness and angiogenesis (Wakisaka et al. 2004). Similarly to KSHV, adenovirus infection also mirrors the Warburg effect, as the glycolytic rate of epithelial cells rises dramatically upon infection. This process was attributed to an increase in MYC activation, a proto-oncogene vastly associated with cancers. MYC expression was driven by a viral polypeptide, E4ORF1, which is essential for optimal adenovirus replication (Thai et al. 2014). Studies with dengue virus (DENV) also demonstrated the importance of glucose metabolism as a crucial hub during infection. Using an intracellular metabolic profiling approach, Fontaine et al. (2015) demonstrated that glucose metabolism was strikingly altered in DENV-infected cells, with several glycolytic metabolites being upregulated at early times postinfection and downregulated at later stages. The utilization of 2DG culminated in the blockage of DENV RNA production, demonstrating the importance of this metabolic pathway for a successful viral infection. Due to the growing interest in the recent outbreaks of fetal microcephaly induced by Zika virus infection, it was proposed that Zika modulates GLUT1 function preventing a normal glucose flux in endothelial cells of the placenta, impacting on normal fetal growth (Blonz 2016). However, additional studies are required to understand if the alterations of host cell metabolism upon Zika virus infection are similar to the ones observed with other virus from the *Flaviviruses* family, such as DENV.

As glucose uptake is important for increasing bioenergetics demand and cellular biomass, it is not surprising that viral modulation of glucose metabolism in host cell impacts viral proliferation. Furthermore, the observed increase in glucose uptake in infected cells may also underlie a requirement to feed other metabolic pathways, such as the glutamine and lipid pathways. These viral-induced changes in the glycolytic machinery are essential for the development of a successful infection and are likely to contribute to the observed pathogenesis among the distinct infections. However, the precise molecular mechanisms that drive this reprogramming remain to be addressed. Further studies will be essential to understand if the pharmacological targeting of specific nodes in the glycolytic pathway can be combined with approved antiviral therapies to achieve a better clinical outcome.

10.3 Glutaminolysis

Glutamine is a nonessential amino acid that plays an important role in several central metabolic pathways. During glutaminolysis, glutamine is converted into glutamate by glutaminase (GLS) and then into α-ketoglutarate (αKG) through glutamate dehydrogenase (GDH) activity. Upon entry in the mitochondria, αKG is capable of replenishing the TCA cycle, thus allowing its continued activity. This amino acid can also be used in purine synthesis, due to its ability to donate nitrogen atoms, and it originates reduced glutathione (GSH), which is vital for cellular detoxification mechanisms.

Despite evidence demonstrating glycolysis as a central mechanism associated with an effective viral replication, additional metabolic pathways have been reported to be hijacked by microbes, namely, glutaminolysis (Fig. 10.2). The pioneer work by Eagle and Habel (1956) showed that glutamine, alongside with glucose, is required for optimal replication of poliovirus in HeLa cells. In vitro supplementation of incomplete growth media with glutamine resulted in the restoration of virion production to normal levels. Furthermore, a similar increase of viral replication was observed when incomplete media were supplemented with glutamic acid, suggesting that glutaminolysis is driving this striking phenotype (Darnell and Eagle 1958).

Infectious spleen and kidney necrosis virus (ISKNV) infects more than 50 marine and freshwater fish and represents an important cause of economic loss. Fu and colleagues showed that ISKNV-infected Chinese perch brain (CPB) cells displayed altered glutamine metabolism, with increased expression of enzymes involved in glutamine metabolism (such as GLS and GDH), TCA cycle, and glutathione metabolism. Furthermore, although glutamine was not essential for maintaining CPB viability, it was required for the anaplerotic flux of carbon into the TCA cycle and to induce ATP and glutathione production, which appears to be essential for ISKNV replication (Fu et al. 2017).

Herpes simplex virus-1 (HSV-1) and human cytomegalovirus (HCMV) are two related viruses from the herpesviruses family. HSV-1 does not alter the host glycolytic rate but instead promotes pyruvate carboxylation and fuels de novo pyrimidine nucleotide synthesis. HCMV-infected fibroblasts display increased uptake of ammonia and glutamine, as well as higher expression of GLS and GDH (Vastag et al. 2011). The relevance of glutamine uptake is demonstrated by the depletion of

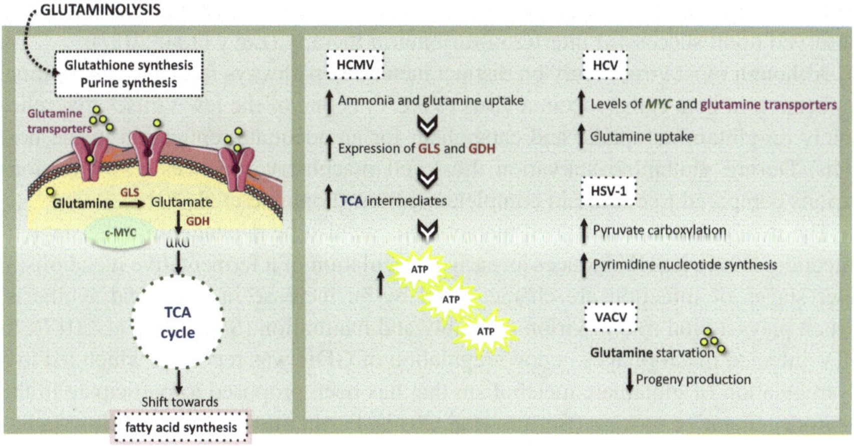

Fig. 10.2 Alterations on host glutaminolysis upon virus infection. Glutaminolysis fuels glutathione synthesis (crucial for cellular detoxification processes), purine synthesis, and the TCA cycle. Upon uptake, glutamine is converted into glutamate, through GLS, and then into α-KG, through GDH. During the TCA cycle, reducing agents and other intermediates are produced and used to provide high quantities of ATP. Citrate is produced in the TCA cycle, and it is an important precursor for fatty acid synthesis, which is crucial during viral infections. Upon infection, glutaminolysis may be enhanced, due to upregulation of glutamine transporters, GLS and GDH. Deletion of glutamine from culture media may decrease intracellular ATP levels and thus impair the production of infectious virions, which is reverted upon supplementation with other TCA intermediates, such as α-KG, oxaloacetate, and pyruvate. *α-KG* α-ketoglutarate, *ATP* adenosine triphosphate, *GDH* glutamate dehydrogenase, *GLS* glutaminase, *HCMV* human cytomegalovirus, *HCV* hepatitis C virus, *HSV-1* herpes simplex virus 1, *KSHV* Kaposi's sarcoma-associated herpesvirus, *TCA* tricarboxylic acid, *VACV* vaccinia virus

glutamine from cell culture that leads to a drop of ATP levels, leading to a diminished viral production. The restoration of the phenotype can be rescued by the exogenous administration of TCA intermediates, such as αKG, oxaloacetate, and pyruvate (Chambers et al. 2010). Furthermore, it has been shown that the marked increase in several metabolites involved in glycolysis during HCMV infection is correlated with an upregulation of phosphofructokinase-1, pyruvate kinase, and pyruvate dehydrogenase (Munger et al. 2006), which suggests that HCMV-infected cells require active glycolysis and nucleotide synthesis, while glutamine is catabolized to replenish the TCA cycle with the necessary intermediates. Similarly, Thai et al. (2015) showed that, besides its role in glucose metabolism, MYC activation is responsible for glutamine catabolism and consequent progeny generation. Recent evidence has indicated an important role for glutamine in the regulation of c-MYC-mediated transcription (Dang 2017), as well as the impact of this transcription factor in mitochondrial biogenesis (Le and Dang 2013). Given the importance of glutamine during carcinogenesis, it was reported that hepatitis C virus (HCV) infection of Huh7.5 cells led to an upregulation of the transcriptional levels of *MYC* and glutamine transporters, which sustained the increase in glutamine catabolism. Furthermore, the same glutaminolytic factors were found to be increased in liver biopsies of chronically infected HCV patients, while a substantial decrease was observed upon successful interferon-α/ribavirin therapy (Levy et al. 2017).

Although most viruses rely on distinct metabolic pathways for achieving optimal replication in host cells, vaccinia virus (VACV) is one of the few viruses that relies solely on glutamine uptake and catabolism for an adequate replication inside host cells. During glutamine starvation, host cell machinery produces less infectious virions compared to culture in complete medium (Fontaine et al. 2014).

The manipulation of host cell metabolism can also be dependent on the stage of infection. While KSHV induces an early upregulation of a fermentative metabolism, later stages of infection are characterized by an increase in fatty acid synthesis, which plays a vital role in virion assembly and maturation (Sanchez et al. 2017). In HIV-infected macrophages, a downregulation of GDH was reported, which led to a dysregulation of glutamate metabolism that has been proposed to participate in the neurodegenerative process (Barrero et al. 2013). Furthermore, metabolic flux studies showed that the viral protein, Vpr, induced increased glutamate release in infected macrophages, as well as accumulation of α-KG and glutamine in the media, which may contribute to neuroexcitotoxic mechanisms in the context of NeuroAIDS (HIV-associated dementia) (Datta et al. 2016).

This cluster of data suggests that targeting glutamine metabolism during microbe infections may offer a therapeutic advantage that leads to a dampening of viral load. Recent evidence in this field has also proposed that viruses induce host cell glutaminolysis when glucose is shunted toward lactate production. However, if this metabolic pathway serves only to produce energy or also to assist during virus replication remains to be elucidated.

10.4 Lipid Metabolism During Viral Infections

Fatty acids are macromolecules of vital importance for cell metabolism and proliferation, representing the main constituents of cellular membranes and a significant reservoir for cellular energy production. These molecules are constituted by a long hydrocarbon chain, attached to a terminal carboxylate group that can be further incorporated in more complex lipids or degraded by fatty acid β-oxidation (FAO). In intracellular compartments, fatty acids can be stored in the form of triglycerides in lipid droplets (LD), which are neutral lipid-rich organelles composed of a lipid ester core and a phospholipid monolayer. Upon lipase degradation and fatty acid entrance into the mitochondrial matrix, fatty acyl-CoA are broken into acetyl-coA during the oxidative process. On the contrary, fatty acid synthesis takes place in the cytosol, and it is highly dependent on nicotinamide adenine dinucleotide phosphate (NADPH) and acetyl-coA, mainly produced by oxidation of pyruvate. In this process, carboxylation of acetyl-coA by acetyl-coA carboxylase (ACC) originates malonyl-coA. The final steps of this anabolic process are catalyzed by the multienzyme complex fatty acid synthase (FASN). The role of lipids during virus infection has long been appreciated, but recent works have started to unveil the molecular and cellular mechanisms associated with alterations in lipid metabolism in infected cells. Lipids have several distinct functions in the life cycle of viruses, as they impact host bioenergetic status, endocytosis, host cell signaling, and virion assembly and budding.

10.4.1 The Role of Lipids during Virus–Host Interactions

Lipids represent a first barrier for infection, and similarly to what happens with other metabolite pathways, viruses have evolved to exploit existing host lipids in order to achieve a successful adhesion and entry. The membrane fluidity conferred by lipids is of crucial importance for fusion between viral envelope and cell membrane, and some lipids may also be used by the viruses to promote infection. Additionally, in some cases, the viral particles directly interact with plasma membrane and circulating lipids (Fig. 10.3).

The polyomavirus family, such as murine polyomavirus and simian virus 40 (SV40), use gangliosides as functional receptors (Campanero-Rhodes et al. 2007; Tsai et al. 2003). Qian and Tsai (2010) showed that while ganglioside GD1a functions as an entry receptor that binds the virus to cell surface, its glycoprotein moiety diverts polyomaviruses away from the endoplasmic reticulum, thus attenuating infection. The presence of glycosphingolipids in the viral surface of HIV has been suggested to promote viral entry in mature dendritic cells (Izquierdo-Useros et al. 2012). Richard and colleagues showed that phosphatidylserine (PS) receptors, including T-cell immunoglobulin mucin domain protein 1 (TIM1), bind to phosphatidylethanolamine (PE), which is present at the surface of virions of flaviviruses and filoviruses, such as West Nile, dengue virus (DENV), and Ebola viruses, allowing

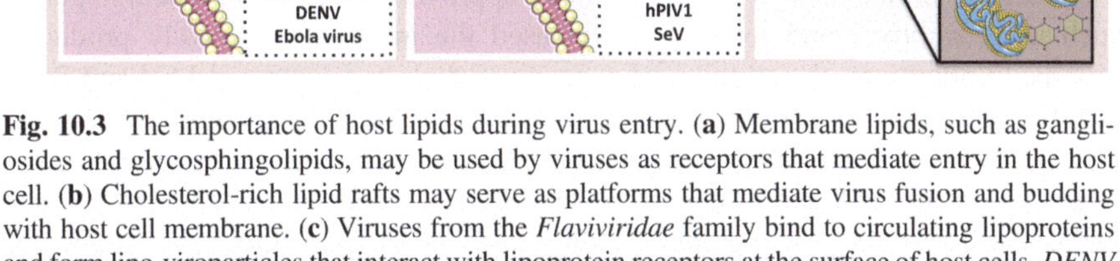

Fig. 10.3 The importance of host lipids during virus entry. (**a**) Membrane lipids, such as gangliosides and glycosphingolipids, may be used by viruses as receptors that mediate entry in the host cell. (**b**) Cholesterol-rich lipid rafts may serve as platforms that mediate virus fusion and budding with host cell membrane. (**c**) Viruses from the *Flaviviridae* family bind to circulating lipoproteins and form lipo-viroparticles that interact with lipoprotein receptors at the surface of host cells. *DENV* dengue virus, *HCV* hepatitis C virus, *HIV* human immunodeficiency virus, *hPIV1* human parainfluenza virus type 1, *IAV* influenza A virus, *RSV* respiratory syncytial virus, *SeV* Sendai virus, *SV40* simian virus 40

viral invasion (Richard et al. 2015). In an alternative fashion, viruses from the *Flaviviridae* family, such as HCV and DENV, have been shown to bind to circulating lipids. HCV interacts directly with low-density lipoprotein (LDL) receptors, through binding to LDL (Agnello et al. 1999; Molina et al. 2007; Targett-Adams et al. 2010), while the viral capsid proteins of DENV bind to very low-density lipoproteins, which suggests the formation of lipo-viroparticles (Faustino et al. 2014).

Distinct viral receptors are present in raft domain allowing viral entry by binding to cell surface (Heaton and Randall 2011). Several viruses require a specific lipid microenvironment to facilitate viral entry. HIV-1 budding depends on the presence of lipid microdomains, highly enriched in cholesterol, sphingolipids, and glycolipids, as cholesterol depletion with β-cyclodextrin reduced HIV-1 binding to primary cells (Liao et al. 2001). Furthermore, CD4 and CCR5 colocalize with lipid rafts, and this interaction is important for virus entry (Popik et al. 2002). Cui et al. (2012) showed that viral accessory protein Nef induces lipid raft formation in RAW 264.7 macrophages, through competition with ABCA1-dependent pathway of cholesterol efflux and mobilization of cholesterol to the plasma membrane. In a molecular basis, del Real et al. (2004) demonstrated that statins, commercially available drugs for hypercholesterolemia, inhibit HIV infection of human peripheral blood mononuclear cells (PBMCs) through reduction of geranylgeranylation and downregulation of Rho GTPase activity. Consistently, it was recently demonstrated, using model systems that mimic HIV envelopes and T-cell membranes, that this virus prefers the edges of cholesterol-rich lipid rafts for docking and entry (Yang et al. 2015).

Filoviruses such as Ebola and Marburg incorporate raft-associated molecules upon release from host cells, suggesting that viral egress occurs in lipid rafts (Bavari et al. 2002). The protein Niemann-Pick C1 (NPC1) involved in cholesterol transport appears to be essential for authentic filovirus infection given that NCP1-mutant cells are more resistant to Ebola and Marburg virus infection, yielding lower viral progeny (Carette et al. 2011). During HHV-6 infection, a significant decrease of viral infectivity was observed upon cholesterol depletion, which was abolished after exogenous cholesterol treatment (Huang et al. 2006).

Finally, the composition of the lipids that constitute viral envelopes is also critical for host cell infection. The envelopes of influenza A (IAV) and respiratory syncytial virus (RSV) contain cholesterol, which is essential for maintaining virion stability. Treatment of human airway cells with FDA-approved cholesterol reducing agents, such as gemfibrozil and lovastatin, increased the instability of newly produced virions. Furthermore, the addition of exogenous cholesterol restored IAV and RSV stability and infectivity (Bajimaya et al. 2017a). Likewise, the same group showed that depletion of cellular cholesterol reduced the association of human parainfluenza virus type 1 (hPIV1) and Sendai virus (SeV) matrix and nucleocapsids with raft membranes, resulting in impaired budding and egress from host cells (Bajimaya et al. 2017b).

Viruses orchestrate host cell metabolism from the initial steps of infection. In particular, the strategies by which lipid-mediated signaling is altered have been explored in the context of distinct viral infections (Heaton and Randall 2011; Mazzon and Mercer 2014). A deeper understanding of the molecular and cellular events that dictate these phenomena is essential for developing new strategies that aim at preventing viral infection.

10.4.2 Modulation of Host Lipid Metabolism upon Infection

Several viruses are known to reprogram host lipid metabolism, by inducing and/or requiring fatty acid synthesis at some point of their maturation in infectious virions (Fig. 10.4). Lipid synthesis and remodeling during infections has proven to be important not only during viral replication but also to favor the egress of newly synthetized virions (Heaton and Randall 2011; Konan and Sanchez-Felipe 2014; Lorizate and Krausslich 2011; Mazzon and Mercer 2014). Additionally, the storage of free fatty acids, triglycerides, and sterol esters in LD may serve as energy reservoirs but also as assembly platforms for new virions. LD formation may be indicative of a higher anabolic activity (increased synthesis of fatty acids) and may prepare the host cell for a rapid membrane generation.

HSV-1 buds at nuclear membranes, thus acquiring an envelope mostly composed of phospholipids. It was suggested that this virus is capable of inducing host cell-derived phospholipid synthesis to promote the maintenance of membrane integrity during cellular expansion and the supply of constituents for envelopment and to provide membranes for the generation of transport vacuoles (Sutter et al. 2012). In accordance with this, FASN has been shown to be crucial for maximal replication of several viruses. Kulkarni et al. (2017) have reported that HIV-1 infection upregulates FASN protein levels in HeLa cells, which correlate with elevated intracellular levels of fatty acids (palmitic, oleic, and staric acids). Inhibition of FASN with a new thiophenopyrimidine molecule (Fasnall) or its knockdown resulted in a reduction of HIV-1 viral production, without interfering with Gag production. This suggests that FASN and de novo lipogenesis are essential during late-stage HIV-1 replication. Similarly to HIV infection, inhibition of FASN by another synthetic compound, TVB-3166, results in decreased production of RSV progeny from infected human lung epithelial cells. Additionally, this compound was shown to reduce at a higher extent the viral titer in the lung of RSV A-infected Balb/c mice when comparing with the current available drug ribavirin (Ohol et al. 2015). Greseth and Traktman (2014) demonstrated that palmitate is required for VACV replication. This is accompanied by an increase in oxygen consumption rate and ATP levels, suggesting a shift toward mitochondrial FAO. Upon etomoxir treatment, which irreversibly inhibits carnitine palmitoyltransferase I (CPTI), ATP levels in infected cells drop with a consequent decrease in viral yield (Greseth and Traktman 2014).

De novo lipogenesis has also been demonstrated to be essential during DENV infection of mosquito cells, as the irreversible FANS inhibitor C75 disrupts viral replication. It has been reported through high-resolution MS that DENV alters the cellular lipid repertoire culminating in significant alterations in membrane

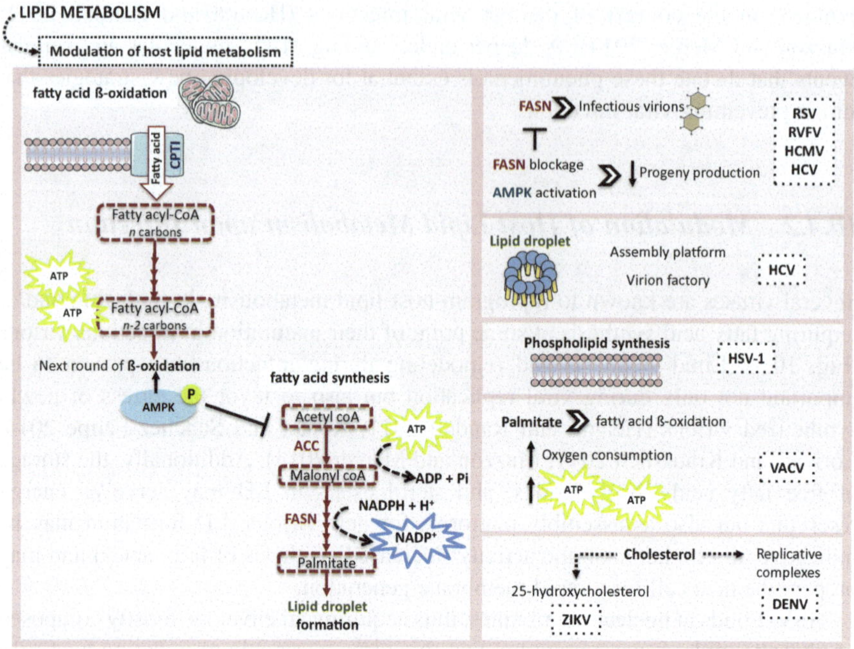

Fig. 10.4 Modulation of host lipid metabolism upon virus infection. Fatty acid β-oxidation of medium to long-chain fatty acids takes place in the mitochondrial matrix, following CPTI transport. Each round of β-oxidation removes two carbon units from the original fatty acyl-coA and produces two molecules of ATP. Fatty acid synthesis starts with carboxylation of acetyl-coA in malonyl-coA, and synthesis culminates with the activity of the multicomplex enzyme FASN. Newly formed fatty acids (such as palmitate) may be further incorporated in lipid droplets. AMPK activation through phosphorylation activates catabolic programs, such as β-oxidation, while blocking anabolic ones, such as fatty acid synthesis (through inactivation of ACC through phosphorylation). CPTI inhibitors (such as etomoxir), AMPK activators (such as A769662), ACC inhibitors (such as TOFA), FASN inhibitors (such as C75), and inhibitors of lipid droplet formation (such as triacsin C) are commercially available and used to modulate lipid metabolism. Upon virus infection, a metabolic reprogramming of host cell lipids occurs, as lipids are crucial during virion replication and assembly. *ACC* acetyl-coA carboxylase, *ADP* adenosine diphosphate, *AMPK* 5' AMP-activated protein kinase, *ATP* adenosine triphosphate, *CPTI* carnitine palmitoyltransferase I, *DENV* dengue virus, *FASN* fatty acid synthase, *HCMV* human cytomegalovirus, *HCV* hepatitis C virus, *HSV-1* herpes simplex virus 1, *NADPH* nicotinamide adenine dinucleotide phosphate, *RSV* respiratory syncytial virus, *RVFV* Rift Valley fever virus, *VACV* vaccinia virus, *ZIKV* Zika virus

permeability and curvature (Perera et al. 2012). Furthermore, by exploiting the action of endosomal lipid contents, DENV promotes fusion with late endosomes in the adequate time frame to persist within the host (Zaitseva et al. 2010). However, recent evidence points toward a requirement of active FAO for DENV replication, which is driven by an increase in lipophagic flux through activation of the 5' AMP-activated protein kinase/mammalian target of rapamycin (AMPK/mTOR) axis (Jordan and Randall 2017).

Host-derived lipogenesis was suggested to be essential for HCMV, as inhibition of FASN or ACC with C75 or TOFA, respectively, arrested the production of infectious virions. By using a flux measurement approach based on LC-MS/MS, Munger et al. (2008) showed that HCMV-infected fibroblasts showed a significant upregulation of the flux from the TCA cycle into the fatty acid biosynthesis. Consistently, Koyuncu et al. (2013) performed a siRNA screen that predicted that multiple long-chain acyl-CoA synthetases and fatty acid elongases are essential during HCMV infection. Pharmacological inhibition of the abovementioned enzymes by triacsin C, Endo 1 k, (S)-1y, and compound 37 reduced the yield of HCMV from infected fibroblasts. Furthermore, MS-based fatty acid analysis with ^{13}C-labeled glucose revealed that the majority of biosynthetic precursors subjected to elongation are preexisting C18 fatty acids.

AMPK is an important inhibitor of lipid synthesis. Upon activation, AMPK inhibits ACC through phosphorylation and drives cellular catabolic programs that result in decreased levels of intracellular lipids. This kinase is thought to be a restrictive factor during Rift Valley Fever Virus (RVFV) infection, given that, in the absence of AMPK, fibroblasts were incapable of controlling infection, which resulted in increased plaque formation and size. Similar results were obtained for other arboviruses, such as the Flavivirus Kunjin virus (KUNV), the Togavirus Sindbis virus (SINV), and the Rhabdovirus Vesicular stomatitis virus (VSV). Additionally, FASN inhibition mediated by cerulenin, C75, or AMPK activation with A769662 decreased the percentage of infected cells, which was bypassed through exogenous supplementation with palmitate. These findings suggest that de novo fatty acid synthesis is crucial for RVFV replication in host cells and AMPK may represent an antiviral mechanism to restrict lipid accumulation (Moser et al. 2012). In contrast Soto-Acosta et al. (2017) showed that upon infection with DENV, Huh7 cells display decreased levels of AMPK phosphorylation and increased activity of 3-hydroxy-3-methyl-glutaryl-CoA reductase (HMG-CoA reductase), which is a rate-limiting enzyme in cholesterol synthesis. Furthermore, they also reported that DENV infection augments cholesterol accumulation at replicative complexes, which is essential for maintaining the architecture of these structures. Using a Zika virus-induced mouse microcephaly model, Li et al. 2017) showed that cholesterol-25-hydroxylase (CH25H) is induced in response to viral infection. Supplementation, both in vivo and in vitro, with its enzymatic product, 25-hydroxycholesterol (25HC), reduced viremia and conferred protection against infection in mice and rhesus macaques.

Other viruses are known to associate with lipid droplets and clinical states of obesity. Human adenovirus 36 is a serotype frequently linked to obesity (Greenway 2006; Whigham et al. 2006). Wang et al. (2010) demonstrated that Adenovirus 36-infected muscle cells display a decrease in fatty acid oxidation and a concomitant increase in lipogenesis, as shown through fat staining with Oil Red O. Furthermore, the accumulation of lipid droplets was mediated by activation of several lipogenic mediators, such as Cidec/fat-specific protein 27, ACC, and sterol regulatory element-binding protein (SREBP) 1c and 2. HCV is one of the most studied viruses associated with host cell lipid modulation (Filipe and McLauchlan 2015). Thus, a growing body of evidence has suggested that alterations in lipid metabolism are associated with disease progression during HCV infection. It is already known that (a) HCV infecting virions circulate in the bloodstream bound to lipoproteins (Thomssen et al. 1992); (b) lipids are essential during HCV life cycle, namely in the assembly stage (Miyanari et al. 2007); and (c) triglyceride accumulation is a common feature of HCV-induced hepatic steatosis (Yoon and Hu 2006). Consistent with the observed decrease of mitochondrial FAO, HCV infection results in increased levels of SREBP cleavage and phosphorylation (Waris et al. 2007), as well as reduced lipid oxidation and efflux (Negro and Sanyal 2009). Additionally, HCV increases the transcriptional levels of FASN, and de novo lipogenesis appears to be essential for viral replication and assembly, as treatment of infected cells with C75 disrupts these processes and results in lower viremias (Yang et al. 2008). HCV-infected Huh-7 cells display increased levels of genes associated with cholesterol biosynthesis, which was consistent with higher intracellular levels of cholesterol and free fatty acids (Woodhouse et al. 2010). Consistently, HCV core protein is known to associate with cytoplasmic LD. The sequential recruitment of nonstructural proteins to the vicinity of LD promotes virion assembly, suggesting that LD may act as assembly platforms for virus production (Miyanari et al. 2007). Similarly to what happens during HCV infection, rotavirus replicates inside cytoplasmic virion factories, namely viroplasms, which colocalize with LD (Crawford and Desselberger 2016; Gaunt et al. 2013). This colocalization appears to be essential for the production of progeny virus, as chemical compounds that interfere with LD biogenesis, such as triacsin C, decrease viral RNA replication as well as viral-induced cytopathicity (Cheung et al. 2010). Recently, Amako et al. (2015) showed that HCV-infected Huh7.5 cells have attenuated levels of FAO as measured by using [^{3}H] palmitate. Regarding hepatitis B virus (HBV) infection, it has been shown that viral X protein (HBx) induces the expression of liver X Receptor (LXR) and consequently activates SREBP-1c, FASN, and peroxisome proliferator-activated receptor (PPAR) culminating in lipid accumulation (Kim et al. 2007, 2008). Additionally, samples from patients with HBV-associated hepatocellular carcinoma display higher levels of LXRβ transcripts, as well as increased transcription of downstream targets FASN and SREBP-1c (Na et al. 2009).

These data suggest that several core lipid metabolic pathways may be hijacked during viral infections to provide the necessary biosynthetic precursors and free energy that will allow maximal viral replication, assembly, and egress. Although the manipulation of host lipids upon viral infections has been vastly studied, it is still unknown if the modulation of lipid metabolism during infection is advantageous to prevent viral multiplication.

10.5 Concluding Remarks

Striking advances have been made in the past few years in the field of metabolomics with the development of highly sensitive and specific techniques that allow a deeper study of the intricate link between metabolic pathways. These sophisticated techniques offered a deeper understanding of the metabolic perturbations of a host cell upon infection and may eventually be advantageous for the development of new prognostic and diagnostic techniques in the field of infectious diseases. A synthesis of the most relevant virus-induced metabolic reprogramming of host cells is presented in Table 10.1.

The study of viral reprogramming of cell metabolism poses a challenge, not only because viruses can hijack several interconnected metabolic pathways to increase their maximal replication rate but also because each virus can induce specific phenotypes. By doing so, viruses oblige host cells to provide the necessary building blocks to ensure their persistence. However, several questions remain, as to which signaling pathways are activated or inhibited during infection as well as the impact of modulating such pathways in the final infection outcome. Additionally, the metabolic dysregulation originated upon viral infection may promote the appearance of other clinical complications, as in the case of lipogenesis and HBV-associated hepatocellular carcinoma. Understanding the targeted metabolic pathways and the main players involved is a crucial step to elucidate how we can envisage specific pathways as antiviral targets. Another vital consideration is that, depending on their life cycle, viruses induce different metabolic and energetic requirements in host cells, which adds a layer of complexity to the system. This results in the increased need for a deeper and individualized study on host cell metabolism during specific viral infections that may unveil therapeutic targets that could be used as possible adjuvants for antiviral therapy.

Further studies regarding host cell bioenergetics, mitochondrial mechanisms associated with apoptosis and antiviral signaling, as well as immune responses associated with viral infections are required to understand if modulating metabolic networks represents a valid attempt to design new therapeutic approaches to fight viral infections.

Funding Statement This work was supported by the Northern Portugal Regional Operational Programme (NORTE 2020), under the Portugal 2020 Partnership Agreement, through the European Regional Development Fund (FEDER) (NORTE-01-0145-FEDER-000013) and the Fundação para a Ciência e Tecnologia (FCT) contract SFRH/BD/120127/2016 to IM. This work was supported by grants to JE from the Agence Nationale de Recherches sur le Sida et les Hépatites Virales (ANRS) and from The Canadian HIV Cure Enterprise Team Grant HIG-13305 from the Canadian Institutes of Health Research (CIHR) in partnership with CANFAR and IAS. JE acknowledges the support of the Canada Research Chair program.

Table 10.1 Virus-induced metabolic reprogramming of host cells

Virus		Molecular and cellular mechanisms	Metabolic reprogramming of host cell	Additional observations	References
Adenoviridae family					
Adenovirus	dsDNA	Viral E4ORF1 induces MYC activation	Increase in glycolytic flux	Increased lactate production	Thai et al. (2014)
		Activation of cidec/fat-specific protein 27, ACC, SREBP-1c/2	Increase in lipogenesis	Adenovirus type 36 associates with obesity	Wang et al. (2010)
			Decrease in FAO	Reduction of AMPK activity and mitochondrial mass	
Bunyaviridae family					
RVFV	ssRNA negative-strand	AMPK activation	Lower lipid accumulation	Restriction of RVFV replication	Moser et al. (2012)
			Decrease in FAS		
Flaviviridae family					
DENV	ssRNA positive-strand	Upregulation of glycolytic enzymes	Increase in glycolytic flux	2DG (glycolysis inhibitor) blocks DENV replication	Fontaine et al. (2015)
		Altered lipid composition in infected cells	Active FAS	C75 (FASN inhibitor) blocks DENV replication	Perera et al. (2012)
		Increase in lipophagic flux via AMPK/mTOR	Active FAO	Lipophagy: mobilization of lipids from LD to mitochondria	Jordan and Randall (2017)
		Decrease in AMPK activation	Increased cholesterol accumulation at replicative centers	Metformin (AMPK activator) has an antiviral effect	Soto-Acosta et al. (2017)
		Increased activity of HMG-CoA Reductase			

HCV	ssRNA positive-strand	Increase MYC and glutamine transporters levels	Glutamine catabolism	*MYC* and *GLS* silencing reduces HCV replication	Levy et al. (2017)
		Increased levels of SREBP cleavage and activation	Active FAS	Enhancement of lipogenic factors	Waris et al. (2007)
		Reduced lipid oxidation and efflux	–	Increased hepatic and serum lipid content in infected patients	Negro and Sanyal (2009)
		Downregulation of mitochondrial tri-functional protein levels	Reduced levels of FAO	Lack of responsiveness to antiviral IFN-α	Amako et al. (2015)
		Increased levels of FASN	Active FAS	C75, orlistat, and FASN knockdown block HCV replication	Yang et al. (2008)
		Increased levels of genes from cholesterol synthesis	Accumulation of cholesterol and fatty acids	Increased production of ROS in the course of infection	Woodhouse et al. (2010)
		Core proteins recruit HCV replication complexes to LD	Increased LD formation	LD as assembly platforms	Miyanari et al. (2007)
Zika virus	ssRNA positive-strand	Induction of CH25H	–	Supplementation with 25HC protects against infection	Li et al. (2017)
Hepadnaviridae family					
HBV	dsDNA-RT	HBx induces LXRα	Lipid accumulation	Activation of SREBP-1c, FASN, and PPAR	Kim et al. (2007, 2008)
		High levels of LXRβ and SREBP-1c transcripts			Na et al. (2009)
Herpesviridae family					
EBV	dsDNA	LMP1 increases HIF-1α synthesis	–	Increase invasiveness and angiogenesis	Wakisaka et al. (2004)

(continued)

Table 10.1 (continued)

Virus		Molecular and cellular mechanisms	Metabolic reprogramming of host cell	Additional observations	References
KSHV	dsDNA	SIRT1-induced epigenetic transactivation of lytic genes	Shift to glucose fermentation	KSHV reactivation in hyperglycemic conditions	Ye et al. (2016)
		AKT activation and GLUT1 trafficking		2DG (glycolysis inhibitor) increases sensitivity to therapy	Gonnella et al. (2013)
		Viral-expressed miRNA induces HIF-1α		Decreased mitochondrial mass	Yogev et al. (2014)
		Increase in GLUT3 and glucose uptake		Cell latency	Delgado et al. (2010)
		–	Active FAS	TOFA (ACC inhibitor) decreases KSHV replication	Sanchez et al. (2017)
		–	Glycolysis is required for maximal replication	Oxamate (LDH inhibitor) decreases KSHV replication	
		–	Glutaminolysis is required for maximal replication	BPTES (GLS inhibitor) decreases KSHV replication	
HCMV	dsDNA	Increase in TCA intermediates	Uptake of ammonia and glutamine	Depends on FAS to produce membranes for progeny generation	Vastag et al. (2011)
		High expression of GLS and GDH			
		High expression of GLS and GDH	Depletion of glutamine leads to drop in ATP	Low viral production (restored by supplementation with TCA intermediates)	Chambers et al. (2010)
		Accumulation of citrate (from TCA cycle)	Active FAS	C75 and TOFA block HCMV replication	Munger et al. (2008)
		Increased levels of VLCFA-associated enzymes		Triacsin C, Endo-1 k, (S)-1y, and compound 37 block HCMV replication	Koyuncu et al. (2013)

HSV	dsDNA	Increase in pyruvate carboxylation	Increased de novo pyrimidine nucleotide synthesis	Knockdown of pyruvate carboxylase and aspartate transaminase 2 inhibits HSV replication	Vastag et al. (2011)
		Increased consumption of aspartate via aspartate transaminase 2			
Iridoviridae family					
ISKNV	dsDNA	Increase in GLS and GDH levels	Glutamine replenishes TCA cycle	Production of ATP and glutathione	Fu et al. (2017)
Picornaviridae family					
Poliovirus	ssRNA positive-strand	–	Uptake of glutamine and glucose	Depletion of these nutrients reduces viral load	Eagle and Habel (1956)
		–			Darnell and Eagle (1958)
Pneumoviridae family					
RSV	ssRNA negative-strand ssRNA negative-strand	–	Active FAS	TVB-3166 (FASN inhibitor) reduces RSV infectivity in vitro	Ohol et al. (2015)
				TVB-3166 reduces viral load in vivo: prophylactic and therapeutic effects	
Retroviridae family					
HIV	ssRNA-RT	Increase in GLUT1 expression and glucose uptake	Active glycolysis	Highly glycolytic effector CD4 T cells produce infectious virions	Loisel-Meyer et al. (2012)
		Polarization toward glycolytic Th1/Th17		HIV persistence during ART	Sun et al. (2015)
		Vpr induces HIF-1α expression	Increase in glycolytic flux	Upregulation of PPP increases nucleotide pool for sustained HIV replication	Barrero et al. (2013)
		Vpr induces HK and PKM2			
		Vpr downregulates GDH	Dysregulation of glutamate metabolism	Increased glutamate release	

(continued)

Table 10.1 (continued)

Virus		Molecular and cellular mechanisms	Metabolic reprogramming of host cell	Additional observations	References
		Upregulation of FASN levels	Active FAS at late stages of infection	Accumulation of fatty acids	Kulkarni et al. (2017)
				FANS inhibition blocks HIV replication	

ACC acetyl-coA carboxylase, *AMP* adenosine monophosphate, *AMPK* 5' AMP-activated protein kinase, *ART* anti-retroviral therapy, *ATP* adenosine triphosphate, *CH25H* 25-Hydroxycholesterol receptor, *CPTI* carnitine palmitoyltransferase I, *DENV* dengue virus, *ds* double stranded, *EBV* Epstein-Barr virus, *FAO* fatty acid oxidation, *FASN* Fatty acid synthase, *GDH* glutamate dehydrogenase, *GLUT* glucose transporter, *GLS* glutaminase, *HBV* hepatitis B virus, *HBx* HBV X protein, *HCMV* human cytomegalovirus, *HCV* hepatitis C virus, *HIF1α* hypoxia-inducible factor 1 alpha, *HIV* human immunodeficiency virus, *HK* hexokinase, *HMG* 3-hydroxy-3-methyl-glutaryl, *HSV* Herpes simplex virus, *IFN* interferon, *ISKNV* infectious spleen and kidney necrosis virus, *KSHV* Kaposi's sarcoma-associated herpesvirus, *LD* lipid droplets, *LMP* latent membrane protein, *LXR* liver X receptor, *MAPK* mitogen-associated protein kinase, *mTOR* mammalian target of rapamycin, *PPAR* peroxisome proliferator-activated receptor, *PPP* pentose phosphate pathway, *ROS* reactive oxygen species, *RT* reverse transcribed, *RVFV* Rift Valley fever virus, *SIRT* sirtuin, *SREBP* sterol regulatory element-binding protein, *ss* single stranded, *TCA* tricarboxylic acid, *Th* T helper, *VACV* vaccinia virus, *VEGF* vascular endothelial growth factor

References

Agnello V, Abel G, Elfahal M, Knight GB, Zhang QX (1999) Hepatitis C virus and other flaviviridae viruses enter cells via low density lipoprotein receptor. Proc Natl Acad Sci USA 96:12766–12771

Amako Y, Munakata T, Kohara M, Siddiqui A, Peers C, Harris M (2015) Hepatitis C virus attenuates mitochondrial lipid beta-oxidation by downregulating mitochondrial trifunctional-protein expression. J Virol 89:4092–4101

Arnoult D, Petit F, Lelièvre JD, Estaquier J (2003) Mitochondria in HIV-1-induced apoptosis. Biochem Biophys Res Commun 304:561–574

Arnoult D, Viollet L, Petit F, Lelièvre JD, Estaquier J (2004) HIV-1 triggers mitochondrion death. Mitochondrion 4:255–269

Bajimaya S, Frankl T, Hayashi T, Takimoto T (2017a) Cholesterol is required for stability and infectivity of influenza a and respiratory syncytial viruses. Virology 510:234–241

Bajimaya S, Hayashi T, Frankl T, Bryk P, Ward B, Takimoto T (2017b) Cholesterol reducing agents inhibit assembly of type I parainfluenza viruses. Virology 501:127–135

Barrero CA, Datta PK, Sen S, Deshmane S, Amini S, Khalili K, Merali S (2013) HIV-1 Vpr modulates macrophage metabolic pathways: a SILAC-based quantitative analysis. PLoS One 8: e68376

Bavari S, Bosio CM, Wiegand E, Ruthel G, Will AB, Geisbert TW, Hevey M, Schmaljohn C, Schmaljohn A, Aman MJ (2002) Lipid raft microdomains: a gateway for compartmentalized trafficking of Ebola and Marburg viruses. J Exp Med 195:593–602

Blonz ER (2016) Zika virus and GLUT1. Lancet Infect Dis 16:642

Buck MD, O'Sullivan D, Pearce EL (2015) T cell metabolism drives immunity. J Exp Med 212:1345–1360

Bukrinsky MI, Stanwick TL, Dempsey MP, Stevenson M (1991) Quiescent T lymphocytes as an inducible virus reservoir in HIV-1 infection. Science 254:423–427

Campanero-Rhodes MA, Smith A, Chai W, Sonnino S, Mauri L, Childs RA, Zhang Y, Ewers H, Helenius A, Imberty A et al (2007) N-glycolyl GM1 ganglioside as a receptor for simian virus 40. J Virol 81:12846–12858

Carette JE, Raaben M, Wong AC, Herbert AS, Obernosterer G, Mulherkar N, Kuehne AI, Kranzusch PJ, Griffin AM, Ruthel G et al (2011) Ebola virus entry requires the cholesterol transporter Niemann–pick C1. Nature 477:340–343

Chambers JW, Maguire TG, Alwine JC (2010) Glutamine metabolism is essential for human cytomegalovirus infection. J Virol 84:1867–1873

Cheung W, Gill M, Esposito A, Kaminski CF, Courousse N, Chwetzoff S, Trugnan G, Keshavan N, Lever A, Desselberger U (2010) Rotaviruses associate with cellular lipid droplet components to replicate in viroplasms, and compounds disrupting or blocking lipid droplets inhibit viroplasm formation and viral replication. J Virol 84:6782–6798

Chun T-W, Carruth L, Finzi D, Shen X, DiGiuseppe JA, Taylor H, Hermankova M, Chadwick K, Margolick J, Quinn TC et al (1997) Quantification of latent tissue reservoirs and total body viral load in HIV-1 infection. Nature 387:183–188

Crawford SE, Desselberger U (2016) Lipid droplets form complexes with viroplasms and are crucial for rotavirus replication. Curr Opin Virol 19:11–15

Cribbs SK, Uppal K, Li S, Jones DP, Huang L, Tipton L, Fitch A, Greenblatt RM, Kingsley L, Guidot DM et al (2016) Correlation of the lung microbiota with metabolic profiles in bronchoalveolar lavage fluid in HIV infection. Microbiome 4:3

Cui HL, Grant A, Mukhamedova N, Pushkarsky T, Jennelle L, Dubrovsky L, Gaus K, Fitzgerald ML, Sviridov D, Bukrinsky M (2012) HIV-1 Nef mobilizes lipid rafts in macrophages through a pathway that competes with ABCA1-dependent cholesterol efflux. J Lipid Res 53:696–708

Cumont MC, Monceaux V, Viollet L, Lay S, Parker R, Hurtrel B, Estaquier J (2007) TGF-beta in intestinal lymphoid organs contributes to the death of armed effector CD8 T cells and is associated with the absence of virus containment in rhesus macaques infected with the simian immunodeficiency virus. Cell Death Differ 14:1747–1758

Dang CV (2017) c-MYC mRNA tail tale about glutamine control of transcription. EMBO J 36:1806–1808

Darnell JE Jr, Eagle H (1958) Glucose and glutamine in poliovirus production by HeLa cells. Virology 6:556–566

Datta PK, Deshmane S, Khalili K, Merali S, Gordon JC, Fecchio C, Barrero CA (2016) Glutamate metabolism in HIV-1 infected macrophages: role of HIV-1 Vpr. Cell Cycle 15:2288–2298

del Real G, Jiménez-Baranda S, Mira E, Lacalle RA, Lucas P, Gómez-Moutón C, Alegret M, Peña JM, Rodríguez-Zapata M, Alvarez-Mon M et al (2004) Statins inhibit HIV-1 infection by down-regulating rho activity. J Exp Med 200:541–547

Delgado T, Carroll PA, Punjabi AS, Margineantu D, Hockenbery DM, Lagunoff M (2010) Induction of the Warburg effect by Kaposi's sarcoma herpesvirus is required for the maintenance of latently infected endothelial cells. Proc Natl Acad Sci USA 107:10696–10701

Eagle H, Habel K (1956) The nutritional requirements for the propagation of poliomyelitis virus by the HeLa cell. J Exp Med 104:271–287

Faustino AF, Carvalho FA, Martins IC, Castanho MARB, Mohana-Borges R, Almeida FCL, Da Poian AT, Santos NC (2014) Dengue virus capsid protein interacts specifically with very low-density lipoproteins. Nanomed: Nanotechnol, Biol Med 10:247–255

Filipe A, McLauchlan J (2015) Hepatitis C virus and lipid droplets: finding a niche. Trends Mol Med 21:34–42

Fontaine KA, Camarda R, Lagunoff M (2014) Vaccinia virus requires glutamine but not glucose for efficient replication. J Virol 88:4366–4374

Fontaine KA, Sanchez EL, Camarda R, Lagunoff M (2015) Dengue virus induces and requires glycolysis for optimal replication. J Virol 89:2358–2366

Fu X, Hu X, Li N, Zheng F, Dong X, Duan J, Lin Q, Tu J, Zhao L, Huang Z et al (2017) Glutamine and glutaminolysis are required for efficient replication of infectious spleen and kidney necrosis virus in Chinese perch brain cells. Oncotarget 8:2400–2412

Gaunt ER, Zhang Q, Cheung W, Wakelam MJ, Lever AM, Desselberger U (2013) Lipidome analysis of rotavirus-infected cells confirms the close interaction of lipid droplets with viroplasms. J Gen Virol 94:1576–1586

Ghannoum MA, Mukherjee PK, Jurevic RJ, Retuerto M, Brown RE, Sikaroodi M, Webster-Cyriaque J, Gillevet PM (2013) Metabolomics reveals differential levels of oral metabolites in HIV-infected patients: toward novel diagnostic targets. OMICS 17:5–15

Gonnella R, Santarelli R, Farina A, Granato M, D'Orazi G, Faggioni A, Cirone M (2013) Kaposi sarcoma associated herpesvirus (KSHV) induces AKT hyperphosphorylation, bortezomib-resistance and GLUT-1 plasma membrane exposure in THP-1 monocytic cell line. J Exp Clin Cancer Res 32:79

Gosselin A, Wiche Salinas TR, Planas D, Wacleche VS, Zhang Y, Fromentin R, Chomont N, Cohen ÉA, Shacklett B, Mehraj V et al (2017) HIV persists in CCR6+CD4+ T cells from colon and blood during antiretroviral therapy. AIDS 31:35–48

Greenway F (2006) Virus-induced obesity. Am J Physiol Regul Integr Comp Physiol 290:R188–R189

Greseth MD, Traktman P (2014) De novo fatty acid biosynthesis contributes significantly to establishment of a bioenergetically favorable environment for vaccinia virus infection. PLoS Pathog 10:e1004021

Heaton NS, Randall G (2011) Multifaceted roles for lipids in viral infection. Trends Microbiol 19:368–375

Hollenbaugh JA, Munger J, Kim B (2011) Metabolite profiles of human immunodeficiency virus infected CD4+ T cells and macrophages using LC-MS/MS analysis. Virology 415:153–159

Huang H, Li Y, Sadaoka T, Tang H, Yamamoto T, Yamanishi K, Mori Y (2006) Human herpesvirus 6 envelope cholesterol is required for virus entry. J Gen Virol 87:277–285

Izquierdo-Useros N, Lorizate M, Contreras FX, Rodriguez-Plata MT, Glass B, Erkizia I, Prado JG, Casas J, Fabrias G, Krausslich HG et al (2012) Sialyllactose in viral membrane gangliosides is a novel molecular recognition pattern for mature dendritic cell capture of HIV-1. PLoS Biol 10: e1001315

Jacobs SR, Herman CE, MacIver NJ, Wofford JA, Wieman HL, Hammen JJ, Rathmell JC (2008) Glucose uptake is limiting in T cell activation and requires CD28-mediated Akt-dependent and independent pathways. J Immunol 180:4476–4486

Jordan TX, Randall G (2017) Dengue virus activates the AMP kinase-mTOR axis to stimulate a proviral lipophagy. J Virol 91:e02020–e02016

Kim KH, Shin HJ, Kim K, Choi HM, Rhee SH, Moon HB, Kim HH, Yang US, Yu DY, Cheong J (2007) Hepatitis B virus X protein induces hepatic steatosis via transcriptional activation of SREBP1 and PPARgamma. Gastroenterology 132:1955–1967

Kim K, Kim KH, Kim HH, Cheong J (2008) Hepatitis B virus X protein induces lipogenic transcription factor SREBP1 and fatty acid synthase through the activation of nuclear receptor LXRalpha. Biochem J 416:219–230

Konan KV, Sanchez-Felipe L (2014) Lipids and RNA virus replication. Curr Opin Virol 9:45–52

Koyuncu E, Purdy JG, Rabinowitz JD, Shenk T (2013) Saturated very long chain fatty acids are required for the production of infectious human cytomegalovirus progeny. PLoS Pathog 9: e1003333

Kulkarni MM, Ratcliff AN, Bhat M, Alwarawrah Y, Hughes P, Arcos J, Loiselle D, Torrelles JB, Funderburg NT, Haystead TA et al (2017) Cellular fatty acid synthase is required for late stages of HIV-1 replication. Retrovirology 14:45

Laforge M, Campillo-Gimenez L, Monceaux V, Cumont MC, Hurtrel B, Corbeil J, Zaunders J, Elbim C, Estaquier J (2011) HIV/SIV infection primes monocytes and dendritic cells for apoptosis. PLoS Pathog 7:e1002087

Le A, Dang CV (2013) Studying Myc's role in metabolism regulation. Methods Mol Biol 1012:213–219

Levy HB, Baron S (1957) The effect of animal viruses on host cell metabolism ii. Effect of poliomyelitis virus on glycolysis and uptake of glycine by monkey kidney tissue cultures. J Infect Dis 100:109–118

Levy PL, Duponchel S, Eischeid H, Molle J, Michelet M, Diserens G, Vermathen M, Vermathen P, Dufour J-F, Dienes H-P et al (2017) Hepatitis C virus infection triggers a tumor-like glutamine metabolism. Hepatology 65:789–803

Li C, Deng YQ, Wang S, Ma F, Aliyari R, Huang XY, Zhang NN, Watanabe M, Dong HL, Liu P et al (2017) 25-Hydroxycholesterol protects host against Zika virus infection and its associated microcephaly in a mouse model. Immunity 46:446–456

Liao Z, Cimakasky LM, Hampton R, Nguyen DH, Hildreth JE (2001) Lipid rafts and HIV pathogenesis: host membrane cholesterol is required for infection by HIV type 1. AIDS Res Hum Retrovir 17:1009–1019

Loisel-Meyer S, Swainson L, Craveiro M, Oburoglu L, Mongellaz C, Costa C, Martinez M, Cosset F-L, Battini J-L, Herzenberg LA et al (2012) Glut1-mediated glucose transport regulates HIV infection. Proc Natl Acad Sci 109:2549–2554

Lorizate M, Krausslich HG (2011) Role of lipids in virus replication. Cold Spring Harb Perspect Biol 3:a004820

Mazzon M, Mercer J (2014) Lipid interactions during virus entry and infection. Cell Microbiol 16:1493–1502

Miyanari Y, Atsuzawa K, Usuda N, Watashi K, Hishiki T, Zayas M, Bartenschlager R, Wakita T, Hijikata M, Shimotohno K (2007) The lipid droplet is an important organelle for hepatitis C virus production. Nat Cell Biol 9:1089–1097

Molina S, Castet V, Fournier-Wirth C, Pichard-Garcia L, Avner R, Harats D, Roitelman J, Barbaras R, Graber P, Ghersa P et al (2007) The low-density lipoprotein receptor plays a role in the infection of primary human hepatocytes by hepatitis C virus. J Hepatol 46:411–419

Moser TS, Schieffer D, Cherry S (2012) AMP-activated kinase restricts Rift Valley fever virus infection by inhibiting fatty acid synthesis. PLoS Pathog 8:e1002661

Munger J, Bajad SU, Coller HA, Shenk T, Rabinowitz JD (2006) Dynamics of the cellular metabolome during human cytomegalovirus infection. PLoS Pathog 2:e132

Munger J, Bennett BD, Parikh A, Feng XJ, McArdle J, Rabitz HA, Shenk T, Rabinowitz JD (2008) Systems-level metabolic flux profiling identifies fatty acid synthesis as a target for antiviral therapy. Nat Biotechnol 26:1179–1186

Na TY, Shin YK, Roh KJ, Kang SA, Hong I, Oh SJ, Seong JK, Park CK, Choi YL, Lee MO (2009) Liver X receptor mediates hepatitis B virus X protein-induced lipogenesis in hepatitis B virus-associated hepatocellular carcinoma. Hepatology 49:1122–1131

Negro F, Sanyal AJ (2009) Hepatitis C virus, steatosis and lipid abnormalities: clinical and pathogenic data. Liver Int 29(Suppl 2):26–37

Ohol YM, Wang Z, Kemble G, Duke G (2015) Direct inhibition of cellular fatty acid synthase impairs replication of respiratory syncytial virus and other respiratory viruses. PLoS One 10: e0144648

Palmer CS, Ostrowski M, Gouillou M, Tsai L, Yu D, Zhou J, Henstridge DC, Maisa A, Hearps AC, Lewin SR et al (2014) Increased glucose metabolic activity is associated with CD4+ T-cell activation and depletion during chronic HIV infection. AIDS 28:297–309

Perera R, Riley C, Isaac G, Hopf-Jannasch AS, Moore RJ, Weitz KW, Pasa-Tolic L, Metz TO, Adamec J, Kuhn RJ (2012) Dengue virus infection perturbs lipid homeostasis in infected mosquito cells. PLoS Pathog 8:e1002584

Physicochemical D, Essex M (1974) Glycolysis during early infection of feline and human cells with feline leukemia virus. Infect Immun 9:824–827

Popik W, Alce TM, Au W-C (2002) Human immunodeficiency virus type 1 uses lipid raft-colocalized CD4 and chemokine receptors for productive entry into CD4+ T cells. J Virol 76:4709–4722

Qian M, Tsai B (2010) Lipids and proteins act in opposing manners to regulate polyomavirus infection. J Virol 84:9840–9852

Richard AS, Zhang A, Park S-J, Farzan M, Zong M, Choe H (2015) Virion-associated phosphatidylethanolamine promotes TIM1-mediated infection by Ebola, dengue, and West Nile viruses. Proc Natl Acad Sci 112:14682–14687

Sanchez EL, Pulliam TH, Dimaio TA, Thalhofer AB, Delgado T, Lagunoff M (2017) Glycolysis, glutaminolysis, and fatty acid synthesis are required for distinct stages of Kaposi's sarcoma-associated herpesvirus lytic replication. J Virol 91(10): e02237–16

Santangelo PJ, Rogers KA, Zurla C, Blanchard EL, Gumber S, Strait K, Connor-Stroud F, Schuster DM, Amancha PK, Hong JJ et al (2015) Whole-body immunoPET reveals active SIV dynamics in viremic and antiretroviral therapy–treated macaques. Nat Methods 12:427–432

Scarpelini B, Zanoni M, Sucupira MCA, Truong HHM, Janini LMR, Segurado IDC, Diaz RS (2016) Plasma metabolomics biosignature according to HIV stage of infection, pace of disease progression, viremia level and immunological response to treatment. PLoS One 11:e0161920

Scharko AM, Perlman SB, Hinds P, Hanson JM, Uno H, Pauza CD (1996) Whole body positron emission tomography imaging of simian immunodeficiency virus-infected rhesus macaques. Proc Natl Acad Sci USA 93:6425–6430

Shi LZ, Wang R, Huang G, Vogel P, Neale G, Green DR, Chi H (2011) HIF1α–dependent glycolytic pathway orchestrates a metabolic checkpoint for the differentiation of T_H 17 and T_{reg} cells. J Exp Med 208:1367–1376

Singh VN, Singh M, August JT, Horecker BL (1974) Alterations in glucose metabolism in chick-embryo cells transformed by Rous sarcoma virus: intracellular levels of glycolytic intermediates. Proc Natl Acad Sci USA 71:4129–4132

Soto-Acosta R, Bautista-Carbajal P, Cervantes-Salazar M, Angel-Ambrocio AH, del Angel RM (2017) DENV up-regulates the HMG-CoA reductase activity through the impairment of AMPK phosphorylation: a potential antiviral target. PLoS Pathog 13:e1006257

Stevenson M, Stanwick TL, Dempsey MP, Lamonica CA (1990) HIV-1 replication is controlled at the level of T cell activation and proviral integration. EMBO J 9:1551–1560

Stieh DJ, Matias E, Xu H, Fought AJ, Blanchard JL, Marx PA, Veazey RS, Hope TJ (2016) Th17 cells are preferentially infected very early after vaginal transmission of SIV in macaques. Cell Host Microbe 19:529–540

Sun H, Kim D, Li X, Kiselinova M, Ouyang Z, Vandekerckhove L, Shang H (2015) Th1/17 polarization of CD4 T cells supports HIV-1 persistence during antiretroviral therapy. J Virol 89:11284–11293

Sutter E, de Oliveira AP, Tobler K, Schraner EM, Sonda S, Kaech A, Lucas MS, Ackermann M, Wild P (2012) Herpes simplex virus 1 induces de novo phospholipid synthesis. Virology 429:124–135

Targett-Adams P, Boulant S, Douglas MW, McLauchlan J (2010) Lipid metabolism and HCV infection. Virus 2:1195–1217

Thai M, Graham NA, Braas D, Nehil M, Komisopoulou E, Kurdistani SK, McCormick F, Graeber TG, Christofk HR (2014) Adenovirus E4ORF1-induced MYC activation promotes host cell anabolic glucose metabolism and virus replication. Cell Metab 19:694–701

Thai M, Thaker SK, Feng J, Du Y, Hu H, Ting Wu T, Graeber TG, Braas D, Christofk HR (2015) MYC-induced reprogramming of glutamine catabolism supports optimal virus replication. Nat Commun 6:8873

Thomssen R, Bonk S, Propfe C, Heermann KH, Kochel HG, Uy A (1992) Association of hepatitis C virus in human sera with beta-lipoprotein. Med Microbiol Immunol 181:293–300

Tsai B, Gilbert JM, Stehle T, Lencer W, Benjamin TL, Rapoport TA (2003) Gangliosides are receptors for murine polyoma virus and SV40. EMBO J 22:4346–4355

Vastag L, Koyuncu E, Grady SL, Shenk TE, Rabinowitz JD (2011) Divergent effects of human cytomegalovirus and herpes simplex virus-1 on cellular metabolism. PLoS Pathog 7:e1002124

Wakisaka N, Kondo S, Yoshizaki T, Murono S, Furukawa M, Joseph S, Pagano JS (2004) Epstein-Barr virus latent membrane protein 1 induces synthesis of hypoxia-inducible factor 1 α. Mol Cell Biol 24:5223–5234

Wang ZQ, Yu Y, Zhang XH, Floyd EZ, Cefalu WT (2010) Human adenovirus 36 decreases fatty acid oxidation and increases de novo lipogenesis in primary cultured human skeletal muscle cells by promoting Cidec/FSP27 expression. Int J Obes 34:1355–1364

Waris G, Felmlee DJ, Negro F, Siddiqui A (2007) Hepatitis C virus induces proteolytic cleavage of sterol regulatory element binding proteins and stimulates their phosphorylation via oxidative stress. J Virol 81:8122–8130

Whigham LD, Israel BA, Atkinson RL (2006) Adipogenic potential of multiple human adenoviruses in vivo and in vitro in animals. Am J Physiol Regul Integr Comp Physiol 290:R190–R194

Woodhouse SD, Narayan R, Latham S, Lee S, Antrobus R, Gangadharan B, Luo S, Schroth GP, Klenerman P, Zitzmann N (2010) Transcriptome sequencing, microarray, and proteomic analyses reveal cellular and metabolic impact of hepatitis C virus infection in vitro. Hepatology 52:443–453

Yang W, Hood BL, Chadwick SL, Liu S, Watkins SC, Luo G, Conrads TP, Wang T (2008) Fatty acid synthase is up-regulated during hepatitis C virus infection and regulates hepatitis C virus entry and production. Hepatology 48:1396–1403

Yang ST, Kiessling V, Simmons JA, White JM, Tamm LK (2015) HIV gp41-mediated membrane fusion occurs at edges of cholesterol-rich lipid domains. Nat Chem Biol 11:424–431

Ye F, Zeng Y, Sha J, Jones T, Kuhne K, Wood C, Gao S-J (2016) High glucose induces reactivation of latent Kaposi's sarcoma-associated Herpesvirus. J Virol 90(21):9654–9663. (JVI.01049-16)

Yogev O, Lagos D, Enver T, Boshoff C (2014) Kaposi's sarcoma herpesvirus microRNAs induce metabolic transformation of infected cells. PLoS Pathog 10:e1004400

Yoon EJ, Hu KQ (2006) Hepatitis C virus (HCV) infection and hepatic steatosis. Int J Med Sci 3:53–56

Zaitseva E, Yang ST, Melikov K, Pourmal S, Chernomordik LV (2010) Dengue virus ensures its fusion in late endosomes using compartment-specific lipids. PLoS Pathog 6:e1001131

Chapter 11
Metabolic Regulation of Innate Immunity to Fungal Infection

Cláudia S. Rodrigues, Cláudia F. Campos, Cristina Cunha, and Agostinho Carvalho

Abstract In recent years, the renewed interest in immune cell metabolism has driven the emergence of a research field aimed at studying the role of metabolic processes during innate and adaptive immune responses. Although the specific requirements of myeloid cells after the canonical lipopolysaccharide/TLR4 stimulation have been extensively addressed, recent evidence suggests that this model may not represent a universally accurate metabolic blueprint. Instead, different microbial stimuli, pathogens, or tissue microenvironments trigger specific and complex metabolic rewiring of myeloid cells. This chapter aims to provide an overview of the metabolic heterogeneity in activated myeloid cells during fungal disease. Directions for future research in dissecting the uniqueness of metabolic signatures during fungal infection are suggested to ultimately provide new tailored diagnostic and therapeutic interventions.

Keywords Fungal disease · Immunometabolism · β-glucan · Trained immunity · Epigenetics

11.1 Introduction

An unparalleled number of fungal and fungal-like diseases have been documented over the past two decades (Fisher et al. 2012). While only a few fungal species can actually cause disease in healthy individuals, opportunistic and otherwise commensal fungi can trigger life-threatening infections in immunosuppressed individuals, ranging from superficial mucocutaneous disease to invasive deep-seated infections (Brown et al. 2012). Although true mortality rates remain unknown because of the lack of adequate epidemiological data, the incidence of invasive fungal diseases is rising as the result of advances in medical care and the increasing numbers among

C. S. Rodrigues · C. F. Campos · C. Cunha · A. Carvalho (✉)
Life and Health Sciences Research Institute (ICVS), School of Medicine, University of Minho, Braga, Portugal

ICVS/3B's - PT Government Associate Laboratory, Braga/Guimarães, Portugal
e-mail: agostinhocarvalho@med.uminho.pt

R. Silvestre, E. Torrado (eds.), *Metabolic Interaction in Infection*, Experientia Supplementum 109, https://doi.org/10.1007/978-3-319-74932-7_11

immunocompromised populations due to acquired (e.g., AIDS) or treatment-induced immunodeficiency in patients receiving hematopoietic stem-cell or solid organ transplants or undergoing anticancer therapy (Pfaller and Diekema 2010). There is, therefore, an unmet clinical need for more rapid and better diagnostics, safer and more effective antifungal drugs, and preventive vaccines.

Many species of fungi are responsible for invasive infections, although more than 90% of all reported fungal-related deaths result from infection with species belonging to one of four genera: *Cryptococcus*, *Candida*, *Aspergillus*, and *Pneumocystis*. Clinical and basic research during the last decade has brought exciting new insights into the pathogenesis of fungal diseases and revealed important molecular and cellular players in host–fungus interactions and host defense. Among these mechanisms, the metabolic profile of immune cells has recently been appreciated as a crucial factor regulated in response to infection. One critical example is provided within the sepsis setting, in which circulating leukocytes undergo an initial hyperinflammatory state fueled by a shift from oxidative phosphorylation (OxPhos) to aerobic glycolysis, which is then followed by a state of immune tolerance or paralysis associated with defective antimicrobial defenses and characterized by a broad defect at the level of both glycolysis and oxidative metabolism (Cheng et al. 2016). In turn, given the immunologically dynamic context, fungi have developed their own strategies to sustain extreme stress conditions, the adaptation to which also relies on profound metabolic changes (Grahl et al. 2011).

In this chapter, we will summarize and review recent advances in our understanding of the different cellular metabolic pathways activated in myeloid cells in response to fungal infection. In addition, we highlight possible approaches targeting host metabolic pathways, whose exploitation may represent a promising novel therapeutic strategy for fungal disease.

11.2 Cellular Immunometabolism During Fungal Infection

Recent discoveries on the role of metabolic pathways in immune cell function have pushed forward the expanding area of immunometabolism to the forefront for many immunologists (O'Neill and Pearce 2016). An infectious challenge triggers a set of signals and alterations in the microenvironment that prompt the immune cells to accommodate their own cellular metabolism. The adaptation of specific metabolic outputs extends beyond the normal cellular energy requirements and is crucial to the regulation of cell differentiation and proliferation, and activation of effector responses (e.g., phagocytosis, production of proinflammatory mediators, antigen presentation, and cell surface receptor expression) (Artyomov et al. 2016; Gleeson and Sheedy 2016).

Over the last years, knowledge of macrophage metabolism has vastly expanded (Galvan-Pena and O'Neill 2014). Macrophage activation results in the differential expression of functional phenotypes characterized by markedly distinct metabolic demands, classically activated (M1) macrophages, and alternatively activated

(M2) macrophages (Olenchock et al. 2017). In general terms, M1 macrophages display an increased glucose metabolism and lactate production via glycolysis, whereas M2 macrophages rely mostly on OxPhos and β-oxidation by fatty acid uptake. In the next sections, we will discuss in more detail the metabolic signatures and their consequences to the function of myeloid cells in response to fungal infection.

11.2.1 Glycolysis

The glycolytic metabolism plays a pivotal role in numerous cellular processes, including the activation and regulation of effector functions of immune cells (Pearce and Pearce 2013) (Fig. 11.1). Following cellular activation, glucose is taken up and converted into pyruvate by a series of enzymatic reactions. Pyruvate can either be converted into lactate by a process called lactic fermentation via anaerobic glycolysis (yielding a total of two molecules of adenosine triphosphate, ATP) or, alternatively, can be converted into acetyl-coenzyme A (acetyl-coA) and enter the tricarboxylic acid (TCA) cycle (yielding 36 ATP by OxPhos). Although glycolysis does not generate large amounts of ATP, its rapid activation allows a faster production comparatively with OxPhos. Therefore, the shift of the basal metabolism from OxPhos to glycolysis and enhanced lactate production allows the rapid acquisition

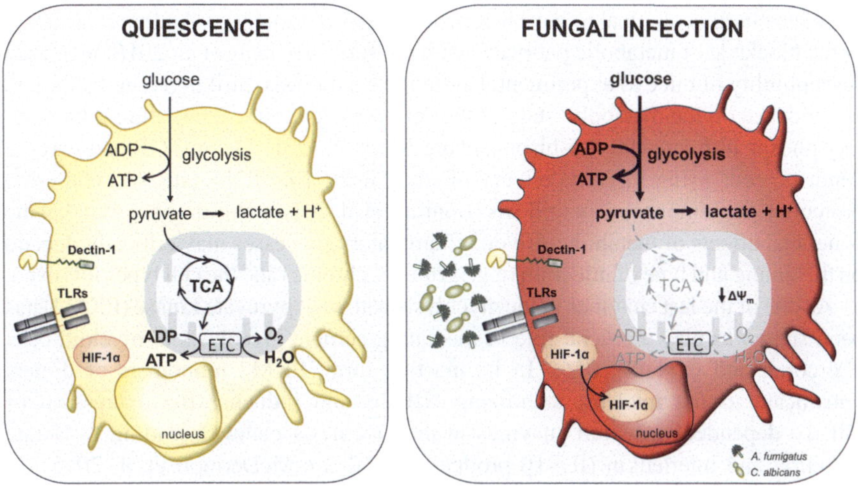

Fig. 11.1 Metabolic reprogramming during fungal infection. Under basal conditions, cellular metabolism of monocytes/macrophages is typically dependent on OxPhos. Upon fungal infection or stimulation, the metabolic profile shifts toward aerobic glycolysis, a state that is associated with the enhancement of antimicrobial effector functions. *ADP* adenosine diphosphate, *ATP* adenosine triphosphate, *TCA* tricarboxylic acid, *ETC* electron transport chain, *HIF-1α*, hypoxia-inducible factor 1α, *TLR* Toll-like receptor

of a highly inflammatory phenotype characterized by the activation of immune effector functions, including production of cytokines and reactive oxygen (ROS) and nitrogen (NO) species (Olenchock et al. 2017). As such, this faster ATP generation is widely exploited by short-lived immune cells and might be favorable in situations that demand energy in a short time frame, such as acute inflammation or infection.

The glycolytic reprogramming of immune cells is typically supported by the activation of the transcription factor hypoxia-inducible factor 1α (HIF-1α) (Corcoran and O'Neill 2016). Upon nuclear translocation, HIF-1α promotes the upregulation of glucose transporters such as glucose transporter 1 (GLUT1) and several glycolytic enzymes. Conversely, HIF-1α inhibits pyruvate oxidation and impairs the activity of pyruvate dehydrogenase (PDH), therefore favoring a skew toward the induction of glycolysis. It is not surprising, therefore, that a pathway involving dectin-1 signaling and activation of serine/threonine-specific protein kinase (Akt)/mammalian target of rapamycin (mTOR)/HIF-1α has been shown to be essentially required for the metabolic reprogramming of monocytes in response to β-glucan from the fungal cell wall (Cheng et al. 2014) (Fig. 11.2). Likewise, HIF-1α was found to be essentially required for the modulation of the metabolic activity and cytokine release by human dendritic cells upon infection with *Aspergillus fumigatus* (Fliesser et al. 2015). Because *Candida albicans* was also demonstrated to induce metabolic reprogramming in human natural killer (NK) cells (Hellwig et al. 2016), the metabolic shift from OxPhos toward aerobic glycolysis appears to be a common process triggered across different immune cells in response to fungal stimuli. More recently, the central role for glucose metabolism in immunity to *C. albicans* was also demonstrated in vivo, since the pharmacological impairment of glycolysis (Dominguez-Andres et al. 2017) or the blockade of metabolic pathways with metformin (Cheng et al. 2016) increased susceptibility of mice to experimental systemic candidiasis. Strikingly, and although no evidence on the metabolic state of the cells was determined, targeting inflammatory phosphatidylinositol-4,5-bisphosphate 3-kinase (PI3K)/Akt/mTOR pathways in dendritic cells using in vivo delivery of small interfering RNA (siRNA) conferred protection to pulmonary aspergillosis (Bonifazi et al. 2010). This may suggest that the beneficial effects of metabolic-driven inflammatory processes may critically depend on the timing and type of infection (e.g., acute vs. chronic) and the cell types involved.

As one of the last enzymes in the glycolysis pathway, pyruvate kinase (PKM2) has been shown to play a crucial role in mediating aerobic glycolysis in myeloid cells (Corcoran and O'Neill 2016). In its inactive form, PKM2 monomers or dimers translocate to the nucleus, stabilizing HIF-1α and inducing the expression of HIF-1α-dependent essential enzymes in the glycolysis pathway leading to lactate secretion and interleukin (IL)-1β production (Palsson-McDermott et al. 2015). In contrast, when PKM2 is enzymatically activated (tetrameric state), it resides in the cytosol, where it is not able to regulate HIF-1α, promoting instead a more anti-inflammatory phenotype. In this regard, PKM2 inhibition results in decreased serum lactate and the levels of the high mobility group box 1 (HMGB1) alarmin, showing promising therapeutic results by protecting mice from lethal endotoxemia

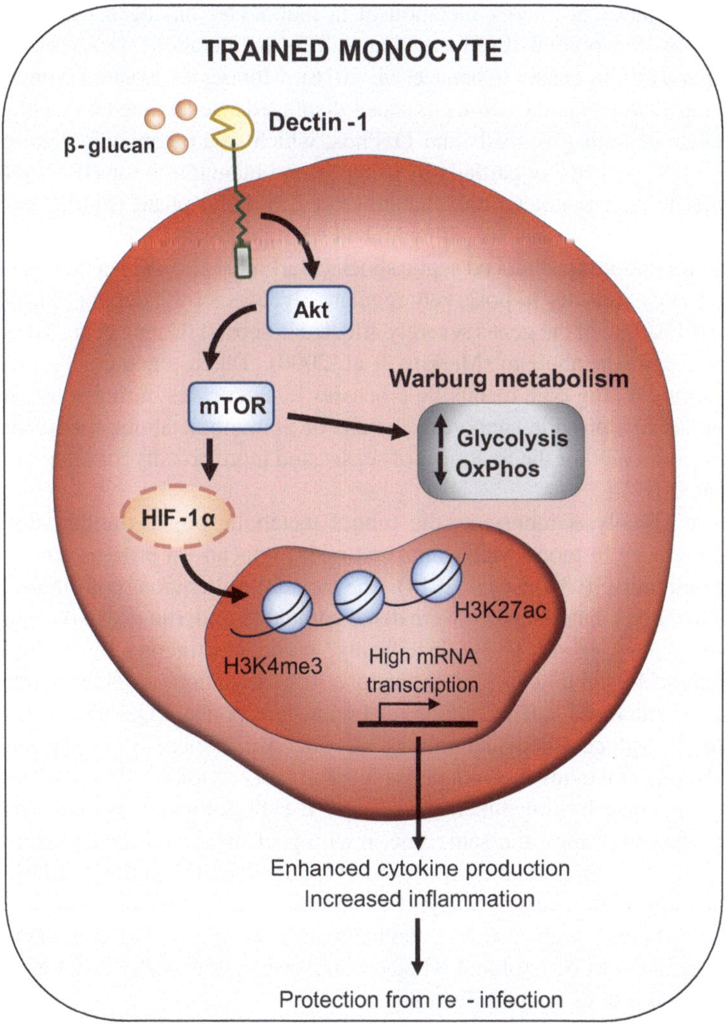

Fig. 11.2 Trained immunity integrates immunological and metabolic pathways. Upon stimulation of dectin-1 by β-glucan, trained cells undergo a shift toward aerobic glycolysis and several epigenetic alterations at the level of histone methylation and acetylation. This functional reprogramming promotes enhanced activity of the immune cell upon secondary stimulation and protection against reinfection from related or unrelated pathogens. *Akt* serine/threonine-specific protein kinase, *mTOR* mammalian target of rapamycin, *HIF-1α* hypoxia-inducible factor 1α

(Yang et al. 2014). Ethyl pyruvate demonstrated similar effects, as it inhibits the nuclear factor (NF)-κB activation and reduces proinflammatory cytokine production and HMGB1 secretion, which might, at least partially, be the result of its inhibitory effect on PKM2 (Ulloa et al. 2002).

The importance of energy metabolism in leukocytes has been recently further illustrated by the pivotal discovery that defects in metabolic processes underlie immunoparalysis in sepsis (Cheng et al. 2016). Monocytes isolated from patients with *Escherichia coli* and *C. albicans* sepsis displayed a generalized metabolic defect at the levels of both glycolysis and OxPhos, which was completely restored after recovery of the patients, or partially by using recombinant interferon (IFN)-γ therapy. Along this line, it is noteworthy that fructose-1,6-diphosphate (FDP), one of the intermediates in the glycolysis pathway, had been proposed in the early 1980s as a treatment for endotoxin-induced septic shock (Markov et al. 1981), a therapeutic role attributed to its capacity to positively regulate glycolysis by inducing phosphofructokinase (PFK, one of the genes severely affected in sepsis) (Cheng et al. 2016), while decreasing lipid metabolism (Markov et al. 2000). Taken together, the promising effects of modulating core metabolic processes in leukocytes during sepsis-induced immunotolerance provide support to the use of immunometabolic modulation as a promising approach for the treatment of sepsis, and more broadly, or other infections (Arts et al. 2017).

Aerobic glycolysis represents the central metabolic process underlying innate immune memory in monocytes stimulated with β-glucan—a process referred to as "trained immunity" (Cheng et al. 2014). The metabolic alterations contributing to the establishment of innate memory were demonstrated to underlie exclusive epigenetic signatures (Saeed et al. 2014). Specifically, monocyte training by β-glucan was associated with stable changes in histone trimethylation at H3K4, ultimately resulting in enhanced innate antifungal immune responses (Quintin et al. 2012). Remarkably, β-glucan-mediated training was not only efficient in vitro by enhancing the ability of monocytes to produce proinflammatory cytokines but also conferred protection to mice lacking functional T- and B-cells following systemic infection with *C. albicans*. Along the same line, rewiring of the T-cell metabolism toward aerobic glycolysis supports the expression of lactate dehydrogenase A (LDHA) that, by maintaining high levels of acetyl-coA, in turn promotes histone acetylation and expression of genes such as IFN-γ, the hallmark of activated Th1 cells (Peng et al. 2016). It remains to be explored whether similar events underlie differentiation of T-cells in response to fungal infection.

11.2.2 Pentose Phosphate Pathway (PPP)

The PPP is a metabolic pathway also emerging from glycolysis. The hexokinase (HK, the first enzyme in glycolysis) converts glucose into glucose-6-phosphate (G6P), which can be metabolized further into fructose-6-phosphate. Alternatively, G6P dehydrogenase (G6PD) can dehydrogenate G6P, which initiates the PPP. The PPP is divided into two branches: the oxidative and nonoxidative branches, which are used to generate nicotinamide adenine dinucleotide phosphate (NADPH) for fatty acid synthesis and ribose-5-phosphate to produce nucleotides. The nonoxidative branch can also shuttle metabolites back to glycolysis. Although the sedoheptulose kinase CARKL, one of the kinases in the PPP, was found to be downregulated upon lipopolysaccharide (LPS) administration in mice and humans, and was associated with a proinflammatory phenotype of macrophages (Haschemi et al. 2012), not much is known about the immunological relevance of this metabolic pathway in response to fungal infection. It has, however, been deemed not essential for trained immunity with β-glucan, since inhibition of the oxidative branch of PPP led to no effect on cytokine production by trained monocytes (Arts et al. 2016). Altogether, and despite an increase in PPP is observed during stimulation with LPS, concomitantly with an increase in glycolysis, no data are available on PPP during the immune response to fungi.

11.2.3 TCA Cycle/OxPhos

The TCA cycle is considered a connecting node between different metabolic pathways. For example, glycolysis fuels the TCA via pyruvate and acetyl-CoA, glutamine can be converted via glutamate into α-ketoglutarate, and β-oxidation of fatty acids also results in acetyl-CoA production. In the TCA cycle, metabolites are reduced by structural modifications, leading to the production of intermediates such as nicotinamide adenine dinucleotide (NADH) and flavin adenine dinucleotide (FAD), which will be used in OxPhos to generate ATP molecules.

The highly inflammatory profile of activated macrophages following stimulation with LPS is known to result from an increase in glycolysis and a decrease in OxPhos (Tannahill et al. 2013; Yang et al. 2014). Although during this metabolic shift, the TCA is globally downregulated, LPS-stimulated cells display nonetheless an increase in the TCA intermediates succinate, malate, and fumarate (Tannahill et al. 2013). Succinate accumulation due to the decreased activity of succinate dehydrogenase (SDH) is particularly relevant, since when it is transported to the cytosol, it can inhibit prolyl hydrolase enzymes (together with fumarate and α-ketoglutarate), which results in the stabilization and activation of HIF-1α and contributes to enhanced IL-1β production. Succinate is partially derived from glutamine via the γ-aminobutyric acid (GABA) shunt, and inhibition of this pathway decreased succinate levels, impaired HIF-1α stabilization, and lowered IL-1β production both in vitro and in murine sepsis (Tannahill et al. 2013). Accordingly, inhibition of the GABA shunt resulted in reduced mortality in murine sepsis, an identical phenotype to that observed in *Hif1a*-deficient mice. Of note, increased succinate levels promote lysine succinylation, which may be regarded as another succinate-driven mechanism of metabolic regulation, since many proteins involved in amino acid metabolism, TCA cycle, fatty acid metabolism, as well as pyruvate dehydrogenase kinase (PDK) are among its targets (Park et al. 2013).

Besides succinate, citrate also accumulates in LPS-stimulated macrophages (Tannahill et al. 2013), and it has been shown to be essential for production of NO, ROS, prostaglandins, and itaconic acid, the latter being synthesized as an antibacterial and antifungal compound from the citric acid cycle intermediate *cis*-aconitic acid through the action of the immune-responsive gene 1 (IRG1) protein

(Lampropoulou et al. 2016; Michelucci et al. 2013). Specifically, itaconic acid targets the glyoxylate cycle in pathogens, which is favored over glycolysis under limited glucose availability (e.g., inside the phagosome) and is essential for growth and survival (Michelucci et al. 2013). The effects of citrate on the effector functions of immune cells are closely related with the activity of the mitochondrial citrate transporter, which carries citrate out of the mitochondrion into the cytosol. Once in the cytosol, citrate can be converted into acetyl-CoA and oxaloacetate, which can then be used for the synthesis of NO, ROS, and arachidonic acid. Moreover, citrate can also be converted into malonyl-CoA, which can be used in fatty acid biosynthesis to support membrane biogenesis. Interestingly, M2 macrophage polarization due to gut dysbiosis has been reported to underlie allergic airway inflammation as the result of fungal-induced production of prostaglandin E2 (Kim et al. 2014). It remains to be assessed whether citrate accumulation plays a role in this metabolic shift underlying the allergic phenotype observed in gut microbiota-depleted mice.

Additional TCA intermediates with regulatory potential have been also identified. Specifically, accumulation of fumarate due to glutamine replenishment of the TCA cycle was found to integrate immune and metabolic circuits to induce monocyte epigenetic reprogramming in response to β-glucan-induced trained immunity (Arts et al. 2016). Fumarate itself was found to induce trained immunity by promoting increased trimethylation of histones at H3K4, but also acetylation at H3K27, linking immunometabolic activation with long-term epigenetic changes. Along this line, inhibition of glutaminolysis blocked fumarate accumulation and prevented the induction of trained immunity by β-glucan. Collectively, these findings provide further evidence of the role of cellular metabolites acting as cofactors for epigenetic enzymes and illustrate the potential of metabolic reorientation to manipulate the epigenetic state of immune cells and confer protection to infection.

Strikingly, whereas OxPhos is typically considered to be reduced following LPS stimulation, conflicting reports have been published regarding sepsis. Indeed, sepsis patients have been reported to either display mitochondrial dysfunction with decreased leukocyte oxygen consumption (Adrie et al. 2001; Brealey et al. 2002; Cheng et al. 2016) or increased respiratory capacity (Belikova et al. 2007; Sjovall et al. 2013). Because the acute phase of sepsis is characterized by a strong inflammatory reaction, and immunoparalysis may instead be encountered at later stages in some patients, both the timing of assessment and the type of cell and tissue are crucial for determining the extent of OxPhos activation in sepsis, which may prove critical in the design of novel therapeutic strategies based on the modulation of immunometabolism.

11.2.4 Amino Acid Metabolism

The availability and metabolism of amino acids are increasingly recognized as crucial regulators of immune function. Nowadays, it is well appreciated that the levels of serum amino acids during acute bacterial and fungal infections are highly dynamic

(Su et al. 2015) and likely pose a considerable impact on the metabolic profiles of immune cells, altering their inflammatory state. One such critical example is arginine which, upon infection, is imported by M1 macrophages to synthesize NO, generating the by-product citrulline (Qualls et al. 2012). In contrast, M2 macrophages upregulate arginase, which provides another competing arginine metabolism pathway, resulting in a relative arginine deficiency for NO production. The importance of this pathway to antifungal immunity was recently illustrated by the demonstration that chitin from the cell wall of *C. albicans* redirects macrophage polarization toward an anti inflammatory phenotype by inducing the substrate-competing host enzyme arginase-1 (ARG1), thereby reducing antimicrobial functions and mediating fungal survival within macrophages (Wagener et al. 2017). This may, however, prove to be a specific trait of this commensal since primary fungal pathogens such as *Blastomyces dermatitidis* have instead developed evading strategies that are not based on limiting the access of inducible NO synthase to its arginine substrate, but rather target the enzyme directly (Rocco et al. 2011). Likewise, granulocyte-mediated clearance of *A. fumigatus* was found to occur independently of arginine availability (Kapp et al. 2014). Along this line, arginine supplementation regimens failed to influence the immune function and development of nosocomial infections in critically ill patients with low NO fractions in the airways (Tadie et al. 2013), which may explain the differential requirements for this amino acid across distinct fungal infections.

The amino acid glutamine is also known to be essential for NO production by macrophages, together with arginine and citrulline. In addition, glutaminolysis was recently found to cooperate with glycolysis and OxPhos toward the production of energy required for adequate innate immune response against yeast from *C. albicans* (Dominguez-Andres et al. 2017). Because this process was not activated by hyphae (responses to which relied primarily on glycolysis), glutamine consumption may be a process that is required early after activation to cope with fast energy-demanding processes required to prevent fungal germination. Trained immunity by β-glucan stimulation was also found to depend on glutamine metabolism, as reflected by the increased production of glutamate, a substrate that can replenish the TCA cycle, during this process (Arts et al. 2016). Interestingly, aspartate was also consumed in large amounts, suggesting that metabolism of amino acids other than glutamine may also be involved in trained immunity. In this regard, it is interesting to note that glutamine supplementation was found to prevent acute respiratory distress syndrome following sepsis by regulating inflammatory cytokine release (Singleton et al. 2005).

Tryptophan is an essential amino acid that cannot be synthesized de novo and is metabolized by the enzyme indoleamine-2,3-dioxygenase 1 (IDO1) (Munn and Mellor 2013). Production of IL-6 by murine macrophages is IDO1-dependent, and tryptophan supplementation in vitro increases IL-6, and partially tumor necrosis factor (TNF)-α and IL-12, production by macrophages (Liu et al. 2014). Because tryptophan degradation via IDO1 activity can develop through the kynurenine pathway, thereby generating a number of immunoregulatory metabolites, a pivotal role for IDO1 in response to infection has been acknowledged. Tryptophan degradation by IDO1 within the airway epithelium is a mechanism by which the number and activation status of CD4+ T cells recruited to the lung during allergic inflammation by *A. fumigatus* are kept under control (Paveglio et al. 2011). Likewise, epithelial cells were found to confer protection from pulmonary aspergillosis via a Toll/IL-1 receptor (TIR) domain-containing adapter-inducing IFN-β (TRIF) pathway downstream Toll-like receptor (TLR)3 converging on IDO1 activation and consequent production of kynurenines (de Luca et al. 2010). IDO1 was also confirmed as an important immunoregulatory enzyme in response to the primary fungal pathogen *Paracoccidioides brasiliensis*, promoting fungal clearance and controlling T-cell immunity and inflammation, particularly in genetically susceptible hosts (Araujo et al. 2014). Likewise, IDO1 was also found to be crucial in balancing resistance and tolerance to *C. albicans*, and its genetic deficiency was found to be a key risk factor for the development of recurrent vulvovaginal candidiasis (De Luca et al. 2013). Taken together, these findings suggest that distinct immune pathways contribute to resistance and tolerance to fungi, and that targeting immunotolerance via IDO1 activity and therapeutic kynurenines may benefit patients suffering from fungal disease. Indeed, a small trial of granulocyte-macrophage colony-stimulating factor (GM-CSF) treatment in sepsis patients has revealed increased IDO1 activity in whole blood, which positively correlated with procalcitonin and negatively with levels of human leukocyte antigen—antigen D related (HLA-DR) in monocytes (a marker for immunotolerance) (Schefold et al. 2010).

11.2.5 Fatty Acid Metabolism

Fatty acids can either be used as an energy source (β-oxidation), or they can be synthesized from acetyl-CoA. β-oxidation mainly starts with triglycerides, either from meals or from the adipose tissue, that are freed from glycerol by lipolysis. This process yields free fatty acids that can be incorporated by immune cells via fatty acid transporters (e.g., CD36). Once inside the cell, the long-chain fatty acids are converted into fatty acyl-adenylates, which are fused with the CoA molecule to yield acyl-CoA. This molecule is then fused with carnitine by carnitine palmitoyltransferase I (CPT-I), and this acyl-carnitine molecule is shuttled into the mitochondrion in exchange for a carnitine molecule by CPT-II. Once inside the mitochondrion, acyl-CoA is freed from carnitine, and β-oxidation can be initiated, with acetate molecules being cut from the fatty acid chain and combined with CoA to form acetyl-CoA, which enters the TCA cycle, while fusing with oxaloacetate to form citrate (Houten and Wanders 2010).

In LPS-stimulated macrophages, increased levels of citrate and fatty acids were observed, suggesting a diversion from the TCA cycle toward fatty acid synthesis during the acute-phase response (Jha et al. 2015; Tannahill et al. 2013). The acute upregulation of glycolysis upon stimulation transits into increased fatty acid β-oxidation, a shift characterized by upregulation of proliferator-activated receptor γ (PPAR-γ) coactivator-1α (PGC-1α) and -β and CD36 and CPT-I. Along this line, a large-scale metabolomics/proteomics study on plasma samples defined fatty acid metabolism as one of the most promising metabolic predictors for survival in sepsis

(Langley et al. 2013). In an animal model of sepsis, PPAR-γ expression was shown to correlate with survival during murine septic shock (Standage et al. 2012). Incidentally, PPAR-γ ligands were found to be required for optimal expression of the β-glucan receptor dectin-1 and the alternative activation of macrophages in response to *C. albicans* (Gales et al. 2010). This may explain, at least partly, the requirement for the cholesterol synthesis pathway to the establishment of trained immunity by β-glucan (Arts et al. 2016). In addition, in human sepsis, transcriptome analysis has revealed a profoundly diminished expression of genes regulating β-oxidation, which was accompanied by decreased CPT-I and CD36 expression in human tolerant monocytes, a phenotype that was restored upon recovery of the patients (Cheng et al. 2016). Taken together, these findings may suggest that β-oxidation is downregulated during severe sepsis, but is instead enhanced in tolerant cells (Arts et al. 2017).

Because of the changes in fatty acid metabolism during sepsis, the therapeutic potential of various fatty acids has been investigated. In a murine cecal ligation model of sepsis or upon intraperitoneal injection with *E. coli*, supplementation with endogenous lysophosphatidylcholine (LPC) decreased deactivation of neutrophils and promoted pathogen clearance, improving survival (Yan et al. 2004). LPC was also demonstrated to decrease HMGB1 release from monocytes and macrophages in murine sepsis (Chen et al. 2005) and to reduce circulating IL-1β, thereby diminishing organ injury and dysfunction (Murch et al. 2006). Of note, omega-3-fatty acids also reduce IL-1β production by suppressing nuclear translocation of NF-κB and enhancing autophagy (Williams-Bey et al. 2014) and by inhibiting caspase-1 activity and, subsequently, nucleotide-binding domain leucine-rich repeat-containing family, pyrin domain-containing 3 (NLRP3) inflammasome activity (Yan et al. 2013). Although the mechanism(s) whereby fatty acids improve outcome after infection may vary, both LPC and omega-3-fatty acid supplements were shown to reduce IL-1β, which suggests this cytokine as a key player in severe inflammation during sepsis, and likely other infectious contexts.

11.3 Targeting Host Immunometabolism in Diagnosis and Treatment

Early diagnosis of fungal disease is critically required to decrease morbidity and mortality, particularly in vulnerable populations of immunocompromised patients, since the delayed initiation of antifungal therapy may contribute to fatal outcomes. Although the incorporation of fungal surrogate markers into clinical algorithms has facilitated diagnostic-driven strategies, there are currently no host-derived biomarkers with clinical application, highlighting therefore a pressing demand for new and improved diagnostic methods for fungal disease (Oliveira-Coelho et al. 2015).

Clinical metabolomics is a new and rapidly expanding field that has become a leading tool in defining disease phenotypes since it can translate the impairment induced by pathogenic mechanism(s) to associated phenotype(s). By providing a time-resolved "snapshot" of all metabolites present in a biological sample, metabolomics is a powerful tool for the detection of temporal physiological changes and to use as a monitoring approach of metabolic processes, contributing to a better understanding of disease mechanisms as well as providing novel diagnostic possibilities (Johnson et al. 2016). As mentioned above, a metabolomics (and proteomics) analysis on plasma samples from sepsis patients revealed differences, especially in fatty acid transport and β-oxidation, gluconeogenesis, and the TCA cycle between survivors and non-survivors (Langley et al. 2013). Therefore, the targeting of metabolic pathways might be a promising therapeutic target and might also serve for diagnosis and prognosis. Because these approaches focused on metabolites in plasma, and given our recent understanding of the metabolic mechanisms underlying immune paralysis in sepsis, assessment of metabolites from the circulating cells that undergo different metabolic stages during infection might have additional predictive value.

Recent insights have revealed that specific metabolites act as cofactors for epigenetic enzymes and, thus, modify the epigenetic and transcriptional landscape of the cell (Buck et al. 2017). To date, little is known about the interaction of metabolism and epigenetics in fungal disease. However, a few interesting observations have already been made. One of the most extensively described roles is that of histone deacetylases (HDACs). This is particularly important since numerous HDAC inhibitors are being tested in clinical trials, primarily for the treatment of cancer. Remarkably, HDAC inhibitors were found to act essentially as negative regulators of critical immune receptors and antimicrobial pathways in innate immune cells (Roger et al. 2011). Accordingly, HDAC inhibitors resulted in impaired phagocytosis and killing by macrophages and increased susceptibility to non-severe bacterial and fungal infections. The important role of HDACs in countering defective host defense mechanisms in sepsis is supported by data showing that inhibition of sirtuin 1 (SIRT1) by the specific inhibitor EX-527 was able to shift tolerant monocytes toward a more proinflammatory phenotype, improving pathogen clearance and, therefore, the outcome of patients (Vachharajani et al. 2014).

In addition to HDACs, adenosine monophosphate-activated protein kinase (AMPK) is activated in sepsis (therefore inhibiting mTOR and HIF-1α), and this has been shown to improve sepsis-induced lung injury and to increase the eradication of *Pseudomonas aeruginosa* by restraining the development of tolerant macrophages. Strikingly, AMPK was reported to activate the host autophagy initiation complex upon infection with *Cryptococcus neoformans*, therefore conferring increased susceptibility to fungal colonization of mice, as the result of an enhanced ability of *C. neoformans* to survive and replicate within macrophages (Pandey et al. 2017). This may also be partially explained by the fact that AMPK activation suppresses NADPH oxidase-derived ROS production and decreases cytokine production, mitogen-activated protein kinase (MAPK) activation, and activating transcription factor 3 (ATF3) induction.

Several other metabolites are known to play a major influence on the epigenetic landscape. Among the most important, acetyl-CoA is a direct acetyl donor for histone acetylation, and the rate of glycolysis has been directly associated with the acetylation of specific histone sites, which appeared to be mostly dependent on acetyl-CoA levels (Cluntun et al. 2015). Furthermore, the TCA metabolite α-ketoglutarate is an essential cofactor for histone demethylases (e.g., the JMJ family of lysine demethylases) and the ten eleven translocation (TET) enzymes, which regulate DNA methylation. Comparable metabolites at the molecular level, such as succinate, fumarate, malate, and 2-hydroxyglutarate, are instead known to antagonize this reaction (Carey et al. 2015; Xu et al. 2011). This is a crucial point since the increased H3K27 trimethylation due to the inhibition of JMD family enzymes may contribute to the suppression of genes associated with M2 phenotypes and sustain a more proinflammatory profile in trained macrophages (Saeed et al. 2014).

All these regulating mechanisms of metabolites on epigenetic modifiers can potentially influence the phenotype of monocytes and macrophages during infection. During the development of monocytes into macrophages, metabolic changes have been demonstrated to result partially from epigenetic changes (Cheng et al. 2014; Saeed et al. 2014), so the effects of metabolism on epigenetics and vice versa are most likely a complex interaction, in which both influence each other. It has, therefore, been proposed that the metabolism–epigenetics interaction is a field that should be the focus of future research, on the one hand for diagnostic and prognostic markers, but on the other hand, also for novel therapeutic approaches (Arts et al. 2017).

11.4 Concluding Remarks

Fungal diseases result from the combination between a dysregulated immune system and the resistance to, or subversion of, host immune mechanisms by the fungus. These two key factors highlight the need to develop novel immunomodulatory therapies, whose effects may contribute to maintain the delicate equilibrium between a successful immune response and an innocuous pathogen behavior. Although different immunotherapeutic approaches including recombinant cytokines, fungal vaccines, and antibodies have been attempted (Armstrong-James et al. 2017), the development of advanced therapies that target specific immune defects or functions remains an ambitious dream. To achieve this purpose, it is essential to improve our understanding of emerging concepts of regulation of host immunity, such as the immunometabolism. Because different metabolic states are associated with different immunologic states during infection, the use of metabolomics for diagnosis and prognosis is a novel field of research but that already demonstrates encouraging results. In addition, the interaction between the metabolic profile and the epigenetic landscape is a new and promising field for the development of novel therapies directed to improve immune responses in sepsis and, more generally, in response to infection.

Acknowledgments This work was supported by the Northern Portugal Regional Operational Programme (NORTE 2020), under the Portugal 2020 Partnership Agreement, through the European Regional Development Fund (FEDER) (NORTE-01-0145-FEDER-000013), and the Fundação para a Ciência e Tecnologia (FCT) (IF/00735/2014 to AC and SFRH/BPD/96176/2013 to CC).

References

Adrie C, Bachelet M, Vayssier-Taussat M, Russo-Marie F, Bouchaert I, Adib-Conquy M, Cavaillon JM, Pinsky MR, Dhainaut JF, Polla BS (2001) Mitochondrial membrane potential and apoptosis peripheral blood monocytes in severe human sepsis. Am J Respir Crit Care Med 164(3):389–395. https://doi.org/10.1164/ajrccm.164.3.2009088

Araujo EF, Loures FV, Bazan SB, Feriotti C, Pina A, Schanoski AS, Costa TA, Calich VL (2014) Indoleamine 2,3-dioxygenase controls fungal loads and immunity in paracoccidioidomicosis but is more important to susceptible than resistant hosts. PLoS Negl Trop Dis 8(11):e3330. https://doi.org/10.1371/journal.pntd.0003330

Armstrong-James D, Brown GD, Netea MG, Zelante T, Gresnigt MS, van de Veerdonk FL, Levitz SM (2017) Immunotherapeutic approaches to treatment of fungal diseases. Lancet Infect Dis 17(12): e393–e402. https://doi.org/10.1016/S1473-3099(17)30442-5

Arts RJ, Novakovic B, Ter Horst R, Carvalho A, Bekkering S, Lachmandas E, Rodrigues F, Silvestre R, Cheng SC, Wang SY, Habibi E, Goncalves LG, Mesquita I, Cunha C, van Laarhoven A, van de Veerdonk FL, Williams DL, van der Meer JW, Logie C, O'Neill LA, Dinarello CA, Riksen NP, van Crevel R, Clish C, Notebaart RA, Joosten LA, Stunnenberg HG, Xavier RJ, Netea MG (2016) Glutaminolysis and fumarate accumulation integrate immunometabolic and epigenetic programs in trained immunity. Cell Metab 24(6):807–819. https://doi.org/10.1016/j.cmet.2016.10.008

Arts RJ, Gresnigt MS, Joosten LA, Netea MG (2017) Cellular metabolism of myeloid cells in sepsis. J Leukoc Biol 101(1):151–164. https://doi.org/10.1189/jlb.4MR0216-066R

Artyomov MN, Sergushichev A, Schilling JD (2016) Integrating immunometabolism and macrophage diversity. Semin Immunol 28(5):417–424. https://doi.org/10.1016/j.smim.2016.10.004

Belikova I, Lukaszewicz AC, Faivre V, Damoisel C, Singer M, Payen D (2007) Oxygen consumption of human peripheral blood mononuclear cells in severe human sepsis. Crit Care Med 35(12):2702–2708

Bonifazi P, D'Angelo C, Zagarella S, Zelante T, Bozza S, De Luca A, Giovannini G, Moretti S, Iannitti RG, Fallarino F, Carvalho A, Cunha C, Bistoni F, Romani L (2010) Intranasally delivered siRNA targeting PI3K/Akt/mTOR inflammatory pathways protects from aspergillosis. Mucosal Immunol 3(2):193–205. https://doi.org/10.1038/mi.2009.130

Brealey D, Brand M, Hargreaves I, Heales S, Land J, Smolenski R, Davies NA, Cooper CE, Singer M (2002) Association between mitochondrial dysfunction and severity and outcome of septic shock. Lancet 360(9328):219–223. https://doi.org/10.1016/S0140-6736(02)09459-X

Brown GD, Denning DW, Gow NA, Levitz SM, Netea MG, White TC (2012) Hidden killers: human fungal infections. Sci Transl Med 4(165):165rv113. https://doi.org/10.1126/scitranslmed.3004404

Buck MD, Sowell RT, Kaech SM, Pearce EL (2017) Metabolic instruction of immunity. Cell 169(4):570–586. https://doi.org/10.1016/j.cell.2017.04.004

Carey BW, Finley LW, Cross JR, Allis CD, Thompson CB (2015) Intracellular alpha-ketoglutarate maintains the pluripotency of embryonic stem cells. Nature 518(7539):413–416. https://doi.org/10.1038/nature13981

Chen G, Li J, Qiang X, Czura CJ, Ochani M, Ochani K, Ulloa L, Yang H, Tracey KJ, Wang P, Sama AE, Wang H (2005) Suppression of HMGB1 release by stearoyl lysophosphatidylcholine: an additional mechanism for its therapeutic effects in experimental sepsis. J Lipid Res 46(4):623–627. https://doi.org/10.1194/jlr.C400018-JLR200

Cheng SC, Quintin J, Cramer RA, Shepardson KM, Saeed S, Kumar V, Giamarellos-Bourboulis EJ, Martens JH, Rao NA, Aghajanirefah A, Manjeri GR, Li Y, Ifrim DC, Arts RJ, van der Veer BM, Deen PM, Logie C, O'Neill LA, Willems P, van de Veerdonk FL, van der Meer JW, Ng A, Joosten LA, Wijmenga C, Stunnenberg HG, Xavier RJ, Netea MG (2014) mTOR- and HIF-1alpha-mediated aerobic glycolysis as metabolic basis for trained immunity. Science 345(6204):1250684. https://doi.org/10.1126/science.1250684

Cheng SC, Scicluna BP, Arts RJ, Gresnigt MS, Lachmandas E, Giamarellos-Bourboulis EJ, Kox M, Manjeri GR, Wagenaars JA, Cremer OL, Leentjens J, van der Meer AJ, van de Veerdonk FL, Bonten MJ, Schultz MJ, Willems PH, Pickkers P, Joosten LA, van der Poll T, Netea MG (2016) Broad defects in the energy metabolism of leukocytes underlie immunoparalysis in sepsis. Nat Immunol 17(4):406–413. https://doi.org/10.1038/ni.3398

Cluntun AA, Huang H, Dai L, Liu X, Zhao Y, Locasale JW (2015) The rate of glycolysis quantitatively mediates specific histone acetylation sites. Cancer Metab 3:10. https://doi.org/10.1186/s40170-015-0135-3

Corcoran SE, O'Neill LA (2016) HIF1alpha and metabolic reprogramming in inflammation. J Clin Investig 126(10):3699–3707. https://doi.org/10.1172/JCI84431

de Luca A, Bozza S, Zelante T, Zagarella S, D'Angelo C, Perruccio K, Vacca C, Carvalho A, Cunha C, Aversa F, Romani L (2010) Non-hematopoietic cells contribute to protective tolerance to *Aspergillus fumigatus* via a TRIF pathway converging on IDO. Cell Mol Immunol 7(6):459–470. https://doi.org/10.1038/cmi.2010.43

De Luca A, Carvalho A, Cunha C, Iannitti RG, Pitzurra L, Giovannini G, Mencacci A, Bartolommei L, Moretti S, Massi-Benedetti C, Fuchs D, De Bernardis F, Puccetti P, Romani L (2013) IL-22 and IDO1 affect immunity and tolerance to murine and human vaginal candidiasis. PLoS Pathog 9(7):e1003486. https://doi.org/10.1371/journal.ppat.1003486

Dominguez-Andres J, Arts RJW, Ter Horst R, Gresnigt MS, Smeekens SP, Ratter JM, Lachmandas E, Boutens L, van de Veerdonk FL, Joosten LAB, Notebaart RA, Ardavin C, Netea MG (2017) Rewiring monocyte glucose metabolism via C-type lectin signaling protects against disseminated candidiasis. PLoS Pathog 13(9):e1006632. https://doi.org/10.1371/journal.ppat.1006632

Fisher MC, Henk DA, Briggs CJ, Brownstein JS, Madoff LC, McCraw SL, Gurr SJ (2012) Emerging fungal threats to animal, plant and ecosystem health. Nature 484(7393):186–194. https://doi.org/10.1038/nature10947

Fliesser M, Morton CO, Bonin M, Ebel F, Hunniger K, Kurzai O, Einsele H, Loffler J (2015) Hypoxia-inducible factor 1alpha modulates metabolic activity and cytokine release in anti-*Aspergillus fumigatus* immune responses initiated by human dendritic cells. Int J Med Microbiol 305(8):865–873. https://doi.org/10.1016/j.ijmm.2015.08.036

Gales A, Conduche A, Bernad J, Lefevre L, Olagnier D, Beraud M, Martin-Blondel G, Linas MD, Auwerx J, Coste A, Pipy B (2010) PPARgamma controls Dectin-1 expression required for host antifungal defense against *Candida albicans*. PLoS Pathog 6(1):e1000714. https://doi.org/10.1371/journal.ppat.1000714

Galvan-Pena S, O'Neill LA (2014) Metabolic reprograming in macrophage polarization. Front Immunol 5:420. https://doi.org/10.3389/fimmu.2014.00420

Gleeson LE, Sheedy FJ (2016) Metabolic reprogramming & inflammation: fuelling the host response to pathogens. Semin Immunol 28(5):450–468. https://doi.org/10.1016/j.smim.2016.10.007

Grahl N, Puttikamonkul S, Macdonald JM, Gamcsik MP, Ngo LY, Hohl TM, Cramer RA (2011) In vivo hypoxia and a fungal alcohol dehydrogenase influence the pathogenesis of invasive pulmonary aspergillosis. PLoS Pathog 7(7):e1002145. https://doi.org/10.1371/journal.ppat.1002145

Haschemi A, Kosma P, Gille L, Evans CR, Burant CF, Starkl P, Knapp B, Haas R, Schmid JA, Jandl C, Amir S, Lubec G, Park J, Esterbauer H, Bilban M, Brizuela L, Pospisilik JA, Otterbein LE, Wagner O (2012) The sedoheptulose kinase CARKL directs macrophage polarization through control of glucose metabolism. Cell Metab 15(6):813–826. https://doi.org/10.1016/j.cmet.2012.04.023

Hellwig D, Voigt J, Bouzani M, Loffler J, Albrecht-Eckardt D, Weber M, Brunke S, Martin R, Kurzai O, Hunniger K (2016) *Candida albicans* induces metabolic reprogramming in human NK cells and responds to perforin with a zinc depletion response. Front Microbiol 7:750. https://doi.org/10.3389/fmicb.2016.00750

Houten SM, Wanders RJ (2010) A general introduction to the biochemistry of mitochondrial fatty acid beta-oxidation. J Inherit Metab Dis 33(5):469–477. https://doi.org/10.1007/s10545-010-9061-2

Jha AK, Huang SC, Sergushichev A, Lampropoulou V, Ivanova Y, Loginicheva E, Chmielewski K, Stewart KM, Ashall J, Everts B, Pearce EJ, Driggers EM, Artyomov MN (2015) Network integration of parallel metabolic and transcriptional data reveals metabolic modules that regulate macrophage polarization. Immunity 42(3):419–430. https://doi.org/10.1016/j.immuni.2015.02.005

Johnson CH, Ivanisevic J, Siuzdak G (2016) Metabolomics: beyond biomarkers and towards mechanisms. Nat Rev Mol Cell Biol 17(7):451–459. https://doi.org/10.1038/nrm.2016.25

Kapp K, Prufer S, Michel CS, Habermeier A, Luckner-Minden C, Giese T, Bomalaski J, Langhans CD, Kropf P, Muller I, Closs EI, Radsak MP, Munder M (2014) Granulocyte functions are independent of arginine availability. J Leukoc Biol 96(6):1047–1053. https://doi.org/10.1189/jlb.3AB0214-082R

Kim YG, Udayanga KG, Totsuka N, Weinberg JB, Nunez G, Shibuya A (2014) Gut dysbiosis promotes M2 macrophage polarization and allergic airway inflammation via fungi-induced PGE (2). Cell Host Microbe 15(1):95–102. https://doi.org/10.1016/j.chom.2013.12.010

Lampropoulou V, Sergushichev A, Bambouskova M, Nair S, Vincent EE, Loginicheva E, Cervantes-Barragan L, Ma X, Huang SC, Griss T, Weinheimer CJ, Khader S, Randolph GJ, Pearce EJ, Jones RG, Diwan A, Diamond MS, Artyomov MN (2016) Itaconate links inhibition of succinate dehydrogenase with macrophage metabolic remodeling and regulation of inflammation. Cell Metab 24(1):158–166. https://doi.org/10.1016/j.cmet.2016.06.004

Langley RJ, Tsalik EL, van Velkinburgh JC, Glickman SW, Rice BJ, Wang C, Chen B, Carin L, Suarez A, Mohney RP, Freeman DH, Wang M, You J, Wulff J, Thompson JW, Moseley MA,Reisinger S, Edmonds BT, Grinnell B, Nelson DR, Dinwiddie DL, Miller NA, Saunders CJ, Soden SS, Rogers AJ, Gazourian L, Fredenburgh LE, Massaro AF, Baron RM, Choi AM, Corey GR, Ginsburg GS, Cairns CB, Otero RM, Fowler VG Jr, Rivers EP, Woods CW, Kingsmore SF (2013) An integrated clinico-metabolomic model improves prediction of death in sepsis. Sci Transl Med 5(195):195ra195. https://doi.org/10.1126/scitranslmed.3005893

Liu H, Huang L, Bradley J, Liu K, Bardhan K, Ron D, Mellor AL, Munn DH, McGaha TL (2014) GCN2-dependent metabolic stress is essential for endotoxemic cytokine induction and pathology. Mol Cell Biol 34(3):428–438. https://doi.org/10.1128/MCB.00946-13

Markov AK, Turner MD, Oglethorpe N, Neely WA, Hellems HK (1981) Fructose-1,6-diphosphate: an agent for treatment of experimental endotoxin shock. Surgery 90(3):482–488

Markov AK, Neely WA, Didlake RH, Terry J 3rd, Causey A, Lehan PH (2000) Metabolic responses to fructose-1,6-diphosphate in healthy subjects. Metabolism 49(6):698–703. https://doi.org/10.1053/meta.2000.6249

Michelucci A, Cordes T, Ghelfi J, Pailot A, Reiling N, Goldmann O, Binz T, Wegner A, Tallam A, Rausell A, Buttini M, Linster CL, Medina E, Balling R, Hiller K (2013) Immune-responsive gene 1 protein links metabolism to immunity by catalyzing itaconic acid production. Proc Natl Acad Sci USA 110(19):7820–7825. https://doi.org/10.1073/pnas.1218599110

Munn DH, Mellor AL (2013) Indoleamine 2,3 dioxygenase and metabolic control of immune responses. Trends Immunol 34(3):137–143. https://doi.org/10.1016/j.it.2012.10.001

Murch O, Collin M, Sepodes B, Foster SJ, Mota-Filipe H, Thiemermann C (2006) Lysophosphatidylcholine reduces the organ injury and dysfunction in rodent models of gram-negative and gram-positive shock. Br J Pharmacol 148(6):769–777. https://doi.org/10.1038/sj.bjp.0706788

O'Neill LA, Pearce EJ (2016) Immunometabolism governs dendritic cell and macrophage function. J Exp Med 213(1):15–23. https://doi.org/10.1084/jem.20151570

Olenchock BA, Rathmell JC, Vander Heiden MG (2017) Biochemical underpinnings of immune cell metabolic phenotypes. Immunity 46(5):703–713. https://doi.org/10.1016/j.immuni.2017.04.013

Oliveira-Coelho A, Rodrigues F, Campos A Jr, Lacerda JF, Carvalho A, Cunha C (2015) Paving the way for predictive diagnostics and personalized treatment of invasive aspergillosis. Front Microbiol 6:411. https://doi.org/10.3389/fmicb.2015.00411

Palsson-McDermott EM, Curtis AM, Goel G, Lauterbach MA, Sheedy FJ, Gleeson LE, van den Bosch MW, Quinn SR, Domingo-Fernandez R, Johnston DG, Jiang JK, Israelsen WJ, Keane J, Thomas C, Clish C, Vander Heiden M, Xavier RJ, O'Neill LA (2015) Pyruvate kinase M2 regulates Hif-1alpha activity and IL-1beta induction and is a critical determinant of the warburg effect in LPS-activated macrophages. Cell Metab 21(1):65–80. https://doi.org/10.1016/j.cmet.2014.12.005

Pandey A, Ding SL, Qin QM, Gupta R, Gomez G, Lin F, Feng X, Fachini da Costa L, Chaki SP, Katepalli M, Case ED, van Schaik EJ, Sidiq T, Khalaf O, Arenas A, Kobayashi KS, Samuel JE, Rivera GM, Alaniz RC, Sze SH, Qian X, Brown WJ, Rice-Ficht A, Russell WK, Ficht TA, de Figueiredo P (2017) Global reprogramming of host kinase signaling in response to fungal infection. Cell Host Microbe 21(5):637–649.e636. https://doi.org/10.1016/j.chom.2017.04.008

Park J, Chen Y, Tishkoff DX, Peng C, Tan M, Dai L, Xie Z, Zhang Y, Zwaans BM, Skinner ME, Lombard DB, Zhao Y (2013) SIRT5-mediated lysine desuccinylation impacts diverse metabolic pathways. Mol Cell 50(6):919–930. https://doi.org/10.1016/j.molcel.2013.06.001

Paveglio SA, Allard J, Foster Hodgkins SR, Ather JL, Bevelander M, Campbell JM, Whittaker LeClair LA, McCarthy SM, van der Vliet A, Suratt BT, Boyson JE, Uematsu S, Akira S, Poynter ME (2011) Airway epithelial indoleamine 2,3-dioxygenase inhibits CD4+ T cells during *Aspergillus fumigatus* antigen exposure. Am J Respir Cell Mol Biol 44(1):11–23. https://doi.org/10.1165/rcmb.2009-0167OC

Pearce EL, Pearce EJ (2013) Metabolic pathways in immune cell activation and quiescence. Immunity 38(4):633–643. https://doi.org/10.1016/j.immuni.2013.04.005

Peng M, Yin N, Chhangawala S, Xu K, Leslie CS, Li MO (2016) Aerobic glycolysis promotes T helper 1 cell differentiation through an epigenetic mechanism. Science 354(6311):481–484. https://doi.org/10.1126/science.aaf6284

Pfaller MA, Diekema DJ (2010) Epidemiology of invasive mycoses in North America. Crit Rev Microbiol 36(1):1–53. https://doi.org/10.3109/10408410903241444

Qualls JE, Subramanian C, Rafi W, Smith AM, Balouzian L, DeFreitas AA, Shirey KA, Reutterer B, Kernbauer E, Stockinger S, Decker T, Miyairi I, Vogel SN, Salgame P, Rock CO, Murray PJ (2012) Sustained generation of nitric oxide and control of mycobacterial infection requires argininosuccinate synthase 1. Cell Host Microbe 12(3):313–323. https://doi.org/10.1016/j.chom.2012.07.012

Quintin J, Saeed S, Martens JH, Giamarellos-Bourboulis EJ, Ifrim DC, Logie C, Jacobs L, Jansen T, Kullberg BJ, Wijmenga C, Joosten LA, Xavier RJ, van der Meer JW, Stunnenberg HG, Netea MG (2012) *Candida albicans* infection affords protection against reinfection via functional reprogramming of monocytes. Cell Host Microbe 12(2):223–232. https://doi.org/10.1016/j.chom.2012.06.006

Rocco NM, Carmen JC, Klein BS (2011) Blastomyces dermatitidis yeast cells inhibit nitric oxide production by alveolar macrophage inducible nitric oxide synthase. Infect Immun 79(6):2385–2395. https://doi.org/10.1128/IAI.01249-10

Roger T, Lugrin J, Le Roy D, Goy G, Mombelli M, Koessler T, Ding XC, Chanson AL, Reymond MK, Miconnet I, Schrenzel J, Francois P, Calandra T (2011) Histone deacetylase inhibitors impair innate immune responses to Toll-like receptor agonists and to infection. Blood 117(4):1205–1217. https://doi.org/10.1182/blood-2010-05-284711

Saeed S, Quintin J, Kerstens HH, Rao NA, Aghajanirefah A, Matarese F, Cheng SC, Ratter J, Berentsen K, van der Ent MA, Sharifi N, Janssen-Megens EM, Ter Huurne M, Mandoli A, van Schaik T, Ng A, Burden F, Downes K, Frontini M, Kumar V, Giamarellos-Bourboulis EJ, Ouwehand WH, van der Meer JW, Joosten LA, Wijmenga C, Martens JH, Xavier RJ, Logie C, Netea MG, Stunnenberg HG (2014) Epigenetic programming of monocyte-to-macrophage differentiation and trained innate immunity. Science 345(6204):1251086. https://doi.org/10.1126/science.1251086

Schefold JC, Zeden JP, Pschowski R, Hammoud B, Fotopoulou C, Hasper D, Fusch G, Von Haehling S, Volk HD, Meisel C, Schutt C, Reinke P (2010) Treatment with granulocyte-macrophage colony-stimulating factor is associated with reduced indoleamine 2,3-dioxygenase activity and kynurenine pathway catabolites in patients with severe sepsis and septic shock. Scand J Infect Dis 42(3):164–171. https://doi.org/10.3109/00365540903405768

Singleton KD, Beckey VE, Wischmeyer PE (2005) Glutamine prevents activation of NF-kappaB and stress kinase pathways, attenuates inflammatory cytokine release, and prevents acute respiratory distress syndrome (ARDS) following sepsis. Shock 24(6):583–589

Sjovall F, Morota S, Persson J, Hansson MJ, Elmer E (2013) Patients with sepsis exhibit increased mitochondrial respiratory capacity in peripheral blood immune cells. Crit Care 17(4):R152. https://doi.org/10.1186/cc12831

Standage SW, Caldwell CC, Zingarelli B, Wong HR (2012) Reduced peroxisome proliferator-activated receptor alpha expression is associated with decreased survival and increased tissue bacterial load in sepsis. Shock 37(2):164–169. https://doi.org/10.1097/SHK.0b013e31823f1a00

Su L, Li H, Xie A, Liu D, Rao W, Lan L, Li X, Li F, Xiao K, Wang H, Yan P, Li X, Xie L (2015) Dynamic changes in amino acid concentration profiles in patients with sepsis. PloS One 10(4): e0121933. https://doi.org/10.1371/journal.pone.0121933

Tadie JM, Cynober L, Peigne V, Caumont-Prim A, Neveux N, Gey A, Guerot E, Diehl JL, Fagon JY, Tartour E, Delclaux C (2013) Arginine administration to critically ill patients with a low nitric oxide fraction in the airways: a pilot study. Intensive Care Med 39(9):1663–1665. https://doi.org/10.1007/s00134-013-2984-y

Tannahill GM, Curtis AM, Adamik J, Palsson-McDermott EM, McGettrick AF, Goel G, Frezza C, Bernard NJ, Kelly B, Foley NH, Zheng L, Gardet A, Tong Z, Jany SS, Corr SC, Haneklaus M, Caffrey BE, Pierce K, Walmsley S, Beasley FC, Cummins E, Nizet V, Whyte M, Taylor CT, Lin H, Masters SL, Gottlieb E, Kelly VP, Clish C, Auron PE, Xavier RJ, O'Neill LA (2013) Succinate is an inflammatory signal that induces IL-1beta through HIF-1alpha. Nature 496 (7444):238–242. https://doi.org/10.1038/nature11986

Ulloa L, Ochani M, Yang H, Tanovic M, Halperin D, Yang R, Czura CJ, Fink MP, Tracey KJ (2002) Ethyl pyruvate prevents lethality in mice with established lethal sepsis and systemic inflammation. Proc Natl Acad Sci USA 99(19):12351–12356. https://doi.org/10.1073/pnas.192222999

Vachharajani VT, Liu T, Brown CM, Wang X, Buechler NL, Wells JD, Yoza BK, McCall CE (2014) SIRT1 inhibition during the hypoinflammatory phenotype of sepsis enhances immunity and improves outcome. J Leukoc Biol 96(5):785–796. https://doi.org/10.1189/jlb.3MA0114-034RR

Wagener J, MacCallum DM, Brown GD, Gow NA (2017) *Candida albicans* chitin increases Arginase-1 activity in human macrophages, with an impact on macrophage antimicrobial functions. mBio 8(1). https://doi.org/10.1128/mBio.01820-16

Williams-Bey Y, Boularan C, Vural A, Huang NN, Hwang IY, Shan-Shi C, Kehrl JH (2014) Omega-3 free fatty acids suppress macrophage inflammasome activation by inhibiting NF-kappaB activation and enhancing autophagy. PloS One 9(6):e97957. https://doi.org/10.1371/journal.pone.0097957

Xu W, Yang H, Liu Y, Yang Y, Wang P, Kim SH, Ito S, Yang C, Wang P, Xiao MT, Liu LX, Jiang WQ, Liu J, Zhang JY, Wang B, Frye S, Zhang Y, Xu YH, Lei QY, Guan KL, Zhao SM, Xiong Y (2011) Oncometabolite 2-hydroxyglutarate is a competitive inhibitor of alpha-ketoglutarate-dependent dioxygenases. Cancer Cell 19(1):17–30. https://doi.org/10.1016/j.ccr.2010.12.014

Yan JJ, Jung JS, Lee JE, Lee J, Huh SO, Kim HS, Jung KC, Cho JY, Nam JS, Suh HW, Kim YH, Song DK (2004) Therapeutic effects of lysophosphatidylcholine in experimental sepsis. Nat Med 10(2):161–167. https://doi.org/10.1038/nm989

Yan Y, Jiang W, Spinetti T, Tardivel A, Castillo R, Bourquin C, Guarda G, Tian Z, Tschopp J, Zhou R (2013) Omega-3 fatty acids prevent inflammation and metabolic disorder through inhibition of NLRP3 inflammasome activation. Immunity 38(6):1154–1163. https://doi.org/10.1016/j.immuni.2013.05.015

Yang L, Xie M, Yang M, Yu Y, Zhu S, Hou W, Kang R, Lotze MT, Billiar TR, Wang H, Cao L, Tang D (2014) PKM2 regulates the Warburg effect and promotes HMGB1 release in sepsis. Nat Commun 5:4436. https://doi.org/10.1038/ncomms5436

Chapter 12
Metabolic Crosstalk Between Host and Parasitic Pathogens

Diana Moreira, Jérôme Estaquier, Anabela Cordeiro-da-Silva, and Ricardo Silvestre

Abstract A complex network that embraces parasite–host intrinsic factors and the microenvironment regulated the interaction between a parasite and its host. Nutritional pressures exerted by both elements of this duet thus dictate this host–parasite niche. To survive and proliferate inside a host and a harsh nutritional environment, the parasites modulate different nutrient sensing pathways to subvert host metabolic pathways. Such mechanism is able to change the flux of distinct nutrients/metabolites diverting them to be used by the parasites. Apart from this nutritional strategy, the scavenging of nutrients, particularly host fatty acids, constitutes a critical mechanism to fulfil parasite nutritional requirements, ultimately defining the host metabolic

D. Moreira
Life and Health Sciences Research Institute (ICVS), School of Medicine, University of Minho, Braga, Portugal

ICVS/3B's-PT Government Associate Laboratory, Braga/Guimarães, Portugal

i3S-Instituto de Investigacão e Inovação em Saúde, Universidade do Porto, Porto, Portugal

IBMC-Instituto de Biologia Molecular e Celular, Universidade do Porto, Porto, Portugal

Departamento de Ciências Biológicas, Faculdade de Farmaácia, Universidade do Porto, Porto, Portugal

J. Estaquier
CNRS FR 3636, Université Paris Descartes, Paris, France

Centre de Recherche du CHU de Québec, Université Laval, Québec, Canada

A. Cordeiro-da-Silva
i3S-Instituto de Investigacão e Inovação em Saúde, Universidade do Porto, Porto, Portugal

IBMC-Instituto de Biologia Molecular e Celular, Universidade do Porto, Porto, Portugal

Departamento de Ciências Biológicas, Faculdade de Farmaácia, Universidade do Porto, Porto, Portugal

R. Silvestre (✉)
Life and Health Sciences Research Institute (ICVS), School of Medicine, University of Minho, Braga, Portugal

ICVS/3B's-PT Government Associate Laboratory, Braga/Guimarães, Portugal
e-mail: ricardosilvestre@med.uminho.pt

R. Silvestre, E. Torrado (eds.), *Metabolic Interaction in Infection*, Experientia Supplementum 109, https://doi.org/10.1007/978-3-319-74932-7_12

landscape. The host metabolic alterations that result from host–parasite metabolic coupling can certainly be considered important targets to improve diagnosis and also for the development of future therapies. Metabolism is in fact considered a key element within this complex interaction, its modulation being crucial to dictate the final infection outcome.

Keywords Nutrient sensing pathways · Host–parasite interaction · Host metabolic pathways · Scavenging of nutrients · Host metabolic landscape

12.1 Host Metabolic Subversion: A Life-Saving Strategy

Parasitism is defined as an intimate, often obligatory and symbiotic relationship between two organisms during which the parasite benefits at the host's expenses. During a parasitic infection, the host cell/organism constitutes both a challenge and a crucial platform to ensure the development of the parasites. Hence, parasitic organisms need to subvert distinct host defence mechanisms for a successful colonization of the host. Most parasite organisms are believed to be metabolically dependent on the host, which has been consequently proven by the experimental demonstration of several nutritional virulence strategies adopted to survive inside the host. It is this dependency that generally makes the parasitic relationship in general obligatory. To fight against the parasites, the host can induce an indirect nutritional scarcity to the parasites through its metabolic dependence on distinct metabolites or can directly impose a metabolic restriction limiting or hijacking the access of the parasites to the host nutritional pool in a process known as nutritional immunity. As counter-attack measures in such harsh metabolic conditions, the parasites need to modulate host nutrient sensing mechanisms, altering host metabolic pathways and ultimately host nutritional requirements to fulfil the parasites' metabolic needs. The scavenging of host nutrients, particularly lipids, by the parasites also helps to outline the host metabolic landscape, being a crucial strategy to sustain parasite survival and proliferation.

This chapter covers the referred topics from a host perspective, highlighting the most significant findings in this exciting area of host–parasite metabolic coupling.

12.2 Host Metabolic Checkpoints Exploited During Infection

The dynamic and complex metabolic coupling established between a parasite and the host counterpart is the result of a long co-evolutionary process. Within such intricate interaction, the parasite and the host were continuously adapting and evolving up to the current host–parasite metabolic interaction, although, due to the constant and sudden alterations imposed by the host microenvironment, such metabolic coupling continues to be dynamically regulated. This constant pursuit and competition for nutrients within the host–parasite duet will alter the host metabolic pathways with major consequences for the host nutritional reserves, affecting eventually the phenotypic and functionality of the host cell. Such host alterations are imposed by the parasite's demand for a suitable niche. The ability to adapt to a host environment, sensing their surroundings, has been attributed to the modulation of host nutrient sensing pathways. This, in fact, constitutes a major strategy to explore host metabolic pathways and manage with metabolic stress. Using such mechanisms, parasites gain access to host nutritional reserves, which are crucial for their survival and persistence inside the host. Host sensing pathways and the mechanisms used by the parasites to explore them will be extensively discussed in the current section.

12.2.1 Nutrient Sensing Pathways

All organisms sense their surroundings, searching for better metabolic conditions to fulfil their nutritional requirements. In fact, nutritional deprivation has been considered the major driver for the establishment of distinct sets of nutrient sensing pathways. Such topic has been quite debated in the scientific community, with several authors proving the major impact of these pathways in times of nutritional scarcity. These pathways are able to detect extracellular and intracellular metabolic cues, switching from anabolism to catabolism in the presence or absence of nutrients availability, respectively (Efeyan 2015). More recently, the role played by nutrient sensing pathways has been addressed in the context of parasitic infections, where two major players stand out, the mechanistic target of rapamycin (mTOR) and AMP-activated protein kinase (AMPK).

12.2.1.1 mTOR Sensing Pathway

mTOR is defined as a protein kinase anchored to a signalling network that by responding to diverse environmental cues will promote anabolic metabolism generating energy and nutrients. mTOR signalling has a broad impact in terms of proliferation and cell growth (mass accumulation). This protein is an atypical serine/threonine kinase that belongs to the phosphoinositide 3-kinase (PI3K)-related kinase family interacting with a distinct set of proteins, forming two major complexes defined as mTOR complex 1 (mTORC1) and mTOR complex 2 (mTORC2). Yet, while mTORC1 and mTORC2 share several components, such as the catalytic mTOR subunit, the mammalian lethal with sec-13 protein 8 (mLST8), DEP domain containing mTOR-interacting protein (DEPTOR) and the Tti/1/Tel2 complex (Jacinto et al. 2004; Kaizuka et al. 2010; Kim et al. 2003; Peterson et al. 2009), the regulatory-associated protein of mechanistic target of rapamycin (raptor) is only present in the mTORC1 subunit (Hara et al. 2002; Kim et al. 2003). Conversely, rapamycin-insensitive companion of mTOR (rictor) is only part of mTORC2 signalling pathway (Jacinto et al. 2004; Sarbassov et al. 2004). The distinct composition between mTORC1 and mTORC2 may explain the different sensibilities to the classical mTOR inhibitor rapamycin (Laplante and Sabatini 2012).

mTORC1 integrates inputs from extracellular and intracellular cues, such as growth factors, stress, energy status, oxygen and amino acids. In that sense, the major metabolic pathways are regulated by mTORC1, inducing and inhibiting anabolic and catabolic processes, respectively, through the enhancement of protein and lipid synthesis while decreasing autophagic processes (Laplante and Sabatini 2012). A heterodimer protein composed of the tuberous sclerosis 1 (TSC1; also known as hamartin) and TSC2 (also known as tuberin) is considered a central regulator of the mTORC1 pathway. This protein complex functions as a GTPase-activating protein (GAP) for the Ras homologue enriched in brain (Rheb) GTPase. Such mechanism impairs the direct interaction between the GTP-bound form of Rheb with mTORC1 that occurs specifically at the lysosomal surface through the conversion of Rheb into its inactive GDP-bound state (Inoki et al. 2002; Tee et al. 2003). The TSC1/2 complex is also regulated upstream by protein kinase B (Akt/PKB), extracellular-signal-regulated kinase 1/2 (ERK1/2) and ribosomal S6 kinase (RS6K), which directly inactivate the TSC1/TSC2 complex by phosphorylation leading to mTORC1 activation (Potter et al. 2002; Roux et al. 2004). Besides these, other factors are involved in mTORC1 activation, such as Akt, TNF-α and Wnt pathway as well as increasing levels of amino acids, specifically leucine and arginine. Both amino acids are mandatory signals; they must be present for any upstream signal, including growth factors, to activate mTORC1 (Laplante and Sabatini 2012). This amino acid-dependent activation mechanism requires the interaction between Rag GTPases with the raptor component of mTORC1. After this complex interplay, mTORC1 is able to translocate from the cytoplasm to the lysosomal surface, leading to the docking of Rag GTPases on a complex known as Ragulator, which is essential for the activation of mTORC1 (Sancak et al. 2010). Ragulator functions as a lysosomal scaffold for other protein complexes such as GATOR1 and GATOR2, which are a negative and positive regulator of mTORC1 pathway, respectively (Bar-Peled et al. 2013) The cytosolic sensor for leucine was recently described as Sestrin2. This protein can interact with GATOR2, consequently inhibiting mTORC1 pathway in the absence of leucine. In terms of arginine, so far two sensors were defined: SLC38A9 and the CASTOR1. The former constitutes a lysosomal sensor, forming a supercomplex with Ragulator that is key for the transmission of arginine levels to mTORC1 pathway. The latter seems to function in parallel with SLC38A9 sensor. CASTOR1 forms a homodimer and heterodimerizes with CASTOR2. Both complexes interact with GATOR2 regulating negatively mTORC1 activity. Interestingly, the disruption of such interaction is accomplished only in the presence of arginine, permitting the direct binding of this amino acid to CASTOR1 and the consequent activation of mTORC1 pathway (Chantranupong et al. 2016; Wolfson et al. 2016). More recently, cholesterol was identified as a nutrient input that drives mTORC1 recruitment and activation at the lysosome surface through the complex SLC38A9-Niemann-Pick C1 (NPC1) (Castellano et al. 2017). SLC38A9 amino acid transporter has a conserved cholesterol-responsive element which may facilitate the mTORC1 activation by cholesterol, independently of its arginine-sensing function. Additionally, NPC1 constitutes the major lysosomal cholesterol transporter, which makes the SLC38A9-NPC1 complex key for the sensing of dietary lipids by mTORC1 pathway. Lysosomes have emerged as an important cellular platform to sustain mTORC1 activation and also as a key nutrient gateway, sensing and communicating cholesterol availability to mTORC1 (Castellano et al. 2017).

Upon activation, mTORC1 induces protein synthesis mainly by direct phosphorylation of the translational regulators eukaryotic translation initiation factor 4E (eIF4E)-binding protein 1 (4E-BP1) and S6 kinase 1 (S6K1). S6K1 is responsible for an increase of mRNA biogenesis and also for the translational initiation and elongation. Such mechanism is indeed highly relevant since the inhibition of mTORC1 significantly reduces the overall rates of protein synthesis in proliferating cells (Laplante and Sabatini 2012). mTORC1 activation also modulates as well the synthesis of lipids acting through the sterol regulatory element-binding protein 1/2 (SREBP1/2) transcription factors that control the expression of numerous genes involved in fatty acid and cholesterol synthesis (Laplante and Sabatini 2009). mTORC1 can regulate SREBP function through S6K1, at least in some cell types, and also by Lipin-1 phosphorylation (Peterson et al. 2011). The expression and activity of peroxisome proliferator-activated receptor γ (PPAR-γ), the master regulator of adipogenesis, is also controlled by the mTORC1 pathway (Zhang et al. 2009). Energy consumption is a hallmark of anabolic metabolism that is in fact regulated by mTORC1 protein. In such context, mTORC1 enhances the glycolytic flux through the increased transcription and translation of hypoxia-inducible factor 1 α (HIF-1α), which is a positive regulator of glycolytic genes (Düvel et al. 2010). Conversely, mTORC1 also promotes growth by negatively regulating autophagy, through direct phosphorylation and suppression of ULK1/Atg13/FIP200 (unc-51-like kinase 1/mammalian autophagy-related gene 13/focal adhesion kinase family-interacting protein of 200 kDa), required to initiate autophagy (Ganley et al. 2009; Hosokawa et al. 2009). Inversely, mTORC2 complex is insensitive to nutrients but does respond to growth factors such as insulin through a mechanism that might be dependent on PI3K signalling. mTORC2 after being activated regulates cytoskeletal organization and cell survival/metabolism. It can also control several members of the AGC subfamily of kinases including Akt, serum- and glucocorticoid-induced protein kinase 1 (SGK1) and protein kinase C-α (PKC-α) (Laplante and Sabatini 2012).

12.2.1.2 AMPK Sensing Pathway

AMPK protein is defined as an energetic sensor, responding to unbalanced energy levels with the upregulation of catabolic and the downregulation of anabolic pathways. In response to an energetic/nutritional stress, AMPK upregulates glucose uptake, mitochondria biogenesis and fatty acid oxidation pathways and simultaneously downregulates protein, cholesterol and fatty acid synthesis. Ultimately, in such context, AMPK will restore metabolic homeostasis ensuring cell survival and growth. AMPK becomes activated when the levels of AMP or ADP increase and subsequently the ATP levels are diminished (Hardie et al. 2012). Many physiological and pathological conditions are capable of activating AMPK such as lack of nutrients, exercise and hormones (adiponectin and ghrelin) besides a diverse set of diseases (Steinberg et al. 2009).

From a structural point of view, the mammalian AMPK is a heterotrimeric protein, composed of a catalytic domain (α) and two regulatory subunits (β and γ). The catalytic domain presents a Thr172 residue that is mainly phosphorylated by the tumour suppressor liver kinase B1 (LKB1), the pseudokinase STRAD and the scaffold protein MO25. Thr172 conserved residue can also be phosphorylated by calcium/calmodulin-dependent protein kinase kinase β (CAMKKβ) that is activated by higher levels of intracellular calcium (Hardie et al. 2012) and also by the transforming growth factor-β-activated kinase 1 (TAK1), while its role still needs to be clearly defined. Moreover, the binding of AMP and/or ADP to the γ subunit is also required for AMPK activation. This alters the structure and subsequent phosphorylation of the conserved residue Thr172, within the activation loop (Moreira et al. 2015a). Interestingly, an intriguing mechanism demonstrated showing that AMP binding causes AMPK-LKB1 interaction at the lysosomal membrane but not only in the cytoplasm, similar to mTORC1 activation (Zhang et al. 2014). Some authors proposed a model of reciprocal regulation between AMPK and mTORC1, which are activated under opposite circumstances, scarcity versus availability of nutrients, respectively (Bar-Peled and Sabatini 2014; Zhang et al. 2014).

To restore the unbalanced energetic status, AMPK upregulates mitochondrial pathways for the recovery of ATP levels (Foretz and Viollet 2011). Among its downstream targets, AMPK activates by phosphorylation of the peroxisome proliferator-activated receptor-gamma coactivator-1α (PGC-1α), a transcription coactivator that induces mitochondria biogenesis (Jäger et al. 2007). In parallel, the AMPK pathway has profound effects on the regulation of lipid metabolism. It induces the phosphorylation and translocation to the membrane of FAT/CD36 that is responsible, a least in part, for the uptake of fatty acids, and controls the activity of acetyl coA carboxylase (ACC) enzyme by phosphorylation. The latter mechanism is involved in the internalization and oxidation of fatty acids into the mitochondria. AMPK also affects glycolysis controlling the transcriptional levels of GLUT4 and HKII enzymes and enhancing the glucose uptake by the modulation of signalling events as Rab-GTPase-activating proteins (TBC1D1 and TBC1D4) (Hardie 2011). Finally, AMPK can also reestablish the nutrient and energetic pool of the cell activating ULK1, a highly conserved autophagy kinase, promoting mitophagy (Egan et al. 2011; Kim et al. 2011). Oppositely, AMPK diminishes biosynthetic pathways to decrease ATP consumption. Fatty acid synthesis is regulated negatively by ACC phosphorylation through the suppression of malonyl-coA formation. Other biosynthetic pathways such as cholesterol synthesis and transcription of genes involved in lipid biosynthesis are also suppressed by AMPK protein (Foretz and Viollet 2011).

12.2.1.3 mTOR–AMPK Interplay

In scarce nutritional conditions or in response to hypoxia, AMPK acts as a metabolic checkpoint, blocking the cellular growth, through mTORC1 inhibition. Such mechanism is achieved by the direct phosphorylation of TSC2, inactivating its GAP activity, and by the phosphorylation of the Raptor component, leading to the allosteric inhibition of mTORC1 (Gwinn et al. 2008; Kalender et al. 2010). The antagonistic interaction between AMPK and mTORC1 is clearly observed in the regulation of autophagy mainly by ULK1 phosphorylation. AMPK can trigger autophagy in a double-prolonged mechanism by directly activating ULK1 and by inhibiting simultaneously the suppressive effect of mTORC1 on ULK1. In fact, some authors were able to find distinct sites in ULK1 target by AMPK and also a direct mTOR phosphorylation site (Shang et al. 2011; Wang et al. 2001). Another convergence point between AMPK and mTOR signalling can be observed in the context of DNA damage. This major event signals through mTORC1 axis, inducing the expression of TSC2 and phosphatase and tensin homologue deleted on chromosome 10 (Pten), causing a downregulation of the entire PI3K-mTORC1 signalling (Stambolic et al. 2001). AMPK is then able to be activated through a mechanism that depends on the induction of sestrin1/2 proteins (Budanov and Karin 2008).

12.2.2 *Subversion of Host Nutrient Sensing Pathways*

The modulation of mTOR and AMPK sensing pathways constitutes an important strategy observed during the infection with apicomplexa parasites, such as *Plasmodium spp.* and *Toxoplasma gondii*, as well as with trypanosomatida parasites, like *Leishmania spp.* and *Trypanosoma cruzi*. The strategies used by these parasites as well as their downstream metabolic effects will be discussed extensively as follows and are synthesized in Box 12.1.

Box 12.1 Why Parasites Hijack Host Metabolic Checkpoints During Infection?

- The modulation of mTOR and AMPK metabolic checkpoints helps the parasites to modulate host metabolic pathways in order to scavenge nutrients/metabolites from the host.
- The nutritional virulence imposed by the parasite may divert those nutrients/metabolites from the host, altering consequently the host immune response against the parasites.
- The modulation of mTOR/AMPK sensing pathways reflects distinct functional/nutritional requirements of the parasites, which are continuously challenged within host–parasite coupling.

(continued)

Box 12.1 (continued)

- AMPK and mTOR metabolic subversion facilitates the adaptation of the parasites, during their life cycle to distinct hosts, which in turn exert different nutritional and immunological pressure.
- The hijacking of host metabolic checkpoints constitutes a successful strategy favouring parasites invasion, survival and eventually proliferation.

12.2.2.1 *Plasmodium spp.*

A possible alteration of mTOR signalling was initially observed during the *P. yoelii* liver stage infection. An increase of p-Akt and p-mTOR and a decrease of p53 expression were detected after performing protein microarray in infected hepatocytes (HepG2 expressing CD81) (Kaushansky et al. 2013). Although the mTOR activation was suggested to play an important role for the survival of *Plasmodium*, other studies generated controversy on the real impact of mTOR during a *Plasmodium* infection. Indeed, the siRNA knockdown of mTORC1 components (mTOR and raptor) in HepG2 cells failed to modify the percentage of infected cells or *P. berghei* development (Hanson et al. 2013). A significant impact of mTOR signalling during *Plasmodium* infection was more recently described. Using an experimental cerebral malaria (ECM) model, mice subjected to brief periods of dietary restriction (10–50%) displayed reduced ECM symptoms associated with a decrease of leptin levels and a reduction of mTORC1 activation in $CD4^+$ and $CD8^+$ T cells. A higher number of active T cells were found in the spleen with less infiltration into the brain tissues, supporting the decreased levels of parasitaemia found in the brain and in the spleen. As a proof of concept, the inhibition of leptin or mTOR signalling in this experimental model of infection can reduce both ECM symptoms and mortality. The authors observed an upregulation of S6 phosphorylation and CXCR3, which is commonly linked to mTORC1 activation, in $CD4^+$ and $CD8^+$ T cells after leptin treatment. Overall, these results establish mTORC1 as a downstream target of leptin in $CD4^+$ and $CD8^+$ T cells (Mejia et al. 2015). The potential for mTOR inhibition in the development of ECM has been further examined. Rapamycin treatment induces a reduction of $CD4^+$ and $CD8^+$ T cells' flux associated with a decreased accumulation of parasitized RBCs in the brain tissues. Concurrently, a decrease of ECM symptoms, with the reduction of blood–brain barrier breakdown and brain haemorrhage and an increase of mice survival, was established. In the light of these results and regarding the recent knowledge concerning the activation of metabolic pathways in T cells, the hypothesis of a possible impact of additional metabolic pathways in the control of this disease, with AMPK protein being one of the most obvious candidates, was raised (Gordon et al. 2015).

The potential impact of AMPK was initially addressed during *P. berghei* infection. A decrease of host p-AMPK and its downstream target p-ACC (S79) protein level was depicted in Huh7-infected cells, until 30 h post-infection. The pharmacological activation of AMPK in vitro and in vivo led to a reduction of *Plasmodium* liver stage infection (Ruivo et al. 2016). The reduction of AMPK activity in this context of infection comes in line with the work from Kashansky and colleagues, showing an increase of p-mTOR, a negative regulator of AMPK signalling, as well as the decrease of p53, which is in fact an AMPK downstream target. In such context, *Plasmodium*-infected cells might have a higher biosynthetic capacity, upregulating the fermentative pathway, the cholesterol and fatty acid synthesis, which is crucial to sustain the higher rate of *Plasmodium* growth (Ruivo et al. 2016). Several authors have previously reported the massive dependency of *Plasmodium* blood and liver stages on host glycolysis to provide the glucose used as the main energy source (Itani et al. 2014; Kirk et al. 1996; Pfaller et al. 1982; Slavic et al. 2011). Recently, Meireles and colleagues described how *Plasmodium* parasites modulate host machinery to acquire host glucose suggesting a possible role for AMPK in such context (Meireles et al. 2017). During the liver stage infection, *P. berghei* parasites lead to a reduction of host glucose intracellular levels associated with a reduction of intracellular ATP and consequently affecting the host energetic pool. The consequent binding of ADP/AMP to the GLUT1 glucose transporters induced a higher translocation of GLUT1 transporters to the surface of infected cells, ultimately leading to a further enhancement of glucose uptake, in a process that was suggested to be AMPK mediated. The relevance of glucose uptake via the GLUT1 transporter for the infection outcome was demonstrated by the chemical inhibition of GLUT1 transporter that impaired *P. berghei* hepatic infection (Meireles et al. 2017). Combining these studies, a restoration/increase of AMPK phosphorylation during *Plasmodium* liver stage infection might be found at later time points of infection, particularly after 30 h post-infection (p.i.). The strategies used by *Plasmodium* to subvert host nutrient sensing pathways are depicted in Fig. 12.1.

12.2.2.2 *Toxoplasma gondii*

T. gondii parasites induce a sustained activation of mTOR nutrient sensing pathway, which is essential for the host cell entering S phase of the growth cycle, although not induced through its classical upstream modulators PI3K-Akt and ERK pathways. Interestingly, only an increase of phosphorylation of ribosomal protein S6 was observed without any modulation of growth-associated mTOR substrates such as 4E-BP1 and S6K1. A striking interaction was noticed between mTOR-bearing vesicles and the parasitophorous vacuole (PV) in infected cells, which is considered the intracellular niche of *T. gondii* parasites, highlighting a possible functional interaction. Overall, these data lead to the suggestion of a potential impact of mTOR signalling in host translational regulation of mRNA, encoding for example mediators of host defence (Wang et al. 2009a). Additionally, due to the known role of autophagy, as a supportive mechanism for *T. gondii* growth during amino acid deprivation, the authors raised a hypothesis based on host–parasite metabolic adaptation (Wang et al. 2009b). It was suggested that *Toxoplasma* might obtain nutritional benefits, guiding host amino acids through a "futile cycle" of protein synthesis (mTOR activation) and degradation (autophagy) (Orlofsky 2009), increasing the

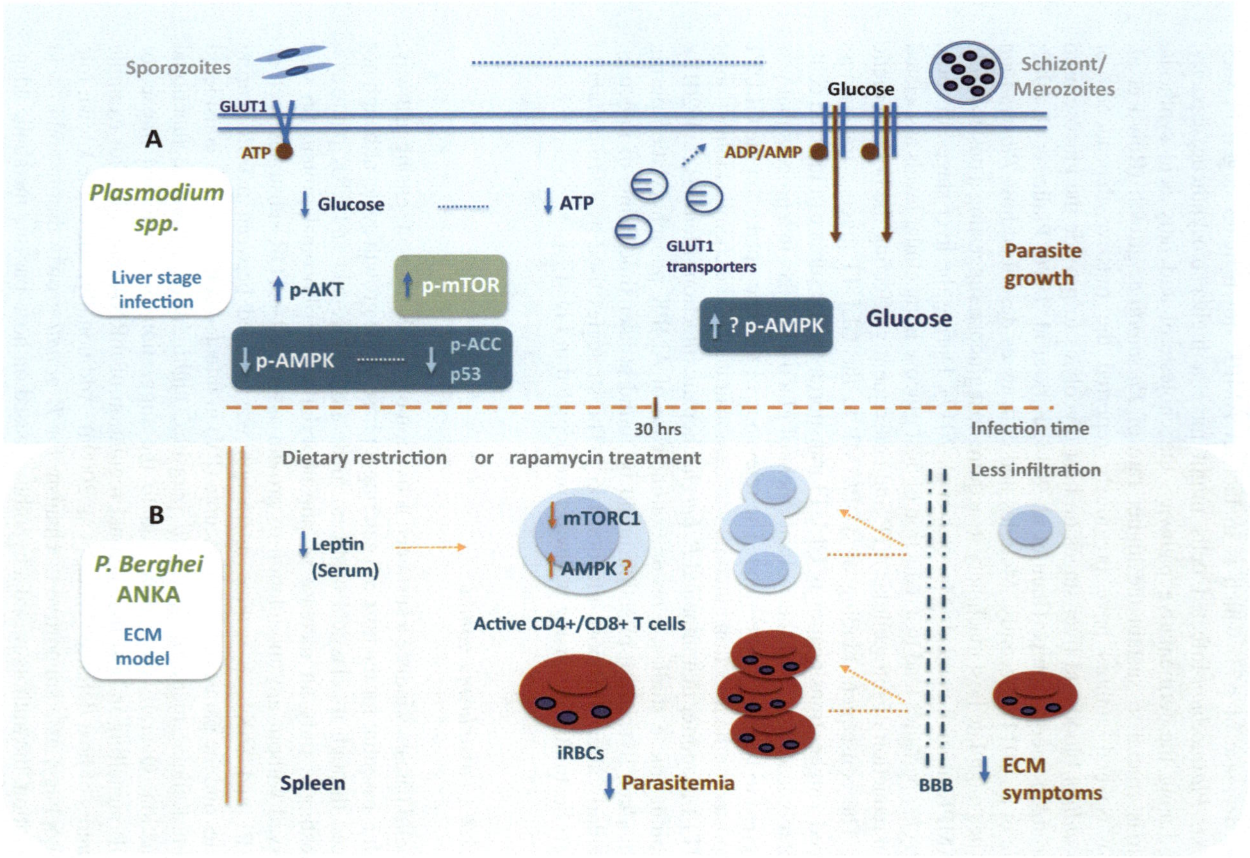

Fig. 12.1 Major strategies used by *Plasmodium* parasites to modulate host mTOR and AMPK sensing pathways. (**a**) Immediately after entry of the host liver cells, sporozoites alter host sensing pathways to accomplish their differentiation into schizont/merozoites forms. At the early interaction phase, early than 30 h p.i., the

amino acid flux through host lysosomes that can be internalized by the PV, eventually becoming available to the parasites (Coppens et al. 2006).

Wang and colleagues defined an interesting phenomenon in another work where a crucial role for mTORC2-AKT-GSK3 pathway in the migration and reorganization of *T. gondii*-infected cell structure was found. A migratory suppression of infected cells in a wounded monolayer model, explained by the inability of centrosomes to reorient in the direction of migration, seems to be governed by mTORC2 pathway (Wang et al. 2010). Moreover, the localization of centrosome and Golgi apparatus to the vacuole and the recruitment of mitochondria, microtubules and endolysosomes, which are homogenously distributed within the vacuole perimeter, constitutes vital mechanisms hijacked by *T. gondii* parasites, and modulated ultimately by mTORC2. The mechanisms used by *T. gondii* to modulate host nutrient sensing pathways are represented in Fig. 12.2. These data suggest that mitochondria and lysosomes can associate with traffic via microtubules with parasites inside the PV (Wang et al. 2010). PV is surrounded by a membrane that is freely permeable to many host metabolites (Schwab et al. 1994), which may support the relevance of homogenous organelle distribution referred earlier, particularly during nutrient starvation, to secure an efficient delivery of host resources to the PV (Wang et al. 2010).

12.2.2.3 *Leishmania spp.*

A global blockage of host translation was described initially on *L. major*-infected B10R macrophages, a BMM cell line, upon 6 h of infection through a mechanism dependent on the action of the glycoprotein GP63. The direct cleavage of mTOR protein by GP63 consequently led to mTORC1 inhibition and the activation of a well-known translational repressor named 4E-BP1 by dephosphorylation. Such mechanism is likely to be important for *Leishmania* survival since the activation

Fig. 12.1 (continued) infected cells exhibit a lower uptake of glucose through GLUT1 transporter. Within this time frame, a significant alteration of host pathways is achieved with the increased expression of p-AKT and p-mTOR and simultaneously a decrease of p-AMPK and its downstream targets, p-ACC and p53. Thus, an energetic deficit is established in such environment, turning possible the interaction between ADP/AMP with GLUT1 transporter (after 30 h p.i.), increasing their translocation to the membranes and subsequently the glucose uptake. In this event, AMPK may play an important role in restoring the energetic levels and increasing the expression of GLUT1 that is important to increase intracellular glucose levels. These early metabolic alterations guide the infected cells to establish a more permissive niche to *Plasmodium* during liver stage infection. (**b**) The modulation of host metabolism plays a crucial role dictating the infection outcome of *P. berghei ANKA* infection, particularly in a model of experimental cerebral malaria (ECM). After dietary restriction or rapamycin treatment, a reduction of leptin levels in the serum of infected mice is obtained, which leads subsequently to the downregulation of mTORC1 and possibly to an increase of AMPK pathway in splenic T cells. T cells become active establishing an environment less permissive to *Plasmodium* in the spleen. A lower infiltration of active T cells and infected RBCs through the blood–brain barrier (BBB) is also acquired, which leads to the reduction of ECM symptoms and to decreased levels of parasite load in the brain

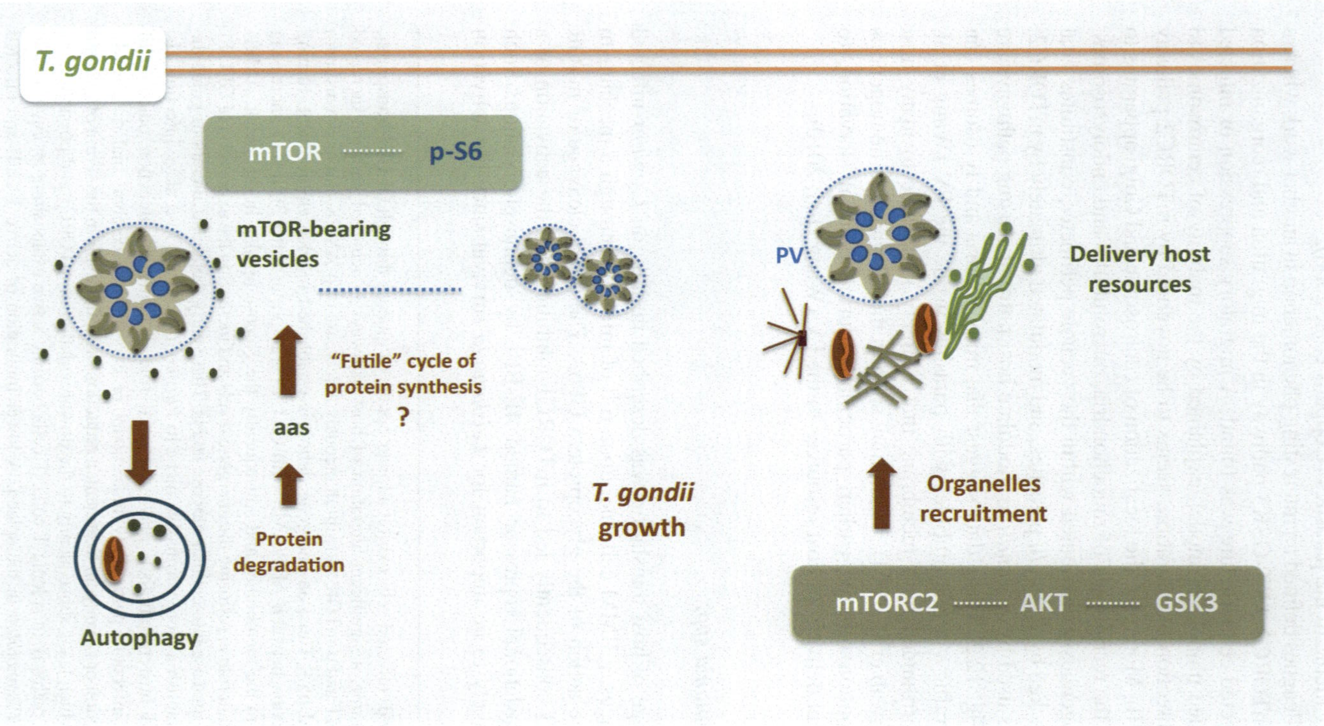

Fig. 12.2 Modulation of host mTOR sensing pathways in the context of *T. gondii* infection. *T. gondii* parasites after interacting with their host cells trigger mTOR pathway leading subsequently to the activation of S6 mTOR substrate and also to the deposition of mTOR-bearing vesicles around the PV structure. In this context was proposed a model, where host proteins degraded by autophagy, a mechanism that is known to be activated during *T. gondii* infection, can be afterwards recycled by mTOR pathway in a continuous cycle defined as futile cycle of protein synthesis. Additionally, *T. gondii* parasites can also hijack mTORC2 nutrient sensing pathway that results in the further activation of AKT and GSK3 signalling components. The activation of this pathway network led to the recruitment of mitochondria, microtubules, Golgi apparatus and centrosomes, among others, to the vacuole perimeter, which is then critical for the delivery of host resources to the parasite. Importantly, all these mechanisms are crucially implicated in the growth of *T. gondii* parasites

of 4E-BP1 through rapamycin treatment increases the parasite load in WT and not in 4EB-P1/2 double knockout (DKO) macrophages. Importantly, the genetic deletion of 4E-BP1/2 reduces parasite load in macrophages ex vivo and decreases the susceptibility to cutaneous leishmaniasis in vivo (Jaramillo et al. 2011). Yet, a contradictory report demonstrated an early hyperphosphorylation of mTOR upon infection with *L. major,* which was interpreted as a counteracting mechanism to prevent autophagic digestion of parasites in the early infection phase, ensuring promastigote differentiation into amastigotes (Frank et al. 2015). Although these data suggested a dynamic process of mTOR activation/deactivation during the initial phase of *L. major* infection, downregulation of mTOR expression by siRNA before infection decreased the infection rate compared to *L. major*-infected BMDM. Thus, mTOR phosphorylation may develop a protective role for the parasite during the early stages of infection. At 24 h p.i., mTOR hyperphosphorylation is already repressed, and an increase of glycolysis, inflammation- and autophagy-related mRNAs and miRNAs was observed, as well as a reduction of *Prkaa2* gene generating a harsh environment to eliminate intracellular amastigotes by an autophagy-dependent mechanism (Frank et al. 2015).

In the same context of infection, Rabhi and colleagues were able to detect an upregulation of a glycolytic and lipid transcriptional signature (Rabhi et al. 2012), which seems to correlate with the reduction of *prkaa2* transcript found earlier, suggesting a decrease of AMPK activity (Frank et al. 2015). Among the glycolytic transcripts, increased levels of mRNA were found for glucose transporters and glycolytic enzymes such as hexokinases (*Hk*), pyruvate kinase M2 (*Pkm2*) and lactate dehydrogenase a (*Ldha*). This phenotype was simultaneously associated with a downregulation of tricarboxylic acid (TCA) and oxidative phosphorylation (OXPHOS)-related genes. This work suggested that upon establishment, *L. major*-infected macrophages rely mainly on increased glycolytic flow for energy production. *L. major* also disturbs cholesterol and triglyceride homeostasis leading to their accumulation through the expression of scavenger receptors involved in the uptake of low-density lipoprotein (LDL), inhibiting cholesterol efflux and increasing the synthesis of triacylglycerides (TAG) (Rabhi et al. 2012).

While exploring the metabolic interactions between macrophages and the visceral *L. infantum*, we identified the activation of host AMPK as a key mechanism favouring parasite survival (Moreira et al. 2015b). We observed that following *L. infantum* infection, macrophages switch from an early glycolytic to an oxidative metabolism, in a process requiring SIRT1 and LKB1/AMPK. In the absence of SIRT1 or LKB1, infected macrophages are not able to induce AMPK activation leading to an impairment of the metabolic switch. In that sense, the AICAR-induced AMPK activation contributes to parasite survival while inhibition of AMPK using compound C resulted in lower parasite numbers in vitro. Interestingly, we demonstrated that AMPK inactivation specifically in the myeloid population led to a reduced *L. infantum* burden in vivo (Moreira et al. 2015b). The mechanisms used to hijack host nutrient sensing during *Leishmania* infection are presented in Fig. 12.3. All these examples highlight the strategies used by *Leishmania* to explore host resources, shedding light on cellular metabolic subversion mechanisms induced by a pathogen that resides strictly within the phagolysosome compartment of the host cell.

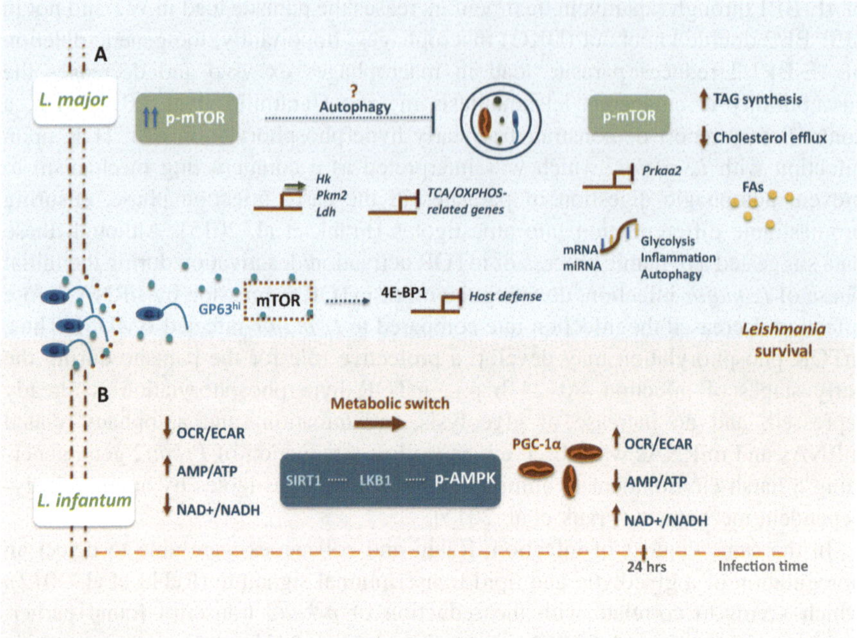

Fig. 12.3 Strategies employed by different species of *Leishmania* to subvert mTOR and AMPK host sensing pathways. (**a**) At early phase of infection (until 24 h p.i.), *L. major* parasites lead to the hyperphosphorylation of mTOR, which has been suggested to inhibit the autophagic machinery. Simultaneously, *Leishmania*-infected cells exhibit an upregulation of glycolytic enzymes and a downregulation of TCA/OXPHOS components at transcriptional levels. At 24 h p.i., a significant reduction of mTOR hyperphosphorylation state is observed with the concomitant reduction of *Prkaa2* gene and the mRNA and miRNA of glycolysis, inflammation and autophagy programs. Finally, a metabolic reprogramming towards the enhancement of TAG synthesis and the reduction of cholesterol efflux then become established. Still in the context of *L. major* infection, a striking cleavage of mTOR can be eventually acquired, being such mechanism regulated by the levels of GP63 secreted by *Leishmania*. The resultant activation of 4E-BP1 protein dampens the transcription of important genes related to host defence. (**b**) At early phase of *L. infantum* infection, a decrease of OCR/ECAR ratio is acquired along with the establishment of an energetic deficit and a reduction of NAD^+/NADH ratio. These metabolic alterations support the induction of a metabolic switch through the activation of SIRT1–LKB1–AMPK axis. A mitochondria-dependent niche is acquired through the increased expression of PGC-1α, a master regulator of mitochondria biogenesis, leading ultimately to the increase of OXPHOS pathway and to the restoration of the energetic and redox levels. Altogether the strategies used are crucial for *Leishmania spp.* survival

12.2.2.4 *Trypanosoma cruzi*

The mammalian invasion strategy used by *T. cruzi* parasite involves the host autophagic machinery, which is crucial for the lysosomal-dependent entry of tissue culture-derived trypomastigotes (TCT). As such, induction of autophagy by starvation or rapamycin-induced mTOR inhibition leads to a significant increase in the number of infected cells, while the opposite holds true as well (Romano et al. 2009).

To clarify the contribution of mTOR signalling during *T. cruzi* host invasion, a comparative analysis was performed of the molecular mechanisms of epithelial cell invasion upon metacyclic trypomastigote (MT) and TCT parasites' entry. MT parasites begin early to explore the host cell through the activation of mTOR and/or PI3K/PKC by the parasite molecule gp82. Cytoskeleton disorganization and lysosomes mobilization to the periphery of infected cells support the induction of parasites' lysosome exocytosis (Martins et al. 2011). The reversion of these effects was established through the inhibition of mTOR, PI3K and PKC pathways, which led to a consequent decrease of MT invasion. Interestingly, TCT parasites also rearrange in a similar manner the host cell structures referred previously, although through a mTOR-independent pathway. As a consequence, rapamycin treatment was responsible for enhancing TCT invasion (Martins et al. 2011). Overall, MT and TCT engage in different sets of pathways interacting differently with the host cell structures.

Recently, the application of a genome-wide RNA interference screen using as a target host gene HeLa cells allowed the identification of cellular processes that fuel *T. cruzi* growth (Caradonna et al. 2013). In this experimental setup, AKT-mTORC1 pathway stands out as a possible regulator of *T. cruzi* growth. The maintenance of intracellular ATP/ADP ratios at higher levels provided a distinct advantage for the parasite, which therefore kept AMPK activity in check. In addition, acute silencing of AMPK catalytic (PRKAA1) or the regulatory subunit (PRKAB1) provides a more favourable growth environment for intracellular *T. cruzi* (Caradonna et al. 2013).

T. cruzi can modulate the type of environment established during infection, leading to a significant inflammatory reaction which is considered one of the major alterations regarding Chagas disease. Strikingly, Nagajyothi and colleagues found that adipocytes constitute an important site of inflammation during the progression of infection. Infected adipocytes exhibited an increase in the transcription of inflammatory cytokines and chemokines, such as IL-1β, IFN-γ, TNF-α, CCL2, CCL5 and CXCL10. A higher expression of PI3K and the consequent activation of AKT, along with the concomitant reduction of adiponectin levels, are key events for the establishment of such inflammatory niche (Nagajyothi et al. 2008). Importantly, the authors suggested an indirect impairment of host AMPK in infected cells, since it constitutes one of the targets of the adiponectin pathway. Adiponectin AMPK interplay may constitute one of the mechanisms behind the anti-inflammatory role of adiponectin (Nagajyothi et al. 2008). The major strategies used by *T. cruzi* to modulate host nutrient sensing pathways are represented in Fig. 12.4. An additional work developed by Vilar-Pereira and colleagues demonstrated that AMPK activity could impair reactive oxygen species (ROS) production, an important inflammatory mediator, improving the cardiac function in established Chagas heart disease. The chronic Chagas cardiomyopathy (CCC) can be developed years after *T. cruzi* infection. Using a mice model that mimics the human CCC, treatment with resveratrol after the onset of CCC led to an activation of the AMPK pathway associated with a decrease of ROS production and heart parasite burden. Moreover, resveratrol could also ameliorate the heart function even when treatment is performed late after infection. The use of metformin, an AMPK activator, or temple

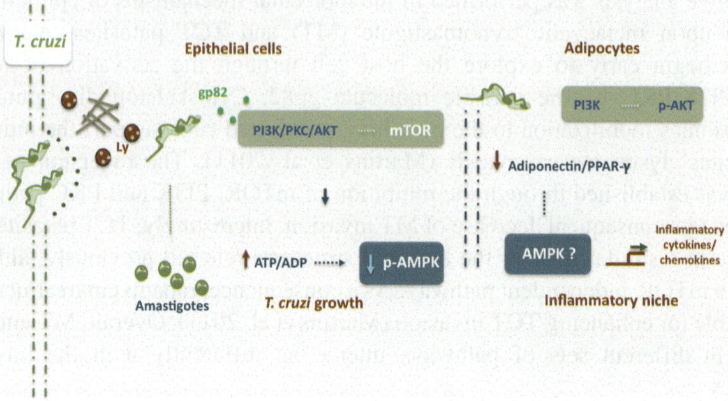

Fig. 12.4 *T. cruzi* virulence mechanisms used to manipulate AMPK/mTOR host sensing pathways. After *T cruzi* internalization, gp82, a secreted virulence factor, triggers the PI3K/PKC/AKT signalling pathways which consequently supports the activation of mTOR protein in *T. cruzi*-infected cells. Importantly, at this early contact with the host, *T. cruzi* parasites are able to alter cytoskeleton organization, recruiting it, as well as lysosomes, to the membrane site of host interaction. This mechanism may facilitate lysosome exocytosis and further *T. cruzi* invasion. The infected cells in such complex niche exhibit a higher energetic level that seems to be crucial to sustain lower levels of AMPK phosphorylation, promoting later on *T. cruzi* growth. In a context of adipocytes infection, *T. cruzi* impose an activation of PI3K-AKT pathway and a simultaneous decrease of adiponectin/PPARγ levels in those cells. All these signalling networks contribute to the establishment of an inflammatory environment with the upregulation of inflammatory cytokines and chemokine transcripts, being such process regulated possibly by AMPK sensing pathway. This local response may support the overall inflammation generated commonly during *T. cruzi* infection

(SOD mimetic) led to similar effects as resveratrol, in terms of heart function improvement and lipid peroxidation reduction, although no modifications were observed on heart parasite burden (Vilar-Pereira et al. 2016).

12.3 Redefining Host Nutritional Content During Parasitic Infections

Apart from the ability to explore host sensing pathways to remodel host metabolic fluxes, parasites can hijack important host reserves, which is considered one of the most crucial parasite nutritional strategies. Despite the variability observed among the distinct parasites regarding their specific niches, nutritional properties and nutritional pressure, a certain degree of conservation is maintained that might account for the large similarities found among them. The specific nutritional requirements of each parasite press them to hijack host nutrient pools with the modulation of host amino acids synthesis, particularly arginine and tryptophan. They can also take advantage of the host iron pool. Apart from these striking strategies that redefine host nutritional content, parasites are also able to alter host fatty acid metabolism, e.g. scavenging fatty acids (FA) such as cholesterol and/or trigger of the flux of lipid droplets (LDs) to the parasite's vicinity. The focus of the next section will be reserved to depict the strategies developed by the parasite to subvert host lipid content that is considered the most broader effect among parasitic infections.

12.3.1 Lipids: A Control Node of Host–Parasite Metabolic Coupling

Lipids constitute the most scavenged class of host nutrients with a relevant impact to the parasite energetic pool. Lipids can be incorporated as fatty acid intermediates in parasite fatty acid metabolic pathways and ultimately define the landscape of host immune response. Despite the fact that the majority of the parasites can perform de novo fatty acid synthesis, these organisms depend on the lipids scavenged from the host to compensate for their rapid replication and their urgent environment adaptation (Ginger 2006). The remodelling of host cholesterol and fatty acid pool and the modulation of lipid droplet biogenesis are strategies used by *Plasmodium spp.*, *Toxoplasma gondii*, *Leishmania spp.*, *Trypanosoma cruzi* and *Trypanosoma brucei* parasites.

12.3.1.1 Cholesterol and Fatty Acids

During the blood stage of *Plasmodium* infection, infected erythrocytes display a significant increase in the phospholipid content alongside a concomitant decrease of polyunsaturated phospholipids. These fatty acids' balance appears to be crucial for the remodelling of parasite membranes during parasite growth and division (Beaumelle and Vial 1988; Simões et al. 1992; Vial et al. 2003). The modulation of lipid metabolism in the host cell is not restricted to blood stage of infection given that a remarkable alteration in the transcription program of host lipid profile was similarly detected in *Plasmodium*-infected hepatocytes (Albuquerque et al. 2009). Such alteration seems to be time dependent, being the metabolism of fatty acids particularly enriched at 12 h post-infection. More recently, an enhancement of the host lipids repertoire, namely in neutral lipids and in phosphatidylcholine (PC), was detected in infected hepatocytes. PC is internalized by the parasite, supporting the correct localization of parasite proteins within the PV membrane (PVM), which was shown to be essential for the parasite survival (Itoe et al. 2014).

During *Plasmodium* infection, a striking interaction was uncovered between the *Plasmodium* protein UIS3, located at the PVM, and the host liver-fatty acid binding protein (L-FABP), which is responsible for the delivery of fatty acids to cytoplasmic compartments (Mikolajczak et al. 2007). This process might support the host fatty acid supply to the parasites, since *Plasmodium* PVM, as well as during *Toxoplasma* infection, is highly permeable to small metabolites forming long tubular extensions that cross the host cytoplasm to facilitate the metabolites interchange (Bano et al. 2007; Mueller et al. 2005). Upon *Plasmodium* invasion, PVM closely associates with the endoplasmic reticulum (ER), which approximates the host lipid biosynthesis machinery to the PV, corroborating the host–parasite fatty acid flow (Bano et al. 2007; Jayabalasingham et al. 2010; Labaied et al. 2011). Besides fatty acids, *Plasmodium* infection also impacts cholesterol metabolism. The expression of the scavenger receptor BI (SR-BI), a major cholesterol provider, is altered in the host membrane during infection. The increased surface expression of the SR-BI receptor supports a higher permissiveness concerning the entry and the development of *Plasmodium* parasite in the host cells (Rodrigues et al. 2008; Yalaoui et al. 2008). In the liver, *Plasmodium* can scavenge continuously cholesterol from hepatocytes until release of merozoites, via host LDL interaction (Labaied et al. 2011). Thus, these authors argue that malaria parasites are moderately dependent on host sterols for optimal development in the liver. Interestingly, they also noticed the plasticity of these parasites to survive in the liver, adapting easily to a cholesterol-deprived environment. *Plasmodium* is able to explore other sources of sterols in order to maintain their proper infectivity (Labaied et al. 2011).

Similarly, *Plasmodium T. gondii* parasites can scavenge host lipids compartmentalized in parasite endomembrane or in the proximity of PV membrane. The ability to use host lipids, such as long-chain fatty acids and cholesterol, may contribute to the biogenesis of parasite membranes (Charron and Sibley 2002). *T. gondii*-infected cells alter host cholesterol metabolism, with increased LDL endocytosis using the LDL receptor (LDLr), inducing as well a higher activity of hydroxymethylglutaryl-CoA (HMG-CoA) reductase (Blader et al. 2001; Coppens et al. 2000, 2006). Both sources are responsible for the lower levels of cholesterol and TAG in serum of infected mice. This reduction could be partially associated with the parasite acquisition of cholesterol and TAG from the host. Yet, under hypercholesterolemia conditions *T. gondii* developed a strategy to uptake lipids via an alternative receptor(s), which maintains intracellular cholesterol at optimal levels, contributing to parasite survival and replication in LDLr-deficient mice (Portugal et al. 2008). The host cell membrane cholesterol constitutes the main source of cholesterol within the PV membrane where *T. gondii* resides. Interestingly, such mechanism is considered a major strategy to support the discharge of organelles of *T. gondii*-infected cells, which is crucial for *Toxoplasma* invasion (Coppens and Joiner 2003).

The remodelling of PV membrane with the intravacuolar network (IVN), which is an extensive collection of elongated nanotubules in the vacuolar space participating in the delivery of nutrients from the PVM to the parasite, is also dependent on lipids. A continuous flow of lipids supports the growth and maturation of PV and IVN, which reflects the dynamic nature of host–parasite lipidome (Caffaro and Boothroyd 2011; Sibley et al. 1995). Additionally, *Toxoplasma* is able to explore host organelles through the recruitment of mitochondria to the cytosolic face of the PV, allowing the scavenging of lipoic acid from the host to fulfil its nutritional requirements (Crawford et al. 2006; Sinai et al. 1997). Finally, *T. gondii* induce the fragmentation of the Golgi apparatus into ministack structures, around the PV, triggering host Golgi-derived vesicles, marked with Rab14, Rab30 or Rab43, within the PV. These mechanisms allow *T. gondii* to scavenge sphingolipids from the host Golgi contributing to the parasite replication (Romano and Coppens 2013).

Several reports have been demonstrating that *Leishmania* parasites are also capable of modulating fatty acid metabolism, particularly regarding the role of cholesterol uptake and metabolism. *L. major* infection induces a transcriptional increase of key enzymes related with cholesterol synthesis, namely HMG-CoA reductase and squalene epoxidase. *L. major* interfere as well with the efflux of cholesterol and phospholipids through the downregulation of ATP binding cassette transporter A1 (ABCA1). The diminished expression of sterol 27-hydroxylase (CYP27) is also crucial to block the prevention of cholesterol accumulation in the membranes (Rabhi et al. 2012). The disruption of cholesterol metabolism in *L. major*-infected cells was attributed to the cleavage of the pre-miRNA processor Dicer enzyme mediated by GP63. The downregulation of miR-122 is coupled with low levels of serum cholesterol in VL mice, leading to disease progression (Ghosh et al. 2013). Interestingly, VL patients present a reduced serum cholesterol that could be correlated with the splenic parasite load (Ghosh et al. 2011; Lal et al. 2010). The depletion of cholesterol from the cytosol or treatment with cholesterol-lowering drugs such as statins reduced the parasite burden demonstrating the impact of cholesterol for the progression of *Leishmania* infection (Dinesh et al. 2014; Kumar et al. 2016; Tewary et al. 2006).

Semini and colleagues recently tried to clarify more recently the complex interaction between host cholesterol and parasite survival. Cholesteryl esters are sequestered by the infected cells forming a halo around the parasites within the PV structure. A reduction of Niemann-Pick Disease type C1 (NPC1) protein, which is responsible for cholesterol efflux from endocytic compartments, supports the intracellular retention of cholesteryl esters as well as the concomitant increase of mRNA-encoding proteins, which are important modulators of cholesterol biosynthesis (Semini et al. 2017). Importantly, during *Leishmania* infection the reduction of cholesterol from the membranes alters the membrane fluidity, impairing the antigen-presenting capacity of the innate immune cells (Chakraborty et al. 2005). In agreement, the replenishment of cholesterol to the membrane increases the resistance in *L. donovani*-infected hamsters (Banerjee et al. 2009).

Some controversy exists among the different authors concerning the content levels of cholesterol following *Leishmania* infection. Mukherjee and colleagues found that at early phase of infection, between 0.5 and 4 h p.i., *L. donovani* lead to an increase of host cell membrane cholesterol through the upregulation and nuclear stabilization of SREBP2. This transcription factor upregulates mitochondrial uncoupling protein 2 (*UCP2*) and *HMGCR* transcripts that are crucial for the cholesterol biosynthesis-mediated intact membrane raft formation (Mukherjee et al. 2014). Such mechanism facilitates the parasite entry and suppresses mitochondrial-ROS production supporting parasite survival.

Within the context of cholesterol trafficking, Rabhi et al. found an enhancement of CD36 receptor, LDLr and low-density lipoprotein receptor-related protein 12 precursors (Lrp12), which are crucial for the uptake of LDL, during *L. major* infection (Rabhi et al. 2012). Additionally, an increase of lipoprotein lipase precursor (Lpl) was detected that has been associated with a hydrolysis of triglycerides. Increased levels of LDL uptake were also noticed during *L. amazonensis* infection, where cholesterol from LDL is esterified and localized in compartments along the parasite body. Interestingly, such mechanism seems to rely on parasite membrane lipid microdomains (De Cicco et al. 2012). Finally, *L. amazonensis* can also alter the lipid metabolism of dendritic cells (DCs), an additional host target for *Leishmania* parasite, with the upregulation of cholesterol and long-chain fatty acids (LVFAs) uptake (Lecoeur et al. 2013). Despite the existence of some antagonism and in the light of serum cholesterol content found in VL patients, *Leishmania* has an impact on the modulation of influx/efflux cholesterol mechanisms, which become accumulated and potentially available to the parasite.

T. cruzi parasites infect a variety of host cells, including adipocytes, and are also dependent on host cholesterol metabolism, which was demonstrated by the impact of LDLr for the parasite invasion and for the fusion of the PV with host-cell lysosomes (Nagajyothi et al. 2011). In fact, the early interaction between the parasite and the host led to an accumulation of LDLr in coated pits, triggering the signalling cascades for lysosome recruitment (Nagajyothi et al. 2011; Rodriguez et al. 1996; Wilkowsky et al. 2001). Strikingly, an increased expression of LDLr was noticed in the heart of *T. cruzi*-infected CD1 mice (Nagajyothi et al. 2011). Recently, an increase in LDL levels and cholesterol was described in host tissues from acute and chronic murine Chagas disease, via LDLr–*T. cruzi* interaction (Johndrow et al. 2014). Interestingly, a direct association between LDL and *T. cruzi* trypomastigotes, but not amastigotes, was also observed, with the LDL covering the parasite cellular surface in close proximity of trans-sialidase localization. Trans-sialidase/glycoprotein 85 (gp85) is a neuraminidase that transfers sialic acid onto parasitic receptors, being involved in trypomastigote cell adhesion and invasion. Prioli and colleagues suggested that LDL inhibition of *T. cruzi* trans-sialidase may facilitate LDL endocytosis favouring *T. cruzi* infection (Prioli and Rosenberg 1990). LDL was also found in *T. cruzi* reservosomes, which are sites to accumulate endocytosed proteins and lipids, supporting a role for LDL in parasite metacyclogenesis from epimastigotes to trypomastigotes (Pereira et al. 2011; Soares and De Souza 1988; Soares et al. 1989). Overall, LDL and LDLr expression has been associated with the pathology and progression of atherosclerosis (Sunnemark et al. 2000). As a proof of concept, *T. cruzi* infection in animals fed with a high cholesterol diet led to early symptoms of atherosclerosis in mice (Nievelstein-Post et al. 1994). Therefore, the interaction between *T. cruzi* and LDL or LDLr may enhance host susceptibility for the development of atherosclerosis disease (Miao and Ndao 2014).

T. cruzi parasites are also able to interact with high-density lipoprotein (HDL), which has major functions such as removing excess cholesterol from the tissues, inhibiting LDL oxidation, inducing endothelial inflammation and promoting endothelial nitric oxide production (Miao and Ndao 2014). HDL inhibits *T. cruzi* trans-sialidase enhancing parasite infection in vitro (Prioli and Rosenberg 1990). Furthermore, it was suggested that such inhibition led to a decreased rate of trypomastigotes escaping from the PV and possibly delaying the process of trypomastigote transformation (Rubin-de-celis et al. 2006). This latter process of trypomastigotes transformation seems to be oppositely regulated by LDL-trans-sialidase inhibition. Also, HDL has been suggested to play a role as a nutritional supply in *T. cruzi* epimastigote form (Prioli and Rosenberg 1990). Finally, the potential modulation of HDL during *T. cruzi* infection was highlighted by the presence of truncated fragments of apolipoprotein A-I (Apo A-I), the major structural component of HDL, in sera of *T. cruzi*-infected patients (Ndao et al. 2010).

12.3.1.2 Lipid Droplets

Lipid droplets (LDs or lipid bodies) are present in almost all cell types, being able to store and supply fatty acids in all eukaryote and some prokaryote cells. Their main role is to secure the balance of lipid availability with metabolic and energetic demand (Pol et al. 2014). These organelles emerge from the ER lipid bilayer or from a subset of ER membranes. In eukaryotes, LDs can form de novo by progressive accumulation of neutral lipids in ER and also through fission processes that have been suggested due to the presence of such mechanism in yeasts (Long et al. 2012). They are composed of a core of neutral lipids, mainly TAG and sterol esters surrounded by a phospholipid hemimembrane associated with proteins (Murphy 2001; Tauchi-Sato et al. 2002). LDs also comprise different types of proteins such as perilipin family proteins perilipin/PLIN1, adipose differentiation-related proteins (ADRP/adipophilin/PLIN2) and tail-interacting protein of 47 kDa (TIP47/PLIN3), which are constitutively associated with the borders of LDs regulating lipid metabolism (Brasaemle et al. 2004; Wolins et al. 2001). In addition, the presence of enzymes of the lipid metabolism and membrane trafficking proteins (GTPases of the Rab family) that are important for vesicular traffic and organelle interaction can also be detected (Wan et al. 2007). Importantly, LDs also contain specific protein markers of other organelles, such as ER luminal chaperones and components of mitochondrial oxidative phosphorylation (Gao and Goodman 2015). During host/pathogen interactions, different pathogens modulate host fatty acid metabolism inducing an enrichment of LDs in proximity or in the PV, where eventually LDs may contact/internalize with the parasites, establishing a protein/nutritional flow.

Two reports found an enhancement of diacylglycerol (DAG) and TAG in *P. falciparum*-infected erythrocytes that accumulated closely to an acidic food vacuole (FV) of the parasite (Jackson et al. 2004; Palacpac et al. 2004). Jackson and colleagues suggested that an organelle storage with lipid intermediates (neutral lipids) might be generated during the digestion of phospholipids in the FV (Jackson et al. 2004), which has also an important role in the digestion of host haemoglobin. The neutral pool accumulated close to the PV might be involved in haem detoxification avoiding the toxic effects resulting from its accumulation such as oxidative damage, membrane disruption and inhibition of enzymatic reactions (Campanale et al. 2003; Loria et al. 1999) (Jackson et al. 2004; Rosenthal and Meshnick 1996). Finally, an enrichment of LDs was also observed in the liver and in the kidney of *Plasmodium*-infected mice, supporting the crucial impact of LDs during *Plasmodium* infection (Dumont et al. 1988; Pulido-Mendez et al. 2006; Rodriguez-Acosta et al. 1998).

The mobilization of host lipid resources for the intracellular *Toxoplasma* parasite was defined as one of the major strategies to sustain the remodelling of parasite membranes (Charron and Sibley 2002). LDs within the parasite may function as a storage of host lipid scavenged from the host, also being a source of lipids for parasite membrane biogenesis and eventually a site of lipid metabolism (Charron and Sibley 2002). In fact, LDs were described as a nutritional source for *T. gondii* parasites with the host lipid content and the lipolytic activity crucial for the parasite's development (Nolan et al. 2017). Upon *T. gondii* infection, there is an association between an upregulation of host LD-associated transcripts adipose differentiation-related protein (*ADRP*), diacylglycerol O-acyltransferase 2 (*DGAT2*) and LD accumulation being the PV become surrounded by LDs. Host LDs were also found to be upregulated in human cells, in a FABP-dependent process (Hu et al. 2017). The LDs enrichment is regulated upstream by two host signalling pathways, namely, c-Jun kinase (JNK) and mTOR. Such mechanism is apparently evolutionarily conserved, being distributed across different *Toxoplasma* strains and in the related parasite *Neospora caninum,* also a member of the apicomplexa phylum (Hu et al. 2017). Importantly, the impairment of LDs or the LD-associated biogenesis mechanisms reduce significantly *Toxoplasma* replication in the host cell. Moreover, *Toxoplasma* has access to host LDs storage that will be used for their own membrane remodelling and LD biogenesis, in a process regulated by Rab 7 protein and IVN structure (Nolan et al. 2017).

An increased transcription of key enzymes involved in the lipid metabolism and modulation of LDs biogenesis was also observed during *L. major* infection. As an example, *L. major* infection enhanced stearoyl-CoA desaturase 2 (*SCD2*) that regulates the ratio of monounsaturated to saturated fatty acids (Rabhi et al. 2016). Infected macrophages shift the transcriptional program towards the synthesis of TAG that is associated with the formation of lipid droplets. As a result, *L. major*-infected macrophages accumulate LDs that have been localized in a close association or even co-localized with the parasites within the PV. Live time lapse imaging allows the observation of a direct cumulative recruitment of LDs into the parasites. This suggests that LDs have a dynamic profile interacting continuously with PV and parasites (Rabhi et al. 2016). Importantly, LD accumulation is independent of parasite viability being detected in uninfected cells, although to a lower extent possibly through paracrine stimulation (Rabhi et al. 2012). LD enrichment is not exclusive of macrophages given that the same phenotype is observed in DCs infected with *L. amazonensis* amastigotes (Lecoeur et al. 2013). Some discrepancies were found concerning LD recruitment during *L. infantum chagasi* infection. Although no significant accumulation of LDs was observed throughout infection, all LDs were detected in infected cells and co-localizing specifically with the parasite in the PV. These LDs could be the result of de novo LDs synthesis performed by the parasite or could be eventually recruited from the host cells (Araújo-Santos et al. 2014).

The work developed by Melo et al. described for the first time an increased number of lipid droplets in infected macrophages during acute *T. cruzi* infection (Melo 1999). The modulation of LDs using C75, an inhibitor of fatty acid synthase, reduced critically the LD formation as well as the parasite division within macrophages highlighting the impact of LDs for a successful infection (D'Avila et al. 2011). LD biogenesis during *T. cruzi* infection is regulated upstream by the activation of toll-like receptor 2 (TLR-2). Yet, the enhanced uptake of apoptotic cells by macrophages also constitutes an important mechanism to trigger LD synthesis. Uptake of apoptotic cells by macrophages together with TLR-2 triggering supports the interaction with $\alpha v\beta 3$ integrin. The activation of TGF-β-dependent lipid droplet formation and PGE_2 synthesis is then established. Overall, this mechanism increases the permissiveness of macrophages to *T. cruzi* infection (D'Avila et al. 2011; Freire-de-lima et al. 2011; Ming et al. 1995; Silva et al. 1991). Most importantly, there is a direct correlation between lipid droplet formation during *T. cruzi* infection and increased parasite load in vivo (D'Avila et al. 2011). Similarly to *Leishmania* infection, LDs accumulate in the vicinity of *T. cruzi*-infected macrophage phagolysosome or eventually within these structures. This suggests again a possible interaction between LDs and the phagolysosome throughout *T. cruzi* infection, which may serve as a platform for lipid exchange (Melo et al. 2003). On the other hand, such interaction might be important to mount a host response. The lipids recruited during LD biogenesis, such as arachidonic acid (AA) an upstream regulator of prostaglandins, can activate actin assembly, phagosome–lysosome fusion, phagosome maturation and ultimately the dinucleotide phosphate (NADPH-) oxidase. The latter structure is implicated in the elimination of intracellular pathogens, contributing to the macrophage microbicidal properties (Anes et al. 2003; D'Avila et al. 2012; Suh et al. 2006). Fig. 12.5 represents the major strategies used by the parasites to modulate host LD biogenesis with severe implications for parasite survival.

Overall, the formation/recruitment of LDs to the vicinity of PV, where the parasites reside, could be an important strategy to obtain a high-energy substrate source, protecting the parasites against toxic intermediates and/or modulating the host immune response.

12.3.1.3 Lipid Droplets: An Immune Signalling Hub

The composition of LDs, as a reflection of their dynamic nature, could vary depending on the cell type and physiological state (Melo and Dvorak 2012). As an example, in immune cells, LDs have the entire enzymatic machinery to produce eicosanoids that are mainly inflammatory lipid mediators (Dvorak et al. 1993; Triggiani et al. 1995). Macrophages, eosinophils and neutrophils can store arachidonic acid that can be associated with phospholipids and/or neutral lipids. The LDs also possess the major enzymes for eicosanoid synthesis, cyclooxygenases (COX), lipoxygenases (LO) and leukotriene C4 (LTC4)-synthase as well as other enzymes such as phospholipase A_2 ($cPLA_2$) and MAPK proteins that are crucial to release arachidonic acid (Bozza et al. 1997, 1998).

LDs also play a crucial role within the immunologic synapse in the process of antigen presentation. LD organelles can regulate the cross-presentation of exogenous antigens to $CD8^+$ T cells in association with MHC class I, given that this process is

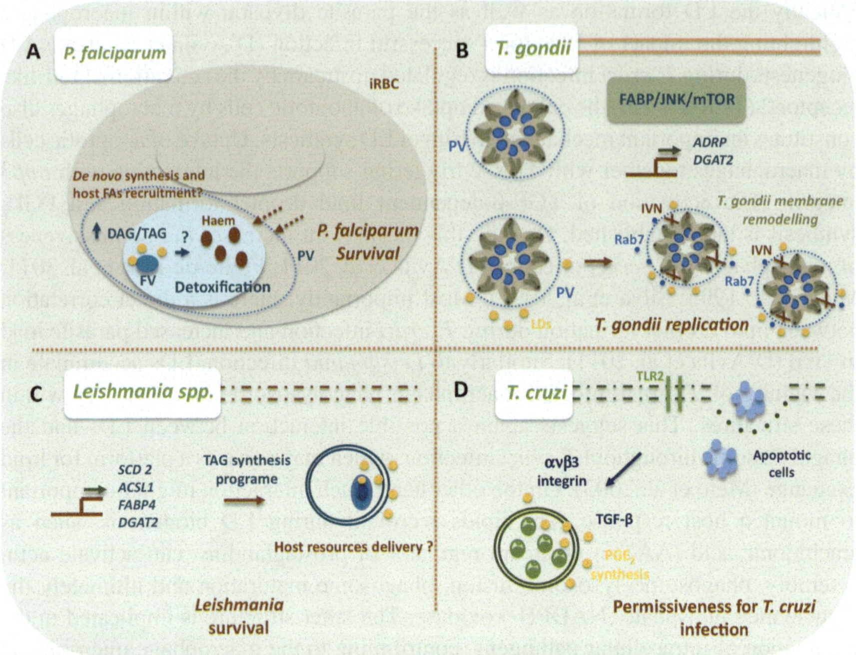

Fig. 12.5 Different mechanisms used by the parasites to hijack host LDs. (**a**) *P. falciparum*-infected red blood cells (iRBCs) exhibit an increase of DAG and TAG synthesis, which are then accumulated closely to the parasite food vacuole (FV) in the form of LDs. LDs may play a critical role in the regulation of parasite oxidative damage via haem detoxification, favouring *Plasmodium* survival. (**b**) During *T. gondii* infection, the upregulation of FABP/JNK/mTOR network together with the increase of *ADRP* and *DGAT2* transcripts guides the infected cells to synthesize LDs. These LDs become closely located to the PV. The presence of Rab7 and the formation of an intravacuolar network (IVN) structure drive the lipids stored in LDs for the remodelling of *T. gondii* membrane. (**c**) *Leishmania spp.* upregulate the transcriptional program of TAG synthesis, potentiating the LD biogenesis in the host cells. These organelles are then found in the vicinity of parasites PV and eventually some of them co-localize with the parasites. Such mechanism may favour host resource delivery and ultimately *Leishmania* survival. (**d**) *T. cruzi* internalization leads to the activation of TLR2 receptor at the surface of host membrane. This mechanism along with the uptake of apoptotic cells by infected macrophages potentiates the program of LD synthesis via $\alpha\nu\beta3$ integrin interaction and TGF-β production. All these events and the subsequent production of PGE_2 create a permissive environment for *T. cruzi* survival

critically reduced when in the presence of defects in LD formation (Bougnères et al. 2009). Recently, LDs have been defined as an important regulator of the activity of saponin-based adjuvants (SBAs) used in animal and human vaccines. The enhancement of antigen cross-presentation induced by SBAs is related to their capacity to induce LDs in the CD11b$^+$ DC in vitro and in vivo. LDs induction results in a saponin-dependent increase in cross-presentation and T cell activation (den Brok et al. 2016).

Several authors have begun to describe a potential role of LDs as crucial intermediates of host immune and microbicidal response. Two reports in the context of *T. gondii* infection described an increase in the number of LDs in the host cytoplasm of macrophages and skeletal muscle cells that fine-tune the capacity to reduce host microbicidal properties (Gomes et al. 2014; Mota et al. 2014). Additionally, a higher production of prostaglandin E2 (PGE_2), an inhibitor of Th1 immune response, eicosanoids, and an impairment of nitric oxide production were found, which benefits ultimately the parasite persistence in the host cell (Gomes et al. 2014; Mota et al. 2014). The correlation between LD formation and enhanced generation of PGE_2 by host cells has been also observed during *T. cruzi* and *L. amazonensis* infection (Melo et al. 2003; Pinheiro et al. 2009). Interestingly, within *T. cruzi*-infected macrophages, COX-2 and PGE2 immunolocalized in LDs indicates that these organelles could be the source of PGE2 in macrophages (D'Avila et al. 2011). Moreover, treatment with nonsteroidal anti-inflammatory drugs (NSAIDs) or COX inhibitors decreases LDs and PGE_2 production, resulting in the reduction of parasite growth (D'Avila et al. 2011; Freire-de-Lima et al. 2000). As such, LDs enrichment may modulate macrophage immune response with the production of PGE_2, becoming the intracellular niche more suitable for parasite survival (Toledo et al. 2016).

12.3.1.4 Is the Adipose Tissue a Safe Niche to the Parasite?

Different sets of parasites can exert a pressure over host lipid metabolism, leading eventually to the accumulation of important lipid reserves in the host cells. Yet, what is the potential role of adipocytes, which are lipid-enriched sites for the success of infection? Can the parasites use this niche to gain access to different sort of nutrients, or to evade the immune response constituting critical reservoirs for chronic infections? So far, parasites such as *T. cruzi*, *T. brucei* and *P. berghei* were observed to survive and persist in the adipose tissue (AT) of infected mice. During *P. berghei* infection, Frank-Fayerd et al. found that sequestration of malaria parasite-infected erythrocytes, also defined as the adherence of *Plasmodium*-infected red blood cells (iRBCs) to host tissue, occurs specifically in AT and lungs in a CD36-dependent mechanism (Franke-Fayard et al. 2005). Postmortem analysis of tissues from individuals who died due to malaria infection confirms that *P. falciparum*-iRBCs are sequestered not only in the lung and spleen but also in AT (Miller 1969; Wilairatana et al. 2000).

T. cruzi was also found in both white (WAT) and brown adipose tissues (BAT) (Shoemaker et al. 1970) as well as adipocytes of infected mice (Andrade and Silva 1995). In fact, both adipose tissues may represent important reservoirs for chronic infections in mice and humans (Combs et al. 2005; Ferreira et al. 2011). Although *T. brucei* are mainly extracellular parasites, residing in the blood and in interstitial spaces of many different tissues of the host, a recent report described the presence of *T. brucei* in mice adipose tissue, more specifically in the interstitial space between the adipocytes (Trindade et al. 2016). Based on scanning electron microscopy, the parasites were shown to interact with subcutaneous adipocytes, with entanglement by reticular fibres and embedment between collagen bundles. However, it is still unknown yet if the parasites can eventually contact directly with the adipocytes (Caljon et al. 2016; Capewell et al. 2016; Trindade et al. 2016). Trindade et al. also found a significant increase in the levels of parasite DNA in adipose tissue compared with the levels in blood, during the chronic stage of infection. The parasites close to the AT modify their genetic profile with the upregulation of genes that encode for fatty acid oxidation enzymes in an adaptive behaviour to its microenvironment allowing the parasitesurvival and replication in close association with that niche. Ultimately, the authors suggested a flow of nutrients between the parasite and AT to explain parasite genetic alteration (Trindade et al. 2016).

AT could represent an advantage reservoir, especially during the chronic infections given the adipocyte large half-life (Spalding et al. 2008; Wang et al. 2013). A striking mechanism can exemplify the major importance of this tissue for chronic infections. *T. cruzi*-infected mice fed with a high-fat diet present low parasitaemia levels with a reduced myocardial pathology, although, despite the parasitaemia reduction, a significant increase of parasites was detected in WAT. Within this tissue, parasites can survive and be maintained without major impact for the host (Nagajyothi et al. 2014). Overall, adipose tissue may constitute an important reservoir, being nutritionally advantageous to the parasites that need to adapt successfully to this niche to overcome the inflammatory pressure exerted within.

12.4 Concluding Remarks

The immunometabolic network established during a host–parasite interaction is one of the most relevant mechanisms dictating the outcome of infection. It can be regulated through the action of distinct parasite factors, which affect different host pathways, and/or even by the nutritional pressure exerted by the parasite and the host nutritional requirements. The parasites' exquisite metabolic demand, to support their growth, occupying either extracellular or intracellular niches, exerts a critical nutritional pressure, which ultimately helps to define host metabolic landscape. In fact, this host–parasite metabolic coupling is dynamically regulated, reflecting different nutritional requirements in each stage of parasite differentiation, a process that is dependent on the host (mammalian/insect vector) to be fully accomplished. In that sense, we may define parasites as the master renovators of their host, becoming eventually impossible to dissociate from their host counterpart, constituting host–parasite interaction a unique entity with their own metabolic requirements.

The modulation of nutrient sensing pathways constitutes one of the major strategies employed by the parasites, especially during the early interaction phase. The redirection of metabolite flux and the dampening of a proper immune response against the parasites serve as critical platforms to support parasite adaptation, survival and proliferation. Interestingly, parasites that live restricted to the PV or to the cytosol, as *Plasmodium*, *Toxoplasma* and *T. cruzi* appear to be dependent on mTOR activation to support their survival. Instead, *Leishmania* parasites that reside inside phagolysosome, despite the early detection of mTOR hyperphosphorylation, become highly dependent on AMPK activation to induce an increase of mitochondria pathways, crucial for *Leishmania* survival. This disparity among the modulation of host sensing pathways might be explained possibly by the nutritional content of phagolysosome intracellular niche. *Leishmania* complex auxotrophic requirement is indeed a tropism for this type of intracellular niche (Naderer and McConville 2011). Other pathogens with similar nutrient auxotrophies, such as the bacteria *Coxiella burnetii,* also display a phagolysosome tropism (Omsland et al. 2013).

The activation of host mTOR or AMPK sensing pathways could be a strategy of the parasites to cope their metabolic phenotype with the metabolic profile of the host cell. *Leishmania* amastigotes inside phagolysosome acquire a stringent metabolic state, with a lower metabolic and growth rate but still dependent on fatty acid oxidation to survive inside that niche (Saunders et al. 2014). The activation of catabolic pathways by AMPK might ensure a flow of tricarboxylic acid (TCA) cycle and fatty acid oxidation (FAO), intermediates and cofactors into the phagolysosome that could be consequently used by the amastigotes to support their own FAO pathway. Oppositely, *T. gondii* and *Plasmodium* parasites use glucose as a carbon source for ATP generation and various anabolic reactions, which might explain the possible modulation of host mTOR sensing pathway (Blume et al. 2009; Itani et al. 2014; Kirk et al. 1996; MacRae et al. 2012; Pfaller et al. 1982; Slavic et al. 2011). The modulation of host sensing pathways during infection might be crucial to define the baseline of the host–parasite metabolic phenotype, tracing the path, for the subsequent activation of further metabolic intermediates/pathways, creating ultimately a permissive niche to the parasites.

The remodelling of host nutrient reserves, particularly through the modulation of cholesterol and fatty acids metabolism, appears to constitute a general marker of a parasite infection. These specific mechanisms may enhance the parasite fatty acids pool to be used in the membrane remodelling, as a source of energy, or to maintain a continuous flow of lipids to specialized vacuoles where parasites reside (Lingelbach and Joiner 1998; Naderer and McConville 2011). Lipids may also be an important pathogenic factor to overcome immune response and manipulate host pathways during parasitic infections. The specialized class of lipids defined as glycosylphosphatidylinositol (GPI) lipids are widely distributed among protozoan parasites (Ramakrishnan et al. 2013). In *Leishmania*, the GPI-anchored lipophosphoglycans constitute an important virulence factor counteracting the host response (Dobson et al. 2010; Sacks 2001). Meanwhile in African trypanosomes, GPI-anchored variant surface glycoproteins are considered the key process to sustain antigenic variation mechanism used to manipulate host immune system. *Plasmodium* and *Toxoplasma* can induce the production of toxins and immune modulators that seems to be dependent on precursor GPI lipids (Debierre-Grockiego and Schwarz 2010; Nebl et al. 2005; Schofield et al. 2002). Even though many of these specialized lipids could be synthesized by the parasites, the host lipids might play a role in this pathogenic process supporting the importance of host lipid manipulation during infection.

The salvage of host lipids through the manipulation of host lipid metabolism indicates the degree of dependence of the parasites on this class of nutrients. It also may explain their tropism for the adipose tissue, a site that constitutes the most important source of lipids. Apart from the nutritional point of view, adipose tissue has also been recently addressed as one of the factors that could account for the fewer efficacies and subsequently for the relapse episodes of some antiparasitic drugs. The hydrophobic nature of adipose tissue might explain the impairment of the activity of some conventional therapeutic drugs, since most drugs are indeed hydrophilic. Such phenomenon was observed recently with posaconazole for *T. cruzi* treatment. The drug failure was directly linked to its inability to eliminate the parasites present in the mice adipose tissue (Francisco et al. 2015). A striking observation from Trindade et al. shows that *T. brucei* can be released from the adipose tissue and be relocated into the blood. This event suggests that adipose tissue might be an important niche for parasites toescape an immunological response, guiding the infection into the chronicity stage, and simultaneously support some relapse episodes that may occur after treatment (Trindade et al. 2016).

The importance of a lipid-enriched environment for the parasite survival may argue for the existence of a possible link between the success of a parasitic infection with host metabolic dysregulation. In fact, a higher susceptibility *L. infantum chagasi* was observed in an obesity mice model (diet-induced obesity), exhibiting an increased parasite burden compared to the standard-diet animals (Sarnáglia et al. 2016). Interestingly, a correlation was made between individuals with higher body mass index, who were diagnosed previously for cutaneous leishmaniasis, compared to the healthy group (Da Cunha et al. 2009). A positive association was also established between *T. gondii* seropositivity and obesity, where higher levels of *T. gondii* IgG titres were detected in obese individuals compared to the non-obese group (Reeves et al. 2013). During *T. cruzi* infection, Nagajyothi and colleagues observed in infected-high fat diet (HFD) mice, a reduction in the mortality, parasitaemia and cardiac parasite load but at the same time an increase of parasite load in adipocytes. Interestingly, during *P. berghei* infection, obese mice, genetically altered (*ob/ob* mice), were proved to be resistant to cerebral malaria (Robert et al. 2008). Additionally, mice subjected to HFD for 2 or 4 days exhibited a clear reduction of *P. berghei* in the liver, due to ROS production, and consequently a delay in the onset and progression of blood stage infection (Zuzarte-Luis et al. 2017). It is important to notice that these two reports regarding *Plasmodium* infection do not show any measurements of the parasite load in AT. And also if the parasites, in this lipid-enriched environment, become more protected from ROS effect since an antioxidant role was recently defined for LDs in a stem cell niche of the *Drosophila* model (Bailey et al. 2015). Finally, in the context of *T. cruzi* infection, two reports described an interesting relationship between diabetes, a metabolic disease and *T. cruzi* infection. Mice with chemical-induced diabetes or with genetically induced diabetes (*db/db* mice) displayed high parasitaemia levels and mortality rate (Tanowitz et al. 1988). Recently, Cabalén et al. described the ability of *T. cruzi* parasites to potentiate adipose tissue macrophage polarization towards an anti-inflammatory profile, favouring diabetes progression in a diet-obesity model

(Cabalén et al. 2016). *T. cruzi* seems to potentiate a dysregulation of host metabolic status, particularly diabetes, in order to create a glucose-rich environment that may constitute an important nutritional source for the parasite.

Metabolism is a key element of the host–parasite duet, which is able to dictate the success of a pathogen infection. Parasites within the host develop sensing and subversion mechanisms to orchestrate host metabolic pathways. Host–parasite interaction has been modulated through evolution, and this may account for the huge plasticity exhibited by the parasite. In fact, such event is clearly observed by the capacity of the parasites to reside in distinct host niches. Taking advantage of the host nutrient resources, parasites manipulate the fluxes of nutrients in accordance with time, space and demand conditions, giving them tremendous adaptation skills.

Funding Statement This work was supported by the Northern Portugal Regional Operational Programme (NORTE 2020), under the Portugal 2020 Partnership Agreement, through the European Regional Development Fund (FEDER) (NORTE-01-0145-FEDER-000013) and the Fundação para a Ciência e Tecnologia (FCT) contracts IF/00021/2014 to RS and INFECT-ERA (Project INLEISH).

References

Albuquerque SS, Carret C, Grosso A, Tarun AS, Peng X, Kappe SH, Prudêncio M, Mota MM (2009) Host cell transcriptional profiling during malaria liver stage infection reveals a coordinated and sequential set of biological events. BMC Genomics 10:270

Andrade ZA, Silva HRR (1995) Parasitism of adipocytes by *Trypanosoma cruzi*. Mem Inst Oswaldo Cruz 90:521–522

Anes E, Kühnel MP, Bos E, Moniz-Pereira J, Habermann A, Griffiths G (2003) Selected lipids activate phagosome actin assembly and maturation resulting in killing of pathogenic mycobacteria. Nat Cell Biol 5:793–802

Araújo-Santos T, Rodríguez NE, Moura-Pontes S, Dixt UG, Abánades DR, Bozza PT, Wilson ME, Borges VM (2014) Role of prostaglandin F2α production in lipid bodies from Leishmania infantum chagasi: insights on virulence. J Infect Dis 210:1951–1961

Bailey AP, Koster G, Guillermier C, Hirst EMA, MacRae JI, Lechene CP, Postle AD, Gould AP (2015) Antioxidant role for lipid droplets in a stem cell niche of *Drosophila*. Cell 163:340–353

Banerjee S, Ghosh J, Sen S, Guha R, Dhar R, Ghosh M, Datta S, Raychaudhury B, Naskar K, Haldar AK, Lal CS, Pandey K, Das VN, Das P, Roy S (2009) Designing therapies against experimental visceral leishmaniasis by modulating the membrane fluidity of antigen-presenting cells. Infect Immun 77(6):2330–2342. https://doi.org/10.1128/IAI.00057-09

Bano N, Romano JD, Jayabalasingham B, Coppens I (2007) Cellular interactions of Plasmodium liver stage with its host mammalian cell. Int J Parasitol 37:1329–1341

Bar-Peled L, Sabatini DM (2014) Regulation of mTORC1 by amino acids. Trends Cell Biol 24:400–406

Bar-Peled L, Chantranupong L, Cherniack AD, Chen WW, Ottina KA, Grabiner BC, Spear ED, Carter SL, Meyerson M, Sabatini DM (2013) A tumor suppressor complex with GAP activity for the rag GTPases that signal amino acid sufficiency to mTORC1. Science 340:1100–1106

Beaumelle BD, Vial HJ (1988) Uninfected red cells from malaria-infected blood: alteration of fatty acid composition involving a serum protein: an in vivo and in vitro study. Vitr Cell Dev Biol 24:711–718

Blader IJ, Manger ID, Boothroyd JC (2001) Microarray analysis reveals previously unknown changes in *Toxoplasma gondii*-infected human cells. J Biol Chem 276:24223–24231

Blume M, Rodriguez-Contreras D, Landfear S, Fleige T, Soldati-Favre D, Lucius R, Gupta N (2009) Host-derived glucose and its transporter in the obligate intracellular pathogen *Toxoplasma gondii* are dispensable by glutaminolysis. Proc Natl Acad Sci USA 106:12998–13003

Bougnères L, Helft J, Tiwari S, Vargas P, Chang BHJ, Chan L, Campisi L, Lauvau G, Hugues S, Kumar P et al (2009) A role for lipid bodies in the cross-presentation of phagocytosed antigens by MHC Class I in dendritic cells. Immunity 31:232–244

Bozza PT, Yu W, Penrose JF, Morgan ES, Dvorak AM, Weller PF (1997) Eosinophil lipid bodies: specific, inducible intracellular sites for enhanced eicosanoid formation. J Exp Med 186:909–920

Bozza PT, Yu W, Cassara J, Weller PF (1998) Pathways for eosinophil lipid body induction: differing signal transduction in cells from normal and hypereosinophilic subjects. J Leukoc Biol 64:563–569

Brasaemle DL, Dolios G, Shapiro L, Wang R (2004) Proteomic analysis of proteins associated with lipid droplets of basal and lipolytically stimulated 3T3-L1 adipocytes. J Biol Chem 279:46835–46842

Budanov AV, Karin M (2008) p53 Target genes Sestrin1 and Sestrin2 connect genotoxic stress and mTOR signaling. Cell 134:451–460

Cabalén ME, Cabral MF, Sanmarco LM, Andrada MC, Onofrio LI, Ponce NE, Aoki MP, Gea S, Cano RC (2016) Chronic *Trypanosoma cruzi* infection potentiates adipose tissue macrophage polarization toward an anti-inflammatory M2 phenotype and contributes to diabetes progression in a diet-induced obesity model. Oncotarget 7:13400–13415

Caffaro CE, Boothroyd JC (2011) Evidence for host cells as the major contributor of lipids in the intravacuolar network of toxoplasma-infected cells. Eukaryot Cell 10:1095–1099

Caljon G, Van Reet N, De Trez C, Vermeersch M, Pérez-Morga D, Van Den Abbeele J (2016) The dermis as a delivery site of *Trypanosoma brucei* for tsetse flies. PLoS Pathog 12. https://doi.org/10.1371/journal.ppat.1005744

Campanale N, Nickel C, Daubenberger CA, Wehlan DA, Gorman JJ, Klonis N, Becker K, Tilley L (2003) Identification and characterization of heme-interacting proteins in the malaria parasite, *Plasmodium falciparum*. J Biol Chem 278:27354–27361

Capewell P, Cren-Travaillé C, Marchesi F, Johnston P, Clucas C, Benson RA, Gorman TA, Calvo-Alvarez E, Crouzols A, Jouvion G et al (2016) The skin is a significant but overlooked anatomical reservoir for vector-borne African trypanosomes. Elife 5. https://doi.org/10.7554/eLife.17716

Caradonna KL, Engel JC, Jacobi D, Lee CH, Burleigh BA (2013) Host metabolism regulates intracellular growth of *Trypanosoma cruzi*. Cell Host Microbe 13:108–117

Castellano BM, Thelen AM, Moldavski O, Feltes M, van der Welle REN, Mydock-McGrane L, Jiang X, van Eijkeren RJ, Davis OB, Louie SM et al (2017) Lysosomal cholesterol activates mTORC1 via an SLC38A9–Niemann-Pick C1 signaling complex. Science 355:1306–1311

Chakraborty D, Banerjee S, Sen A, Banerjee KK, Das P, Roy S (2005) Leishmania donovani affects antigen presentation of macrophage by disrupting lipid rafts. J Immunol 175(5):3214–3224

Chantranupong L, Scaria SM, Saxton RA, Gygi MP, Shen K, Wyant GA, Wang T, Harper JW, Gygi SP, Sabatini DM (2016) The CASTOR proteins are arginine sensors for the mTORC1 pathway. Cell 165:153–164

Charron AJ, Sibley LD (2002) Host cells: mobilizable lipid resources for the intracellular parasite *Toxoplasma gondii*. J Cell Sci 115:3049–3059

Combs TP, Nagajyothi, Mukherjee S, De Almeida CJG, Jelicks LA, Schubert W, Lin Y, Jayabalan DS, Zhao D, Braunstein VL et al (2005) The adipocyte as an important target cell for *Trypanosoma cruzi* infection. J Biol Chem 280:24085–24094

Coppens I, Joiner KA (2003) Host but not parasite cholesterol controls toxoplasma cell entry by modulating organelle discharge. Mol Biol Cell 14:3804–3820

Coppens I, Sinai AP, Joiner KA (2000) *Toxoplasma gondii* exploits host low-density lipoprotein receptor-mediated endocytosis for cholesterol acquisition. J Cell Biol 149:167–180

Coppens I, Dunn JD, Romano JD, Pypaert M, Zhang H, Boothroyd JC, Joiner KA (2006) *Toxoplasma gondii* sequesters lysosomes from mammalian hosts in the vacuolar space. Cell 125:261–274

Crawford MJ, Thomsen-Zieger N, Ray M, Schachtner J, Roos DS, Seeber F (2006) *Toxoplasma gondii* scavenges host-derived lipoic acid despite its de novo synthesis in the apicoplast. EMBO J 25:3214–3222

D'Avila H, Freire-de-Lima CG, Roque NR, Teixeira L, Barja-Fidalgo C, Silva AR, Melo RCN, DosReis GA, Castro-Faria-Neto HC, Bozza PT (2011) Host cell lipid bodies triggered by *Trypanosoma cruzi* infection and enhanced by the uptake of apoptotic cells are associated with prostaglandin E2 generation and increased parasite growth. J Infect Dis 204:951–961

D'Avila H, Toledo DAM, Melo RCN (2012) Lipid bodies: inflammatory organelles implicated in host- *Trypanosoma cruzi* interplay during innate immune responses. Mediat Inflamm 2012:1–11

Da Cunha DF, Da Cunha SFDC, Nunes AG, Silva-Vergara ML (2009) Is an increased body mass index associated with a risk of cutaneous leishmaniasis? Rev Soc Bras Med Trop 42:494–495

De Cicco NNT, Pereira MG, Corrêa JR, Andrade-Neto VV, Saraiva FB, Chagas-Lima AC, Gondim KC, Torres-Santos EC, Folly E, Saraiva EM et al (2012) LDL uptake by leishmania amazonensis: involvement of membrane lipid microdomains. Exp Parasitol 130:330–340

Debierre-Grockiego F, Schwarz RT (2010) Immunological reactions in response to apicomplexan glycosylphosphatidylinositols. Glycobiology 20:801–811

den Brok MH, Büll C, Wassink M, de Graaf AM, Wagenaars JA, Minderman M, Thakur M, Amigorena S, Rijke EO, Schrier CC et al (2016) Saponin-based adjuvants induce cross-presentation in dendritic cells by intracellular lipid body formation. Nat Commun 7:13324

Dinesh N, Pallerla DSR, Kaur PK, Kishore Babu N, Singh S (2014) Exploring Leishmania donovani 3-hydroxy-3-methylglutaryl coenzyme A reductase (HMGR) as a potential drug target by biochemical, biophysical and inhibition studies. Microb Pathog 66:14–23

Dobson DE, Kamhawi S, Lawyer P, Turco SJ, Beverley SM, Sacks DL (2010) Leishmania major survival in selective Phlebotomus papatasi sand fly vector requires a specific SCG-encoded lipophosphoglycan galactosylation pattern. PLoS Pathog 6:e1001185. https://doi.org/10.1371/journal.ppat.1001185

Dumont ME, Dei-Cas E, Maurois P, Slomianny C, Prensier G, Houcke-Lecomte M, Vernes A, Camus D (1988) Histopathology of the liver and kidney during malaria: relation to malaria-induced dyslipoproteinemia. Ann Parasitol Hum Comp 63:171–183

Düvel K, Yecies JL, Menon S, Raman P, Lipovsky AI, Souza AL, Triantafellow E, Ma Q, Gorski R, Cleaver S et al (2010) Activation of a metabolic gene regulatory network downstream of mTOR complex 1. Mol Cell 39:171–183

Dvorak AM, Weller PF, Harvey VS, Morgan ES, Dvorak HF (1993) Ultrastructural localization of prostaglandin endoperoxide synthase (Cyclooxygenase) to isolated, purified fractions of guinea pig peritoneal macrophage and line 10 hepatocarcinoma cell lipid bodies. Int Arch Allergy Immunol 101:136–142

Efeyan ACWCSDM (2015) Nutrient-sensing mechanisms and pathways. Nature 517:302–310

Egan DF, Shackelford DB, Mihaylova MM, Gelino S, Kohnz RA, Mair W, Vasquez DS, Joshi A, Gwinn DM, Taylor R et al (2011) Phosphorylation of ULK1 (hATG1) by AMP-activated protein kinase connects energy sensing to mitophagy. Science 331:456–461

Ferreira AVM, Segatto M, Menezes Z, Macedo AM, Gelape C, de Oliveira Andrade L, Nagajyothi F, Scherer PE, Teixeira MM, Tanowitz HB (2011) Evidence for *Trypanosoma cruzi* in adipose tissue in human chronic chagas disease. Microbes Infect 13:1002–1005

Foretz M, Viollet B (2011) Regulation of hepatic metabolism by AMPK. J Hepatol *54*:827–829

Francisco AF, Lewis MD, Jayawardhana S, Taylor MC, Chatelain E, Kelly JM (2015) Limited ability of posaconazole to cure both acute and chronic *Trypanosoma cruzi* infections revealed by highly sensitive in vivo imaging. Antimicrob Agents Chemother 59:4653–4661

Frank B, Marcu A, de Oliveira Almeida Petersen AL, Weber H, Stigloher C, Mottram JC, Scholz CJ, Schurigt U (2015) Autophagic digestion of Leishmania major by host macrophages is associated with differential expression of BNIP3, CTSE, and the miRNAs miR-101c, miR-129, and miR-210. Parasit Vectors 8:404

Franke-Fayard B, Janse CJ, Cunha-Rodrigues M, Ramesar J, Büscher P, Que I, Löwik C, Voshol PJ, den Boer MA, van Duinen SG et al (2005) Murine malaria parasite sequestration: CD36 is the major receptor, but cerebral pathology is unlinked to sequestration. Proc Natl Acad Sci USA 102:11468–11473

Freire-de-Lima CG, Nascimento DO, Soares MB, Bozza PT, Castro-Faria-Neto HC, de Mello FG, DosReis G a, Lopes MF (2000) Uptake of apoptotic cells drives the growth of a pathogenic trypanosome in macrophages. Nature 403:199–203

Freire-de-lima G, Roque NR, Teixeira L, Barja-fidalgo C, Avila HD, Silva AR, Melo RCN, Dosreis GA, Castro-faria-neto HC (2011) Host cell lipid bodies triggered by *Trypanosoma cruzi* infection and enhanced by the uptake of apoptotic cells are associated with Prostaglandin E 2 generation and increased parasite growth. J Infect Dis 204:951–961

Ganley IG, Lam DH, Wang J, Ding X, Chen S, Jiang X (2009) ULK1·ATG13·FIP200 complex mediates mTOR signaling and is essential for autophagy. J Biol Chem 284:12297–12305

Gao Q, Goodman JM (2015) The lipid droplet—a well-connected organelle. Front Cell Dev Biol 3:1–12

Ghosh J, Lal CS, Pandey K, Das VN, Das P, Roychoudhury K, Roy S (2011) Human visceral leishmaniasis: decrease in serum cholesterol as a function of splenic parasite load. Ann Trop Med Parasitol 105(3):267–271. https://doi.org/10.1179/136485911X12899838683566

Ghosh J, Bose M, Roy S, Bhattacharyya SN (2013) Leishmania donovani targets dicer1 to downregulate miR-122, lower serum cholesterol, and facilitate murine liver infection. Cell Host Microbe 13:277–288

Ginger ML (2006) Niche metabolism in parasitic protozoa. Philos Trans R Soc B Biol Sci 361:101–118

Gomes AF, Magalhães KG, Rodrigues RM, de Carvalho L, Molinaro R, Bozza PT, Barbosa HS (2014) *Toxoplasma gondii*-skeletal muscle cells interaction increases lipid droplet biogenesis and positively modulates the production of IL-12, IFN-g and PGE2. Parasit Vectors 7:47

Gordon EB, Hart GT, Tran TM, Waisberg M, Akkaya M, Skinner J, Zinöcker S, Pena M, Yazew T, Qi C et al (2015) Inhibiting the mammalian target of rapamycin blocks the development of experimental cerebral malaria. mBio 6:1–17

Gwinn DM, Shackelford DB, Egan DF, Mihaylova MM, Mery A, Vasquez DS, Turk BE, Shaw RJ (2008) AMPK phosphorylation of raptor mediates a metabolic checkpoint. Mol Cell 30:214–226

Hanson KK, Ressurreicao AS, Buchholz K, Prudencio M, Herman-Ornelas JD, Rebelo M, Beatty WL, Wirth DF, Hanscheid T, Moreira R et al (2013) Torins are potent antimalarials that block replenishment of Plasmodium liver stage parasitophorous vacuole membrane proteins. Proc Natl Acad Sci 110:E2838–E2847

Hara K, Maruki Y, Long X, Yoshino K, Oshiro N, Hidayat S, Tokunaga C, Avruch J, Yonezawa K (2002) Raptor, a binding partner of target of rapamycin (TOR), mediates TOR action. Cell 110:177–189

Hardie DG (2011) AMP-activated protein kinase-an energy sensor that regulates all aspects of cell function. Genes Dev 25:1895–1908

Hardie DG, Ross FA, Hawley SA (2012) AMPK: a nutrient and energy sensor that maintains energy homeostasis. Nat Rev Mol Cell Biol 13:251–262

Hosokawa N, Hara T, Kaizuka T, Kishi C, Takamura A, Miura Y, Iemura S, Natsume T, Takehana K, Yamada N et al (2009) Nutrient-dependent mTORC1 association with the ULK1 – Atg13 – FIP200 complex required for autophagy. Mol Biol Cell 20:1981–1991

Hu X, Binns D, Reese ML (2017) The coccidian parasites Toxoplasma and Neospora dysregulate mammalian lipid droplet biogenesis. J Biol Chem 292:11009–11020

Inoki K, Li Y, Zhu T, Wu J, Guan K-L (2002) TSC2 is phosphorylated and inhibited by Akt and suppresses mTOR signalling. Nat Cell Biol 4:648–657

Itani S, Torii M, Ishino T (2014) D-Glucose concentration is the key factor facilitating liver stage maturation of Plasmodium. Parasitol Int 63:584–590

Itoe MA, Sampaio JL, Cabal GG, Real E, Zuzarte-Luis V, March S, Bhatia SN, Frischknecht F, Thiele C, Shevchenko A et al (2014) Host cell phosphatidylcholine is a key mediator of malaria parasite survival during liver stage infection. Cell Host Microbe 16:778–786

Jacinto E, Loewith R, Schmidt A, Lin S, Rüegg MA, Hall A, Hall MN (2004) Mammalian TOR complex 2 controls the actin cytoskeleton and is rapamycin insensitive. Nat Cell Biol 6:1122–1128

Jackson KE, Klonis N, Ferguson DJP, Adisa A, Dogovski C, Tilley L (2004) Food vacuole-associated lipid bodies and heterogeneous lipid environments in the malaria parasite, *Plasmodium falciparum*. Mol Microbiol 54:109–122

Jäger SS, Handschin CC, St-Pierre JJ, Spiegelman BMBM (2007) AMP-activated protein kinase (AMPK) action in skeletal muscle via direct phosphorylation of PGC-1alpha. PNAS 104:12017–12022

Jaramillo M, Gomez MA, Larsson O, Shio MT, Topisirovic I, Contreras I, Luxenburg R, Rosenfeld A, Colina R, McMaster RW et al (2011) Leishmania repression of host translation through mTOR cleavage is required for parasite survival and infection. Cell Host Microbe 9:331–341

Jayabalasingham B, Bano N, Coppens I (2010) Metamorphosis of the malaria parasite in the liver is associated with organelle clearance. Cell Res 20:1043–1059

Johndrow C, Nelson R, Tanowitz H, Weiss LM, Nagajyothi F (2014) *Trypanosoma cruzi* infection results in an increase in intracellular cholesterol. Microbes Infect *16*:337–344

Kaizuka T, Hara T, Oshiro N, Kikkawa U, Yonezawa K, Takehana K, Iemura SI, Natsume T, Mizushima N (2010) Tti1 and Tel2 are critical factors in mammalian target of rapamycin complex assembly. J Biol Chem 285:20109–20116

Kalender A, Selvaraj A, Kim SY, Gulati P, Brûlé S, Viollet B, Kemp BE, Bardeesy N, Dennis P, Schlager JJ et al (2010) Metformin, independent of AMPK, inhibits mTORC1 in a rag GTPase-dependent manner. Cell Metab 11:390–401

Kaushansky A, Ye AS, Austin LS, Mikolajczak SA, Vaughan AM, Camargo N, Metzger PG, Douglass AN, MacBeath G, Kappe SHI (2013) Suppression of host p53 is critical for plasmodium liver-stage infection. Cell Rep 3:630–637

Kim D, Sarbassov DD, Ali SM, Latek RR, Guntur KVP, Erdjument-bromage H, Tempst P, Sabatini DM (2003) GbetaL, a positive regulator of the rapamycin- sensitive pathway required for the nutrient- sensitive interaction between raptor and mTOR. Mol Cell 11:895–904

Kim J, Kundu M, Viollet B, Guan K-L (2011) AMPK and mTOR regulate autophagy through direct phosphorylation of Ulk1. Nat Cell Biol 13:132–141

Kirk K, Horner HA, Kirk J (1996) Glucose uptake in *Plasmodium falciparum*-infected erythrocytes is an equilibrative not an active process. Mol Biochem Parasitol 82:195–205

Kumar GA, Roy S, Jafurulla M, Mandal C, Chattopadhyay A (2016) Statin-induced chronic cholesterol depletion inhibits Leishmania donovani infection: relevance of optimum host membrane cholesterol. Biochim Biophys Acta—Biomembr 1858:2088–2096

Labaied M, Jayabalasingham B, Bano N, Cha SJ, Sandoval J, Guan G, Coppens I (2011) Plasmodium salvages cholesterol internalized by LDL and synthesized de novo in the liver. Cell Microbiol 13:569–586

Lal CS, Verma N, Rabidas VN, Ranjan A, Pandey K, Verma RB, Singh D, Kumar S, Das P (2010) Total serum cholesterol determination can provide understanding of parasite burden in patients with visceral leishmaniasis infection. Clin Chim Acta 411(23–24):2112–2113. https://doi.org/10.1016/j.cca.2010.08.041

Laplante M, Sabatini DM (2009) An emerging role of mTOR in lipid biosynthesis. Curr Biol 19: R1046–R1052

Laplante M, Sabatini DM (2012) MTOR signaling in growth control and disease. Cell 149:274–293

Lecoeur H, Giraud E, Prévost M-C, Milon G, Lang T (2013) Reprogramming neutral lipid metabolism in mouse dendritic leucocytes hosting live leishmania amazonensis amastigotes. PLoS Negl Trop Dis 7:e2276

Lingelbach K, Joiner K (1998) The parasitophorous vacuole membrane surrounding Plasmodium and Toxoplasma: an unusual compartment in infected cells. J Cell Sci 111(Pt 1):1467–1475

Long AP, Manneschmidt AK, Verbrugge B, Dortch MR, Minkin SC, Prater KE, Biggerstaff JP, Dunlap JR, Dalhaimer P (2012) Lipid droplet de novo formation and fission are linked to the cell cycle in fission yeast. Traffic 13:705–714

Loria P, Miller S, Foley M, Tilley L (1999) Inhibition of the peroxidative degradation of haem as the basis of action of chloroquine and other quinoline antimalarials. Biochem J 339:363

MacRae JI, Sheiner L, Nahid A, Tonkin C, Striepen B, McConville MJ (2012) Mitochondrial metabolism of glucose and glutamine is required for intracellular growth of *Toxoplasma gondii*. Cell Host Microbe 12:682–692

Martins RM, Alves RM, Macedo S, Yoshida N (2011) Starvation and rapamycin differentially regulate host cell lysosome exocytosis and invasion by *Trypanosoma cruzi* metacyclic forms. Cell Microbiol 13:943–954

Meireles P, Sales-dias J, Andrade CM, Mello-vieira J, Mancio-silva L, Simas JP, Staines HM, Prudêncio M (2017) GLUT1-mediated glucose uptake plays a crucial role during plasmodium hepatic infection. Cell Biol 19. https://doi.org/10.1111/cmi.12646

Mejia P, Treviño-Villarreal JH, Hine C, Harputlugil E, Lang S, Calay E, Rogers R, Wirth D, Duraisingh MT, Mitchell JR (2015) Dietary restriction protects against experimental cerebral malaria via leptin modulation and T-cell mTORC1 suppression. Nat Commun 6:6050

Melo RC (1999) Depletion of immune effector cells induces myocardial damage in the acute experimental *Trypanosoma cruzi* infection: ultrastructural study in rats. Tissue Cell 31:281–290

Melo RCN, Dvorak AM (2012) Lipid body–phagosome interaction in macrophages during infectious diseases: host defense or pathogen survival strategy? PLoS Pathog 8:e1002729

Melo RCN, Ávila HD, Fabrino DL, Almeida PE, Bozza PT (2003) Macrophage lipid body induction by Chagas disease in vivo: putative intracellular domains for eicosanoid formation during infection. Tissue Cell 35:59–67

Miao Q, Ndao M (2014) *Trypanosoma cruzi* infection and host lipid metabolism. Mediators Inflamm 2014:902038. https://doi.org/10.1155/2014/902038

Mikolajczak SA, Jacobs-Lorena V, MacKellar DC, Camargo N, Kappe SHI (2007) L-FABP is a critical host factor for successful malaria liver stage development. Int J Parasitol 37:483–489

Miller LH (1969) Distribution of mature trophozoites and schizonts of *Plasmodium falciparum* in the organs of Aotus trivirgatus, the night monkey. Am J Trop Med Hyg 18:860–865

Ming M, Ewen ME, Pereira ME (1995) Trypanosome invasion of mammalian cells requires activation of the TGF beta signaling pathway. Cell 82:287–296

Moreira D, Silvestre R, Cordeiro da Silva A, Estaquier J, Foretz M, Viollet B (2015a) AMP-activated protein kinase as a target for pathogens: friends or foes? Curr Drug Targets 17:163–166

Moreira D, Rodrigues V, Abengozar M, Rivas L, Rial E, Laforge M, Li X, Foretz M, Viollet B, Estaquier J et al (2015b) Leishmania infantum Modulates Host Macrophage Mitochondrial Metabolism by Hijacking the SIRT1-AMPK Axis. PLoS Pathog. 11:1–24

Mota LAM, Roberto Neto J, Monteiro VG, Lobato CSS, de Oliveira MA, da Cunha M, D'Ávila H, Seabra SH, Bozza PT, DaMatta RA (2014) Culture of mouse peritoneal macrophages with mouse serum induces lipid bodies that associate with the parasitophorous vacuole and decrease their microbicidal capacity against *Toxoplasma gondii*. Mem Inst Oswaldo Cruz 109:767–774

Mueller A-K, Camargo N, Kaiser K, Andorfer C, Frevert U, Matuschewski K, Kappe SHI (2005) Plasmodium liver stage developmental arrest by depletion of a protein at the parasite-host interface. Proc Natl Acad Sci USA 102:3022–3027

Mukherjee M, Basu Ball W, Das PK (2014) Leishmania donovani activates SREBP2 to modulate macrophage membrane cholesterol and mitochondrial oxidants for establishment of infection. Int J Biochem Cell Biol 55:196–208

Murphy DJ (2001) The biogenesis and functions of lipid bodies in animals, plants and microorganisms. Prog Lipid Res 40:325–438

Naderer T, McConville MJ (2011) Intracellular growth and pathogenesis of *Leishmania* parasites. Essays Biochem 51:81–95

Nagajyothi F, Desruisseaux MS, Thiruvur N, Weiss LM, Braunstein VL, Albanese C, Teixeira MM, de Almeida CJ, Lisanti MP, Scherer PE et al (2008) *Trypanosoma cruzi* infection of cultured adipocytes results in an inflammatory phenotype. Obesity (Silver Spring) 16:1992–1997

Nagajyothi F, Weiss LM, Silver DL, Desruisseaux MS, Scherer PE, Herz J, Tanowitz HB (2011) *Trypanosoma cruzi* utilizes the host low density lipoprotein receptor in invasion. PLoS Negl Trop Dis 5:e953

Nagajyothi F, Weiss LM, Zhao D, Koba W, Jelicks LA, Cui M, Factor SM, Scherer PE, Tanowitz HB (2014) High fat diet modulates *Trypanosoma cruzi* infection associated myocarditis. PLoS Negl Trop Dis 8:e3118

Ndao M, Spithill TW, Caffrey R, Li H, Podust VN, Perichon R, Santamaria C, Ache A, Duncan M, Powell MR et al (2010) Identification of novel diagnostic serum biomarkers for chagas' disease in asymptomatic subjects by mass spectrometric profiling. J Clin Microbiol 48:1139–1149

Nebl T, De Veer MJ, Schofield L (2005) Stimulation of innate immune responses by malarial glycosylphosphatidylinositol via pattern recognition receptors. Parasitology 130(Suppl):S45–S62

Nievelstein-Post P, Mottino G, Fogelman A, Frank J (1994) An ultrastructural study of lipoprotein accumulation in cardiac valves of the rabbit. Arterioscler Thromb 14:1151–1161

Nolan SJ, Romano JD, Coppens I (2017) Host lipid droplets: an important source of lipids salvaged by the intracellular parasite *Toxoplasma gondii*. PLOS Pathog 13:e1006362

Omsland A, Hackstadt T, Heinzen RA (2013) Bringing culture to the uncultured: *Coxiella burnetii* and lessons for obligate intracellular bacterial pathogens. PLoS Pathog 9:e1003540

Orlofsky A (2009) Toxoplasma-induced autophagy: a window into nutritional futile cycles in mammalian cells? Autophagy 5:404–406

Palacpac NMQ, Hiramine Y, Mi-ichi F, Torii M, Kita K, Hiramatsu R, Horii T, Mitamura T (2004) Developmental-stage-specific triacylglycerol biosynthesis, degradation and trafficking as lipid bodies in *Plasmodium falciparum*-infected erythrocytes. J Cell Sci 117:1469–1480

Pereira MG, Nakayasu ES, Sant'Anna C, de Cicco NNT, Atella GC, de Souza W, Almeida IC, Cunha-e-Silva N (2011) *Trypanosoma cruzi* epimastigotes are able to store and mobilize high amounts of cholesterol in reservosome lipid inclusions. PLoS One 6. https://doi.org/10.1371/journal.pone.0022359

Peterson TR, Laplante M, Thoreen CC, Sancak Y, Kang SA, Kuehl WM, Gray NS, Sabatini DM (2009) DEPTOR is an mTOR inhibitor frequently overexpressed in multiple myeloma cells and required for their survival. Cell 137:873–886

Peterson TR, Sengupta SS, Harris TE, Carmack AE, Kang SA, Balderas E, Guertin DA, Madden KL, Carpenter AE, Finck BN et al (2011) MTOR complex 1 regulates lipin 1 localization to control the srebp pathway. Cell 146:408–420

Pfaller MA, Krogstad DJ, Parquette AR, Nguyen-Dinh P (1982) *Plasmodium falciparum*: stage-specific lactate production in synchronized cultures. Exp Parasitol 54:391–396

Pinheiro RO, Nunes MP, Pinheiro CS, D'Avila H, Bozza PT, Takiya CM, Côrte-Real S, Freire-de-Lima CG, DosReis GA (2009) Induction of autophagy correlates with increased parasite load of Leishmania amazonensis in BALB/c but not C57BL/6 macrophages. Microbes Infect 11:181–190

Pol A, Gross SP, Parton RG (2014) Biogenesis of the multifunctional lipid droplet: lipids, proteins, and sites. J Cell Biol 204:635–646

Portugal LR, Fernandes LR, Pietra Pedroso VS, Santiago HC, Gazzinelli RT, Alvarez-Leite JI (2008) Influence of low-density lipoprotein (LDL) receptor on lipid composition, inflammation and parasitism during *Toxoplasma gondii* infection. Microbes Infect 10:276–284

Potter CJ, Pedraza LG, Xu T (2002) Akt regulates growth by directly phosphorylating Tsc2. Nat Cell Biol 4:658–665

Prioli P, Rosenberg I (1990) High- and low-density lipoproteins enhance infection of *Trypanosoma cruzi* in vitro. Mol Biochem Parasitol 38:191–198

Pulido-Mendez M, Finol HJ, Giron ME, Aguilar I (2006) Ultrastructural pathological changes in mice kidney caused by Plasmodium berghei infection. J Submicrosc Cytol Pathol 38:143–148

Rabhi I, Rabhi S, Ben-Othman R, Rasche A, Consortium S, Daskalaki A, Trentin B, Piquemal D, Regnault B, Descoteaux A et al (2012) Transcriptomic signature of leishmania infected mice macrophages: a metabolic point of view. PLoS Negl Trop Dis 6:e1763

Rabhi S, Rabhi I, Trentin B, Piquemal D, Regnault B, Goyard S, Lang T, Descoteaux A, Enninga J, Guizani-Tabbane L (2016) Lipid droplet formation, their localization and dynamics during leishmania major macrophage infection. PLoS One 11:1–19

Ramakrishnan S, Serricchio M, Striepen B, Bütikofer P (2013) Lipid synthesis in protozoan parasites: a comparison between kinetoplastids and apicomplexans. Prog Lipid Res 52:488–512

Reeves GM, Mazaheri S, Snitker S, Langenberg P, Giegling I, Hartmann AM, Konte B, Friedl M, Okusaga O, Groer MW et al (2013) A positive association between *T. gondii* seropositivity and obesity. Front Public Heal 1:73

Robert V, Bourgouin C, Depoix D, Thouvenot C, Lombard M-N, Grellier P (2008) Malaria and obesity: obese mice are resistant to cerebral malaria. Malar J 7:81

Rodrigues CD, Hannus M, Prudêncio M, Martin C, Gonçalves LA, Portugal S, Epiphanio S, Akinc A, Hadwiger P, Jahn-Hofmann K et al (2008) Host scavenger receptor SR-BI plays a dual role in the establishment of malaria parasite liver infection. Cell Host Microbe 4:271–282

Rodriguez A, Samoff E, Rioult MG, Chung A (1996) Host cell invasion by trypanosomes requires lysosomes and microtubule/kinesin-mediated transport. J Cell Biol 134:349–362

Rodriguez-Acosta A, Finol HJ, Pulido-Mendez M, Marquez A, Andrade G, Gonzalez N, Aguilar I, Giron ME, Pinto A (1998) Liver ultrastructural pathology in mice infected with Plasmodium berghei. J Submicrosc Cytol Pathol 30:299–307

Romano JD, Coppens I (2013) Host Organelle Hijackers: a similar modus operandi for *Toxoplasma gondii* and *Chlamydia trachomatis*: co-infection model as a tool to investigate pathogenesis. Pathog Dis 69:72–86

Romano PS, Arboit MA, Vázquez CL, Colombo MI (2009) The autophagic pathway is a key component in the lysosomal dependent entry of *Trypanosoma cruzi* into the host cell. Autophagy 5:6–18

Rosenthal PJ, Meshnick SR (1996) Hemoglobin catabolism and iron utilization by malaria parasites. Mol Biochem Parasitol 83:131–139

Roux PP, Ballif BA, Anjum R, Gygi SP, Blenis J (2004) Tumor-promoting phorbol esters and activated Ras inactivate the tuberous sclerosis tumor suppressor complex via p90 ribosomal S6 kinase. Proc Natl Acad Sci USA 101:13489–13494

Rubin-de-celis SSC, Uemura H, Yoshida N, Schenkman S (2006) Expression of trypomastigote trans -sialidase in metacyclic forms of *Trypanosoma cruzi* increases parasite escape from its parasitophorous vacuole. Cell Microbiol 8:1888–1898

Ruivo MTG, Vera IM, Sales-Dias J, Meireles P, Gural N, Bhatia SN, Mota MM, Mancio-Silva L (2016) Host AMPK is a modulator of plasmodium liver infection. Cell Rep 16:2539–2545

Sacks DL (2001) Leishmania-sand fly interactions controlling species-specific vector competence. Cell Microbiol 3:189–196

Sancak Y, Bar-Peled L, Zoncu R, Markhard AL, Nada S, Sabatini DM (2010) Ragulator-rag complex targets mTORC1 to the lysosomal surface and is necessary for its activation by amino acids. Cell 141:290–303

Sarbassov DD, Ali SM, Kim D-H, Guertin DA, Latek RR, Erdjument-Bromage H, Tempst P, Sabatini DM (2004) Rictor, a novel binding partner of mTOR, defines a rapamycin-insensitive and raptor-independent pathway that regulates the cytoskeleton. Curr Biol 14:1296–1302

Sarnáglia G, Covre L, Pereira F, DE Matos Guedes H, Faria A, Dietza R, Rodrigues R, Maioli T, Gomes D (2016) Diet-induced obesity promotes systemic inflammation and increased susceptibility to murine visceral leishmaniasis. Parasitology 143:1647–1655

Saunders EC, Ng WW, Kloehn J, Chambers JM, Ng M, McConville MJ (2014) Induction of a stringent metabolic response in intracellular stages of *Leishmania mexicana* leads to increased dependence on mitochondrial metabolism. PLoS Pathog 10:e1003888

Schofield L, Hewitt MC, Evans K, Siomos M-A, Seeberger PH (2002) Synthetic GPI as a candidate anti-toxic vaccine in a model of malaria. Nature 418:785–789

Schwab JC, Beckers CJ, Joiner KA (1994) The parasitophorous vacuole membrane surrounding intracellular *Toxoplasma gondii* functions as a molecular sieve. Proc Natl Acad Sci USA 91:509–513

Semini G, Paape D, Paterou A, Schroeder J, Barrios-Llerena M, Aebischer T (2017) Changes to cholesterol trafficking in macrophages by Leishmania parasites infection. Microbiologyopen 6:1–13

Shang L, Chen S, Du F, Li S, Zhao L, Wang X (2011) Nutrient starvation elicits an acute autophagic response mediated by Ulk1 dephosphorylation and its subsequent dissociation from AMPK. Proc Natl Acad Sci 108:4788–4793

Shoemaker JP, Hoffman RV, Huffman DG (1970) *Trypanosoma cruzi*: preference for brown adipose tissue in mice by the Tulahuen strain. Exp Parasitol 27:403–407

Sibley LD, Niesman IR, Parmley SF (1995) Regulated secretion of multi-lamellar vesicles leads to formation of a tubulo- vesicular network in host-cell vacuoles occupied by *Toxoplasma gondii*. J Cell Sci 1677:1669–1677

Silva JS, Twardzik DR, Reed SG (1991) Regulation of *Trypanosoma cruzi* infections in vitro and in vivo by transforming growth factor beta (TGF-beta). J Exp Med 174:539–545

Simões AP, Roelofsen B, Op den Kamp J (1992) Incorporation of free fatty acids can explain alterations in the molecular species composition of phosphatidylcholine and phosphatidyethanolamine in human erythrocytes as induced by *Plasmodium falciparum*. Cell Biol Int Rep 16:533–545

Sinai AP, Webster P, Joiner KA (1997) Association of host cell endoplasmic reticulum and mitochondria with the *Toxoplasma gondii* parasitophorous vacuole membrane: a high affinity interaction. J Cell Sci 2128:2117–2128

Slavic K, Delves MJ, Prudêncio M, Talman AM, Straschil U, Derbyshire ET, Xu Z, Sinden RE, Mota MM, Morin C et al (2011) Use of a selective inhibitor to define the chemotherapeutic potential of the plasmodial hexose transporter in different stages of the parasite's life cycle. Antimicrob Agents Chemother 55:2824–2830

Soares MJ, De Souza W (1988) Cytoplasmic organelles of trypanosomatids: a cytochemical and stereological study. J Submicrosc Cytol Pathol 20:349–361

Soares MJ, Souto-Padrón T, Bonaldo MC, Goldenberg S, de Souza W (1989) A stereological study of the differentiation process in *Trypanosoma cruzi*. Parasitol Res 75:522–527

Spalding KL, Arner E, Westermark PO, Bernard S, Buchholz BA, Bergmann O, Blomqvist L, Hoffstedt J, Näslund E, Britton T et al (2008) Dynamics of fat cell turnover in humans. Obstet Gynecol Surv 63:577–578

Stambolic V, MacPherson D, Sas D, Lin Y, Snow B, Jang Y, Benchimol S, Mak TW (2001) Regulation of PTEN transcription by p53. Mol Cell 8:317–325

Steinberg GR, Kemp BE (2009) AMPK in health and disease. Physiol Rev 89(3):1025–1078. https://doi.org/10.1152/physrev.00011.2008

Suh C-I, Stull ND, Li XJ, Tian W, Price MO, Grinstein S, Yaffe MB, Atkinson S, Dinauer MC (2006) The phosphoinositide-binding protein p40phox activates the NADPH oxidase during FcgammaIIA receptor-induced phagocytosis. J Exp Med 203:1915–1925

Sunnemark D, Harris RA, Frostega J, Anders O (2000) Induction of early atherosclerosis in CBA/J mice by combination of *Trypanosoma cruzi* infection and a high cholesterol diet. Atherosclerosis 153:273–282

Tanowitz HB, Amole B, Hewlett D, Wittner M (1988) *Trypanosoma cruzi* infection in diabetic mice. Trans R Soc Trop Med Hyg 82:90–93

Tauchi-Sato K, Ozeki S, Houjou T, Taguchi R, Fujimoto T (2002) The surface of lipid droplets is a phospholipid monolayer with a unique fatty acid composition. J Biol Chem 277:44507–44512

Tee AR, Manning BD, Roux PP, Cantley LC, Blenis J (2003) Tuberous sclerosis complex gene products, Tuberin and Hamartin, control mTOR signaling by acting as a GTPase-activating protein complex toward Rheb. Curr Biol 13:1259–1268

Tewary P, Veena K, Pucadyil TJ, Chattopadhyay A, Madhubala R (2006) The sterol-binding antibiotic nystatin inhibits entry of non-opsonized Leishmania donovani into macrophages. Biochem Biophys Res Commun 339:661–666

Toledo DAM, D'Avila H, Melo RCN (2016) Host lipid bodies as platforms for intracellular survival of protozoan parasites. Front Immunol 7:1–6

Triggiani M, Oriente A, Seeds MC, Bass DA, Marone G, Chilton FH (1995) Migration of human inflammatory cells into the lung results in the remodeling of arachidonic acid into a triglyceride pool. J Exp Med 182:1181–1190

Trindade S, Rijo-Ferreira F, Carvalho T, Pinto-Neves D, Guegan F, Aresta-Branco F, Bento F, Young SA, Pinto A, Van Den Abbeele J et al (2016) Trypanosoma brucei parasites occupy and functionally adapt to the adipose tissue in mice. Cell Host Microbe 19:837–848

Vial HJ, Eldin P, Tielens AGM, Van Hellemond JJ (2003) Phospholipids in parasitic protozoa. Mol Biochem Parasitol 126:143–154

Vilar-Pereira G, Carneiro VC, Mata-Santos H, Vicentino ARR, Ramos IP, Giarola NLL, Feijó DF, Meyer-Fernandes JR, Paula-Neto HA, Medei E et al (2016) Resveratrol reverses functional chagas heart disease in mice. PLOS Pathog 12:e1005947

Wan H-C, Melo RCN, Jin Z, Dvorak AM, Weller PF (2007) Roles and origins of leukocyte lipid bodies: proteomic and ultrastructural studies. FASEB J 21:167–178

Wang Z, Wilson WA, Fujino MA, Roach PJ (2001) Antagonistic controls of autophagy and glycogen accumulation by Snf1p, the yeast homolog of AMP-activated protein kinase, and the cyclin-dependent kinase Pho85p. Mol Cell Biol 21:5742–5752

Wang Y, Weiss LM, Orlofsky A (2009a) Intracellular parasitism with *Toxoplasma gondii* stimulates mammalian-target-of-rapamycin-dependent host cell growth despite impaired signalling to S6K1 and 4E-BP1. Cell Microbiol 11:983–1000

Wang Y, Weiss LM, Orlofsky A (2009b) Host cell autophagy is induced by *Toxoplasma gondii* and contributes to parasite growth. J Biol Chem 284:1694–1701

Wang Y, Weiss LM, Orlofsky A (2010) Coordinate control of host centrosome position, organelle distribution, and migratory response by *Toxoplasma gondii* via host mTORC2. J Biol Chem 285:15611–15618

Wang QA, Tao C, Gupta RK, Scherer PE (2013) Tracking adipogenesis during white adipose tissue development, expansion and regeneration. Nat Med 19:1338–1344

Wilairatana P, Riganti M, Puchadapirom P, Punpoowong B, Vannaphan S, Udomsangpetch R, Krudsood S, Brittenham GM, Looareesuwan S (2000) Prognostic significance of skin and subcutaneous fat sequestration of parasites in severe falciparum malaria. Southeast Asian J Trop Med Public Health 31:203–212

Wilkowsky SE, Barbieri MA, Stahl P, Isola ELD (2001) *Trypanosoma cruzi*: phosphatidylinositol 3-kinase and protein kinase B activation is associated with parasite invasion. Exp Cell Res 218:211–218

Wolfson RL, Chantranupong L, Saxton RA, Shen K, Scaria SM, Cantor JR, Sabatini DM (2016) Sestrin2 is a leucine sensor for the mTORC1 pathway. Science 351:43–48

Wolins NE, Rubin B, Brasaemle DL (2001) TIP47 associates with lipid droplets. J Biol Chem 276:5101–5108

Yalaoui S, Huby T, Gego A, Rametti A, Moreau M, Collet X, Siau A, Van Gemert G, Sauerwein RW, Luty AJF et al (2008) Article scavenger receptor BI boosts hepatocyte permissiveness to plasmodium infection. Cell Host Microbe 4:283–292

Zhang HH, Huang J, Düvel K, Boback B, Wu S, Squillance RM, Wu CL, Manning BD (2009) Insulin stimulates adipogenesis through the Akt-TSC2-mTORC1 pathway. PLoS One 4:e6189

Zhang CS, Jiang B, Li M, Zhu M, Peng Y, Zhang YL, Wu YQ, Li TY, Liang Y, Lu Z et al (2014) The lysosomal v-ATPase-ragulator complex is a common activator for AMPK and mTORC1, acting as a switch between catabolism and anabolism. Cell Metab 20:526–540

Zuzarte-Luis V, Mello-Vieira J, Marreiros IM, Liehl P, Chora AF, Carret C, Carvalho T, Mota MM (2017) Dietary alterations modulate susceptibility to Plasmodium infection. Nat Microbiol 2(12):1600–1607

Chapter 13
Microbiome and Gut Dysbiosis

José E. Belizário and Joel Faintuch

Abstract The gastrointestinal (GI) tract is the residence of trillions of microorganisms that include bacteria, archaea, fungi and viruses. The collective genomes of whole microbial communities (microbiota) integrate the gut microbiome. Up to 100 genera and 1000 distinct bacterial species were identified in digestive tube niches. Gut microbiomes exert permanent pivotal functions by promoting food digestion, xenobiotic metabolism and regulation of innate and adaptive immunological processes. Proteins, peptides and metabolites released locally and at distant sites trigger many cell signalling and pathways. This intense crosstalk maintains the host-microbial homeostasis. Diet, age, diet, stress and diseases cause increases or decreases in relative abundance and diversity bacterial specie of GI and other body sites. Studies in animal models and humans have shown that a persistent imbalance of gut's microbial community, named dysbiosis, relates to inflammatory bowel diseases (IBD), irritable bowel syndrome (IBS), diabetes, obesity, cancer, cardiovascular and central nervous system disorders. Notably specific bacterial communities are promising clinical target to treat inflammatory and infectious diseases. In this context, intestinal microbiota transplantation (IMT) is one optional treatment for IBD, in particular to patients with recurrent *Clostridium difficile*-induced pseudo-membrane colitis. Here we discuss on recent discoveries linking whole gut microbiome dysbiosis to metabolic and inflammatory diseases and potential prophylactic and therapeutic applications of faecal and phage therapy, probiotic and prebiotic diets.

Keywords Microbiomes · Metabolic and gastrointestinal diseases · Probiotics · Prebiotics · Faecal therapy

J. E. Belizário (✉)
Department of Pharmacology, Institute of Biomedical Sciences, University of São Paulo, São Paulo, Brazil
e-mail: jebeliza@usp.br

J. Faintuch
Department of Gastroenterology of Medical School, University of Sao Paulo, São Paulo, Brazil

R. Silvestre, E. Torrado (eds.), *Metabolic Interaction in Infection*, Experientia Supplementum 109, https://doi.org/10.1007/978-3-319-74932-7_13

13.1 Introduction

The evolution of *Homo sapiens* is linked to a mutualistic partnership among bacteria, archaea, viruses, protozoans and fungi that cohabit many sites of the human body (Backhed et al. 2012). The number and diversity of the microorganisms are modulated by physiological and environmental factors, such as host genotype, habitat and mainly diet. The total number of bacteria in a standard adult man of 70 kg is estimated to be 3.8×10^{13}, which is approximately the number of cells in the human body (Sender et al. 2016). Bacteria are classified morphologically and biochemically based on cell wall type, cell shape, oxygen requirements, endospore production, motility and energy requirements. They are also classified phylogenetically based on the analysis of nucleotide sequences of variable regions of small subunit ribosomal RNA operons or 16S and 18S rRNA genes (Belizario and Napolitano 2015). The next generation sequencing (NGS) metagenomic method has been applied to genomic profiling of known and unknown microbes present in diverse human body sites. The phylogenetic clustering of all bacterial genomes, collectively named microbiomes, into taxonomic groups (domain, kingdom, phyla, class, order, family, genus and species) have generated complex microbiome datasets (Qin et al. 2010; Gevers et al. 2012; Flintoft 2012; Li et al. 2014; The Human Microbiome Project Consortium 2012).

The human microbiomes are divided in four major phyla: Firmicutes (65%), Bacteroidetes (16%), Actinobacteria (9%) and Proteobacteria (5%). Bacteroidetes and Firmicutes represent more than 90% of the relative abundance of gut microbes (Arumugam et al. 2011). The phylum Bacteroidetes is composed of Gram-negative, not spores forming, anaerobic bacteria that tolerate the presence of oxygen but cannot use it for growth. Firmicutes are a diverse phylum of gut microbiota composed mainly of Bacilli and Clostridia classes. They are Gram-positive, anaerobic (Clostridia) and obligate or facultative aerobes (Bacilli) characterized by a low GC content. *Clostridium* species produce endospores to survive adverse conditions, allowing them to return only when the ecosystem is favourable for their growth. In contrast, Actinobacteria, for example *Bifidobacterium*, are Gram-positive, multiple branching rods, non-motile, non-spore-forming and anaerobic. Proteobacteria (e.g. *Escherichia*, *Klebsiella*, *Enterobacter*) are aerobic or facultative anaerobic, Gram-negative, non-spore-forming rods, which inhabit the intestinal tract of all vertebrates. While *Lactobacillus*, *Staphylococcus* and *Escherichia coli* are examples of facultative anaerobic bacteria, *Bacteroides* are obligate anaerobes (Belizário and Napolitano 2015).

Microbiome studies have identified a large number of microbial compositions across a range of people from European, South Americans, Indians and Africans (Qin et al. 2010; Gevers et al. 2012; Flintoft 2012). Phylogenic approaches have been used to compare modern human microbiomes from those of ancestral population (Moeller et al. 2014; Clemente et al. 2013). European children had higher Firmicutes and Proteobacteria and lower levels depletion of Bacteroidetes, which contain the fibre degradation species such as *Prevotella* and *Xylanibacter*, in the gut flora (De Filippo et al. 2010). West Africa children had a higher prevalence of

Bacteroidetes and absence of Firmicutes (De Filippo et al. 2010). The diversity of bacterial composition, caloric load and nutrient absorption has served to clustering species-level phylotypes into enterotypes (Arumugam et al. 2011; Koren et al. 2013). A personal enterotype is based on presence in the gut of healthy and disease-associated microbial strains/species, and it determines the physiopathological state of each individual. The human skin, lung, oral, genital and gastrointestinal tract microbiomes are influenced by environment and microbe colonization and represent a key step in the dynamic evolution of *Homo sapiens* (Ley et al. 2008; Qin et al. 2010; Abubucker et al. 2012; Flintoft 2012; Moeller et al. 2014).

The deeper taxonomic resolution (strains and subspecies) and functional analyses of single nucleotide polymorphisms (SNPs) and other variant sequences in microbiome datasets are still being defined (Ley et al. 2008; Qin et al. 2010; Abubucker et al. 2012; Flintoft 2012; Li et al. 2014). At least as important is the analysis of the metabolome, which mirrors the multiple functional activities of the microbiome, along with the corresponding transcriptome and proteome. Human gut tissue-specific genome-scale metabolic models displaying metabolome maps and networks have been developed for this particular purpose (Pornputtapong et al. 2015).

The investigation of such numerous and diverse microbiome datasets is a complex task. It requires biological knowledge and skills on a set of tools to accurately predict the influence of perturbed microbe–microbe interaction network on health and disease outcomes. Here, we will present and discuss studies exploring the role of gut microbiomes on aetiology of gastroenterology and metabolic diseases. We also discuss on emerging nutritional and pharmacological strategies including faecal transplantation, probiotics, prebiotics and phage therapy to modulate species-level phylotypes, as well as general bacterial richness and diversity, of the gut microbiota.

13.2 Gut Microbiota Dictates Host's Immunological Functions

The gut is colonized by 100–400 trillion of microorganisms that live in a close symbiotic relationship (Sender et al. 2016). Most of bacteria are tightly adherent to mucus which forms the outer and inner physical barrier of over 7 m long. Many of these organisms are transmitted at early life to babies, mainly via mother's milk, which contains higher numbers of *Bifidobacteria* and *Lactobacillus* (Palmer et al. 2012). In adults, the majority of the bacteria found in the gut belong to the genera *Bacteroides*, *Parabacteroides* (Bacteroidetes) and *Clostridium* (Firmicutes) (Eckburg et al. 2005; Lepage et al. 2013). The gut is an anaerobic environment and aerobic pathogenic species cannot invade and colonize it; however, anaerobic and facultative pathogenic species can invade it causing disease. Each site of the GI tract has an unique microbiota; bacteria density increases in the jejunum/ileum and in the large intestine in comparison with the stomach and duodenum. The highest cell density is present in the colon, approximately 10^{12} colony per units/mL (99%). Colon microbiota population is mostly composed of anaerobes such *Bacteroides*, *Porphyromonas*, *Bifidobacterium*, *Lactobacillus* and *Clostridium* (genera that belong to the most abundant phyla: Bacteroidetes, Actinobacteria and Firmicutes), with anaerobic bacteria outnumbering aerobic bacteria by a factor of 100 to 1000:1.

Host and microbial communities have evolved strategies that promote a steady state for their mutual benefit. Dysbiosis refers to an imbalance in microbial species abundance, which is commonly linked to impaired gut barrier function and inflammatory cell activation (Backhed et al. 2012). It is likely that failure to adequately regulate the composition (microbial diversity) is at the onset and chronicity of many diseases, including inflammatory bowel diseases (IBD), irritable bowel syndrome (IBS), diabetes, obesity and cancer (Clemente et al. 2013; Brown et al. 2013a, b; Chang and Lin 2016; Thaiss et al. 2016). The potential association between intestinal dysbiosis and certain human diseases can be assigned, for example, by the presence of anti-inflammatory species, such as *Faecalibacterium prausnitzii*, which predominate in healthy individuals, or the presence of potentially pro-inflammatory bacteria *Bacteroides* and *Ruminococcus gnavus*, which are associated with IBD (Swidsinski et al. 2005). A decreased Bacteroidetes/Firmicutes ratio and low abundance of *Bacteroides* species have been associated with obesity in animal model studies (Verdam et al. 2013). Dysbiosis is also known in other microbiomes, such as in the skin, vagina, oral cavity, stomach and particularly in the form of small intestine bacterial overgrowth.

The gut microbiota plays a critical role in the development of gut-associated lymphoid tissues including Peyer's patches, isolated lymphoid follicles and mesenteric lymph nodes, which are responsible for a fully functional immune system (Round and Mazmanian 2009; Brown et al. 2013a, b). At early life, interactions of microbes and their products in the gut (intestine) prepare the immune system to distinguish self from non-self (invaders) (Round and Mazmanian 2009; Hooper et al. 2012; Honda and Littman 2016). The innate immune response is mediated by innate lymphoid cells (ILCs) formed by natural killer (NK) cells, cytotoxic and non-cytotoxic and helper lymphoid cell subsets that include ILC1, ILC2 and ILC3. NK cells and ILC1 produce large amounts of IFN-γ, killer antimicrobial peptides (AMPs), such as granulysins, cathelicins and defensins, and chemokines (Ostaff et al. 2013). Together, they exert critical functions on the regulation of microbial ecology and immune tolerance (Thaiss et al. 2016). Imbalance of either NK activity or AMP production can lead to the development of local inflammatory diseases as well as systemic infectious (Fuchs and Colonna 2011; Ostaff et al. 2013).

The intestinal epithelial cells (IECs) and immune cells express a variety of receptors called pattern recognition receptors (PRR) that mediate the interactions between the immune system and the commensal microbiota. Toll-like receptors (TLRs) and nucleotide-binding and oligomerization domain (NOD)-like receptors (NLRs) are examples of PRR that recognize unique microbial molecules named pathogen-associated molecular patterns (PAMPs) including lipopolysaccharides, lipid A, peptidoglycans, flagella and microbial RNA/DNA (Sangiuliano et al. 2014). These receptors activate inflammasomes, which in turn activates pro-inflammatory caspases, leading to the production of cytokines interleukins IL-1β and IL-18 (Maynard et al. 2012; Brown et al. 2013a, b; Sangiuliano et al. 2014). The different inflammasomes have redundant roles during infection, and their activation can lead to cell death by pyroptosis, which protects against infection (Jorgensen et al. 2017).

Firmicutes are positively associated with TNF-α and IL-1β production, which are critical cytokines involved in the regulation of immune cells (Maynard et al. 2012; Clemente et al. 2013; Honda and Littman 2016; Webb et al. 2016). Different inflammatory diseases are characterized by mutations or loss of some innate response genes in lymphoid tissues, smaller Peyer's patches and mesenteric lymph nodes (Hooper et al. 2012; Honda and Littman 2016). The ability to control systemic infection is markedly reduced in $NOD1^{-}/^{-}$ as well as $NOD2^{-}/^{-}$ (Maynard et al. 2012; Brown et al. 2013a, b; Sangiuliano et al. 2014). Experiments with $NOD2^{-}/^{-}$ mice revealed its participation in type 1 diabetes, which is triggered by products from gut bacterial species not yet identified (Costa et al. 2016). Patients' immunity status and genotypes are directly associated with overgrowth of pathogenic microbes.

Luminal bacteria of gut epithelium induce the synthesis and secretion of approximately 70% of bacterial-specific IgA that prevents invasion of the microorganisms into the gut tissue (Brown et al. 2013a, b). IgA class switching by mucosal B cells is maintained by retinoic acid and transforming growth factor-β (TGF-β), which are critical immune suppressive processes mediated by T helper (Th) 17 cells and regulatory T cells (Treg) (Honda and Littman 2016). The use of germ-free (GF) mice and gnotobiotic mouse models have demonstrated that immunological functions can be restored by the development of bacterial communities after GF are colonizing with microorganisms (Honda and Littman 2016; Webb et al. 2016). GF mice display underdeveloped lymphoid tissues, defective T and B cell function and low numbers of circulating $CD4^{+}$ T cells and antibody production. The equilibrium between the whole body immune system and immune regulatory functions of gut bacteria appears to be delicate. Loss of a specific species can lead to an overreaction or suppression of the innate immune system (Round and Mazmanian 2009; Webb et al. 2016).

13.3 Gut Microbiota Control the Host's Metabolic States

Microorganisms perform their functions largely through enzyme pathways to digest complex dietary carbohydrates and proteins. The synthesis of a wide range of low-molecular weight signalling molecules, including methane, hydrogen sulfide and metabolites, play either harmful or beneficial effects to the host (Wikoff et al. 2009; Tremaroli and Bäckhed 2012). Those products turn on or off both host genes and genes from other microbes. Thousands of microbiota-derived metabolites have been identified as components of the human metabolome (Overbeek et al. 2005). The maintenance of a stable, fermentative gut microbiota requires diets rich in whole plant foods particularly high in dietary fibre and polyphenols, which are processed by the enzymes from intestinal microbiota such as glycoside hydrolases and polysaccharide lyases. Under anaerobic conditions, species belonging to the Bacteroides family, and to the Clostridiaceae and Lactobacillaceae families, particularly *Citrobacter* and Serratia strains, produce short-chain fatty acids (SCFAs). SCFAs produced by microbiota are acetate (with two carbons), propionate (with three carbons) and butyrate (with four carbons). Butyrate provides energy for cellular metabolism.

Acetate and propionate pass into the bloodstream, and they are taken by liver and peripheral organs, where they can act as substrates for gluconeogenesis and lipogenesis, which, in turn, control body weight gain. SCFAs are preferred source of energy used by colonocytes (colon) (Meijer et al. 2010). SCFAs also bind to the G-protein-coupled receptors (GPRs), among them GPR41 and 43, which are present in a variety of tissues, including adipose, gut enteroendocrine cells and inflammatory cells. Acting in the gut, SCFAs induces the secretion of glucagon-like peptide (GLP-1) and peptide YY (PYY). PYY inhibits nutrient absorption from the intestinal lumen and appetite, whereas GLP1 stimulate β-cells of the pancreas to produce insulin. The gut microbiota contributes to fat deposition through the regulation of the farnesoid X receptor (FXR), the bile acid receptor that is responsible for the regulation of bile acid synthesis and hepatic triglyceride accumulation. Moreover, microbiota converts choline to trimethylamine and thus regulates directly the bioavailability of choline and indirectly the accumulation of triglycerides in the liver. Therefore, targeting FXR signalling by bile acid derivatives could help control the gut bacterial community.

SCFAs play a key role in controlling mucosal proliferation, differentiation and maintenance of mucosal integrity. Impairment in the production of SCFAs, tryptophan metabolites, GABA, noradrenaline, dopamine, acetylcholine and 5-HT (5-hydroxytryptamine or serotonine), all related to microbial metabolism, is involved in GI abnormalities, metabolic diseases and neuropsychiatric derangements (Round and Mazmanian 2009; Tremaroli and Bäckhed 2012; Gilbert et al. 2016). Indeed, several of these molecules are major neurotransmitters. Serotonin, in particular, has been vigorously investigated in connection with its central roles in appetite, sexuality, substance addiction, emotions and stress response. There are reasons to believe that most of the brain serotonin either originates from the gut microbiome or has its synthesis modulated by the brain–gut axis (O'Mahony et al. 2015).

Furthermore, the microbiota produces large quantities of epigenetically active metabolites, such as folate and vitamin A. Therefore, changes in gut microbiota (dysbiosis) can result in the epigenome changes not only directly in adjacent intestinal cells, but also distant cell populations, such as hepatocytes and adipocytes (Honda and Littman 2016). SCFA generated by microbial fermentation are critical for generation of Treg. Finally, bacteria can inhibit the growth of their competitors by specific microbial communication, cell signalling through cell-to-cell contact, metabolites and quorum sensing peptides (Miller and Bassler 2001; Ostaff et al. 2013). These phenomena are relevant for the prevention of intestinal infection and colonization by pathobionts (deleterious microorganisms).

13.4 Gut Dysbiosis Induces Inflammatory and Metabolic Diseases

Elevated intestinal barrier permeability and translocation of bacteria or endotoxin are associated with GI diseases. Certain strains are sometimes mentioned such as *E. coli*, *Klebsiella*, *Proteus*, *Enterobacter*, *Shigella*, *Salmonella* and *Serratia*. However, it is still not clear whether these changes are a contributing factor or a consequence of the disease (Harley and Karp 2012). Microbiota composition varies between different locations in the gastrointestinal tract (Eckburg et al. 2005; Arumugam et al. 2011; Cucchiara et al. 2012; Lepage et al. 2013). However, most microbiome studies reported in the literature have focused on only distal large intestine or faecal microbiota. The faecal samples contain over 1000 bacterial species, and up to 50% are uncultivable and thus uncharacterized molecularly (Qin et al. 2010; Zhou et al. 2014; Li et al. 2014). 16S rRNA gene sequencing was used to precisely identify 28 clinically relevant species and genera in stool samples of 897 healthy individuals (Almonacid et al. 2017). This curated database was used to define a 99% healthy confidence interval for each target (14 species and 14 genera). Then, the authors used this healthy reference range as a clinical tool to evaluate for positive and negative associations in people who have one of various chronic disease and medical conditions, including diarrhoea, constipation, indigestion, inflammatory bowel syndrome, ulcerative colitis, Crohn's disease, obesity, type II diabetes and other conditions. Table 13.1 presents the taxon associated and inversely associated with one or more condition.

The ratio of potentially pathogenic to beneficial commensal microbes seems to be more crucial for disease development (Baumler and Sparandio 2016). *Clostridium difficile* infection is one of the most common infectious that cause of nosocomial diarrhoea and colitis, associated with antibiotic use (Antharam et al. 2013). The vast majority of Firmicutes sequences (70%) are assigned into the Clostridia class, which is divided into 19 clusters on the basis of growth, metabolic and morphological parameters (Antharam et al. 2013). *Clostridium difficile* overgrowth is associated with the depletion of *Ruminococcaceae*, *Lachnospiraceae* and butyrogenic bacteria belonging to Clostridium cluster IV or XIVa.

IBD is a chronic relapsing inflammatory disorder affecting the GI tract, and its two main clinical phenotypes are Crohn's disease (CD) and ulcerative colitis (UC). IBDs are characterized by the infiltration of neutrophils, monocytes and lymphocytes into the intestinal lamina propria causing tissue injury, loss of goblet cells, fibrosis, erosions and ulcerations. These diseases are associated with Firmicutes and Bacteroidetes, and increases in Proteobacteria, Actinobacteria, in particular, within the families *Pasteurellaceae*, *Veillonellaceae*, *Fusobacteriaceae*, *Enterobacteriaceae*, and the adherent-invasive *E. coli* strains. IBS is a heterogeneous clinical condition diagnosed by abdominal pain, bowel movement irregularity and bloating. Patient's faecal samples have decreased *Bifidobacterium Lactobacillus*, Bacteroidetes and Actinobacteria and increased Firmicutes and Proteobacteria. The methane production, which is sometimes elevated, can be measured by breath testing. This is caused by the archaebacteria *Methanobrevibacter smithii.*

Obesity, insulin resistance and the metabolic syndromes result from a complex interaction of genetic, immune and microbial factors produced by associated gut microbiota (Ley et al. 2006; Cani et al. 2007; Harley and Karp 2012). Metabolic syndrome is a chronic inflammatory disease characterized by translocation of bacteria into blood causing metabolic endotoxemia. The clinical manifestations of metabolic syndrome include cardiovascular disease, insulin resistance and type 2 diabetes, hypertension and fatty liver disease. Individuals colonized by

Table 13.1 Gastrointestinal pathogens with positive and negative association with metabolic and inflammatory diseases

Taxon	Associated	Inversely associated
Akkermansia muciniphila	Type II Diabetes	Obesity, ulcerative colitis Crohn's disease
Anaerotruncus colihominis	Bloating	Obesity
Bacteroides fragilis	Flatulence, Diarrhoea	
Butyrivibrio crossotus		Obesity
Clostridium difficile	Diarrhoea	
Collinsella aerofaciens		IBS[a]
Desulfovibrio piger	IBD[b]	
Dialister invisus	Crohn's disease	
Methanobrevibacter smithii	Abdominal tenderness	Obesity, Constipation
Oxalobacter formigenes	Kidney stones	
Ruminococcus albus	Ulcerative colitis	
Salmonella enterica	IBS, Diarrhoea	
Streptococcus sanguinis	Kawasaki disease	
Vibrio cholerae	Diarrhoea	
Alistipes		Obesity, IBS
Barnesiella		Obesity
Bifidobacterium		Constipation , Crohn's disease, IBS
Campylobacter	IBS, Diarrhoea	
Clostridium	Diarrhoea	
Escherichia/Shigella	Crohn's disease, IBS	
Fusobacterium	IBD	
Lactobacillus	Obesity	Obesity, Type II Diabetes IBS, Diarrhea
Odoribacter		Ulcerative colitis, Crohn's disease
Prevotella		Ulcerative colitis
Roseburia		Type II Diabetes
Ruminococcus	Ulcerative colitis	
Salmonella	Diarrhoea	
Veillonella	IBS	

Associated: taxa in stool microbiome data more abundant in people who have this condition
Inversely Associated: taxa in stool microbiome less abundant in people who have this condition
Adapted from Almonacid et al. (2017)
[a]Irritable Bowel Syndrome (IBS)
[b]Inflammatory Bowel Disease (IBD)

bacteria of genera *Faecalibacterium*, *Bifidobacterium*, *Lactobacillus*, *Coprococcus* and *Methanobrevibacter* have significantly less tendency to develop metabolic disturbances and inflammation that lead to type-2-diabetes and ischemic cardiovascular disorders (Ley et al. 2006; Chatelier et al. 2013). These species are characterized by their high production of SCFAs as well as hydrogen peroxides, which are known to inhibit biofilm formation and activity of pathogenic species including *Staphylococcus aureus* (Chatelier et al. 2013). Genetic and diet-induced mouse models of obesity have shown that Firmicutes/Bacteroidetes ratio is increased in the obese animals as compared to non-obese animals (Ley et al. 2006; De Filippo et al. 2010; Chatelier et al. 2013; Verdam et al. 2013). However, conflicting results exist regarding the overall gut microbiome taxonomic differences (Harley and Karp 2012; Debelius et al. 2016; Almonacid et al. 2017). Metabolic disturbances in human and animal models of obesity may be caused by increased intestinal permeability and diffusion of LPS through circulation, which in turn promotes low-grade inflammation and insulin resistance (Cani et al. 2007; Cani and Delzenne 2011, Kootte et al. 2012). Recently, one study described the effects of high-fed diet on acetate produced by intestinal microbiota using a rat model (Perry et al. 2016). Through its effects on both digestive tract and parasympathetic system, acetate plasma levels may be responsible to increased glucose-stimulated insulin secretion, increased ghrelin secretion, hyperphagia and obesity (Perry et al. 2016). As stated by the authors, variation of acetate levels could explain why consumption of calorically concentrated foods in abundance promotes obesity and insulin resistance.

The gut–brain axis may be crucial for controlling the animal metabolic states and metabolic diseases (Seeley et al. 2015). Gastric bypass surgery named Roux-en-Y Gastric Bypass (RYGB) was applied to study food preference in rats (Hankir et al. 2017). Experimentally, RYGB induces anatomic changes, and the enteroendocrine adaptations modify intestinal gene-encoding hormone expression. Normally, a fatty-acid derivative named oleoylethanolamide (OEA) is produced when fat is ingested and it directly activates PPAR-α receptors, which is responsible to promote satiety via vague nerve that controls dopamine release in the brain (Hankir et al. 2017). Dopamine-suppressed obese animals have low levels of OEA and dopamine. By means of surgically increasing the levels of OEA and dopamine 1 receptor (D1R) expression, a shift in GI–brain axis signalling is speculated to altering animal behaviour and leading them to prefer low fat food. An ongoing human pan-omics protocol is exploring RYGB-induced genetic and enteroendocrine changes in the GI mucosa and the impact of gut microbiome metabolites in humans (Sala et al. 2016). There has been a lot of expectation that dysbiosis associated with gastrointestinal diseases, obesity and diabetes could be treated by targeted modulation of some bacteria species and/or phyla in intestinal tract.

13.5 Targeting Gut Microbiota Dysbiosis

13.5.1 Faecal Microbial Transplant (FMT)

Antibiotics have been used to treat infectious diseases over the past century. However, it is clear that antibiotic treatment can change gut microbial community significantly (Dethlefsen et al. 2008; Forslund et al. 2013). High doses and frequent use of antibiotics particularly against anaerobes, such as vancomycin, can disrupt and destabilize the normal bowel microbiome (Andersson and Hughes 2014). One of the consequences could be *Clostridium difficile* colitis. The chronic recurrent nosocomial pattern of this colitis is associated with two toxins, TcdA and TcdB, produced by

C. difficile (Baumler and Sparandio 2016). Antibiotic resistant infections, which are not uncommon, are associated with pathogens named "ESKAPE" which include *Enterococcus faecium, Staphylococcus aureus, Klebsiella pneumoniae, Acinetobacter baumanii, Pseudomonas aeruginosa* and *Enterobacter species* (Baumler and Sparandio 2016).

Faecal bacteriotherapy refers to transference of microbiota of unprocessed stool, via the upper or lower GI tract. FMT has been used as a clinical procedure, in which a liquid suspension of stool from a human (a family member, or a disease-free, screened donor) is inoculated, in order to treat refractory cases of *C. difficile* colorectal infection (Gough et al. 2011; Rupnik 2015). The faecal preparation contains from 3.0% to 10% of viable and dead bacteria and colonic cells.

It is a matter of discussion whether bacteria and cell's components affect transplant outcomes, by enhancing ILCs immunostimulatory functions (Bojanova and Bordenstein 2016). Among over 150 on-going clinical trials involving FMT, most of them– in the USA, only few target kidney- and liver-transplanted patients, however, with antibiotic-resistant bacterial colonization or infection (https://clinicaltrials.gov).

About 90% of the patients that receive microbiota transplantation for *C. difficile* infection are cured, versus 31% of those receiving vancomycin (Van Nood et al. 2013). A few species, including *Lactobacillus, Enterococcus, Bifidobacterium* and *Bacteroides* species, have inhibitory activities against *C. difficile*. Some problems remain to be solved before FMT becomes widespread (Brandt and Reddy 2011). It is known that a variety of viruses, archaea, fungi, parasitic species and metabolites are transferred together with commensal bacteria (Bojanova and Bordenstein 2016). One of the pending questions is their role, potentially both positive and negative, in intestinal immunity and restoration of microbiome normality.

Pediatric Crohn's disease patients were treated with FMT from suitable donors (Suskind et al. 2015). In this study, 31 species in the faecal microbiome following FMT at two weeks were highlighted, including *Ruminococcus torques, Odoribacter splanchnicus, Bacteroides caccae, Alistipes shahii, A. putredinis and Parabacteroides merdae. E. coli* expansion contributed to disease outcome and may be a helpful predictor of efficacy of faecal therapy (Suskind et al. 2015).

In animals, the gut microbiome plays a crucial role in weight gain and obesity (Ley et al. 2006; Harley and Karp 2012; Tremaroli and Bäckhed 2012). In ob/ob mice, there is a 50% reduction in Bacteroidetes, with proportional increases in Firmicutes and Archaea in obese mice (Turnbaugh et al. 2006). Transferring the microbiota from fat mice to germ-free hosts induces greater weight gain than in those receiving the microbiota from lean donors (Ley et al. 2006). High-fat intake increases intestinal permeability. In contrast, the administration of *Bifidobaterium infantis* reduced the production of proinflammatory cytokines and white adipose tissue gain in mice (Cani et al. 2007).

Results in human metabolic disorders have not been conclusive. Transfer of intestinal microbiota from lean donors mildly increased insulin sensitivity in subjects with metabolic syndrome (Vrieze et al. 2012). So far, however, these results have not been confirmed in larger protocols. In 75 obese and prediabetic volunteers that underwent an 8-day antibiotic treatment (amoxicillin and vancomycin), dysbiosis did not alter gut permeability (Reijnders et al. 2016). Neither did LPS blood levels, used as biomarker of systemic inflammation, nor enzymes involved in the regulation of hepatic metabolism of lipids and cholesterol, despite tremendous alterations in gut microbiota (Reijnders et al. 2016).

The use of microbiome-modulating therapies to treat obesity is still a big challenge (Debelius et al. 2016). Novel preventive and therapeutic approaches to obesity depend on better experimental techniques. In this way, recently scientific communities have initiated global efforts to standardization, which has the support of the Unified Microbiome Initiative consortium and the Microbiome Quality Control Project (Alivisatos et al. 2015, http://www.mbqc.org/).

13.5.2 Probiotics and Prebiotics

Probiotics are defined as live microorganisms that can survive and temporarily colonize the gut. The gastric juice pH ranges from 3 to 5, and this is the first major obstacle that one probiotic must overcome in clinical practice to recover the gut microbiomic dysbiosis. *Lactobacillus plantaraum* and *Bifidobacterium* are probiotic bacteria capable of modulating negative effects of high fat diets and even to manage immunological reactions mediated by inflammatory diseases (Maynard et al. 2012; Hooper et al. 2012). Non-pathogenic lactic acid producer such as bioengineered *E. coli* strains named Nissle 1917 (Ukena et al. 2007) act also as probiotics. *Lactobacillus rhamnosus* and *gasseri* and *Bifidobacterium lactis* may reduce adiposity, body weight, and weight gain in humans (Mekkes et al. 2014). Furthermore, these probiotics can prevent or ameliorate clinical symptoms of irritable bowels syndrome, inflammatory and necrotizing enterocolitis and acute diarrhoea (Whelan and Quigley 2013).

Dietary prebiotics are selectively fermented ingredients that can change the composition and activity of the gastrointestinal microbiota in beneficial ways (Roberfroid 2007). Dietary prebiotic components include poorly digestible carbohydrates, such as non-starch polysaccharides, resistant starch, non-digestible oligosaccharides and polyphenols. Colonic bacteria produce carbohydrate hydrolysing enzymes and via fermentation of prebiotics produce hydrogen, methane, carbon dioxide and SCFAs, which can affect host energy levels and gut hormone regulation. Mixtures of probiotic and prebiotic ingredients can selectively stimulate growth or activity of health promoting bacteria (Roberfroid 2007). Large-scale metagenomics studies have confirmed that synergic association of pro- and prebiotics really change the gut microbial specie dynamics at the genomic level.

13.5.3 Phage Therapy and CRISPR/Cas9 System

Phages have been used as antimicrobials in human medicine in Eastern Europe as replacement of antibiotics for over 90 years (Abedon 2014). Phage therapy consists of using bacterial viruses (known as phages) as vector for controlling microbial populations (Sulakvelidze et al. 2001; Abedon 2014). Bacteriophages attach to specific receptors in the host membrane and then inject their genetic material into bacteria. Phage infection can result in lysis, lysogeny or resistance. Lytic bacteriophages induce host cell death. Lysogenic (or temperate) phages insert their genome into the host DNA bacteria as a prophage that can transfer it to another during excision-infection process. Temperate and lytic bacteriophages can be programmed to sensitize and kill antibiotic-resistant bacteria.

The use of phages as antibacterial therapeutics has gained new frontier with discoveries of single targeted bacterial specific phage therapies to *Pseudomonas aeruginosa*, *Salmonella*, *Klebsiella pneumonia*, *Shigela dysenteriae*, *Vibrio cholerae*, Vancomycin-resistant *Enterococcus*, methicillin-resistant *Staphylococcus aureus* (MRSA) strains and *Escherichia coli* referred to as extended-spectrum beta-lactamase producing *E. coli* resistant to several classes of antibiotics (Rhoads et al. 2009; Kutter et al. 2015). Clinical studies have confirmed the effectiveness of large spectrum commercial bacteriophage cocktails, such as Enko-phage, SES-bacteriophage, Pyo-bacteriophage and Intesti-bacteriophage. However, the uses of these generic phage cocktails have been approved only in few Eastern European countries (Pelfrene et al. 2016).

Bacteria and archaea have evolved various mechanisms of defence against phage infections (Sampson et al. 2010). Clustered Regularly Interspaced Short Palindromic Repeats (CRISPR) and CRISPR associated proteins (Cas) system is an RNA-mediated DNA interference mechanism of defence identified in both Archaea and Bacteria domain (Garneau et al. 2010; Zhang et al. 2013; Haeussler and Concordet 2016). Cas genes encode a heterogeneous family of nucleases, helicases, polymerases, and polynucleotide-binding proteins. This adaptive immune system is triggered upon phage invasion. Bacteria integrate in its genome short fragments of the foreign DNA, called protospacers. These sequences are integrated into specific regions of the host chromosome known as "CRISPR arrays" that are rich in short palindromic invariable repeats. The protospacer becomes a spacer sequence, which is flanked by the repeated regions. The spacer and the invariable repeats are then transcribed into a CRISPR-RNA (crRNA), which guides specific endonuclease Cas9 to a target DNA containing regions complementary to the spacer fragment of the crRNA. Upon recognition, Cas9 cleaves invasive DNA preventing phage infection.

CRISPR/Cas9 system has served as tool for many genetic studies, particularly genomic editing (Jiang et al. 2013; Haeussler and Concordet 2016). One of its enormous potential is the manipulation of bacterial genome (Garneau et al. 2010; Vercoe et al. 2013; Selle and Barrangou 2015; Yosef et al. 2015; Marraffini and Sontheimer 2008). Engineering of commensal bacteria with CRISPR/Cas9 system can be used to introduce specific mutations into essential, antibiotic resistance, virulence genes encode resistance genes and lytic phages. CRISPR/Cas system may constitute an effective vaccination tool in public health for prevention of diseases. However, much work is needed to be done in order to anticipate undesired off-target effects and efficiency of delivery system.

A healthy human gut contains viral communities—the human gut "virome" (over 10^{12} per person) that infect host cells and other types of microorganisms that inhabit us (Virgin 2014). The viral communities, many of which were unknown so far, may affect human health and disease as do bacteria and archaea. Metagenomic DNA sequencing remains the unique way to identify the growing number of bacteriophages in commensal and pathogenic bacteria present in human body. Lytic phages have also been identified in environmental reservoirs such as zoo composting and human and domestic animal faecal sewages (Amgarten et al. 2017). These are promising tools for the development of more adequate phage-based therapies.

Finally, the functional exploration of pathogen-specific bacteriophages and gene therapy depends on development of in vitro microbiome-mediated disease models that are not so easy to standardize (Fritz et al. 2013). Next, proof-of-concept for phage therapy approaches needs to be addressed in well-designed human clinical trials.

13.6 Conclusions and Perspectives

Metagenomic DNA sequencing of human microbiomes has revealed the high complexity and diversity of microbial communities living in various sites of the human body. These human–microbiota ecosystems are critical in health-promoting processes, and thus clinical studies have highlighted the unexpected and wide-ranging consequences of killing bacteria with antibiotics. Therefore, bacteriotherapy has been explored as a strategy to the steady-state control of bacterial community in disease-causing processes, although yet to be fully determined. The annotation and analyses of many human microbiomes have shown that absence of specific microbial species facilitate the breakdown of epithelial-immune cell barrier and dissemination of pathogenic species into the distant sites. Cultivated and uncultivated microbes may contribute to aetiology of human disorders. Emerging new biomarkers for better discriminatory phyla and species will contribute to diagnosis of type II diabetes, obesity, metabolic disorders, inflammatory bowel diseases and even certain cancers (Fig. 13.1).

The use of antibiotics compromises genome defence and increases bacteria's ability to acquire antibiotic resistance. Prebiotics, probiotics, synbiotics, phage therapy and CRISPR/cas systems are emerging as tools to eradicate and modulate pathogenic bacterial species in microbial communities. Genetic engineering and synthetic biology offer alternative and complementary approaches to the production of quorum sensing natural products and their analogues. Engineering of multidrug resistant bacteria-specific bacteriophages will be possible and fully explored in near future. The use of novel pharmaceuticals and nutraceuticals to manage microbial colonization and development of healthy gut microbial community at early childhood will help the adult human body functions and prevent the occurrence of several pathologies.

Funding Statement This work was supported by grants from Coordenação de Aperfeiçoamento de Pessoal de Nível Superior (CAPES/PNPD proc 2011/188-37/2011), Conselho Nacional de Desenvolvimento Científico e Tecnológico (CNPq, proc 2011/486048) and Fundação de Amparo a Pesquisa do Estado de Sao Paulo (2012/02497-7, 2014/20847-0).

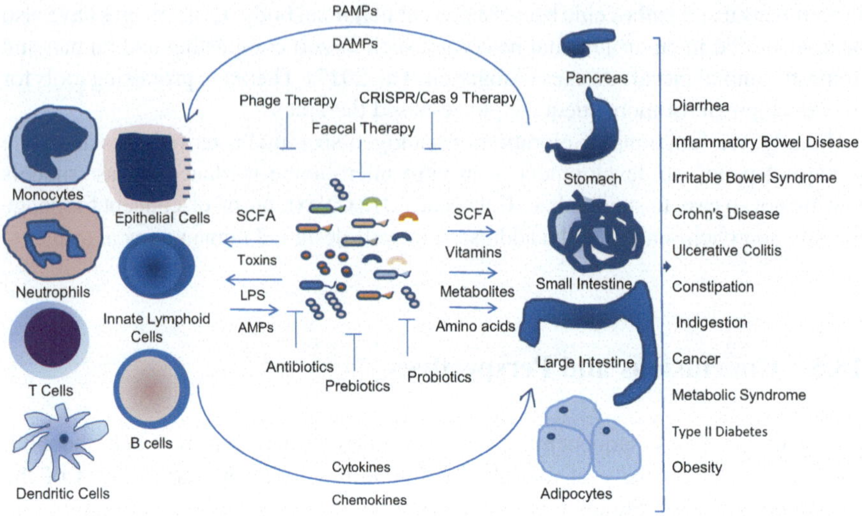

Fig. 13.1 The ecological community of commensal, symbiotic and pathogenic microorganisms play a central role in host metabolism, energy homeostasis and the innate and adaptive immune response. A shift in the gut microbial ecosystem (dysbiosis) results from a decrease in symbiont and/or an increase in pathobiont bacteria. The complex interaction between diet, age, host genetic and environmental factors (antibiotics and infections) and products released by gut microbiota such as SCFA, tyrosine, tryptophan and phenylalanine and their derivatives affect host signalling pathways and thus host metabolism and immunity. Increases in gut permeability and in the systemic levels of bacterial products such as LPS, a condition called metabolic endotoxemia, cause systemic inflammation. Low grade inflammation leads to impaired insulin action, insulin resistance, fat deposition and obesity and the metabolic syndrome. Pharmaceutical and nutraceutical agents targeting specific bacterial species or phylum/class may help to re-establish homeostasis and microbiome balance. SCFA, short-chain fatty acid, PAMP (Pathogen-Associated Molecular Patterns), DAMP (danger-associated molecular patterns and danger signals), AMPs (antimicrobial peptides)

Acknowledgements We thank Aline Maria da Silva, João Carlos Setubal, Dan Waitzberg and colleagues of University of Sao Paulo and Clinical Hospital of Medical School for insights and productive discussions.

References

Abedon ST (2014) Phage therapy: eco-physiological pharmacology. Scientifica 214:581639

Abubucker S, Segata N, Goll J, Schubert AM, Izard J et al (2012) Metabolic reconstruction for metagenomic data and its application to the human microbiome. PLoS Comput Biol 8(6): e1002358

Alivisatos AP, Blaser MJ, Brodie EL, Chun M, Dangl JL, Donohue TJ, Dorrestein PC, Gilbert JA, Green JL, Jansson JK, Knight R, Maxon ME, McFall-Ngai MJ, Miller MF, Pollard KS, Ruby EG, Taha SA, Unified Microbiome Initiative Consortium (2015) A unified initiative to harness Earth's microbiomes. Science 350(6260):507–508

Almonacid DE, Kraal L, Ossandon FJ, Budovskaya YV, Cardenas JP, Bik EM, Goddard AD, Richman J, Zachary S, Apte ZS (2017) 16S rRNA gene sequencing and healthy reference ranges for 28 clinically relevant microbial taxa from the human gut microbiome. PLoS One 12(5): e0176555

Amgarten D, Martins LG, Lombardi KC, Antunes LP, Souza APS et al (2017) Three novel *Pseudomonas* phages isolated from composting provide insights into the evolution and diversity of tailed phages. BMC Genomics 18:346

Andersson DI, Hughes D (2014) Microbiological effects of sublethal levels of antibiotics. Nat Rev Microbiol 12:465–478

Antharam VC, Li EC, Ishmael A, Sharma A, Mai V, Rand KH, Wang GP (2013) Intestinal dysbiosis and depletion of butyrogenic bacteria in Clostridium difficile infection and nosocomial diarrhea. J Clin Microbiol 51(9):2884–2892

Arumugam M, Raes J, Pelletier E, Le Paslier D, Yamada T, Mende DR, Fernandes GR, Tap J, Bruls T et al (2011) Enterotypes of the human gut microbiome. Nature 473(7346):174–180

Backhed F, Fraser CM, Ringel Y, Sanders ME, Sartor B, Sherman PM, Versalovic J, Young V, Finlay BB (2012) Defining a healthy human gut microbiome: current concepts, future directions, and clinical applications. Cell Host Microbe 12:611–622

Baumler AJ, Sparandio V (2016) Interactions between the microbiota and pathogenic bacteria in the gut. Nature 535(7610):85–93

Belizario JE, Napolitano M (2015) Human microbiomes and their role in dysbiosis, common diseases and novel therapeutic approaches. Front Microbiol 6:1050

Bojanova DP, Bordenstein SR (2016) Faecal transplants: what is being transferred? PLoS Biol 13(7):e1002503

Brandt LJ, Reddy SS (2011) Faecal microbiota transplantation for recurrent *Clostridium difficile* infection. J Clin Gastroenterol 45(suppl):S159–S167

Brown EM, Sadarangani M, Finlay BB (2013a) The role of the immune system in governing host–microbe interactions in the intestine. Nat Immunol 14(2013):660–667

Brown CT, Sharon I, Thomas BC, Castelle CJ, Morowitz MJ, Banfield JF (2013b) Genome resolved analysis of a premature infant gut microbial community reveals a *Varibaculum cambriense* genome and a shift towards fermentation-based metabolism during the third week of life. Microbiome 1(1):30. https://doi.org/10.1186/2049-2618-1-30

Cani PD, Delzenne NM (2011) The gut microbiome as therapeutic target. Pharmacol Ther 130:202–212

Cani DP, Neyrinck AM, Fava F, Knauf C, Burcelin RG, Tuohy KM, Gibson GR, Delzenne NM (2007) Selective increases of bifidobacteria in gut microflora improve high-fat-diet-induced diabetes in mice through a mechanism associated with endotoxaemia. Diabetologia 50:2374–2383

Chang C, Lin H (2016) Dysbiosis in gastrointestinal disorders. Best Pract Res Clin Gastroenterol 30:3–15

Chatelier EL, Nielsen T, Qin J, Prifti E, Hildebrand F, Falony G, Almeida M, Arumugam M et al (2013) Richness of human gut microbiome correlates with metabolic markers. Nature 500:541–546

Clemente JC, Ursell LK, Parfrey LW, Knight R (2013) The impact of the gut microbiota on human health: an integrative view. Cell 148(6):1258–1270

Costa FRC, Françozo MCS, Oliveira GG, Ignacio A, Castoldi A, Zamboni DS, Ramos SG, Câmara NO, Zoete MR, Palm NW, Flavell RA, Silva JS, Carlos D (2016) Gut microbiota translocation to the pancreatic lymph nodes triggers NOD2 activation and contributes to T1D onset. J Exp Med 213:1223–1239

Cucchiara S, Stronati L, Aloi M (2012) Interactions between intestinal microbiota and innate immune system in pediatric inflammatory bowel disease. J Clin Gastroenterol 46(Suppl):S64–S66

De Filippo C, Cavalieri D, Di Paola M, Ramazzotti M, Poullet JB, Massart S, Collini S, Pieraccini G, Lionetti P (2010) Impact of diet in shaping gut microbiota revealed by a comparative study in children from Europe and rural Africa. Proc Natl Acad Sci USA 107:14691–14696

Debelius J, Song SJ, Vazquez-Baeza Y, Xu ZZ, Gonzalez A, Knight R (2016) Tiny microbes, enormous impacts: what matters in gut microbiome studies? Genome Biol 17:217
Dethlefsen L, Huse S, Sogin ML, Relman DA (2008) The pervasive effects of an antibiotic on the human gut microbiota, as revealed by deep 16S rRNA sequencing. PLoS Biol 6:e280
Eckburg PB, Bik EM, Bernstein CN, Purdom E, Dethlefsen L et al (2005) Diversity of the human intestinal microbial flora. Science 308:1635–1638
Flintoft L (2012) Disease genomics: associations go metagenome-wide. Nat Rev Genet 13 (11):756–757. https://doi.org/10.1038/nrg3347
Forslund K, Sunagawa S, Kultima JR, Mende DR, Arumugam M, Typas A, Bork P (2013) Country-specific antibiotic use practices impact the human gut resistome. Genome Res 23(7):31–39
Fritz JV, Desai MS, Shah P, Schneider JG, Wilmes P (2013) From meta-omics to causality: experimental models for microbiome research. Microbiome 1:14
Fuchs A, Colonna A (2011) Natural killer (NK) and NK-like cells at mucosal epithelia: mediators of anti-microbial defense and maintenance of tissue integrity. Eur J Microbiol Immunol 1:257–266
Garneau JE, Dupuis ME, Villion M, Romero DA, Barrangou R et al (2010) The CRISPR/Cas bacterial immune system cleaves bacteriophage and plasmid DNA. Nature 468:67–71
Gevers D, Po M, Schloss PD, Huttenhower C (2012) Bioinformatics for the human microbiome project. PLoS Comput Biol 8(11):e1002779
Gilbert JA, Quinn RA, Debelius J, Xu ZZ, Morton J, Garg N, Jansson JK, Dorrestein PC, Knight R (2016) Microbiome-wide association studies link dynamic microbial consortia to disease. Nature 535:94–103
Gough N, Shaikh H, Manges AR (2011) Systematic review of intestinal microbiota transplantation (Faecal Bacteriotherapy) for recurrent *Clostridium difficile* infection. Clin Infect Dis 53 (10):994–1002
Haeussler M, Concordet J-P (2016) Genome editing with CRISPR-Cas9: Can it get any better? J Genet Genomic 43:239–250
Hankir MK, Seyfried F, Hintschich CA, Diep TA, Kleberg K, Kranz M, Deuther-Conrad W et al (2017) Gastric bypass surgery recruits a gut PPAR-α-striatal D1R pathway to reduce fat appetite in obese rats. Cell Metab 25(2):335–344
Harley ITW, Karp CL (2012) Obesity and the gut microbiome: striving for causality. Mol Metab 1(1–2):21–31
Honda K, Littman DR (2016) The microbiota in adaptive immune homeostasis and disease. Nature 535:75–84
Hooper LV, Littman DR, Macpherson AJ (2012) Interactions between the microbiota and the immune system. Science 336:1268–1273
Jiang W, Bikard D, Cox D, Zhang F, Marraffini LA (2013) RNA-guided editing of bacterial genomes using CRISPR-Cas systems. Nat Biotech 31:233–239
Jorgensen I, Rayamajhi M, Miao EA (2017) Programmed cell death as a defence against infection. Nat Rev Immunol 17:151–164
Kootte RS, Vrieze A, Holleman F, Dallinga-Thie GM, Zoetendal EG, de Vos WM et al (2012) The therapeutic potential of manipulating gut microbiota in obesity and type2 diabetes mellitus. Diabetes Obes Metab 14:112–120
Koren O, Knights D, Gonzalez A, Waldron L, Segata N, Knight R, Huttenhower C, Ley RE (2013) A guide to enterotypes across the human body: meta-analysis of microbial community structures in human microbiome datasets. PLoS Comput Biol 9(1):e1002863
Kutter EM, Kuhl SJ, Abedon ST (2015) Re-establishing a place for phage therapy in western medicine. Future Microbiol 10:685–688
Lepage P, Leclerc MC, Joossens M, Mondot S, Blottière HM, Raes J, Ehrlich D, Doré J (2013) A metagenomic insight into our gut's microbiome. Gut 62(1):146–158
Ley RE, Turnbaugh PJ, Klein S, Gordon JI (2006) Microbial ecology: human gut microbes associated with obesity. Nature 444:1022–1023
Ley RE, Lozupone CA, Hamady M, Knight R, Gordon JI (2008) Worlds within worlds: evolution of the vertebrate gut microbiota. Nat Rev Microbiol 6(10):776–788

Li J, Jia H, Cai X, Zhong H, Feng Q, Sunagawa S et al (2014) An integrated catalogue of reference genes in the human gut microbiome. Nat Biotechnol 32:834–841

Marraffini LA, Sontheimer EJ (2008) CRISPR interference limits horizontal gene transfer in 623 Staphylococci by targeting DNA. Science 322(5909):1843–184

Maynard CL, Elson CO, Hatton RD, Weaver CT (2012) Reciprocal interactions of the intestinal microbiota and immune system. Nature 489:231–241

Meijer K, de Vos P, Priebe MG (2010) Butyrate and other short-chain fatty acids as modulators of immunity: what relevance for health? Curr Opin Clin Nutr Metab Care 13:715–721

Mekkes MC, Weenen TC, Brummer RJ, Claassen E (2014) The development of probiotic treatment in obesity: a review. Benef Microbes 5(1):19–28

Miller MB, Bassler BL (2001) Quorum sensing in bacteria. Annu Rev Microbiol 55:165–199

Moeller AH, Li Y, Mpoudi Ngole E, Ahuka-Mundeke S, Lonsdorf EV, Pusey AE, Peeters M, Hahn BH, Ochman H (2014) Rapid changes in the gut microbiome during human evolution. Proc Natl Acad Sci USA 111(46):16431–16435

O'Mahony SM, Clarke G, Borre YE, Dinan TG, Cryan JF (2015) Serotonin, tryptophan metabolism and the brain-gut-microbiome axis. Behav Brain Res 277:32–48

Ostaff MJ, Stange EF, Wehkamp J (2013) Antimicrobial peptides and gut microbiota in homeostasis and pathology. EMBO Mol Med 5:1465–1483

Overbeek R, Begley T, Butler RM, Choudhuri JV, Chuang HY, Cohoon M, de Crécy-Lagard V et al (2005) The subsystems approach to genome annotation and its use in the project to annotate 1000 genomes. Nucleic Acids Res 33(17):5691–5702

Palmer DJ, Metcalfe J, Prescott SL (2012) Preventing disease in the 21st century: the importance of maternal and early infant diet and nutrition. J Allergy Clin Immunol 130(3):733–734

Pelfrene E, Willebrand E, Sanches AC, Sebris Z, Cavaleri M (2016) Bacteriophage therapy: a regulatory perspective. J Antimicrob Chemother 71:2071–2207

Perry RJ, Peng L, Barry NA, Cline GW, Zhang D, Cardone RL, Petersen KF, Kibbey RG, Goodman AL, Shulman GI (2016) Acetate mediates a microbiome-brain-β-cell axis to promote metabolic syndrome. Nature 534(7606):213–217

Pornputtapong N, Nookaew I, Nielsen J (2015) Human metabolic atlas: an online resource for human metabolism. Database (Oxford) 2015:bav068. https://doi.org/10.1093/database/bav068

Qin J, Li R, Raes J, Arumugam M, Burgdorf KS, Manichanh C et al (2010) A human gut microbial gene catalogue established by metagenomic sequencing. Nature 464:59–65

Reijnders D, Goossens GH, Hermes GDA, Neis EPJG, van der Beek CM, Most J et al (2016) Effects of gut microbiota manipulation by antibiotics on host metabolism in obese humans: a randomized double-blind placebo-controlled trial. Cell Metabol 24:63–74

Rhoads DD, Wolcott RD, Kuskowski MA, Wolcott BM, Ward LS, Sulakvelidze A (2009) Bacteriophage therapy of venous leg ulcers in humans: results of a phase I safety trial. J Wound Care 18(6):237–244

Roberfroid MB (2007) Prebiotics: the concept revisited. J Nutr 137(3):830s–837s

Round JL, Mazmanian SK (2009) The gut microbiota shapes intestinal immune responses during health and disease. Nat Rev Immunol 9(5):313–323

Rupnik M (2015) Toward a true bacteriotherapy for *Clostridium difficile* infection. N Engl J Med 372:1566–1568

Sala P, Belarmino G, Machado NM, Cardinell CS et al (2016) The SURMetaGIT study: design and rationale for a prospective pan-omics examination of the gastrointestional response to Roux-em-Y gastric bypass surgery. J Intl Med Res 44(6):1359–1375

Sampson TR, Saroj SD, Llewellyn AC, Tzeng YL, Weiss DS (2010) A CRISPR/Cas system mediates bacterial innate immune evasion and virulence. Nature 479:254–257

Sangiuliano B, Perez M, Moreira D, Belizário J (2014) Cell death associated molecular-pattern molecules: inflammatory signalling and control. Mediators Inflamm 2014:249784

Seeley RJ, Chambers AP, Sandoval DA (2015) The role of gut adaptation in the potent effects of multiple bariatric surgeries on obesity and diabetes. Cell Metab 21:369–378

Selle K, Barrangou R (2015) Harnessing CRISPR-Cas systems for bacterial genome editing. Trends Microbiol 23:225–232

Sender R, Fuchs S, Milo R (2016) Revised estimates for the number of human and bacteria cells in the body. PLoS Biol 14(8):e1002533
Sulakvelidze A, Alavidze Z, Morris JC Jr (2001) Bacteriophage therapy. Antimicrob Agents Chemother 45(3):649–659
Suskind DL, Brittnacher MJ, Wahbeh G, Shaffer ML, Hayden HS, Qin X, Singh N, Damman CJ, Hager KR, Nielson H, Miller SI (2015) Faecal microbial transplant effect on clinical outcomes and faecal microbiome in active Crohn's disease. Inflamm Bowel Dis 21(3):556–563
Swidsinski A, Weber J, Loening-Baucke V, Hale LP, Lochs H (2005) Spatial organization and composition of the mucosal flora in patients with inflammatory bowel disease. J Clin Microbiol 43:3380–3389
Thaiss CA, Zmora N, Levy M, Elinav E (2016) The microbiome and innate immunity. Nature 535:65–74
The Human Microbiome Project Consortium, Huttenhower C, Gevers D, Knight R, Abubucker S, Badger JH et al (2012) Structure, function and diversity of the healthy human microbiome. Nature 486:207–214
Tremaroli V, Bäckhed F (2012) Functional interactions between the gut microbiota and host metabolism. Nature 489:242–249
Turnbaugh PJ, Ley RE, Mahowald MA, Magrini V, Mardis ER, Gordon JI (2006) An obesity-associated gut microbiome with increased capacity for energy harvest. Nature 444:1027–1031
Ukena SN, Singh A, Dringenberg U, Engelhardt R, Seidler U et al (2007) Probiotic *Escherichia coli* Nissle 1917 inhibits leaky gut by enhancing mucosal integrity. PLoS One 2(12):e1308
Van Nood E, Vrieze A, Nieuwdorp M, Fuentes S, Zoetendal EG, de Vos WM, Visser CE, Kuijper EJ, Bartelsman JB, Tijssen JG, Speelman P, Dijkgraaf MG, Keller JJ (2013) Duodenal infusion of donor faeces for recurrent *Clostridium difficile*. N Engl J Med 368:407–415
Vercoe RB, Chang JT, Dy RL, Taylor C, Gristwood T, Clulow JS et al (2013) Cytotoxic chromosomal targeting by CRISPR/Cas systems can reshape bacterial genomes and expel or remodel pathogenicity islands. PLoS Genet 9:e1003454
Verdam FJ, Fuentes S, de Jonge C, Zoetendal EG, Erbil R, Greve JW, Buurman WA, de Vos WM, Rensen SS (2013) Human intestinal microbiota composition is associated with local and systemic inflammation in obesity. Obesity (Silver Spring) 21(12):E607–E615
Virgin HW (2014) The virome in mammalian physiology and disease. Cell 157(1):142–150
Vrieze A, van Nood E, Holleman F, Salojärvi J, Kootte RS, Bartelsman JF, Dallinga-Thie GM, Ackermans MT, Serlie MJ, Oozeer R, Derrien M, Druesne A, van Hylckama Vlieg JE, Bloks VW, Groen AK, Heilig HG, Zoetendal EG, Stroes ES, de Vos WM, Hoekstra JB, Nieuwdorp M (2012) Transfer of intestinal microbiota from lean donors increases insulin sensitivity in subjects with metabolic syndrome. Gastroenterology (4):913–916
Webb CR, Koboziev I, Furr KL, Grisham MB (2016) Protective and pro-inflammatory roles of intestinal bacteria. Pathophysiology 23:67–80
Whelan K, Quigley EM (2013) Probiotics in the management of irritable bowel syndrome and inflammatory bowel disease. Curr Opin Gastroenterol 29(2):184–189
Wikoff WR, Anfora AT, Liu J, Schultz PG, Lesley SA, Peters EC, Siuzdaka G (2009) Metabolomics analysis reveals large effects of gut microflora on mammalian blood metabolites. Proc Natl Acad Sci USA 106(10):3698–3703
Yosef MM, Kiro R, Qimron U (2015) Temperate and lytic bacteriophages programmed to sensitize and kill antibiotic-resistant bacteria. Proc Natl Acad Sci USA 112(23):7267–7272
Zhang Q, Rho M, Tang H, Doak TG, Ye Y (2013) CRISPR-Cas systems target a diverse collection of invasive mobile genetic elements in human microbiomes. Genome Biol 14:R40
Zhou Y, Mihindukulasuriya KA, Gao H, La Rosa P, Wylie KM, Martin JC et al (2014) Exploration of bacterial community classes in major human habitats. Genome Biol 15(5):R66

The manufacturer's authorised representative in the EU is Springer Nature Customer Service Centre GmbH, Europaplatz 3, 69115 Heidelberg, Germany. If you have any concerns regarding our products, please contact ProductSafety@springernature.com

Printed and bound by CPI Group (UK) Ltd, Croydon, CR0 4YY
05/10/2026
02228138-0019